STATISTICS
For
MANAGEMENT

STATISTICS For MANAGEMENT

[For MBA Course, Anna University, Chennai, Trichy, Tirunelveli, Coimbatore and Other Indian Universities]

Dr. P.N. ARORA
M.A., Ph.D. (Delhi Univ.)
Reader, Dyal Singh College
University of Delhi, Delhi.

S. ARORA
M.A., (Eco.), B.Ed.

THOROUGHLY REVISED EDITION

S. Chand And Company Limited
(ISO 9001 Certified Company)

S. Chand And Company Limited

(ISO 9001 Certified Company)

Head Office: D-92, Sector–2, Noida – 201301, U.P. (India), Ph. 91-120-4682700

Registered Office: A-27, 2nd Floor, Mohan Co-operative Industrial Estate, New Delhi – 110 044, Phone: 011-49731800

www.**schandpublishing.com;** e-mail: **info@schandpublishing.com**

Marketing Offices:

Chennai	:	Ph: 23632120; chennai@schandpublishing.com
Guwahati	:	Ph: 2738811, 2735640; guwahati@schandpublishing.com
Hyderabad	:	Ph: 40186018; hyderabad@schandpublishing.com
Jalandhar	:	Ph: 4645630; jalandhar@schandpublishing.com
Kolkata	:	Ph: 23357458, 23353914; kolkata@schandpublishing.com
Lucknow	:	Ph: 4003633; lucknow@schandpublishing.com
Mumbai	:	Ph: 25000297; mumbai@schandpublishing.com
Patna	:	Ph: 2260011; patna@schandpublishing.com

First Edition 2003
Subsequent Editions and Reprints 2003, 2006, 2007, 2008, 2009
Fifth Revised Edition 2010; Reprints 2012, 2013, 2015, 2016, 2017, 2020, 2022 (Twice)

Reprint 2024

ISBN: 978-81-219-2285-2 **Product Code:** H8STA60STAT10ENAE10O

PRINTED IN INDIA

By Vikas Publishing House Private Limited, Plot 20/4, Site-IV, Industrial Area Sahibabad, Ghaziabad – 201 010 and Published by S. Chand And Company Limited, A-27, 2nd Floor, Mohan Co-operative Industrial Estate, New Delhi – 110 044.

Dedication

This book is dedicated to the Loving,
Encouraging and Everlasting
memories of our esteemed Parents
who were a permanent source
of inspirations to us

Dr. P.N. Arora & S. Arora

PREFACE TO THE FIFTH EDITION

The present edition of the book has been thorougly revised in the light of the revised syllabus of M.B.A. program of Anna University, Chennai, Tiruchi, Tirunelveli and Combatore for their paper "Statistics for Management".

The present edition also contains the following four chapters which form a part of the syllabus of Anna University and are necessary tools for the study of statistics in every field.

Chapter 1 : ***Introduction to Statistics***

Chapter 2 : ***Frequency Distributions***

Chapter 3 : ***Measures of Central Tendency or Averages***

Chapter 4 : ***Measures of Dispersion or Variation***

The present book is written for M.B.A. program of Indian Universities and other Indian B-Schools. Since their programs are already overloaded, they need a book which is self contained and brief. This book is designed to meet the above criterion. The purpose of this book is to teach the language, principles and practice of Statistics and see to it that the students are able to handle many statistical problems themselves.

A mathematical back ground is not a pre-requisite for this book and the students will not be intimidated by complex mathematical notations. Each chapter is supported by exercises and copiously worked out examples. **Each chapter of the book reveals a refreshing new touch with scores of valuable points the student is unlikely to find in any other book on Statistics.**

Some of the exclusive features of the book are :

* ***Every concept has been explained with the help of solved examples.***
* ***Working rules showing the various steps for the applications of formulae have been given.***
* ***The diagrams and graphs have been neatly and correctly drawn in such a way that the students have the complete understanding of the problem by simply looking at them.***
* ***It is a user-friendly book containing a large number of solved problems.***
* ***Efforts have been made to make the subject thoroughly exhaustive and nothing important has been omitted.***
* ***Efforts have been made to make the book error free.***

We wish to express our sincere thanks to Mrs. Nirmala Gupta, Chairperson-Managing Director: Mr. Himansu Gupta, Joint-Managing Diiector, S. Chand & Company Ltd., who have taken keen interest and untiring efforts in the publication of the book. Thanks are also due to Mr. Navin Joshi (Vice President– Publishing), Mr. R.S. Saxena (Advisor – Publishing), Mr. Shishir Bhatnagar (Manager-Pre-press & Editorial) for their constant encouragement and guidance.

Criticism and suggestions are invited for the improvement of the book.

Dr. P.N. ARORA
S. ARORA

ANNA UNIVERSITY – MBA – SYLLABUS (REVISED) 2009

(This Syllabus is Common for Chennai, Tiruchi and Tirunelveli)

1. Introduction to Statistics & Probability **12**

Statistics–Definition, Types. Types of variables – Organising data – Descriptive measures. Basic definitions and rules for probability, conditional probability independence for events, Beye's theorem, and random variables, Probability distributions : Binominal, Poisson, Uniform and Normal distributions.

2. Sampling Distribution and Estimation : **12**

Introduction to sampling distributions, sampling distribution of mean and proportion, application of central limit theorem, sampling techniques. Estimation : Point and interval, estimation for population parameters of large samples and small samples, determining the sample size.

3. Testing of Hypothesis : **12**

Hypothesis testing : one sample and two sample tests for means and proportions of large samples (Z-test), one sample and two sample tests for means of small samples (t-test), F-test for two sample standard ANOVA one and two way – design of experiments.

4. Non-parametric Methods : **12**

Chi-square test for single sample standard deviation. Chi-square tests for independence of attributes and goodness of fit; Yates correction for continuity. Sign test for paired data. Rank sun test. Kolmogorov-Smirnon-test for goodness of fit, comparing two populations. Mann–Whitney U Test and Kruskal Walli's test. One sample run test, rank correlation.

5. Correlation, Regression and Time Series Annalysis **12**

Correlation analysis, Estimation, Regression lines. Time series analysis : Variations in time series, trend analysis, cyclical variations, seasonal variations and irregular variations. Index Numbers – Laspeyre's, Paasche's and Fisher's Ideal index.

TOTAL : 60 PERIODS

QUANTITATIVE METHODS FOR MANAGEMENT

(Syllabus Anna, Coimbatore)

Unit	Topics	No. of Hours
1.	**Introduction of Statistics–** Types of statistics-Importance of statistics in business-Discribing and displaying data using Frequence Tables, Frequency distributions, diagrammatic and graphic presentations.	07
2.	Measures of central tendency and dispersion Measures of central tendency-Mean, median, mode, quartiles, deciles, percentiles, Measures of dispersion-Range, quartile deviation, Mean deviation, standard deviation, coefficient of variation. Estimation & confidence Intervals-Types of estimates-Point estimates and confidence intervals for a mean and proportion	08
3.	**Testing of Hypothesis I** Procedure for testing of hypothesis-Hypothesis testing for large sample and small sample and small samples (using z-test, 't' test)–F-Test.	09
4.	**Testing of Hypothesis II** Non-parametric tests : chi-square tests Sign test, Mann Whitney U-Test, Kruskal-Wallis Test, Spearman's Rank correlation Test	09
5.	**Correlation and Regression** Concepts of correlation-Types of correlation-Kari Pearson's coefficient of correlation, Simple Regression-Regression Coefficients-Method of least squares.	07
	Total	07

CONTENTS

1

CHAPTER

INTRODUCTION TO STATISTICS

1.1 INTRODUCTION

The word **'Statistics'** has been derived from the *Latin* word **'Status'** which means a political state. It has also its root either to the Italian word **'Statista'** or the German word **'Statistik'** each one of which means a **political state.** For several decades, the word **'statistics'** was associated solely with the display of facts and figures pertaining to the economic, demographic and political situations prevailing in a country, usually, collected and brought out by the local governments.

Statistics is a tool in the hands of mankind to translate complex facts into simple and understandable statements of facts.

1.2 MEANING AND DEFINITION OF STATISTICS

Meaning of Statistics. The word *Statistics* is used in two different senses — **Plural and Singular**. In its **plural form,** *it refers to the numerical data collected in a systematic manner with some definite aim or object in view* such as the number of persons suffering from malaria in different colonies of Delhi or number of unemployed girls in different states of India and so on. In **singular form**, *the word statistics means the science of statistics that deals with the principles, devices or statistical methods of collecting, analysing and interpreting numerical data.*

Thus, **'statistics' when used in singular refers to that branch of knowledge which implies Applied Mathematics.**

The science of statistics is an old science and it has developed through ages. This science has been defined in different ways by different authors and even the same author has defined it in different ways on different occasions.

It is impossible to enumerate all the definitions given to statistics both as "**Numerical Data,** *i.e.,* **Plural Form**" and "**Statistical Methods,** *i.e.,* **Singular Form**". However, we have give below some selected definitions of both the forms.

1.2.1 Definitions of "Statistics in Plural Form or Numerical Data"

Different authors have given different definitions of statistics. Some of the definitions of statistics describing it quantitatively or in plural form are:

"Statistics are the classified facts representing the conditions of the people in a state especially those facts which can be stated in number or in a table of numbers or in any tabular or classified arrangement."

— Webster.

This definition is narrow as it is confined only to the collection of the people in a state.

But the following definition given by **Secrist** is modern and convincing. It also brings out the major characteristics of statistical data.

> ***"By statistics we mean the aggregate of facts affected to a marked extent by multiplicity of causes, numerically expressed, enumerated or estimated according to reasonable standards of accuracy collected in a systematic manner for a pre-determined purpose and placed in relation to each other."*** ***Secrist.***

This definition makes it clear that **statistics (*in plural form or numerical data*)** should possess the following characteristics.

I. Statistics are aggregate of facts

II. Statistics are affected by a large number of causes

III. Statistics are always numerically expressed

IV. Statistics should be enumerated or estimated

V. Statistics should be collected in a systematic manner

VI. Statistics should be collected for a pre-determined purpose

VII. Statistics should be placed in relation to each other

1.2.2 Statistics as Statistical Methods or Statistics in Singular Sense

We give below the definitions of **statistics** used in **singular sense, *i.e.*, statistics as statistical methods.**

Statistical methods provide a set of tools which can be profitably used by different sciences in the manner they deam fit. The term statistics in this context has been defined differently by different authors. A few definitions are given below:

"Statistics may be called the science of counting."

—A.L. Bowley.

This definition covers only one aspect, *i.e.*, counting, but the other aspects such as classification, tabulation, etc., have been ignored. As such, the definition is inadequate and incomplete.

"Statistics may be defined as the collection, presentation, analysis and interpretation of numerical data."

—Croton and Cowden.

This definition given by **Croxton and Cowden is simple,** clear and concise.

According to this definition, there are four stages — collection of data, and presentation of data, analysis of data, and interpretation of data. However, one more stage may be added and that is the organisation of data. Thus, there are four stages:

1. Collection and Organisation of data. Three are various methods for collecting the data such as census, sampling, primary and secondary data etc.

2. Presentation of data. The mass data collected should be presented in a suitable, concise form as the mass data collected is difficult to understand and analyse.

3. Analysis of data. The mass data collected should be presented in a suitable, concise form for further analysis. Analysis includes condensation, summarisation, conclusion, etc., through the means of measures of central tendencies, dispersion, skewness, kurtosis, correlation, regression, etc.

4. Interpretation of data. The last step is the drawing conclusions from the data collected as the figures do not speak for themselves.

Having briefly discussed some of the definitions of the term statistics and having seen their drawbacks we are now in a position to give a simple and complete definition of the **'statistics'** in the following words:

Statistics ***(as used in the sense of data) are numerical statements of facts capable of analysis and interpretation and the science of statistics is a study of the principles and methods used in the collection, presentation, analysis and interpretation of numerical data in any sphere of enquiry.***

1.3 IMPORTANCE AND SCOPE OF STATISTICS IN BUSINESS

I. Statistics and Economics

According to ***Prof. Alfred Marshall*, "Statistics are the straws out of which I, like every other economist, have to make bricks."** The following are some of the fields of economics where statistics is extensively used.

(*a*) **Consumption.** Statistical data of consumption enable us to find out the ways in which people in different strata of society spend their incomes.

(*b*) **Production.** The statistics of production describe the total productivity in the country. This enables us to compare ourselves with other countries of the world.

(*c*) **Exchange.** In the field of exchange, an economist studies markets, laws of prices are determined by the forces of demand and supply, cost of production, monopoly, competition, banking etc. A systematic study of all these can be made only with the help of statistics.

(*d*) **Econometrics.** With the help of econometrics, economics has become exact science. Econometrics is the combination of economics, mathematics and statistics.

(*e*) **Public Finance.** Public finance studies the revenue and expenditure activities of a country. ***Budget,*** (a statistical document), ***fiscal policy, deficit financing,*** etc., are the concepts of economics which are based on statistics.

(*f*) **Input-Output Analysis.** The input-output analysis is based on statistical data which explain the relationship between the input and the output. ***Sampling, Time series, Index numbers, Probability, Correlation and Regressions*** are some other concepts which are used in economic analysis.

II. Statistics and Commerce

Statistical methods are widely applied in the solution of most of the business and trade activities such as ***production, financial analysis, costing, manpower, planning, business, market research, distribution and forecasting etc.*** A shrewd businessman always makes a proper and scientific analysis of the past records in order to predict the future course of the business conditions. **Index numbers** help in predicting the future course of business and economic events. *Statistics or statistical methods help the business establishments in analysing the business activities* such as:

(*a*) **Organisation of Business.** Businessman makes extensive use of statistical data to arrive at the conclusion which guides him in establishing a new firm or business house.

(*b*) **Production.** The production department of an organisation prepares the forecast regarding the production of the commodity with the help of statistical tools.

(*c*) **Scientific Management and Business Forecasting.** Better and efficient control of a business can be achieved by scientific management with the help of statistical data. *"The success of businessman lies on the accuracy of forecast made."* ***"The successful businessman is one who estimate most closely approaches accuracy,"*** *said* **Prof. Boddington.**

(*d*) **Purchase.** The price statistics of different markets help the businessman in arriving at the correct decisions. Raw material is purchased from those markets only where the prices are low.

III. Statistics and 'Auditing and Accounting'

Statistics is widely used in accounting and auditing.

IV. Statistics and Economic Planning

According to **Prof. Dickinson,**

"Economic Planning is making of major decisions—what and how much is to be produced, and to whom it is to be allocated—by the conscious decisions of a determinate authority on the basis of a comprehensive survey of economy as a whole." The various documents accompanying, preceding and following each of the eight **Five Year Plans of India** *are a standing testimony to the fact that statistics is an indispensable tool in economic planning.*

V. Statistics and Business

Statistics is an indispensable tool in all aspects of business. When a man enters business he enters the profession of forecasting because success in business is always the result of precision in forecasting and failure in business is very often due to wrong expectations, which arise in turn due to faulty reasoning and inaccurate analysis of various causes affecting a particular phenomenon. **Boddington** observes, "*The successful businessman is the one, whose estimate most closely approaches the accuracy.*"

1.4 LIMITATIONS OF STATISTICS

Statistics and its techniques are widely used in every branch of knowledge. **W.I. King** rightly says: ***"Science of statistics is the most useful servant, but only of great value to those who understand its proper use."*** The scope of statistics is very wide and it has great utility; but these are restricted by its limitations. Following are the important limitations of statistics:

1. Statistics does not deal with individual item. **King** says, ***"Statistics from the very nature of the subject cannot and never will be able to take into account individual cases."*** Statistics proves inadequate, where one wants to study individual cases. Thus, it fails to reveal the true position.

2. Statistics deals with quantitative data. According to **Prof. Horace Secrist, *"Some phenomenon cannot be quantitatively measured; honesty, resourcefulness, integrity, goodwill, all important in industry as well as in life, are generally not susceptible to direct statistical measurement."***

3. Statistical laws are true only on averages. According to **W.I. King, *"Statistics largely deals with averages and these may be made up of individual items radically different from each other."*** Statistics are the means and not a solution to a problem.

4. Statistics does not reveal the entire story. According to **Marshall, *"Statistics are the straws, out of which, I like every other economist to have to make bricks."*** **Croxton** says: ***"It must not be assumed that statistical method is the only method for use in research; neither should this method be considered the best attack for every problem."***

5. Statistics is liable to be misused. According to **Bowley, *"Statistics only furnishes a tool though imperfect, which is dangerous in the hands of those who do not know its use and deficiencies."*** **W.I. King** states, ***"Statistics are like clay of which you can make a God or Devil as you please."*** He remarks, ***"Science of Statistics is the useful servant, but only of great value to those who understand its proper use."***

6. Statistical data should be uniform and homogeneous.

1.5 STATISTICAL TOOLS USED IN ECONOMIC ANALYSIS

The following are some of the important statistical techniques which are applied in economic analysis.

(*a*) Collection of data ; **(*b*) Tabulation ;**
(*c*) Measures of Central Tendency ; **(*d*) Measures of Dispersion ;**
(*e*) Time Series ; **(*f*) Probability ;**
(*g*) Index Numbers ; **(*h*) Sampling and its uses ;**
(*i*) Business Forecasting ;
(*j*) Tests of Significance and Analysis of Variance ; **(*k*) Statistical Quality Control.**

1.6 COLLECTION OF DATA

Data *The information collected through censuses and surveys or in a routine manner or other sources is called a* **raw data.** The word **data** means **information** (its literary meaning is given as facts). The adjective **raw** attached to data indicates that the information thus collected and recorded cannot be put to any use immediately and directly. It has to be converted into more suitable form or processed before it begins to make sense to be utilised gainfully. **A raw data is a statistical data in original form before any statistical techniques are used to redefine, process or summarize it.**

There are two types of statistical data :

(*i*) Primary data **(*ii*) Secondary data**

1. **Primary data.** *It is the data collected by a particular person or organisation for his own use from the primary source.*
2. **Secondary data.** *It is the data collected by some other person or organisation for their own use but the investigator also gets it for his use.*

In other words, the **primary** data are those data which are collected by you to meet your own specific purpose, whereas the **secondary** data are those data which are collected by somebody else.

A data can be primary for one person and secondary for the other.

1.6.1 Methods of Collecting Primary Data

The primary data can be collected by the following methods :

1. **Direct personal observation.** In this method, the investigator collects the data personally and, therefore, it gives reliable and correct information.
2. **Indirect oral investigation.** In this method, a third person is contacted who is expected to know the necessary details about the persons for whom the enquiry is meant.
3. **Estimates from the local sources and correspondence.** Here the investigator appoints agents and correspondents to collect the data.
4. **Data through questionnaires.** The data can be collected by preparing a questionnaire and getting it filled by the persons concerned.
5. **Investigations through enumerators.** This method is generally employed by the Government for population census, etc.

1.6.2 Methods of Collecting Secondary Data

The secondary data can be collected from the following sources :

1. ***Information collected through newspapers and periodicals.***
2. ***Information obtained from the publications of trade associations.***
3. ***Information obtained from the research papers published by University departments or research bureaus or U.G.C.***
4. ***Information obtained from the official publications of the central, state, and the local governments dealing with crop statistics- industrial statistics, trade and transport statistics, etc.***
5. ***Information obtained from the official publications of the foreign governments for international organisations.***

1.6.3 Distinction of Between Primary Data and Secondary Data

Primary Data	*Secondary Data*
1. Primary data is collection of original data for the first time.	1. Secondary data is basically compilation of existing data.
2. It is collected by the investigator or his agents.	2. It is compiled by persons other than who collected the primary data.
3. Its collection is relatively more costly.	3. Its collection is relatively less costly.
4. It is usually directly suitable to the purpose of enquiry.	4. It may or may not be directly suitable to the purpose of enquiry.
5. It may be used as it is for the purpose of enquiry.	5. It may require certain adjustments to be made to suit the purpose of enquiry.
6. There is possibility of a personal prejudice in its collection.	6. There is no possibility of a personal prejudice in compilation since such data are already collected.

1.7 CLASSIFICATION OF DATA

The process of arranging things in groups or classes according to their common characteristics and affinities is called the **classification of data.**

"*Classification is the process of arranging data into sequences and groups according to their common characteristics or separating them into different but related parts.*" – **Secrist.**

Thus classification is the process of arranging the available data into various homogenous classes and sub-classes according to some common characteristic or attribute or objective of investigation.

1.8 REQUISITES OF A GOOD CLASSIFICATION

The main characteristics of a good classification are :

1. It should be exhaustive.
2. It should be unambiguous .
3. It should be mutually exclusive.
4. It should be stable.
5. It should be flexible.
6. It should have suitability.
7. It should be homogeneous.
8. It should be a revealing classification.

1.9 ADVANTAGES OF CLASSIFICATION OF DATA

(i) It condenses the data and ignores unnecessary details.

(ii) It facilitates comparison of data.

(iii) It helps in studying the relationships between several characteristics.

(iv) It facilitates further statistical treatments.

1.10 TYPES OF CLASSIFICATION OF DATA

There are four types of classification of data:

(i) Quantitative Classification ;
(ii) Temporal Classification ;
(iii) Spatial Classification and
(iv) Qualitative Classification.

***(i)* Quantitative Classification :** When the basis of classification is according to differences in quantity, the classification is called **quantitative.**

A quantitative classification refers to classification that is based on figures. In other words, it is a classification which is based on such characteristics which are capable of quantitative measurement such as height, weight, number of marks obtained by students of a class.

(ii) Temporal Classification : When the basis of classification is according to differences in time, the classification is called **temporal** or **chronological classification.**

(iii) Spatial or Geographical Classification : When the basis of classification is according to geographical location or place, the classification is called **spatial** or **geographical**.

(iv) Qualitative Classification : When the basis of classification is according to characteristics or attributes like social status etc. is called **qualitative classification.**

Classification according to attributes is a method in which the data are divided on the basis of qualities. (i.e. married or single; honest or dishonest; beautiful or ugly; on the basis of religion, viz., Hindu, Muslim, Sikh, Christian, etc. known as **attributes**), which cannot be measured quantitatively.

Classification of this nature is of two types :

(i) *Simple Classification or Two-Fold Classification.*

(ii) *Manifold Classification.*

I. Simple Classification or Two-fold Classification

If the data are classified only into two categories according to the presence or absence of only one attribute, the classification is known as **simple or two–fold classification or Dichotomous** *For example*, the population of India may be divided into males and females; literate and illiterate etc.

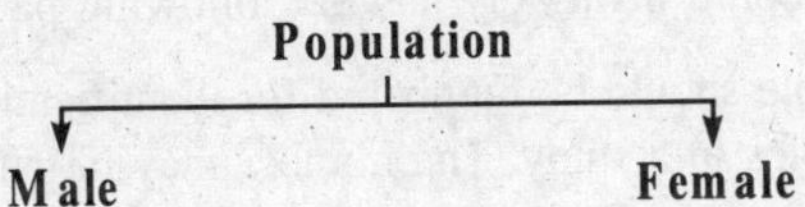

Moreover, if the classification is done according to a single attribute it is also known as **One-way Classification.**

II. Manifold Classification : *It is a classification where more than one attributes are involved.*

1.11 MODE OF PRESENTATION OF DATA

In this section we shall consider the following three modes of presentation of data.

(a) Textual presentation;

(b) Tabular presentation or Tabulation;

(c) Diagrammatic presentation.

1.12 TEXTUAL PRESENTATION

This method comprises presenting data with the help of a paragraph or a number of paragraphs. The official report of an enquiry commission is usually made by textual presentation. Following the examples of textual presentation.

Example 1. *In 1995, out of total of 2,000 students in a college, 1,400 were for graduation and the rest for post-graduation (P.G.). Out of 1,400 Graduate students 100 were girls, however, in all there were 600 girls in the college. In 2000, number of graduate students increased to 1,700 out of which 250 were girls, but the number of P.G. students fall to 500 of which only 50 were boys. In 2005, out of 800 girls 650 were for graduation, whereas the total number of gradurtes was 2,200. The number of boys and girls in P.G. classes was equal.*

1.12.1 Merits and Demerits of Texual Presentation

The merit of this mode of presentation lies in its simplicity and even a layman can present and understand the data by this method. The observations with exact magnitude can be presented with the help of textual presentation. This type of presentation can be taken as the first step towards the other methods of presentation.

Textual presentation, however, is not preferred by a statistician simply because, it is dull, monotonous and comparison between different observations is not possible in this method. For manifold classification, this method cannot be recommended.

1.13 TABULAR PRESENTATION OR TABULATION OF DATA

Tabulation **is a *scientific process used in setting out the collected data in an understandable form.***

Tabulation may be defined **as logical and systematic arrangement of statistical data in rows and columns. It is designed to simplify the presentation of data for the purposes of analysis and statistical inferences.**

"Tabulation is the process of condensing classified data in the form of a table so that it may be more easily understood and so that any comparison involved may be more readily made."
— D. Gregory and H. Ward.

Thus tabulation is one of the most important and ingenious devices of presenting the data in a condensed and readily comprehensible form. It attempts to furnish the maximum information in the minimum possible space, without sacrificing the quality and usefulness of the data.

1.13.1 Essential Parts of a Statistical Table

A good statistical table should invariably has the following parts :

1. **Table Number.** A table should be numbered for identification, especially, when there are a large number of tables in a study. The number may be put at the centre, above the title or at the bottom of the table.
2. **Title of the table.** Every table should have a title. It should be clear, brief and self-explanatory. The title should be set in bold type so as to give it prominence.
3. **Date.** The date of preparation of table should always be written on the table. It enables to recollect the chronological order of the table prepared.
4. **Stubs or Row designations.** Each row of the table must have *a heading*. The designations of the rows are called *stubs or stub items.* Stubs clarify the figures in the rows. As far as possible, the items should be condensed so that they can be included in a single row.
5. **Captions or Column headings.** A table has many *columns.* Sub-headings of the columns are called ***captions or headings.*** They should be well-defined and brief.
6. **Body of the Table.** It is the most vital part of the table. It contains the numerical information. It should be made as comprehensive as possible. The actual data should be arranged in such a manner that any figure may be readily located. Different categories of numerical variables should be set out in an ascending order, from left to right in rows and in the same fashion in the columns, from top downwards.
7. **Unit of Measurements.** The unit of measurement should always be stated along with the title, if this is uniform throughout. If different units have been adopted, then they should be stated along the stubs or captions.
8. **Source Notes.** A note at the bottom of the table should always be given to indicate the primary source as well as the secondary source from where the data has been taken, particularly, when there is more than one source.
9. **Foot Notes and References.** It is always placed at the bottom of the table. It is a statement which contains explanation of some specific items, which cannot be understood by the reader from the title, or captions and stubs.

1.13.2 Format of a Table

The following is the specimen or format of a table indicating the above parts.

Format of a Blank Table

Title : [Head Note or Prefatory Note (if any)]

Stub Heading	*Caption*					*Total*
	Sub-Heads		*Sub-Heads*			
	Column Head **(1)**	*Column Head* **(2)**	*Column Head* **(3)**	*Column Head* **(4)**	*Column Head* **(5)**	**(6)**
			Body			
Total						

Foot Note: **Source Note:**

Number and Date : **No. DC 1172/25/2/2006**

Example 2. *In 1995, out of total of 2,000 students in a college, 1,400 were for graduation and the rest for Post-graduation (P.G.). Out of 1,400 Graduate students 100 were girls, however, in all there were 600 girls in the college. In 2000, number of graduate students increased to 1,700 out of which 250 were girls, but the number of P.G. students fall to 500 of which only 50 were boys. In 2005, out of 800 girls 650 were for graduation, whereas the total number of graduates was 2,200. The number of boys and girls in P.G. classes was equal.*

Represent the above information in tabular form. Also calculate the percentage increase in the number of graduate students in 2005 as compared to 1995.

Solution.

Table No. 1.1

Title : Students of a College according to Degree and Sex (1995 – 2005)

Degrees / *Year*	*Graduation*		*Total* *(a)*	*Post - graduation*		*Total* *(b)*	*Total* *(a + b)*
	Boys	*Girls*		*Boys*	*Girls*		
1995	1300	100	1400	100	500	600	2000
2000	1450	250	1700	50	450	500	2200
2005	1550	650	2200	150	150	300	2500
Total	**4300**	**1000**	**5300**	**300**	**1100**	**1400**	**6700**

Source : University Paper.
No. PC 293 / 5 /3 / 2006.

Percentage increase in graduation in 2000 as compared to 1990

$$= \frac{(2200-1400)}{1400} \times 100 = \mathbf{57.14\ \%}.$$

1.14 DIFFERENCE BETWEEN TEXUAL AND TABULAR PRESENTATION

The tabulation method is usually preferred to textual presentation as :

(*i*) ***It facilitates comparison between rows and columns.***

(*ii*) ***Complicated data can be represented using tabulation.***

(*iii*) ***Without tabulation, statistical analysis of data is not possible.***

(*iv*) ***It is a must for diagrammatic representation.***

1.15 DIAGRAMMATIC PRESENTATION OF DATA

The representation of statistical data through charts, diagrams and picture is another attractive and alternative method. Unlike the first two methods of representation of data, diagrammatic representation can be used for both the educated section and uneducated section of the society. Furthermore, any hidden trend presented in the given data can be noticed only in this mode of representation. However, compared to tabulation, this is less accurate. So if there is a priority for accuracy, we have to recommend tabulation.

In this chapter we shall consider the following three types of diagrams :

I. Line diagram or Historiagram;

II. Bar diagram;

III. Pie chart.

1.15.1 Rules For Construction of a Diagram

1. **Title:** Each diagram should be given a suitable title, either at the top or at the bottom. It should clearly convey the main theme which the diagram intends to portray. The title should be brief, self-explanatory, clear and un-ambiguous.
2. **Foot Notes:** Foot notes are given at the bottom of each diagram in order to clarify some points relating to the diagram.
3. **Source Note:** Source note, wherever possible, should be appended at the bottom of the diagram. It is given to indicate the source from where the data have been taken.
4. **Neat and Attractive:** Diagrams should be made very neat, clean and attractive by proper size and lettering.
5. **Size of diagram:** The size of diagram would depend on the magnitude of data which is to be shown. The size should be such that all the important characteristics of the data are properly emphasised and can be understood by a look at the diagram.
6. **Selection of scale:** Scale of presentation should be consistent with the size of the paper and the size of the observations to be displayed, so that the diagram obtained is neither too small nor too big. The lettering of the scale must be very clear and legible.
7. **Index:** A brief index explaining various types of shades, colours, lines and designs used in the diagram should be given for clear understanding of the diagram.
8. **Simplicity:** The diagrams should be as simple as possible so that they are easily understood even by a lay-man who does not have any mathematical background.
9. **Proportion between length and breadth:** The length and breadth of the diagram should be consistent with the space available.
10. **Choice of a diagram:** The choice of a diagram for the given data (or data under consideration) be made with utmost caution and care.

1.16 LINE DIAGRAM OR HISTORIGRAM

It is the simplest of all the diagrams. Line graphs are simple mathematical graphs that are drawn on the graph paper by plotting the data concerning one variable on the horizontal known as *x*-axis and other variable of data on the vertical known as *y*-axis. With the help of such graphs, the effect of one variable upon another variable during an experimental or normative study may be clearly demonstrated. The *x* values are presented along *x*-axis and the corresponding frequencies are presented along *y*-axis. This diagram facilitates comparison. Even a time series data can be presented by a line diagram. The distance between the lines is, generally, kept uniform.

When we observe the values of a variable at different period of time, the series so formed is known as **Time series**. We take the time on *X*-axis and value of the variable on *Y*-axis. Join the points by straight lines. This is known as **Line graph** or the graph of **Time series. A time series graph is also known as Historigram.**

This is the simplest and easiest graph, and no technical skill is needed in its construction. More over, it enables an individual to present more information in a more precise form than any other kind of chart can do. Two or more variables can be shown on the same graph and thus comparison becomes easy. ***Graphs of Time series can be constructed either on a natural scale or on a ratio scale. In natural scale, absolute changes from one period to another are shown, whereas in a ratio scale the relative changes are shown.***

Example 3. *A word-nonsense syllables association test was administered on students of a class X to demonstrate the effect of practice on learnings. The data so obtained may be studied from the following table:*

Trial No. :	*1*	*2*	*3*	*4*	*5*	*6*	*7*	*8*	*9*	*10*	*11*	*12*
Score :	*4*	*5*	*8*	*8*	*10*	*13*	*12*	*12*	*14*	*16*	*16*	*16*

Draw a line graph for the representation and interpretation of the above data.

Solution. Plot the points (1, 4), (2, 5), (3, 8), (4, 8), (5, 10), (6, 13), (7, 12), (8, 12), (9, 14) and (10, 16). Join them to get the required line graph as given by **figure 1.1 on page 1.14.**

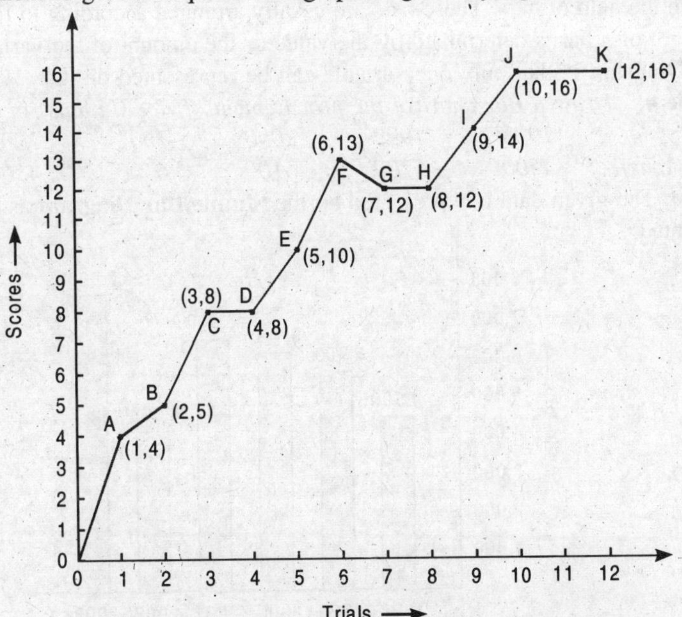

Fig. 1.1. Line Graph—The effect of practice on learning.

1.17 BAR DIAGRAM

The simplest type of graph is the bar diagram. It is especially useful in comparing qualitative data or quantitative data of discrete type. ***A bar diagram is a graph on which the data are represented in the form of bars. It consists of a number of bars or rectangles which are of uniform width with equal space between them on the x-axis. The length of the bar is proportional to the value it represents.*** It should be seen that the bars are neither too short nor too long. The scale should be clearly indicated and base line be clearly shown.

Bars may be drawn either horizontally or vertically. A good rule to use in determining the direction is that if the legend describing the bar can be written under the bars when drawn vertically, vertical bars should be used; when it cannot be, horizontal ones must be used. In this way, the legends can be read without turning the graph. The descriptive legend should not be written at the ends of the bars or within the bars, since such writing may distort the comparison. Usually the diagram will be more attractive if the bars are wider than the spaces between them.

The width of bars is not governed by any set rules. It is an arbitrary factor. Regarding the space between two bars, it is conventional to have a space about one half of the width of a bar.

The data capable of representation through bar diagrams, may be in the form of row scores, or total scores, or frequencies, or computed statistics and summarised figures like percentages and averages etc.

The bar diagram is generally used for comparison of quantitative data. It is also used in presenting data involving time factor. When two or more sets of data over a certain period of time are to be compared a group bar diagram is prepared by placing the related data side by side in the shape of bars. The bars may be vertical or horizontal in a bar diagram. If the bars are placed horizontally, it is called a **Horizontal Bar Diagram.** When the bars are placed vertically, it is called a **Vertical Bar Diagram.**

There are six types of Bar diagram: **(*i*) Simple Bar Diagram; (*ii*) Multiple or Grouped Bar Diagram; (*iii*) Subdivided or Component Bar Diagram; (*iv*) Percentage Subdivided Bar Diagram; (*v*) Deviation or Bilateral Bar Diagram; (*vi*) Broken Bars.**

1.17.1 Simple Bar Diagram

It is used to compare two or more items related to a variable. In this case, the data are presented with the help of bars. These bars are usually arranged according to relative magnitude of bars. The length of a bar is determined by the value or the amount of the variable. A limitation of **Simple Bar Diagram** is that only one variable can be represented on it.

Example 4. *Draw a Bar chart of the procurement of rice (in tons) in an Indian state:*

Year	:	*1998*	*1999*	*2000*	*2001*	*2002*	*2003*
Rice (in tons)	:	*4500*	*5700*	*6100*	*6500*	*4300*	*7800*

Solution. The given data is represented by the **Simple Bar Diagram** as only one variable is to be represented.

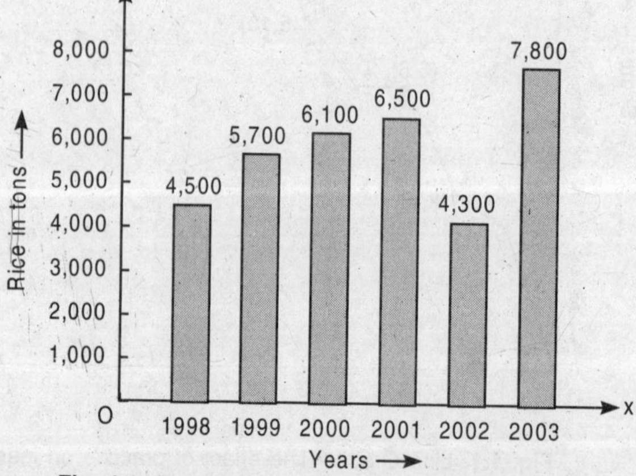

Fig. 1.2. Bar diagram of the procurement of rice.

Here we represent years on the x-axis and procurement of rice on the y-axis.

1.17.2 Multiple or Grouped Bar Diagram

A multiple or grouped bar diagram is used when a number of items are to be compared in respect of two, three or more values. In this case, the numerical values of major categories are arranged in ascending or descending order so that the categories can be readily distinguished. Different shades or colours are used for each category.

Example 5. *Represent the following data by a suitable diagram showing the difference between proceeds and costs:*

Proceeds and costs of a firm (in thousands of rupees).

Year	*Total Proceeds*	*Total costs*
1999	*22.0*	*19.5*
2000	*27.3*	*21.7*
2001	*28.2*	*30.0*
2002	*30.3*	*25.6*
2003	*32.7*	*26.1*
2004	*33.3*	*34.2*

Solution. In this case, the two types of information: proceeds and costs are shown on a diagram indicating the difference between them. The two types of information for each year are placed in such a way that a comparison may be made between them. So a grouped bar diagram is drawn in order to represent the given data.

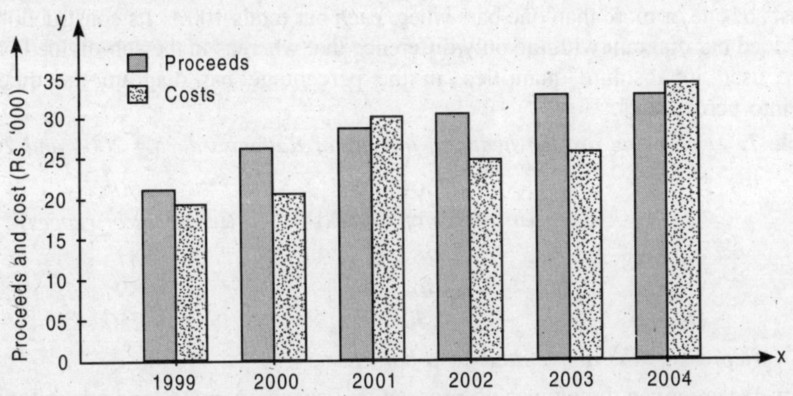

Fig. 1.3. Grouped bar diagram showing proceeds and costs.

1.17.3 Sub-divided or Component Bar Diagram

A component bar diagram is one which is formed by dividing a single bar into several component parts. A single bar represents the aggregate value whereas the component parts represent the component values of the aggregate value. It shows the relationship among the different parts and also between the different parts and the main bar.

Example 6. *Represent the following data of the development expenditure of Central Governments in India during 1997 – 98, 1998 – 99, 1999 – 2000 by bar diagram.*

Year	*Loans and advances*	*Capital*	*Revenue*	*Total*
1997 – 98	*8,601*	*3,787*	*3,477*	*15,865*
1998 – 99	*10,335*	*4,456*	*4,036*	*18,827*
1999 – 2000	*11,549*	*4,803*	*3,709*	*20,061*

Solution. In this case, we use subdivided bar diagram. Here the data for different years is represented by various parts of the single bar for the year. Different parts are indicating loans and advances capital and revenues for each year as shown in the figure.

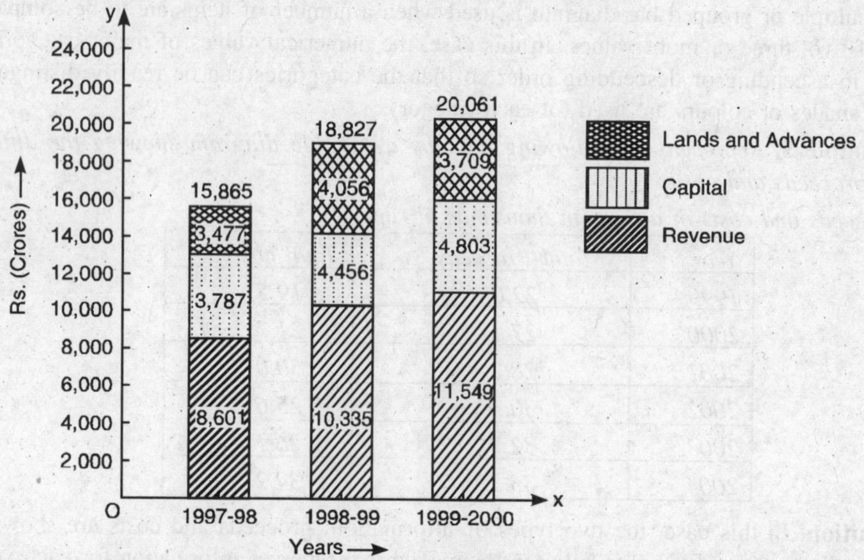

Fig. 1.4. Development expenditure of central and state governments in India

1.17.4 Percentage Sub-divided Bar Diagram

It consists of one or more than one bars where each bar totals 100%. Its construction is similar to the sub-divided bar diagram with the only difference that whereas in the sub-divided bar diagram segments are used in absolute quantities, in the percentage bar diagram the quantities are transformed into percentages.

Example 7. *Following are the heads of income of Railways during 2004 and 2005.*

	2004 (in crores of rupees)	*2005 (in crores of rupees)*
Coaching	*26*	*31*
Goods	*40*	*39*
Others	*4.50*	*3.50*

Represents the above data by a bar chart.

Solution. We are given three types of information : coaching, goods and others for two years. In order to facilitate comparison among them and also between the two years, a percentage sub-divided bar chart is drawn to represent the given data. The graph has been drawn on percentage figures. Percentage of each item has been expressed on the total income of respective years by the formula.

$$\textbf{Percentage} = \frac{\textbf{Income of the item}}{\textbf{Total income of respective year}} \times 100$$

Calculation of Percentage

Items	*2004*		*2005*	
	Income	*Percentage*	*Income*	*Percentage*
Coaching	**26**	**36.88**	**31**	**42.18**
Goods	**40**	**56.74**	**39**	**53.06**
Other	**4.50**	**6.38**	**3.50**	**4.76**
Total	**70.50**	**100.00**	**73.50**	**100.00**

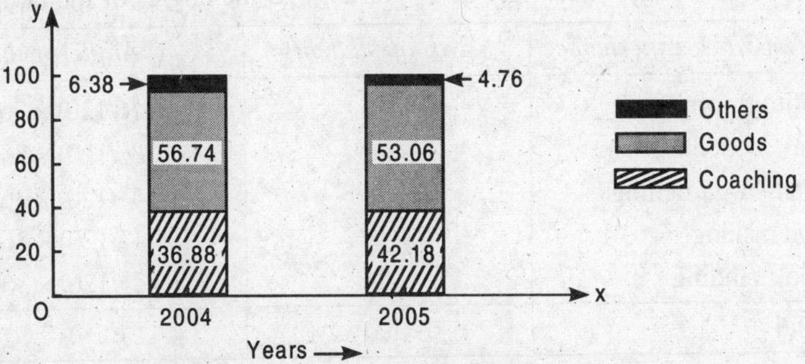

Fig. 1.5. Percentage sub-divided bar chart showing incomes of Railways.

1.17.5 Deviation Bars

Deviation bars are popularly used for graphic presentation of net deviations in different values which have both positive and negative values. The positive values are shown above the baseline (or x-axis) and the negative values below it.

1.17.6 Broken Bars

In certain cases we may come across data which contain very wide variations in values, *i.e.*, very small or very large. In order to provide adequate and reasonable shape to the smaller bars, the larger bars may be broken at the top. The value of each bar is written at the top of the bar.

1.18 PIE DIAGRAM OR ANGULAR DIAGRAM

A pie diagram is a circular graph which represents the total value with its components. The area of a circle represents the total value and the different sectors of the circle represent the different parts. The circle is divided into sectors by radii and the areas of the sectors are proportional to the angles at the centre. It is generally used for comparing the relation between various components of a value and between components and the total value. In ***pie diagram, the data are expressed as percentages.*** Each component is expressed as percentage of the total value. A pie diagram is also known as **angular diagram**.

The name pie diagram is given to a circle diagram because in determining the circumference of a circle we have to take into consideration a quantity known as 'pie' (written as π).

Method of construction. The surface area of a circle is known to cover 2π radians or 360 degrees. The data to be represented through a circle diagram may therefore be presented through 360 degrees, parts or sections of a circle. The total frequencies or value is equated to 360° and then the angles corresponding to component parts are calculated (or the component parts are expressed as percentages of the total and then multiplied by 360/100 or 3.6). After determining these angles, the required sectors in the circle are drawn. Different shades or colours of designs or different types of cross-hatchings are used to distinguish the various sectors of the circle.

Example 8. *120 students of a college were asked to opt for different work experiences. The details of these options are as under.*

Areas of work experience	*No. of students*
Photography	6
Clay modelling	30
Kitchen gardening	48
Doll making	12
Book binding	24

Represent the above data through a pie diagram.

Solution. The numerical data may be converted into the angles of the circle as given below:

Areas of work experience	*No. of students*	*Angle of the circle*
Photography	6	(6/120) × 360 = 18°
Clay modelling	30	(30/120) × 360 = 90°
Kitchen gardening	48	(48/120) × 360 = 144°
Doll making	12	(12/120) × 360 = 36°
Book binding	24	(24/120) × 360 = 72°
Total	120	360°

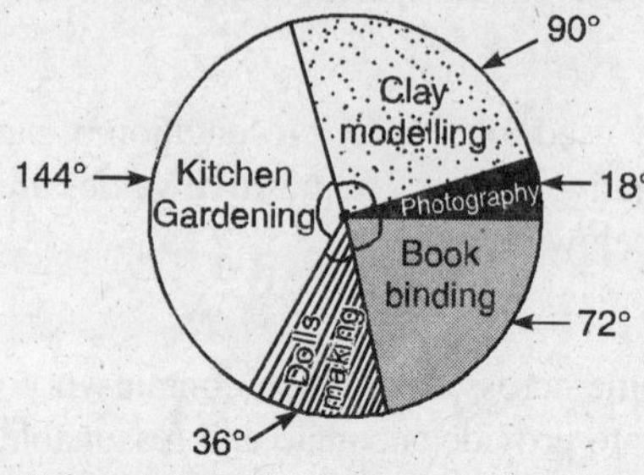

Fig. 1.6. Pie diagram: Areas of work experience opted by the students.

Example 9. *The following figures relate to the cost of construction of a house in Delhi:*

Item	: *Cement*	*Steel*	*Bricks*	*Timber*	*Labour*	*Miscellaneous*
Expenditure	: *20%*	*18%*	*10%*	*15%*	*25%*	*12%*

Represent the data by a suitable diagram.

Solution.

Item	*Expenditure*	*Cenral Angles (Degrees)*
Cement	20%	$\frac{20}{100} \times 360 = 72°$
Steel	18%	$\frac{18}{100} \times 360 = 64.8°$
Bricks	10%	$\frac{10}{100} \times 360 = 36.0°$
Timber	15%	$\frac{15}{100} \times 360 = 54.0°$
Labour	25%	$\frac{25}{100} \times 360 = 90.0°$
Misc.	12%	$\frac{12}{100} \times 360 = 43.2°$
Total	**100%**	**360°**

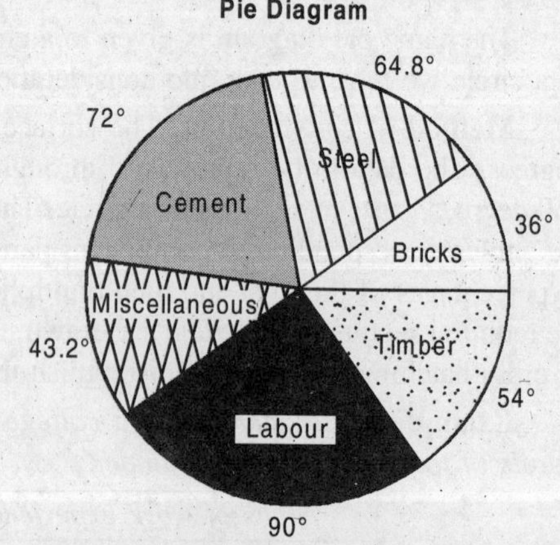

Fig. 1.7. Pie diagram showing the cost of construction of a house in Delhi.

Example 10. *Draw a two-dimensional diagram to represent the following data:*

Item of expenditure	*Expenditure in Rupees*	
	Family A	*Family B*
1. Food	*2000*	*3000*
2. Clothing	*480*	*750*
3. Education	*320*	*400*
4. House rent	*400*	*750*
5. Miscellaneous	*800*	*1100*
Total	*4000*	*6000*

Solution. The total expenditure is taken as 100. The expenditure on each item will be expressed in percentage. The width of the two rectangles will be in proportion to the total expenditure of the two families ; *i.e.*, 4000 : 6000 or 2 : 3. The height of each rectangle will be the same as it represents 100 per cent.

Items of expenditure	*Monthly Expenditure*					
	Family A: (Rs. 4000)			*Family B: (Rs. 6000)*		
	Rs.	*%*	*Cumulative %*	*Rs.*	*%*	*Cumulative %*
1. Food	2000	50	50	3000	50	50
2. Clothing	480	12	62	750	12.5	62.5
3. Education	320	8	70	400	6.67	69.17
4. House rent	400	10	80	750	12.5	81.67
5. Miscellaneous	800	20	100	1100	18.33	100
Total	**4000**	**100**		**6000**	**100**	

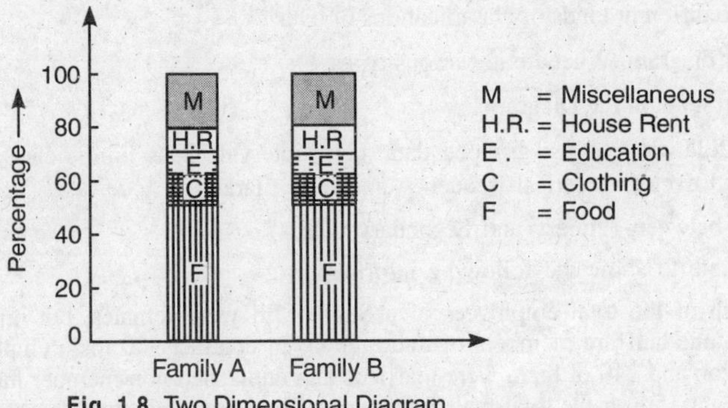

Fig. 1.8. Two Dimensional Diagram

Example 14. *Draw a suitable diagram for the following:*

Expenditure on item	*Percentage of Total expenditure*
Food	*65*
Clothing	*10*
Housing	*12*
Fuel and lighting	*5*
Miscellaneous	*8*

Solution.

Item	*Expenditure*	*Central Angle (Degrees)*
Food	65%	$\frac{65}{100} \times 360 = 234°$
Clothing	10%	$\frac{10}{100} \times 360 = 36°$
Housing	12%	$\frac{12}{100} \times 360 = 43.2°$
Fuel and Lighting	5%	$\frac{5}{100} \times 360 = 18°$
Miscellaneous	8%	$\frac{8}{100} \times 360 = 28.8°$
Total	**100%**	**360°**

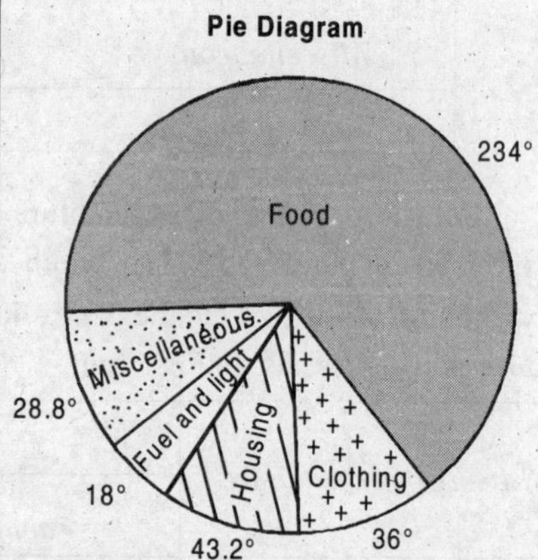

Fig. 1.9. Pie diagram showing expenditure.

EXERCISE 1.1

1. Define the term statistics and discuss its functions and limitations.
2. Discuss the importance of statistics in business and management planning.
3. Explain the origin and development of science of statistics.
4. Define classification of data and state four primary rules that should be observed while classifying a data.
5. What is a tabulation? What are its uses ?
6. What are the advantages of data classification ?
6. What are the different kinds of classifications of data ?
7. Define a bar diagram. What are its various types ?
8. Write a short note on Pie Diagram.
9. Define the following terms: grouped data, class interval, class limits, class boundaries, class mark, inclusive and exclusive series, frequency and tally marks.
10. Distinguish between Primary and Secondary data.
11. Prepare a table to show the following information :

    Exactly 1/4th of the total employees of a Sugar Mill were females, but only 1/10th of those were married and one-half are members of trade union. In contrast 500 male employees were members of the trade union and 240 of them were married. The unmarried non-member male employees were only 40, this being 1/15th of the total male employees. The unmarried employees in the mill, who were not members of the trade union were 132.

12. The budgets of two families are given below. Represent the data by percentage diagrams.

Items of Expenditure	*Family A* *Rs.*	*Family B* *Rs.*
Food	1,600	1,200
Clothing	800	320
Rent	600	480
Light and Fuel	200	160
Miscellaneous	800	240
	4,000	2,400

13. Represent the following data by a suitable diagram:

Items of Expenditure	*Family A* *Rs.*	*Family B* *Rs.*
Food	200	250
Clothing	100	200
House Rent	80	100
Fuel and Light	30	40
Education	90	210

Construct an appropriate diagram to represent the above data.

14. The following data relate to the monthly expenditure of a family. Represent this data by angular diagram :

Item	:	Food	Clothing	Rent	Recreation	Miscellaneous
Rupees	:	160	110	80	60	40

15. Draw a pie diagram to represent the following population in a town :

Males	*Females*	*Girls*	*Boys*	*Total*
2000	1800	4200	2000	10000

[**Hint.**	Population	:	Males	Females	Girls	Boys
	Degree	:	72°	65°	151°	72°]

2
CHAPTER

FREQUENCY DISTRIBUTIONS

2.1 PRESENTATION OF DATA BY FREQUENCY DISTRIBUTION

Statistical data are collected in order to serve a purpose. It should be presented in such a manner, so that it may be easily understood and grasped, and the conclusion may be drawn promptly from the data presented. The data can be presented in a tabular or graphical or diagramatical form. When the data are presented in tables or charts in order to bring out their salient features, it is called the **presentation of data.**

In other words, the methods to condense the data (collected by an investigator) in the tabular form to study their salient features is called the **presentation of data.**

The raw data can be arranged in any one of the following ways :

(*i*) **Serial order** (*ii*) **Ascending or Descending order**

(*iii*) **Tables or Charts** (*iv*) **Groups or Class intervals.**

Illustration 1. The marks obtained by 30 students in a class test, out of 50 marks, according to their roll numbers are :

41, 25, 5, 33, 12, 21, 19, 39, 19, 21, 12, 1, 19, 12, 19, 17, 12, 17, 17, 41, 41, 19, 41, 33, 12, 21, 33, 5, 1, 21.

The data in this form is called a **raw data.** The above data can be arranged in **serial order.**

2.1.1 Data in Serial Order

The above raw data can be arranged in serial order as follows :

Roll No.	*Marks*	*Roll No.*	*Marks*	*Roll No.*	*Marks*
1	41	11	12	21	41
2	25	12	1	22	19
3	5	13	19	23	41
4	33	14	12	24	33
5	12	15	19	25	12
6	21	16	17	26	21
7	19	17	12	27	33
8	39	18	17	28	5
9	19	19	17	29	1
10	21	20	41	30	21

The above arrangement is called the data is arranged in **'Serial Order'.**

The data in this form do not gives us clear picture of the class.

2.1.2 Data in Ascending or Descending Order

Now suppose we wish to judge the standard of achievement of the students. The data in this form do not give us a clear picture of the group. If we arrange them in ascending or descending order, it gives us a slightly better picture. In **ascending order**, the data of illustration 1 will look as:

1, 1, 5, 5, 12, 12, 12, 12, 12, 17, 17, 17, 19, 19, 19, 19, 19, 21, 21, 21, 21, 25, 33, 33, 33, 39, 41, 41, 41, 41.

In **descending order,** the data look as follows :

41, 41, 41, 41, 39, 33, 33, 33, 25, 21, 21, 21, 21, 19, 19, 19, 19, 19, 17, 17, 17, 12, 12, 12, 12, 12, 5, 5, 1, 1.

The raw **data when put in ascending or descending order of magnitude is called an array or arrayed data.**

Array. ***An array*** *is an arrangement of data in order of magnitude either in ascending order or in descending order.*

The **arrayed data** *do not tell us much except the* maximum(*s*) *or* minimum(*s*) *of data.*

Moreover, if the observations are large, then arranging the data in ascending or descending or serial order is a difficult job.

2.2 TALLY BARS AND FREQUENCY

In order to make the data easily understandable, we tabulate the data in the form of ***tables or charts.*** **A table has three columns.**

(*i*) **Variable ;** (*ii*) **Tally marks ;** (*iii*) **Frequency.**

2.2.1 Variable

Any character which can vary from one individual to another is called a **variable** or a **variate.** For example, age, income, height, intelligence, colour etc. are variates. Some variates are measurable and others are not directly measurable. The examples of measurable variates are age, height, temperature, etc., whereas colour and intelligence are the examples of those variates which cannot be measured numerically. Variables or observations with numbers as possible values are called **quantitative variables,** whereas those with names of places, quality, things etc., as possible values are called **qualitative variables or attributes.**

Variables are of two types **(*i*) Continuous; (*ii*) Discontinuous** or **Discrete.** *Quantities which can take all numerical values within a certain interval are called* **continuous variables;** *But those variables which can take only a finite of values are called* **discrete variables;** For example, number of students in a particular class, number of sections in a school etc.

2.2.2 Tally

It is a method of keeping count in blocks of five. For example: 1 = 1; || = 2; ||| = 3; |||| = 4; ~~||||~~ = 5; ~~||||~~ | = 6 and so on.

Tally Bars. These are the straight bars used in the Tally.

Each item falling in the class interval, a stroke (vertical Bar) is marked against it. This **stroke (vertical Bar)** is called the **Tally Bar.** Usually, after every four strokes **(Tally Bar),** in a class, **the fifth item is marked by a horizontal or slanted line across** the Tally Bars (strokes). For example, the frequency 5, 6, 7 is represented by, ~~||||~~, ~~||||~~ |, ~~||||~~ || respectively.

The above method of presentation of data is known as **'Frequency Distribution'.** Marks are called **variates**. The number of students who have secured a particular number of marks is called **Frequency of that variate.**

In the first column of the table, we write all marks from lowest to highest. We now look at the first mark or value in the given raw data and put a bar (vertical line) in the second column opposite to it. We then, see the second mark or value in the given raw data and put a bar opposite to it in the second column. This **process is repeated till all the observations in the given raw data are exhausted.** The bars drawn in the second column are known as *tally marks and to facilitate we record tally marks in bunches of five, the fifth tally marks is drawn diagonally across the first four.* For example, |||| ||| = 8. We finally count the number of tally marks corresponding to each observation and write in the third column headed by frequency or number of students.

2.2.3 Frequency

The number of times an observation occurs in the given data is called the frequency of the observation.

Frequency Distribution. A frequency distribution is *the arrangement of the given data in the form of a table showing frequency with which each variable occurs. In other words,* **frequency** ***distribution*** **of** a variable *x* **is the ordered set** $\{x, f\}$**, where** f **is the frequency.** It shows all scores in a set of data together with the frequency of each score.

The data of illustration 1 can be put in the tabular form as given below:

Marks (*Variate*)	*Tally Marks*	*No. of Students* (*Frequency*)
1	||	2
5	||	2
12	𝍸	5
17	|||	3
19	𝍸	5
21	||||	4
25	|	1
33	|||	3
39	|	1
41	||||	4
Total		**30**

The presentation of data can be further condensed into groups or classes. In this presentation all observations are divided into groups. These groups are called **classes** or **class intervals.**

We can arrange the above data into classes as follows :

Marks obtained	*Tally Marks*	*No. of Students* (*Frequency*)
1–10	||||	4
11–20	𝍸 𝍸 |||	13
21–30	𝍸	5
31– 40	||||	4
41–50	||||	4
Total		**30**

The class 11–20 means the marks obtained between 11 and 20 including both 11 and 20. **The number of observations falling in a particular class is called the frequency of that class or class frequency.** The frequency of the class 1–10 is 4, 11–20 is 13, 21–30 is 5 and so on.

In the class 11–20, 11 is called the ***lower limit*** and 20 is called the **upper limit of the class. The presentation of data in the form of above table is called Grouped Frequency Distribution.**

2.3 TYPES OF FREQUENCY DISTRIBUTIONS

Frequency distribution are of two types :

(*i*) Discrete Frequency Distribution

(*ii*) Grouped (or Continuous) Frequency Distribution

2.3.1 Discrete Frequency Distribution

The construction of discrete frequency distribution from the given raw data is done by the method of tally marks as explained earlier.

2.3.2 Construction of Discrete Frequency Distribution Table

The **frequency distribution table** has three columns headed by

1. Variables (or Classes) **2. Tally Marks or Bars** **3. Frequency.**

The table is constructed by the following steps :

Step 1. *Prepare three columns, viz., one for the* **variable** *(or classes), another for* **tally marks** *and the* **third** *for the* **frequency** *corresponding the variable (or class).*

Step 2. *Arrange the given data (or values) from the lowest to the highest in the first column under the heading* **variable (or classes).**

Step 3. *Take the first observation in the raw data and put a bar (or vertical line |) in the second column under* **Tally Marks** *opposite to it. Then take a second observation and put a* **tally mark** *opposite to it. Continue this process till all the observations of the given raw data are exhausted. For the sake of convenience, record the tally marks in* ***bunches of five*****, the fifth bar is placed diagonally crossing the other four (5 is represented by** ~~||||~~) *leave some space between each block of bars.*

Step 4. *Count the* **tally marks** *of column 2 and place this number opposite to the value of the variable in the third column headed by* **Frequency.**

Step 5. *Give a suitable title to the frequency distribution table so that it exactly conveys the information contained in the table.*

Example 1. *Form a discrete frequency distribution from the following scores :*

15, 18, 16, 20, 25, 24, 25, 20, 16, 15, 18, 18, 16, 24, 15, 20, 28, 30, 27, 16, 24, 25, 20, 18, 28, 27, 25, 24, 24, 18, 18, 25, 20, 16, 15, 20, 27, 28, 29, 16.

Solution.

Frequency Distribution of Scores

Variate	*Tally Marks*	*Frequency*	*Variate*	*Tally Marks*	*Frequency*
15	||||	4	25	~~||||~~	5
16	~~||||~~ |	6	27	|||	3
18	~~||||~~ |	6	28	|||	3
20	~~||||~~ |	6	29	|	1
24	~~||||~~	5	30	|	1
Total					**40**

2.3.3 Continuous or Grouped Frequency Distribution

The method of presenting the raw data in **Discrete Frequency Distribution** form is convenient only when the values in the raw data are largely repeating and the difference between the greatest and the largest observation is not very large. If the number of observations in the raw data is large and the difference between the greatest and the smallest observation is large, we condense the data in classes or groups.

We now define certain statistical terms required for continuous frequency distribution.

2.4 SOME STATISTICAL TERMS

2.4.1. Data or Raw Data. *A raw data is a statistical data in original form before any statistical technique is applied to redefine, process or summarize it.*

2.4.2. Variable or Variate. *Any character which can vary from one individual to another is called variable or variate.* For example, age, income, height, intelligence, colour, etc., are variates. Some variates are measurable and others are not directly measurable. The examples of measurable variates are : age, height, temperature, etc., whereas colour and intelligence are the examples of those variates which cannot be measured numerically. **Variables or observations with numbers as possible values are called quantitative variables,** whereas **those with names of places, attributes, things etc., as possible values are called qualitative variables.**

Types of Variables :

Variables are of two types :

(*i*) **Continuous ;** (*ii*) **Discontinuous or Discrete**

Continuous Variable. ***A continuous variable is capable of assuming any value within a certain range or interval.*** The height, weight, age and temperature of any person can be expressed not only in integral part but also in fractions of any part. For example, the weight of a boy may 44.0 kg or 44.6 kg or 44.65 kg, similarly, his height may be 56 inches or 56.4 inches and age may be 10 years or 10.5 years. Thus, the height, weight, age or temperature etc. are continuous variables.

Discrete Variable. ***A discrete variable can assume only integral values and is capable of exact measurement.*** In other words, **those variables which can take only a finite set of values are called discrete variables.** For example : the number of students in a particular class, or the number of sections in a school, etc. are the examples of discrete variables. **Discrete variables are also known as discontinuous variables.**

2.4.3. Continuous Series. *When the continuous variables are arranged in the form of a series, it is called continuous series or exclusive series.*

2.4.4. Discrete or Discontinuous Series. *When the discrete variables are arranged in the form of a series, it is called a discrete or discontinuous series.*

2.4.5. Array. *An array is an arrangement of data in order of magnitude either in descending or ascending order.*

2.4.6. Descending Order. *When data is arranged from the highest value to the lowest value, the array so formed is in descending order.*

2.4.7. Ascending order. *When the data is arranged from the lowest value to the highest value, the array so formed is in ascending order.*

Illustration 2. If the given data is 17, 7, 11, 5, 13, 9 then

Array in Ascending order : 5, 7, 9, 11, 13, 17.

Array in Descending order : 17, 13, 11, 9, 7, 5

2.4.8. Range. *It is the difference between the largest and the smallest number in the given data.*

The range of the data given in illustration 2 is 17 – 5 = **12.**

2.4.9. Class, Class Interval and Class limits. If the observations of a series are divided into groups and the groups are bounded by limits, then each group is called a *class.* The end values of a class are called the *class limits.* In other words, *the two end limits of each class are called* class limits. The smaller value of the two limits is called the **lower limit** and the higher value of the same is called the **upper limit** of the class. These two **class limits** are sometimes called the **stated class limits.**

2.4.10. Class Interval. *The difference between the lower limit (L) and the upper limit (U) of the class is known as class interval* (I). Thus : **I = U – L.**

In other words, the range of a class is called its **Class Interval.**

Illustration 3. The given data is

Marks obtained	*Tally Marks*	*No. of Students*
1 – 10	𝍸 |	6
11 – 20	|||	3
21 – 30	𝍸 ||	7
31 – 40	||	2
Total		**16**

In the above data the classes are : **1 – 10, 11 – 20, 21 – 30, 31– 40.**

2.4.11. Class Interval. The range of the marks from 1 to 40 is grouped into four classes or groups, *viz.* : 1 – 10, 11 – 20, 21 – 30, 31 – 40. Each group is known as **class interval.** The interval between one class and its adjacent class being 9.

[as: 10 – 1 = **9,** 20 – 11= **9,** 30 – 21 = **9,** etc.]

2.4.12. Class Limits. In the first class 1 – 10, its lower limit is 1 and upper limit is 10. Similarly, **31** is the lower limit and 40 is the upper limit of the class interval **31 – 40.**

2.4.13. Actual Class Limit or Class Boundaries. In the illustration 3, there is a gap of 1 mark between the limits of any two adjacent classes. This gap may be filled up by extending the two limits of each class by half of the value of the gap. Thus

$$\textbf{Lower class boundary} = \textbf{Lower class limit} - \frac{1}{2} \textbf{ of the gap}$$

$$\textbf{Upper class boundary} = \textbf{Upper class limit} + \frac{1}{2} \textbf{ of the gap}$$

The class boundary of the class 11 to 20 are

$$\textbf{Lower class boundary} = 11 - \frac{1}{2} \text{ of } 1 = 11 - 0.5 = \mathbf{10.5}$$

$$\textbf{Upper class boundary} = 20 + \frac{1}{2} \text{ of } 1 = 20 + 0.5 = \mathbf{20.5}$$

In other words, **the class boundaries are the limits up to which the two limits, (actual) of each class may be extended to fill up the gap that exists between the classes.** The class boundaries of each class, so obtained are called the **Actual class limits** or **True class limits.**

$$\textbf{True lower class limit} = \textbf{Lower class limit} - \frac{1}{2} \textbf{ of the gap}$$

$$\therefore \quad \textbf{True upper class limit} = \textbf{Upper class limit} + \frac{1}{2} \textbf{ of the gap}$$

Note: 1. *In the case of exclusive series True class limits are the same as class limits.*

Illustration 4.

Class Interval	*Class Boundaries*
11 – 20	10.5 – 20.5
21 – 30	20.5 – 30.5
31 – 40	30.5 – 40.5

2.4.14. Class-mark or Mid-point or Mid-value. *The central value of the class interval is called the mid-point or mid-value or class mark.* It is the arithmetic mean of the lower class and upper class limit of the same class.

$$\textbf{Mid-value of Class} = \frac{\textbf{Lower class limit + Upper class limit}}{\mathbf{2}}$$

or

$$\textbf{Class mark} = \frac{\textbf{True upper class limit + True lower class limit}}{\mathbf{2}}$$

The class mark of the class 11 – 20 is $\frac{11+20}{2} = 15.5.$

2.4.15. Class Magnitude. *It is the difference between the upper class boundary and the lower class boundary of the class.* In the illustration 4 : the class magnitude of the class, **20.5 – 30.5** is (30.5 – 20.5) = **10.**

2.4.16. Inclusive and Exclusive Series. In the above illustration, all the marks we considered were integers. Hence, it was possible for us to choose classes 11 to 20,

21 to 30 etc. There is a gap of 1 between the upper limit of a class and the lower limit of its next consecutive class, which has not created any difficulty. But there can be situations where the raw data is not in integers. For example, in the information regarding maximum temperature of a city or time required to solve a statistical problem is recorded in the data, it may contain fractions as well. In such cases, the consecutive classes have to be necessarily continuous. We have the following :

Inclusive Series. *When the class-intervals are so fixed that the upper limit of the class is included in that class, it is known as inclusive method of classification, e.g.,* 0 – 5, 6 – 10, 11– 15, 16 – 20.

In the inclusive series, the upper limit and lower limit are included in that class interval. For example, in illustration 3, the marks 11 and 20 are included in the class 11 – 20. It is a **discontinuous series or inclusive series.** In order to make it a continuous one, some adjustment with the class limits is necessary. The **class limits are extended to class boundaries by the adjusting adjustment factor, which is equal to half of the difference between the upper limit of the one class and lower limit of the next class.** The series so obtained is **continuous and** is known **as exclusive series.**

Exclusive or Continuous Series. In this ***series the upper limit of one class is the lower limit of the other class. the common point of the two classes is included in the higher class.*** For example, 10 – 15, 15 – 20, 20 – 25, ..., represent a continuous series or the exclusive series. In this series, **15** is included in the class **15 – 20** and **20** is included in **20 – 30.** Here the class intervals overlap and the ***upper limit of each class is treated as less than that limit and lower limit of each class actually represents exact value.*** Thus

When the class-intervals are so fixed that the upper limit of one class is the lower limit of the next class, *it is known as* **Exclusive method of classification.**

2.5 CONVERSION OF A DISCONTINUOUS SERIES INTO CONTINUOUS SERIES (INCLUSIVE SERIES TO EXCLUSIVE SERIES)

Step 1. Find the adjustment factor by the following formula :

$$\textbf{Adjustment factor} = \frac{1}{2} \times \textbf{[Lower limit of second class – Upper limit of the first class]}$$

Step 2. Subtract the adjustment factor from the lower limit and add it to the upper limit of each class. We will get the required exclusive series.

For Example, to convert the Discontinuous series (or Inclusive series) 11 – 20, 21 – 30, 31 – 40, we have the following method.

Step 1.Adjustment factor. The difference between the upper limit of first class and the lower limit of the second class is 1.

$$\therefore \textbf{ Adjustment factor} = \frac{1}{2} \times [21 - 20] = \frac{1}{2} = \mathbf{0.5}$$

Step 2. The continuous series (or exclusive series) of the given discontinuous series is

11 – 0.5 to 20 + 0.5, *i.e.,* **10.5 – 20.5**
21 – 0.5 to 30 + 0.5, *i.e.,* **20.5 – 30.5**
31 – 0.5 to 40 + 0.5, *i.e.* **30.5 – 40.5**

Inclusive Form	*Exclusive Form*
11 – 20	10.5 – 20.5
21 – 30	20.5 – 30.5
31 – 40	30.5 – 40.5

Example 2. *Construct a continuous frequency distribution table* (**inclusive type**) *of the following marks scored by a class of 40 students.*

20, 11, 11, 37, 15, 40, 31, 29, 38, 27, 13, 7, 29, 25, 37, 42, 30, 10, 9, 27, 25, 18, 2, 9, 47, 17, 11, 32, 41, 6, 29, 15, 13, 39, 21, 40, 10, 15, 3, 4.

Solution. Here the minimum marks scored by a student = **2.**

Maximum marks scored = **47.**

∴ **Range = (Maximum marks – Minimum marks)** = 47 – 2 = **45.**

Suppose, we want to represent the given data by **5 classes,** then the each class interval should be $\frac{45}{5}$ = **9.**

Thus, the classes are 0 – 9, 10 – 19, 20 – 29, 30 – 39, 40 – 49. We have the following Table:

Discrete Frequency Distribution Table

Class	*Tally*	*Frequency*
0 – 9	𝍸 II	7
10 – 19	𝍸 𝍸 II	12
20 – 29	𝍸 IIII	9
30 – 39	𝍸 II	7
40 – 49	𝍸	5
	Total	**40**

Example 3. *The class marks of a frequency distribution are 6, 10, 14, 18, 22, 26, 30. Find the class size and the class intervals.*

Solution. Class size = difference between two successive class marks = (10 – 6) = **4.**

Let the lower limit of the first class interval be *a*.

Then, its upper limit = $(a + 4)$. ∴ $\frac{a+(a+4)}{2} = 6 \Rightarrow 2a = 8 \Rightarrow a = 4.$

∴ **The first class interval is: 4 – 8.**

Let the lower limit of the last class interval be *b*. Then, its upper limit = $(b + 4)$.

∴ $\frac{b+(b+4)}{2} = 30 \Rightarrow 2b = 56 \Rightarrow b = 28.$

∴ **The last class interval is: 28 – 32.**

Hence, the required class intervals are : **4 – 8, 8 – 12, 12 – 16, 16 – 20, 20 – 24, 24 – 28 and 28 – 32.**

Example 4. *The class marks of a distribution are: 26, 31, 36, 41, 46, 51, 56, 61, 66. 71. Find the true class limits.*

Solution. As the class marks are uniformly spaced, so *the class size is the difference between any two consecutive class marks.*

∴ Class size = 31 – 26 = 5.

If a is the class mark of a class interval of size h, then the lower and upper limits of the class interval are $a-\frac{h}{2}$ and $a+\frac{h}{2}$ respectively. Here $h = 5$.

∴ **Lower limit of first class interval** = $26 - \frac{5}{2}$ = **23.5.**

And, **Upper limit of first class interval** = $26 + \frac{5}{2}$ = **28.5.**

∴ **First class interval is : 23.5 – 28.5.**

Thus, the class intervals are:

23.5 – 28.5, 28.5 – 33.5, 33.5 – 38.5, 38.5 – 43.5, 43.5 – 48.5, 48.5 – 53.5

Since, **the classes are formed by exclusive method, therefore, these class limits are true class limits.**

2.6 CUMULATIVE FREQUENCY DISTRIBUTION

Cumulative frequency *corresponding to a class is the sum of all the frequencies upto and including that class.* In cumulative frequency distribution (or series), ***the frequency of a particular class is obtained by adding to the frequency of that class all the frequencies of its previous classes.*** Thus, the cumulative frequency table is obtained from the ordinary frequency table by successively adding the several frequencies.

Cumulative frequency series are of two types.

(*i*) Less than Series **(*ii*) More than Series.**

Suppose, we are given the following discrete series of marks obtained by 100 students. With the help of this series, we shall form the **'Less than' and 'More than' series.**

Marks obtained :	30 – 40	40 – 50	50 – 60	60 – 70	70 – 80	80 – 90
No. of Students :	8	12	20	25	18	17

Less than Cumulative Series

Marks obtained	*No. of Students*
Less than 40	**8**
Less than 50	**20** (= 8 + 12)
Less than 60	**40** (= 20 + 20)
Less than 70	**65** (= 40 + 25)
Less than 80	**83** (= 65 + 18)
Less than 90	**100** (= 83 + 17)

More than Cumulative Series

Marks obtained	*No. of Students*
More than 30	**100**
More than 40	**92** (= 100 – 8)
More than 50	**80** (= 92 – 12)
More than 60	**60** (= 80 – 20)
More than 70	**35** (= 60 – 25)
More than 80	**17** (= 35 – 18)

Example 5. *The distances (in km) covered by 24 cars in 2 hours are given below* :

125, 140, 128, 108, 96, 149, 136, 112, 84, 123, 130, 120, 103, 89, 65, 103, 145, 97, 102, 87, 67, 78, 98, 126

Represent them as a cumulative frequency table using 60 as the lower limit of the first group and all the classes having the class size of 15.

Solution. **Class size = 15.**

Maximum distance covered = 149 km.

Minimum distance covered = 65 km.

∴ **Range = (maximum – minimum)** = (149 – 65) km = 84 km.

$$\textbf{Number of classes} = \frac{\textbf{Range}}{\textbf{Class size}} = \frac{84}{15} = 5.6 \text{ or } 6.$$

Thus, the class intervals are 60 – 75, 75 – 90, 90 – 105, 105 – 120, 120 – 135, 135 – 150.

Cumulative frequency distribution of distances covered

Class interval	*Tally marks*	*Frequency*	*Cumulative frequency*
60 – 75	II	2	**2**
75 – 90	IIII	4	**6**
90 – 105	卌 I	6	**12**
105 – 120	II	2	**14**
120 – 135	卌 I	6	**20**
135 – 150	IIII	4	**24**

Example 6. *The heights (in centimetres) of 40 persons are :*

110, 112, 125, 135, 150, 155, 152, 150, 159, 130, 128, 138, 133, 143, 147, 151, 154, 156, 112, 116, 119, 111, 113, 115, 118, 121, 123, 120, 125, 121, 110, 113, 114, 149, 153, 155, 150, 155, 152, 111.

Array the data and form a cumulative frequency table with class interval of 10.

Solution. Arranging the given data in ascending order of magnitude, we get

110, 110, 111, 111, 112, 112, 113, 113, 114, 115, 116, 118, 119, 120, 121, 121, 123, 125, 125, 128, 130, 133, 135, 138, 143, 147, 149, 150, 150, 150, 151, 152, 152, 153, 154, 155, 155, 155, 156, 159.

Class interval	*Tally (ii)*	*Frequency*	*Cumulative Frequency*
110 – 120	卌 卌 III	13	13
120 – 130	卌 II	7	20 (= 13 + 7)
130 – 140	IIII	4	24 (= 20 + 4)
140 – 150	III	3	27 (= 24 + 3)
150 – 160	卌III	13	40 (= 27 + 13)

Example 7. *Form a frequency table from the following data :*

Age (in years)	*No. of Persons*	*Age (in years)*	*No. of Persons*
under 5	*0*	*under 40*	*22*
under 10	*7*	*under 45*	*29*
under 15	*10*	*under 50*	*35*
under 20	*14*	*under 55*	*41*
under 25	*16*	*under 60*	*47*
under 30	*17*	*under 65*	*50*
under 35	*20*		

Solution. Here cumulative frequencies of the variates have been given. So to get the frequency of a particular class-interval, subtract the cumulative frequency of the lower limit from that of the upper limit.

Age (in years)	*No. of Persons*	*Age (in years)*	*No. of Persons*
5 – 10	7	35 – 40	2
10 – 15	*3*	*40 – 45*	*7*
15 – 20	*4*	*45 – 50*	*6*
20 – 25	*2*	*50 – 55*	*6*
25 – 30	*1*	*55 – 60*	*6*
30 – 35	*3*	*60 – 65*	*3*

2.7 PRACTICAL PROBLEMS IN FORMING A FREQUENCY DISTRIBUTION

The following practical problems are, generally, faced while forming a frequency distribution :

1. ***The number of classes should neither be too small nor too large. It should preferably lie between 5 and 15. As far as possible, open-ended classes should be avoided as these create problems in analysing and interpretation.***
2. ***Class interval is obtained by dividing the range with the number of classes. Again this may create problem unless the classes are of equal width. Sometimes, however, unequal class intervals become necessary.***
3. ***While recording the data in frequency tally, one has to be careful in putting the tally marks, especially, when the number of observations is large.***
4. ***Once tally marks are obtained, the frequencies are obtained by counting the tallies. One should be careful while counting the tallies.***

2.8 RELATIVE FREQUENCY AND PERCENTAGE FREQUENCY OF A CLASS INTERVAL

Relative Frequency: Frequency of each class can also be expressed as **a fraction or percentage terms.** These are known as **relative frequencies.** In other words, a **relative frequency** *is the class frequency expressed as a ratio of the total frequency, i.e.,*

$$\textbf{Relative frequency} = \frac{\textbf{Class frequency}}{\textbf{Total frequency}}$$

Percentage Frequency: ***Percentage frequency of a class interval may be defined as the ratio of the class frequency to the total frequency expressed as a percentage.***

$$\textbf{Percentage frequency} = \frac{\textbf{Class frequency}}{\textbf{Total frequency}} \times \textbf{100.}$$

Relative frequency distribution

Class (Marks)	*Frequency (No. of Students)*	*Relative frequency*
20 – 40	6	6/25 = 0.24 or 24 %
40 – 60	12	12/25 = 0.48 or 48 %
60 – 80	4	4/25 = 0.16 or 16 %
80 – 100	3	3/25 = 0.12 or 12 %
Total	**25**	**1.00 or 100 %**

It may be observed that the sum of all relative frequencies is 1.00 or 100 per cent because the frequency of each class has been expressed as a percentage of the total frequencies (or data).

The relative frequencies express the frequency of any class as a percentage of the total frequency. It is obtained by the formula :

$$\textbf{Relative frequency} = \frac{\textbf{Frequency of item in a group}}{\textbf{Total frequency of the group}}$$

The relative frequencies are used to compare two or more frequency distributions or two or more items in the same frequency distribution.

In the above example, we can compare the number of students securing the marks 20 – 40 and 80 – 100. As there are 24% students who secured marks between 20 and 40 and 12 % students have secured marks between 80 and 100. In the same manner, we can compare the production of rice in Punjab and U.P. in 2006.

It is important to note that for comparing two frequency distributions by means of relative frequencies, the variates as well as the division of the values of the variate or variable into classes must be same for both the distributions. Relative frequencies are used in Percentage, sub-divided Bar diagrams and Pie diagrams.

2.9 BIVARIATE OR TWO-WAY FREQUENCY DISTRIBUTION

So far we studied the frequency distribution of one variable only. Such frequency distribution is called a **univariate frequency distribution.** In many situations simultaneous study of two variables becomes necessary. For example, we may study the heights and weights of a group of students, or the marks obtained by a group of students in Statistics and Economics; or income and expenditure of a group of individuals etc.. The frequency distribution so obtained as a result of this cross classification of two variables give rise to **bivariate frequency distribution** and this can be summarised in the form of a **two-way table** called **Bivariate Frequency Table.**

While preparing a bivariate frequency distribution, the same consideration of classification are applied as for univariate frequency distribution. If the data corresponding to one variable, say X, is grouped into m classes and the data corresponding to the other variable, say Y, is grouped into n classes, then the bivariate table will consist of $m \times n$ cells. By going through the different pairs of the values (X, Y) of the variables and using tally marks we can find the frequency of each cell and thus form bivariate frequency distribution.

The *frequency distribution of the values of the variable X together with their frequency total is* called ***the marginal distribution of X***. *The frequency distribution of the values of the variable Y together with the total frequencies is* known as the ***marginal frequency distribution of Y***.

Example 8. *Prepare a bivariate frequency distribution for the following data for 20 students :*

Marks in Law (X) :	*10*	*11*	*10*	*11*	*11*	*14*	*12*	*12*	*13*	*10*
Marks in Statistics (Y) :	*20*	*21*	*22*	*21*	*23*	*23*	*22*	*21*	*24*	*23*
Marks in Law (X) :	*13*	*12*	*11*	*12*	*10*	*14*	*14*	*12*	*13*	*10*
Marks in Statistics (Y) :	*24*	*23*	*22*	*23*	*22*	*22*	*24*	*20*	*24*	*23.*

Solution.

The bivariate table will have 25 cells (5×5).

Let the marks in Law be denoted by X (values 10 to 14 *i.e.*, 5 values).

Let the marks in Statistics be denoted by Y (values 20 to 24 *i.e.*, 5 values).

Bivariate Frequency Table Showing Marks in Law and Statistics

Marks in Law (X) → *Marks in Statistics (Y)* ↓	10	11	12	13	14	Total
20	| (1)		| (1)			**2**
21		|| (2)	| (1)			**3**
22	|| (2)	| (1)	| (1)		| (1)	**5**
23	|| (2)	| (1)	|| (2)		| (1)	**6**
24				||| (3)	| (1)	**4**
Total	**5**	**4**	**5**	**3**	**3**	**20**

Note : The figures within brackets show the frequency corresponding to each cell.

2.10 GRAPH OF A FREQUENCY DISTRIBUTION

The graphs of a frequency distribution is designed to present the characteristic features of a frequency data. They facilitate comparative study of two or more frequency distributions regarding their shape and pattern.

The most commonly used graphs are :

1. Histogram **2. Frequency Polygon**

3. Frequency Curve **4. Cumulative Frequency Curve or Ogive**

2.10.1 Histogram

A *Histogram* is a graph containing a set of rectangles, each being constructed to represent the size of the class interval by its width and the frequency in each class-interval by its height. The area of each rectangle is proportional to the frequency in the respective class-interval and the total area of the histogram is proportional to the total frequency. *A histogram is used to depict a frequency distribution.*

In constructing a histogram, the class-boundaries are considered to be very important; and they are located on the horizontal axis of the graph. The vertical rectangles are erected side by side on the basis of the frequencies over the class-intervals which are extended to their class-boundaries.

When the class-intervals are unequal, difficulty arises, because the rectangles so formed will also of unequal width. So in order to remove this difficulty, the heights of rectangles are made proportional not to the class frequencies, but to the frequency densities.

This type of diagram is used exclusively for showing frequency distributions of quantitative data that are continuous in nature. It is essentially an area diagram composed of a series of adjacent rectangles. Hence, the areas used to represent the frequency in groups or class intervals when added together will give the composite area for the entire group.

REMARKS :

1. It is not necessary that the scale on x-axis and y-axis be the same. **Different scales may be taken on the two axes.**
2. The position of origin on the y-axis is according to scale, which may not be so on the x-axis. If **class intervals are not starting from 0, then it is indicated by drawing a kink mark (-⋀⋀-) on the x-axis near the origin. If necessary the kink mark (-⋀⋀-) may be made on y-axis or on both the axes.**
3. A histogram may not be confused with a bar diagram. In a bar diagram, the height of each bar matters and not its width, whereas in a histogram height as well as the width of each rectangle matters.

2.10.2 Types of Histograms

TYPE I. HISTOGRAM WITH EQUAL CLASS INTERVALS

The sizes of class intervals are drawn on x-axis with equal distances and their respective frequencies on y-axis. Class and its frequency taken together form a rectangle. The graph of rectangles is known as histogram.

Each class has lower and upper values. This gives us two equal lines representing the frequencies. Upper ends of the lines are joined together. This process will give us rectangles. The heights of the rectangles will be proportional to their frequencies.

Example 9. *Draw the histogram for the following data :*

Marks	*0 – 10*	*10 – 20*	*20 – 30*	*30 – 40*	*40 – 50*	*50 – 60*	*60 – 70*
No. of Students	*20*	*30*	*70*	*50*	*40*	*50*	*40*

Solution. We represent the class limits along x-axis and frequencies along y-axis. Taking class intervals as bases and the corresponding frequencies (No. of students) as heights, we construct the rectangles to get the histogram of the given frequency distribution as shown in Fig. 1.1.

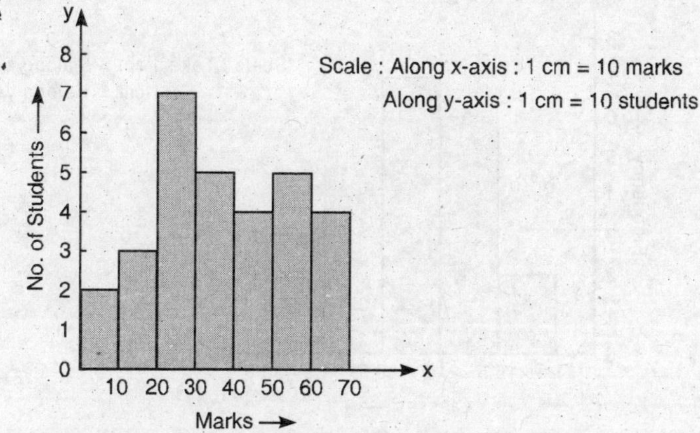

Fig. 2.1. Histogram showing marks obtained by students.

TYPE II. HISTOGRAM WHEN CLASS INTERVALS ARE GIVEN IN EXCLUSIVE FORM, *i.e.*, WHEN CLASS INTERVALS ARE NOT CONTINUOUS

In a histogram, it is necessary that the adjacent rectangles be attached to each other. If we try to represent the given data (in exclusive form) as such we shall get a bar diagram as there will be gaps in between the classes. **But in a histogram the bars are continuous without any gaps.** Therefore, we shall convert the **exclusive series** into an **inclusive series**, *i.e.*, make the classes continuous by taking class limits.

STEPS FOR CONVERTING AN EXCLUSIVE SERIES INTO AN INCLUSIVE SERIES

Step I. *Find the difference between the lower limit of second class interval and the upper limit of the first class interval. Let the difference be h.*

Step II. *Subtract h/2 from the lower limit of each and every class, i.e., find (lower limit – h/2) and add (h/2) to the upper limits of every class, i.e., find (upper limit + h/2).*

Step III. *Now write down the frequency distribution with continuous class intervals.*

Step IV. *Draw the histogram of the new distribution.*

The above method is illustrated by the following examples :

Example 10. *Draw the histogram of the following frequency distribution :*

Class Interval :	*12 – 16*	*17 – 21*	*22 – 26*	*27 – 31*	*32 – 36*
Frequency :	*2*	*6*	*7*	*5*	*3*

Solution. The given distribution is not continuous. If we represent the given data by a graph; we shall **not get a histogram**. But we will **get a bar diagram** as there will be gaps in between the classes, *i.e.*, we shall get isolated rectangles. But in a histogram the bars or rectangles are continuous without gaps. Therefore, we shall have to make the classes continuous by taking actual class limits.

Here : **Lower limit of 2nd class interval = 5; Upper limit of 1st class interval = 4.**

h **= difference between the lower limit of second class interval and upper limit of the first class interval = 5 – 4 = 1.**

To convert the given frequency distribution, we subtract $\frac{h}{2} = \frac{1}{2}$ **= 0.5 from each lower limit and add** $\frac{h}{2}$ **= 0.5 to each upper limit.** The new distribution obtained is given below :

Class Interval :	11.5 – 16.5	16.5 – 21.5	21.5 – 26.5	26.5 – 31.5	31.5 – 36.5
Frequency :	2	6	7	5	3

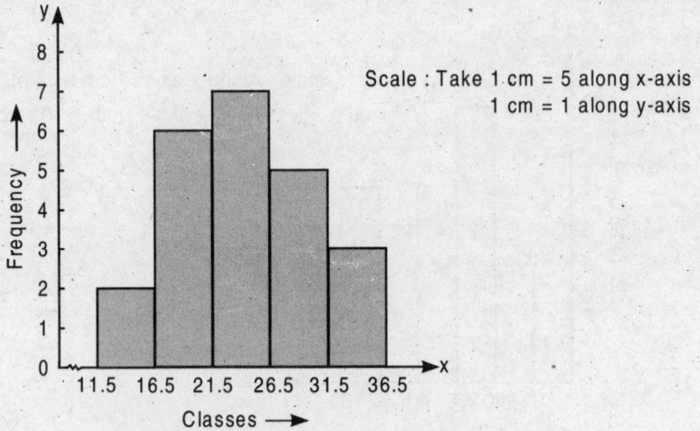

Fig. 2.2. Histogram of frequency distribution.

Take class intervals along x-axis. The kink mark (–ᴧᴧ–) on x-axis means that owing to lack of space the small distance represents 0 – 11.5. Take frequencies along y-axis.

Construct rectangles having base equal to class intervals and the corresponding frequencies as heights. Join these rectangles to get the histogram given by Fig. 1.2.

TYPE III. HISTOGRAM WHEN MID-POINTS OF CLASS INTERVALS ARE GIVEN

Working Rule

Step I. ***Find the difference between the second and first mid-point. Let it be h.***

Step II. ***To get the lower limit of first class, subtract (h/2) from the mid-points.***

To get the upper limit of first class, add (h/2) to the mid-point.

i.e., First Class Interval = (mid-point – h/2, mid-point + h/2).

Step III. ***Find all other class intervals by repeating the Step I and Step II.***

Now write the given data in the class intervals with frequency form and draw **histogram of this frequency distribution.**

Example 11. *Construct a histogram from the following distribution of total marks of 30 students of a class :*

Marks (mid-points) :	*100*	*120*	*140*	*160*	*180*	*200*
No. of Students :	*5*	*6*	*4*	*7*	*5*	*3*

Solution. We shall first convert the given data of mid-points into class interval form.

Here $\boldsymbol{h}$ = Difference between the second and first mid-point.

$$= 120 - 100 = \mathbf{20}. \quad \therefore \quad \frac{h}{2} = \frac{20}{2} = \mathbf{10}.$$

$$\therefore \textbf{ First class interval} = \left(100 - \frac{h}{2}\right) - \left(100 + \frac{h}{2}\right) = (100 - 10) - (100 + 10) \text{ i.e., } \mathbf{90 - 110}.$$

Using the above procedure, we get the other class intervals as under :

Marks :	**90 – 110**	**110 – 130**	**130 – 150**	**150 – 170**	**170 – 190**	**190 – 210**
No. of Students :	**5**	**6**	**4**	**6**	**5**	**4**

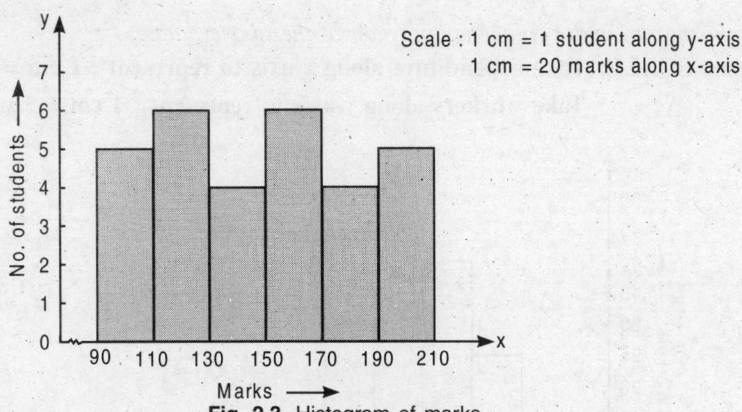

Fig. 2.3. Histogram of marks.

The histogram of the above frequency distribution is given by Fig. 1.3.

Note: The crank mark or kink (—ʌʌʌ—) in the curve on the horizontal axis means that owing to lack of space this small distance represents **0 to 90.**

2.12 FREQUENCY POLYGON

A frequency polygon is a line graph for the graphical representation of the frequency distribution. I*f we mark the mid-points of the top horizontal sides of the rectangles in a histogram and join them by straight lines, the figure so formed is called a frequency polygon.* It is assumed that the frequencies in a class interval are evenly distributed throughout the classes and consequently the mid-points are representative. A frequency polygon is useful in comparing two or more frequency distributions. It is called **a polygon as it consists of a number of lines as the sides of a polygon.** It has no special significance but it gives a fair idea about shape of frequency distribution. In order *to complete the drawing of a* ***frequency polygon****, the two end points of it are joined to the base line at the mid points of the empty classes at both ends of the frequency distribution.* A frequency polygon for a grouped frequency distribution with equal class intervals, may be constructed in two ways.

CASE 1. WHEN A HISTOGRAM AND A FREQUENCY POLYGON ARE DRAWN ON THE SAME SCALE

Working Rule

Step I. ***Draw a histogram from the given data.***

Step II. ***Join the consecutive mid-points of the upper sides (tops) of the adjacent rectangle of the histogram by means of line segments.***

Step III. ***It is assumed that the class preceding the first class and the class succeeding the last class in the classification exists and the frequency of each extreme class is zero. Join the mid-points of these extreme classes with the mid-points of upper horizontal sides of the extreme (first and last) rectangles of the histogram.***

Example 13. *The following is the distribution of total household expenditure (in Rs.) of 202 workers in a city.*

Expenditure (in Rs.)	*Number of Workers*	*Expenditure (in Rs.)*	*Number of Workers*
100 – 150	*25*	*300 – 350*	*30*
150 – 200	*40*	*350 – 400*	*22*
200 – 250	*33*	*400 – 450*	*16*
250 – 300	*28*	*450 – 500*	*8*

Draw a histogram and a frequency polygon of the above data.

Solution. SCALE. Take expenditure along *x*-axis to represent : 1 cm = Rs. 50

Take workers along *y*-axis to represent : 1 cm = 5 workers

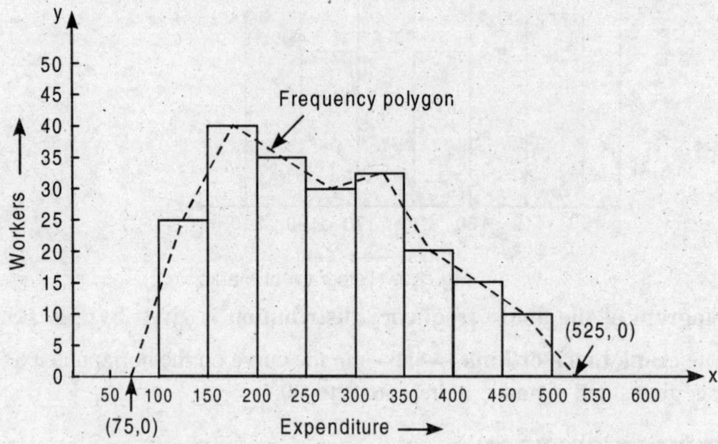

Fig. 2.4. Histogram and frequency polygon of wage earners.

Also the two **extreme classes are** 50 – 100 and 500 – 550 **whose frequencies are taken 'zero' in each case.**

The histogram and a frequency polygon of the given data are given by Fig. 1.4.

Example 12. *Draw, on the same diagram, a histogram and a frequency polygon to represent the following data which shows the monthly cost of living index of a city in a period of 2 years.*

Cost of living index	*Number of months*	*Cost of living index*	*Number of months*
440 – 460	*2*	*520 – 540*	*3*
460 – 480	*4*	*540 – 560*	*2*
480 – 500	*3*	*560 – 580*	*1*
500 – 520	*5*	*580 – 600*	*4*
Total			*24*

Solution. Take the following scale :

SCALE: 1 cm = 20 units along *x*-axis. 1 cm = 1 month along *y*-axis.

Also take the extreme classes as 420 – 440 and 600 – 620 with zero frequency in each case.

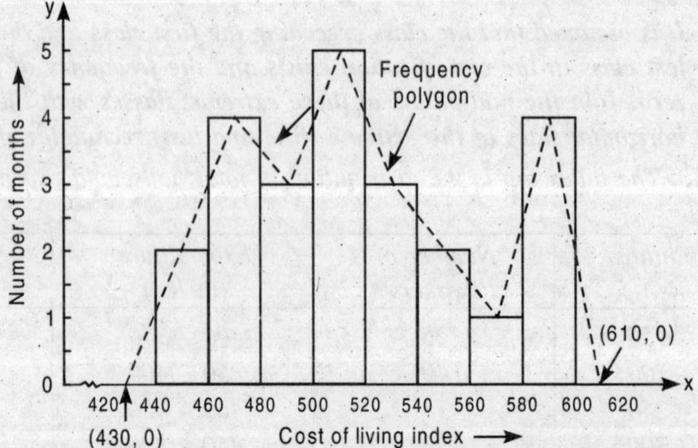

Fig. 2.5. Histogram and frequency polygon graphs.

CASE II. WHEN ONLY THE FREQUENCY POLYGON IS TO BE DRAWN.

Working Rule

Step I. ***Represent the mid-points of the class intervals along x-axis and the frequencies along the y-axis.***

Step II. ***Let x_i and f_i respectively be the mid-points and corresponding frequencies of the class intervals. Find all the points (x_i, f_i).***

Step III. ***Plot the point (x_i, f_i).***

Step IV. ***Join all these points (x_i, f_i) by straight lines taken in order to get a frequency polygon.***

Step V. ***The polygon obtained in Step IV is enclosed by joining the first and last point to the corresponding points on the x-axis representing the mid-values of the hypothetical classes, i.e., the class just before the first class or class just after the last class.***

Example 14. *The distribution of monthly wages (in Rs.) of 40 workers in a factory is given below:*

Monthly wages	*No of workers*	*Monthly wages*	*No. of workers*
210 – 230	*4*	*290 – 310*	*5*
230 – 250	*7*	*310 – 330*	*6*
250 – 270	*5*	*330 – 350*	*4*
270 – 290	*9*		

Draw a frequency polygon.

Solution. The two hypothetical intervals; one in the beginning and the other at the end with zero frequency are 190 – 210 and 350 – 370 respectively.

Step I. ***Represent monthly wages along x-axis with scale 1 cm = Rs. 20 and the number of workers along y-axis with scale 1 cm = 1 labour.***

Step II. ***We prepare the following frequency table:***

Monthly wages	*Class marks*	*No. of workers*
190 – 210	200	**0**
210 – 230	220	4
230 – 250	240	7
250 – 270	260	5
270 – 290	280	9
290 – 310	300	5
310 – 330	320	6
330 – 350	340	4
350 – 370	**360**	**0**

Step III. ***Plot the points (220, 4) (240, 7) (260, 5), (280, 9), (300, 5), (320, 6) and (340, 4). Also plot the two hypothetical points (200, 0) and (360, 0).***

Step IV. ***Join all these points to get the required frequency polygon.***

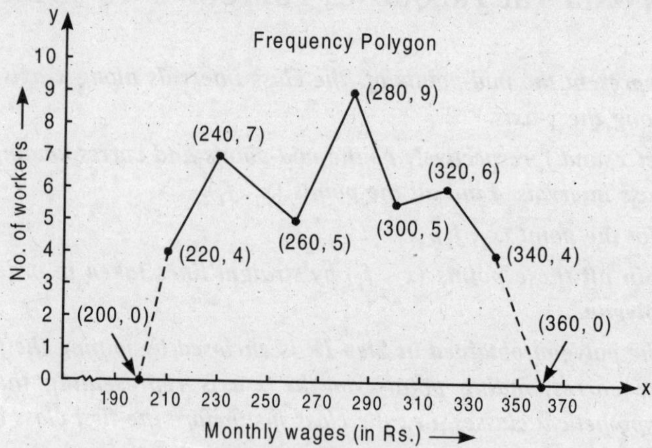

Fig. 2.6. Frequency polygon of monthly wages of workers.

The frequency polygon of the given data is given by Figure 1.6.

2.12 FREQUENCY CURVE

A *frequency curve is drawn by smoothing the frequency polygon. It is smoothed in such a way that the sharp turns are avoided.*

The curve is drawn free hand in such a way that the area included under the curve is approximately the same as that of the frequency polygon.

A frequency polygon, if smoothed further, so as to minimise sudden changes, results into a continuous smooth curve known as **frequency curve or smooth frequency curve.** The curve should begin and end at the base line. ***To draw the frequency curve, firstly draw a histogram, then draw a frequency polygon and lastly smooth it to get frequency curve.*** The basic object of drawing a frequency curve is to present graphically, the area covered by histogram in a more symmetrical manner. It is also known as **smooth frequency curve.**

Example 15. *Draw a histogram, a frequency polygon and a smoothed frequency curve from the following data relating to marks secured in Economics by students of a school in a house-test.*

Marks	*No. of students*	*Marks*	*No. of students*
0 – 5	*3*	*25 – 30*	*14*
5 – 10	*7*	*30 – 35*	*10*
10 – 15	*13*	*35 – 40*	*7*
15 – 20	*25*	*40 – 45*	*4*
20 – 25	*40*	*45 – 50*	*2*

Solution. The required curves are shown in Fig 1.7.

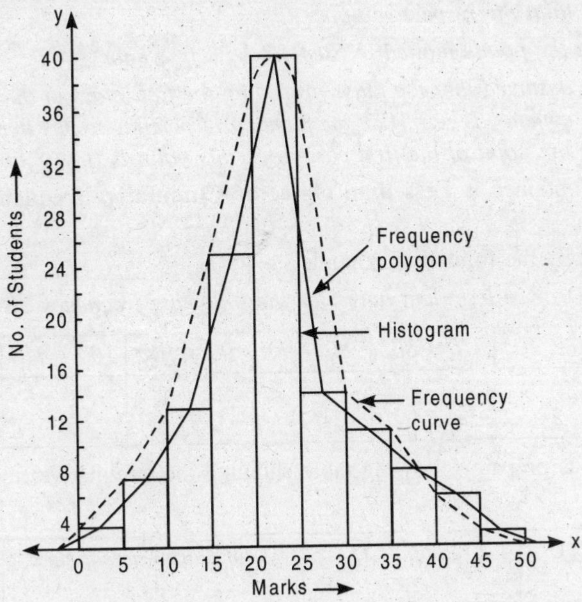

Fig. 2.7. Histogram, frequency polygon and smoothed frequency curve.

2.13 CUMULATIVE FREQUENCY CURVE OR OGIVE

It is a graph which represent the data of a cumulative frequency distribution. An ***Ogive, when it is drawn, is in the form of a curve. The curve drawn from the data cumulated downward is known as less than ogive and the curve drawn from the data cumulated upward as more than ogive. An ogive is used to find median, quartites, deciles and percentiles etc.*** It is also used to find the number of observations which are expected to lie between two given values.

A cumulative frequency curve or ogive is obtained by plotting the successive cumulative frequencies on the graph paper and joining them by a free hand. A cumulative frequency curve or ogive represents the cumulative frequency distribution by plotting actual upper limits of the class intervals on the x-axis and the representing a cumulative frequency of these class intervals on the y-axis.

In drawing an ogive, the cumulative frequency is plotted at the upper limit of the class interval. The successive points are later joined together to get an ogive curve. There are two types of ogives:

(*i*) **Less Than Ogive** and

(*ii*) **More Than Ogive**.

In drawing an ogive, ***if the cumulative frequencies are plotted at the upper limit of the class interval, if it is a less than ogive.***

But in the case of ***more than ogive, the cumulative frequencies are plotted against the lower class boundaries of the respective class intervals.***

Less Than Ogive or

Cumulative Frequency Curve Less than the upper limit

Working Rule

Step I. *Represent the upper limits of the classes along x-axis.*

Step II. *Represent the cumulative frequency of respective frequency along y-axis.*

Step III. *Plot the points corresponding to upper class limits and the cumulative frequencies less than the respective upper limits.*

Step IV. *Join the points plotted in* **Step III** *by a free hand.*

Step V. *It is assumed that the class preceding the first class in the classification exists and its frequency is zero. Plot the point corresponding to this* **hypothetical point** *and join it to the point of the first class, i.e., this point is (lower limit of the first class, 0).*

The curve so obtained is **Less than Ogive or Cumulative frequency curve less than the upper class limit.**

It is illustrated by the following example.

Example 16. *Draw a less than ogive for the following frequency distribution:*

I.Q. :	*60 – 70*	*70 – 80*	*80 – 90*	*90 – 100*	*100 – 110*	*110 – 120*	*120 – 130*
No. of students :	*2*	*5*	*12*	*31*	*39*	*10*	*4*

Solution. Let us prepare following table showing the cumulative frequencies more than the upper limit.

Class Interval (I.Q.)	*Frequency (f)*	*Cumulative Frequency (c.f.)*	*Plots to be plotted*
50 – 60	**0**	**0**	**(60, 0)**
60 – 70	2	2	**(70, 2)**
70 – 80	5	2 + 5 = **7**	**(80, 7)**
80 – 90	12	7 + 12 = **19**	**(90, 19)**
90 – 100	31	19 + 31 = **50**	**(100, 50)**
100 – 110	39	50 + 39 = **89**	**(110, 89)**
110 – 120	10	89 + 10 = **99**	**(120, 99)**
120 – 130	4	99 + 4 = **103**	**(30, 103)**

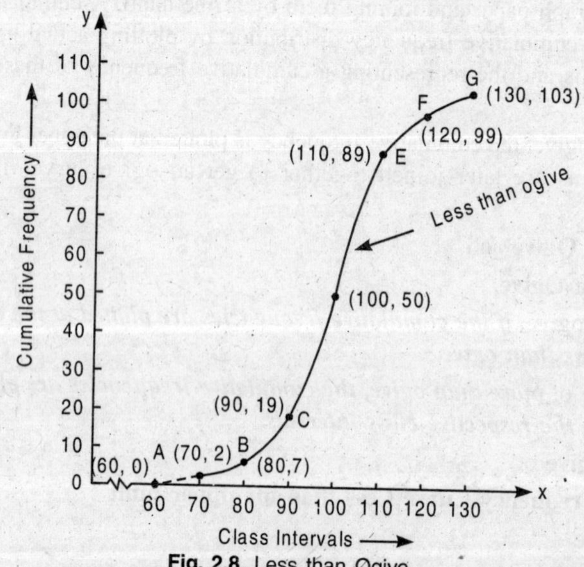

Fig. 2.8. Less than Ogive.

From this table, we get the pairs of numbers corresponding to **the upper class limits** and their corresponding cumulative frequencies as **(70, 2), (80, 7), (90, 19), (100, 50), (110, 89), (120, 99) and (130, 103).** Plot these points and join them by a free hand curve to get the ogive as shown in the Fig. 1.8. To complete it, **we join the curve to the hypothetical point (Lower limit of first interval, 0)**, *i.e.* **(60, 0)** ***(by means of broken lines).***

Example 17. *Draw the ogive for the following distribution:*

Age (in years) :	*0 – 9*	*10 – 19*	*20 – 29*	*30 – 39*	*40 – 49*	*50 – 59*	*60 – 69*
No. of persons :	*5*	*15*	*20*	*25*	*15*	*12*	*8*

Solution. The given **frequency distribution is not continuous.** We first **make it continuous** and prepare the cumulative frequency distribution given by the following table.

Age (in years)	*Actual limits*	*No. of persons (Frequency)*	*Cumulative Frequency*	*Points to be plotted*
0 – 9	– 0.5 – 9.5	5	5	**(9.5, 5)**
10 – 19	9.5 – 19.5	15	20	**(19.5, 20)**
20 – 29	19.5 – 29.5	20	40	**(29.5, 40)**
30 – 39	29.5 – 39.5	25	65	**(39.5, 65)**
40 – 49	39.5 - 49.5	15	80	**(49.5, 80)**
50 – 59	49.5 – 59.5	12	92	**(59.5, 92)**
60 – 69	59.5 – 69.5	8	100	**(69.5, 100)**

Plot the points **(9.5, 5), (19.5, 20), (29.5, 40), (39.5, 65), (49.5, 80), (59.5, 92), and (69.5, 100)** and joined them by a free hand smooth curve. To complete it, **the curve is joined to the hypothetical point (lower limit of the first class interval, 0), *i.e.*, the point (– 0.5, 0). This portion is always shown by dotted line.**

Class	*Frequency*	*Cumulative frequency (More than the lower limits)*	*Points to be plotted*
40 – 44	7	128 + 7 = **135**	(40, 135)
44 – 48	12	116 + 12 = **128**	(44, 128)
48 – 52	33	83 + 33 = **116**	(48, 116)
52 – 56	47	36 + 47 = **83**	(52, 83)
56 – 60	20	16 + 20 = **36**	(56, 36)
60 – 64	11	5 + 11 = **16**	(60, 16)
64 – 68	5	5 = **5**	(64, 5)

From the table we get the points of numbers corresponding to the lower class limits and the respective cumulative frequency, as : **(40, 135), (44, 128), (48, 116) (52, 83), (56, 36), (60, 16) and (64, 5).**

Plot these points and join them by a free hand curve to get the **More Than Ogive** as shown in the Fig. 1.9. In order to complete it, we join the curve to the **hypothetical point (upper limit of last interval, 0),** *i.e.,* **(68, 0) (shown by dotted lines).**

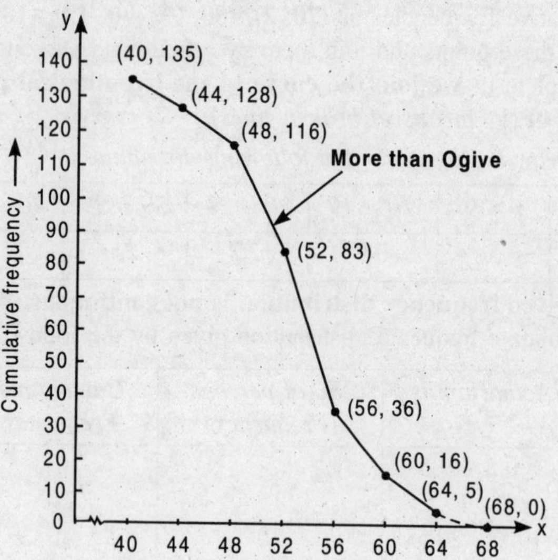

Fig. 2.9. More than Ogive.

The More Than Ogive of the given data is given by Fig. 2.9.

Example 18. *Plot Less than Ogive and More than Ogive for the following data:*

Cost of production	*4 – 6*	*6 – 8*	*8 – 10*	*10 – 12*	*12 – 14*	*14 – 16*
No. of farms	*13*	*111*	*182*	*105*	*19*	*7*

Solution. We prepare a cumulative frequency distribution table showing cumulative frequency 'less than Ogive' and 'more than Ogive' as shown below:

Class interval (Cost of production)	*Class boundary*	*Frequency (No. of farms)*	*Cumulative Frequency*	
			(Less than Ogive)	*(More than Ogive)*
2 – 4	**4**	**0**	**0**	**437**
4 – 6	6	13	13	424
6 – 8	8	111	124	313
8 – 10	10	182	306	131
10 – 12	12	105	411	26
12 – 14	14	19	430	7
14 – 16	16	7	437	0

By plotting the points (**4, 0**), (6, 13), (8, 124), (10, 306), (12, 411), (14, 430), (16, 437) and joining them by a free hand, we get the **less than Ogive**. Again by plotting the points (16, 0), (14, 7), (12, 26), (10, 131), (8, 313), (6, 424), (3, 37) and (**16, 0**) then by a free hand, we get the **more than Ogive**. Both the curves are given by Figure 2.10.

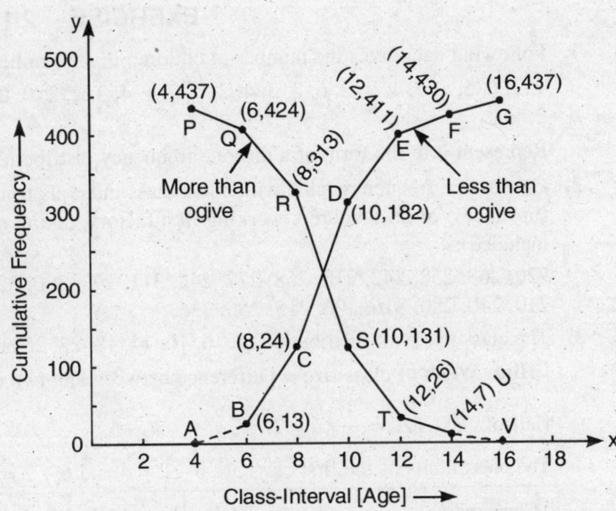

Fig. 2.10. Less than Ogive and More than Ogive.

Example 19. *Convert the following distribution into more than frequency distribution.*

Weekly Wages less than (Rs.) :	*20*	*40*	*60*	*80*	*100*
No. of workers :	*41*	*92*	*156*	*194*	*201*

For the data given above, draw 'less than' and 'more than' ogive

Solution. **Table: Computation of less than' and more than Ogive**

Weekly Wages	*N. of Workers(f)*	*Cumulative frequencies*	
		Less than	*More than*
0 – 20	41	41	201
20 – 40	92 – 41 = 51	92	160
40 – 60	156 – 92 = 64	156	109
60 – 80	194 – 156 = 38	194	45
80 – 100	201 – 194 = 7	201	7

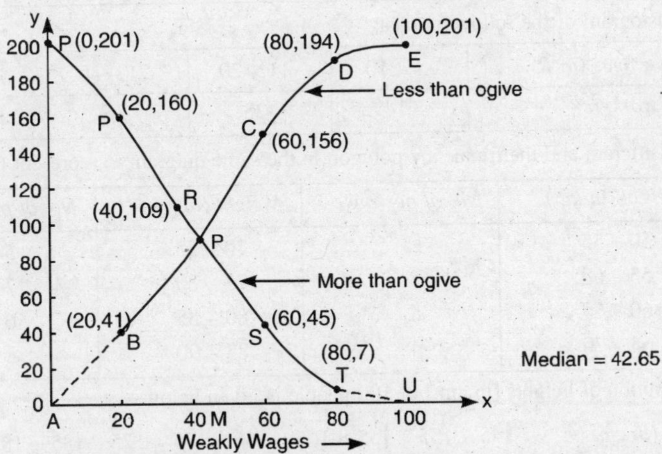

Fig. 2.11 'More than' and Less than' Ogive.

EXERCISE – 2.1

1. Following data gives the number of children in 40 families :

1, 2, 6, 5, 1, 5, 1, 3, 2, 6, 2, 3, 4, 2, 0, 0, 4, 4, 3, 2, 2, 0, 0, 1, 2, 2, 4, 3, 2, 10, 5, 1, 2, 4, 3, 4, 1, 6, 2, 2.

Represent it in the form of a discrete frequency distribution.

2. Construct a frequency table with equal class intervals from the following data on the monthly wages (in rupees) of 28 labourers working in a factory, taking one of the class intervals as 210– 230 (230 included):

220, 268, 258, 242, 210, 268, 272, 242, 311, 290, 300, 320, 319, 304, 302, 318, 306, 292, 254, 278, 210, 240, 280, 316, 306, 215, 256, 236.

3. The class marks of distribution are: 6, 10, 14, 18, 22, 26 and 30. Find the class size and class interval.

[**Hint.** Width of class size = Difference between any two class marks, *i.e.*, 10 – 6 = 4.]

Half of class size = $\frac{1}{2} \times 4 = 2.$

The class limits of the first class are 6 – 2 = 4 and 6 + 2 = 8.

Class mark :	6	10	14	18	22	26	30
Class Interval :	4 – 8	8 – 12	12 – 16	16 – 20	20 – 24	24 – 28	28 – 32

4. Following are the prices of the shoes of a particular company in a stock market recorded for 30 days. Tabulate the data in the form of a frequency distribution showing frequencies of days against interval of prices (in Rs.)

256, 221, 232, 241, 250, 246, 232, 240, 249, 253, 229, 236, 239, 234, 238, 246, 249, 255, 250, 251 253, 251, 249, 248, 247, 231, 239, 243, 229, 232.

Also form the cumulative frequency distribution table.

[**Hint.**

Table: Frequency Distribution

Class Interval	*Tally Marks*	*Frequency*	*Cumulative Frequency*
220-230	|||	3	3
230-240	~~||||~~ ~~||||~~	10	13
240-250	~~||||~~ ~~||||~~ |	11	24
250-260	~~||||~~ |	6	30

5. Draw a histogram of the following data:

Weekly wages (in Rs.) :	1 – 10	11 – 20	21 – 30	31 – 40	41 – 50
No. of workers :	14	28	36	12	10

6. Draw a histogram and the frequency polygon in the same diagram to represent the following data:

Weight (in kg.)	*No. of persons*	*Weight (in kg.)*	*No. of persons*
50 – 55	12	70 – 75	5
55 – 60	8	75 – 80	7
60 – 65	5	80 – 85	6
65 – 70	4	85 – 90	3

7. The distribution of heights (in cm.) of 100 people is given below:

Height (in cm) :	145 – 155	155 – 165	165 – 175	175 – 185	185 – 195	195 – 205
Frequency :	3	35	25	15	20	2

Construct a histogram and polygon in the same diagram.

8. Draw histogram of the following distribution :

Life of electric lamps (in hours) mid-value	2020	2060	2100	2140	2180
Firm A	10	130	480	340	20
Firm B	300	100	80	200	500

9. The following is the age distribution of 542 Members of Parliament. On the basis of this information construct a histogram and convert it into a frequency polygon.

Age	*No. of Members*	*Age*	*No. of Members*
20 – 30	3	60 – 70	140
30 – 40	61	60 – 70	51
40 – 50	132	70 – 80	2
50 – 60	153		

10. What is an 'Ogive'? Point out its advantages over an ordinary frequency curve, particularly for the purposes of interpolation. Represent the following data by an ogive curve and obtain the percentage of holdings with size equal to or less than 7 acres.

Size groups in acres	*Percentage of holdings*	*Size groups in acres*	*Percentage of holdings*
Up to 1 acre	7.6	5.1 to 10 acres	25.6
1.1 to 2.5 acres	18.0	10.1 to 25 acres	15.2
2.6 to 5 acres	29.3	Above 25 acres	4.3

11. The following table gives the marks of students in statistics:

Marks :	0-10	10-20	20-30	30-40	40-50
No. of students :	5	10	10	20	5

Draw an Ogive of less than type and find out the percentage of students with marks between 20 and 40.

12. Draw less than and more than Ogives from the data given below:

Profits (Rs. Lakhs)	*No. of Cos.*	*Profits (Rs. Lakhs)*	*No. of Cos.*
10 – 20	6	60 – 70	16
20 – 30	8	70 – 80	8
30 – 40	12	80 – 90	5
40 – 50	18	90 – 100	2
50 – 60	25		

13. The following table gives the distribution monthly income of 600 middle class families in a certain city:

Monthly income (in Rs.)	*Frequency*	*Monthly income (in Rs.)*	*Frequency*
Below 75	69	300 – 375	58
76 – 150	167	375 – 450	25
150 – 225	207	450 and over	10
225 – 300	65		

Draw an Ogive for the above data. Obtain the limits of income of central 50% of the observed families.

14. Draw a 'less than' and 'more than' Ogive curve for the following data and find median value:

No. of Children :	0	1	2	3	4	5	6
No. of Families :	150	72	50	28	12	8	5

3
CHAPTER

MEASURES OF CENTRAL TENDENCY OR AVERAGES

3.1 MEASURES OF CENTRAL TENDENCY OR LOCATION

A single expression, representing the whole group, is selected which may convey a fairly adequate idea about the whole group. This single expression in statistics is known as the **average.** Averages are, generally, **the central part of the distribution and, therefore, they are also called the measures of central tendency.**

The different measures of central tendency are shown below:

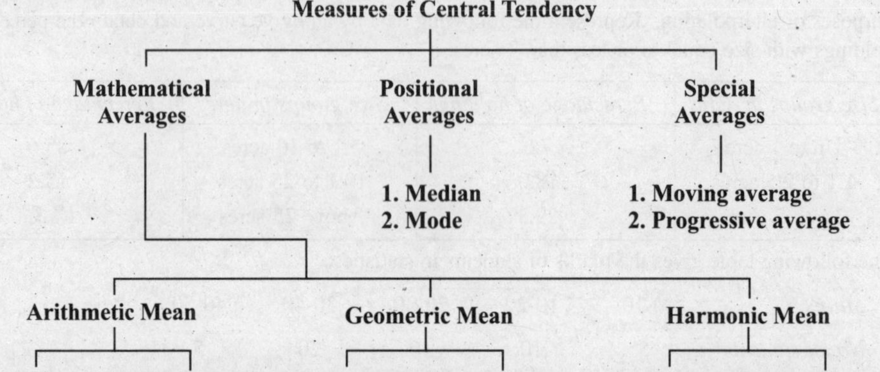

3.2 FEATURES OF A GOOD AVERAGE

According to **Professor G.U. Yule,** a good average must have the following characteristics:

1. *It should be rigidly defined so that different persons may not interpret it differently.*
2. *It should be easy to understand and easy to calculate.*
3. *It should be based on all the observations of the data.*
4. *It should be easily subjected to further mathematical calculations.*
5. *It should be least affected by the fluctuations of the sampling.*
6. *It should not be unduly affected by the extreme values.*
7. *It should be easy to interpret.*
8. *It should have sampling stability. It means that if the average is computed for similar groups, the result should also be similar.*

3.3 ARITHMETIC MEAN

3.3.1 Arithmetic Mean of Raw Data

We know that in an ungrouped or raw data, we are given individual items. We also know that the average of n numbers is obtained by finding their sum (by adding) and then dividing it by n. **Let $x_1, x_2, ..., x_n$ be n numbers, then their average or arithmetic mean is given by**

$$\bar{x} = \frac{x_1 + x_2 + x_3 + ... + x_n}{n} = \frac{\sum_{i=1}^{n} x_i}{n}.$$

Example 1. *Find the arithmetic mean of the marks obtained by 10 students of class X in Mathematics in a certain examination. The marks obtained are:*

25, 30, 21, 55, 47, 10, 15, 17, 45, 35.

Solution. Let $\bar{x}$ be the average marks.

∴ Sum of all the observations = 25 + 30 + 21 + 55 + 47 + 10 + 15 + 17 + 45 + 35 = **300.**

Number of students = **10.**

$$\therefore \quad \textbf{Arithmetic mean} = \frac{\Sigma x_i}{n} = \frac{300}{10} = \mathbf{30.}$$

3.4 PROPERTIES OF ARITHMETIC MEAN

Property 1. *If the mean of n observations $x_1, x_2, \ldots\ldots, x_n$ is $\bar{x}$, then*

$$(x_1 - \bar{x}) + (x_2 - \bar{x}) + \ldots\ldots + (x_n - \bar{x}) = 0. \quad \textit{i.e.,} \sum_{i=1}^{n} (x - \bar{x}) = 0$$

Proof. $(x_1 - \bar{x}) + (x_2 - \bar{x}) + \ldots\ldots + (x_n - \bar{x}) = (x_1 + x_2 + \ldots\ldots + x_n) - n\bar{x} = n\bar{x} - n\bar{x} = \mathbf{0}.$

Thus, the algebraic sum of deviations from mean is zero.

Property II. *If the mean of n observations, $x_1, x_2, \ldots\ldots, x_n$ is $\bar{x}$, then the mean of the observations $(x_1 + a), (x_2 + a), (x_3 + a), \ldots\ldots, (x_n + a)$, is $\bar{x} + a$.*

i.e., **if each observation is increased by '*a*', then the mean is also increased by *a*.**

Property III. *If the mean of n observations $x_1, x_2, x_3, \ldots\ldots, x_n$ is $\bar{x}$, then the mean of the observations $(x_1 - a), (x_2 - a), (x_3 - a), \ldots\ldots, (x_n - a)$ is $\bar{x} - a$.*

i.e., **if each observation is decreased by '*a*', then the new mean is also decreased by *a*.**

Property IV. *The mean of n observations $x_1, x_2, x_3, \ldots. x_n$ is $\bar{x}$. If each observation is multiplied by p, $p \neq 0$, then the mean of new observations is $\mathbf{p\bar{x}}$.*

Property V. *The mean of n observations $x_1, x_2, x_3, \ldots\ldots, x_n$ is $\bar{x}$. If each observation is divided by p, then the mean of new observations is $\frac{\bar{x}}{p}$.*

3.5 MERITS, DEMERITS AND USES OF ARITHMETIC MEAN

Merits

1. *It can be easily calculated.*
2. *Its calculations are based on all the observations.*
3. *It is easy to understand.*
4. *It is rigidly defined by the mathematical formula.*
5. *It is least affected by sampling fluctuations.*
6. *It is the best measure to compare two or more series (data).*
7. *It is the average obtained by calculations and it does not depend upon any position.*

Demerits

1. *It may not be represented in actual data and so it is theoretical.*
2. *The extreme values have greater effect on mean.*
3. *It cannot be calculated if all the values are not known.*
4. *It cannot be determined for the qualitative data such as love, beauty, honesty, etc.*
5. *Mean may lead to fallacious conclusions in the absence of original observations.*

Uses of Arithmetic Mean

1. *A common man uses mean for calculating average marks obtained by a student.*
2. *It is extensively used in practical statistics.*
3. *Estimates are always obtained by mean.*
4. *Businessman uses it to find out the operation cost, profit per unit of article, output per man and per machine, average monthly income and expenditure, etc.*

Example 2. *The mean of 68 numbers is 18. If each number is divided by 6, find the new mean. Find the relation between the new mean and old mean.*

Solution. Mean of 68 numbers = 18. $\therefore$ Sum of 68 numbers = 68 × 18.

If each number is divided by 6, the sum is also divided by 6.

$$\therefore \text{ New sum} = \frac{68 \times 18}{6}$$

$$\textbf{New Mean} = \frac{68 \times 18}{6 \times 68} = \frac{18}{6} = \mathbf{3}. \qquad \therefore \quad \textbf{New Mean} = \frac{\textbf{Original mean}}{\mathbf{6}}.$$

Example 3. *The mean weight of 150 students in a class is 60 kg. The mean weight of the boys is 70 kg while that of the girls is 55 kg. Find the number of boys and that of the girls in the class.*

Solution. Let the number of boys be x. Then, the number of girls = $(150 - x)$.

$\therefore$ **Total weight of 150 students = (total weight of x boys) + [total weight of $(150 - x)$ girls]**

= 150 × 60 = $70x + 55 \times (150 - x)$. This gives, $15x = 750$ **or $x = 50$.**

$\therefore$ **The number of boys = 50 and the number of girls =** (150 – 50) **= 100.**

Example 4. *In a school 85 boys and 35 girls appeared in a public examination. The mean marks of boys was found to be 40 %, whereas the mean marks of girls was 60 %. Determine the average marks percentage of the school.*

Solution. Total score of 85 boys = 85 × 40 = **3400** **[Total = Number × Mean]**

Total score of 35 girls = 35 × 60 = 2100.

Total score of 120 students (85 boys and 35 girls) = 3400 + 2100 = **5400.**

$$\textbf{Average marks percentage of the school} = \frac{\mathbf{5400}}{\mathbf{120}} = \mathbf{45\%}.$$

3.6 METHODS TO CALCULATE ARITHMETIC MEAN

3.6.1 Arithmetic Mean of an Ungrouped Data

Direct Method. If n observations in the raw data consist of n distinct values denoted by x_1, x_2, x_3, ..., x_n of the observed variable x occurring with frequencies f_1, f_2, f_3, ..., f_n respectively, then the arithmetic mean of the variate x is given by $\bar{x}$, where

$$\bar{x} = \frac{f_1x_1 + f_2x_2 + f_3x_3 + ... + f_nx_n}{f_1 + f_2 + f_3 + ... + f_n} = \frac{\sum_{i=1}^{n} f_ix_i}{\Sigma f_i} = \frac{\sum_{i=1}^{n} f_ix_i}{N}$$

where $$N \doteq \sum_{i=1}^{n} f_i = f_1 + f_2 + ... + f_n = \textbf{Sum of frequencies}$$

Example 5. *Find the Arithmetic Mean from the following frequency table:*

Marks	52	58	60	65	68	70	75
No. of Students	7	5	4	6	3	3	2

Solution. Let x be the marks and f be the frequency, so that we have the following table:

Computation Table

x :	52	58	60	65	68	70	75	**Total**
f :	7	5	4	6	3	3	2	**30**
$f \times x$:	360	290	240	390	204	210	150	**1848**

Here $$N = \Sigma f = 30 \quad \text{and} \quad \Sigma fx = 1848.$$

$$\therefore \textbf{ Mean : } \bar{x} = \frac{\Sigma fx}{N} = \frac{1848}{30} = \frac{616}{10} = 61.6.$$

3.6.2 Concealed Frequency

Example 6. *The following table gives the distribution of 100 accidents in New Delhi during seven days of a week of a given month. During that month there were 5 Mondays, 5 Tuesdays and 5 Wednesdays and only four each of the other days. Calculate the number of accidents per day.*

Day:	*Sunday*	*Monday*	*Tuesday*	*Wednesday*	*Thursday*	*Friday*	*Saturday*
No. of Accidents:	*26*	*16*	*12*	*10*	*8*	*10*	*18*

Solution. Let us prepare the following table.

Table : Computation of Mean

Day	*No. of Accidents* (x)	*Number* (f)	$f \times x$
Sunday	26	4	104
Monday	16	5	80
Tuesday	12	5	60
Wednesday	10	5	50
Thursday	8	4	32
Friday	10	4	40
Saturday	18	4	72
		$\Sigma f = 31$	$\Sigma fx = 438$

Now **Mean:** $$\bar{x} = \frac{\Sigma fx}{\Sigma f} = \frac{438}{31} = 14.13.$$

3.6.3 Arithmetic Mean of a Grouped Data or Continuous Series

Direct Method. In this case the raw data is presented in the form of a **Frequency Distribution with Class Intervals.** The arithmetic mean is calculated by the **Direct Method.** The **arithmetic mean** is defined as:

Mean: $\bar{x} = \dfrac{\sum_{i=1}^{k} f_i x_i}{n}$, **where** $n = \sum_{i=1}^{k} f_i$,

and x_i = **class mark** (or **mid-value or mid-point**) **of the** ***i*th class interval** and is given by:

$$x_i = \frac{\textbf{lower limit of } i\textbf{ th class interval} + \textbf{upper limit of } i\textbf{ th class interval}}{2}$$

Example 7. *The data on number of patients attending a hospital in a month are given below. Find the average number of patients attending the hospital in a day.*

Number of patients	0 – 10	10 – 20	20 – 30	30 – 40	40 – 50	50 – 60
Number days attending the hospital	2	6	9	7	4	2

Solution.

TABLE: CALCULATION OF MEAN

No. of patients (*Class Interval*)	*Frequency* f_i	*Mid-value* x_i	$f_i \times x_i$
0 – 10	2	5	2 × 5 = 10
10 – 20	6	15	6 × 15 = 90
20 – 30	9	25	9 × 25 = 225
30 – 40	7	35	7 × 35 = 245
40 – 50	4	45	4 × 45 = 180
50 – 60	2	55	2 × 54 = 110
Total	$\Sigma f_i = 30$		$\Sigma f_i \times x_i = 860$

Average = $\bar{x} = \dfrac{\Sigma f_i x_i}{\Sigma f_i} = \dfrac{860}{30}$ = **28.67 ≈ 29 patients per day.**

3.6.4 Arithmetic Mean by Short-Cut Method

CASE I. Short-cut Method for Ungrouped Data

For the mean of ungrouped data, the short-cut method is applied when the frequencies and the values of the variables are quite large and it becomes very difficult to compute the arithmetic mean as in the case of above example. In a frequency table of such a type, the provisional mean '*a*' is taken as that value of *x* (mid-value of the class interval) which corresponds to the highest frequency or which comes near the middle value of the frequency distribution. This number is called the **Provisional Mean** or **Assumed Mean.** Also find the deviations of the variates from this provisional mean. Then the arithmetic mean is given by the formula.

(*i*) In the case of **ungrouped data**

$\bar{x} = a + \dfrac{\Sigma d}{n}$, where **$a$ = assumed mean, n = number of items,**

$d = x - a$ = deviations of the variate 'x' from a.

CASE II. Short-cut Method for Grouped Data when Deviations are Taken from Assumed Mean

In the case of **grouped data**, the mean $\bar{x}$ is given by

$\bar{x} = a + \dfrac{\Sigma fd}{N}$, where **$a$ = assumed mean; $d = x - A$;**

fd = product of the frequency and the corresponding deviation,

$N = \Sigma f$ = the sum of all the frequencies.

Example 8. *Ten coins were tossed together and the number of the resulting from them were observed. The operation was performed 1050 times and the frequencies thus obtained for different number of tails (x) are shown in the following table. Calculate the arithmetic mean by the short-cut method.*

x :	*0*	*1*	*2*	*3*	*4*	*5*	*6*	*7*	*8*	*9*	*10*
f :	*2*	*8*	*43*	*133*	*207*	*260*	*213*	*120*	*54*	*9*	*1*

Solution. Let **5** be the **assumed mean,** *i.e.,* $a = 5$. Let us prepare the following table in order to calculate the arithmetic mean:

x	f	$d = x - a = x - 5$	fd
0	2	– 5	– 10
1	8	– 4	– 32
2	43	– 3	– 129
3	133	– 2	– 266
4	207	– 1	– 207
5	260	0	0
6	213	1	+ 213
7	120	2	+ 240
8	54	3	+ 162
9	9	4	+ 36
10	1	5	+ 5
		$\Sigma f = 1050$	$\Sigma fd = 12$

$\therefore$ **Arithmetic Mean:** $\bar{x} = a + \dfrac{\Sigma fd}{\Sigma f} = 5 + \dfrac{12}{1050} = 5 + 0.0114 = \mathbf{5.0114.}$

Example 9. *For the following frequency table, find the mean.*

Class :	*100 – 120*	*120 – 140*	*140 – 160*	*160 – 180*	*180 – 200*	*200 – 220*	*220 – 240*
Frequency :	*10*	*8*	*4*	*4*	*3*	*1*	*2.*

Solution. Let $a = 170$, then we have the following table:

Table: Computation of Mean

Class	*Frequency* (f)	*Mid-value* (x)	$d = x - a$ $= x - 170$	fd
100 – 120	10	110	– 60	– 600
120 – 140	8	130	– 40	– 320
140 – 160	4	150	– 20	– 80
160 – 180	4	170	0	0
180 – 200	3	190	20	60
200 – 220	1	210	40	40
220 – 240	2	230	60	120
	$\Sigma f = 32$			$\Sigma fd = -780$

Now **Mean** $= a + \dfrac{\Sigma fd}{\Sigma f} = 170 - \dfrac{780}{32} = 170 - 24.375 = \mathbf{145.625.}$

3.6.5 Arithmetic Mean by Step-Deviation Method for Continuous Series

When the class intervals in a grouped data are equal, then the calculation can be simplified further by taking out the common factor from the deviations. This common factor is equal to the width of the class-interval. In such cases, the deviation of variate x from the assumed mean 'a' (*i.e.*, $d = x - a$) are divided by the common factor. The arithmetic mean is then obtained by the following formula:

$$\bar{x} = a + \frac{\Sigma fd}{\Sigma f} \times i,$$

where **a = assumed mean or provisional mean,**

$d = \frac{x-a}{i}$ = the deviation of the variate x from a,

i = the width of the class-interval ; N = the number of observations.

Example 10. *Calculate,* ***by step deviation method,*** *the arithmetic mean of the following marks obtained by students in English.*

Marks :	5	10	15	20	25	30	35	40	45	50
No. of Students :	20	43	75	67	72	45	39	9	8	6.

Solution. Let **a = 30 and i = 5.**

Table: Computation of Mean

x	f	$dx = x - 30$	$d'x = dx/i$	$fd'x$
5	20	– 25	– 5	– 100
10	43	– 20	– 4	– 172
15	75	– 15	– 3	– 225
20	67	– 10	– 2	– 134
25	72	–5	– 1	– 72
30	45	0	0	0
35	39	5	1	39
40	9	10	2	18
45	8	15	3	24
50	6	20	4	24
	$\Sigma f = 384$			**$\Sigma fd'x = -598$**

Now $$\bar{x} = a + \frac{\Sigma fd'x}{\Sigma f} \times i$$

$$= 30 - \frac{598}{384} \times 5 = 30 - 1.56 \times 5 = 30 - 7.80 = \mathbf{22.2.}$$

Example 11. *In a study on patients, the following data were obtained. Find the arithmetic mean.*

Age (in years) :	*10 – 19*	*20 – 29*	*30 – 39*	*40 – 49*	*50 – 59*	*60 – 69*	*70 – 79*	*80 – 89*
No. of cases :	*1*	*0*	*1*	*10*	*17*	*38*	*9*	*3*

Solution. The data is presented in the form of an inclusive series. We have to **transform, the inclusive series into exclusive series.** It can be transformed as follows:

We find the distance between the lower limit of the second class interval and the upper limit of the first class-interval. This is equal to 20 – 19 = 1. We subtract $\frac{1}{2}$ of this distance (*i.e.*, 0.5) from the lower limit and add it to the upper limit. The new classes will be formed as follows:

The new data is 10 – 0.5 = 9.5; 19 + 0.5 = 19.5 and so on.

Let $a = 44.5$, and $d = (x - a)/i = (x - 44.5)/10$, where $i = 10$.

Table: Computation of Mean

Age (in years)	*No. of cases (f)*	*Mid-value (x)*	$d = \frac{x - 44.5}{10}$	*fa*
9.5 – 19.5	1	14.5	– 3	– 3
19.5 – 29.5	0	24.5	– 2	0
29.5 – 39.5	1	34.5	– 1	– 1
39.5 – 49.5	10	44.5	0	0
49.5 – 59.5	17	54.5	1	17
59.5 – 69.5	38	64.5	2	76
69.5 – 79.5	9	74.5	3	27
79.5 – 89.5	3	84.5	4	12
	N = 79			**Σ fd = 128**

Now $\bar{x} = a + \frac{\Sigma fd}{N} \times i.$ Here $a = 44.5$; $i = 10$; $N = 79$.

$\therefore \quad \bar{x} = 44.5 + \frac{128}{79} \times 10 = 44.5 + 16.2 = \mathbf{60.7}.$

Example 12. *Following is the distribution of marks obtained by 60 students in Economics test.*

Marks	*No. of students*	*Marks*	*No. of students*
More than 0	*60*	*More than 30*	*20*
More than 10	*56*	*More than 40*	*10*
More than 20	*40*	*More than 50*	*3*

Calculate the arithmetic mean.

Solution. We are given the cumulative frequency series. All the 60 students have secured more than 0 marks; but 56 students have secured more than 10 marks and so on. It means 4 students have secured marks between 0 – 10. We can similarly rearrange the distribution as follows:

Table: Computation of Mean

Marks	*No. of Students*	*Mid-value* ***x***	$d = \frac{x - 35}{10}$	*fd*
0 – 10	4	5	– 3	– 12
10 – 20	16	15	– 2	– 32
20 – 30	20	25	– 1	– 20
30 – 40	10	35	0	0
40 – 50	7	45	1	7
50 – 60	3	55	2	6
	N = 60			**Σ fd = – 51**

Here $a = 35,\ N = 60,\ \Sigma fd = -51,\ i = 10.$

Now $$\bar{x} = a + \frac{\Sigma fd}{N} \times i$$

Hence, $$\bar{x} = 35 + \frac{(-51)}{60} \times 10 = 35 - 8.5 = \mathbf{26.5.}$$

3.6.6 Missing Frequencies

Example 13. *Find the value of p for the following distribution whose mean is 16.6.*

f :	12	16	20	24	16	8	4.
x :	8	12	15	p	20	25	30

Solution. Here $\Sigma f_i x_i = 8 \times 12 + 12 \times 16 + 15 \times 20 + p \times 24 + 20 \times 16 + 25 \times 8 + 30 \times 4$
$= 96 + 192 + 300 + 24p + 320 + 200 + 120 = \mathbf{24p + 1228}$

Also $\Sigma f_i = 12 + 16 + 20 + 24 + 16 + 8 + 4 = \mathbf{100.}$

$\therefore$ $$\textbf{Mean} = \frac{\Sigma f_i x_i}{\Sigma f_i} = \frac{24p + 1228}{100} = \mathbf{16.6} \qquad \textbf{[Given]}$$

$\Rightarrow$ $24\,p = 1660 - 1228 = 432 \ \Rightarrow\ \mathbf{p = 18.}$

3.7 WEIGHTED ARITHMETIC MEAN

If $w_1, w_2, w_3, ..., w_n$ are the weights assigned to the values $x_1, x_2, x_3, ..., x_n$ respectively, then the weighted average is defined as:

$$\textbf{Weighted Arithmetic Mean} = \frac{w_1 x_1 + w_2 x_2 + ... + w_n x_n}{w_1 + w_2 + ... + w_n} = \frac{\Sigma wx}{\Sigma w}.$$

It is, generally, denoted by $\bar{x}_w$ or $\bar{X}_w$.

Example 14. *A candidate obtains the following percentages in an examination. English 46%, Mathematics 67%, Sanskrit 72%, Economics 58%, Political Science 53%. It is agreed to give double weights to marks in English and Mathematics as compared to other subjects. What is the weighted mean?*

Solution. Since it is agreed to give double weights to marks in English and Mathematics, we have the following data:

Subjects	*Marks* (x)	*Weights* (w)	*wx*
English	46	2	92
Mathematics	67	2	134
Sanskrit	72	1	72
Economics	58	1	58
Political Science	53	1	53
		$\Sigma w = 7$	$\Sigma wx = 409$

$$\textbf{Weighted mean} = \frac{\Sigma wx}{\Sigma w} = \frac{409}{7} = \mathbf{58.43.}$$

3.8 CORRECTED MEAN

FORMULA: $$\textbf{Correct Mean} = \frac{\textbf{Incorrect } \Sigma x - \textbf{Incorrect item} + \textbf{Correct item}}{n}$$

Example 15. *The mean marks scored by 100 students was found to be 40. Later on, it was discovered that a score of 53 was misread as 83. Find the correct mean.*

Solution. Here $\bar{X} = 40, n = 100$. Also $\bar{X} = \frac{\Sigma x}{n} \Rightarrow 40 = \frac{\Sigma x}{10} \Rightarrow \Sigma x = 4000$

or **Incorrect value of Σx = 4000.**

∴ **Correct value of Σx = Incorrect value of Σx – Incorrect item + correct item**

$= 4000 - 83 + 53 = \mathbf{3970.}$

$$\therefore \textbf{ Correct mean} = \frac{\textbf{Correct value of } \Sigma x}{n} = \frac{3970}{100} = \mathbf{397.}$$

Example 16. *Mean of 25 observations was found to be 78.4. But later on it was found that 96 was misread as 69. Find the correct mean.*

Solution. We know that the mean is given by: $\bar{x} = \frac{\Sigma x}{n}$ or $\Sigma x = n\bar{x}$.

Here $\bar{x} = 78.4, n = 25$. ∴ $\Sigma x = 25 \times 78.4 = 1960$.

But this Σx is incorrect as 96 was misread as 69.

∴ **Correct** $\Sigma x = 1960 - 69 + 96 = \mathbf{1987.}$

$$\textbf{Correct mean} = \frac{\textbf{Corrected value of } \Sigma x}{n} = \frac{1987}{25} = \mathbf{79.48.}$$

3.9 COMBINED MEAN

If we are given the mean of two series and their sizes, then the combined mean for the resultant series can be obtained by the formula:

$$\bar{X}_{12} \text{ or } \bar{X} = \frac{n_1\bar{x}_1 + n_2\bar{x}_2}{n_1 + n_2},$$

where $\bar{X}_{12}$ or $\bar{X}$ = **Combined mean of the two series**

$\bar{x}_1$ = **Mean of the first series**

$\bar{x}_2$ = **Mean of the second series**

n_1 = **Size of the first series**

n_2 = **Size of the second series**

Example 17. *A firm of readymade garments make both men's and women's shirts. Its profit average is 6% of sales. Its profits in men's shirts average 8% of sales; and women's shirts comprise 60% of output. What is the average profit per sales rupee in women's shirts.*

Solution. Here $\bar{X} = 6, \bar{x}_1 = 8, n_1 = 40, n_2 = 60$. Assuming that the total output is 100, we are required to find out $\bar{x}_2$. We know that

$$\bar{X} = \frac{n_1\bar{x}_1 + n_2\bar{x}_2}{n_1 + n_2} = \frac{40 \times 8 + 60\bar{x}_2}{40 + 60}$$

$$\Rightarrow \quad 6 = \frac{320 + 60\bar{x}_2}{100}$$

$$\Rightarrow \quad \bar{x}_2 = \frac{600 - 320}{60} = \frac{280}{60} = \frac{14}{3} = \mathbf{4.67.}$$

Thus, **the average profits in women's shirt is 4.66 per cent of sales, or Re. 0.0467 per sale rupee.**

Example 18. *The average score of girls in class X examination in a school is 67 and that of boys is 63. The average score for the whole class is 64.5, find the percentage of girls and boys in the class.*

Solution. Let n_1 and n_2 respectively be the number of girls and boys.

$\overline{X}_1$ = Average score of girls = 67; $\overline{X}_2$ = Average score of boys = 63;

$\overline{X}$ = Average score of the whole class = 64.5;

$$\therefore \quad \overline{X} = \frac{n_1\overline{X}_1 + n_2\overline{X}_2}{n_1 + n_2} \quad \Rightarrow \quad 64.5 = \frac{67n_1 + 63n_2}{n_1 + n_2}$$

$\Rightarrow 64.5n_1 + 64.5n_2 = 67n_1 + 63n_2 \quad \Rightarrow \quad 2.5n_1 = 1.5n_2 \quad \Rightarrow \quad 25n_1 = 15n_2 \quad \Rightarrow \quad 5n_1 = 3n_2.$

Total number of students in the class = $n_1 + n_2$.

$$\therefore \quad \textbf{Percentage of girls} = \frac{n_1}{n_1 + n_2} \times 100 = \frac{n_1}{n_1 + (5/3)n_1} \times 100 \qquad [\because \; 5n_1 = 3n_2]$$

$$= \frac{3n_1}{3n_1 + 5n_1} \times 100 = \frac{3}{8} \times 100 = \textbf{37.5\%}$$

and
$$\textbf{Percentage of boys} = \frac{n_2}{n_1 + n_2} \times 100 = \frac{n_2}{(3/5)n_2 + n_2} \times 100 = \frac{5n_2}{3n_2 + 5n_2} \times 100 = \textbf{62.5\%}.$$

Hence, **there are 37.5% girls and 62.5% boys in the class.**

Example 19. *There are 50 students in a class of which 40 are boys and rest girls. The average weight of the class is 44 kg and the average weight of the girls is 40 kg. Find the average weight of the boys.*

Solution. Here: n = Number of students in a class = 50

n_1 = Number of boys in a class = 40; n_2 = Number of girls in a class = 10

$\overline{X}_1$ = Average weight of boys = ?; $\overline{X}_2$ = Average weight of girls = 40 kg.

$$\therefore \text{ Using} \quad \overline{X} = \frac{n_1\overline{X}_1 + n_2\overline{X}_2}{n_1 + n_2}, \text{ we get:}$$

$$44 = \frac{40\overline{X}_1 + 10 \times 40}{40 + 10} \quad \Rightarrow \quad 50 \times 44 = 40\,\overline{X}_1 + 400 \quad \Rightarrow \quad \overline{X}_1 = \textbf{45.}$$

Example 20. *The mean annual salary of all employees in a company is Rs. 25,000. The mean salary of male and female employees is Rs.27,000 and Rs.17,000 respectively. Find the percentage of males and females employed by the company.*

Solution. The combined mean is given by:

$$\overline{X}_{12} = \frac{N_1\overline{X}_1 + N_2\overline{X}_2}{N_1 + N_2} \qquad \text{... (1)}$$

Here $\overline{X}_{12}$ = 25,000, X_1 = 27,000, $\overline{X}_2$ = 17,000.

Let $N_1 + N_2 = 100$, where N_1 denote males and N_2 denote females.

$\therefore \; N_2 = 1000 - N_1$

Substituting the values in the formula (1), we get:

$$25000 = \frac{N_1(27{,}000) + 17000\,(100 - N_1)}{100}$$

$2500000 = 27000N_1 + 1700000 - 17000N_1 \Rightarrow 10000N_1 = 2500000 - 1700000 = 800000$

or $N_1 = \mathbf{80}$. $\therefore N_2 = 100 - 80 = 20$.

Hence, percentage of males = 80 and females = 20.

EXERCISE – 3.1

1. Five coins were simultaneously tossed 1000 times and at each toss, the number of heads was observed. The number of tosses during which 0, 1, 2, 4 and 5 heads were obtained are shown in the table below. Find the mean number of heads per toss

No. of heads per toss :	0	1	2	3	4	5
No. of tosses :	38	144	342	287	164	25.

[**Hint.** $\bar{x} = \frac{\Sigma f_i x_i}{\Sigma f_i} = \frac{0\times38+1\times144+2\times342+3\times287+4\times164+5\times25}{\Sigma f_i = 1000}$ = **2.47.**]

2. Find the mean of the following data:

x_i	19	21	23	25	27	29	31
f_i	13	15	16	18	16	15	13.

[**Hint. Mean** = $\frac{19\times13+21\times15+23\times16+25\times18+27\times16+29\times15+31\times13}{13+15+16+18+16+15+13}$ = **25**]

3. Find the missing frequency (p) for the following distribution whose mean is 50.

x :	10	30	50	70	90
f :	17	p	32	24	19.

[**Hint.** $\Sigma fx = 10 \times 17 + 30 \times p + 50 \times 32 + 70 \times 24 + 90 \times 19 = \mathbf{30p + 5160}$.

Also $\mathbf{\Sigma f = p + 92}$. **Mean** = $\frac{\Sigma fx}{\Sigma f} = \frac{30p + 5160}{p + 92}$ = **50** **(given)**

$\Rightarrow 30p + 5160 = 50p + 4600 \Rightarrow 20p = 560$ or $\mathbf{p = 28}$.]

4. Find the missing frequency (p) for the following distribution whose mean is 7.68.

x	3	5	7	9	11	13
f	6	8	15	p	8	4.

[**Hint.** $\Sigma fx = 18 + 40 + 105 + 9p + 88 + 52 = \mathbf{9p + 303}$. Also $\mathbf{\Sigma f = 41 + p}$.

$\therefore$ **Mean** = $\frac{9p + 303}{41 + p} = 7.68 \Rightarrow 9p + 303 = 314.88 + 7.68p \Rightarrow 1.32 = 11.88 \Rightarrow p = \mathbf{9}$.]

Find the mean of the following frequency Distribution

5.

Class Interval	0 – 8	8 – 16	16 – 24	24 – 32	32 – 40
Frequency	6	7	10	8	9

[**Hint.**

Mid-point	4	12	20	28	36
f_i	6	7	10	8	9

Mean = $\frac{\Sigma f_i x_i}{\Sigma f_i} = \frac{4\times6+12\times7+20\times10+28\times8+36\times9}{6+7+10+8+9} = \frac{856}{40}$ = **21.4.**]

6. The mean of the following frequency table is 50. But the frequencies f_1 and f_2 in classes 20 – 40 and 60 – 80 are missing. Find the missing frequencies.

Class :	0 – 20	20 – 40	40 – 60	60 – 80	80 – 100	Total
Frequency :	17	f_1	32	f_2	19	120

[**Hint.** $\Sigma f_i x_i = 3480 + 30f_1 + 70f_2$

$68 + f_1 + f_2 = 120 \Rightarrow f_1 + f_2 = 52.$ $[\because \Sigma f_i = 120]$ ... (1)

Also the mean = 50. $\therefore \bar{x} = \dfrac{3480 + 30f_1 + 70f_2}{120} = 50$

$\Rightarrow 3480 + 30f_1 + 70f_2 = 6000 \Rightarrow 30f_1 + 70f_2 = 2520 \Rightarrow 3f_1 + 7f_2 = 252.$ ... (2)

Multiplying (1) by 3, and subtracting it from (2), we have: $4f_2 = 96 \Rightarrow f_2 = 24.$ ... (3)

Now (1) and (3) $\Rightarrow f_1 + 24 = 52 \Rightarrow \boldsymbol{f_1 = 28.}$ Hence, $\boldsymbol{f_1 = 28,}$ $\boldsymbol{f_2 = 24.}$]

7. The weight in kilograms of 60 workers in a factory are given below. Find the mean weight of a worker by short-cut method.

Weight (in kg.)	60	61	62	63	64	65
No. of Workers	5	8	14	16	10	7

[**Hint.** Take A = 63, so that $d_i = x_i - 63$, $\Sigma f_i d_i = -21$.

$$\therefore \quad \textbf{Mean} = \textbf{A} + \frac{\Sigma f_i d_i}{\Sigma f_i} = 63 + \left(-\frac{21}{60}\right) = \textbf{62.65 kg.}]$$

8. The following table gives the distribution of total household expenditure (in rupees) of manual workers in a city.

Expenditure in (Rs.)	100–150	150–200	200–250	250–300	300–350	350–400	400–450	450–500
Frequency	24	40	33	28	30	22	16	7

Find, by step deviation method, the average expenditure (in Rs.) per household.

[**Hint.** Take assumed mean A = 325, and c = 50, then

$$u_i = (x_i - \text{A})/c = \frac{x_i - 325}{50} \text{ and } \Sigma f_i u_i = -235.$$

$$\therefore \textbf{Mean} = \textbf{A} + \boldsymbol{c}\left(\frac{\Sigma f_i u_i}{\Sigma f_i}\right) = 325 + 50 \times \frac{(-235)}{200} = \textbf{Rs. 266.25.}]$$

ANSWERS

1. 2.47. **2.** 25. **3.** 28. **4.** 9.

5. 21.4. **6.** $f_1 = 28, f_2 = 24.$ **7.** 62.65 kg. **8.** Rs 266.25.

3.10 MEDIAN

Median ***is defined as the middle most or the central value of the variable in a set of observations, when the observations are arranged either in ascending or in descending order of their magnitudes. It divides the arranged series in two equal parts. Median is a position average, whereas the arithmetic mean is a calculated average.*** When a series consists of an even number of terms, median is the arithmetic mean of the two central items. **It is generally denoted by** $\boldsymbol{M}$ **or** $\boldsymbol{M_d}$**.**

3.11 CALCULATION OF MEDIAN

3.11.1 When the Data is Ungrouped

Arrange the n values of the given variable in ascending (or descending) order of magnitudes.

Case I. When n is odd. In this case $\frac{n+1}{2}$ th term is the median,

$$\textbf{Median:} \qquad M_d \text{ or } M = \frac{n+1}{2} \text{ th term.}$$

Example 21. *The number of runs scored by 11 players of a cricket team of a school are*

5, 19, 42, 11, 50, 30, 21, 0, 52, 36, 27.

Find the median.

Solution. Let us arrange the values in **ascending order :**

0, 5, 11, 19, 21, 27, 30, 36, 42, 50, 52. ... (*i*)

In this case the number of items is 11, which is an odd number.

$$\therefore \qquad \textbf{Median: } M = \left(\frac{n+1}{2}\right)\textbf{th value} = \left(\frac{11+1}{2}\right)\text{th value} = \textbf{6th value.}$$

$\because n = 11$

Now, the **6th value in the data is 27.**

$\therefore$ **Median = 27 runs.**

Case II. When n is even. In this case, there are two middle terms $(n/2)$th and $(n/2 + 1)$th. **The median is the average of these two terms,** *i.e.,*

$$\textbf{Median:} \qquad M_d \text{ or } M = \frac{(n/2)\text{th term} + [(n/2)+1]\text{th term}}{2}$$

Example 22. *Find the median of the following items:*

6, 10, 4, 3, 9, 11, 22, 18.

Solution. Let us arrange the items in ascending order :

3, 4, 6, 9, 10, 11, 18, 22.

In this data, the number of items is **n = 8, which is even.**

$$\therefore \qquad \textbf{Median: } M = \textbf{average of } \left\{\left(\frac{n}{2}\right)\textbf{th and } \left(\frac{n}{2}+1\right)\textbf{th}\right\} \textbf{ terms}$$

$$= \text{Average of } \left(\frac{8}{2}\right)\text{th and } \left(\frac{8}{2}+1\right)\text{th terms}$$

$$= \text{average of 4th and 5th terms}$$

$$= \frac{9+10}{2} = \frac{19}{2} = \mathbf{9.5.}$$

Example 23. *The weights (in kilogram) of 15 students are as follows:*

31, 35, 27, 29, 32, 43, 37, 41, 34, 28, 36, 44, 45, 42, 30.

Find the median. If the weight 44 kg is replaced by 46 kg and 27 kg by 25 kg, find the new median.

Solution. Weights of 15 students in **ascending order is**

27, 28, 29, 30, 31, 32, 34, 35, 36, 37, 41, 42, 43, 44, 45.

$$\textbf{Median} = \left(\frac{n+1}{2}\right)^{th} \textbf{ item} = \frac{15+1}{2} = \mathbf{8^{th}} \textbf{ item} = \mathbf{35} \textbf{ kg.}$$

When 44 is replaced by 46 and 27 by 25, then the new data in ascending order is

25, 28, 29, 30, 31, 32, 34, 35, 36, 37, 41, 42, 43, 46, 45.

New median = 8th item = 35 kg.

Example 24. *The following table represents the marks obtained by a batch of 12 students in certain class tests in Statistics and Physics.*

Sr. No.	*1*	*2*	*3*	*4*	*5*	*6*	*7*	*8*	*9*	*10*	*11*	*12*
Marks (Statistics) :	*53*	*54*	*32*	*30*	*60*	*46*	*28*	*25*	*48*	*72*	*33*	*65*
Marks (Physics) :	*55*	*41*	*48*	*49*	*27*	*25*	*23*	*20*	*28*	*60*	*43*	*67*

Indicate in which subject is the level of achievement higher?

Solution. ***The level of achievement is higher in that subject for which the median marks are more.***

Let us arrange the marks in the two subjects in ascending order,

Sr. No. :	1	2	3	4	5	6	7	8	9	10	11	12
Marks (Statistics) :	25	28	30	32	33	46	48	53	54	60	65	72
Marks (Physics) :	20	23	25	27	28	41	43	48	49	55	60	67

Since the **number of students is 12, the marks of the middle-most student are the mean of the marks of 6th and 7th students.**

$$\therefore \qquad \textbf{Median marks in Statistics} = \frac{46+48}{2} = \mathbf{47.}$$

$$\textbf{Median marks in Physics} = \frac{41+43}{2} = \mathbf{42.}$$

Here the median marks in Statistics are greater than the median marks in Physics, therefore, the level of achievement of the students is higher in Statistics.

3.11.2 Calculation of Median for Grouped Data

CASE I. When the series is discrete.

In this case, the values of the variable are arranged in ascending or descending order of magnitudes. A table is prepared showing the corresponding frequencies and cumulative frequencies.

$$\textbf{Median: } M = \left(\frac{n+1}{2}\right)\textbf{th value}$$

Example 25. *Calculate median for the following data:*

No. of Students :	*6*	*4*	*16*	*7*	*8*	*2*
Marks :	*20*	*9*	*25*	*50*	*40*	*80*

Solution. Arranging the marks in ascending order and preparing the following table:

Table: Computation of Median

Marks	*Frequency*	*Cumulative Frequency*
9	4	4
20	6	10
25	16	26
40	8	34
50	7	41
80	2	
	$n = \Sigma f = 43$	

Here $n = 43$. $\therefore$ **Median :** $M = \left(\frac{n+1}{2}\right)$**th value** $= \left(\frac{43+1}{2}\right)$th value = **22nd value.**

The above table shows that all items from 11 to 26 have their values 25. Since **22nd item lies in this interval, therefore, its value is 25.**

Hence, Median = 25 marks.

Example 26. *Find the median of the following frequency distribution:*

x :	*5*	*7*	*9*	*12*	*14*	*17*	*19*	*21*
f :	*6*	*5*	*3*	*6*	*5*	*3*	*2*	*4*

Solution.

x :	5	7	9	12	14	17	19	21
f :	6	5	3	6	5	3	2	4
c.f. :	6	11	14	20	25	28	30	34

Median = Average of $\frac{n}{2}$ **and** $\left(\frac{n}{2}+1\right)$ **items.** ($\because$ Here $n = 34$ is an even number)

= Average of 17th and 18th items = $\frac{12+12}{2}$ **= 12.**

The above table shows that all items between 12 to 14 have their values 9. **Since 12th item lies in this interval therefore its valus is 9.**

Hence, Median = 9.

Example 27. *The median of the observations 8, 11, 13, 15, x + 1, x + 3, 30, 35, 40, 43 arranged in ascending order is 22. Find x.*

Solution. The observations arranged in ascending order are:

8, 11, 13, 15, $x + 1$, $x + 3$, 30, 35, 40, 43.

Total number of observations = 10 **(even)**

$$\therefore \text{ Median} = \frac{\frac{n}{2}\text{th item} + \left(\frac{n}{2}+1\right)\text{th item}}{2} = \frac{5\text{th item} + 6\text{th item}}{2}$$

$$\Rightarrow 22 = \frac{(x+1)+(x+3)}{2} \quad \Rightarrow \quad 22 = \frac{2x+4}{2} \quad \Rightarrow \quad 22x + 4 \quad \Rightarrow \quad x = 20.$$

CASE II. When the series is continuous.

In this case the data is given in the form of a frequency table with class-interval, etc., and the following formula is used to calculate the Median.

Median: $$M = L + \frac{(n/2) - C}{f} \times i, \text{ where}$$

L = **lower limit of the class in which the median lies,**

n = **total number of frequencies, *i.e.*, $n = \Sigma f$,**

f = **frequency of the class in which the median lies,**

C = **cumulative frequency of the class preceding the median class,**

i = **width of the class-interval of the class in which the median lies.**

Example 28. *The following table gives the weekly expenditure of 100 families. Find the median weekly expenditure.*

Weekly Expenditure : (in Rs.)	0 – 10	10 – 20	20 – 30	39 – 40	40 – 50
Number of Families :	14	23	27	21	15

Solution. Let us prepare a table which gives the frequencies and cumulative frequencies.

Table: Computation of Median

Weekly Expenditure (in Rs.)	*Number of families (frequency) f*	*Cumulative frequency*
0 – 10	14	14
10 – 20	23	37
20 – 30	27	64
30 – 40	21	85
40 – 50	15	100
		Here $n = \Sigma f = 100$

$\therefore$ **Median** = $\left(\frac{n}{2}\right)$**th value** = $\left(\frac{100}{2}\right)$th value = **50th value.**

Median class = 20 – 30.

Here $\frac{n}{2} = 50, \quad L = 20, \quad C = 37, \quad i = 10.$

$$\therefore \textbf{ Median} = L + \frac{(n/2) - C}{f} \times i = 20 + \frac{50 - 37}{27} \times 10 = 20 + \frac{13}{27} \times 10$$

$$= \mathbf{20 + 4.815 = 24.815.}$$

Hence, **median = 24.815.**

Example 29. *Calculate the mean and median for the following data:*

Height (in cm)	*No. of boys*	*Height (in cm)*	*No. of boys*
135 – 140	*4*	*155 – 160*	*24*
140 – 145	*9*	*160 – 165*	*10*
145 – 150	*18*	*165 – 170*	*5*
150 – 155	*28*	*170 – 175*	*2*

Solution. Let us prepare the following table by taking 157.5 assumed mean.

Table: Computation of Median

Class	x_i	f_i	*c.f.*	$d'_i = \dfrac{x_i - 157.5}{5}$	$f_i d'_i$
135 – 140	137.5	4	4	– 4	– 16
140 – 145	142.5	9	13	– 3	– 27
145 – 150	147.5	18	31	– 2	– 36
150 – 155	152.5	28	59	– 1	– 28
155 – 160	157.5	24	83	0	0
160 – 165	162.5	10	93	1	10
165 – 170	167.5	5	98	2	10
170 – 175	172.5	2	100	3	6
		$\Sigma f_i = 100$			**– 81**

Now **Arithmetic Mean** $= a + \dfrac{\sum_{i=1}^{n} f_i d'_i}{\sum_{i=1}^{n} f_i} \times i = 157.5 + \dfrac{(-81)}{100} \times 5$

$$= 157.5 - \frac{405}{100} = 157.5 - 4.05 = \mathbf{153.45.} \qquad (\because\ a = 157.5,\ i = 5)$$

For median: $\dfrac{N}{2} = \dfrac{100}{2} = 50.$ $\therefore$ **Median class:** 150 – 155, $L = 150$; $C = 31$; $i = 5$.

$$\therefore\ \textbf{Median} = L + \frac{\left(\frac{N}{2}\right) - C}{f} \times i = 150 + \frac{50 - 31}{28} \times 5 = 150 + \left(\frac{95}{28}\right) = 150 + 3.39 = \mathbf{153.39.}$$

3.12 PROPERTIES OF MEDIAN

1. ***If the two variables X and Y are related by Y = ax + b, where a and b are constants. Also the value of the median of X is given, then*** $Y_{md} = a + bX_{md}$.
2. ***For a given set of observations, the sum of deviations is minimum when the deviations and taken from the median i.e., if A is the median then*** $\Sigma \mid X_i - A \mid$ ***is minimum.***

Example 30. *If x and y are related by 7x + 2y = 15 and median of x = 3, then find the median of y.*

Solution. $7x + 2y = 5 \Rightarrow y = (-7/2)\,x + 15$

$\Rightarrow \quad y_{md} = (-7/2)\, x_{md} + 5 = (-7/2) \times 3 + 15 = \mathbf{4.5.}$

3.13 MERITS, DEMERITS AND USES OF MEDIAN

Merits:

1. *It is easily understood.*
2. *It is not affected by extreme values.*
3. *It can be located graphically.*
4. *It is the best measure for qualitative data such as beauty, intelligence, honesty, etc.*
5. *It can be easily located even if the class-intervals in the series are unequal.*
6. *It can be determined even by inspection in many cases.*

Demerits:

1. *It is not subject to algebraic treatments.*
2. *It cannot represent the irregular distribution series.*
3. *It is a positional average and is based on the middle item.*
4. *It does not have sampling stability.*
5. *It is an estimate in case of a series containing even number of items.*
6. *It does not take into account the values of all the items in the series.*
7. *It is not suitable in those cases where due importance and weight should be given to extreme values.*

Uses:

1. *It is useful in those cases where numerical measurements are not possible.*
2. *It is also useful in those cases where mathematical calculations cannot be made in order to obtain the mean.*
3. *It is generally used in studying phenomena like skill, honesty, intelligence, etc.*

Example 30. *Compute median from the following data:*

Mid values :	*115*	*125*	*135*	*145*	*155*	*165*	*175*	*185*	*195*
Frequency :	*6*	*25*	*48*	*72*	*116*	*60*	*38*	*22*	*3*

Solution. The given problem is a continuous frequency distribution. Mid-values of the class-intervals are given. The difference between two mid-values is 10. Therefore, 10/2 or 5 is reduced from each mid-value to get the lower limit and 5 is added to get the upper limit of a class.

Table: Computation of Median

Class-Intervals	*Frequency*	*c.f.*
110 – 120	6	6
120 – 130	25	31
130 – 140	48	79
140 – 150	72	151
150 – 160	116	267
160 – 170	60	327
170 – 180	38	365
180 – 190	22	387
190 – 200	3	390
Total	**390**	

$$\textbf{Median} = \textbf{Size of } \frac{N}{2} \textbf{ th item}$$

$$= \text{Size of } \frac{390}{2} = 195\text{th item, which lies in the class } 150 - 160$$

Here $L = 150$, $N/2 = 195$, $i = 10$.

$$\textbf{Median} = L + \frac{\frac{N}{2} - c.f.}{f} \times i = 150 + \frac{195 - 151}{116} \times 10 = 150 + 3.79 = \mathbf{153.79.}$$

Example 31. *Following is the distribution of marks obtained by 50 students in Mercantile Law:*

Marks more than:	*0*	*10*	*20*	*30*	*40*	*50*
No. of students:	*50*	*46*	*40*	*20*	*10*	*3*

Calculate the median marks.

Solution. We first convert 'more than series' into continuous series and get the following frequency table

Table: Frequency Table

Marks more than	*Cumulative 'f'*	*Single 'f'*	*Class-Interval*	*c.f.*
0	50 = N	50 – 46 = 4	0 – 10	4
10	46	46 – 40 = 6	10 – 20	10
20	40	40 – 20 = 20	20 – 30	30
30	20	20 – 10 = 10	30 – 40	40
40	10	10 – 3 = 87	40 – 50	47
50	3	3	50 and above	50

$$\textbf{Median = size of } \frac{N}{2} \textbf{ th item = Size of } \frac{50}{2} \textbf{ th item.}$$

= 25th item, which lies in the class 20 – 30.

Here $L = 20$; $f = 20$; $c.f. = 10$; $i = 10$.

$$\textbf{Median} = L + \frac{\frac{N}{2} - c.f.}{f} \times i = 20 + \frac{25-10}{20} \times 10 = 20 + 7.5 = \textbf{27.5 marks.}$$

3.14 GRAPHIC LOCATION OF MEDIAN

There are two methods of locating median graphically:

1. By one Ogive

An ogive is a cumulative frequency curve which shows a rising trend. We put the **variables** on the X-axis and the **cumulative frequencies** on the Y-axis. The middle item is then marked on vertical scale by the formula $N/2$. A line parallel to the base is drawn, cutting the ogive at a point. Draw a line perpendicular to the base and the median is read off. Similarly, quartiles, deciles etc. are calculated with the help of a graph.

2. By Two Ogives

One ogive is drawn by "more than cumulative frequency method" and the other is drawn by "less than cumulative frequency type". From the intersection point of these two ogives, a perpendicular is drawn on the X-axis. The point where the perpendicular touches the X-axis, the median value will be read.

EXERCISE – 3.2

1. The heights (in cm) of 15 students of class X are: 152, 147, 156, 149, 151, 159, 148, 160, 153, 154, 150, 143, 155, 157, 161. Find the median.

2. Find the median of the following observations: 46, 64, 87, 41, 58, 77, 35, 90, 55, 92, 33. If 92 is replaced by 99 and 41 by 43 in the above data, find the new median.

3. The median of the following observations arranged in the ascending order: 11, 12, 14, 18, $x + 2$, $x + 4$, 30, 32, 35, 41 is 24. Find x.

4. Find the median of the following distribution.

x :	8	5	6	10	9	4	7
y :	6	4	5	8	9	6	4

5. Find the median of the following frequency distribution.

x :	3	6	10	12	7	15
y :	3	4	2	8	13	10

[**Hint.** Here $N = 40$, Median = Average of $\left(\frac{40}{2}\right)$th and $\left(\frac{40}{2}+1\right)$th value.]

6. The marks (out of 75) obtained by 60 students in a certain examination are given below. Calculate the median.

Marks	15–20	20–25	25–30	30–35	35–40	40–45	45–50	50–55	55–60	60–65
No. of students	4	5	11	6	5	8	9	6	4	2

7. Find the median of the following frequency distribution

Wages (in Rs.)	20 – 30	30 – 40	40 – 50	50 – 60	60 – 70
No. of labourers	3	5	20	10	5.

[**Hint.**

Class-Interval	*Frequency f*	*c.f.*
20 – 30	3	3
30 – 40	5	**8**
40 – 50	**20**	28
50 – 60	10	38
60 – 70	5	43

Here. $N = 43$, which is an odd number.

$\therefore$ Median = size of $\left(\frac{N}{2}\right)$th item = Size of $\frac{43}{2}$ = (21.5)th item which lies in the class 40 – 50

$\therefore$ Median $= L \times \frac{N/2 - C}{f} \times i = 40 + \frac{21.5 - 8}{20} \times 10 = \mathbf{46.75.}$]

ANSWERS

1. 153 cm. **2.** 58, 58. **3.** $x = 21$. **4.** 8. **5.** 8.5.
6. 39. **7.** Rs 46.75.

3.15 MODE

Mode is that value in a series which occurs most frequently. In a frequency distribution mode is that variate which has the maximum frequency. In other words, **mode represents that value which is most frequent or typical or predominant.** Mode is also known as **Norm.**

For example, in the series 6, 5, 3, 4, 3, 7, 8, 5, 9, 5, 4; we notice that 5 occurs most frequently, therefore, **5 is the mode.**

Example 32. *A shoe shop in Delhi had sold 100 pairs of shoes of a particular brand on a, certain day with the following distribution:*

Size of Shoe .	*4*	*5*	*6*	*7*	*8*	*9*	*10*
No. of Pairs :	*10*	*15*	*20*	*35*	*16*	*3*	*1*

Find the mode of the distribution.

Solution. Let us prepare the table showing the frequency.

Size of Shoe :	4	5	6	7	8	9	10
Frequency :	10	15	20	35	16	3	1

In the above table, we notice that the size 7 has the maximum frequency, *viz.*, 35.

Therefore, **7 is the mode of the distribution.**

3.15.1 Types of Model Series

A series of observations may have one or more modes.

Unimodal series. ***The series of observations which contains only one mode, is called unimodal series.***

Bimodal series. ***The series of observations which contains two modes is called a bimodal series. In this series, the two modes are of same value of greatest density.***

Trimodal series. ***The series of observations which contains three modes is called a trimodal series. In this series, the three modes are of same value of greatest density and highest concentration of observations.***

Ill-defined Mode. ***If a series of observations has more than one mode then the mode is said to be ill-defined.***

3.16 COMPUTATION OF MODE

3.16.1 Simple Series

In the case of simple series, the value which is repeated maximum number of times is the mode of the series.

Example 33. *In Rajdhani Rubber Industry, Tilak Nagar, New Delhi seven labourers are receiving the daily wages of Rs. 5, 6, 6, 8, 8, 8 and 10. Find the modal wage.*

Solution. In the series 5, 6, 6, 8, 8, 8, 10; since 8 occurs thrice and no other item occurs three times or more than three times and hence the modal wage is **Rs. 8.**

3.16.2 Discrete frequency distribution series

In the case of discrete frequency distribution, mode is the value of the variable corresponding to the maximum frequency.

Example 34. *A set of numbers consists of four 4's, five 5's, six 6's and nine 9's. What is the mode?*

Solution. Let us prepare the following frequency table.

Size of item:	4	5	6	9
Frequency:	4	5	6	9

Since 9 has the maximum frequency, *viz.*, 9, therefore, **9 is the mode.**

3.16.3 Computation of Mode in a Continuous Frequency Distribution

(*i*) **Modal class. It is that class in a grouped frequency distribution in which the mode lies.** The modal class can be determined either by inspection or with the help of grouping table. After finding the modal class, we calculate the mode by the following formula:

$$\textbf{Mode} = l + \frac{f_m - f_1}{2f_m - f_1 - f_2} \times i,$$

where l = **the lower limit of the modal class**
i = **the width of the modal class**
f_1 = **the frequency of the class preceding the modal class**
f_m = **the frequency of the modal class**
f_2 = **the frequency of the class succeeding the modal class.**

Sometimes, it so happened that the above formula fails to give the mode. In this case, **the modal value lies in a class other than the one containing maximum frequency.** In such cases, we take the help of the following formula:

$$\textbf{Mode} = l + \frac{\Delta_1}{\Delta_1 + \Delta_2} \times i, \text{ where } \Delta_1 = f_m - f_1, \Delta_2 = f_m - f_2,$$

where l, f_1, f_2, f_m, and l have usual meanings.

The precedure of finding the mode by the above method is called **Method of Interpolation.**

Example 35. *Find the mode for the following data:*

Marks :	*1 – 5*	*6 – 10*	*11 – 15*	*16 – 20*	*21 – 25*
No. of Students :	*7*	*10*	*16*	*32*	*24*

Solution. From the above table, it is clear that the maximum frequency is **32** and it lies in the **class** 16 – 20. Thus, the **modal class is 16 – 20.**

Here $l = 16,\ f_m = 32,\ f_1 = 16,\ f_2 = 24,\ i = 5.$

$$\textbf{Mode} = l + \frac{f_m - f_1}{2f_m - f_2 - f_1} \times i = 16 + \frac{32 - 16}{64 - 24 - 16} \times 5$$

$$= 16 + \frac{16}{24} \times 5 = 16 + \frac{10}{3} = 16 + 3.33 = \mathbf{19.33.}$$

Example 36. *Calculate Median and Mode for the following distribution:*

Production per day (in Tons) :	*21 – 22*	*23 – 24*	*25 – 26*	*27 – 28*	*29 – 30*
No. of days :	*7*	*13*	*22*	*10*	*8*

Solution. We shall first convert the given date into a continuous series.

Table: Calculations of Median and Mode

Class-boundaries	*No. of days (f)*	*Cumulative Frequency (c.f.)*
20.5 – 22.5	7	7
22.5 – 24.5	13	20
24.5 – 26.5	22	42
26.5 – 28.5	10	52
28.5 – 30.5	8	60
Total	$N = \Sigma f = 60$	

Since $\frac{1}{2}N = 30 \Rightarrow$ Median lies in the class **24.5 – 26.5.**

$$\therefore \textbf{Median} = 24.5 + \frac{30 - 20}{22} \times 2 = 24.5 + 0.91 = \mathbf{25.41\ tonnes.}$$

Since the maximum frequency is 20 so the modal class is **24.5 – 20.5.**

$$\text{Mode} = l + \frac{f_m - f_1}{2f_m - f_2 - f_1} \times i = 24.5 + \frac{22 - 13}{2 \times 22 - 13 - 10} \times 2 = \mathbf{25.36\ tonnes.}$$

3.16.4 Computation of Mode by Grouping Method

We notice that, in discrete series, mode is determined by inspection and, therefore, an error of judgement is possible in those cases where the **difference between the maximum frequency and the frequency preceding or succeeding it is very small and the items are heavily concentrated on either side.** Under such circumstances the value of mode is determined by preparing **a grouping table and analysis table.**

GROUPING TABLE

A grouping table has the following six columns.

Column I. *It has original frequencies and the maximum frequency is marked by bold type.*

Column II. *In this column the frequencies of column 1 are combined 'two by two'. (1 and 2; 3 and 4; 5 and 6; and so on). Here also the maximum frequency is marked by bold type.*

Column III. *Here, we leave the first frequency of the column I and combine the others in 'two by two' (2 and 3; 4 and 5; 6 and 7; and so on). Again the maximum frequency is marked by bold type.*

Column IV. *In this column the frequencies of the column I are combined (grouped) in 'three by three' (1, 2 and 3; 4, 5 and 6; 7, 8 and 9; and so on). And again the maximum frequency is marked by bold type.*

Column V. *Here we leave the first frequency of the column I and group the others in 'three by three' (2, 3 and 4; 5, 6 and 7; 8, 9 and 10; and so on). Again mark the maximum frequency by bold type.*

Column VI. *Now leave the first two frequencies of column I and combine the others in 'three by three'. (3, 4 and 5; 6, 7 and 8; 9, 10 and 11; and so on). Mark the maximum frequency by bold type.*

ANALYSIS TABLE

After preparing the grouping table, we prepare the **analysis table.** While preparing this table we put the ***column numbers on the left-hand side and the various probable values of the mode on the right-hand side. The values against which frequencies are maximum marked in the grouping table and are entered by means of a bar in the relevant 'box' corresponding to the values they represent.***

The procedure of preparing a grouping table and analysis table shall be clear from the following example:

Example 37. *Calculate the mode from the following frequency distribution:*

Size (x)	4	5	6	7	8	9	10	11	12	13
Frequency (f)	2	5	8	9	12	14	14	15	11	13

Solution. We find that the value 11 of the variable x occurs maximum numbers of times, *i.e.*, 15. But we also notice that the difference between the frequencies of the values of the variable, on both sides of 15, which are very closed to 11 is very small. This shows that the values of the variable x are heavily concentrated on either side of 11. Therefore, if we find mode just by inspection, **an error is possible.**

This problem is solved by **the method of grouping as it is an irregular distribution** in the sense that the difference between maximum frequency 15 and frequency preceding it is very small. Let us prepare the grouping and analysis table.

GROUPING TABLE

Size (x)	(I) Frequency (f)	Grouping (II) Col. of two	(III) Col. of two leaving the first	(IV) Col. of three	(V) Col. of three leaving the first	(VI) Col. of three leaving the first two
4	2	7		15		
5	5		13		22	
6	8	17				29
7	9		21	35		
8	12	26			**40**	
9	14		**28**			**43**
10	14	**29**		**40**		
11	**15**		26		39	
12	11	24				
13	13					

Let us now prepare the **Analysis Table.**

ANALYSIS TABLE

X → / ↓ Col. No.	4	5	6	7	8	9	10	11	12	13
I								I		
II							I	I		
III						I	I			
IV							I	I	I	
V					I	I	I			
VI						I	I	I		
Total Frequency					**1**	**3**	**5**	**4**	**1**	

From the analysis table, it is clear that the **value 10 has the maximum number of bars, *i.e.*, maximum frequency, *viz.*, 5.** Hence, **the modal value is 10.**

Remark. *But by inspection one is likely to say that the modal value is 11, since it occurs the maximum number of times, i.e., 15, which is incorrect as revealed by analysis and grouping tables which give the correct modal value as 10 (though it occurs 15 times).*

3.17 PROPERTY OF MODE

1. ***If X and Y are two variates related by : $Y = aX + b$, where a and b are real constants and variates also the value of mode of X is given, then the mode of Y is given by : $Y_{mo} = a + b_{mo}$.***

3.18 MERITS, DEMERITS AND USES OF MODE

Merits:

1. *It can be easily understood.*
2. *It can be located in some cases by inspection.*
3. *It is capable of being ascertained graphically.*
4. *It is not affected by extreme values.*
5. *It represents the most frequent value and hence it is very often in practice.*
6. *The arrangement of data is not necessary if the items are a few.*

Demerits:

1. *There are different formulae for its calculations which ordinarily give different answers.*
2. *Mode is determinate. Some series have two or more than two modes.*
3. *It cannot be subjected to algebraic treatments. For example, the combined mode cannot be calculated for the modes of two series.*
4. *It is an unsuitable measure as it is affected more by sampling fluctuations.*
5. *Mode for the series with unequal class-intervals cannot be calculated.*

Uses:

1. *It is used for the study of most popular fashion.*
2. *It is extensively used by businessmen and commercial managements.*

3.19 EMPIRICAL RELATION BETWEEN MEAN, MEDIAN AND MODE

A distribution in which mean, median and mode coincide is called asymmetrical distribution. If the distribution is **moderately asymmetrical,** then mean, median and mode are connected by the formula:

$$\textbf{Mode} = \textbf{3 Median} - \textbf{2 Mean}$$

or

$$\textbf{Median} = \frac{1}{3}\,(\textbf{Mode} + \textbf{2 Mean})$$

Example 38. *If the value of mode and mean is 60 and 66 respectively, find the value of median.*

Solution. We know that:

$$\textbf{Median} = \frac{1}{3}\,(\textbf{Mode} + \textbf{2 Mean}) = \frac{1}{3}\,(60 + 2 \times 66) = \mathbf{64}.$$

3.20 GEOMETRIC MEAN

Geometric Mean. If $x_1, x_2, x_3, \ldots\ldots, x_n$ are n values of a variate x, none of them being zero, then the geometric mean G is defined as:

$$G = (x_1 \cdot x_2 \cdot x_3 \ldots\ldots x_n)^{1/n}. \qquad \ldots (1)$$

In particular, the **geometric mean** of 3, 9 and 27 is $= (3 \cdot 9 \cdot 27)^{1/3} = \mathbf{9}$.

The difficulty of calculating the nth root is overpowered with the help of logarithms. Now taking logarithms of both sides of (1), we get:

$$\log G = \log (x_1 \cdot x_2 \cdot x_3 \ldots\ldots x_n)^{1/n} = \frac{1}{n} \log (x_1 \cdot x_2 \ldots\ldots x_n)$$

$$= \frac{\log x_1 + \log x_2 + \ldots\ldots + \log x_n}{n}$$

$$\therefore \; G = \text{antilog} \left(\frac{\log x_1 + \log x_2 + \ldots\ldots + \log x_n}{n} \right).$$

In the case of a **frequency distribution, the geometric mean of n values $x_1, x_2, \ldots\ldots x_n$ of a variate x occurring with frequency $f_1, f_2, \ldots\ldots, f_n$ respectively is given by**

$$G = \left[x_1^{f_1} \cdot x_2^{f_2} \cdot x_3^{f_3} \ldots\ldots x_n^{f_n} \right]^{1/n}$$

$$\Rightarrow \log G = \frac{f_1 \log x_1 + f_2 \log x_2 + \ldots\ldots + f_n \log x_n}{n} \quad \text{or} \quad G = \text{antilog} \left[\frac{\sum_{i=1}^{n} f_i \log x_i}{n} \right].$$

Thus, Geometric mean is the anti log of weighted mean of the different values of log x_i, whose weights are their respective frequencies f_i.

In the case of continuous or grouped frequency distribution, the values of the variate x are taken to be the values corresponding to the mid-points of the class-intervals.

3.21 PROPERTIES OF GEOMETRIC MEAN

1. ***The logarithm of G for a set of observations x, x_2, ...; x_n is the arithmetic mean of the logarithm of the observations, i.e., log G =*** $\dfrac{\log x_1 + \log x_2 + \ldots + \log x_n}{n}$
2. ***If all the observations of a variable assumes the same constant value K, i.e., $x_1 = x_2 = \ldots x_n = K$, then the Geometric mean these observations is also K.***
3. ***The G.M. of the product of two variables is the product of their G.M. i.e., if Z = XY, then G.M. of Z = (G.M. of X) × (G.M. of Y).***
4. ***The G.M. of the ratio of two variables is the ratio of the G.M. of these two variables i.e., if Z = X/Y. Then G.M. of Z = (G.M. of X)/(G.M. of Y).***

3.22 MERITS, DEMERITS AND USES OF GEOMETRIC MEAN

Merits:

1. *It is the only average that can be used to indicate the rate of change or ratios.*
2. *It is very simple and lends itself to algebraic treatments.*
3. *It is useful in the construction of index numbers.*
4. *It is not much affected by the fluctuations of sampling.*
5. *It is based on all the observations.*
6. *It gives less weightage to large items and large weightage to small items.*

Demerits:

1. *It cannot be easily understood.*
2. *It is relatively difficult to compute as it requires some special knowledge of logarithms.*
3. *It cannot be calculated when any item or value is zero or negative.*
4. *It may be a value which does not correspond to actual value.*
5. *It cannot be obtained by inspection.*

Uses:

1. *It is used in those cases where it is necessary to average ratios which express rate of change.*
2. *It is also used for the construction of index numbers.*

Example 39. *Compute the geometric mean for the following data:*

10, 110, 120, 50, 52, 80, 37, 60.

Solution. Let us prepare the following table:

Size of item (x)	*log x*	*Value of log x*
10	log 10	1.000
110	log 110	2.0414
120	log 120	2.0792
50	log 50	1.6990
52	log 52	1.7160
80	log 80	1.9031
37	log 37	1.5682
60	log 60	1.7782
n = 8		**Σ log x = 13.7851**

Now $$\log G = \frac{1}{n}\Sigma \log x = \frac{13.7851}{8} = \mathbf{1.723.}$$

$$\therefore \quad G = \text{antilog } 1.723 = \mathbf{52.84.}$$

Example 40. *Calculate the geometric mean for the following data:*

x :	*12*	*13*	*14*	*15*	*16*	*17*
f :	*5*	*4*	*4*	*3*	*2*	*1*

Solution. Let us prepare the following table in order to calculate the geometric mean for the given data:

x	*log x*	*f*	*f × log x*
12	1.0792	5	5.3960
13	1.1139	4	4.4556
14	1.1461	4	4.5844
15	1.1761	3	3.5283
16	1.2041	2	2.4092
17	1.2304	1	1.2304
		n = Σf = 19	**Σf log x = 21.6029**

Here $n = 19$

$$\therefore \quad \log G = \frac{\Sigma f \log x}{n} = \frac{21.6029}{19} = 1.137.$$

$$\therefore \quad G = \textbf{Antilog (1.137) = 13.71.}$$

Example 41. *Find the geometric mean for the following data:*

Marks :	*0 – 10*	*10 – 20*	*20 – 30*	*30 – 40*	*40 – 50*
No. of Students :	*4*	*8*	*10*	*6*	*7*

Solution. Let us prepare the following table in order to calculate the geometric mean for the given data:

Marks	*Mid-value* (x)	*Frequency* (f)	*log x*	$f \times \log x$
0 – 10	5	4	0.6990	2.7960
10 – 20	15	8	1.1761	9.4088
20 – 30	25	10	1.3979	13.9790
30 – 40	35	6	1.5441	9.2646
40 – 50	45	7	1.6532	11.5724
	Total	$\boldsymbol{\Sigma f = n = 35}$		$\boldsymbol{\Sigma f \log x = 47.0208}$

Now
$$\log G = \frac{\Sigma f \log x}{n} = \frac{47.0208}{35} = 1.3435.$$

$$\therefore \quad G = \text{antilog } (1.3435) = \mathbf{22.055 \text{ marks.}}$$

3.23 HARMONIC MEAN

Definition. The harmonic mean of n items $x_1, x_2, x_3, ..., x_n$ is defined as;

$$\textbf{Harmonic Mean} = \frac{n}{\dfrac{1}{x_1}+\dfrac{1}{x_2}+\dfrac{1}{x_3}+......+\dfrac{1}{x_n}}$$

For example, the harmonic mean of 2, 4 and 5 is $= \dfrac{3}{\dfrac{1}{2}+\dfrac{1}{4}+\dfrac{1}{5}} = \dfrac{60}{19} = \mathbf{3.16.}$

Harmonic Mean of Frequency Distribution Data. Let $x_1, x_2, x_3, ..., x_n$ be n items which occur with frequencies $f_1, f_2, f_3, ..., f_n$ respectively. Then their Harmonic Mean is given by:

$$\textbf{Harmonic Mean} = \frac{f_1+f_2+f_3+......+f_n}{\left(\dfrac{f_1}{x_1}+\dfrac{f_2}{x_2}+\dfrac{f_3}{x_3}+......+\dfrac{f_n}{x_n}\right)} = \frac{\Sigma f_i}{\Sigma f_i \times \dfrac{1}{x_i}}.$$

3.24 PROPERTIES OF HARMONIC MEAN

1. ***If all the observations of a variable takes the same constant value K, then their H.M. is also K.***
2. ***If there are two groups with n_1 and n_2 observations and H_1 and H_2 as their respective H.M., then their combined H.M.*** $= \dfrac{n_1+n_2}{(n_1/H_1)+(n_2/H_2)}.$

3.25 MERITS, DEMERITS AND USES OF HARMONIC MEAN

Merits:

1. *It is easy to calculate.*
2. *It is rigidly defined.*
3. *It gives largest weight to the smallest items and can be used whenever so desired.*
4. *It is a useful average when we deal with* ***average of rates.***

Demerits:

It cannot be located by inspection.

Uses:

1. *The harmonic mean is especially useful in averaging time rates, in finding the average price per unit when the data gives the amount of commodity for a given price and in the development of index numbers.*
2. *It is used when rates are expressed as x per y, where x is constant.*

3.26 WEIGHTED AVERAGE

When the observations under consideration have a hierarchical order of importance, we take recourse to computing weighted average, which could be either weighted A.M. or weighted G.M. or weighted HM.

Let $x_1, x_2, x_3, ..., x_n$ be n observations with their respective weights, $w_1, w_2, ..., w_n$, then

$$\textbf{Weighted A.M.} = \frac{\Sigma w_1 x_1}{\Sigma w_1};$$

$$\textbf{Weighted G.M.} = \text{Ante-log}\left(\frac{\Sigma w_1 \log x_1}{\Sigma w_1}\right); \textbf{ Weighted H.M.} = \frac{\Sigma w_1}{\Sigma\left(\frac{w_1}{x_1}\right)}.$$

Example. 42. *Find the H.M. of 4, 6 and 10.*

Solution. $\textbf{H. M.} = \frac{n}{\Sigma(1/x_i)} = \frac{3}{(1/4)+(1/6)+(1/10)} = \frac{3}{0.25+0.17+0.10} = \textbf{5.77.}$

Example 43. *The interest paid on each of three different sums of money yielding 3%, 4% and 5% simple interest p.a. respectively is the same. What is the average yield per cent on the total sum invested.*

Solution. $\textbf{H.M.} = \frac{3}{\frac{1}{3}+\frac{1}{4}+\frac{1}{5}} = \frac{3}{\frac{20+15+12}{60}} = \frac{3\times 60}{47} = \frac{180}{47} = \textbf{3.83\%}.$

Example 44. *A train travels first 300 km. at an average rate of 30 k.p.h. and further travels the same distance at an average rate of 40 k.p.h. What is the average speed over the whole distance?*

Solution. The average speed of train over the whole distance shall be the weighted harmonic mean of the speeds of 30 kilometres per hour over 300 kilometres and 40 kilometres per hour over 300 km.

Here the weights $w_1 = 300$, $w_2 = 300$.

We know that: $\frac{\mathbf{1}}{\mathbf{H.M.}} = \frac{\Sigma(w/x)}{\Sigma w}$.

or $$= \frac{1}{\text{average speed}} = \left(\frac{300}{30}+\frac{300}{40}\right)\times\frac{1}{(300+300)}.$$

or $$\textbf{Average speed} = \frac{600}{10+7.5} = \frac{600}{17.5} = \textbf{34.286 k.p.h.}$$

Example 45. *Point out the mistake or ambiguity in the following statement, "A person goes X to Y on cycle at 20 k.p.h. and returns at 24 k.p.h. His average speed was 22 kph".*

Solution. ***It is a question on rate where the variable is time but the constant is distance. We shall apply harmonic mean in calculating the average speed.***

$$\textbf{Average speed} = \text{A.M. of 20 and 24} = \frac{20+24}{2} = \textbf{22 k.p.h.}$$

So the average speed would not be 22 k.p.h. which is based on arithmetic mean.

The appropriate average speed is the harmonic mean of 20 and 24.

Here $$\Sigma \frac{1}{x} = \frac{1}{20} + \frac{1}{24} = \frac{11}{120}$$

$$\therefore \quad \textbf{Harmonic Mean} = \frac{N}{\Sigma(1/x)} = 2 \times \frac{120}{11} = \frac{240}{11} = \textbf{21.82 k.p.h.}$$

Example 46. *If two grades of oranges sell at 10 for Re. 1 and 20 for Re. 1, respectively. Calculate the average price per orange, stating your assumptions explicitly.*

Solution. It is a question based on rate. Here the variable is orange and constant is rupee. So harmonic mean is applicable in calculating the average rate.

Here $$\Sigma\left(\frac{1}{x}\right) = \frac{1}{10} + \frac{1}{20} = \frac{3}{20}.$$

N = number of items = 2.

$$\therefore \textbf{Harmonic Mean} = \frac{N}{\Sigma(1/x)} = 2 \times \frac{20}{3} = \textbf{13.33.}$$

Average rate of orange sold per Re. = **13.33.**

$$\text{Average price per orange} = \frac{1}{13.33} = \textbf{7.5 paise.}$$

Example 47. *Find the H.M. for the following data :*

x	:	2	4	8	16
f	:	2	3	3	2.

Solution. $$\textbf{H. M.} = \frac{N}{\Sigma(f_i/x_i)} = \frac{10}{(2/2)+(3/4)+(3/8)+(2/16)} \quad \textbf{[Here N} = \Sigma f_i\textbf{]}$$

$$= \frac{10}{1+0.75+0.37+0.125} = \frac{10}{2.25} = \textbf{4.44.}$$

Example 48. *Find the weighted AM and weighted HM of first n natural weights being equal to the squares of the corresponding numbers.*

Solution. As given,

x	:	1	2	3	...	n
w	:	1^2	2^2	3^2	...	n^2

$$\textbf{Weighted A.M.} = \frac{\Sigma w_i x_i}{\Sigma w_i} = \frac{1\times1^2 + 2\times2^2 + 3\times3^2 + \ldots n\times n^2}{1^2+2^2+3^2+\ldots+n^2}$$

$$= \frac{1^3+2^3+3^3+\ldots+n^3}{1^2+2^2+3^2+\ldots+n^2} = \frac{[n(n+1)/2]^2}{\{n(n+1)(2n+1)\}/6} = \frac{\mathbf{3n(n+1)}}{\mathbf{2(2n+1)}}$$

$$\textbf{Weighted H.M.} = \frac{\Sigma w_i}{\Sigma\left(\frac{w_i}{x_i}\right)} = \frac{1^2+2^2+3^2+\ldots n^2}{\frac{1^2}{1}+\frac{2^2}{2}+\frac{3^2}{3}+\ldots\frac{n^2}{n}} = \frac{1^2+2^2+3^2+\ldots+n^2}{1+2+3+\ldots+n}$$

$$= \frac{\{n(n+1)(2n+1)\}/6}{\{n(n+1)\}/2} = \frac{\mathbf{(2n+1)}}{\mathbf{3}}.$$

3.27 RELATION BETWEEN ARITHMETIC MEAN, GEOMETRIC MEAN AND HARMONIC MEAN

The arithmetic mean (A.M.), Geometric mean (G.M.) and Harmonic Mean (H.M.) for a given set of observations of a series are related as:

$$\textbf{A.M.} \geq \textbf{G. M.} \geq \textbf{H. M.}$$

Theorem 1: *Given two positive numbers a and b, prove that A.H. = G^2. Does the result hold for any set of observations ?*

Proof. For two positive numbers a and b, we have :

$$\text{A.M.} = \frac{a+b}{2}\ ; \qquad \text{G.M.} = \sqrt{ab}\ ;$$

$$\text{H.M.} = \frac{2}{(1/a)+(1/b)} = \frac{2ab}{a+b}.$$

$$(\text{A.M}) \times (\text{H.M}) = \frac{a+b}{2} \times \frac{2ab}{a+b} = ab = (\text{G.M})^2.$$

This result holds for only two positive observations or if the observations are in arithmetical progression.

Example 49. *If the arithmetic mean of two numbers is 10 and their geometric mean is 8, find their harmonic mean. Find the numbers also.*

Solution. We are given that: A.M. = 10, G.M. = 8.

Also we know that **(A.M.) (H.M.) = (G.M.)²**

or $\qquad (10)\ (\text{H.M.}) = 8^2 = 64 \quad \text{or} \quad \textbf{H.M.} = \frac{64}{10} = \mathbf{6.4}.$

Let the **two numbers be *a* and *b*.**

Now $\qquad \text{A.M.} = \frac{a+b}{2} = 10 \text{ or } a + b = 20.$

$\qquad \text{G.M.} = \sqrt{ab} = 8 \text{ or } ab = 64.$

But $\quad (a-b)^2 = (a+b)^2 - 4ab = 202 - 4 \times 64 = 144.$

$\therefore \qquad (a-b) = \sqrt{144} = 12.$

Again, $\quad a + b = 20$ and $a - b = 12$ implies that

$2a = 32$ or $a = 16$

and $\quad b = a - 12 = 16 - 12 = 4.$

Hence, **the numbers are 16, 4.**

EXERCISE – 3.3

1. A set of numbers consisting of four 4's, five 5's and six 6's and nine 9's. What is the mode ?
2. (a) The size of ten shirts in centimetres are given below: 90, 95, 90, 105, 100, 95, 90, 95, 100, 90. Find the modal size. If the size of one shirt is measured as 90 instead of 95, find the correct modal size.
   (b) When is mode said to be ill-defined ?
3. Compute the mode from the following data:

x:	6	12	18	24	30	36
f:	12	24	36	38	37	6.

4. Calculate the mode for the following frequency distribution:

Class-Interval	0 – 4	4 – 8	8 – 12	12 – 16
Frequency	4	8	5	6

5. Find out the mode from the following data showing frequency with which profits are made:

Profits (in '000 rupees)	3 – 4	4 – 5	5 – 6	6 – 7	7 – 8	8 – 9	9 – 10
Frequency	83	27	25	50	75	38	18.

6. Calculate the Geometric and Harmonic means of the following series of monthly expenditure of a batch of students:

Rs. 125 130 75 10 45 0.5 0.4 500 1505.

7. An auto ride in Delhi costs Re. one for the first kilometer and sixty paise for each additional kilometer. The cost for each kilometer is incurred at the beginning of kilometer so that the rider pays for a whole kilometer. What is the average cost for 2.75 kilometers?

8. An investor buys Rs. 1200 worth of shares in a company each month. During the first 5 months he bought the shares at a price of Rs. 10, Rs. 12, Rs. 15, Rs. 20 and Rs. 24 per share. After 5 months what is the average price paid for the shares.

[**Hint.** Use Harmonic mean = $\dfrac{5}{\dfrac{1}{10}+\dfrac{1}{12}+\dfrac{1}{15}+\dfrac{1}{20}+\dfrac{1}{24}}$ = **Rs. 14.63**]

9. A cyclist covers first three kms. at an average speed of 8 k.p.h. Another two kms. at 3 k.p.h. and the last two kms. at 2 k.p.h. Find the average speed for the entire journey.

[**Hint.** Average speed = Weighted H.M. = $\dfrac{3+2+2}{\left(\dfrac{1}{8}\times 3\right)+\left(\dfrac{1}{3}\times 2\right)+\left(\dfrac{1}{2}\times 2\right)}$ = **3.43 k.p.h.**]

10. The number of telephone calls received in 245 successive one minute intervals at an exchange are shown in the following frequency distribution.

No. of calls	0	1	2	3	4	5	6	7
Frequency	14	21	25	43	51	40	39	12

Evaluate mean, median and mode.

11. "The H.M., A.M., and G.M. of a set of 5 observations are 10.2, 16 and 14 respectively." Comment.
[**Hint.** A.M. > G.M. > H.M.]

12. A cyclist pedals from his house to his college at a speed of 10 k.p.h. and back from the college to his house at 15 k.p.h. Find his average speed.

[**Hint.** Average speed = $\dfrac{2x}{\dfrac{x}{10}+\dfrac{x}{15}}$ = **12 k.p.h.** where x is the distance of the college from the house.]

13. The number of workers in each section of a factory and their average monthly salaries are given below. Find the combined average salary of all workers in the factory.

Section	*No. of workers*	*Average of Monthly Salary*
A	50	113
B	60	120
C	70	115

[**Hint.** Combined mean of three series is given by : $\overline{X} = \dfrac{n_1\bar{x}_1 + n_2\bar{x}_2 + n_3\bar{x}_3}{n_1+n_2+n_3}$.]

14. A person walks 9 hours at a speed of 3 kms per hour and again walks 8 hours at a speed of 4 kms per hour. Find the average speed per hour by using appropriate average.

[Hint. Average Speed $= \dfrac{\textbf{Total distance}}{\textbf{Total time taken}} = \dfrac{9\times3+8\times4}{9+8} = \dfrac{59}{17}$ **= 3.47 km / hr.]**

15. The mean age of a combined group of men and women is 30 years. If the mean age of the group of men is 32 and that of the group of women is 27, find out the percentage of men and women in the group.

[Hint. $\overline{X}_{12} = \dfrac{N_1\overline{X_1} + N_2\overline{X_2}}{N_1 + N_2} \Rightarrow \dfrac{N_1 \times 32 + 400 - N_1 \times 27}{100} \Rightarrow N_1 = 60\%$ **...(1)**

$N_2 = 100 - N_1 = 100 - 60 = 40\%$.]

16. *The mean annual salaries paid to all employees of a company was Rs. 500. The mean annual salaries paid to male and female employees were Rs. 520 and Rs. 420 respectively. Determine the percentage of males and females employed by the company.*

[Hint. $\overline{X}_{12} = \dfrac{N_1\overline{X_1} + N_2\overline{X_2}}{N_1 + N_2}$.

$\therefore \quad 500 = \dfrac{N_1(520) + N_2(420)}{N_1 + N_2} \Rightarrow 500 = \dfrac{N_1(520) + N_2(420)}{100}$

$\Rightarrow \quad 100N_1 = 8000 \Rightarrow N_1 = 80\%. \therefore \quad N_2 = 100 - N_1 = 100 - 80 = 20\%$.]

17. The sum of deviations of a certain number of observations measured from 4 is 72 and the sum of observations of the same value measured from 7 is –3. Find the number of observations and their mean.

[Hint. $4 + \dfrac{72}{N} = 7 - \dfrac{3}{N} \Rightarrow \dfrac{72+3}{N} = 3 \Rightarrow \mathbf{N = 25}$.

Putting N = 25 we have : $\overline{X} = 4 + \dfrac{72}{25} = 4 + 2.88 = \mathbf{6.88}$.

Hence the number of observations is 25 and their mean is 6.88.]

18. Average marks in Statistics of 10 students of a class was 68. A new student took admisstion with 72 marks whereas two existing students left the college. If the marks of these student were 40 and 39. find the average marks of the remaining students.

[Hint. $\bar{X} = \dfrac{\textbf{Total marks of remaining students}}{\textbf{Number of remaining students}} = \dfrac{(68\times10)+72-(40+39)}{10+1-2}$ **= 74.77].**

19. Shri Naredra Kumar has increased his capital in three secruities, namely Reliance Ltd. Tisco Ltd. and Statyam Rs. 40.000, Rs 50.000 and 80.000 respectively. If he collect dividends of Rs. 10,000 from each company, compute his average return from three secruities.

[Hint. Average return from the three securities $= \dfrac{25+20+12.5}{3} = \dfrac{57.5}{3} = \mathbf{19.167\%}$.]

20. A candidate obtained the following perecentaage of marks in different subjects in the half yearly examination.

English	Statistics	Cost Accountancy	Economics	Income Tax
46%	67%	72%	58%	53%

It is agreed to give double weights to marks in English and Statistics as compared to other object. What is the simple and weighted arithmetic mean?

[Hint. Here $\Sigma x = 46 + 67 + 72 + 58 + 53 = 296$.

$$\Sigma WX \equiv 2 \times 46 + 4 \times 67 + 3 \times 72 + 4 \times 88 + 5 \times 53 = \mathbf{1073.}$$

$$\textbf{Simple Arithmetic Mean} = \frac{\Sigma X}{N} = \frac{296}{5} = \mathbf{5.92\%.}$$

$$\textbf{Weighted Arithmetic Mean} = \frac{\Sigma WX}{\Sigma W} = \frac{1073}{18} = \mathbf{59.61\%.]}$$

21. Given mean = 70.2 and mode = 70.5. Find median using empirical relationship among them.
[Hint. Mode = 3 Median – 2 Mean
Substituting the given values of mean and mode, we get:
$\Rightarrow$ 70.5 = 3 Median – 2 (70.2) $\Rightarrow$ 3 Median – 140.4 = 70.5.
or 3 Median = 70.5 + 140.4 = 210.9 $\Rightarrow$ **Median = 70.3.**

22. A person purchase one kilogram of cabbage from each of the four place at the rate of 20 kg., 16 kg., 12 kg. and 10 kg. per rupee respectively. On the average how many kg. of cabbage has he purchased per rupee?

$$\textbf{[Hint. H.M.} = \frac{N}{(1/x_1)+(1/x_2)+(1/x_3)+(1/x_4)} = \frac{4}{(1/20)+(1/16)+(1/12)+(1/10)}$$

$$= \frac{4 \times 240}{12+15+20+24} = \frac{960}{71} = \mathbf{13.52}\boldsymbol{kg}.\,]$$

23. In a certain office a letter is typed by A in 4 minutes. The same letter is typed by B, C and D in 5, 6 and 10 minutes respectively. What is the average time taken in completing one letter? How many letters do you expect to be typed in one day comprising eight working hours?

$$\textbf{[Hint. H.M.} = \frac{4}{\frac{1}{4}+\frac{1}{5}+\frac{1}{6}+\frac{1}{10}} = \frac{4}{\frac{30+24+20+12}{120}} = \frac{4 \times 120}{86} = \textbf{5.58 minutes per letter.}$$

Hence, the average time taken for typing one letter is 5.58 minutes.
Number of letters typed in one day comprising eight hours, *i.e.,*

$$\textbf{(8 × 60 = 480 minutes)} = \frac{\mathbf{480}}{\mathbf{5.58}} = \textbf{86.02 or 86 letters per day.]}$$

24. A man travelled by car for 3 days. He covered 480 miles each day. On the first day he drove for 10 hours at 48 miles an hour, on the second day he drove for 12 hours at 40 miles an hour and on last day he drove for 15 hours at 32 miles per hour. What was his average speed?

$$\textbf{[Hint. Harmonic Mean} = \frac{N}{\frac{1}{X_1}+\frac{1}{X_2}+\frac{1}{X_3}} = \frac{3}{\frac{1}{48}+\frac{1}{40}+\frac{1}{32}}$$

$$= 3 \times \frac{480}{37} = \textbf{38.92 m.p.h.]}$$

ANSWERS

1. 9. **2.** 90; 95. **3.** 24. **4.** 6.28. **5.** 7.40..
6. G.M. = Rs. 22.98, H.M. = Rs. 2.06. **7.** Rs. 0.80. **8.** Rs. 14.63.
9. 3.43 k.p.h. **10.** mean = 3.76, median = 4, mode = 4. **11.** The statement is correct.
12. 12 k.p.h. **13.** Rs. 116. **14.** 3.47 km/hr.. **15.** $N_1 = 40\%$, $N_2 = 60\%$
16. $N_1 = 20\%$, $N_2 = 80\%$ **17.** N = 25, Meen = 6.88. **18.** 74.77
19. 19.16%. **20.** 5.92%, 59.61%. **21.** 70.3 **22.** 13.52 Kg.
23. 5.58 minutes/letter, 86 letters/day. **24.** 38.92 m.p.h.

3.28 PARTITION VALUES

Median divides an arrayed series in ascending or descending series into two equal parts. When we are requires to divide an arrayed series into than two equal parts, the dividing places are known as partition values.

We know that one point divide a series into two equal parts called halves. Similarly, three points(3) divides the arrayed series into four parts called **quartiles, nine points** (9) divide it into ten parts called **deciles** and **ninety-nine (99)** point divide it into one hundred parts called **percentile.**

Quartiles, Deciles and Percentiles are called Partition Values.

For getting partition values, the most important rule is that the values must be arranged in ascending order only. In case of finding the medium, we could arrange the data either in ascending or descending order **but here there is no choice - only ascending order is possible for calculating partition values.**

3.29 DIFFERENCE BETWEEN AVERAGES AND PARTITION VALUES

An average (mean, median, mode, G.M., H.M.) ***is the representative of whole series,*** **but quartiles, deciles, percentiles** ***are averages of parts of the distribution (series).*** For example, third quartile (D_3) is the average of the second half of the series, first decile (D_1) is the average of first ten observations of the series and 35th percentile (P_{35}) is the average of first 35 observations of the series.

Thus, quartiles, deciles, percentiles are not averages like, mean, median and mode. Partition values help us in understanding how various values are scattered around median.

Thus, **partition values are used to study the scatteredness of the values of the variable in relation to the median.** Therefore, the special use of partition values is to study **dispersion of items in relation to the median,** ***i.e.,*** **it helps us in understanding the composition of a series.**

3.30 QUARTILES

Quartiles. ***Quartiles are those values of variate which divides the series, when arranged in ascending order, into four equal parts.***

Each portion contains equal number of items. The first, second and third points are termed as first quartile (Q_1), second quartile (better named as median) and third quartile (Q_3). The first quartile (Q_1) or lower quartile, has 25% of the items of the distribution below it and 75% of the items are greater than it. Q_2 (Median), the second quartile or median has 50% of the observations above it and 50% of the observations below it. The upper quartile or third quartile (Q_3) has 75% of the items of the distribution below it and 25% of the items are above it.

Note. It must be **noted that $Q_1 < Q_2 < Q_3$.**

3.30.1 Computation of Quartiles

CASE I. Computation of Quartile for Individual Series

Let $x_1, x_2, x_3, ..., x_n$ be n values of a variate X. We can compute ***quartiles*** of these values by the following method.

METHOD

Step I. *Arrange the given data in ascending order of magnitude.*

Step II. *Find the total number of observations. Let it be N.*

Step III. *Calculate the three quartiles Q_1, Q_2, Q_3 by the following formulae.*

$$Q_1 = \text{Value of } \left(\frac{N+1}{4}\right)^{\text{th}} \text{ observation in the arrayed series}$$

$$Q_2 = \text{Value of } \left(\frac{N+1}{2}\right)^{\text{th}} \text{ observation in the arrayed series if n is odd.}$$

$$Q_2 = \text{Mean of the values of } \left(\frac{N}{2}\right)^{\text{th}} \text{ and } \left(\frac{N}{2}+1\right)^{\text{th}} \text{ observations if n is even.}$$

$$Q_3 = \text{Value of } 3\left(\frac{N+1}{2}\right)^{\text{th}} \text{ observation in the arrayed series.}$$

Example 50. *Compute Q_1, Q_2 and Q_3 of the following data:*

13, 14, 7, 12, 17, 8, 10, 6, 15, 18, 21, 20

Solution. Arranging the given data in **ascending order**, we get:

6, 7, 8, 9, 10, 12, 13, 14, 15, 17, 18, 20, 21

Here n = 13.

$$Q_1 = \text{Value of } \left(\frac{N+1}{4}\right)^{\text{th}} \text{ term} = \text{Value of } \left(\frac{13+1}{4}\right)^{\text{th}}, \textit{ i.e., } \text{3.5th term} \quad (\because\ n = 13)$$

$$= \text{Value of 3rd term} + \frac{1}{2} \text{ (value of 4th term – 3rd term)}$$

$$= 8 + \frac{1}{2}(9-8) = 8 + \frac{1}{2} = \mathbf{8.5.}$$

$$Q_2 = \text{Median} = \text{Value of } \left(\frac{N+1}{2}\right)^{\text{tn}} \text{ term}$$

$$= \text{Value of } \left(\frac{13+1}{2}\right)^{\text{th}}, \textit{ i.e., } \text{7th term} = \mathbf{13.}$$

$$Q_3 = \text{Value of } 3\left(\frac{N+1}{4}\right)^{\text{th}} = \text{Value of } 3\left(\frac{13+1}{4}\right)^{\text{th}}, \textit{ i.e., } \text{10.5th term}$$

$$= \text{Value of 10th term} + \frac{1}{2} \text{ (value of 11th term – value of 10th term)}$$

$$= 17 + \frac{1}{2}(18-17) = 17 + 0.5 = \mathbf{17.5.}$$

Hence, Q_1 = **8.5**, Q_2 = **13** and Q_3 = **17.5.**

CASE II. Computation of Quartiles for a Discrete Frequency Distribution

Let the variate X takes n values $x_1, x_2, x_3, ..., x_n$ with frequencies $f_1, f_2, f_3, ..., f_n$ respectively. Then we calculate quartiles Q_1, Q_2, Q_3 by the following method.

METHOD

Step I. *Compute the cumulative frequencies.*

Step II. *Find* $N = \sum_{i=1}^{n} f_i$

Step III. Calculate the quartiles as under:

Q_1 = *Find the cumulative frequency just greater than* $\frac{N}{4}$. *Determine the corresponding value of the variable X. This value is the* ***lower quartile*** $\boldsymbol{Q_1}$.

Q_2 = *Find the cumulative frequency just greater than* $\frac{N}{2}$. *Determine the corresponding value of the variable X. This value is the* ***middle quartile*** $\boldsymbol{Q_2}$, ***i.e., the median.***

Q_3 = *Find the cumulative frequency just greater than* $\frac{3N}{4}$. *Determine the value of the variable X. This value is the* ***third quartile*** $\boldsymbol{Q_3}$.

Example 51. *The following data relate to the sizes of shoes sold at a store during a given week. Find upper and lower quartile from the given data.*

Find Q_1 *and* Q_3 *of the following series:*

Size of shoes	*Frequency*	*Size of shoes*	*Frequency*
4.0	*10*	*6.0*	*40*
4.5	*18*	*6.5*	*15*
5.0	*22*	*7.0*	*10*
5.5	*25*	*7.5*	
		8.0	*7*

Solution. Let us arrange the values of the variate in ascending order and prepare the cumulative frequency table.

Table: Calculation of Q_1 and Q_3

Size of shoes	*Frequency*	*c.f.*
4.0	10	10
4.5	18	28
5.0	22	**50**
5.5	25	75
6.0	40	115
6.5	15	**130**
7.0	10	140
7.5	8	148
8.0	7	155

Here N = 155.

Lower Quartile Q_1 = Size of $\frac{(N+1)}{4}$ th item = Size of $\frac{(155+1)}{4}$ th item

= Size of 39th item = **5** [$\because$ **39 lies in the row with c.f. 50]**

Upper Quartile Q_3 = Size of $\frac{3(N+1)}{4}$ th item = Size of $3\frac{(155+1)}{4}$ th item

= Size of 117th item = **6.5** **[∵ 117 lies in the row with c.f. 130]**

Hence, Q_1 **= 5 and** Q_3 **= 6.5.**

CASE III. Computation of Quartiles for a Frequency Distribution with Class-Intervals

The following method is used to compute quartiles Q_1, Q_2, Q_3 for a continuous distribution.

METHOD

Step I. *Compute the cumulative frequency of the given distribution. Let* $N = \sum_{i=1}^{n} f_i$

Step II. *Compute* $\frac{iN}{4}$, *where i = 1 for lower quartile* Q_1, *i = 2 for middle quartile* Q_2 *(or median), i = 3 for upper quartile* Q_3.

Step III. *Find the cumulative frequency just greater than* $\frac{iN}{4}$ *and the corresponding class. This class is called the* ***quartile class.***

Step IV. *Use the following formula to calculate* Q_1, Q_2 *or* Q_3.

$$Q_i = L + \frac{(i \times N/4) - C}{f} \times h, \; i = 1, 2, 3;$$

where L = **Lower limit of the class in which a particular quartile lies,**

f = **Frequency of the class-interval in which a particular quartile lies**

h = **Width of the class-interval of the class in which a particular quartile lies,**

C = **Cumulative frequency of the class preceding the class in which the quartile lies.**

Example 52. *Calculate median and the three quartiles from the following data:*

Marks :	*0 – 10*	*10 – 20*	*20 – 30*	*30 – 40*	*40 – 50*	*50 – 60*	*60 – 70*	*70 – 80*
No. of Students :	*5*	*7*	*8*	*12*	*28*	*22*	*10*	*8.*

Solution.

Size of shoes	*Frequency*	*c.f.*
0 – 10	5	5
10 – 20	7	12
20 – 30	8	20
30 – 40	12	32
40 – 50	28	60
50 – 60	22	82
60 – 70	10	92
70 – 80	8	100

Median = Size of $\left(\frac{n}{2}\right)$**th item** = $\frac{100}{2}$ th item = Size of 50th item. **[Here** n **= 100]**

Now, **the median lies in the group 40 – 50 and** $L = 40$; $\frac{n}{2} = 50$; $C = 32$; $h = 10$; $f = 28$.

$\therefore$ **Median** $(Q_2) = L + \frac{(n/2) - C}{f} \times h = 40 + \frac{50 - 32}{28} \times 10$

or **Median** $= 40 + \frac{18 \times 10}{28} = 40 + 6.43 =$ **46.43.**

Again Q_1 **= size of** $\left(\frac{n}{4}\right)$ **th item** = $\frac{100}{4}$ th, *i.e.*, size of 25th item.

Q_1 lies in the group 30 – 40 and $L = 30$, $C = 20$, $h = 10$, $f = 12$

$\therefore$ $Q_1 = L + \frac{(n/4) - C}{f} \times h = 30 + \frac{25 - 20}{12} \times 10 =$ **34.17.**

Also Q_3 **= size of** $\left(\frac{3n}{4}\right)$ **th item**, *i.e.*, size of 7th item, *i.e.*,

Q_3 lies in the group 50 – 60 and $L = 50$, $C = 60$, $h = 10$ and $f = 22$.

$Q_3 = L + \frac{(3n/4) - C}{f} \times h = 50 + \frac{75 - 60}{22} \times 10 = 50 + \frac{15}{22} \times 10 =$ **56.81.**

Hence, $Q_1 = 34.17$, Median = $Q_2 = 46.43$ and $Q_3 = 56.81$.

3.32 DECILES

***The value of the variable which divides the series, when arranged in ascending order, into 10 equal parts is called a decile. The nine points which divide the given series (when arranged in ascending order) into ten parts are called* deciles.**

Deciles are denoted by $D_1, D_2, D_3, ..., D_9$. The first decile, D_1, is the value of the variable such that it exceeds 10% of the observations and is exceeded by 90% of the observations. Similarly, D_6, the sixth decile, has 60% observations before it and 40% observations after it.

The fifth decile, D_5, is the median of the given data.

2.32.1 Computation of Deciles

CASE I. Computation of Deciles for Individual Series

In this case, the ***k*th decile is given by**

$$D_k = \text{Value of } k\left(\frac{n+1}{10}\right)\text{th term, } k = 1, 2, 3, 4, ..., 9,$$

when the series is arranged in ascending order.

Example 53. *Compute D_3 of the following data:*

13, 14, 7, 12, 9, 17, 8, 10, 6, 15, 18, 21, 20

Solution. Arranging the given data in **ascending order,** we get:

6, 7, 8, 9, 10, 12, 13, 14, 15, 17, 18, 20, 21.

Here $n = 13$.

D_3 **= Value of** $3\left(\frac{n+1}{10}\right)^{\text{th}}$ **term** = Value of $3\left(\frac{13+1}{10}\right)^{\text{th}}$ term

= Value of 4.2th term = Value of 4th term + $\frac{1}{5}$ **(value of 5th term – value of 4th term)**

$= 9 + \frac{1}{5}(10 - 9) = 9.2$

Thus, **3rd decile = D_3 = 9.2.**

CASE II. Computation of Deciles for a Discrete Frequency Distribution

METHOD

Step I. *Arrange the given data in ascending order.*

Step II. *Compute the cumulative frequencies*

Step III. *Find* $N = \Sigma f_i$

Step IV. *Find* $\frac{iN}{10}$ *to compute* D_i, *the ith decile,* $i = 1, 2, 3, ..., 9$.

Step V. *Find the cumulative frequency just greater than* $\frac{iN}{10}$. *Then the corresponding value of variable is the ith decile* D_i, $i = 1, 2, 3, ..., 9$.

Example 54. *Compute* D_3 D_6 *and* D_7 *from the following data:*

Marks :	10	20	30	40	50	80
No. of Students :	4	7	15	18	7	2.

Solution. The cumulative frequency distribution is given below:

Marks	*No. of Students* (f)	*Cumulative Frequency (c.f.)*
10	4	4
20	7	11
30	**15**	**26**
40	**18**	**44**
50	7	51
80	2	53
	$N = \Sigma f_i = 53$	

Third Decile = D_3. Here $\frac{3N}{10} = \frac{3 \times 53}{10} = \frac{159}{10}$ **= 15.9.**

The cumulative frequency just greater than $\frac{3N}{10}$ **= 15.9 is 26 and the corresponding value of the variable is 30. Thus** D_3 **= 30.**

Sixth Decile = D_6. Here $\frac{6N}{10} = \frac{6 \times 53}{10}$ **= 31.8.**

The cumulative frequency just greater than $\frac{6N}{10}$ **= 31.8 is 44 and the corresponding value of the variable is 40. Thus** D_6 **= 40.**

Seventh Decile = D_7. Here $\frac{7N}{10} = \frac{7 \times 53}{10}$ **= 37.1.**

The cumulative frequency just greater than $\frac{7N}{10}$ **= 37.1 is 44 and the corresponding value of the variable is 40. Thus** D_7 **= 40.**

Hence, D_3 **= 30,** D_6 **= 40 and** D_7 **= 40.**

CASE III. Computation of Deciles for a Frequency Distribution with Class-Intervals.

METHOD

Step I. *Compute the cumulative frequency table. Let* $N = \Sigma f_i$.

Step II. *Compute* $\frac{iN}{10}$ *to find* D_i, *the ith decile i = 1, 2, 3, ..., 9.*

Step III. *Find the cumulative frequency just greater than* $\frac{iN}{10}$ *and the corresponding class. This* ***class is called the decile class.***

Step IV. *Use the formula*

$$D_i = L + \frac{\frac{iN}{10} - C}{f} \times h,$$

where **L = Lower limit of the decile class**

C = Cumulative frequency of the class preceding the decile class

f = frequency of the ith decile class

h = width of the ith decile class.

Example 55. *Compute* Q_3 *and* D_7 *for the following frequency distribution:*

Marks :	*0 – 10*	*10 – 20*	*20 – 30*	*30 – 40*	*40 – 50*	*50 – 60*	*60 – 70*	*70 – 80*
No. of Students :	*3*	*10*	*17*	*7*	*6*	*4*	*2*	*1*

Solution. The cumulative frequency distribution is given by

Marks	*No. of Students (f)*	*Cumulative Frequency (c.f.)*
0 – 10	3	3
10 – 20	10	13
20 – 30	17	30
30 – 40	7	37
40 – 50	6	43
50 – 60	4	47
60 – 70	2	49
70 – 80	1	50
	$N = \Sigma f_i = 50$	

Third Quartile (Q_3) : $N = 50 \Rightarrow \frac{3N}{4} = \frac{3 \times 50}{4} = 37.5.$

The cumulative frequency just greater than $\frac{3N}{4}$ = 37.5 is 43, so 40 – 50 is the upper quartile class such that: $L = 40$, $h = 10$, $f = 6, f = 37$

$$\therefore \quad Q_3 = L + \frac{\frac{3N}{4} - C}{f} \times h = 40 + \frac{37.5 - 37}{6} \times 10 = \mathbf{40.83.}$$

Seventh Decile = D_7. Here $N = 50$; $\frac{7N}{10} = \frac{7 \times 50}{10} = 35.$

The cumulative frequency just greater than 35 is 37 and the corresponding class is 30 – 40. Thus, the 7th decile class is 30 – 40, such that $L = 30$, $h = 10$, $f = 7$, $C = 30$.

$$\therefore \quad D_7 = L + \frac{\frac{7N}{10} - C}{f} \times h = 30 + \frac{35 - 30}{7} \times 10 = \mathbf{27.14.}$$

Hence, Q_3 = **40.83 marks and** D_7 = **24.14 marks.**

3.33 PERCENTILES

The value of the variables which divides the series, when arranged in ascending order, into 100 equal parts are called percentiles. There are 99 percentiles denoted by P_1, P_2, P_3, P_4, ..., P_{99} respectively.

The ninety-nine points which divide the given data, when arranged in ascending order, into hundred equal parts are called percentiles of the data.

P_{30} is the value of variable such that it exceeds 30% of the observations and is exceeded by 70%. Similarly, P_{50} is the median.

2.36.1 Computation of Percentiles

CASE I. Computation of Percentile of Individual Series.

In this case, **the *k*th percentile** is given by

$$P_k = \text{value}\left[k\left(\frac{n+1}{100}\right)\right]^{\text{th}} \text{ term, when}$$

arranged in ascending order, k = 1, 2, 3, ..., 100.

CASE II. Computation of Percentiles for Discrete Frequency Distribution

Step I. *Arrange the given data in ascending order.*

Step II. *Compute the cumulative frequencies.*

Step III. *Find $N = \Sigma f_i$.*

Step IV. *Compute $\frac{i \times N}{100}$ to compute P_i, the ith percentile, i = 1, 2, 3, ..., 100.*

Step V. *Find the cumulative frequency just greater than $\frac{iN}{100}$ and the corresponding value of the variable. This value is the D_i, the ith percentile of the given data.*

Example 56. *Calculate D_4 and P_{30} from the following data.*

Marks :	20	25	30	35	45
No. of Students :	5	14	20	11	10.

Solution. **Table: Cumulative Frequency Table**

Marks	*Frequency (f)*	*Cumulative Frequency (c.f.)*
20	5	5
25	14	19
30	20	39
35	11	50
40	10	60
	Total: $N = \Sigma f = 60$	

$$D_4 = \textbf{Value of} \left(\frac{4N}{10}\right)^{\text{th}} \textbf{item} = \text{Value of} \left(\frac{4 \times 60}{10}\right)^{\text{th}} \text{item}$$

= Value of 24th item = **30.** **[∵ 19 < 24 < 39]**

[$\because$ The cumulative frequency just greater than $\frac{4N}{10}$ = 24 is 39 and the corresponding value of marks is 30]

$$P_{30} = \textbf{Value of} \left(\frac{30N}{100}\right)^{\text{th}} \textbf{item} = \text{Value of} \left(\frac{30 \times 60}{100}\right)^{\text{th}} \text{item}$$

= Value of 18th item = **25.** **[$\because$ 5 < 18 < 19]**

[$\because$ The cumulative frequency just greater than $\frac{30N}{100}$ = 18 is 19 and the corresponding value of marks is 25]

Case III. Computation of Percentiles for a Frequency Distribution with Class-Intervals.

Step I. *Compute the cumulative frequency table. Let* $N = \Sigma f_i$.

Step II. *Compute* $\frac{i \times N}{100}$ *for* P_i, *the ith percentile,* $i = 1, 2, 3, ..., 99$.

Step III. *Find the cumulative frequency just greater than* $\frac{i \times N}{100}$ *and the corresponding class. This class is called* ***ith percentile class.***

Step IV. *Use the formula:* $P_i = L + \dfrac{\frac{i \times N}{100} - C}{f} \times h,$

where **L = lower limit of percentile class**

C = cumulative frequency of the class precending the percentile class

f = frequency of the percentile class

h = width of the percentile class.

Example 57. *Calculate* $Q_1, Q_3, D_2, P_5, P_{90}$ *from the following data:*

Marks :	0 – 10	10 – 20	20 – 30	40 – 60	60 – 80	80 – 100
Number of Students :	8	10	22	25	10	5

Solution. Let us prepare the following table showing the frequency and cumulative frequency.

Marks	*Number of Students (f)*	*Cumulative Frequency (c.f.)*
0 – 10	8	8
10 – 20	10	18
20 – 40	22	40
40 – 60	25	65
60 – 80	10	75
80 – 200	5	80
	Total = Σf = 80.	

Here ***n*** **= 80.**

Now $Q_1 = \left(\frac{n}{4}\right)$ **th value** = $\left(\frac{80}{4}\right)$th value = **20th value.**

Here Q_1 **- class = 20 – 40,** $L = 20,\quad \frac{n}{4} = 20;\quad C = 18, f = 22, i = 20.$

$$\therefore \quad Q_1 = L + \frac{(n/4) - C}{f} \times i = 20 + \frac{20 - 18}{22} \times 20 = 20 + 1.8 = \textbf{21.8 marks.}$$

Again $Q_3 = \left(\frac{3 \times n}{4}\right)$**th value** = $\left(\frac{3 \times 80}{4}\right)$th value = **60th value.**

Here Q_3 **- class = 40 – 60,** $L = 40,$ $\frac{3 \times n}{4} = 60,$ $C = 40, f = 25, i = 20.$

$$\therefore \quad Q_3 = L + \frac{(3 \times n)/4 - C}{f} \times i = 40 + \frac{60 - 40}{25} \times 20 = 40 + 16 = \textbf{56 marks.}$$

Again $D_2 = \frac{2 \times n}{10}$ **th value** = $\frac{2 \times 80}{10}$ th value = **16th value.**

∴ Here D_2 **– class = 10 – 20**; $L = 10,$ $\frac{2 \times n}{10} = 16,$ $C = 8, f = 10, i = 10.$

$$\therefore \quad D_2 = L + \frac{(2 \times n)/10 - C}{f} \times i = 10 + \frac{16 - 8}{10} \times 10 = 10 + 8 = \textbf{18 marks.}$$

Again $P_5 = \left(\frac{5 \times n}{100}\right)$**th value** = $\frac{5 \times 80}{100}$ = 4th value.

Here P_5 **- class = 0 – 10**; $L = 0,$ $\frac{5 \times n}{100} = 4,$ $C = 0, f = 8, i = 10.$

$$\therefore \quad P_5 = L + \frac{(5 \times n)/100 - C}{f} \times i = 0 + \frac{4 - 0}{8} \times 10 = \textbf{5 marks.}$$

Again $P_{90} = \left(\frac{90 \times n}{100}\right)$**th value** = $\frac{90 \times 80}{100}$ th value = 72nd value.

Here P_{90} **- class = 60 – 80**; $L = 60,$ $\frac{20 \times n}{100} = 72,$ $C = 65, f = 10, i = 20$

$$\therefore \quad P_{90} = L + \frac{(90 \times n)/100 - C}{f} \times i = 60 + \frac{72 - 65}{10} \times 20 = 60 + 14 = \textbf{74 marks.}$$

Example 58. *Given below are the marks obtained by 50 students appearing for an admission test:*

Marks :	*0 – 10*	*10 – 20*	*20 – 30*	*30 – 40*	*40 – 50*
No. of Students :	*6*	*8*	*20*	*9*	*7*

If cut off point was 34, find the percentage of students scoring more than 34 marks.

Solution. Suppose x % students scored 34 marks. Then, number of students scoring class 34 is x ⇒ x th percentile = 34 or $P_x = 34$.

Table: Cumulative Frequency

Marks	*Number of Students* (f)	*Cumulative Frequency* ($c.f$)
0 – 10	6	6
10 – 20	8	14
20 – 30	20	34
30 – 40	9	43
40 – 50	7	50
	$N = \Sigma f_i = 50$	

Now, $P_x = 34$ which lies in the interval 30 – 40, so, the xth **percentile class is 30 – 40** such that $L = 30$, $h = 10$, $f = 9$, $N = 50$ and $C = 34$.

$$\therefore\ P_x = L + \frac{\frac{xN}{100} - C}{f} \times h \quad \Rightarrow \quad 34 = 30 + \frac{\frac{x \times 50}{100} - 34}{9} \times 10$$

$$\Rightarrow 4 \times 9 = \left(\frac{x}{2} - 34\right) \times 10 \quad \Rightarrow \quad 5x = 376 \quad \Rightarrow \quad x = 75.2.$$

Thus, **75.2% students scored upto 34 marks and hence the remaining 24.8% students scored more than 34 marks.**

Example 59. *In a small branch of a bank in a rural area the following is the average deposit balance of current accounts (x) during a month. Calculate median, seventh decile and 85th percentile.*

Deposit balance less than	*No. of deposits*	*Deposit balance less than*	*No. of deposits*
1000	*500*	*500*	*374*
900	*498*	*400*	*300*
800	*480*	*250*	*125*
600	*475*	*100*	*25.*
550	*440*		

Solution.

Table : Computation of Median, Decilies and Percentiles

x	f	$c.f.$
0 – 100	25	25
100 – 250	125 – 25 = 100	125
250 – 400	300 – 125 = 175	300
400 – 500	374 – 300 = 74	374
500 – 550	440 – 374 = 66	440
550 – 600	475 – 440 = 35	475
600 – 800	480 – 475 = 5	480
800 – 900	498 – 480 = 18	498
900 – 1000	500 – 498 = 2	500

MEDIAN: **Median** = **Size of $N/2^{th}$ item** = Size of $(500/2)^{th}$ item

= **250th item, which lies in the class 250 – 400.**

$$\textbf{Median} = L_1 + \frac{\frac{N}{2} - c.f.}{f} \times i.$$

Here $L_1 = 250$; $N/2 = 250$; $c.f. = 125$; $f = 175$; $i = 150$.

$$\therefore \quad \textbf{Median} = 250 + \frac{250 - 125}{175} \times 150 = 250 + 107.14 = \mathbf{357.14}.$$

DECILES: D_7 = **Size of** $\frac{7N}{10}$ **th item** = Size of $\left(\frac{7 \times 500}{10}\right)^{th}$ item

= **350th item, which lies in the class interval 400 – 500.**

$$D_7 = L_1 + \frac{\frac{7N}{10} - c.f.}{f} \times i.$$

Here $L_1 = 400; \quad \frac{7N}{10} = 350; \quad c.f. = 300; \quad f = 74; \quad i = 100.$

$\therefore \quad D_7 = 400 + \frac{350 - 300}{74} \times 100 = 400 + 67.57 = \mathbf{467.57.}$

PERCENTILES: P_{85} = **Size of 85 ×** $\frac{N}{100}$ **th item** = Size of 85 × $\frac{500}{100}$ th item

= **425th item, which lies in the class interval 500 – 550.**

Here $L_1 = 500; \quad \frac{85N}{100} = 425; \quad c.f. = 374; \quad f = 66; \quad i = 50.$

$$P_{85} = L_1 + \frac{\frac{85N}{100} - c.f.}{f} \times i$$

$$= 500 + \frac{425 - 374}{66} \times 50 = 500 + 38.64 = \mathbf{538.64.}$$

Hence, Median = 357.14; D_7 = 467.57 and P_{85} = 538.64.

Example 60. *Find the 45th and 57th percentiles of the following data on marks obtained by 100 students.*

Marks :	*20 – 25*	*25 – 30*	*30 – 35*	*35 – 40*	*40 – 45*	*45 – 50*
No. of Students :	*10*	*20*	*20*	*15*	*15*	*20*

Solution. *The value of the variable which divide the series, when arranged in ascending or descending order into 100 equal parts is called a* **percentile.** In other words, when a series is divided into 100 equal parts, it gives 99 positions in the series and they are called **percentiles.**

Table : Cumulative Frequency Table

Marks	*Number of Students (f)*	*Cumulative Frequency (c.f.)*
20 – 25	10	**10**
25 – 30	20	10 + 20 = **30**
30 – 35	20	30 + 20 = **50**
35 – 40	15	50 + 15 = **65**
40 – 45	15	65 + 15 = **80**
45– 50	20	80 + 20 = **100**

$P_{45} = \frac{\mathbf{45 \times n}}{\mathbf{100}} = \frac{45 \times 100}{100}$ = **45.** Similarly, P_{57} = **57.**

45th Percentile: P_{45} class is 30 – 35. $\quad \therefore L = 30, \; C = 30, \; i = 5, \; f = \mathbf{20}$

$\therefore \; P_{45} = 30 + \frac{45 - 30}{20} \times 5 = 30 - 1.25 = \mathbf{28.75.}$

57th Percentile:

In this case P_{57} = 57, which lies in the class 35 – 40. $\quad \therefore \quad C = 50, \; i = 5, \; f = 15, \; L = 35$

$P_{57} = L + \frac{(57n/100) - C}{f} \times i = 35 + \frac{57 - 50}{15} \times 15 = 35 - 2.33 = \mathbf{32.6\%}.$

Example 61. *Following is the distribution of marks obtained by 100 students in a class test:*

Marks more than	*0*	*10*	*20*	*30*	*40*	*50*	*60*	*70*
Number of Students	*100*	*92*	*77*	*57*	*40*	*25*	*15*	*5*

If 70% of the students passed the class test, find the minimum pass marks by a pass student.

Solution. Here 70% of the students passed the class test and 30% failed. Thus 30% are below pass mark and 70% are above the pass marks. The minimum pass marks are just greater than P_{30}.

Table: Cumulative Frequency

Marks	*Number of Students (f)*	*Cumulative Frequency (c.f.)*
0 – 10	8	8
10 – 20	15	23
20 – 30	20	**43**
30 – 40	17	60
40 – 50	15	75
50 – 60	10	85
60 – 70	10	95
70 and above	5	100
	$N = \Sigma f_i = 100$	

Here $N = 100$. $\therefore \quad \dfrac{30N}{100} = \dfrac{30 \times 100}{100} = 30.$

The cumulative frequency just greater than 30 is 43 and the corresponding class is 20 – 30. Thus, 20 – 30 is **30th percentile class** such that

$L = 20, h = 10, f = 20$ and $C = 23$.

$$\therefore \; P_{30} = L + \frac{\frac{30N}{100} - C}{f} \times h = 20 + \frac{30 - 23}{20} \times 10 = 20 + \frac{7}{2} = 23.5.$$

Hence, **the minimum marks to pass the test are 24.**

3.34 MERITS AND DEMERITS OF PARTITION VALUES

Merits:

1. *Partition values can be computed for distributions with open ended class-intervals.*
2. *The partition values are not affected by extreme values.*
3. *One of the partition values, viz., the median, can be obtained graphically.*

Demerits:

1. *While computing partition values, the data have to be arranged in ascending order, whereas it is not so in case of other measures of central tendency.*
2. *Partition values do not depend on all the observations.*

3.35 GENERAL REVIEW OF DIFFERENT MEASURES OF CENTRAL TENDENCY

The **best measure of central tendency is the Arithmetic Mean (A.M.).** It is rigidly defined and is based on all the observations. It is easy to comprehend, simple to calculate and amenable to mathematical properties. However, A.M. has one drawback in the sense that it is very much affected by sampling fluctuations. In case of frequency distribution, mean cannot be calculated for open end classification.

Median, like A.M., is also rigidly defined and easy to comprehend and compute. But median is not based on all the observations. It does not allow itself to mathematical treatment. However, median is not much affected by sampling fluctuations. It is the most appropriate measure of central tendency for an open-end classifications.

Mode is the most popular measure of central tendency but there are cases when mode remains undefined or **the distribution has more than one mode.** Unlike mean, it has no mathematical property. Mode is also affected by sampling fluctuations.

G.M. and H.M. possess some mathematical properties. They are rigidly defined and based on all the observations. But they are difficult to comprehend and compute and, as such, have limited applications for the computation of average rates and ratios and such like things.

3.36 CHOICE OF AN AVERAGE FOR DECISION MAKING

We have studied the various kinds of averages such as: **Arithmetic mean, Median, Mode, Geometric mean and Harmonic mean.** It is important to know **when and how to use which average?** Thus averages cannot be used indiscriminately. A judicial selection of averages for sound statistical analysis depends upon the following factors:

(*i*) *The nature of the variable involved.*

(*ii*) *The purpose of analysis.*

(*iii*) *The system of classification adopted.*

(*iv*) *The quality, nature and availability of data.*

(*v*) *The study of average for further statistical computation required for the enquiry in mind.*

We give below the suitability of some of the averages.

Arithmetic Mean. It is, generally, used in business.Whenever, we talk of average cost of production or sale or average wages, we use arithmetic mean. It is also used for further statistical calculations such as **standard deviation.** Arithmetic mean is not recommended while dealing with frequency distribution with extreme observations or open end classes.2.21 Partition Values

Median divides an arrayed series in ascending or descending series into two equal parts. When we are required to divide an arrayed series into more than two equal parts, the dividing places are known as **partition values.**

We know that one point divide a series into two equal parts called **halves.** Similarly, **three points (3)** divides the arrayed series into four parts called **quartiles**, **nine points** *(9)* divide it into ten parts called **deciles** and **ninety-nine (99)** points divide it into one hundred parts called **percentile.**

Quartiles, Deciles and Percentiles are called Partition values.

For getting partition values, the most important rule is that the values must be arranged in ascending order only. In case of finding the median, we could arrange the data either in ascending or descending order **but here there is no choice — only ascending order is possible for calculating partition values.Median.** It is to be used for finding the average when the data is qualitative, *i.e.*, for finding average of intelligence, honesty, beauty etc., median is the only average to be used. It is a **positional average** as it divides the entire series in two equal parts, 50% of actual values will be below and 50% will be above it. It is suitable when there are **open extreme classes** or where there are **extreme values.** It is commonly used for **average wages of worker** as it would avoid the influence of a few very high or very low wage rates.

Mode. It is a **positional average**. It is to be used while dealing with **open end classes.** It is particularly used in **business,** when the businessman is not interested in the magnitude but only in the most common or fashionable value.

Geometric and Harmonic Means. Geometric mean and Harmonic mean are known as **ratio averages** as they are most appropriate where the data comprise **rates, ratios or percentages instead of actual quantities.** *Geometric mean is to be used while dealing with rates and ratios. Harmonic mean is to be used in compiling special types of average rates or ratios, where time factor is variable and the act being performed, e.g., distance is constant.*

EXERCISE – 3.4

1. Explain the difference between averages of a distribution and its partition values.

2. Following are the scores of 12 students in a class test: 15, 9, 18, 20, 21, 26, 14, 13, 27, 22, 16, 28 Find D_7 and P_{33}.

3. Calculate Q_1, Q_3, D_8, P_{35} and P_{80} from the following data;

Marks :	20	30	40	50	60
No. of Students :	4	16	20	18	11.

4. Find first quartile, third quartile, median and third decile of the following distribution:

Height (in inches) :	58	59	60	61	62	63	64	65	66
No. of Persons :	2	3	6	15	10	5	4	3	1.

5. Calculate the median, 3rd decile and 20th percentile from the following data:

x :	0 – 5	5 – 10	10 – 15	15 – 20	20 – 25
f :	7	18	25	30	20.

6. Compute Q_1, Q_3, D_6, P_8 and P_3 from the following data:

Marks :	0 – 10	10 – 20	20 – 40	40 – 60	60 – 80	80 – 100
No. of Students :	8	10	22	25	10	5.

7. Calculate the median, Q_3, D_7 and P_{70} of the following data:

Marks :	0 – 10	10 – 20	20 – 30	30 – 40	40 – 50	50 – 60	60 – 70	70 – 80
No. of Students :	3	10	17	7	6	4	2	1.

8. Find Q_1, Q_3, P_{40} and P_{84} from the following data:

Marks :	0 – 7	7 – 14	14 – 21	21 – 28	28 – 35	35 – 42	42 – 49
No. of Students :	3	4	7	11	2	14	9.

ANSWERS

1. $Q_1 = 11.5$, $Q_3 = 31.5$ and median = 18.

2. $D_7 = 24$, $P_{33} = 15.5$.

3. $Q_1 = 30$, $Q_3 = 50$, $D_8 = 50$, $P_{35} = 40$, $P_{80} = 50$.

4. $Q_1 = 61$, $Q_3 = 63$, Median = 61, $D_3 = 61$.

5. Median = 15, $D_3 = 11$, $P_{20} = 8.6$

6. $Q_1 = 21.8$, $Q_3 = 56$, $D_6 = 46.4$, $P_8 = 8$, $P_{30} = 25.45$.

7. Median = 20.07 marks, $Q_3 = 40.83$ marks, $D_7 = 37.14$ marks, $P_{70} = 37.14$ marks.

8. $Q_1 = 19.5$, $Q_3 = 40.5$, $P_{40} = 25$, $P_{64} = 37.5$.

MEASURES OF DISPERSION OR VARIATION

4.1 DISPERSION OR VARIATION.

The **meaning of variation or dispersion is scatteredness.** It helps in finding out the **variability of the data** or scatteredness of individual items from an appropriate measure of central tendency in a given distribution. In other words, **the degree to which numerical data tend to spread about an average value is called the dispersion of the data.**

These are two types of measures of dispersion or variation :

1. Absolute measures of dispersion **2. Relative measures of dispersion**

I. Absolute measures of dispersion. The absolute measures of dispersion are :

(*i*) **Range** (*ii*) **Quartile Deviation**

(*iii*) **Mean Deviation** (*iv*) **Standard Deviation**

II. Relative measures of dispersion. Corresponding to these four absolute measures of dispersion, we have the following four relative measures of dispersions or variations :

(*i*) **Coefficient of Range** (*ii*) **Coefficient of Quartile**

(*iii*) **Coefficient of Mean Deviation** (*iv*) **Coefficient of Variation or Coefficient of Standard Deviation**

4.1.1 Difference between Absolute and Relative Measures of Dispersion

The following are the points of distinction between the absolute and relative measures of dispersion :

I. ***Absolute measures are dependent on the units of the variable under consideration whereas the relative measures of dispersion are unit free.***

II. ***The relative measures are used only for the purpose of comparison between two or more series with varying size or the number of items or varying central values or varying units of calculations.***

III. ***Compared to absolute measures of dispersion, relative measures of dispersion are difficult to compute and comprehend.***

4.1.2 Characteristics for an Ideal Measure of Dispersion or Variation

An ideal measure of dispersion should be properly defined, and easy to comprehend, simple to compute, based on all the observations and unaffected by sampling fluctuations. It should be amenable to some desirable mathematical treatment.

4.2 RANGE

It is the simplest measure of dispersion. ***It is the difference between the minimum and maximum items of the series.***

For example, in the series 20, 21, 22, 25, 30, 32, 37, 47, 65 the range is 65 – 20 = 45.

Absolute Range or Range = $x_{max} - x_{min}$ **or Range =** $L - S$

where L **is the largest value and** S **is the smallest value of the term.**

$$\textbf{Coefficient of Range or Relative Range} = \frac{\textbf{Absolute Range}}{\textbf{Sum of the two extremes values}} = \frac{L-S}{L+S}$$

4.2.1 Range for Continuous Frequency Distribution

Range in continuous series is determined by the following two methods :

Method I. *Find the differnce between the upper limit of the highest wage class and the lower limit of the lowest wage class.*

Method II. *Subtract the mid-point of the lowest wage class from the mid-point of the highest wage class.*

4.2.2 Properties of Range

1. ***Range remains unaffected due to a change of origin but is affected in the same ratio due to a change in scale, i.e., if for any two constants a and b, the two variables x and y are related by y = a + bx, then the range of y is given by :*** $R_y = |b| \times R_x$.
2. ***If all the observations assumed by a variable are constant, say k, then the range is zero.***

4.2.3 Merits, Demerits and Uses of Range

Merits:

1. *It can be easily understood.*
2. *It is easy to calculate and it is the simplest method of measuring dispersion.*
3. *It lends itself to algebraic treatments.*
4. *It is an absolute measure of dispersion.*

Demerits:

1. *It is too indefinite to be used as a practical measure of dispersion because it depends entirely upon the extreme values.*
2. *It is not based on all the observations.*
3. *It is affected by sampling fluctuations.*

Uses:

It is used in quality control.

Example 1. *Find the range and coefficient of range of the weights of 10 students from the following data:*

41 20 15 65 73 84 53 35 71 55

Solution. Arranging the data in the **ascending order**, we get:

15 20 35 41 53 55 65 71 73 84

Here **largest value : L = 84,** **Smallest value : S = 15**

$$\therefore \quad \textbf{Range} = \mathbf{L - S} = 84 - 15 = \mathbf{69.}$$

$$\textbf{Coefficient of Range} = \frac{\mathbf{L-S}}{\mathbf{L+S}} = \frac{84-15}{84+15} = \frac{69}{96} = \mathbf{0.696.}$$

Example 2. *Find the range and the coefficient of range of the marks obtained by 100 students given below:*

Marks	*No. of Students*	*Marks*	*No. of Students*
0 – 10	*13*	*40 – 50*	*11*
10 – 20	*8*	*50 – 60*	*23*
20 – 30	*7*	*60 – 70*	*18*
30 – 40	*10*	*70 – 80*	*10*

Solution. Here $L = 80$, $S = 0$.

$$\therefore \quad \textbf{Range} = \mathbf{L - S} = 80 - 0 = \mathbf{80}.$$

$$\textbf{Coefficient of Range} = \frac{\mathbf{L-S}}{\mathbf{L+S}} = \frac{80}{80} = \mathbf{1}.$$

Example 3. *If the relationship between x and y is given by 3x + 2y = 13 and the range of x is 12, what would be the range of y.*

Solution. $3x + 2y = 13 \Rightarrow y = (-3/2)\,x + (13/3)$

We know that if $y = a + bx$, then range of $y = \boldsymbol{R_y} = |\,\boldsymbol{b}\,| \times \boldsymbol{R_x}$.

$$\therefore \quad R_y = |-3/2|\,R_x = (3/2) \times 12 = \mathbf{18}.$$

4.3 QUARTILE DEVIATION OR SEMI-INTER QUARTILE RANGE

Quartile Deviation ***is a measure of dispersion based on the Upper Quartile (Q_3) and Lower Quartile (Q_1) of a series. It is half of the difference between the upper quartile and the lower quartile. Since the difference is the range between these two quartiles so it is also called inter-quartile range. The half of this range is*** **semi-inter-quartile range.** The **quartile deviation** is also known as **Semi-inter-quartile Range.**

$$\therefore \quad \textbf{Quartile Deviation} = \frac{Q_3 - Q_1}{2},$$

where $\quad Q_1$ **= first or lower quartile;** $\quad Q_3$ **= third or upper quartile.**

In most cases, the central 50% observations of a series tend to be fairly typical. Thus quartile deviation can be used as a suitable measure of dispersion in such cases. It can also be computed from a frequency distribution with open-end classes at both ends. *It is important to note that range cannot be computed from a frequency distribution with open end classes at both ends, so,* ***quartile deviation is a more suitable measure of dispersion than range.***

$$\textbf{Coefficient of Quartile Deviation}: \frac{Q_3 - Q_1}{Q_3 + Q_1}.$$

4.3.1 Coefficient of Variation

There is another relative measures of dispersion where the coefficient of respective absolute measure is multiplied by 100 to convert the figures into percentages or in relation to 100.

4.3.2 Coefficient of Quartile Variation = Coefficient of Quartile Deviation × 100

$$= \frac{Q_3 - Q_1}{Q_3 + Q_1} \times \mathbf{100}.$$

4.3.3 Properties of Quartile Deviation

1. *It is the best measure of dispersion for open-end distribution.*

2. *It is less affected due to sampling fluctuations.*

3. *Quartile deviation remains unaffected due to a change of origin but it is affected in the same ratio due to change in scale, i.e., if the two variables x and y are related as y = a + bx, where a and b are constants, then*

$$\textbf{Q.D. of } y = |\,b\,| \times (\textbf{Q.D. of } x)$$

4.3.4 Merits, Demerits and Uses of Quartile Deviation

Merits:

1. *It is easy to calculate.*
2. *It can be easily understood.*

3. *It is not affected by the extreme values.*
4. *It has a special utility in measuring variation in case of frequency distribution with open end classes at both ends.*

Demerits:

1. *It is not based on all the observations.*
2. *It is not the representative value of data.*
3. *It is not capable of algebraic treatments.*
4. *It is affected by sampling fluctuations.*
5. *It cannot be regarded as measure of dispersion as it really does not show the scatteredness around an average but rather a distance on a scale.*

Uses:

It can be used only for descriptive statistics.

Example 4. *Calculate the quartile deviation and coefficient of quartile deviation and the coefficient of quartile variation from the following data :*

Age in years :	*20*	*30*	*40*	*50*	*60*	*70*	*80*
No. of Members :	*3*	*61*	*132*	*153*	*140*	*51*	*3*

Solution. We have the following table :

Age in years (x)	*No. of Members (f)*	*Cumulative Frequency (c.f)*
20	3	3
30	61	64
40	**132**	**196**
50	153	349
60	**140**	**489**
70	51	540
80	3	543

Here N = **543.**

Q_1 **= Value of** $\left(\frac{N+1}{4}\right)$**th** item $= \frac{543+1}{4}$ = 136th item = **40 years.**

Q_3 **= Value of** $3\left(\frac{N+1}{4}\right)$**th** item = 3 × 136 = 408th item = **60 years.**

$\therefore$ **Quartile Deviation** $= \frac{Q_3 - Q_1}{2} = \frac{60-40}{2}$ = **10 years**.

Coefficient of Q.D. $= \frac{Q_3 - Q_1}{Q_3 + Q_1} = \frac{60-40}{60+40} = \frac{20}{100}$ = **0.2.**

Coefficient of Quartile Variation $= \frac{Q_3 - Q_1}{Q_3 + Q_1} \times 100 = 0.2 \times 100$ = **20%.**

Example 5. *If the first quartile is 104 and quartile deviation is 18, find the third quartile.*

Solution. Here $Q_1 = 104, \quad Q_3 = ?$

Quartile Deviation $= \frac{Q_3 - Q_1}{2} = 18$ **[Given]**

$\Rightarrow \quad Q_3 = 2 \times 18 + Q_1 = 36 + 104 = \mathbf{140.}$

Third Quartile: $Q_3 = \mathbf{140.}$

Example 6. *From the following data, calculate the quartile deviation and the coefficient of quartile variation :*

Wages :	*0 – 10*	*10 – 20*	*20 – 30*	*30 – 40*	*40 – 50*	*50 – 60*	*60 – 70*	*70 – 80*
No. of workers :	*20*	*45*	*85*	*160*	*70*	*55*	*35*	*30*

Solution. We have the following table :

Wages: (x)	*No. of Workers: (f)*	*Cumulative Frequency: (c.f.)*
0 – 10	20	20
10 – 20	45	65
20 – 30	85	150
30 – 40	160	310
40 – 50	70	380
50 – 60	55	435
60 – 70	35	470
70 – 80	30	500

Here, N = Total number of workers = **500.**

COMPUTATION of Q_1 and Q_3

For Q_1, we have $\frac{N}{4} = 125$, cumulative frequency just greater than 125 is **150**, and therefore, it lies in the class 20 – 30, whose frequency $f = 85$. Also $C = 65$ and $i = 10$

$$\therefore \; Q_1 = L + \frac{(N/4) - C}{f} \times i = 20 + \frac{125 - 65}{85} \times 10 = 20 + 7.06 = \mathbf{27.06.}$$

Similarly, for Q_3, we have: $\frac{3N}{4} = 375$, $L = 40$, $C = 310$, $f = 70$ and $i = 10$.

$$\therefore \quad Q_3 = L + \frac{(3N/4) - C}{f} \times i = 40 + \frac{375 - 310}{70} \times 10 = 40 + 9.29 = \mathbf{49.29.}$$

$$\therefore \; \textbf{Quartile Deviation} = \frac{Q_3 - Q_1}{2} = \frac{49.29 - 27.06}{2} = \frac{22.23}{2} = \mathbf{11.115.}$$

$$\textbf{Coefficient of Quartile Variation} = \frac{Q_3 - Q_1}{Q_3 + Q_1} \times \mathbf{100}$$

$$= \frac{49.29 - 27.06}{49.29 + 27.06} = \frac{22.23}{76.35} \times 100 = \mathbf{29.12\%}.$$

Example 7. *Calculate the appropriate measure of dispersion from the following data:*

Income :	*Less than 35*	*35 – 37*	*38 – 48*	*41 – 43*	*over 43*
No. of Persons :	*14*	*62*	*99*	*18*	*7*

Solution. ***Since the given data has frequency distribution with open end classes at the both ends, so quartile deviation is the appropriate measure of dispersion.*** Making the given series as inclusive series, we have the following table :

Class-Interval	*Frequency: (f)*	*Cumulative Frequency: (c.f.)*
Less than 34.5	14	14
34.5 – 37.5	**62**	**76**
37.5 – 40.5	**99**	**175**
40.5 – 43.5	18	193
Over 43.5	7	200

Here $N = \mathbf{200}$, $\dfrac{N}{4} = \dfrac{200}{4} = \mathbf{50}$, $\dfrac{3N}{4} = \mathbf{150}$.

Since, the cumulative frequency just greater than 50 is 76, so **Q_1 lies in the class 34.5 – 37.5.** Similarly, cumulative frequency just greater than 150 is 175, so **Q_3 lies in the class 37.5 – 40.5.**

Here $L = 34.5$, $N/4 = 50$, $C = 14$, $i = 3$, $f = 62$.

Now, $Q_1 = L + \dfrac{(N/4) - C}{f} \times i = 34.5 + \dfrac{50 - 14}{62} \times 3 = 34.5 + 1.74 =$ **Rs. 36.24.**

Similarly, for Q_3, we have :

$$L = 37.5,\ \frac{3N}{4} = 150,\ C = 76,\ i = 3,\ f = 99.$$

$$\therefore\ Q_3 = L + \frac{(3N/4) - C}{f} \times i = 37.5 + \frac{150 - 76}{99} \times 3 = 37.5 + 2.24 = \mathbf{39.74.}$$

$$\therefore\ \textbf{Quartile Deviation} = \frac{Q_3 - Q_1}{2} = \frac{39.74 - 36.24}{2} = \frac{3.5}{2} = \textbf{Rs. 1.75.}$$

Example 8. *If the quartile deviation of x is 8 and 3x + 6y = 15, what is the quartile deviation of y.*

Solution. $3x + 6y = 15 \Rightarrow y = (15/6) - (3/6)\,x.$

$$\textbf{Q.D. of } y = |-3/6| \times (\textbf{Q.D. of } x) = (1/2) \times 8 = \mathbf{4}.$$

4.4 MEAN DEVIATION OR AVERAGE DEVIATION

Mean deviation ***of a set of observations of a series is the arithmetic mean of all the deviations, (without their algebraic signs), taken from its central value (mean or median or mode).*** In other words, ***it is the average of the modulus of the deviations of the observations in a series taken from mean or median or mode.*** Mean deviation is one of the calculated measure in which all the values are considered in its calculations.

4.4.1 Coefficient of Mean Deviation $= \dfrac{\textbf{Mean deviation about } A}{A}$,

where we take A as mean or median or mode and accordingly, we get the mean deviation about mean or median or mode.

4.4.2 Coefficient of Mean Variation $= \dfrac{\textbf{Mean Deviation about } A}{A} \times \mathbf{100}$

where A = mean or median or mode.

4.4.3 Properties of Mean Deviation

1. ***If all the values taken by a variable x is a constant k, then mean deviation is zero.***
2. ***Mean deviation takes minimum values when the deviations are taken from the median.***
3. ***Mean deviation remains unchanged due to change of origin but changes in the same ratio due to change in scale, i.e., if y = a + bx, a and b being constants, then***

$$\textbf{M.D. of } y = |b| \times (\textbf{M.D. of } x)$$

Example 9. *If a relationship between x and y is given by : 2x + 3y = 10 and M.D. of x is 15, what is the M.D. of y.*

Solution. $2x + 3y = 10 \Rightarrow y = (-2/3)x + (10/3)$

If $y = a + bx$, then **M.D. of** $y = |b| \times$ **(M.D. of** x**)**

$\therefore$ **M.D. of** $y = |-2/3| \times$ M.D. of $x = (2/3) \times 15 =$ **10.**

4.4.4 Mean Deviation for Discrete Series

While calculating the mean deviations, the algebraic sign of the deviation ***is always taken as positive, because the sum of deviations with their algebraic signs, + and –, from the arithmetic mean is always zero.*** We have the following rule :

(*i*) **Signs (plus and minus) of deviations are disregarded and absolute values of the deviations are summed up. Symbolically, we use $|x_i - \bar{x}|$, which means the deviation of the *i*th observation of *x* from the central value $\bar{x}$, (which may be mean or median or mode) with positive sign. Here the vertical lines stand for positive value. Now add up all *n* observations to get** $\sum_{i=1}^{n}|x_i - \bar{x}|$. Then

$$\textbf{Mean Deviation: } MD_{\bar{x}} = \frac{1}{n}\sum_{i=1}^{n}|x_i - \bar{x}|, \textbf{ where } \bar{x} \textbf{ is the arithmetic mean.}$$

Similarly, **Mean deviation from the median M_d for ungrouped data is** denoted by $\mathbf{MD_{med}}$ or $\mathbf{d_{med}}$, where

$$\mathrm{MD_{med}} = \frac{1}{n}\Sigma|x_i - \mathrm{M}_d|,$$

(*a*) **Mean deviation from the mode Z for ungrouped data**

$$\mathrm{d_v} \text{ or } \mathrm{MD_v} = \frac{1}{n}\Sigma|x_i - \mathrm{Z}|$$

But in practice, **we generally use arithmetic mean as it is amenable to algebraic treatments. Median at times is not an actual quantity, while mode of a series may not exist, *i.e.*, it may be ill defined.**

Example 10. *Calculate the mean deviation about the mean for the following series :*

15, 20, 17, 19, 21, 13, 12, 10, 17, 9, 12.

Solution. Here $n = 11$, and therefore

$$\text{Mean} = \frac{15+20+17+19+21+13+12+10+17+9+12}{11} = \frac{165}{11} = \mathbf{15 = M} \text{ (say)}.$$

$$\Sigma|d| = \Sigma|\mathrm{X} - \mathrm{M}| = \Sigma|\mathrm{X} - 15| = 0 + 5 + 2 + 4 + 6 + 2 + 3 + 5 + 2 + 6 + 3 = \mathbf{38}.$$

$$\therefore \quad \textbf{Mean deviation} = \frac{\Sigma|d|}{N} = \frac{38}{11} = \mathbf{3.455}.$$

Example 11. *The marks obtained by 10 students in an examination were as follows:*

70, 65, 68, 70, 75, 73, 80, 70, 83, 86.

Find the mean deviation about the mean and coefficient of variation.

Solution. Mean $= \dfrac{\Sigma X}{n} = \dfrac{65+68+70+70+70+73+75+80+83+86}{10} = \dfrac{740}{10} = \mathbf{74}.$

Mean Deviation about mean $= \dfrac{\Sigma|X - \bar{X}|}{n}$

$$= \frac{4+9+6+4+1+1+6+4+9+12}{10} = \frac{56}{10} = \mathbf{5.6}.$$

Coefficient of Variation $= \dfrac{\textbf{Mean Deviation}}{\textbf{Mean}} \times \mathbf{100} = \dfrac{5.6}{74} \times 100 = \mathbf{7.57\%}.$

4.4.5 Mean Deviation for Grouped data

Let $x_1, x_2, x_3, \ldots\ldots, x_n$ occur with frequencies $f_1, f_2, f_3, \ldots\ldots, f_n$ respectively and let $\Sigma f = n$ and M can be either Mean or Median or Mode, then the mean deviation is given by the formula :

$$\textbf{Mean Deviation} = \frac{\Sigma f|x - M|}{\Sigma f} = \frac{\Sigma f|d|}{n}, \text{ where } d = |x - M| \text{ and } \Sigma f = n.$$

Example 12. *Find the mean deviation from* (*i*) *mean and* (*ii*) *median for the following data:*

Marks :	*20*	*18*	*16*	*14*	*12*	*10*	*8*	*6*
No. of Students :	*2*	*4*	*9*	*18*	*27*	*25*	*14*	*1*

Also calculate coefficient of variation.

Solution. (*i*) **Mean deviation from mean.**

Here Σf = 1 + 14 + 25 + 27 + 18 + 9 + 4 + 2 = **100.**

$\Sigma f \times x$ = 6 + 112 + 250 + 324 + 252 + 144 + 72 + 40 = **1200.**

$\therefore$ **Arithmetic mean :** $\bar{X} = \dfrac{\Sigma fx}{\Sigma f} = \dfrac{1200}{100} = \mathbf{12}.$

Also $\Sigma f|d|$ = 6 + 56 + 50 + 0 + 36 + 36 + 24 + 16 = **224.**

$\therefore$ **Mean deviation (about mean)** $= \dfrac{\Sigma f|d|}{n} = \dfrac{224}{100} = \mathbf{2.24}.$

Coefficient of Variation $= \dfrac{\textbf{M.D.}}{\bar{X}} \times 100 = \dfrac{2.24}{12} \times 100 = \mathbf{18.67\%}.$

(*ii*) **Mean deviation about median.** Let us prepare the following table in order to calculate the median :

Table: Computation of Mean Deviation about Median

Marks: (*x*)	*Frequency : f*	*Cumulative frequency : c.f.*	$\lvert d\rvert = \lvert x - 12\rvert$
6	1	1	6
8	14	15	56
10	25	40	50
12	27	67	0
14	18	85	36
16	9	94	36
18	4	98	24
20	2	100	16
	Σf = **100**		$\Sigma f\lvert d\rvert$ = **224**

Median = Average of $\frac{n}{2}$th and $\left(\frac{n}{2}+1\right)$th item. = Average of 50th and 51st item = **12.**

Mean Deviation (about median) = $\frac{\Sigma f|d|}{n} = \frac{224}{100} = \mathbf{2.24}$.

Coefficient of Variation (mean deviation) = $\frac{\textbf{M.D.}}{\textbf{Median}} \times \mathbf{100} = \frac{2.24}{12} \times 100 = \mathbf{18.67\%}$.

Example 13. *Calculate the mean deviation about the mean for the following data:*

x:	5	15	25	35	45	55	65
f:	8	12	10	8	3	2	7

Also find the coefficient of Mean Variation.

Let the **assumed mean be A = 35.** **Let** = $d = \frac{x - \text{A}}{i} = \frac{x - 35}{10}$.

Here $\Sigma fd = -24 - 24 - 10 + 0 + 3 + 4 + 21 = \mathbf{-30}$.

Mean : $\bar{x} = \text{A} + \frac{\Sigma fd}{\Sigma f} \times i = 35 + \left(\frac{-30}{50}\right) \times 10 = \mathbf{29}.$

$\Sigma f\,(x - \bar{x}) = \Sigma f(x - 29) = 192 + 168 + 40 + 48 + 48 + 52 + 252 = \mathbf{800}.$

Mean Deviation = $\frac{1}{\text{N}} \Sigma f\,|x - \bar{x}| = \frac{1}{50}(800) = \mathbf{16}.$

Coefficient of Mean Variation = $\frac{\textbf{M.D.}}{\bar{x}} \times 100 = \frac{16}{29} \times 100 = \mathbf{55.17\%}$.

4.4.6 Mean Deviation for Continuous Series

Mean Deviation = $\frac{\Sigma fd}{n}$, where

d = deviation of the mid-points of the classes from mean or median

f = frequency of the corresponding mid-point (or class)

$n = \Sigma f$ = number of total frequencies

Example 14. *Calculate the mean deviation from the mean for the following data:*

Class Interval:	*0 – 4*	*4 – 8*	*8 – 12*	*12 – 16*	*16 – 20*
Frequency :	*4*	*6*	*8*	*5*	*2*

Solution. Let us prepare the following table **by assuming that the frequencies in each class are centred at its mid-value.**

Table: Computation of Mean Deviation

Class Interval	*Mid-value: (x)*	*f*	*fx*	$\lvert d \rvert = \lvert x - 9.2 \rvert$	$f\lvert d \rvert$
0 – 4	2	4	8	7.2	28.8
4 – 8	6	6	36	3.2	19.2
8 – 12	10	8	80	0.8	6.4
12 – 16	14	5	70	4.8	24.0
16 – 20	18	2	36	8.8	17.6
		$\mathbf{\Sigma f = 25}$	$\mathbf{\Sigma fx = 230}$		$\mathbf{\Sigma f\lvert d \rvert = 96.0}$

Now **Arithmetic Mean:** $\bar{x} = \frac{\Sigma f x}{\Sigma f} = \frac{230}{25} = \mathbf{9.2.}$

Mean Deviation: $MD_x = \frac{\Sigma f |d|}{\Sigma f} = \frac{96}{25} = \mathbf{3.84.}$

4.4.6 Merits, Demerits and Uses of Mean Deviation

Merits:
1. *It is easy to understand and compute.*
2. *Mean deviation is less affected by the extreme values as compared to standard deviation.*
3. *Mean deviation about an arbitrary point is least when the point is median.*

Demerits:
1. *In mean deviation the signs of all deviations are taken as positive and therefore, it is not suitable for further algebraic treatments.*
2. *It is rarely used in social sciences.*
3. *It does not give accurate results because the mean deviation from the median is least but median itself is not considered a satisfactory average when the variation in the series is large.*
4. *It is often not useful for statistical inferences.*

Uses: *Mean deviation and its coefficient are used in studying economic problems such as distribution of income and wealth in a society.*

4.5 STANDARD DEVIATION

Standard deviation is the most important and commonly used measure of dispersion. ***It measures the absolute dispersion or variability of a distribution. A small standard deviation means a high degree of uniformity in the observations as well as homogeneity of the series.*** It is extremely useful in judging the representativeness of the mean.

***Standard deviation is the positive square root of the average of squared deviations taken from arithmetic mean.* It is, generally, denoted by the Greek alphabet σ or by S.D. or s.d.** Let x be a random variate which takes on n values, *viz.*, $x_1, x_2, x_3, \ldots, x_n$, then the standard deviation of these n observations is given by :

Standard Deviation: $\sigma = \sqrt{\Sigma \frac{(x-\bar{x})^2}{n}}$, **where**

$\bar{x} = \frac{\Sigma x}{n}$ **is the mean of these observations**

ALTERNATIVELY

$$\sigma = \sqrt{\frac{\Sigma x^2}{n} - \left(\frac{\Sigma x}{n}\right)^2}$$

Also $\sigma = \sqrt{\frac{\Sigma(x-x)^2}{n}} = \sqrt{\frac{\Sigma x^2}{n} - (\bar{x})^2}$ $\left[\because \bar{x} = \frac{\Sigma x}{n}\right]$

But the formula $\sigma = \sqrt{\frac{\Sigma x^2}{n} - \left(\frac{\Sigma x}{n}\right)^2}$,

is used when the items are very small. On the other hand the relevance of this method is particularly useful when computers are used whereby the values, even when they are of high magnitude, can be used directly for calculating σ.

Standard deviation is also known as **Root Mean Square Deviation.**

4.5.1 Variance

The **variance** ***is the square of standard deviation and is denoted by*** σ^2.

4.5.2 Coefficient of Standard Deviation = $\dfrac{\sigma}{\bar{x}}$,

where $\bar{x}$ **is the arithmetic mean of the given series. It is a relative measure of standard deviation.**

4.5.3 Coefficient of Variation

Coefficient of Variation. It is a **relative measure of dispersion.** It is, generally, denoted by **C.V.** and is given by the formula:

$$\textbf{Coefficient of Variation : C.V.} = \frac{\sigma}{\bar{x}} \times 100,$$

where σ **is the standard deviation and** $\bar{x}$ **is the mean of the given series.**

It is important to note that the **coefficient of variation is always a percentage.**

The coefficient of variation has a great practical significance and is the best measure for comparing the variability of the two series. The series or group for which the coefficient of variation is greater is said to be more variable (or less consistent). On the other hand, the series for which the variation is less is said to be less variable (or more consistent).

4.5.4 Properties of Standard Deviation

1. ***If all the observations assumed by a variable are constant, i.e., equal, then the S.D. is zero. This means that if all the values taken by a variable x is a constant k, then the standard deviation is zero.***
2. ***S.D. remains unaffected due to a change of origin but is affected in the same ratio due to a change of scale, i.e., if there are two variables x and y related as y = a + bx for any two constants a and b, then S.D. of y is given by*** $\sigma_y = |b|\,\sigma_x$.

Example 15. *If the two variables x and y are related as : 2x + 3y = 10 and the standard deviation of x is 15, then what is the standard deviation of y.*

Solution. $2x + 3y = 10 \Rightarrow y = 10 + (-2/3)\,x.$

Also, **S.D. of** y = $|b|$ × **(S.D. of** x**);** **where** $y = a + bx$.

$\therefore$ **S.D. of** y = $|-2/3|$ × (S.D. of x) $\Rightarrow$ (2/3) × 15 = **10.**

Example 16. *If A.M. and the coefficient of variation are 10 and 50 respectively, what is the variance of (15 – 2x) ?*

Solution. $\text{C.V. }(x) = \dfrac{\sigma_x}{\bar{x}} \times 100 \Rightarrow 50 = \dfrac{\sigma_x}{10} \times 100 \Rightarrow \sigma_x = 5.$

$\therefore$ **Variance of** x = $\sigma_x^2 = 25.$

S.D. of (15 – 2x) = $|-2|$ (S.D. of x) = 2 × 5 = **10** = s_y **(say)**

$\therefore$ **Variance of (15 – 2x)** = $s_y^2 = 10^2 = $ **100.**

Example 17. *Find the standard deviation of 3, 4, 5, 6.*

Solution. Here $n = 4$, $\Sigma x = 3 + 4 + 5 + 6 = $ **18.**

$\Sigma x^2 = 3^2 + 4^2 + 5^2 + 6^2 = 9 + 16 + 25 + 36 = $ **86.**

$$\therefore \quad \sigma = \sqrt{\frac{\Sigma x^2}{n} - \left(\frac{\Sigma x}{n}\right)^2} = \sqrt{\frac{86}{4} - \left(\frac{18}{4}\right)^2}$$

$$= \sqrt{21.5 - (4.5)^2} = \sqrt{21.5 - 20.25} = \sqrt{1.25} = \textbf{1.12} \text{ nearly.}$$

4.5.5 Merits, Demerits and Uses of Standard Deviation

Merits:

1. *It is based on all the observations.*
2. *It is rigidly defined.*
3. *It has a greater mathematical significance and is capable of further mathematical treatments.*
4. *It represents the true measurement of dispersion of a series.*
5. *It is least affected by fluctuation of sampling.*
6. *It is not reliable and dependable measure of dispersion.*
7. *It is extremely useful in correlation etc.*

Demerits:

1. *It is difficult to compute unlike other measures of dispersion.*
2. *It is not simple to understand.*
3. *It gives more weightage to extreme values.*
4. *It consumes much time and labour while computing it.*

Uses:

1. *It is widely used in biological studies.*
2. *It is used in fitting a normal curve to a frequency distribution.*
3. *It is most widely used as a measure of dispersion.*

4.6 CALCULATION OF STANDARD DEVIATION – INDIVIDUAL OBSERVATIONS

When the data under consideration consists of individual observations, the standard deviation may be computed by any of the following two methods :

(*a*) **By taking deviations of the items from the actual mean.**

(*b*) **By taking deviations of the items from an assumed mean.**

4.6.1 When the Deviations are Taken from the Actual Arithmetic Mean

This method is known as **Direct Method.**

Direct Method

In case of simple series, the standard deviation can be obtained by the formula:

$$\sigma = \sqrt{\frac{\Sigma (x_i - \bar{x})^2}{n}} \quad \text{or} \quad \sigma = \sqrt{\frac{\Sigma d^2}{n}}, \text{ where } d = x_i - \bar{x}$$

and x_i = value of the variable or observation, $\bar{x}$ = arithmetic mean, and

n = total number of observations.

Example 18. *Find the standard deviation of 16, 13, 17, 22.*

Solution. Here A.M. = $\bar{x} = \frac{16+13+17+22}{4} = \frac{68}{4} = \mathbf{17}$. Here $d = x - \bar{x} = x - 17$.

Let us prepare the following table in order to calculate the standard deviation.

x :	16	13	17	22
d :	– 1	– 4	0	5
d^2 :	1	16	0	25

$\therefore \qquad \Sigma d^2 = 1 + 16 + 25 = 42.$

Now $$\sigma = \sqrt{\frac{\Sigma(x-\bar{x})^2}{n}} = \sqrt{\frac{\Sigma d^2}{n}} = \sqrt{\frac{42}{4}} = \mathbf{3.24.}$$

4.6.2 When the Deviations are Taken from the Assumed Mean

This method is also known as **Short-cut Method.**

Short-cut Method

This method is applied to calculate standard deviation, when the mean of the data comes out to be a fraction. In that case, it is very difficult and tedious to find the deviations of all observations from the **actual mean** by the above method. The formula used is:

$$\sigma = \sqrt{\frac{\Sigma d^2}{n} - \left(\frac{\Sigma d}{n}\right)^2},$$

where $d = x - A$, A = assumed mean, n = total number of observations.

Example 19. *Find the standard deviation of the following data :*

48, 43, 65, 57, 31, 60, 37, 48, 59, 78.

Solution. Let us prepare the following table in order to calculate the value of S.D. by assuming $A = 50$. Also, $d = x - A = x - 50$ and $n = 10$.

x :	48	43	65	57	31	60	37	48	59	78
d :	– 2	– 7	15	7	– 19	10	– 13	– 2	9	28
d^2 :	4	49	225	49	361	100	169	4	81	784

$$\therefore \quad \mathbf{\Sigma d = 26, \quad \Sigma d^2 = 1826.}$$

Here $$\bar{x} = A + \frac{\Sigma d}{n} = 50 + \frac{26}{10} = \mathbf{52.6,}$$

which is a fraction. Let us **apply the short-cut formula in order to calculate S.D.**

$$\therefore \text{ S.D.} : \sigma = \sqrt{\frac{\Sigma d^2}{n} - \left(\frac{\Sigma d}{n}\right)^2} = \sqrt{\frac{1826}{10} - \left(\frac{26}{10}\right)^2}$$

$$= \sqrt{182.60 - 6.76} = \sqrt{175.84} = \mathbf{13.26.}$$

4.7 CALCULATION OF STANDARD DEVIATION – DISCRETE SERIES OR GROUPED DATA

The standard deviation of a discrete series or grouped data can be calculated by any one of the following three methods.

(a) Actual Mean Method or Direct Method

(b) Assumed Mean Method or Short-cut Method

(c) Step Deviation Method

4.7.1 Actual Mean Method or Direct Method

The standard deviation for the discrete series is given by the formula :

$$\sigma = \sqrt{\frac{\Sigma f(x-\bar{x})^2}{n}},$$

where $\bar{x}$ is the arithmetic mean, x is the size of item, f is the corresponding frequency and $n = \Sigma f$.

However, in practice, this method is rarely used because if the arithmetic mean is a **fraction,** the calculations take a lot of time and are cumbersome.

4.7.2 Assumed Mean Method or Short-cut Method

In this method we use the following formula to calculate the standard deviation σ:

$$\sigma = \sqrt{\frac{\Sigma fd^2}{n} - \left(\frac{\Sigma fd}{n}\right)^2},$$

where A = is the assumed mean, $d = x - A$, and $n = \Sigma f$.

Example 20. *Find the standard deviation from the following data by assumed mean method:*

Size of the item :	*10*	*11*	*12*	*13*	*14*	*15*	*16*
Frequency :	2	7	*11*	*15*	*10*	4	*1*

Also find the coefficient of variation.

Solution.

Table: Computation of Standard Deviation

Size of the item: (x)	*Frequency:* (f)	$d = x - A$, $A = 13$	fd	d^2	fd^2
10	2	– 3	– 6	9	18
11	7	– 2	– 14	4	28
12	11	– 1	– 11	1	11
13	15	0	0	0	0
14	10	1	10	1	10
15	4	2	8	4	16
16	1	3	3	9	9
Total	$n = \Sigma f = 50$		$\Sigma fd = -10$		$\Sigma fd^2 = 92$

Now **Mean :** $\bar{x} = A + \frac{\Sigma fd}{n} = 13 + \frac{(-10)}{50} = \mathbf{12.8.}$

Here $\bar{x}$ **= 12.8, is a fraction, so we use Assumed Mean Method**

$$\therefore \quad \textbf{S.D.} = \sigma = \sqrt{\frac{\Sigma fd^2}{n} - \left(\frac{\Sigma fd}{n}\right)^2} = \sqrt{\frac{92}{50} - \left(\frac{-10}{50}\right)^2} = \sqrt{1.84 - 0.04} = \sqrt{1.80} = \mathbf{1.342.}$$

$\therefore$ **Coefficient of Variation** = $\frac{\sigma}{\bar{x}} \times 100 = \frac{1.342}{12.8} \times 100 = \mathbf{10.4.}$

4.7.3 Step Deviation Method

In this method we divide the deviations by a common class interval and use the following formula for computing standard deviation:

$$\sigma = \sqrt{\frac{\Sigma fd^2}{n} - \left(\frac{\Sigma fd}{n}\right)^2} \times i,$$

where i = common class interval, $d = \frac{x - A}{i}$, A is assumed mean, f is the respective frequency

Example 21. *Daily sales (recorded in rupees) of a retail shop are given below :*

Daily sales in Rs. (X-Mid-point of interval) :	*102*	*106*	*110*	*114*	*118*	*122*	*126*
No. of days (Frequency-f) :	*3*	*9*	*25*	*35*	*17*	*10*	*1*

Calculate the mean and the standard deviation of the above data. Also calculate the coefficient of variation.

Solution. Let us take the assumed mean A = 114.

Then $d = (x - A) / i = (x - 114) / 4$, where i = 4.

Table: Computation of Mean and Standard Deviation

No. of days: x	*f*	$d = \frac{x - 114}{4}$	*fd*	fd^2
102	3	– 3	– 9	27
106	9	– 2	– 18	36
110	25	– 1	– 25	25
114	35	0	0	0
118	17	1	17	17
122	10	2	20	40
126	1	3	3	9
	N = 100		**Σ fd = – 12**	**Σ fd = 154**

Arithmetic Mean : $\bar{x} = A + \frac{\Sigma fd}{N} \times i.$

Here A =114 ; Σfd = – 12 ; n = 100 ; i = 4.

$\therefore \quad \bar{x} = 114 - \frac{12}{100} \times 4 = 114 - 0.48 = \mathbf{113.52.}$

$\therefore \quad \sigma = \sqrt{\frac{\Sigma fd^2}{n} - \left(\frac{fd}{n}\right)^2} \times i = \sqrt{\frac{154}{100} - \left(\frac{-12}{100}\right)^2} \times 4 = \sqrt{1.54 - (-0.12)^2} \times 4$

$= \sqrt{1.54 \quad 00144} \times 4 = 1.235 \times 4 = \mathbf{4.94.}$

Coefficient of Variation : C.V. $= \frac{\sigma}{\bar{x}} \times 100 = \frac{4.94}{113.52} \times 100 =$ **4.35%.**

4.8 CALCULATION OF STANDARD DEVIATION – CONTINUOUS SERIES

The standard deviation of a continuous series can be calculated by any one of the methods discussed for discrete frequency distribution. However, in practice **either assumed means or Step Deviation Method is mostly used.**

Assumed Mean Method:

$$\sigma = \sqrt{\frac{\Sigma fd^2}{n} - \left(\frac{\Sigma fd}{n}\right)^2},$$

where $d = n - A$, A = assumed mean, m = mid-value, f = frequency.

Step Deviation Method:

$$\sigma = \sqrt{\frac{\Sigma fd^2}{n} - \left(\frac{\Sigma fd}{n}\right)^2} \times i,$$

where $d = \frac{m - A}{i}$, i = **class interval (or the common factor in case the class intervals are unequal),** m **is the mid-value of the interval,** A **is the assumed mean.**

Example 22. *Calculate Standard Deviation. Coefficient of Variation and Variance from the following data by assumed mean method.*

Marks	*0 – 10*	*10 – 20*	*20 – 30*	*30 – 40*	*40 – 50*	*50 – 60*
No. of Students	*10*	*20*	*30*	*50*	*40*	*30*

Solution. Let us take Assumed mean = 40. Then $d = x - 40$, m = mid-value of the interval.

Table : Computation of Standard Deviation

Marks (x)	m	*No. of Students* f	$(X - 40) = d$	fd	fd^2
0 – 10	5	10	– 35	– 350	12250
10 – 20	15	20	– 25	– 500	12500
20 – 30	25	30	– 15	– 450	6750
30 – 40	35	50	– 5	250	1250
40 – 50	45	40	5	200	1000
50 – 60	55	30	15	450	6750
Total		**N = 180**	$\Sigma d = -60$	$\Sigma fd = -900$	$\Sigma fd^2 = 40500$

Mean $(\bar{X}) = A + \frac{\Sigma fd}{N} = 40 + \frac{-900}{180} =$ **35.**

Standard Deviation $(\sigma) = \sqrt{\frac{\Sigma fd^2}{N} - \left(\frac{\Sigma fd}{N}\right)^2} = \sqrt{\frac{40500}{180} - \left(\frac{-900}{180}\right)^2} = \sqrt{225 - 25} =$ **14.142.**

Coefficient of Variation (C.V.) $= \frac{\sigma}{\textbf{Mean}} \times 100 = \frac{14.142}{35} \times 100 =$ **400.4061%..**

Variance = (Standard Deviation)^2 = (σ^2) = $(14.142)^2$ = **200.**

Example 23. *Find the standard deviation for the following distribution by Step deviation Method :*

Marks	*10 – 20*	*20 – 30*	*30 – 40*	*40 – 50*	*50 – 60*	*60 – 70*	*70 – 80*
No. of Students	*5*	*12*	*15*	*20*	*10*	*4*	*2*

Solution. Let us prepare the following table in order to calculate the standard deviation, by assuming $A = 45$. Here $d = (m - 45)/10$, $i = 10$.

Table: Computation of Standard Deviation

Marks (Class interval)	*No. of Students (f)*	*Mid-value (m)*	$d = \frac{m-45}{10}$	*fd*	fd^2
10 – 20	5	15	– 3	– 15	45
20 – 30	12	25	– 2	– 24	48
30 – 40	15	35	– 1	– 15	15
40 – 50	20	45	0	0	0
50 – 60	10	55	1	10	10
60 – 70	4	65	2	8	16
70 – 80	2	75	3	6	18
Total	$\boldsymbol{\Sigma f = n = 68}$			$\boldsymbol{\Sigma fd = -30}$	$\boldsymbol{\Sigma fd^2 = 152}$

$$\therefore\ \sigma = i \times \sqrt{\frac{\Sigma fd^2}{n} - \left(\frac{\Sigma fd}{n}\right)^2} = 10 \times \sqrt{\frac{152}{68} - \left(\frac{-30}{68}\right)^2} = 10 \times \sqrt{(2.2352 - 0.1946)}$$

$$= 10 \times \sqrt{2.0406} = \textbf{14.3 Approx.}$$

Example 24. *The arithmetic mean and the standard deviation of a set of 9 items are 43 and 5 respectively. If an item of value 63 is added to the set, find the mean and standard deviation of all the 10 items.*

Solution. Here it is given that :

FIRST CASE. $n = 9$, $\bar{x} = 43$, $\sigma = 5$.

But $$\bar{x} = \frac{\Sigma x}{n} \Rightarrow \Sigma x = n\bar{x} = 9 \times 43 = \mathbf{387.}$$

Also $$\sigma^2 = \frac{\Sigma x^2}{n} - \left(\frac{\Sigma x}{n}\right)^2 = \frac{\Sigma x^2}{n} - (\bar{x})^2$$

or $$\Sigma x^2 = n(\sigma^2 + \bar{x}^2) = 9[25 + (43)^2] = 9\,(25 + 1849) = \mathbf{16866.}$$

SECOND CASE.

If a new item 63 is added then the new number of terms becomes 10.

$$\therefore \quad \text{New } \Sigma x = (\text{old } \Sigma x) + 63 = 387 + 63 = \mathbf{450.}$$

New Mean: $$\bar{x} = \frac{\text{New } \Sigma x}{10} = \frac{450}{10} = \mathbf{45.}$$

$$\therefore \quad \Sigma x^2 = \Sigma x^2 + (63)^2 = 16866 + 3969 = \mathbf{20835.}$$

$$\therefore \text{ New S.D.} = \sqrt{\frac{\Sigma x^2}{10} - (\bar{x})^2} = \sqrt{\frac{20835}{10} - (45)^2} = \sqrt{2083.5 - 2025} = \sqrt{58.5} = \mathbf{7.65.}$$

Example 25. *A student obtained the mean and standard deviation of 100 observations as 40 and 5.1 respectively. It was later discovered that he had wrongly copied down an observation 50 instead of 40. Calculate the correct mean and standard deviation.*

Solution. Here $N = 100$, $\bar{X} = 40$, $\sigma = 5.1$

Also $\bar{X} = \frac{\Sigma x}{N} \Rightarrow 40 = \frac{\Sigma x}{100}$

or $\Sigma x = 40 \times 100 = \mathbf{4000}$ **(Uncorrected).**

$\therefore$ **Correct** Σx = **uncorrected** Σx **– wrong item + correct item**

$= 4000 - 50 + 40 = \mathbf{3990.}$

$\therefore$ **Corrected** $\bar{X} = \frac{\textbf{Correct } \Sigma x}{N} = \frac{3990}{100} = \mathbf{39.9.}$

Uncorrected $\Sigma x^2 = N(\sigma^2 + \bar{X}^2) = 100\,(5.1)^2 + (40)^2] = \mathbf{1,62601.}$

Correct $\Sigma x^2 = 162601 - 50^2 + 40^2 = \mathbf{161701.}$

$\therefore$ **Corrected** $\sigma = \sqrt{\frac{\textbf{Corrected } \Sigma x^2}{N} - (\textbf{Corrected } \bar{X})^2}$

$= \sqrt{\frac{161701}{100} - (39.9)^2} = \sqrt{1617.01 - 1592.01} = \sqrt{25} = \mathbf{5}.$

Example 26. *The quarterly profits (in lakhs of Rs.) of three companies are given below. Calculate their standard deviation and coefficient of variation which company is more solid ?*

Quarters	*Company A*	*Company B*
Jan-March	*9.5*	*15.5*
April-June	*13.7*	*21.2*
July-September	*10.4*	*23.4*
October-December	*8.6*	*18.8*

Solution.

COMPANY A

Here $\Sigma x = 9.5 + 13.7 + 10.4 + 8.6 = \mathbf{42.2.}$ $\therefore \bar{x}_A = \frac{\Sigma x}{n} = \frac{42.2}{4} = \mathbf{10.55}$

$\Sigma x^2 = (9.5)^2 + (13.7)^2 + (10.4)^2 + (8.6)^2 = 90.25 + 187.69 + 108.16 + 73.96 = \mathbf{460.06.}$

Now $\sigma_A = \sqrt{\frac{\Sigma x^2}{n} - \left(\frac{\Sigma x}{n}\right)^2} = \sqrt{\frac{460.06}{4} - \left(\frac{42.2}{4}\right)^2} = \sqrt{115 - 111.3} = \sqrt{3.7} = \mathbf{1.92}$

$\text{C.V. (A)} = \frac{\sigma_A}{\bar{x}_A} \times \mathbf{100} = \frac{1.92}{10.55} \times 100 = \mathbf{18.23\%}$

COMPANY B

$\Sigma x = 15.5 + 21.2 + 23.4 + 18.8 = \mathbf{78.9.}$ $\therefore \bar{x}_B = \frac{78.9}{4} = \mathbf{19.72.}$

$\Sigma x^2 = (15.5)^2 + (21.2)^2 + (23.4)^2 + (18.8)^2 = 240.25 + 449.44 + 547.56 + 353.44. = \mathbf{1590.69.}$

$\therefore \sigma_B = \sqrt{\frac{1590.69}{4} - \left(\frac{78.9}{4}\right)^2} = \sqrt{397.67 - 388.87} = \sqrt{8.8} = \mathbf{2.96.}$

$$\therefore \quad \text{C.V. (B)} = \frac{\sigma_B}{\bar{x}_B}\times 100 = \frac{2.96}{19.72}\times 100 = \mathbf{15\%}.$$

Since C.V. (B) is less than the C.V. (A), so the company B is more solid in profits.

Example 27. *For a set of 100 observations, taking assumed mean as 4, the sum of the deviations is – 11 cm. and the sum of the squares of these deviations is 257 cm². Find the coefficient of variation.*

Solution. In order to find the coefficient of variation, we require mean and standard deviations. They are calculated as follows :

$$\bar{x} = A + \frac{\Sigma fd}{n} = 4 - \frac{11}{100} = \mathbf{3.89.}$$

Also
$$\sigma = \sqrt{\frac{\Sigma d^2}{n} - \left(\frac{\Sigma d}{n}\right)^2} = \sqrt{\frac{257}{100} - \left(\frac{-11}{100}\right)^2} = \sqrt{2.57 - 0.012} = \sqrt{2.558} = \mathbf{1.6.}$$

$$\therefore \quad \text{Coefficient of Variation} = \frac{\sigma}{\bar{x}}\times 100 = \frac{1.6}{3.89}\times 100 = \mathbf{41.13\%}.$$

Example 28. *Comment on the statement "after settlement the average weekly wage in a factory has increased from Rs. 8 to Rs. 12 and standard deviation has increased from 2 to 2.5. After the settlement, the wage has become higher and more uniform".*

Solution. It is clear that after the settlement, the weekly wage has increased from Rs. 8, Rs. 12, so the total wages received per week by all the workers have increased. Thus **the wages have become higher.**

$$\text{C.V. before settlement} = \frac{\sigma_{\text{Before}}}{\bar{x}_{\text{Before}}}\times 100 = \frac{2}{8}\times 100 = \mathbf{25\%}.$$

$$\text{C.V. after settlement} = \frac{\sigma_{\text{After}}}{\bar{x}_{\text{After}}}\times 100 = \frac{2.5}{12}\times 100 = \mathbf{20.83\%}.$$

Thus after the settlement C.V. is less, so the variation is also less.

Example 29. *If $n = 10$, $\bar{x} = 12$, $\Sigma x^2 = 1530$, find the coefficient of variation.*

Solution. We know that:

$$\sigma = \sqrt{\frac{\Sigma x^2}{n} - (\bar{x})^2} = \sqrt{\frac{1530}{10} - (12)^2} = \sqrt{9} = \mathbf{3.}$$

$$\therefore \quad \text{Coefficient of variation} = \frac{\sigma}{\bar{x}}\times 100 = \frac{3\times 100}{12} = \mathbf{25\%}.$$

Example 30. *The arithmetic mean of the runs scored by three batsman, Amit, Sumeet and Kapil in the series are 50, 48 and 12 respectively. The standard deviations of their runs are respectively 15, 12 and 2. Who is the most consistent of the three? If one of the three is to be selected, who will be selected ?*

Solution. Let $\bar{x}_1, \bar{x}_2, \bar{x}_3$ be the means and $\sigma_1, \sigma_2, \sigma_3$ be the standard deviations of the runs scored by Amit, Sumeet and Kapil. Here $\bar{x}_1 = 50, \bar{x}_2 = 48, \bar{x}_3 = 12; \sigma_1 = 15, \sigma_2 = 12, \sigma_3 = 2.$

$$\text{Coefficient of variation of runs scored by Amit} = \frac{\sigma_1}{\bar{x}_1}\times 100 = \frac{15\times 100}{50} = \mathbf{30\%}$$

$$\text{Coefficient of variation of runs scored by Sumeet} = \frac{\sigma_2}{\bar{x}_2}\times 100 = \frac{12\times 100}{48} = \mathbf{25\%}$$

Coefficient of variation of runs scored by Kapil $= \frac{\sigma_3}{\bar{x}_3} \times 100 = \frac{2}{12} \times 100 = 16.67\%$

The most consistent among the three is one whose C.V. is least. Here C.V. of Kapil is least, so he is most consistent. He is to be **selected.**

4.9 COMBINED STANDARD DEVIATION

It is possible to find out the standard deviation of a composite series from the standard deviations of its component parts. Let A_1 and A_2 be two groups having N_1 and N_2 items respectively. Let their means be $\bar{X}_1$ and $\bar{X}_2$ and their respective standard deviations be σ_1 and σ_2. Then the **combined standard deviation σ_{12} of two groups A_1 and A_2 is given by the formula:**

$$\sigma_{12} = \sqrt{\frac{N_1\sigma_1^2 + N_2\sigma_2^2 + N_1 d_1^2 + N_2 d_2^2}{N_1 + N_2}} = \sqrt{\frac{N_1(\sigma_1^2 + d_1^2) + N_2(\sigma_2^2 + d_2^2)}{N_1 + N_2}}$$

where $d_1 = (\bar{X}_1 - \bar{X})$, $\quad d_2 = (\bar{X}_2 - \bar{X})$, **and**

$$\bar{X} = \frac{N_1\bar{X}_1 + N_2\bar{X}_2}{N_1 + N_2}$$ **is their combined mean.**

Similarly, for three groups *A*, *B*, *C* with means $\bar{X}_1, \bar{X}_2, \bar{X}_3$, and S.D. $\sigma_1, \sigma_2, \sigma_3$; having number of items N_1, N_2, N_3, their combined Standard Deviation is given by:

$$\sigma_{123} = \sqrt{\frac{N_1(\sigma_1^2 + d_1^2) + N_2(\sigma_2^2 + d_2^2) + N_3(\sigma_3^2 + d_3^2)}{N_1 + N_2 + N_3}}$$

where $d_1 = \bar{X}_1 - \bar{X}, d_2 = \bar{X}_2 - \bar{X}, d_3 = \bar{X}_3 - \bar{X}, \bar{X}$ **is the combined mean of *A*, *B* and *C*.**

Example 31. *The mean and standard deviation of the marks obtained by two groups of students, consisting of 50 each, are given below. Calculate the mean and standard deviation of the marks obtained by all the in 100 students:*

Group	*Mean*	*Standard Deviation*
1	*60*	*8*
2	*55*	*7*

Solution. Here

For Group 1: $\bar{x}_1 = 60$; $\sigma_1 = 8$; $n_1 = 50$;

For Group 2: $\bar{x}_2 = 55$; $\sigma_2 = 7$; $n_2 = 50$.

Combined Mean: $$\bar{x}_{12} = \frac{\bar{x}_1 \times n_1 + \bar{x}_2 \times n_2}{n_1 + n_2} = \frac{60 \times 50 + 55 \times 50}{50 + 50} = \frac{5750}{100} = 57.5$$

Combined Standard Deviation : $$\sigma_{12} = \sqrt{\frac{\sigma_1^2 n_1 + \sigma_2^2 n_2 + d_1^2 n_1 + d_2^2 n_2}{n_1 + n_2}}.$$

$[\because d_1 = \bar{x}_{12} - \bar{x}_1 = 57.5 - 60 = -2.5$ and $d_2 = \bar{x}_{12} - \bar{x}_2 = 57.5 - 55 = 2.5]$

$$\sigma_{12} = \sqrt{\frac{8^2 \times 50 + 7^2 \times 50 + (-2.5)^2 \times 50 + (2.5)^2 \times 50}{50 + 50}}$$

$$= \sqrt{\frac{3200 + 2450 + 312.5 + 312.5}{100}} = \sqrt{62.75} = \mathbf{7.92.}$$

Example 32. *If the range of x is 2, what would be the range of – 3x + 50 ?*

Solution. **If $y = a + bx$, then $R_y = |b| R_x$.**

$\therefore$ Range $(-3x + 50) = |-3| R_x = 3 \times 2 = \mathbf{6}$.

Example 33. *If two variables x and y are related by 2x + 3y – 7 = 0 and the mean and mean deviation about mean of x are 1 and 0.3 respectively. What is the coefficient of variance of mean deviation of y about mean ?*

Solution. If $y = a + bx$, a, and b being constant, then **M.D. of $y = |b|$ (M.D. of x)**

Here $y = -\frac{2}{3}x + \frac{7}{3}$. $\therefore$ **M.D. of y =** $\left|\frac{2}{3}\right|$ **(M.D. of x)** $= (2/3) \times 0.3 = \mathbf{0.2.}$

Also $2x + 3y - 7 = 0 \quad \Rightarrow 2\bar{x} + 3\bar{y} - 7 = 0 \quad \Rightarrow \bar{y} = \frac{7}{3} - \frac{2}{3}\bar{x} = \mathbf{\frac{5}{3}}$.

Coefficient of Variance(about Mean deviation of y) :

$$= \frac{\textbf{M.D. of } y}{\bar{y}} \times \mathbf{100} = \frac{0.2}{(5/3)} \times 100 = \mathbf{12.}$$

Example 34. *If x and y are related as 3x + 4y = 20 and the quartile deviation of x is 12, then find the quartile deviation of y.*

Solution. If $y = ax + b$, then **Q.D of $y = |a| \times$ (Q.D. of x)**

Here $3x + 4y = 20 \quad \Rightarrow \quad y = \frac{-3}{4}x + 5$.

Q.D. of y $= |3/4|$ (Q.D. of x) $= |3/4|\ 12 = \mathbf{9}$.

Example 35. *If x and y are related by y = 2x + 5 and the S.D. and A.M. of x are known to be 5 and 10 respectively, then find the coefficient of variation of y.*

Solution. $y = 2x + 5 \quad \Rightarrow \quad \sigma_y = |2|\sigma_x = 2 \times 5 = \mathbf{10}$.

Also $\bar{y} = 2\bar{x} + 5 = 20 + 5 = 25$. $\quad [\because \bar{x} = 10]$

Coefficient of Variation of y $= \frac{\sigma_y}{\bar{y}} \times \mathbf{100} = \frac{10}{25} \times 100 = \mathbf{40\%}$.

Example 36. *If the S.D. of x is 3, what is the variance of (5 – 2x) ?*

Solution. If $y = a + bx$, then $\sigma_y = |b|\sigma_x$

Let $y = 5 - 2x$, $\therefore \sigma_y = |-2|\sigma_x = 2 \times 3 = \mathbf{6}$.

$\therefore$ **Variance $(5 - 2x) = 6^2 = 36$.**

Example 37. *Find the variance of the first n natural numbers.*

Solution. S.D. $= \sqrt{\frac{\Sigma n^2}{n} - \left(\frac{\Sigma n}{n}\right)^2} = \sqrt{\frac{n(n+1)(2n+1)}{6n} - \left(\frac{n(n+1)}{2n}\right)^2}$

$$= \sqrt{\frac{(n+1)(2n+1)}{6} - \left(\frac{(n+1)}{2}\right)^2} = \sqrt{\frac{n+1}{2}\left(\frac{2n+1}{3} - \frac{n+1}{2}\right)}$$

$$= \sqrt{\frac{n+1}{2}\left(\frac{n-1}{6}\right)} = \sqrt{\frac{\mathbf{n^2 - 1}}{\mathbf{12}}}.$$

Variance = (S.D)^2 = $\frac{\mathbf{n^2 - 1}}{\mathbf{12}}$.

Example 38. *A sample of 35 observations has the mean 80 and S.D. as 4. A second sample of 65 observations from the same population has mean 70 and S.D. 3. What is the S.D. of the combined sample ?*

Solution. Here $n_1 = 35,\ \bar{x}_1 = 80,\ \sigma_1 = 4,\ n_2 = 65,\ \bar{x}_2 = 70,\ \sigma_2 = 3$.

$$\therefore \quad \bar{X} = \frac{35 \times 80 + 65 \times 70}{35 + 65} = \frac{2800 + 4550}{100} = \frac{7350}{100} = \mathbf{73.5;}$$

$$d_1 = 73.5 - 80 = -6.5,\ d_2 = 73.5 - 70 = 3.5.$$

$$\sigma_{12} = \sqrt{\left[\frac{35(16 + 42 \cdot 25) + 65(9 + 12 \cdot 25)}{100}\right]} = \sqrt{\frac{35 \times 58.25 + 65(21.25)}{100}} = \sqrt{34.21} = \mathbf{5.85.}$$

Example 39. *The mean of 5 observations is 4.4 and their variance is 8.24. If three observations are 1, 2 and 6, find the other two observations.*

Solution. Let the two unknown items be x and y, then their mean = 4.4

$$\textbf{Mean :} \quad \frac{1 + 2 + 6 + x + y}{5} = 4.4 \Rightarrow \mathbf{x + y = 13.} \qquad \text{... (1)}$$

Also variance = 8.24.

$$\textbf{Variance} = \frac{1^2 + 2^2 + 6^2 + x^2 + y^2}{5} - (\text{mean})^2 = 8.24 \Rightarrow 41 + x^2 + y^2 = 5\{(4.4)^2 + 8.24\}$$

$$\Rightarrow x^2 + y^2 = 138 - 41 \Rightarrow \mathbf{x^2 + y^2 = 97.} \qquad \text{... (2)}$$

Solving (1) and (2), for x and y, we get : $\mathbf{x = 9, y = 4}$ **or** $\mathbf{x = 4, y = 9.}$

Example 40. *The mean and S.D. for group of 100 observations are 65 and 7.03 respectively. If 60 of these observations have mean and S.D. as 70 and 3 respectively, what is the S.D. for the group comprising 40 observation ?*

Solution. $N_1 = 60,\ \bar{x}_1 = 70,\ \sigma_1 = 3;\ N_2 = 40,\ \bar{x}_2 = ?;\ \sigma_2 = ?;\ N = 100,\ \bar{x}_{12} = 65,\ \sigma_{12} = 7.03.$

$$\bar{x}_{12} = \frac{N_1 \bar{x}_1 + N_2 \bar{x}_2}{N_1 + N_2} \Rightarrow \frac{60 \times 170 + 40\bar{x}_2}{100} = 65 \Rightarrow \mathbf{\bar{x}_2 = 57.5.}$$

Now $d_1^2 = (\bar{x}_1 - \bar{x}_{12})^2 = (70 - 65)^2 = \mathbf{25;}$

$d_2^2 = (\bar{x}_2 - \bar{x}_{12})^2 = (57.5 - 65)^2 = \mathbf{56.25;}$

$$\sigma_{12}^2 = \frac{[N_1 \sigma_1^2 + N_2 \sigma^2 + N_1 d_1^2 + N_2 d_2^2]}{N_1 + N_2}$$

$$\Rightarrow \quad (7.03)^2 = \frac{[60 \times 9 + 40\sigma_2^2 + 60 \times 25 + 40 \times 56.25]}{100}$$

$$\Rightarrow \quad \sigma_2^2 = 16.3 \text{ or } \sigma_2 = \sqrt{16.3} = \mathbf{4.04.}$$

Example 41. *Find the value of third quartile if the values of first quartile and quartile deviation are 80 and 10 respectively.*

Solution. Here $Q_1 = 80,\ Q.D. = 10.$

$$\textbf{Quartile Deviation} = \frac{Q_3 - Q_1}{2} \Rightarrow 10 = \frac{Q_3 - 80}{2}$$

$\Rightarrow \quad Q_3 = 20 + 80 =$ **100. Hence the third quartile is 100.**

Example 42. *From the following information calculate combined standard deviation:*
$N_1 = 90$, $\bar{X}_1 = 20$, $\sigma_1 = 8$: $N_2 = 60$, $\bar{X}_2 = 15$, $\sigma_2 = 6$.

Solution. In order to calculate the combined standard deviation, we first calculate combined mean.

Combined Mean : $$\bar{X}_{12} = \frac{N_1\bar{X}_1 + N_2\bar{X}_2}{N_1 + N_2}.$$

Here, $N_1 = 90$, $\bar{X}_1 = 20$, $\bar{X}_2 = 15$ and $N_2 = 60$.

$$\bar{X}_{12} = \frac{(90\times20)+(60\times15)}{90+60} = \frac{1800+900}{150} = \frac{2700}{150} = \mathbf{18.}$$

CALCULATION OF COMBINED STANDARD DEVIATION

Combined S.D. : $$\sigma_{12} = \sqrt{\frac{N_1\sigma_1^2 + N_2\sigma_2^2 + N_1d_1^2 + N_2d_2^2}{N_1 + N_2}}$$

Here, $N_1 = 90$, $\sigma_1 = 8$, $N_2 = 60$, $\sigma_2 = 6$, $d_1 = |\bar{X}_1 - \bar{X}_{12}| = |20 - 18| = 2$;

$d_2 = |\bar{X}_2 - \bar{X}_{12}| = |15 - 18| = 3$.

$$\therefore \quad \sigma_{12} = \sqrt{\frac{90(8)^2 + 60(6)^2 + 90(2)^2 + 60(3)^2}{90+60}}$$

$$= \sqrt{\frac{5760+2160+360+540}{150}} = \sqrt{\frac{8820}{150}} = \mathbf{7.67.}$$

Hence the combined standard deviation is 7.67.

Example 43. *The mean and standard deviation of marks obtained by 40 students of a class in Statistics are 55 and 8 respectively. If there are only 5 girls in the class and their respective marks are 40, 55, 63, 75 and 87, find mean and standard deviation of the marks obtained by the boys.*

Solution. Let $\bar{X}_1$ be mean of the marks secured by Girls.

$$\bar{X}_1 = \frac{40+55+63+75+87}{5} = \frac{320}{5} = \mathbf{64.}$$

Llet σ_1 be the standard deviation of the marks secured by Girls.

$$\sigma_1^2 = \frac{1}{N}\Sigma|X - \bar{X}_G|^2 = \frac{1}{5}\left[24^2 + 9^2 + (-1)^2 + 11^2 + 23^2\right]$$

$$= \frac{1}{5}[576+81+1+121+529] = \frac{1308}{5} = \mathbf{26.16.}$$

Here $N_1 = 5$, $N_2 = 40 - 5 = 35$.

$\bar{X}_2$ = mean of the marks secured by boys.

σ_2 = standard deviation of the marks secured by boys.

Also it is given that : $\bar{X}_{12} = 55$, $\sigma_{12} = 8$.

$$\therefore \quad \bar{X}_{12} = \frac{N_1\bar{X}_1 + N_2\bar{X}_2}{N_1 + N_2} \Rightarrow 55 = \frac{5\times 64 + 35\times \bar{X}_2}{5+35}$$

$$\Rightarrow \quad \bar{X}_2 = (2200 - 320) / 35 = \mathbf{53.72.}$$

Now, $d_1^2 = \left(\bar{X}_1 - \bar{X}_{12}\right)^2 = (64-55)^2 = \mathbf{81.}$

$$d_2^2 = \left(\bar{X}_2 - \bar{X}_{12}\right)^2 = (53.72 - 55)^2 = \mathbf{1.63.}$$

We know that:

$$\sigma_{12}^2 = \frac{N_1\sigma_1^2 + N_2\sigma_2^2 + N_1 d_1^2 + N_2 d_2^2}{N_1 + N_2} \Rightarrow 8^2 = \frac{5\times 26.16 + 35\times\sigma_2^2 + 5\times 81 + 35\times 1.63}{5+35}$$

$$\Rightarrow 35\ \sigma_2^2 = 64 \times 40 - 130.8 - 405 - 57.4 = 2560 - 593.20 = 1966.80$$

or $\sigma_2^2 = 1966.80 / 35 = 56.20 \Rightarrow \sigma_2 \sqrt{\mathbf{56.20}}$ or $\sigma_2 = \mathbf{7.496}$

Example 44. *Mean and standard deviation of a series are 40 and 5 respectively. Number of observations is 100. Later on, it has been discovered that one item value 25 was taken as 15. Calculate the correct mean and standard deviation.*

Solution. Calculation of Correct Mean :

$$\bar{X} = \frac{\Sigma X}{N} \text{ or } N\bar{X} = \Sigma X$$

Here, $N = 100,\ \bar{X} = 40$ or $\Sigma X = N\bar{X} = 100 \times 40 = 4000.$

Correct $\Sigma X = 4000 - 15 + 25 = 4010.$

$$\text{Correct } \bar{X} = \frac{\text{Correct } \Sigma X}{N} = \frac{4010}{100} = \mathbf{40.1.}$$

Calculation of Correct Standard Deviation :

$$\sigma^2 = \frac{\Sigma X^2}{N} - (\bar{X})^2$$

$$\text{or} \quad 5^2 = \frac{\Sigma X^2}{100} - (40)^2$$

$25 \times 100 = \Sigma X^2 - 1600 \times 100$ or ΣX^2
$= \mathbf{162500.}$

Correct $\Sigma X^2 = 62500 - (15)^2 + (25)^2$
$= 162500 - 225 + 625 = \mathbf{162900.}$

$$\text{Correct } \sigma = \sqrt{\frac{\text{Correct } \Sigma X^2}{N} - (\text{Correct } \bar{X})^2}$$

$$= \sqrt{\frac{162900}{100} - (40.1)^2}$$

$$= \sqrt{1629 - 1608.01}$$

$$= \sqrt{20.9} = \mathbf{4.57.}$$

Hence the correct mean is 40.1 and correct standard deviation is 4.57.

Example 45. *Share prices of two companies A Ltd. and B Ltd. were recorded as follows:*

A Ltd. :	*12*	*13*	*15*	*14*	*14*	*14*	*13*	*17*
B Ltd. :	*113*	*114*	*113*	*115*	*117*	*114*	*112*	*114 .*

Which company's share price are more variable ?

Solution. In order to find out which company's share price is more variable, we compare their coefficient of variations.

Table : Computation of Coefficient of Variation

X (A Ltd.)	(X – 14) = x	x^2	Y (B Ltd.)	(Y – 114) = y	y^2
12	–2	4	113	–1	1
13	–1	1	114	0	0
15	+1	1	113	–1	1
14	0	0	115	+1	1
14	0	0	117	+3	9
14	0	0	114	0	0
13	–1	1	112	–2	4
17	+3	9	114	0	0
Σ X = 112	**Σ X = 0**	**$\Sigma x^2 = 16$**	**ΣY = 912**	**Σ y = 0**	**$\Sigma y^2 = 16$**

Company *A* Ltd.	Company *B* Ltd.
Mean : $\bar{X}_A = \frac{\Sigma X}{N} = \frac{112}{8} = 14.$	Mean : $\bar{X}_B = \frac{\Sigma Y}{N} = \frac{912}{8} = 114.$
$\sigma_A = \sqrt{\frac{\Sigma x^2}{N}} = \sqrt{\frac{16}{8}} = 1.41.$	$\sigma_B = \sqrt{\frac{\Sigma y^2}{N}} = \sqrt{\frac{16}{8}} = 1.41.$
Coefficient of Variation :	Coefficient of Variation :
C.V. $(A) = \frac{\sigma_A}{\bar{X}_A} \times 100$	C.V. $(B) = \frac{\sigma_B}{\bar{X}_B} \times 100$
$= \frac{1.41}{14} \times 100 = \mathbf{10.07}$ **%.**	$= \frac{1.41}{114} \times 100 = \mathbf{1.24}$ **%.**

Since C.V (*A*). > C.V (*B*), so the share price of *A* Ltd. is more variable than the share price of *B* Ltd.

Example 46. *Find the standard deviation of the weights of the 100 male students of ABC University*

Weights (in kg) :	*60 – 62*	*63 – 65*	*66 – 68*	*69 – 71*	*72 – 74*
No. of students :	*5*	*18*	*42*	*27*	*8*.

Solution. The given data is in inclusive form, we first convert it into exclusive form

Table : Computation of Standard Deviation

Weight (in kg)	*mid-point (m)*	*f*	*(m – 67) / 3*	*fd*	*fd* ²
59.5 – 62.5	61	5	–2	–10	20
62.5 – 65.5	64	18	–1	–18	18
65.5 – 68.5	67	42	0	0	0
68.5 – 71.5	70	27	+1	+27	27
71.5 – 74.5	73	8	+2	+16	32
		***N* = 100**		**Σ *fd* = 15**	**$\Sigma fd^2 = 97$**

Here $\Sigma fd^2 = 97$, $N = 100$, $\Sigma fd = 15$, $i = 3$

$$\therefore \sigma = \sqrt{\frac{\Sigma fd^2}{N} - \left(\frac{\Sigma fd}{N}\right)^2} \times i = \sqrt{\frac{97}{100} - \left(\frac{15}{100}\right)^2} \times 3$$

$$= \sqrt{0.97 - .0225} \times 3$$

$$= 0.9734 \times 3 = \mathbf{2.92.}$$

Example 47. *For two firms A and B, the following details are available:*

	A	*B*
No. of Employees	*100*	*200*
Average Salary (Rs.)	*1,600*	*1,800*
S.D. of Salary	*16*	*18*

Compute the following :

(i) Which firms pays larger package of salary ?

(ii) Which firm shows greater variation in the distribution of salary ?

(iii)Compute the combined average salary and combined variance of both the firms.

Solution.

(i) Firm *A* pays = 100 × 1600 = Rs. 1,60,000.

Firm *B* pays = 200 × 1800 = Rs. 3,60,000.

Hence the firm *B* pays larger salary package.

(ii) **For finding out which firm shows greater variation, we compare their coefficient of variation :**

$$\text{C.V. } (A) = \frac{\sigma}{\bar{X}} \times 100 = \frac{16}{1600} \times 100 = 1\%.$$

$$\text{C.V. } (B) = \frac{\sigma}{\bar{X}} \times 100 = \frac{18}{1800} \times 100 = 1\%.$$

Hence the variation is same in both the firms.

(iii) **Combined Average Salary:**

$$\bar{X}_{12} = \frac{N_1\bar{X}_1 + N_2\bar{X}_2}{N_1 + N_2} = \frac{100(1600) + 200(1800)}{100 + 200}$$

$$= \frac{1,60,000 + 3,60,000}{300} = \frac{5,20,000}{300} = \text{Rs. } 1,733.33.$$

$$\textbf{Combined Variance : } \sigma_{12} = \frac{N_1\sigma_1^2 + N_2\sigma_2^2 + N_1 d_1{}^2 + N_2 d_2{}^2}{N_1 + N_2}$$

$$= \frac{100(16)^2 + 200(18)^2 + 100(1733.33 - 1600)^2 + 200(1733.33 - 1800)^2}{300}$$

$$= \frac{25,600 + 64,800 + 17,77,688.89 + 8,88,977.8}{300} = \frac{27,57,066.67}{300} = 9190.22.$$

Example 48. *Mean and Standard Deviation of 100 items are found to be 40 and 10. At the time of calculation two items are wrongly taken as 30 and 70 instead of 3 and 27. Find the correct mean and correct standard deviation.*

Solution. We know that: $\bar{X} = \frac{\Sigma X}{N} \Rightarrow \Sigma X = N\bar{X}$.

Calculation of Correct Mean

Here $N = 100$, $\bar{X} = 40$ or $\Sigma X = 100 \times 40 =$ **4,000.**

Correct $\Sigma X = 4000 - 30 - 70 + 3 + 27 =$ **3,930.**

$$\text{Correct } \bar{X} = \frac{\text{Correct } \Sigma X}{N} = \frac{3930}{100} = \mathbf{39.3.}$$

Calculation of Correct Standard Deviation

$$\sigma^2 = \frac{\Sigma X^2}{N} - (\bar{X})^2 \text{ or } \sigma^2 = \frac{\Sigma X^2}{N} - (40)^2$$

$\Rightarrow \Sigma X^2 = 100 \times 100 + 1600 \times 100$ or $\Sigma X^2 =$ **1,70,000**

Correct $\Sigma X^2 = 1,70,000 - (30)^2 - (70)^2 + (3)^2 + (27)^2$

$= 1,70,000 - 900 - 4,900 + 9 + 729 =$ **1,64,938**

$$\text{Correct } \sigma = \sqrt{\frac{\text{Correct } \Sigma X^2}{N} - (\text{Correct } \bar{X})^2}$$

$$= \sqrt{\frac{164938}{100} - (39.3)^2} = \sqrt{1649.38 - 1544.49} = \sqrt{104.89} = \mathbf{10.24.}$$

Hence, the correct mean is 39.3 and correct standard deviation is 10.24.

Example 49. *The mean and standard deviation of monthly salary (in Rs.) are given below for two factories A and B.*

Factory	*No. of employees*	*Average monthly salary*	*Standard deviation of monthly salary*
A	*100*	*2,500*	*200*
B	*120*	*2,800*	*250*

(*i*) *Find the pooled mean and pooled standard deviation.*

(*ii*) *Which factory has better consistent salary structure ?*

Solution.

(*i*) $\bar{X}$ = **Combined Mean** $= \frac{n_1\bar{X}_1 + n_2\bar{X}_2}{n_1 + n_2} = \frac{100 \times 2500 + 120 \times 2800}{100 + 120} = \mathbf{2663.63.}$

Pooled Standard Deviation: $\sigma_{12} = \sqrt{\frac{n_1\sigma_1^2 + n_2\sigma_2^2 + n_1 d_1^2 + n_2 d_2^2}{n_1 + n_2}}$,

where $d_1 = \bar{x}_1 - \bar{x}$ **and** $d_2 = \bar{x}_2 - \bar{x}$. Here $d_1 = 163.63$ and $d_2 = 136.37$.

$$\sigma_{12} = \sqrt{\frac{100 \times (200)^2 + 120 \times (250)^2 + 100 \times (163.63)^2 \times 120 \times (136.37)^2}{100 + 120}} = \sqrt{\frac{16409090.92}{220}}$$

$= \mathbf{273.106.}$

(*ii*) **Coefficient of Variation for Factory** $A = \frac{\sigma_1}{\bar{X}_1} \times 100 = \frac{200}{2500} \times 100 = \mathbf{8\%.}$

Coefficient of Variation for Factory $B = \frac{\sigma_2}{\bar{X}_2} \times 100 = \frac{250}{2800} \times 100 = 8.9\%$.

Hence **Factory *A* is more consistent.**

Example 50. *The coefficient of variations of two series are 58% and 69%. Their standard deviations are 21.2 and 15.6. What are their arithmetic mean?*

Solution. We know that

$$\text{C.V. = Coefficient of Variation} = \frac{\sigma \times 100}{\bar{X}} \quad \text{or} \quad \bar{X} = \frac{\sigma}{\text{C.V.}} \times 100.$$

Mean of First Series: $\bar{X}_1 = \frac{\sigma_1}{\text{C.V.}} \times 100 = \frac{21.2}{58} \times 100 = 36.6.$

Mean of Second Series: $\bar{X}_2 = \frac{\sigma_2}{\text{C.V.}} \times 100 = \frac{15.6}{69} \times 100 = 22.6.$

4.10 LIMITS OF VARIABILITY

Standard deviation shows the limits of variability by which the individual item in a distribution will very from the mean. **For a symmetrical distribution with mean $\bar{X}$, the following area relationship holds good:**

(*i*) $\bar{X} \pm \sigma$ **covers 68.27% items.**

(*ii*) $\bar{X} \pm 2\sigma$ **covers 95.54% items.**

(*iii*) $\bar{X} \pm 3\sigma$ **covers 99.73% items.**

These limits are illustrated by the following curve known as Normal Curve.

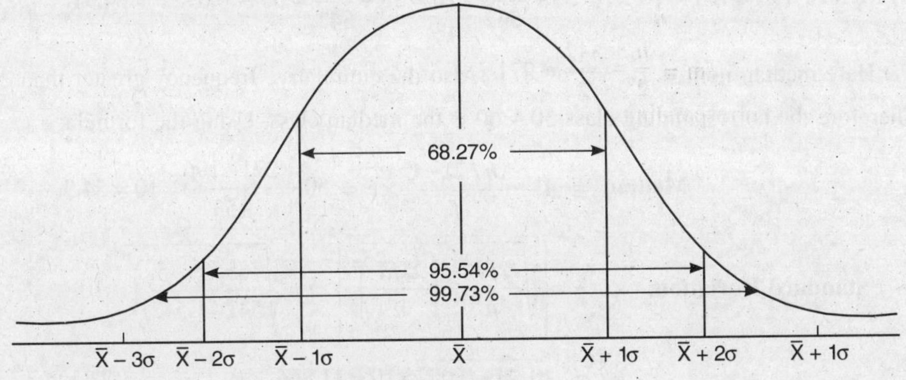

Fig. 4.1. Normal Curve.

4.11 EMPIRICAL RELATIONSHIPS

If the data is **moderately non-symmetrical,** then we have the following empirical relationships.

Mean Deviation = $\frac{4}{5} \times$ **Standard Deviation** = $\frac{4}{5}\sigma$

Semi-Interquartile Range = $\frac{2}{3} \times$ **Standard Deviation** = $\frac{2}{3}\sigma$.

Probable Error of Standard Deviation = $\frac{2}{3}\sigma$ = **Semi-interquartile Range.**

Quartile Deviation = $\frac{5}{6}$ **of Mean Deviation.**

From these relationships, we have

4 Standard Deviation = 5 Mean Deviation = 6 Quartile Deviation.

MISCELLANEOUS EXAMPLES

Example 51. *Calculate the mean, median and standard deviation of the following distribution:*

Age in years :	*20 – 30*	*30 – 40*	*40 – 50*	*50 – 60*	*60 – 70*	*70 – 80*	*80 – 90*
Number of Marks:	*3*	*61*	*132*	*153*	*140*	*51*	*2*

Solution. Let the assumed mean $A = 55$. $i = 10$. Then we have the following table:

Table: Computation of Mean, Median and Standard Deviation

Class Interval	*Mid-value (x)*	*Frequency (f)*	*Cumulative (c.f.)*	$d = \frac{x-A}{10}$	*fd*	*fd*^{2}
20 – 30	25	3	3	– 3	– 9	27
30 – 40	35	61	64	– 2	– 122	244
40 – 50	45	132	196	– 1	– 132	132
50 – 60	**55**	**153**	**349**	0	0	0
60 – 70	65	140	489	1	140	140
70 – 80	75	51	540	2	102	204
80 – 90	85	2	542	3	6	18
Total		$n = \Sigma f = 542$			$\Sigma fd = -5$	$\Sigma fd^2 = 765$

(*i*) **Mean :** $(\bar{x}) = A + \frac{\Sigma fd}{n} \times i = 55 + \frac{(-5)}{542} \times 10 = 55 - \frac{50}{542} = 55 - 0.0922 = \mathbf{54.91}.$

(*ii*) Here median item = $\frac{n}{2} = \frac{542}{2} = 271.$ Also the cumulative frequency greater than 271 is 349. Therefore the corresponding class 50 – 60 is the median class. Using the formula:

$$\textbf{Median} = L + \frac{(n/2) - C}{f} \times i = 50 + \frac{271 - 196}{153} \times 10 = \mathbf{54.9}.$$

$$(iii)\ \textbf{Standard Deviation:}\ \sigma = \left\{\sqrt{\frac{\Sigma fd^2}{n} - \left(\frac{\Sigma fd}{n}\right)^2}\right\} \times i = \sqrt{\frac{765}{542} - \left(\frac{-5}{542}\right)^2} \times 10$$

$$= \sqrt{1.41 - 0.002} \times 10 = \mathbf{11.866}.$$

Example 52. *The arithmetic mean and standard deviation of a series of 20 items were calculated by a student as 20 cm and 5 cm respectively. But while calculating them an item 13 was misread as 30. Find the correct standard deviation.*

Solution. We know that: $\bar{X} = \frac{\Sigma X}{N}$

or $\quad$ **Wrong** $= \Sigma X = \bar{X} N = \mathbf{20 \times 20 = 400}$

Correct $\Sigma X = 400 - 30 + 13 = \mathbf{383}$

Correct $\bar{X} = \frac{383}{20} = \mathbf{19.15}$

Also $\quad$ **Variance** $(\sigma^2) = \frac{\Sigma X^2}{N} - (\bar{X})^2 \Rightarrow 25 = \frac{\Sigma X^2}{20} - (20)^2$

$\Rightarrow \quad 500 = \Sigma X^2 - (20)^2 \times 20 \Rightarrow 500 = \Sigma X^2 - 8{,}000 \Rightarrow 500 + 8{,}000 = \Sigma X^2$

$\therefore \quad$ **Corrected** $\Sigma X^2 = 8{,}500 - (30)^2 + (13)^2 = 7769$

$$\textbf{Correct variance} = \frac{\textbf{Corrected } \Sigma X^2}{N} - (\textbf{Corrected } \bar{X})^2$$

$$= 388.5 - 366.6 = 21.9 \Rightarrow \sigma = \sqrt{21.9} = 4.68.$$

Hence the correct Standard Deviation is 4.68.

Example 53. *The prices of shares X and Y are given below, state which share is more stable in value ?*

X :	*55*	*54*	*52*	*53*	*56*	*58*	*52*	*50*	*51*	*49*
Y :	*108*	*107*	*105*	*105*	*106*	*107*	*104*	*103*	*104*	*101*

Solution. The share which has less coefficient of variation (C.V.) is more stable.

Here, $\bar{X} = \frac{\Sigma X}{n} = \frac{530}{10} = \mathbf{53}$, $\bar{Y} = \frac{\Sigma Y}{n} = \frac{1050}{10} = \mathbf{105}.$

Table: Computation of Mean and S.D. of Shares

Share X			*Share Y*		
X	$X-\bar{X}=X-53$	$(X-\bar{X})^2$	Y	$Y-\bar{Y}=Y-105$	$(Y-\bar{Y})^2$
55	2	4	108	3	9
54	1	1	107	2	4
52	– 1	1	105	0	0
53	0	0	105	0	0
56	3	9	106	1	1
58	5	25	107	2	4
52	– 1	1	104	– 1	1
50	– 3	9	103	– 2	4
51	– 2	4	104	– 1	1
49	– 4	16	101	– 4	16
$\Sigma X = 530$, $\Sigma(X-\bar{X}) = 0$, $\Sigma(X-\bar{X})^2 = 70.$			$\Sigma Y = 1050$, $\Sigma(Y-\bar{Y}) = 0$, Σ	$(Y-\bar{Y}^2) = 40$	

$$\therefore \quad \sigma_x = \sqrt{\frac{1}{n}\Sigma(X-\bar{X})^2} = \sqrt{\frac{70}{10}} = \sqrt{7} = \mathbf{2.646.}$$

$$\sigma_y = \sqrt{\frac{1}{n}\Sigma(Y-\bar{Y})^2} = \sqrt{\frac{40}{10}} = \sqrt{4} = \mathbf{2.}$$

$$\textbf{C.V. of } X \textbf{ series} = \frac{\sigma_x}{\bar{X}} \times 100 = \frac{2.646}{53} \times 100 = \mathbf{4.99\%}.$$

$$\textbf{C.V. of } Y \textbf{ series} = \frac{\sigma_y}{\bar{Y}} \times 100 = \frac{2}{105} \times 100 = \mathbf{1.90\%}.$$

Hence C.V. of *Y* series is less than C.V. of *X* series so the shares of *Y* are more stable.

Example 54. *Comment on the followings:*

"For a set of 10 observations, mean = 5, S.D. = 2 and C.V. = 60%."

Solution. We are given that: $n = 10$, $\bar{x} = 5$, $\sigma = 2$

$$\therefore \text{ C.V.} = \frac{\sigma}{\bar{x}} \times \mathbf{100} = \frac{2}{5} \times 100 = \mathbf{40\%}$$

But we are given that C.V. = 60%.

Hence the given statement is wrong.

Example 55. *The runs scored by two batsmen A and B in 9 consecutive matches are given below:*

A :	*85*	*20*	*62*	*28*	*74*	*5*	*69*	*4*	*13*
B :	*72*	*4*	*15*	*30*	*59*	*15*	*49*	*27*	*26*

Which of the batsmen is more consistent ?

Solution. For Batsman *A* :

$$\textbf{Mean score:} \quad \bar{x}_A = \frac{85+20+62+28+74+5+69+4+13}{9} = 40.$$

$$\textbf{Variance in score:} \sigma_A^2 = \frac{\Sigma(X - \bar{X}_A)^2}{N} = \frac{45^2+20^2+22^2+12^2+34^2+35^2+29^2+36^2+27^2}{9}$$

$$= \frac{8300}{9} = \mathbf{922.22.}$$

$$\textbf{Coefficient of Variation : C.V}_A = \frac{\sigma_A}{\bar{x}_A} \times \mathbf{100} = \frac{\sqrt{922.22}}{40} \times 100 = \mathbf{75.92}\ \%.$$

For Batsman *B*,

$$\textbf{Mean score:} \quad \bar{x}_B = \frac{72+4+15+30+59+15+49+27+26}{9} = \mathbf{33.}$$

$$\textbf{Variance in score:} \sigma_B^2 = \frac{\Sigma(X - \bar{X}_B)^2}{N} = \frac{39^2+29^2+18^2+3^2+26^2+18^2+16^2+6^2+7^2}{9}$$

$$= \frac{4036}{9} = \mathbf{448.44}.$$

$$\textbf{Coefficient of Variation: C.V}_B = \frac{\sigma_B}{\bar{x}_B} \times \mathbf{100} = \frac{\sqrt{448.44}}{33} \times 100 = \mathbf{64.17}\ \%.$$

Since $C.V_B < C.V_A$, so batsman *B* is more consistent.

Example 56. *The mean and standard deviation of distributions of 100 and 150 items are 50, 5 and 40, 6, respectively. Find the mean and standard deviation of all the 250 items taken together.*

Solution. Here $N_1 = 100$, $\bar{X}_1 = 50$, $\sigma_1 = 5$,

$N_2 = 150$, $\bar{X}_2 = 40$, $\sigma_2 = 6$

$$\therefore \quad \textbf{Combined Mean: } \bar{X} = \frac{N_1\bar{X}_1 + N_1\bar{X}_2}{N_1 + N_2} = \frac{100 \times 50 + 150 \times 40}{100+150}$$

$$= \frac{5000+6000}{250} = \mathbf{44.}$$

Also $d_1 = \bar{X}_1 - \bar{X} = 50 - 44 = 6$, $d_2 = \bar{X}_2 - \bar{X} = 40 - 44 = -4$.

$\therefore$ **Combined Standard Deviation:** $\sigma = \sqrt{\dfrac{N_1\sigma_1^2 + N_2\sigma_2^2 + N_1 d_1^2 + N_2 d_2^2}{N_1 + N_2}}$

$$= \sqrt{\frac{100\times(5)^2 + 150\times(6)^2 + 100\,(6)^2 + 150\,(-4)^2}{250}}$$

$$= \sqrt{\frac{2500 + 5400 + 3600 + 2400}{250}} = \sqrt{\frac{13900}{250}} = \sqrt{55.6} = \mathbf{7.46.}$$

Example 57. *Mean and coefficient of standard deviation of 100 items are found by a student as 50 and 5. If at the time of calculation two items are wrongly taken as 40 and 50 instead of 60 and 30, find the correct mean and standard deviation.*

Solution. Coefficient of standard deviation = $\dfrac{\sigma}{\bar{X}} \Rightarrow 0.1 = \dfrac{\sigma}{50} \Rightarrow \sigma = 5.$

Also $\bar{X} = \dfrac{\Sigma X}{N} \Rightarrow 50 = \dfrac{\Sigma X}{100}$ or $\Sigma X = 5{,}000$.

Correct ΣX **= Incorrect** ΣX **– Wrong items + Correct items**

$= 5{,}000 - 40 - 50 + 60 + 30 = \mathbf{5{,}000.}$

Correct $\bar{X} = \dfrac{\text{Correct } \Sigma X}{N} = \dfrac{5{,}000}{100} = \mathbf{50.}$

Also $\sigma = \sqrt{\dfrac{\Sigma X^2}{N} - (\bar{X})^2} \Rightarrow 5 = \sqrt{\dfrac{\Sigma X^2}{100} - (50)^2} \Rightarrow 25 = \dfrac{\Sigma X^2}{100} - 2500$

$\Rightarrow$ $2525 \times 100 = \Sigma X^2 \Rightarrow \Sigma X^2 = \mathbf{2{,}52{,}500.}$

Correct ΣX^2 **= Incorrect** ΣX^2 **– Wrong items' square + Correct items' square**

$= 2{,}52{,}500 - (40)^2 - (50)^2 + (60)^2 + (30)^2$

$= 2{,}52{,}500 - 1{,}600 - 2{,}500 + 3{,}600 + 900 = \mathbf{2{,}52{,}900.}$

Correct $\sigma = \sqrt{\dfrac{\text{Correct } \Sigma X^2}{N} - (\text{Correct } \bar{X})^2} = \sqrt{\dfrac{2{,}52{,}900}{100} - (50)^2}$

$= \sqrt{2{,}529 - 2{,}500} = \sqrt{29} = \mathbf{5.385.}$

Hence the correct mean = 50; and the correct standard deviation = 5.385.

Example 58. *The mean and standard deviation of 200 items are found to be 60 and 20 respectively. If at the time of calculation, two items were wrongly taken as 3 and 67 instead of 13 and 17, find the correct mean and standard deviation. What is the correct coefficient of variation?*

Solution.

CORRECT MEAN:

We are given $\bar{X} = 60$; $\sigma = 20$; $N = 200$.

Also $\bar{X} = \dfrac{\Sigma X}{N} \Rightarrow 60 = \dfrac{\Sigma X}{100}$ or $\Sigma X = 12{,}000$.

But, **Correct ΣX = ΣX – Wrong items + Correct items**

$$= 12{,}000 - 3 - 67 + 13 + 17 = \mathbf{11{,}960.}$$

$$\therefore \quad \textbf{Correct Mean} = \frac{\textbf{Correct } \Sigma X}{N} = \frac{11{,}960}{200} = \mathbf{59.8.}$$

CORRECT STANDARD DEVIATION:

$$\sigma = \sqrt{\frac{\Sigma X^2}{N} - (\overline{X})^2} \Rightarrow 20 = \sqrt{\frac{\Sigma X^2}{200} - (60)^2}$$

$$\Rightarrow \quad 400 = \frac{\Sigma X^2}{200} - 3{,}600 \qquad \textbf{[Squaring both sides]}$$

$$\Rightarrow \quad 400 \times 200 = \Sigma X^2 - 3{,}600 \times 200 \Rightarrow 80{,}000 = \Sigma X^2 - 7{,}20{,}000$$

$$\Rightarrow \quad \Sigma X^2 = 80{,}000 + 7{,}20{,}000 = \mathbf{8{,}00{,}000.}$$

Now **Correct ΣX^2 = Incorrect ΣX^2 – Wrong items square + Correct items square**

$$\therefore \quad \textbf{Correct } \Sigma X^2 = 8{,}00{,}000 - (3)^2 - (67)^2 + (13)^2 + (17)^2$$

$$= 8{,}00{,}000 - 9 - 4{,}489 + 169 + 289 = \mathbf{7{,}95{,}960.}$$

Again $$\textbf{Correct } \sigma = \sqrt{\frac{\textbf{Correct } \Sigma X^2}{N} - (\textbf{Correct } \overline{X})^2} = \sqrt{\frac{7{,}95{,}960}{200} - (59.8)^2}$$

$$= \sqrt{3979.8 - 3576.04} = \sqrt{403.76} = \mathbf{20.09.}$$

$$\textbf{Correct Coefficient of Variation} = \frac{\sigma}{\overline{X}} \times 100 = \frac{20.09}{59.8} \times 100 = \mathbf{33.595.}$$

Hence the correct mean = 59.8; the correct standard deviation = 20.09; and the correct coefficient of variation = 33.595.

Example 59. *For a group containing 100 observations, the arithmetic mean and standard deviation are 8 and* $\sqrt{10.5}$ *. For 50 observations selected from these 100 observations the mean and the standard deviation are 10 and 2 respectively. Find the arithmetic mean and the standard deviation of the other half.*

Solution. Here $N_1 + N_2 = 100, \quad \overline{X}_{12} = 8, \quad \sigma_{12} = \sqrt{10.5}$,

$$N_1 = 50, \quad \overline{X}_1 = 10, \quad \sigma_1 = 2; \; N_2 = 100 - N_1 = 50.$$

We have to find the mean and standard deviation of the other half. We know that:

$$\mathbf{X_{12}} = \frac{\mathbf{N_1 \overline{X}_1 + N_2 \overline{X}_2}}{\mathbf{N_1 + N_2}} \Rightarrow 8 = \frac{50 \times 10 + 50\overline{X}_2}{100}$$

$$\Rightarrow \quad 800 = 500 + 50\overline{X}_2 \Rightarrow 50\overline{X}_2 = 300 \text{ or } \mathbf{\overline{X}_2 = 6.} \qquad [\because N_2 = 50]$$

Also, $$\sigma_{12}^2 = \frac{N_1\sigma_1^2 + N_2\sigma_2^2 + N_1 d_1^2 + N_2 d_2^2}{N_1 + N_2}. \qquad ...(1)$$

Here $d_1 = (\overline{X}_1 - \overline{X}_{12}) = 10 - 8 = \mathbf{2}$; $d_2 = (\overline{X}_2 - \overline{X}_{12}) = 6 - 8 = \mathbf{-2}$.

$$\therefore \quad 10.5 = \frac{50(2)^2 + 50\sigma_2^2 + 50(2)^2 + 50(-2)^2}{100}$$

$$\Rightarrow \quad 10.5 \times 100 = 50 \times 4 + 50\sigma_2^2 + 200 + 200$$

$$\Rightarrow \quad 50\sigma_2^2 = 1050 - 600 = 450 \Rightarrow \sigma_2^2 = \frac{450}{50} = 9 \Rightarrow \boldsymbol{\sigma_2 = 3.}$$

Hence, the mean of the remaining 50 items is 8 and its standard deviation is 3.

Example 60. *The first of two sub-groups has 100 items with mean 15 and standard deviation 3. If the whole group has 250 items with mean 15.6 and standard deviation* $\sqrt{13.44}$, *find the standard deviation of the second sub-group.*

Solution. Here $N_1 = 100$; $\overline{X}_1 = 15$; $\sigma_1 = 3$;

$$N_2 = 250 - N_1 = (250 - 100) = 150; \quad \overline{X}_{12} = 15.6; \quad \sigma_{12} = \sqrt{13.44}$$

Also $$\overline{X}_{12} = \frac{N_1\overline{X}_1 + N_2\overline{X}_2}{N_1 + N_2} \Rightarrow 15.6 = \frac{100 \times 15 + 150\overline{X}_2}{250}$$

$$\therefore \Rightarrow \quad 3{,}900 = 1500 + 150\overline{X}_2 \Rightarrow -150\overline{X}_2 = -3{,}900 + 1{,}500$$

$$\Rightarrow \quad -150\overline{X}_2 = -2{,}400 \Rightarrow \boldsymbol{\overline{X}_2 = 16.}$$

$\therefore$ $d_1 = (\overline{X}_1 - \overline{X}_{12}) = 15 - 15.6 = \mathbf{-0.6.}$; $d_2 = (\overline{X}_2 - \overline{X}_{12}) = 16 - 15.6 = \mathbf{0.4.}$

$$\sigma_{12}^2 = \frac{N_1\sigma_1^2 + N_2\sigma_2^2 + N_1 d_1^2 + N_2 d_2^2}{N_1 + N_2}$$

$$13.44 = \frac{100(3)^2 + 150\sigma_2^2 + 100(-0.6)^2 + 150(0.4)^2}{250}$$

$$\Rightarrow 3{,}360 = 900 + 150\sigma_2^2 + 36 + 24 \Rightarrow 3{,}360 - 960 = 150\sigma_2^2 \text{ or } 2{,}400 = 150\sigma_2^2$$

$$\Rightarrow \quad \sigma_2^2 = \frac{2{,}400}{150} = \mathbf{16} \Rightarrow \boldsymbol{\sigma_2 = 4.}$$

Hence, the standard deviation of the second sub-group = 4.

Example 61. *Two cricketers scored the following runs in the several innings. Find who is a better run-getter and who is more consistent player.*

A :	*42*	*17*	*83*	*59*	*72*	*76*	*64*	*45*	*40*	*32*
B :	*28*	*70*	*31*	*0*	*59*	*108*	*82*	*14*	*3*	*95.*

Solution.

Table: Computation of Mean and Coefficient of Variation

Cricketer A			*Cricketer B*		
X	$(X-\overline{X})$ $= (X-53)$	$(X-\overline{X})^2$	Y	$(Y-\overline{Y})$ $= (Y-49)$	$(Y-\overline{Y})^2$
42	−11	121	28	−21	441
17	−36	1296	70	+21	441
83	+30	900	31	−18	324
59	+6	36	0	−49	2401
72	+19	361	59	+10	100
76	+23	529	108	+59	3481
64	+11	121	82	+33	1089
45	−8	84	14	−35	1225
40	−13	169	3	−46	2116
32	−21	441	95	+46	2116
$\Sigma X = 530$	$\Sigma(X-\overline{X}) = 0$	$(X-\overline{X})^2 = 4{,}058$	$\Sigma Y = 490$	$\Sigma(Y-\overline{Y}) = 0$	$\Sigma(Y-\overline{Y})^2 = 13{,}734$

Cricketer A

$$\overline{X} = \frac{\Sigma X}{N} = \frac{530}{10} = \mathbf{53.}$$

$$\sigma = \sqrt{\frac{\Sigma(X-\overline{X})^2}{N}} = \sqrt{\frac{4{,}058}{10}}$$

$$= \sqrt{405.8} = \mathbf{20.14.}$$

Cricketer B

$$\overline{Y} = \frac{\Sigma Y}{N} = \frac{490}{10} = \mathbf{49.}$$

$$\sigma = \sqrt{\frac{\Sigma(Y-\overline{Y}^2)}{N}} = \sqrt{\frac{13{,}734}{10}}$$

$$= \sqrt{1373.4} = \mathbf{37.06.}$$

$$\text{C.V}_A. = \frac{\sigma}{\overline{X}} \times 100 = \frac{20.14}{53} \times 100 = \mathbf{38\%}. \quad \text{C.V}_B. = \frac{\sigma}{\overline{Y}} \times 100 = \frac{37.06}{49} \times 100 = \mathbf{75.6\%}.$$

A is better run getter as average is higher than B.

A is also more consistent player as the C.V$_A$. is less than C.V$_B$.

Example 62. *The arithmetic means of runs scored by the three batsmen, X, Y and Z, in a series of 10 innings are 50, 48 and 12 respectively. The standard deviations of their runs are 15, 12 and 2 respectively. Who is the most consistent of the three?*

Solution. To compare the variability of the runs scored by the batsmen X, Y and Z, we shall find their **coefficient of variation (C.V.)** given by:

$$\text{C.V.} = \frac{100 \times \sigma}{\text{Mean}}; \qquad \therefore \text{C.V.(X)} = \frac{100 \times 15}{50} = \mathbf{30\%};$$

$$\text{C.V.(Y)} = \frac{100 \times 12}{48} = \mathbf{25\%}; \qquad \text{C.V.(Z)} = \frac{100 \times 2}{12} = \mathbf{16.67\%}.$$

Since C.V.(Z) is the least, so the batsman 'Z' is most consistent.

Example 63. *The coefficients of variation of wages of male workers and female workers are 55 per cent and 70 per cent respectively, while the standard deviations are 22.0 and 15.4 respectively. Calculate the overall average wages of all workers given that 80 per cent of the workers are male.*

Solution. For calculating overall average wages of all workers, we can calculate the average of each group.

The data given in the question is:

	Males	***Females***
C.V.	*55*	*70*
σ	*22.0*	*15.4*

AVERAGE: Male Workers

$$\textbf{C.V.} = \frac{\sigma}{\overline{X}} \times \mathbf{100} \text{ or } \overline{X} = \frac{\sigma}{\text{C.V.}} \times \mathbf{100} \Rightarrow \overline{X} = \frac{22}{55} \times 100 = \mathbf{40.}$$

AVERAGE: Female Workers

$$\textbf{C.V.} = \frac{\sigma}{\overline{X}} \times \mathbf{100} \text{ or } \overline{X} = \frac{\sigma}{\text{C.V.}} \times 100 \Rightarrow \overline{X} = \frac{15.4}{70} \times 100 = \mathbf{22.}$$

EXERCISE – 4.1

1. Explain the meaning of dispersion. What purpose does a measure of dispersion serve ?

2. What do you mean by the dispersion of a distribution? Discuss relative merits and demerits of various measures of dispersion.

3. What are the requisites of a good measure of dispersion? In the light of these comment on some of the well known measures of dispersion.

4. What do you understand by coefficient of variation ? Discuss the relative advantages of coefficient of variation and standard deviation as a measure of variability.

5. Calculate the range from the following data:

12 20 15 22 16 14 21 17.

6. Calculate the range from the following data:

Classes :	10 – 20	20 – 30	30 – 40	40 – 50	50 – 60
Frequency :	5	8	12	7	4

7. The first quartile derived from a set of observations is 25 and its quartile deviation is 15, find the third quartile.

8. Calculate the quartile deviation and its coefficient from the following data:

Marks obtained :	10	20	30	40	50	60
No. of Students :	4	7	15	8	7	2

[**Hint.** Q_1 = value of $\left(\frac{N+1}{4}\right)$th item = $\left(\frac{43+1}{4} = 11\right)$th item = **20.**

Q_3 = value of $\left(\frac{3 \times 44}{4}\right)$th item = 33rd item is **40.**

$\therefore$ $Q.D. = \frac{40-20}{2}$ = **10; Coefficient of** $Q.D. = \frac{Q_3 - Q_1}{Q_3 + Q_1} = \frac{20}{60}$ = **0.333.**]

9. If the first quartile is 142 and the semi-interquartile range is 18, find the median (assuming the distribution to be symmetrical.

10. If the first quartile is 104 and quartile deviation is 8, find the third quartile.

11. Calculate mean deviation from the following data:

Variable	: 0	1	2	3	4	5	6	7	8	9
Frequency	: 15	46	91	162	110	95	82	26	13	2.

12. Calculate standard devation of the following marks obtained by 5 students in a turorial group:

Marks obtained :	8	12	13	15	22 .

13. Co-efficient of variation of two series is 60% and 80%. Their standard deviation are 24 and 20. What are their arithmetic means ?

14. You are given the following raw sums in a statistical survey of two variables *X* and *Y*:

$\Sigma X = 240, \Sigma Y = 250, \Sigma X^2 = 6400$ and $\Sigma Y^2 = 7060$.

Ten items are included in each survey. Compute Standard Deviation of the *X* and *Y* variables.

15. Calculate the Combined Standard Deviation from the following data :

	No. of Students	Average Marks	Student Deviation
Section *A* :	60	72	8
Section *B* :	40	68	11

[**Hint. Combined Mean :** $\bar{X}_{12} = \dfrac{N_1\bar{X}_1 + N_2\bar{X}_2}{N_1 + N_2} = \dfrac{(60\times 72)+(40\times 68)}{60+40} = \mathbf{70.4.}$

$d_1 = |X_1 - X_{12}| = |72 - 70.4| = 1.6;\ d_2 = |\bar{X}_2 - \bar{X}_{12}| = |68 - 70.4| = 2.4.$

$$\sigma_{12} = \sqrt{\frac{N_1\sigma_1^2 + N_2\sigma_2^2 + N_1d_1^2 + N_2d_2^2}{N_1+N_2}} = \sqrt{\frac{60(8)^2 + 40(11)^2 + 60(1.6)^2 + 40(2.4)^2}{60+40}} = \mathbf{9.52.}]$$

16. Find the standard deviation from the following data:

Marks	: 10 – 20	20 – 30	30 – 40	40 – 50	50 – 60
No. of students	: 8	10	12	8	4

[**Hint.** Mid–point (*m*): 15, 25, 35, 45, 55

f : 8 10 12 8 4.

Take $d = (m - 35) / 10$, then $\Sigma fd = -10, \Sigma fd^2 = 60, N = 42$.

$$\sigma = \sqrt{\frac{\Sigma fd^2}{N} - \left(\frac{\Sigma fd}{N}\right)^2} \times i = \sqrt{\frac{66}{42} - \left(\frac{-10}{42}\right)^2} \times 10 = \mathbf{12.31.}]$$

17. From the prices of shares of *X* and *Y* given below, state which share is more stable in value.

X	: 41	44	43	48	45	46	49	50	42	40
Y	: 91	93	96	92	90	97	99	94	98	95

Given that $d_x = (x - 45), d_y = y - 95$, then, $\Sigma d_x = -2, \Sigma d_x^2 = 106, \Sigma d_y = -5, \Sigma d_y^2 = 85, N = 10$.

[**Hint.** $\bar{X} = A + \dfrac{\Sigma d_x}{N} = 45 - \dfrac{2}{10} = \mathbf{44.8.}$; $\bar{Y} = B + \dfrac{\Sigma d_y}{N} = 95 - \dfrac{5}{10} = \mathbf{94.5.}$

$$\sigma_x = \sqrt{\frac{\Sigma d_x^2}{N} - \left(\frac{\Sigma d_x}{N}\right)^2} = \sqrt{\frac{106}{10} - \left(\frac{-2}{10}\right)^2} = \sqrt{10.6 - 0.04} = \mathbf{3.25.}$$

$$C.V.\ (X) = \frac{\sigma_x}{\bar{X}} \times \mathbf{100} = \frac{3.25}{44.8} \times 100 = \mathbf{7.25\%.}$$

$$\sigma_y = \sqrt{\frac{\Sigma d_y^2}{N} - \left(\frac{\Sigma d_y}{N}\right)^2} = \sqrt{\frac{85}{10} - \left(\frac{-5}{10}\right)^2} = \sqrt{8.5 - 0.25} = \mathbf{2.87.}$$

$$C.V.(Y) = \frac{\sigma_y}{\bar{Y}} \times 100 = \frac{2.87}{94.5} \times 100 = \mathbf{3.04\%.}$$

Since the coefficient of variation is much less in case of shares Y, hence it is more stable in value.]

18. The marks of the students are given in the sutdents are given in statistics (out of 50).

8, 10, 12, 15, 18, 20, 14, 16, 24, 32, 28, 24,
20, 16, 16, 16, 16, 16, 34, 36, 28, 42, 46, 46,
44, 42, 18, 20, 24, 25, 26, 26, 26, 26, 23, 15.

Put the information in the form of class intervals by taking the classes as 0 – 10, 10 – 20, 20 – 30, 30 – 40, 40 – 50. Find coefficient of variation ?

[**Hint.**

Interval	:	0 – 10	10 – 20	20 – 30	30 – 40	40 – 50
Mid-point(m)	:	5	15	25	35	45
f	:	1	13	15	2	5.

Take $d = (m-25) / 10$, then $\Sigma d = 0$, $\Sigma fd = -3$, $\Sigma fd^2 = 39$.

$$\therefore \quad \bar{X} = A + \frac{\Sigma fd}{N} \times i = 25 - \frac{3}{36} \times 10 = \mathbf{24.17.}$$

$$\sigma = \sqrt{\frac{\Sigma fd^2}{N} - \left(\frac{\Sigma fd}{N}\right)^2} \times i = \sqrt{\frac{39}{36} - \left(\frac{-3}{36}\right)^2} \times 10 = \mathbf{10.36.}$$

$$C.V. = \frac{\sigma}{\bar{X}} \times 100 = \frac{10.36}{24.17} \times 100 = \mathbf{42.86\%.}\ ]$$

19. Calculate the mean deviation about the mean for the following data:

x	:	5	15	25	35	45	55	65
f	:	8	12	10	8	3	2	7 .

Given mean $\bar{x} = 29$.

$$[\ \textbf{Hint. } M.D. = \frac{1}{N} \Sigma f\,|x - \bar{x}| = \frac{[8\times24 + 12\times14 + 10\times4 + 8\times6 + 3\times16 + 2\times26 + 7\times36]}{50} = \mathbf{16.}\ ]$$

20. Find the mean deviation from the mean for the following data :

Class Interval	:	0 – 10	10 – 20	20 – 30	30 – 40	40 – 50	50 – 60	60 – 70
Frequency	:	8	12	10	8	3	2	7

Given mean $\bar{X} = 29$.

[**Hint.**

mid - value – X	:	5	15	25	35	45	55	65
f	:	8	12	10	8	3	2	1
$f\,\|X - \bar{X}\|$	:	24	14	4	6	16	26	36.

$$\Sigma f\,|X - \bar{X}| = 800; \quad M.D. = \frac{1}{N} \Sigma f\,|x - \bar{x}| = \frac{800}{50} = \mathbf{16.}\ ]$$

21. For team A, the average number of goals scored per match was 2 with $S.D.$ of 1.09 goals. For team B, the average number of goals scored per match was 2.5 with $S.D.$ of 1.25 goals. Find which team may be considered more consistent.

[**Hint.** Coefficient of Variation for team $A = \frac{S.D.}{\text{Mean}} \times 100 = \frac{1.09}{2} \times 100 = \mathbf{54.5\%.}$

$$\text{Coefficient of Variation for team } B = \frac{S.D.}{\overline{X}} \times 100 = \frac{1.25}{2.5} \times 100 = \mathbf{50\%}.$$

Team *B* is better in average scores as well as in consistency of performance because the coefficient of variation of team *B* is lower.]

22. Given the following data for prices of shares *X* and *Y* :

$\overline{X} = 53$, $\sigma_x = 2.646$, $\overline{Y} = 105$, $\sigma_y = 2$. Which share is more stable ?

23. Given the following data for the runs scored by two batsman *A* and *B* in ten innings:

$\overline{X}_A = 39$, $\sigma_A = 32.26$, $\overline{X}_B = 43$, $\sigma_B = 35.72$. Which batsman is more consistent ?

ANSWERS

5. 10.	**6.** 50.	**7.** 55.	**8.** 0.333.	**9.** 160.	**10.** 120.
11. 1.47	**12.** 5.617.	**13.** 25.	**14.** $\sigma_x = 8$, $\sigma_y = 9$.		**15.** 9.52.
16. 12.31.	**17.** Share *Y*.	**18.** 42.86%.	**19.** 16.	**20.** 16.	**21.** Team *B*.
22. Share *Y*.	**23.** Batsman *A*.				

5
CHAPTER

PROBABILITY AND BAYES' THEOREM

5.1 INTRODUCTION

Probability theory was originated from gambling theory. A large number of problems exist even today which are based on the game of chance, such as coin tossing, dice throwing and playing cards. The utility of probability in business and economics is most emphatically revealed in the field of predictions for future. We have to anticipate consequences of the each of these plans and finally we compare the results. *Uncertainty plays an important role in business and probability is a concept which measures the degree of uncertainty and that of certainty also as a corollary. The probability when defined in the simplest way is chance of occurrence of a certain event when expressed quantitatively*. The probability is defined in two different ways:

(*i*) **Mathematical (or *a priori*) definition**

(*ii*) **Statistical (or *empirical*) definition.**

Before we study the probability theory in details, it is appropriate to give the definition of certain terms, which are essential for the study of *Probability Theory*.

5.2 SOME IMPORTANT TERMS AND CONCEPTS

5.2.1 Random Experiment or Trial

An experiment is characterised by the property, that its observations under a given set of circumstances do not always lead to the same observed outcome but rather to different outcomes which follows a sort of statistical regularity. It is also called a **Trial.**

For example, tossing a coin, or throwing a dice.

5.2.2 Sample Space

A set of all possible outcomes from an experiment is called a **sample space**. Let us toss a coin, the result is either head or tail. Let 1 denote head and 0 denote tail. Mark the point 0, 1 on a straight line. These points are called **sample points or event points**. For a given experiment there are different possible outcomes and hence different sample points. The collection of all such sample points is a **Sample Space.** Toss two coins simultaneously, then the possible results are four pairs (0, 0), (1, 0), (0, 1) (1, 1) and they can be represented as points in a coordinate plane. These points constitute a sample space of four sample points. Similarly, by tossing three coins simultaneously, we get eight points which can be denoted by the triplets (0, 0, 0), (0, 1, 0), (0, 0, 1), (1, 0, 0), (1, 1, 0), (0, 1, 1), (1, 0, 1), (1, 1, 1). These 8 points form a sample space of 3-dimension. In general, *in tossing n coins simultaneously we can have an n-dimensional sample space consisting of 2^n sample points.*

5.2.3. Discrete Sample Space

A sample space whose elements are finite or infinite but countable is called a **discrete sample space.**

For example, if we toss a coin as many times as we require for turning up one head, then the sequence of points $S_1 = (1)$, $S_2 = (0, 1)$, $S_3 = (1, 0, 0)$, $S_4 = (0, 0, 0, 1)$ etc., is a discrete sample space.

5.2.4. Continuous Sample Space

A sample space whose elements are infinite and uncountable or assume all the values on a real line R or on an interval of R is called a **continuous sample space.** In this case the sample points build up a continuum, and the **sample space** is said **to be continuous.**

For example, all the points on a line or all points on a plane is a sample space.

5.2.5. Event

A sub-collection of a number of sample points under a definite rule or law is called an **event.**

For example, let us take a dice. Let its faces 1, 2, 3, 4, 5, 6 be represented by E_1, E_2, E_3, E_4, E_5, E_6 respectively. Then all the E_i's are sample points. Let E be the event of getting an even number on the dice. Obviously, $E = \{E_2, E_4, E_6\}$, which is a **subset** of the set $\{E_1, E_2, E_3, E_4, E_5, E_6\}$.

5.2.6. Null Event

An event having no sample point is called a **null event** *and is denoted by* **ϕ.**

5.2.7. Simple event

An event consisting of only one sample point of a sample space is called a **simple event.**

For example, let a dice be rolled once and A be the event that face number 5 is turned up, then A is a **simple event.**

5.2.8. Compound Events

When an event is decomposable into a number of simple events, then it is called a **compound event.**

For example, the sum of the two numbers shown by the upper faces of the two dice is seven in the simultaneous throw of the two unbiased dice, is a **compound event** as it can be decomposable.

5.2.9. Exhaustive Cases or Events

It is the total number of all the possible outcomes of an experiment.

For example, when we throw a dice, then any one of the six faces (1, 2, 3, 4, 5, 6) may turn up and, therefore, there are six possible outcomes. Hence there are six **exhaustive cases or events** in throwing a dice.

5.2.10. Mutually Exclusive Events

If in an experiment the occurrence of an event precludes or prevents or rules out the happening of all other events in the same experiment, then these events are said to be **mutually exclusive events.**

For example, in tossing a coin, the events head and tail are **mutually exclusive**, because if the outcome is head, then the possibility of getting a tail in the same trial is ruled out.

5.2.11. Equally likely Events

Events are said to be **equally likely** *if there is no reason to expect any one in preference to other.*

For example, in throwing a dice, all the six faces (1, 2, 3, 4, 5, 6) are equally likely to occur.

5.2.12. Collectively Exhaustive Events

The total number of events in a population exhausts the population. So they are known as **collectively exhaustive events.**

5.2.13. Equally Probable Events

If in an experience all possible outcomes have equal chances of occurrence, then such events are said to be **equally probable events.**

For example, in throwing a coin, the events head and tail have equal chances of occurrence, therefore, they are **equally probable events.**

5.2.14. Favourable Cases

The cases which ensure the occurrence of an event are said to be **favourable** *to the event.*

5.2.15. Independent and Dependent Events

When the experiments are conducted in such a way that the occurrence of an event in one trial does not have any effect on the occurrence of this or other events at a subsequent experiment, then the events are said to be **independent.** In other words, two or more events are said to be **independent** if the happening of any one does not depend on the happening of the other. *Events which are not independent are called* **dependent events.**

Illustration 1. If we draw a card in a pack of well shuffled cards and again draw a card from the rest of pack of cards (containing 51 cards), then the second draw is **dependent** on the first. But if on the other hand, we draw a second card from the pack by replacing the first card drawn, the second draw is known as **independent** of the first.

5.3. DEFINITIONS OF PROBABILITY

5.3.1. Classical Definition of Probability

If an experiment has n mutually exclusive, equally likely and exhaustive cases, out of which m are favourable to the happening of the event A, then the probability of the happening of A is denoted by $P(A)$ and is defined as:

$$P(A) = \frac{m}{n} = \frac{\textbf{No. of cases favourable to } A}{\textbf{Total (Exhaustive) number of cases}}$$

Notes 1. ***Probability of an event which is certain to occur is 1 and the probability of an impossible event is zero.***

2. ***The probability of occurrence of any event lies between 0 and 1, both inclusive.***

Example 1. *What is the probability of getting an even number in a single throw with a dice?*

Solution. The possible cases in the throw of a dice are six, *viz.*, 1, 2, 3, 4, 5, 6.

Favourable cases are those which are marked with 2, 4, 6 and these are three in number.

$\therefore$ **Probability of getting an even number** $= \frac{3}{6} = \frac{1}{2}$.

Example 2. *What is the probability of getting tail in a throw of a coin ?*

Solution. When we toss a coin, there are two possible outcomes, *viz.*, Head or Tail. In this case the number of possible cases $n = 2$.

No. of favourable cases $= m = 1$ ($\because$ **The outcome of tail is a favourable event**)

$\therefore$ **Probability of getting a tail** $= \frac{m}{n} = \frac{1}{2}$.

Example 3. *A bag contains 6 white balls, 9 black ball. What is the probability of drawing a black ball ?*

Solution. The total number of equally likely and exhaustive cases $= n = 6 + 9 = 15$.

Number of favourable cases = $m = 9$. (∵ **Number of black balls = 9**)

∴ **Probability of drawing a black ball** $= \frac{9}{15} = \frac{3}{5}$.

Example 4. *What is the probability that if a card is drawn at random from an ordinary pack of cards, it is (i) a red card, (ii) a club, (iii) one of the court cards (Jack or Queen or King).*

Solution. No of exhaustive cases = 52.

(*i*) There are 26 red cards and 26 black cards in an ordinary pack.

∴ Favourable cases = $n = 26$ **(number of red cards)**

∴ **Probability of getting a red card** $= \frac{26}{52} = \frac{1}{2}$.

(*ii*) Number of clubs in a pack = 13.

∴ Favourable cases = 13.

∴ **Probability of getting a club** $= \frac{13}{52} = \frac{1}{4}$.

(*iii*) There are 4 × 3 = 12 court cards in a pack of cards.

∴ Number of favourable cases = $m = 12$.

Number of exhaustive cases = $n = 52$.

∴ **Probability of getting a court card** $= \frac{12}{52} = \frac{3}{13}$.

Example 5. *What is the probability of throwing a number greater than 3 with an ordinary dice ?*

Solution. The six faces of a dice are marked with numbers 1, 2, 3, 4, 5, 6 and thus there are only 6 exhaustive cases, *i.e.*, $n = 6$.

There are three numbers 4, 5, 6 which are greater than 3 in a dice.

∴ Favourable cases = $m = 3$.

∴ **Probability of getting a number greater than** $3 = \frac{3}{6} = \frac{1}{2}$.

Example 6. *What is the probability that a leap year, selected at random, will have 53* **Sundays** ?

Solution. There are 366 days in a leap year and it has 52 complete weeks and 2 days over. These two extra days can occur in following possible ways.

(*i*) Sunday and Monday; (*ii*) Monday and Tuesday;

(*iii*) Tuesday and Wednesday; (*iv*) Wednesday and Thursday:

(*v*) Thursday and Friday (*vi*) Friday and Saturday;

(*vii*) Saturday and Sunday.

∴ No. of exhaustive cases = 7.

No. of favourable cases = 2. {∵ **There are two cases which have Sunday [(*i*) and (*vii*)]**}

∴ **Probability** $= \frac{2}{7}$.

Example 7. *What is the probability of getting a total of more than 10 in a single throw with two dice ?*

Solution. The number of exhaustive cases when two dice are thrown simultaneously $= 6^2$
$= 36.$

Out of these 36 cases, there will be only three favourable cases. These are (5, 6), (6, 5), (6, 6).

$\therefore$ **Probability** $= \dfrac{3}{36} = \dfrac{1}{12}.$

Example 8. *A card is drawn from an ordinary pack of playing cards and a person bets that it is a spade or an ace. What are the odds against his winning this bet ?*

Solution. In a pack of 52 cards, 1 card can be drawn in 52 ways. Since there are 13 spades and 3 aces (one ace is also present in spade), therefore, the **favourable** cases $= m = 13 + 3 = 16$.

No. of exhaustive cases $= n = 52$.

Probability of getting a spade or an ace $= \dfrac{16}{52} = \dfrac{4}{13} = \dfrac{4}{9+4}$

$\therefore$ **Odds against winning the bet are 9 to 4.**

5.3.2. Statistical or Empirical Definition of Probability

Von Mises has given the following statistical or empirical definition of probability.

"*If the experiment be repeated a large number of times under essentially identical conditions, the limiting value of the ratio of the number of times the event A happens to the total number of trials of the experiment as the number of trials increases indefinitely is called the probability of happening of the event A.*"

Symbolically. Let $P(A)$ denote the probability of the occurrence of A. Let m be the number of times in which an event A occurs in a series of n trials, then

$$\mathbf{P(A) = \lim_{n \to \infty} \frac{m}{n}}$$, **privided the limit is finite and unique.**

5.4. THEOREMS ON PROBABILITY

There are two important theorems of probability, namely

1. **Addition Theorem, or Theorem on Total Probability.**
2. **Multiplication Theorem or Theorem on Compound Probability.**

5.4.1. Addition Theorem or Theorem on Total Probability

Statement. *If n events are mutually exclusive, then the probability of happening of any one of them is equal to the sum of the probabilities of the happening of the separate events, i.e., in other words, if,* $E_1, E_2, E_3, ..., E_n$ *be n events and* $P(E_1), P(E_2), ..., P(E_n)$, *be their respective probabilities, then*

$$P(E_1 + E_2 + E_3 + ... + E_n) = P(E_1) + P(E_2) + ... P(E_n).$$

Proof. Let $E_1, E_2, E_3, ..., E_n$ be n mutually exclusive events with probabilities $P(E_1), P(E_2), ... P(E_n)$ respectively. Let N be the total number of trials. Let $m_1, m_2, m_3, ..., m_n$ be the cases favourable to events $E_1, E_2, E_3, ..., E_n$ respectively, then

$$\mathbf{P(E_i) = \frac{m_i}{N}, i = 1, 2, ..., n.}$$

Since the events $E_1, E_2, ..., E_n$ are mutually exclusive, therefore, the cases favourable to the happening of any one of the events $E_1, E_2, ..., E_n$ are $m_1 + m_2 + ... + m_n$. Hence the probability of happening of any one of the events $E_1, E_2, ..., E_n$ is

$$\mathbf{P(E_1 + E_2 + ... + E_n)} = \frac{m_1 + m_2 + ... + m_n}{N} = \frac{m_1}{N} + \frac{m_2}{N} + \frac{m_3}{N} + ... + \frac{m_n}{n}$$
$$= \mathbf{P(E_1) + P(E_2) + P(E_3) + ... + P(E_n)}.$$

Example 9. *A dice is rolled. What is the probability that a number 1 or 6 may appear on the upper face ?*

Solution. The probability of appearing the number 1 on the upper face $= \frac{1}{6}$.

The probability of appearing 6 on the upper face $= \frac{1}{6}$.

∴ **The probability of 1 or 6 appearing on the face** $= \frac{1}{6} + \frac{1}{6} = \frac{1}{3}$.

Example 10. *If the probability of the horse A winning the race is* $\frac{1}{5}$ *and the probability of the horse B winning the same race is* $\frac{1}{6}$, *what is the probability that one of the horses will win the race ?*

Solution. Probability of winning of the horse $A = \frac{1}{5}$.

Probability of winning of the horse $B = \frac{1}{6}$.

$$\therefore \quad P(A + B) = P(A) + P(B) = \frac{1}{5} + \frac{1}{6} = \frac{11}{30}.$$

5.4.2. Multiplicative Theorem or Theorem on Compound Probability

Before we proceed further in stating and proving this theorem, we must learn the following definitions.

Simple and compound events. *A single event is called a* **simple event**. *When two or more than simple events occur in connection with each other, then their simultaneous occurrence is called a* **compound event**. If *A* and *B* are two **simple events**, the simultaneous occurrence of *A* and *B* is called a **compound event and is denoted by AB.**

Conditional Probability. *The probability of the happening of an event B, when it is known that A has already happened, is called the* **conditional probability** of *B* and is denoted by **P (B/A)**.

i.e., $P(B/A) \Rightarrow$ **conditional probability of *B* given tthat *A* has already occurred.**

Similarly, $P(A/B) \Rightarrow$ **conditional probability of *A* given that *B* has already happened.**

Mutually Independent Events. *An event A is said to be independent of the event B if* **P (*A*/*B*) = P (*A*)**, *i.e., the probability of the happening of A is independent of the happening of B.*

THEOREM ON COMPOUND PROBABILITY

Statement : *The probability of the simultaneous occurrence of the two events A and B is equal to the probability of one of the events multiplied by the conditional probability of other, given the occurrence of the first, i.e.,*

$$\mathbf{P(AB) = P(A).P(B/A) = P(B).P(A/B).}$$

Proof. Let an experiment have *N* mutualy exclusive, exhaustive and equally likely cases out of which *M* are favourable to the happening of the event *A*. Let the cases which are favourable to the simultaneous occurrence of the events *A* and *B* be *m*. Obviously, *m* cases are included in *M* cases which are favourable to *A*. Now

$$P(AB) = \frac{m}{N} = \frac{m}{M}.\frac{M}{N}. \qquad ... (1)$$

We know that the cases favourable to the event *B* have to come out of *M* cases favourable to the event A, therefore, the exhaustive number of cases for B is M. Thus the ratio $\frac{m}{M}$ represents the happening of B given that A has allready occurred. Also $\frac{M}{N} = P(A)$, so from (1) we have

$$P(AB) = \frac{M}{N} \times \frac{m}{M} = P(A).P(B/A).$$

Cor. 1. If the events A and B are statistically independent then $P(B/A) = P(B)$ and $P(A/B) = P(A)$.

$\therefore$ $\quad P(AB) = P(A)\,P(B/A) = P(A) = P(B)$ $\quad [\because P(B/A) = P(B)]$

Example 11. *If five coins are tossed, what is the probability that all will show a head ?*

Solution. Let $P(A)$ denote the probability of showing a head.

$$\therefore \qquad P(A) = \frac{1}{2}.$$

Now Probability of showing a head when 5 coins are tossed

$$= P(A) \times P(A)\,P(A) \times P(A) \times P(A) = \frac{1}{2} \times \frac{1}{2} \times \frac{1}{2} \times \frac{1}{2} \times \frac{1}{2} = \frac{1}{32}.$$

Example 12. *A card is drawn from a pack of 52 cards and then a second card is drawn. What is the probability that both the cards drawn are queen ?*

Solution. First draw. Probability of getting a queen $= \frac{4}{52} = \frac{1}{13}$

Second draw. After drawing the first queen, we are left with 51 cards having 3 queens.

$\therefore$ Probability of getting a queen in second draw $= \frac{3}{51} = \frac{1}{17}$.

$\therefore$ **Probability that both the cards are queen** $= \frac{1}{13} \times \frac{1}{17} = \frac{1}{221}$.

Example 13. *A bag contains 5 white and 3 black balls. Two balls are drawn at random one after the other without replacement. Find the probability that both the balls drawn are black.*

Solution. Let $P(A)$, $P(B/A)$ denote the probability of drawing a black ball in the first and second attempt respectively.

$\therefore$ Probability of drawing a black ball in the first attempt is $P(A)$,

where $$P(A) = \frac{\text{favourable cases}}{\text{Total exhaustive cases}} = \frac{3}{5+3} = \frac{3}{8}.$$

Probability of drawing the second black ball given the first ball drawn is black is

$$P(B/A) = \frac{\text{favourable cases}}{\text{Total number of exhaustive cases}} = \frac{2}{5+2} = \frac{2}{7}.$$

Probability that both the balls drawn are black is = $P(AB) = P(A)\,P(B/A)$

$$= \frac{3}{8} \times \frac{2}{7} = \frac{3}{28}.$$

5.5. COMPLEMENTARY EVENTS

The event **'A occurs'** *and the event* **'A does not occur'** *are called* ***complementary events.*** The **"event A does not occur" is denoted by A^C or $\overline{A}$ and is read as complementary of A.** It is important to note that

$$P(A) + P(A^C) = 1$$

$$\therefore \quad P(A^C) = 1 - P(A).$$

Proof : Let there be n exhaustive, mutually exclusive outcomes of an experiment. Let m of these outcomes be favourable to the **happening of the event A.** Thus the event **'*A* does not occur' in remaining (*n* – *m*) outcomes.** Thus these **(*n* – *m*)** outcomes are favourable to the event **"*A* does not occur"** *i.e.*, to the event A^C.

$$\therefore P(A^C) = \frac{n-m}{n} = \frac{n}{n} - \frac{m}{n} = 1 - \frac{m}{n} = 1 - P(A). \qquad \left[\because P(A) = \frac{m}{n}\right]$$

Thus $\quad \mathbf{P(A^C) = 1 - P(A)}$ **or** $\mathbf{P(A) + P(A^C) = 1.}$

Example 14. *Find the chance of throwing at least one up in a single throw with three dices.*

Solution. When we throw a dice, it results in six equally likely, mutually exclusive and exhaustive cases. Therefore the total number of outcomes of throwing three dices is $6 \times 6 \times 6 =$ **216 (By Fundamental Principle).**

Let **A be the event that at least 1 appears on a dice.** Then the complementary event A^C **is 1 does not appear on the dice.**

There are 5 outcomes on each dice when 1 does not appear, *i.e.*, occurrence of 2 or 3 or 4 or 5 or 6. Hence, the number of cases favourable to $A^C = 5 \times 5 \times 5 =$ **125.**

$$\therefore \quad P(A^C) = \frac{125}{216}.$$

Now $\quad P(A) = 1 - P(A^C) = 1 - \frac{125}{216} = \frac{91}{216}.$

5.6 HAPPENING OF AT LEAST ONE EVENT

Let A and B be the independent events and p_1 and p_2 be the probabilities of their happening, then the chance that **both *A* and *B* happen is** $p_1 \times p_2$. Also the probabilities of not happening of A and B are $1 - p_1$, $1 - p_2$ respectively.

The probability that both do not happen is $(1 - p_1) \times (1 - p_2)$.

Now the chance that at least one of them happens :

$$p_1 p_2 + p_1(1 - p_2) + p_2(1 - p_1) = 1 - (1 - p_1)(1 - p_2)$$

Thus the above result can be generalised for n events $E_1, E_2, ..., E_n$ with $p_1, p_2, ..., p_n$ respective probabilities.

Then the probability that at least one of them happens is

$$= 1 - (1 - p_1)(1 - p_2) \ldots (1 - p_n)$$

Example 15. *4 coins are tossed. Find the probability that at least one head turns up.*

Solution. Probability of getting a head in a toss of a coin = 1/2.

Probability of getting a tail in each case = 1/2

Probability of getting a tail in all the four cases $= \frac{1}{2} \times \frac{1}{2} \times \frac{1}{2} \times \frac{1}{2} = \frac{1}{16}$

$\therefore$ **Probability of getting at least one head** $= 1 - \frac{1}{16} = \frac{15}{16}.$

Example 16. *A problem in Mathematics is given to three students Dayanand, Ramesh and Naresh whose chances of solving it are* $\frac{1}{2}, \frac{1}{3}, \frac{1}{4}$, *respectively. What is the probability that the problem will be solved ?*

Solution. The probabilities of Dayanand, Ramesh and Naresh **solving** the problem are $\frac{1}{2}, \frac{1}{3}, \frac{1}{4}$ respectively.

$\therefore$ The probabilities of all the three, *i.e.*, Dayanand, Ramesh and Naresh **not solving** the problem $= \frac{1}{2} \times \frac{2}{3} \times \frac{3}{4} = \frac{1}{4}$.

The probability that the problem will be solved by at least one of them $= 1 - \frac{1}{4} = \mathbf{\frac{3}{4}}$.

Example 17. *A card is drawn at random from a well shuffled pack of 52 cards. Find the probability of getting a two of heart or a diamond.*

Solution. Since there is only one two of hearts and only one two of diamonds, therefore, the probability of getting a two of hearts $= \frac{1}{52}$ and the probability of getting a two of diamonds $= \frac{1}{52}$.

Since these are mutually exclusive cases, therefore, **the probability of getting a two of heart or a two of diamond** $= \frac{1}{52} + \frac{1}{52} = \frac{2}{52} = \mathbf{\frac{1}{26}}$.

Example 18. *A man and his wife appear for an interview for two posts. The probability of the husbands selection is $\frac{1}{7}$ and that of the wife's selection is $\frac{1}{5}$. What is the probability that only one of them will be selected ?*

Solution. The probability that husband is not selected $= 1 - \frac{1}{7} = \mathbf{\frac{6}{7}}$.

The probability that wife is not selected $= 1 - \frac{1}{5} = \mathbf{\frac{4}{5}}$.

Probability that only husband is selected $= \frac{1}{7} \times \frac{4}{5} = \mathbf{\frac{4}{35}}$.

Probability that only wife is selected $= \frac{1}{5} \times \frac{6}{7} = \mathbf{\frac{6}{35}}$.

Probability that only one of them is selected $= \frac{4}{35} + \frac{6}{35} = \frac{10}{35} = \mathbf{\frac{2}{7}}$.

Example 19. *A salesman has a 60% chance of making a sale to each customer. The behaviour of successive customers is independent. If two customers A and B enter, what is the probability that the salesman will make a sale to A or B.*

Solution. Probability making of sale to a customer $= \frac{60}{100} = \mathbf{0.6}$.

Probability of making no sale to a customer = 1 – 0.6 = **0.4.**

Probability of making no sale to $A = P(A^C) = \mathbf{0.4}$

Probability of making no sale to $B = P(B^C) = \mathbf{0.4}$.

Probability of making no sale to A and no sale to $B = P(A^C) \times P(B^C) = 0.4 \times 0.4 = \mathbf{0.16}$.

Probability of making a sale to A or B $= 1 - P(A^C) \times P(B^C)$.

$= 1 - 0.16 = 0.84 = \mathbf{84\%}$.

5.7 ADDITION THEOREM FOR COMPATIBLE EVENTS

Theorem. *The probability of the occurrence of at least one of the events A and B (not mutually exclusive) is given by*

$$\mathbf{P(A+B) = P(A) + P(B) - P(AB).}$$

Proof. If A and B are two events, then the event $A + B$ means the occurrence of at least one of the two events A and B. It is also written as (A or B). It is the union of the following mutually exclusive events.

(*i*) **$A\overline{B}$ which means that A happens but B does not happen.**

(*ii*) **$\overline{A}B$ which means B happens but A does not happen.**

(*iii*) **AB which means both A and B occur. AB is also written as A or B.**

$$\therefore \quad A + B = A\overline{B} + \overline{A}B + AB.$$

Applying the theorem of addition of probabilities, we have

$$P(A+B) = P(A\overline{B}) + P(\overline{A}B) + (AB) \qquad \text{... (i)}$$

But $A = A\overline{B} + AB$

$$\therefore \quad P(A) = P(A\overline{B}) + P(AB)$$

or $$P(A\overline{B}) = P(A) - P(AB) \qquad \text{... (ii)}$$

Similarly, $$P(B) = P(\overline{A}B) + P(AB)$$

or $$P(\overline{A}B) = P(B) - P(AB) \qquad \text{... (iii)}$$

Now (*i*), (*ii*) and (*iii*) imply

$$P(A+B) = P(A) - P(AB) + P(B) - P(AB) + P(AB)$$
$$= P(A) + P(B) - P(AB).$$

Hence $$\mathbf{P(A+B) = P(A) + P(B) - P(AB)}$$

It is also written as **$P(A \text{ or } B) = P(A) + P(B) - P(A \text{ and } B)$.**

Example 20. *Two students X and Y work independently on a problem. The probability that X will solve it is (3/4) and the probability that Y will solve it is (2/3). What is the probability that the problem will be solved.*

Solution. Let $P(X)$ and $P(Y)$ denote the probability that X and Y respectively will solve the problem.

$$\therefore \quad P(X) = \frac{3}{4}, \qquad P(Y) = \frac{2}{3}.$$

The events X and Y are mutually exclusive as both of them may solve the problem.

$$\therefore \quad P(X \text{ and } Y) = \frac{3}{4} \times \frac{2}{3} = \frac{1}{2}$$

∴ The **probability of solving the problem is :**

$$P(X \text{ or } Y) = P(X) + P(Y) - P(X \text{ and } Y)$$

$$= \frac{3}{4} + \frac{2}{3} - \frac{1}{2} = \frac{9+8-6}{12} = \frac{\mathbf{11}}{\mathbf{12}}.$$

Second Method.

$$P(X^C) = 1 - \frac{3}{4} = \frac{1}{4}; \qquad P(Y^C) = 1 - \frac{2}{3} = \frac{1}{3}.$$

$$P(X^C \text{ and } Y^C) = \frac{1}{4} \times \frac{1}{3} = \frac{\mathbf{1}}{\mathbf{12}}.$$

$\therefore$ **Probability that problem will be solved** $= \mathbf{1 - P(X^C \text{ and } Y^C)} = 1 - \frac{1}{12} = \mathbf{\frac{11}{12}}.$

Example 21. *The probability that a student passes a Physics test is (2/3) and the probability that he passes both Physics and English test is (14/45). The probability that he passes at least one test is (4/5). What is the probability that the student passes the English test.*

Solution. Let A denote that the student passes Physics test :

$$\therefore \qquad P(A) = \frac{2}{3}.$$

Let B denote that the student passes English test: then **$P(B)$ = ?**

It is given that the probability that the student passes both Physics and English test is $\frac{4}{5}$.

$$\therefore \qquad P\ (A \text{ and } B) = \frac{14}{45}$$

Also the probability that he passes at least one test is $\frac{4}{5}$: $\quad \therefore$ **$P(A$ or $B) = \frac{4}{5}$.**

Now using the addition theorem

$$\mathbf{P(A \text{ or } B) = P(A) + P\ (B) - P(A \text{ and } B),}$$

we get

$$\frac{4}{5} = \frac{2}{3} + P(B) - \frac{14}{45}$$

or

$$P(B) = \frac{4}{5} + \frac{14}{45} - \frac{2}{3} = \frac{36 + 14 - 30}{45} = \frac{20}{45} = \mathbf{\frac{4}{9}}.$$

Hence the probability that the passes English test is $\frac{4}{9}$.

5.8 DEFINITION OF PROBABILITY IN TERMS OF ODD IN FAVOUR OR ODDS AGAINST THE EVENT

Odds in Favour of an event. If the odds in favour of an event **A** are in the ratio $x : y$ (or x to y), then

$$P(A) = \frac{x}{x + y} \qquad \text{and} \qquad P(A^C) = \frac{y}{x + y}. \qquad \text{... (1)}$$

It is important to note that $P(A^C) = 1 - P(A) = 1 - \frac{x}{x + y} = \frac{y}{x + y}$,

which is same as given by **(1)**.

Odds against an event. If the odds against an event **B** are $m : n$, (or **m to n**) then

$$P(B) = \frac{n}{m + n} \qquad \text{and} \qquad P(B^C) = \frac{m}{m + n} \qquad \text{... (2)}$$

It is important to note that

$$P(B^C) = 1 - P(B) = 1 - \frac{n}{m + n} = \frac{m + n - n}{m + n} = \frac{m}{m + n},$$

which is the same as given by (2).

This concept is illustrated by the following examples.

Example 22. *Find the probability of the event A :*

(a) if the odds in favour are 3 : 2.

(b) if the odd against it are 1 : 4.

Solution.

(*a*) **Required** $P(A) = \frac{3}{3+2} = \mathbf{\frac{3}{5}}.$

(*b*) **Required** $P(A) = \frac{4}{1+4} = \mathbf{\frac{4}{5}}$

Example 23. *A problem in statistics is given to two students A and B. The odds in favour of A solving the problem are 6 to 9 and against B solving the problem are 12 to 10. If A and B attempt, find the probability of the problem being solved.*

Solution.

Probability of A solving the problem $P(A) = \frac{6}{6+9} = \mathbf{\frac{6}{15}}$ **(odds in favour)**

Probability of B solving the problem $P(B) = \frac{10}{12+10} = \mathbf{\frac{5}{11}}$ **(odds against)**

$$P(A^C) = 1 - P(A) = 1 - \frac{6}{15} = \mathbf{\frac{9}{15}}.$$

$$P(B^C) = 1 - P(B) = 1 - \frac{5}{11} = \mathbf{\frac{6}{11}}.$$

Probability that both of them will fail to solve the problem

$$\mathbf{P(A^C \text{ and } B^C) = P(A^C) \times P(B^C)} = \frac{9}{15} \times \frac{6}{11} = \frac{18}{55}.$$

∴ **Probability that the problem will be solved = 1 –** $\mathbf{P(A^C \text{ and } B^C)}$

$$= 1 - \frac{18}{55} = \mathbf{\frac{37}{55}}.$$

Second Method.

$P(A \text{ and } B)$ = Probability that both will solve it $= \frac{6}{15} \times \frac{5}{11}.$

∴ $\mathbf{P(A \text{ or } B) = P(A) + P(B) - P(A \text{ and } B)}.$

$$= \frac{6}{15} + \frac{5}{11} - \frac{6}{15} \times \frac{5}{11} = \frac{66 + 75 - 30}{165} = \frac{141 - 30}{165} = \frac{111}{165} = \mathbf{\frac{37}{55}}.$$

Example 24. *The odds against a certain event are 5 to 3 and the adds in favour of another event, independent of the former are 7 to 5. Find the chance that at least one of the events will happen.*

Solution. Let the two events be A and B.

The probability of the first event A not taking place $P(A^C) = \frac{5}{5+3} = \mathbf{\frac{5}{8}}$

The probability of second event not taking place $P(B^C) = \frac{5}{7+5} = \mathbf{\frac{5}{12}}.$

The probability that neither of the two events takes place is :

$$\mathbf{P(A^C \text{ and } B^C)} = P(A^C) \times P(B^C) = \frac{5}{8} \times \frac{5}{12} = \mathbf{\frac{25}{96}}.$$

Probability of happening of at least one of the events = 1 – P $\mathbf{(A^C \text{ and } B^C)}$

$$= 1 - \frac{25}{96} = \mathbf{\frac{71}{96}}.$$

Example 25. *Suppose that it is 11 to 5 against to a person who is now 38 years of age living till he is 73 and 5 to 3 against B now 43 living till he is 78 years. Find the chance that at least one of these persons will be alive 35 years hance.*

Solution. The probability that A will die within 35 year

$$= \frac{11}{11+5} = \frac{11}{16} \quad i.e., \quad \mathbf{P(A^C) = \frac{11}{16}}.$$

The probability that B will die within 35 years

$$= \frac{5}{5+3} = \frac{5}{8} \quad i.e., \quad \mathbf{P(B^C) = \frac{5}{8}}.$$

The probability that both will die within 35 years $= \frac{11}{16} \times \frac{5}{8} = \frac{55}{128}$

i.e., $\quad P(A^C \text{ and } B^C) = \mathbf{\frac{55}{128}}$

∴ **Probability that both of them will not die *i.e.*,** or **at least one of them will be alive**

$$\mathbf{= 1 - P(A^C \text{ and } B^C)} = 1 - \frac{55}{128} = \mathbf{\frac{73}{128}} \text{ or } \frac{73}{128} \times 100 = 57\%$$

Example 26. *The chances of living a person who is now 35 years old till he is 75 are 8 : 6 and of living another person now 40 years old till he is 80 are 4 : 5. Find the probability that at least one of these persons would dice before completing 40 years hence.*

Solution.

Probability that the first person lives till he is 75 years $= \mathbf{\frac{8}{14}}$.

Probability that the second person lives till he is 80 years $= \mathbf{\frac{4}{9}}$.

Probability of the compound event that both the persons live 40 years

$$= \frac{8}{14} \times \frac{4}{9} = \frac{32}{126} = \mathbf{\frac{16}{63}}$$

Probability that at least one of them would die without living 40 years

$$= 1 - \frac{16}{63} = \mathbf{\frac{47}{63}}.$$

5.9 APPLICATION OF PERMUTATION AND COMBINATION

Example 27. *From a pack of 52 cards, two are drawn at random. Find the chance that one is a king and the other is a queen.*

Solution. The total number of ways in which two cards can be drawn out of 52 cards **is** $^{52}C_2$.

Since in a pack of cards there are 4 kings and 4 queens, therefore, the number of **favourable cases are** $^4C_1 \times {}^4C_1$.

Thus required probability is $= \dfrac{^4C_1 \times {}^4C_1}{^{52}C_2} = \dfrac{4 \times 4}{\frac{52 \times 51}{2}} = \dfrac{4 \times 4 \times 2}{52 \times 51} \mathbf{\dfrac{8}{663}}.$

Example 28. *There are two bags. One bag contains 4 white and 2 black balls. Second bag contains 5 white and 4 black balls. Two balls are transferred from first bag to second bag. Then one ball is taken from the second bag. Find the probability that it is a white ball.*

Solution. There are 3 mutually exclusive and exhaustive ways in which 2 balls can be transferred from first bag to second bag.

First Case. Two white balls have been transferred from first bag to the second bag so that probability for that is $\frac{^4C_2}{^6C_2} = \mathbf{\frac{2}{5}}$

In the second bag we have 7 white and 4 black balls and the probability of getting a white ball is $= \frac{7}{11}$.

$$\therefore \quad \textbf{Required probability} = \frac{^2C_2}{^6C_2} \times \frac{7}{11} = \frac{2}{5} \times \frac{7}{11} = \mathbf{\frac{14}{55}}.$$

Second Case. Two black balls have been transferred from first bag to the second bag so that probability for that is $\frac{^2C_2}{^4C_2} = \mathbf{\frac{1}{15}}$.

In the second second bag we have five white and 6 black balls and probability of getting a white ball is $\frac{5}{11}$.

$$\therefore \quad \textbf{Required probability} = \frac{1}{15} \times \frac{5}{11} = \mathbf{\frac{1}{33}}.$$

Third Case. One black and one white ball have been transferred from first bag to the second so that the probability for this is

$$= \frac{^4C_1 \times {^2C_1}}{^4C_1} = \mathbf{\frac{8}{15}}$$

In the second bag we have 6 white and 5 black balls and the probability of drawing a white ball is $\frac{6}{11}$.

$$\therefore \quad \textbf{Required probability} = \frac{8}{15} \times \frac{6}{11} = \mathbf{\frac{16}{55}}.$$

Since these three cases are mutually exclusive, therefore, the required probability of drawing a white ball

$$= \frac{14}{55} + \frac{1}{33} + \frac{16}{55} = \frac{42+5+48}{165} = \frac{95}{165} = \mathbf{\frac{19}{33}}.$$

Example 29. *Two cards are drawn from a well shuffled pack of playing cards. Determine the probability that both are aces.*

Solution. The total number of ways in which 2 cards can be drawn out of 52 is

$$^{52}C_2 = \frac{52!}{2! \times 50!} = \frac{52 \times 51}{2} = 26 \times 51 = \mathbf{1326}.$$

Since there are 4 aces in a pack of cards, so the number of favourable cases is $^4C_2$.

$$\therefore \quad \textbf{Required probability} = \frac{^4C_2}{^{52}C_2} = \frac{4!}{2! \times 2!} \times \frac{2! \times 50!}{52!}$$

$$= \frac{4 \times 3 \times 2 \times 1}{2 \times 2} \times \frac{2}{52 \times 51} = \frac{12}{52 \times 51} = \mathbf{\frac{1}{221}}.$$

Example 30. *Four persons are chosen at random from a group containing 3 men, 2 women and 4 children. Show that the chance that exactly two of them will be children is* $\frac{10}{21}$.

Solution. The total number of ways in which four persons can be selected out of 9 persons is

$$^9C_4.$$

For favourable cases, we want that 2 out of the four selected should be children. Two children can be selected out of 4 in $^4C_2$ **ways.** The other two are to be selected out of 5 persons (3 men and 2 women). Two persons can be selected out of 5 in $^5C_2$ **ways.**

∴ **The number of favourable cases =** $^4C_2 \times {}^5C_2$.

$$\therefore \textbf{ Required probability} = \frac{^4C_2 \times {}^5C_2}{^9C_4} = \frac{4!}{2!\times 2!} \times \frac{5!}{2!\times 3!} / \frac{9!}{4!\times 5!} = \frac{\mathbf{10}}{\mathbf{21}}.$$

Example 31. *In a factory, there are 6 skilled workers and 4 unskiled workers. What is the probability that (a) worker selected is skilled worker; (b) the two workers selected are skilled.*

Solution. (*a*) Total number of favourable cases = $^6C_1 = \mathbf{6}$

Total number of equally likely cases = $^{10}C_1 = \mathbf{10}$

$$\therefore \quad \textbf{Required probability} = \frac{\mathbf{6}}{\mathbf{10}}.$$

(*b*) Number of favourable cases = Selection of two workers out of 6 skilled workers. This is done in $^6C_2 = \frac{6\times 5}{1\times 2}$ ways.

∴ Total number of equally likely cases = selection of 2 workers out of a total of 10 workers $= {}^{10}C_2$.

$$\therefore \quad \textbf{Required Probability} = \frac{^6C_2}{^{10}C_2} = \frac{6\times 5}{1\times 2} \times \frac{1\times 2}{10\times 9} = \frac{\mathbf{1}}{\mathbf{3}}.$$

Example 32. *Four cards are drawn from a full pack of cards. Find the probability that two are spades and two are hearts.*

Solution. Total number of possible selections = $^{52}C_4$.

Number of ways of selecting 2 spades cards = $^{13}C_2$.

Number of ways of selecting 2 hearts cards = $^{13}C_2$.

Total favourable cases = $^{13}C_2 \times {}^{13}C_2$.

$$\therefore \quad \textbf{Required Probability} = \frac{^{13}C_2 \times {}^{13}C_2}{^{52}C_4}$$

$$= \frac{13\times 12}{1\times 2} \times \frac{13\times 12}{1\times 2} \times \frac{1\times 2\times 3\times 4}{52\times 51\times 50\times 49} = \frac{\mathbf{468}}{\mathbf{20825}}.$$

Example 33. *Ten students are seated at random in a row. Find the probability that two particular students are not seated side by side.*

Solution. Let us tie these two particular students into one. Then it amounts to selection of two students out of 9. This is done in $^9C_2$ **ways.**

But the two students can be permuted in **2 ways.**

∴ Total number of ways for selection of two students such that two particular students are not seated side by side = 2 × $^9C_2$ ways.

∴ **Favourable cases** = **2 × $^9C_2$.**

Total exhaustive cases = **$^{10}C_2$.**

$$\textbf{Required Probability} = \frac{2\times {}^9C_2}{{}^{10}C_2} = \frac{2\times 9!}{7!\times 2!}\times\frac{8!\times 2!}{10!} = \frac{4}{5}.$$

Example 34. *A committee of three is to be chosen from a group consisting of 4 men and 5 women. If the selection is made at random, find the probability that (a) all three are men; (b) two are men.*

Solution. ***Exhaustive cases.*** It is the selection of 3 out of (4 + 5) = 9 = **$^9C_3$.**

(***a***) ***Favourable cases*** : selection of 3 men out of 4. = $^4C_3$.

$$\therefore \quad \textbf{Required Probability} = \frac{{}^4C_3}{{}^9C_3} = \frac{4!}{3!}\times\frac{3!\times 6!}{9!} = \frac{1}{21}.$$

(***b***) In this case two are men and 1 is woman. It amounts to selection of 2 men out of 4 men and 1 woman out 5 women = **$^4C_2$ × $^5C_1$.**

$$\therefore \quad \textbf{Required Probability} = \frac{{}^4C_2\times {}^5C_1}{{}^9C_3} = \frac{4!}{2!\times 2!}\times\frac{5!}{4!}\times\frac{3!\times 6!}{9!} = \frac{5}{84}.$$

Example 35. *A bag contains 5 white and 8 red balls. Two drawings of 3 balls are made such that (a) balls are replaced before the second trial and (b) the balls are not replaced before the second trial. Find the probability that the first drawing will give 3 white and the second 3 red balls in each case.*

Solution.

(*a*) When The Balls Are Replaced

Possible cases : Selection of 3 balls out of 13 = **$^{13}C_3$ ways.**

Favourable cases : Selection of 3 white balls out of 5 white balls in first draw = **$^5C_3$.**

$$\textbf{Probability of 3 white in first trial} = \frac{{}^5C_3}{{}^{13}C_3} = \frac{5}{143}.$$

$$\textbf{Probability of 3 red in second trial} = \frac{{}^8C_3}{{}^{13}C_3} = \frac{28}{143}.$$

$$\textbf{Probability of 3 white in first trial × Probability of 3 red in second trial} = \frac{5}{143}\times\frac{28}{143} = \frac{140}{20449}.$$

(*b*) When The Balls Are Not Replaced

In this case the probability of drawing 3 white in first draw $= \frac{5}{143}$.

When the balls are not replaced, after the first draw, the total number of balls left = 2 white + 8 red = **10 balls.**

$$\therefore \textbf{ In this case Probability of drawing 3 red balls in the second draw} = \frac{{}^8C_3}{{}^{10}C_3} = \frac{7}{15}.$$

$$\textbf{Required Compound Probability} = \frac{5}{143}\times\frac{7}{15} = \frac{7}{429}.$$

Example 36. *An urn contains 25 balls numbered 1 through 25. Two balls are drawn from the urn with replacement. Find the probability of selecting :*

(*a*) *both odd numbers.*

(*b*) *one odd and one even number.*

(*c*) *at least one odd number.*

(*d*) *no odd number.*

(*e*) *both even number.*

Solution. Total number of cases = 25.

Let E and D denote the events of drawing 'even' and 'odd' numbered ball. There are **13 odd numbered balls and 12 even numbered balls.**

Then $$P(E) = \frac{12}{25}, \qquad P(D) = \frac{13}{25}.$$

(*a*) When the balls are replaced back in the urn.

Let $P(DD)$ be the probability of getting both odd numbered balls, then

$$\therefore \quad P(DD) = P(D) \,.\, P(D) = \frac{13}{25} \times \frac{13}{25} = \frac{169}{625}.$$

(*b*) The total number of such cases consists of (**first odd and second even**) + (**first even and second odd.**)

$$\therefore \quad P(\text{one odd and one even}) = P(ED) = P(E) \,.\, P(D) + P(D).\, P(E)$$

$$= \frac{12}{25} \times \frac{13}{25} + \frac{13}{25} \times \frac{12}{25} = \frac{312}{625}.$$

(*c*) At least one odd

Probability of getting both the balls even numbered

$$P(EE) = P(E) \,.\, P(E) = \frac{12}{25} \times \frac{12}{25} = \frac{144}{625}.$$

$$P(\text{at least one odd}) = 1 - P(EE) = 1 - \frac{144}{625} = \frac{481}{625}.$$

(*d*) No odd number

$$P(\text{no odd number}) = 1 - P\text{ (at least one odd)} = 1 - \frac{481}{625} = \frac{144}{625}.$$

(*e*) **Probability of (both even numbered)** $= \dfrac{12}{25} \times \dfrac{12}{25} = \dfrac{144}{625}.$

Example 37. *n cadets have to stand in a row. If all possible permutations are equally likely, find the probability that two particular cadets stand side by side.*

Solution. The number of ways in which n cadets can stand in a row is = $n!$

Combine the two particular cadets together and regard them as one cadet.

The number of ways new $(n - 1)$ cadets stand in a row = $(n - 1)!$

Now the two cadets tied together can be arranged in 2! = 2 ways.

∴ Number of ways in which n cadets can stand in a row where two of them stand together

$$= 2 \times (n - 1)!$$

$\therefore$ **Required probability** $= \dfrac{2 \times (n-1)!}{n!} = \dfrac{2 \times (n-1)}{n(n-1)} = \dfrac{2}{n}$.

Example 38. *One bag contains five white and four black balls. Another bag contains seven white and nine black balls. A ball is transferred from the first bag to the second and then a ball is drawn from the second. Find the probability that the ball is white.*

Solution. Case I. ***Let a white ball be transferred from the first bag to the second bag.***

The probability of selecting white ball in the first bag is $p_1 = \dfrac{5}{9}$.

Now the second bag has 8 white and 9 black balls.

The probability of selecting a white ball from the second bag is

$$p_2 = \frac{8}{8+9} = \frac{8}{17}.$$

The probability that both these events take place simultaneously is

$$= p_1 \times p_2 = \frac{5}{9} \times \frac{8}{17} = \frac{40}{153}.$$

Case II. ***Let a black ball be transferred from the first bag to the second bag.***

Its probability is $p_3 = \dfrac{4}{9}$.

Now the second bag contains 7 white and 10 black balls.

The probability of drawing a white ball from the second bag is

$$p_4 = \frac{7}{7+10} = \frac{7}{17}.$$

$\therefore$ The probability of both these events taking place simultaneously is

$$= p_3 \times p_4 = \frac{4}{9} \times \frac{7}{17} = \frac{28}{135}.$$

$\therefore$ **The required probability** $= p_1 p_2 + p_3 p_4 = \dfrac{40}{153} + \dfrac{28}{153} = \dfrac{68}{153}$.

Example 39. *A speaks truth in 75% and B in 80% of the cases. In what percentage of cases are they likely to contradict each other narrating the same incident ?*

Solution. Let *P(A)*, *P(B)* be the probability of A and B speaking the truth, then

$$P(A) = \frac{75}{100} = \frac{3}{4}, \qquad P(B) = \frac{80}{100} = \frac{4}{5}$$

$$P(\bar{A}) = P(A \text{ tells a lie}) = 1 - P(A) = 1 - \frac{3}{4} = \frac{1}{4}.$$

$$P(\bar{B}) = P(B \text{ tell a lie}) = 1 - P(B) = 1 - \frac{4}{5} = \frac{1}{5}$$

Now $P(A \text{ and } B \text{ will contradict}) = P(A)\,P(\bar{B}) + P(B)\,P(\bar{A})$

$$= \frac{3}{4} \times \frac{1}{5} + \frac{4}{5} \times \frac{1}{4} = \frac{7}{20} \text{ or } 35\%.$$

Example 40. *Two dice are tossed. What is the probability that total is divisible by 3 or 4 ?*

Solution. A dice has **6 six faces,** *viz.*, **1, 2, 3, 4, 5, 6.**

Total number of possible cases of 2 dice for the event = 6 × 6 = **36.**

Number of outcomes which are multiple of 3 = **12.**

(*viz.*, 3, 6, 9, 12, 15, 18, 21, 24, 27, 30, 33, 36)

Number of outcomes which are multiple of 4 = **9.**

(*viz.*, 4, 8, 12, 16, 20, 24, 28, 32, 36)

Number of outcomes which are multiple of both 3 and 4 *i.e.*, 12 = 3.

(*viz.*, 12, 24, 36)

Let A = the event that the outcome is a multiple of 3. Then $P(A) = \frac{12}{36} = \frac{1}{3}$;

B = the event that the outcomes, is a multiple of 4. Then $P(B) = \frac{\mathbf{9}}{\mathbf{36}} = \frac{1}{4}$

Also $\quad P(A \cap B) = \frac{\mathbf{3}}{\mathbf{36}}$

Probability that the total is divisible by 3 or 4.

$$P(A \cup B) = P(A) + P(B) - P(A \cap B)$$

$$= \frac{12}{36} + \frac{9}{36} - \frac{3}{36} = \frac{12+9-3}{36} = \frac{18}{36} = \frac{\mathbf{1}}{\mathbf{2}}.$$

Example 41. *The probability that a person A who is now 25 years old, lives for another 30 years is* $\frac{2}{5}$; *and the probability that the person B who is now 45 years old lives for another 30 years is 7/16. Find the probability that at least one of these persons will be alive 30 years hence.*

Solution. Let X, Y be the events that the persons A and B respectively live for another 30 years.

$$P(X) = \frac{2}{5}; \qquad P(Y) = \frac{7}{16}.$$

Probability of living at least one of them live for another 30 years

$$P(X \cup Y) = P(X) + P(Y) - P(X \cap Y)$$

$$= P(X) + P(Y) - P(X).P(Y) = \frac{2}{5} + \frac{7}{16} - \frac{2}{5} \times \frac{7}{16} = \frac{67}{80} - \frac{14}{80} = \frac{\mathbf{53}}{\mathbf{80}}.$$

Hence the Probability that at least one of the two persons will be alive for another 30 years is $\frac{53}{80}$.

5.10. CONDITIONAL PROBABILITY

Let A be any event in the sample space S, and $P(A) > 0$. The probability that an event B occurs subject to the condition that A has already occurred is called the Conditional Probability of B given that A has already occured. It is denoted by $P(B/A)$.

EXPRESSION FOR $P(B/A)$:

Let S be a finite equiprobable space. Let $n(S)$ = total number of sample points of S. Let A be any event with sample points $n(A)$. Let B be another event such that the sample points in $A \cap B$ is $n(A \cap B)$. Then we know that

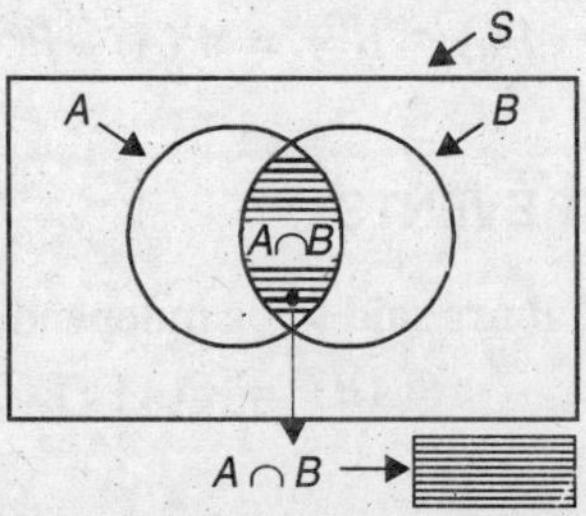

$$P(A) = \frac{\textbf{Sample points in } (A)}{\textbf{Sample points in } S} = \frac{n(A)}{n(S)}$$

Also, $$P(A \cap B) = \frac{n(A \cap B)}{n(S)}$$

Under the assumption that A has occurred, the sample space is restricted to the points lying inside the region A. (see figure). It has $n(A)$ sample points. This new sample space is called the **Reduced Sample Space.**

Under this assumption the occurrence of the event B is restricted to the shaded region in figure 0.9, which is $A \cap B$ having $n(A \cap B)$ sample points.

$\therefore$ ($P(B/A)$ = conditional probability of B given that A has occurred $= \frac{n(A \cap B)}{n(A)}$ = the ratio of the sample points of the events ($A \cap B$) to the reduced sample space consisting of equiprobable sample points.

$$\textbf{Ratio} = \frac{n(A \cap B)/n(S)}{n(A)/n(S)} = \frac{P(A \cap B)}{P(A)}$$

Hence $$P(B/A) = \frac{P(A \cap B)}{P(A)}.$$

or $$P(A \cap B) = P(A) \times P(B/A).$$

It is obvious that $P(B/A)$ is defined if and only if $P(A) \neq 0$. Hence the condition $P(A) > 0$. Similarly, the conditional probability of A given that B has occurred is $P(A/B)$, where

$$P(A/B) = \frac{P(A \cap B)}{P(B)}, P(B) > 0.$$

$$\Rightarrow P(A \cap B) = P(B) \times P(A/B)$$

Example 42. *Let A and B be events with* $P(A) = \frac{1}{3}$, $P(B) = \frac{1}{4}$, $P(A \cap B) = \frac{1}{12}$.

Find (i) P(A/B), (ii), P(B/A), (iii) P(B/A^c), (iv) P(A $\cap$ B^c)

Solution.

(*i*) Here $$P(A/B) = \frac{P(A \cap B)}{P(B)} = \frac{1/12}{1/4} = \frac{\mathbf{1}}{\mathbf{3}}.$$

(*ii*) $$P(B/A) = \frac{P(A \cap B)}{P(A)} = \frac{1/12}{1/3} = \frac{1}{12} \times \frac{3}{1} = \frac{\mathbf{1}}{\mathbf{4}}.$$

(*iii*) Now, $$P(B/A^c) = \frac{P(B \cap A^c)}{P(A^c)} = \frac{P(B) - P(B \cap A)}{1 - P(A)}$$

$$= \frac{(1/4) - (1/12)}{(1 - 1/3)} = \frac{1}{6} \times \frac{3}{2} = \frac{\mathbf{1}}{\mathbf{4}}.$$

(*iv*) $$P(A \cap B^C) = P(A) - P(A \cap B) = \frac{1}{3} - \frac{1}{12} = \frac{3}{12} = \frac{\mathbf{1}}{\mathbf{4}}.$$

5.11. INDEPENDENT EVENTS

The events A and B are said to be independent if

$$P(AB) = P(A) \cdot P(B).$$

Thus in case of independent events the above multiplication theorem becomes

or $$P(A \cap B) = P(A) \times P(B) \qquad [\because P(A/B) = P(A) \text{ and } P(B/A) = P(B)]$$

Example 43. *Given that* $P(A) = \frac{3}{8}$, $P(B) = \frac{5}{8}$ *and* $P(A \cup B) = \frac{3}{4}$, *find P(A/B) and P(B/A). Show whether A and B are independent.*

Solution. We have

$$P(A \cup B) = P(A) + P(B) - P(A \cap B).$$

$$\Rightarrow \quad \frac{3}{4} = \frac{3}{8} + \frac{5}{8} - P(A \cap B) \quad \Rightarrow \quad P(A \cap B) = \frac{3}{8} + \frac{5}{8} - \frac{3}{4} = \frac{1}{4}.$$

Now, $$P(A/B) = \frac{P(A \cap B)}{P(B)} = \frac{1/4}{5/8} = \frac{1}{4} \times \frac{8}{5} = \frac{2}{5}$$

and $$P(B/A) = \frac{P(A \cap B)}{P(A)} = \frac{1/4}{3/8} = \frac{1}{4} \times \frac{8}{3} = \frac{2}{3}.$$

Again, $$P(A \cap B) = \frac{1}{4} \quad \text{and} \quad P(A) \,.\, P(B) = \frac{3}{8} \times \frac{5}{8} = \frac{15}{64}$$

A and *B* will be **independent events, if** $P(A \cap B) = P(A) \,.\, P(B)$.

Thus we notice that $P(A \cap B) \neq P(A) \,.\, P(B)$, **so the events *A* and *B* are not independent.**

Example 44. *The odds against a student X solving a business statistics problem are 8 to 6 and odds in favour of the student Y solving the problem are 14 to 16.*

(*a*) *What is the chance that the problem will be solved if they both try independent of each other.*

(*b*) *What is the probability that none of them is able to solve the problem.*

Solution. Let *A* = **the event that the student *X* solves the problem.**

$$\therefore \quad P(A) = \frac{6}{8+6} = \frac{6}{14}$$

Let *B* = **the event that the student *Y* solves the problem.**

$$\therefore \quad P(B) = \frac{14}{14+16} = \frac{14}{30}.$$

(***a***) Probability that the problem will be solved

= ***P*(at least one of them solve the problem)**

$= P(A \cup B) = P(A) + P(B) - P(A \cap B) = P(A) + P(B) - P(A) \,.\, P(B)$

$$= \frac{6}{14} + \frac{14}{30} - \frac{6}{14} \times \frac{14}{30} = \frac{73}{105}.$$

(***b***) **Probability that none will solve the problem.**

$$= P(A^c) \times P(B^c) = \{1 - P(A)\}\{1 - P(B)\} = \frac{8}{14} \times \frac{16}{30} = \frac{32}{105}.$$

Example 45. *Show that for the three events A, B, C, the probability that at least one of the events will occur is*

$$P(A \cup B \cup C) = P(A) + P(B) + P(C) - P(A \cap B) - P(B \cap C) - P(A \cap C) + P(A \cap B \cap C)$$

Solution. Let us put $B \cup C = D$, so that

$$P(D) = P(B \cup C) = P(B) + P(C) - P(B \cap C) \quad \text{... (1)}$$

$$\therefore P(A \cup B \cup C) = P(A \cup D) = P(A) + P(D) - P(A \cap D)$$

$$= P(A) + P(B) + P(C) - P(B \cap C) - (A \cap D) \quad \text{... (2)}$$

But $P(A \cap D)$
$$= P[A \cap (B \cup C)] = P[(A \cap B) \cup (A \cap C)]$$

$$= P(A \cap B) + P(A \cap C) - P[(A \cap B) \cap (A \cap C)]$$

$$= P(A \cap B) + P(A \cap C) - P(A \cap B \cap C) \quad \text{... (3)}$$

Using (3) in (2), we get

$$P(A \cup B \cup C) = P(A) + P(B) + P(C) - P(A \cap B) - P(A \cap C) - P(B \cap C) + P(A \cap B \cap C).$$

Example 46. *The probability that A can solve the problem is* $\frac{4}{5}$*, B can solve it is* $\frac{2}{3}$*, and C can solve it is* $\frac{3}{7}$*. It all of them try independently, find the probability that the problem will be solved.*

Solution. Here $P(A) = \frac{4}{5}$, $P(B) = \frac{2}{3}$ and $P(C) = \frac{3}{7}$.

$$P(A \cup B \cup C) = P(A) + P(B) + P(C) - P(A \cap B) - P(A \cap C) - P(B \cap C) + P(A \cap B \cap C)$$

$$= P(A) + P(B) \div P(C) - P(A).P(B) - P(A).P(C) - P(B).P(C) + P(A) \,.\, P(B) \,.\, P(C)$$

$$= \frac{4}{5} + \frac{2}{3} + \frac{3}{7} - \left(\frac{4}{5} \times \frac{2}{3}\right) - \left(\frac{4}{5} \times \frac{3}{7}\right) - \left(\frac{2}{3} \times \frac{3}{7}\right) + \frac{4}{5} \times \frac{2}{3} \times \frac{3}{7}.$$

$$= \frac{4}{5} + \frac{2}{3} + \frac{3}{7} - \frac{8}{15} - \frac{12}{35} - \frac{2}{7} + \frac{8}{35} = \mathbf{\frac{101}{105}}.$$

Example 47. *A bag contains 6 red and 5 blue balls and another bag contains 5 red and 8 blue balls. A ball is drawn from the first bag and without noticing its colour is put in the second. a ball is then drawn from the second bag. Find the probability that the ball drawn is blue in colour.*

Solution. There are two possibilities:

Case 1. ***When the ball drawn from the first bag is red.***

$$p_1 = \textbf{P(red ball from 1st bag)} = \frac{6}{11}.$$

Now the second bag contains 6 red and 8 blue balls.

$$p_2 = \textbf{P(blue ball from second bag)} = \frac{8}{14}.$$

Therefore, **the probability of both these events taking place simultaneously** is:

$$= p_1 \times p_2 = \frac{6}{11} \times \frac{8}{14} = \frac{48}{154}$$

Case 2. ***When the ball drawn from first bag is blue.***

$$p_3 = \textbf{P(blue ball from 1st bag)} = \frac{5}{11}.$$

Now the second bag has 5 red and 9 blue balls.

$$p_4 = \textbf{P(blue ball from second bag)} = \frac{9}{14}.$$

Probability of both these events taking place simultaneous is :

$$= p_3 \times p_4 = \frac{5}{11} \cdot \frac{9}{14} = \mathbf{\frac{45}{154}}.$$

But the **Case 1 and Case 2 are mutually exclusive, therefore,**

$$\textbf{the required probability} = p_1p_2 + p_3p_4 = \mathbf{\frac{48}{154} + \frac{45}{154} = \frac{93}{154}}.$$

Example 48. *Two cards are drawn from a pack of 52 cards at random and kept out. Then one card is drawn from the remaining 50 cards. Find the probability that it is an ace.*

Solution. ***Let A denote the event 'getting 2 aces in the first draw of 2 cards. 'B' denote the event getting 1 ace and 1 non-ace in the first draw of 2 cards. Let 'C' denote the event getting 2 non-aces in the first draw of 2 cards.***

Then

$$P(A) = \frac{^4C_2}{^{52}C_2} = \frac{4 \times 3}{52 \times 51} = \mathbf{\frac{1}{221}}.$$

$$P(B) = \frac{^4C_1 \times {}^{48}C_1}{^{52}C_2} = \mathbf{\frac{32}{221}}.$$

$$P(C) = \frac{^{48}C_2}{^{52}C_2} = \mathbf{\frac{188}{221}}.$$

Let *E* be the event of drawing an ace in the second draw of one card.

Required Probability = ***P (E) = P [(A ∩ E) ∪ (B ∩ E) ∪ (C ∩ E)]***
= P(A ∩ E) + P(B ∩ E) + P(C ∩ E) = P(A) P(E/A) + P(B) P(E/B) + P(C) P(E/C)

$$= \frac{1}{221} \times \frac{2}{50} + \frac{32}{221} \times \frac{3}{50} + \frac{188}{221} \times \frac{4}{50} = \frac{2 + 96 + 752}{221 \times 50} = \frac{850}{11050} = \mathbf{\frac{1}{13}}.$$

EXERCISE – 5.1

1. List the reasons for classical definition of probability not being very satisfactory. Give the modern definition of probability.
2. State the addition and multiplication theorem on probability.
3. Give the mathematical and statestical definition of probability.
4. State the addition and multiplication rules of probability giving one example of each case.
5. The probability that machine *A* will be performing a usual function in 5 years time is (1/4), while the probability that machine *B* will still be operating usefully at the end of same period is (1/3). Find the probability that in five years time

   (*a*) both machines will be performing a usual function.

   (*b*) at least one of the machines will be operating.

   [**Hint.** (*a*) Required Probability $= \frac{1}{4} \times \frac{1}{3} = \mathbf{\frac{1}{12}}$.

   (*b*) Required Probability $= \frac{1}{4} + \frac{1}{3} - \frac{1}{12} = \mathbf{\frac{1}{2}}$.]

6. The probability that (*i*) *A* can solve a problem in statistics is 4/5; (*ii*) *B* can solve it is 2/3; and (*iii*) *C* can solve it is 3/7. If all the them try independently, find the probability that the problem will be solved.

   [**Hint.** Required Probability $= 1 - \frac{4}{5} \times \frac{2}{3} \times \frac{3}{7} = \mathbf{\frac{81}{105}}$.]

7. The odds against X solving a problem are 8 to 6, and odds in favour of Y solving the same problem are 14 to 16.

(*a*) What is the chance that the problem will be solved if both try ?

(*b*) What is the probability that neither solves the problem ?

[**Hint.** Here $P(X) = \frac{6}{8+6} = \frac{6}{4}$, $\therefore$ $P(X^C) = 1 - \frac{6}{14} = \frac{8}{14}$,

$$P(Y) = \frac{14}{14+16} = \frac{14}{30}, \quad P(Y^C) = \frac{16}{30}.$$

(*a*) Probability that the problem is not solved $= \frac{8}{14} \times \frac{16}{30} = \mathbf{\frac{32}{105}}$.

Probability that the Problem is solved $= 1 - \frac{32}{105} = \mathbf{\frac{73}{105}}$.

(*b*) Probability that None will solve the problem $= \mathbf{\frac{32}{105}}$.]

8. The probability that a boy will get a scholarship is 0.9 and that a girl will get is 0.8. What is the probability that at least one of them will get the scholarship?

[**Hint.** Required Probability = 1 – (1 – 0.9) × (1 – 0.8) = 1 – 0.02 = **0.98**].

9. Which event has the greatest probability (*i*) getting 4 with one dice; (*ii*) getting 8 with two dice; and (*iii*) getting 12 with three dice.

[Hint. (*i*) P (getting a four) $= \frac{1}{6}$

(*ii*) P (getting 8 with two dice) $= \frac{\text{Possible pairs are (6, 2), (4, 4), (5, 3), (3, 5), (2, 6)}}{36} = \mathbf{\frac{5}{36}}$.

(*iii*) P (getting 12 with 3 dice) $= \frac{25}{216}$

Hence getting 4 with one dice has the greatest probability.]

10. A bag contains 10 white, 15 red and 8 green balls. a single draw of 3 balls is made.

(*a*) What is the probability that a white, a red and green balls are drawn?

(*b*) What would be the probability of getting all the three white balls?

[**Hint.** (*a*) $\frac{{}^{10}C_1 \times {}^{10}C_1 \times {}^{10}C_1}{{}^{33}C_1} = \mathbf{\frac{75}{341}}$. (*b*) $\frac{{}^{10}C_3}{{}^{33}C_3} = \mathbf{\frac{15}{642}}$.]

11. A bag contains 7 red balls and 5 white balls. 4 balls are drawn at random. What is the probability that (*i*) all of them are red; (*ii*) two of them are red and two white ?

[**Hint.** (*i*) $\frac{C(7,4)}{C(12,4)}$, (*ii*) $\frac{C(7,2) \times C(5,2)}{C(12,4)}$.]

12. The odds against A solving a problem are 8 : 6 and odds in favour of B solving it are 9 : 12. What is the probability that if both of them try it, it will be solved ?

13. A piece of equipment will function only when all the three components A, B and C are working. The probability of A failing during one year is 0.15, that of B failing is 0.05 and that of C failing is 0.10. What is the probability that the euqipment will fail before the end of the year ?

[**Hint.** Probability that all components will work = (1 – 0.15) (1 – 0.05) (1 – 0.10)

= 0.85 × 0.95 × 0.90 = 0.727.

Probability that equipment fails = 1 – 0.727 = **0.273.**]

14. There are 3 economists, 4 engineers, 2 statisticians and 1 doctor. A committee of 4 from among them is to be formed. Find the probability that the committee.

(*i*) consists of one each kind.

(*ii*) has at least one economist.

(*iii*) has the doctor as a member and three others.

15. A candidate is selected for interview for 3 posts. For the first there are 3 candidates, for the second there are 4 and for the third there are 2. What are the chances of his getting at least one?

[**Hint.** Let A, B, C be the posts, then $P(A) = \frac{1}{3} \Rightarrow P(\bar{A}) = 1-\frac{1}{3}=\frac{2}{3}.$

Similarly $P(\bar{B}) = 1-\frac{1}{4}=\frac{3}{4}, \quad P(\bar{C})=1-\frac{1}{2}=\frac{1}{2};$

$$P(\bar{A}\bar{B}\bar{C}) = P(\bar{A})P(\bar{B})P(\bar{C})=\frac{2}{3}\times\frac{3}{4}\times\frac{1}{2}=\mathbf{\frac{1}{4}}.]$$

16. There are four hotels in a certain city. If 3 men check into hotels in a day, what is the probability they each are into a different hotel ?

[**Hint.** 3 men can check into a hotel in = 4 × 4 × 4 = 64 ways.

They can check into a different hotel in only 4 × 3 × 2 = 24 ways.

Required Probability $=\frac{24}{64}=$ **0.375.**]

17. What is the probability that a leap year selected at random will contain either 53 Thursdays or 53 Fridays ?

[**Hint.** Leap year has 366 days out of which 52 complete weeks and 2 days. These two days can occur in any of the following combinations:

(*i*) Sunday and Monday (*ii*) Monday and Tuesday

(*iii*) Tuesday and Wednesday. (*iv*) Wednesday and Thursday

(*v*) Thursday and Friday (*vi*) Friday and Saturday.

(*vii*) Saturday and Sunday

Let $P(A)$ = Probability that leap year will contain 53 Thursdays $=\frac{2}{7}.$

$P(B)$ = Probability that leap year will contain 53 Fridays $=\frac{2}{7}.$

∴ $P(A \text{ and } B) = 1/7.$

∴ $P(A \text{ or } B) = P(A) + P(B) - P(A \text{ and } B) = \frac{2}{7}+\frac{2}{7}-\frac{1}{7}=\mathbf{\frac{3}{7}}.]$

18. If a pair of dice is thrown, find the probability that the sum is neither 7 nor 11.

19. Three horses *A*, *B* and *C* are in a race; *A* is twice as likely to win as *B* and *B* is twice as likely to win as *C*. What are there respective probabilities of winning?

20. An investment consultant predicts that the odds against the price of certain stock going up during the next week are 2 : 1 and the odds in favour of price remaining the same are 1 : 3. What is the probability that the price of stock will go down during the next week ?

21. A pack of 50 tickets numbered 1 to 50 is shuffled and then two tickets are drawn. Find the probability that

(*i*) both the tickets drawn have prime numbers.

(*ii*) none of the tickets drawn has prime numbers.

[**Hint.** (*i*) Required probability = 15C2 / 50C2 = $\mathbf{\frac{3}{35}}$

(*ii*) Required probability = 35C2 / 50C2 = $\frac{\mathbf{17}}{\mathbf{35}}$.]

22. There are three urns *A*, *B* and *C*. Urn *A* contains 4 red balls and 3 black balls. Urn *B* contains 5 red balls and 4 black balls. Urn *C* contains 4 red balls and 4 black balls. One ball is drawn from each of these urns. What is the probability that the three balls drawn consist of a red ball and two black balls?

[**Hints.** *P* (1 red and 2 black) = *P* (red from *A*, black from *B*, black from *C*) + *P* (black from *A*, red from *B*, black from *C*) + *P* (black from *A*, black from *B*, red from *C*)

$$= \frac{4}{7}\times\frac{4}{9}\times\frac{4}{8}+\frac{3}{7}\times\frac{5}{9}\times\frac{4}{8}+\frac{3}{7}\times\frac{4}{9}\times\frac{4}{8}=\frac{\mathbf{43}}{\mathbf{126}}.]$$

23. A coin is tossed six times. What is the probability of getting at least two heads?

[**Hint.** Let X denotes the number of heads, then the required probability
$= P(X = 2) + P(X = 3) + P(X = 4) + P(X = 5) + P(X = 6)$.

Required probability = $1 - P(X = 0) - P(X = 1) = \frac{\mathbf{57}}{\mathbf{64}}$.]

24. Two persons throw a dice alternately, till one of them gets a multiple of three and wins the game. Find their respective probabilities of winning.

[**Hint.** *P* (multiple of 3) = *P* (getting 3 or 6) = $\frac{2}{6}=\frac{1}{3}$. Here $p = \frac{1}{3}$, $q=1-\frac{1}{3}=\frac{2}{3}$.

Let the two persons be *A* and *B*. Let *A* start throwing the dice. He wins in the first throw, 3rd, 5th, 7th and so on. Let *P*(*A*) = probability of *A*'s winning, $P(\bar{A})$ = Probability of *A*'s not winning. Similarly $P(B)$ and $P(\bar{B})$ denote the probability of *B*'s winning and not winning respectively.

$$\text{Probability of } A\text{'s winning} = P(A) + P(\bar{A})P(\bar{B})P(A) + P(\bar{A})P(\bar{B})P(\bar{A})P(\bar{B})P(A) + \ldots\ldots$$

$$= \frac{1}{3}+\frac{2}{3}\times\frac{2}{3}\times\frac{1}{3}+\frac{2}{3}\times\frac{2}{3}\times\frac{2}{3}\times\frac{2}{3}\times\frac{1}{3}+\ldots\ldots$$

$$= \frac{1}{3}\left[1+\left(\frac{2}{3}\right)^2+\left(\frac{2}{3}\right)^4+\ldots\ldots\right]=\frac{1}{3}\cdot\frac{1}{1-(2/3)^2}=\frac{1}{3}\times\frac{9}{5}=\frac{3}{5}$$

Probability of *B*'s winning the first game $=1-\frac{3}{2}=\frac{\mathbf{2}}{\mathbf{5}}$.]

25. A problem in statistics is given to three students *A*, *B* and *C* whose chances of solving it are 1/2, 3/4 and 1/4 respectively. What is the probability that the problem is solved?

[**Hint.** $P(\bar{A}) = \frac{1}{2}$, $P(\bar{B})=\frac{1}{4}$, $P(\bar{C})=\left(\frac{3}{4}\right)$. $\therefore P(\overline{ABC}) = \frac{1}{2}\times\frac{1}{4}\times\frac{3}{4}=\frac{3}{32}$.

Probability that the problem is solved $=1-\frac{3}{32}=\frac{\mathbf{29}}{\mathbf{32}}$.]

26. Find the chance of drawing an ace, a queen, a king, and a knave, in this order from an ordinary pack of four consecutive draws, the card drawn not being replaced.

27. A dice has two of its sides painted red, two black and two yellow. If the dice is rolled twice, what is the probability that the same colour appears both the times.

28. A bag contains *n* distinct white and *n* distinct red balls. Pairs of balls are drawn without replacement until the bag is empty. Show that the probability that each pair consists of one white and one red ball is $2^n/C(2n, n)$.

29. A pack of 50 tickets numbered 1 to 50 is shuffled and two tickets are drawn at random without replacement. Find the probability that at least one of the tickets drawn has a prime number.

[**Hint.** There 15 prime numbers and 35 no-prime numbers between 1 and 50.

Let $P(A) = P$ (None is prime number) $= \dfrac{C(35, 2)}{C(50, 2)} = \dfrac{17}{35}$.

P (at least one prime) $= 1 - P(A) =$ **18/25.**]

30. The face cards are removed from a full pack of 52 playing cards. Out of the remaining 40 cards are drawn at random. What is the probability that they belong to different suit?

31. In a group there are 2 men and 3 women. Three persons are selected at random from the group. Find the probability that 1 man and 2 women or 2 men and 1 women are selected.

[**Hint.** Required Probability = P (1 M and 2 W) + P (2 M + 100).

$$= \frac{C(2,1)\times C(3,2)}{C(5,3)} + \frac{C(2,2)\times C(3,1)}{C(5,3)} = \frac{\mathbf{9}}{\mathbf{10}}.]$$

32. There are 3 red and 5 black balls in bag A; and 2 red and 3 black balls in bag B. One ball is drawn from bag A and two from bag B. Find the probability that out of the 3 balls drawn. 1 is red and 2 are black.

33. A man alternatively tosses a coin and throws a die beginning with the coin. What is the probability that he will get a head before he gets '5 or 6' on the die ?

[**Hints.** Let S = the event of getting a '5 or 6'.

$$\therefore\ P(S) = \frac{1}{3} \text{ and } P(\bar{S}) = \frac{2}{3}. \text{ Also } P(H) = \frac{1}{2},\ P(T) = \frac{1}{2}.$$

$\therefore$ P (head before '5 or 6') = P [H or $T\bar{S}H, T\bar{S}T\bar{S}H$ and so on]

$$= \frac{1}{2} P(H) + P(T)\,P(\bar{S})\,P(H) + P(T)\,P(\bar{S})\,P(T)\,P(\bar{S})\,P(H) + \ldots..$$

$$= \frac{1}{2} + \frac{1}{2}\times\frac{2}{3}\times\frac{1}{2} + \frac{1}{2}\times\frac{2}{3}\times\frac{1}{2}\times\frac{2}{3}\times\frac{1}{2} + \ldots\ldots. \text{ to } \infty = \frac{(1/2)}{1-(1/3)} = \frac{3}{4}.]$$

34. A bag contains 4 red and 5 black balls. Another bag contains 3 red and 6 black balls. One ball is drawn at random from the first bag and two balls are drawn from the second bag. Find the probability that out of the three, two are black and one is red.

[**Hint. Case I.** p_1 = P(one Red from bag I and 2 Blacks from bag II)

$$= \frac{C(4,1)}{C(9,1)} \times \frac{C(6,2)}{C(9,2)} = \frac{\mathbf{60}}{\mathbf{324}}.$$

Case II. p_2 = (1 Black from bag I, 1 Black and one Red from bag II)

$$= \frac{C(5,1)}{C(9,1)} \times \frac{C(3,1)\times C(6,1)}{C(9,2)} = \frac{\mathbf{90}}{\mathbf{324}}.$$

Required Probability $= p_1 + p_2 = \dfrac{6}{324} + \dfrac{90}{324} = \dfrac{\mathbf{25}}{\mathbf{54}}$.]

35. A bag contains 8 red, 3 white and 9 blue balls. Three balls are drawn at random from the bag. Determine the probability that none of the balls drawn is white.

[**Hint.** Required Probability $= \dfrac{C(17,3)}{C(20,3)} = \dfrac{\mathbf{34}}{\mathbf{57}}$.]

36. A, B and C play a game and chances of their winning it in an attempt are 2/3, 1/2 and 1/4 respectively. A has the first chance, followed by B and then by C. This cycle is repeated till one of them wins the game. Find their respective chances of winning the game.

37. In a class there are 3 boys and 2 girls. 3 students are selected at random from the class. Find the probability that 2 boys and 1 girl or 1 boy and 2 girls are selected.

[**Hint.** Required Probability $= \dfrac{C(3,1) \times C(2,2)}{C(5,3)} + \dfrac{C(3,2) \times C(2,1)}{C(5,3)} = \dfrac{6}{10} + \dfrac{3}{10} = \mathbf{\dfrac{9}{10}}$.]

38. A card is drawn from a well shuffled pack of 52 cards. The events A and B are:

A : getting a card of spade; B : getting an ace. Determine whether A and B are independent or not.

[**Hint.** $P(A) = \dfrac{13}{52}$, $P(B) = \dfrac{4}{52}$, $P(A \cap B) = \dfrac{1}{52}$; $P(A \cap B) = P(A) \,.\, P(B)$

$\Rightarrow$ A and B are independent.]

39. Out of the numbers 1 to 120, one is selected at random. What is the probability that it is divisible by 8 or 10?

[**Hint.** Let A : the event that the number is divisible by 8 and B : the event that the number is divisible by 10.

$\therefore$ A = {8, 16, 24, 32, 40, 48, 56, 64, 72, 80, 88, 96, 104, 112, 120.}

B = {10, 20, 30, 40, 50, 60, 70, 80, 90, 100, 110, 120.}

$\therefore$ $A \cap B$ = {30, 40, 80, 120.}

$\therefore$ $n(A) = 15$, $n(B) = (12)$, $n(A \cap B) = 4$.

Now $\mathbf{P(A \text{ or } B) = P(A) + P(B) - P(A \cap B)} = \dfrac{15}{120} + \dfrac{12}{120} - \dfrac{4}{120} = \mathbf{\dfrac{23}{120}}$.]

40. Probability that a trainee will remain with company is 0.8. The probability that an employee earns more than Rs. 20,000 per year is 0.4. The probability that an employee, who was a trainee and remained with the company or who earns more than 20,000 per year is 0.9. What is the probability that an employee earns more than Rs. 20,000 per year given that he is a trainee, who stayed with the company?

[**Hint.** A = a trainee who remain with company $\Rightarrow P(A) = 0.8$.

B = employee who earn more than Rs. 20,000 $\Rightarrow P(B) = 0.4$

$\therefore$ $P(A$ or $B) = 0.9$

$\therefore$ $P(A$ or $B) = P(A) + P(B) - P(A$ Ç B)

$\Rightarrow$ 0.9 = 0.8 + 0.04 – $P(A$ Ç $B)$ Þ $P(A$ Ç $B)$ = 0.3.

$\therefore$ Required probability $= P(B/A) = \dfrac{P(A \cap B)}{P(A)} = \dfrac{0.3}{0.8} = \mathbf{0.375}$.]

41. Three machines A, B and C produce respectively 50%, 30% and 20% of the total number of items of a factory. The percentage of defective output of these machines are 3%, 4% and 5%. If an item is selected at random, find the probability that the item is defective.

[**Hint.** P = Prob. (defective item from A) + Prob. (defective item from B)

+ Prob. (defective item from C).

= 50% × 3% + 30% × 4% + 20% × 5% = 0.5 × 0.03 + 0.3 × 0.04 + 0.2 × 0.05

= 0.15 + 0.012 + 0.010 = **0.037**].

42. A box contains three coins; one coin is fair, one coin is two-headed, and one is weighted so that the probability of heads appearing is 1/3. A coin is selected at random and tossed. Find the probability P that head appears.

[**Hint.** Required probability $=\left(\frac{1}{3}\times\frac{1}{2}\right)+\left(\frac{1}{3}\times 1\right)+\frac{1}{3}\times\frac{1}{3}=\mathbf{\frac{11}{18}}$.]

43. The odds that a book will be favourably received by 3 independent critics are 5 to 2, 4 to 3 and 3 to 4 respectively. What is the probability that of the three reviews, a majority will be favourable?

[**Hint.** Let A, B, C denote the events of favouring the book by the first, the second and the third critic respectively, then, $P(A)=\frac{5}{7}$, $P(B)=\frac{4}{7}$ and $P(C)=\frac{3}{7}$.

$\therefore$ Required probability = P(two favour the book or three favour the book)

= P(two favour the book) + P(three favour the book)

$= P(AB\overline{C} \text{ or } A\overline{B}C \text{ or } \overline{A}BC) + P(ABC)$

$= P[A\cap B\cap \overline{C}]\cup(A\cap\overline{B}\cap C)(\overline{A}\cap B\cap C)] + P(A\cap B\cap C)]$.

44. A students takes his examination in four subjects a, b, g, d. He estimates his chance of passing in α as $\frac{4}{5}$, in β as $\frac{3}{4}$, in γ as $\frac{5}{6}$ and δ as $\frac{2}{3}$. To qualify he must pass in a and at least two other subjects. What is the possibility that the qualifies?

[**Hint.** Required probability $= P(\alpha\,\beta\,\gamma\,\overline{\delta}) + P(\alpha\,\beta\,\delta\,\overline{\gamma}) + P(\alpha\,\gamma\,\delta\,\overline{\beta}) + P(\alpha\,\beta\,\gamma\,\delta)$]

45. Two cards are drawn at random from a pack of 52 cards. What is the probability that it will be (*a*) a club and a heart; (*b*) a king and a queen.

[**Hint.** (*a*) $\frac{C(13,1)\,C(13,1)}{C(52,2)}=\mathbf{\frac{13}{102}}$; (*b*) $\frac{C(4,1)\times C(4,1)}{C(52,2)}=\mathbf{\frac{8}{663}}$.]

46. A salesman is known to sell a product in 3 out of 5 attempts while another salesman in 2 out of 5 attempts. Find the probability that

(*i*) no sale will be effected when both try to sell th $P(A)=\frac{3}{5}$, $P(\overline{A})=\frac{2}{5}$; $P(B)=\frac{2}{5}$ and $P(\overline{B})=\frac{3}{5}$.

(*i*) Probability that both will not be able to sell is : $P(\overline{A}\cap\overline{B}) = P(\overline{A})\,P(\overline{B}) = \frac{2}{5}\times\frac{3}{5}=\frac{6}{25}=\mathbf{0.24}$.

(*ii*) Probability that either A or B will succeed is :

$P(A\cup B) = P(A)+P(B)-P(A\cap B)=\frac{3}{5}+\frac{2}{5}-\frac{6}{25}=\mathbf{0.76}$.]

47. Three balls are drawn successively from an urn containing 6 red balls, 4 white balls and 5 blue balls. Find the probability that these are drawn in order as red, white and blue if each ball is not replaced.

[**Hints.** Let A = the event first ball is red; B = the event the second ball is white and C = the event the third ball is blue.

Now $P(ABC) = P(A)\,P(B/A)\,P(C/AB) = \frac{6}{15}\times\frac{4}{14}\times\frac{5}{13}=\mathbf{\frac{4}{91}}$.]

48. A bag contains 7 green, 4 white and 5 red balls. If four balls are drawn one by one without replacement, what is the probability that none is red?

49. A bag contains 4 white and 2 black balls. Another contains 3 white and 5 black balls. If one ball is drawn from each bag, find the probability that (*i*) both are white; (*ii*) both are black; (*iii*) one is white and one is black.

50. There are two bags containing 5 red and 7 white; 3 red and 12 white balls respectively. A ball is drawn from one of the two bags. Find the probability of drawing a red balls.

51. Bag A contains 5 red and 3 green balls, and bag B contains 3 red and 5 green balls. One ball is drawn from bag A and two from bag B. Find the probability that the balls are green.

52. A bag contains 5 white and 3 black balls. Four balls are drawn one at a time without replacement. Find the probability that the balls are alternatively of different colours.

53. An article manufactured by a company consists of two parts A and B. In the process of manufacturing of the parts A, 9 out of 100 parts may be defective. Similarly 5 out of 100 are likely to be defective in the manufacture of part B. Find the probability that the product will not be defective.

54. Arun and Tarun appeared for an interview for two vacancies. The probability of Arun's selection is 1/4 and that of Tarun's rejection of 2/3. Find the probability that at least one of them will be selected.

[**Hint.** Required Probability $= 1 - \left(\frac{3}{4} \times \frac{2}{3}\right) = \frac{1}{2}$.]

55. A speaks truth in 60% of the cases and B in 90% of the cases. In what percentage of cases are likely to contradict each other in stating the same fact?

56. Find the probability of drawing a one rupee coin from a purse with two compartments. One of which contain 3 fifty-paise coins and 2 one rupee coins; and the other contains 2 fifty-paise coins and 3 one-rupee coins.

57. Events E and F are known to be independent. Examine if the following

(*i*) $\bar{E}$ and $\bar{F}$; (*ii*) $\bar{E}$ and F ; (*iii*) E and $\bar{F}$ are independent.

58. A lot contains 50 defective and 50 non-defective bulbs. Two bulbs are drawn at random, one at a time with replacement. The events A, B and C are defined as:

A = {the first bulb is defective}; B = {the second bulb is defective}

C = {the two bulbs are both defective or non-defective}.

Show that (*a*) A, B and C are pairwise independent;

(*b*) A, B and C are not independent.

59. Tickets are numbered 1 to 10. Two tickets are drawn one after the other with replacement. Find the probability that the number on one of the tickets is a multiple of 5 and on the other is a multiple of 4.

[**Hint.** Let A = {(5, 4), (5, 8), (10, 4), (10, 8), (4, 5), (4, 10), (8, 5), (8, 10)}.

$\therefore \quad P(A) = \frac{8}{10 \times 10} = \frac{2}{25}$.]

60. Out of 9 outstanding students in a school, there are 4 boys and 5 girls. A team of 4 students is to be selected for a quiz programme. Find the probability that 2 are girls and 2 are boys.

ANSWERS

5. (*a*) $\frac{1}{12}$; (*b*) $\frac{1}{2}$.
6. $\frac{101}{105}$.
7. (*a*) $\frac{73}{105}$; (*b*) $\frac{32}{105}$.
8. 0.98.

9. Throwing a four with one dice.
10. (*a*) $\frac{75}{341}$; (*b*) $\frac{15}{682}$.

11. (*i*) $\frac{C(7,4)}{C(12,4)}$; (*ii*) $\frac{C(7,2) \times C(5,2)}{C(12,4)}$.
12. $\frac{37}{49}$.
13. 0.273

14. (*i*) $\frac{4}{35}$; (*ii*) $\frac{5}{6}$; (*iii*) $\frac{2}{5}$.
15. $\frac{1}{4}$.
16. 0.375.
17. $\frac{3}{7}$.

18. $\frac{7}{9}$.
19. $\frac{1}{7}$.
20. $\frac{5}{12}$.

21. (*i*) $\frac{3}{35}$, (*ii*) $\frac{17}{35}$. **22.** $\frac{43}{126}$. **23.** $\frac{57}{64}$. **24.** $\frac{2}{5}$.

25. $\frac{29}{32}$. **26.** $\frac{32}{812175}$. **27.** $\frac{1}{3}$. **29.** $\frac{18}{35}$.

30. $\frac{1000}{9139}$. **31.** $\frac{9}{10}$. **32.** $\frac{39}{80}$. **33.** $\frac{3}{4}$.

34. $\frac{25}{54}$. **35.** $\frac{34}{57}$. **36.** $\frac{16}{21}, \frac{4}{21}, \frac{1}{21}$. **37.** $\frac{9}{10}$.

38. Independent **39.** $\frac{23}{120}$ **40.** 0.375 **41.** 0.037

42. $\frac{11}{18}$ **43.** $\frac{209}{343}$ **44.** $\frac{61}{90}$ **45.** $\frac{8}{663}$

46. (*i*) 0.24; (*ii*) 0.76. **47.** $\frac{4}{91}$ **48.** $\frac{33}{182}$

49. (*i*) $\frac{1}{4}$; (*ii*) $\frac{5}{24}$; (*iii*) $\frac{13}{24}$ **50.** $\frac{37}{120}$

51. $\frac{15}{112}$ **52.** $\frac{1}{7}$ **53.** 0.8645. **54.** $\frac{1}{2}$

55. 42% **56.** $\frac{1}{2}$

57. (*i*) Independent; (*ii*) Independent; (*iii*) Independent **59.** $\frac{2}{25}$ **60.** $\frac{10}{21}$

BAYES' THEOREM

5.12 BAYES' THEOREM

Bayes' Theorem ***is based on the concept that probabilities should be revised when some new information is available.*** The idea of revising the probabilities is used by all of us in daily life even though we may not be knowing anything about probability.

Illustration 7: A student while going to a college may start without a rain coat, but as soon as he comes out of his home and sees a thick mass of clouds in the sky he may decide to take a rain coat with him. Thus, he has revised his earlier decision of going to a college without a rain coat. In the same way **the probabilities are revised as soon as some new information is available about the problem concerned.**

The necessity of revising probabilities arises from a need to make better use of available information and, thereby, reduce the element of risk involved in decision making. The idea of revising probabilities on the bases of new information was given by British Mathematician **Thomas Bayes (1702–61)**. His theorem known as **Bayes' Theorem or Rule**, was actually published in 1763 after his death in the form of a small paper.

Bayes' Theorem offers a powerful statistical method of evaluating new information and revising our prior estimates (based on the limited information) of the probability. Modern decision theory is often called Bayesian Decision Theory.

5.13 PRIOR PROBABILITIES (OR PRIORI) AND POSTERIOR PROBABILITIES

Prior Probabilities or Priori

Probabilities before revision, by Bayes' Rule, are called Prior probabilities or simply Priori, because they are determined before the sample (or new) information is taken into account.

Posterior Probabilities

The psoterior probability is the revision of probability with added information. A probability which has been revised in the light of sample information (by Bayes' Rule) is called a **Posterior probability** *since it represents a probability calculated after the additional information is taken into account.* Posterior probabilities are also called **Revised Probabilities** because they are obtained by revising the prior probabilities in the light of the additional information gained. ***Posterior probabilities are always conditional probabilities, the conditional event being the sample information.*** Thus, ***a prior probability which is unconditional probability becomes a posterior probability, which is a conditional probability, by using Bayes' Rule.***

5.14 INVERSE PROBABILITY

The concept of conditional probability discussed in earlier section takes into account the information about the occurrence of one event to predict the probability of another event. In other words, in conditional probability we studied only those problems on probability in which our knowledge of ***factors affecting the event was sufficient to enable us to determine the chances of happening of the event. But now we shall discuss the problems of reverse nature. By this we mean that if an event has happened as a result of one of the several causes, we may be interested to find out the probability of a particular cause which really affected the event to happen. The problem of such a type are called the problems of*** **Inverse Probability**.

Illustration 1: Suppose a **bag A** contains 5 white and 3 green balls. Another **bag B** contains 4 white and 6 green balls. A white ball has been drawn from one of the two bags. A white ball must have been drawn either from the bag A or from bag B. To find out the probability that it came from the **bag B**, (or the white ball has come from **Bag A**) is an example of **Inverse Probability.**

In order to study these types of problems, British Mathematician **Thomas Bayes (1702–61)**, gave a theorem known as **Bayes' Theorem**. It is also known as **Inverse Probability Theorem**.

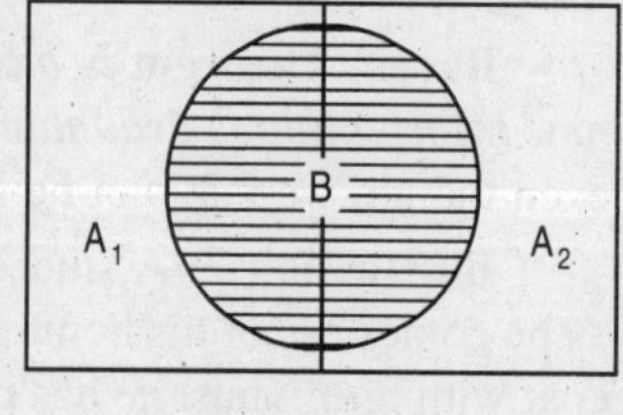

Illustration 2: Consider a sample output of 5 defectives in 100 trials (event A) might be used to estimate the probability that a machine is not working correctly (event B).

Let A_1 and A_2 be the set of events which are mutually exclusive (the two events cannot occur together) and exhaustive (the combination of the two events is the entire experiment), *i.e.*, $A = A_1 \cup A_2$ and $A_1 \cap A_2 = \phi$.

Let B = a simple event which intersects each of the events A_1 and A_2 as shown in the figure. We observe the following:

(*i*) The part of B which is in A_1 represents "the area (A_1 and B)" or $A_1 \cap B$.

(*ii*) The part of B which is in A_2, represents the "the area (A_2 and B)" or $A_2 \cap B$.

(*iii*) The event $\mathbf{B = (A_1 \cap B) \cup (A_2 \cap B)}$

$$\therefore \quad P(B) = P[(A_1 \cap B) \cup (A_2 \cap B)]$$

$$= P(A_1 \cap B) + P(A_2 \cap B) = P(A_1 \text{ and } B) + P(A_2 \text{ and } B)$$

or $$\mathbf{P(B)} = P(A_1)\, P(B/A_1) + P(A_2)\, P(B/A_2). \text{ Also } P(B) \neq 0.$$

Then, **the probability of event A_1 given the event B, is**

$$P(A_1/B) = \frac{P(A_1 \text{ and } B)}{P(B)} = \frac{P(A_1 \cap B)}{P(B)}.$$

Similarly, the probability of event A_2, given the B, is

$$P(A_2/B) = \frac{P(A_2 \text{ and } B)}{P(B)} = \frac{P(A_2 \cap B)}{P(B)}.$$

The conditional probabilities $P(A_1/B)$ and $P(A_2/B)$ (which are computed after getting the sample information B) are called posterior (or inverse) probabilities of event A_1 and A_2.

5.15 STATEMENT OF BAYES' THEOREM

Theorem: ***Let $E_1, E_2, \ldots\ldots, E_n$ be the set of n mutually exclusive and exhaustive events whose union is the random sample space S, of an experiment. If A be any arbitrary event of the sample space of the above experiment with $P(A) \neq 0$, then the probability of the event E_i, when the event A has actually occurred is given by $P(E_i/A)$, where***

$$P(E_i/A) = \frac{P(A \cap E_i)}{P(A \cap E_1) + P(A \cap E_2) + \ldots\ldots + P(A \cap E_n)}$$

[FIRST FORM]

We know that $\quad P(A \cap E_i) = P(E_i)\, P(A/E_i)$

$$\therefore \quad P(E_i/A) = \frac{P(E_i)\, P(A/E_i)}{\Sigma P(E_i)\, P(A/E_i)} \qquad \textbf{[SECOND FORM]}$$

In this theorem, we come across three types of probabilities **$P(E_i)$, $P(E_i/A)$ and $P(A/E_i)$.** We shall explain each one of them in the following manner:

$P(E_i)$ = It is the probability associated with the happening of the event E_i, $i = 1, 2, \ldots\ldots, n$; such that $P(E_1) + P(E_2) + \ldots\ldots + P_n(E_n) = 1$. $P(E_i)$ are known as Priori or a Prior probabilities.

$P(A/E_i)$ = It is the Conditional probability of the event A given that E_i has already occurred.

$P(E_i/A)$ = The probabilities $P(E_i/A)$ are called the Posterior probabilities. The information that A has occurred allow us to re-assess the probability $P(E_i)$ assigned to the event E_i.

Example 49. *A new pregnancy test was given to 100 pregnant women and 100 non-pregnant women. The test indicated pregnancy of 92 of 100 pregnant and to 12 of the 100 non-pregnant women. If a randomly selected woman takes this test and the test indicates that she is pregnant. What is the probability that she was not pregnant?*

Solution: Let **E_1: the event that a pregnant women is selected;**

E_2: the event that a non-pregnant women is selected.

A: the event that test shows that the selected woman is pregnant.

$$\therefore\ P(E_1) = \frac{100}{200} = \frac{1}{2}, \quad P(E_2) = \frac{100}{200} = \frac{1}{2}\ ; \ P(A/E_1) = \frac{92}{100},\ P(A/E_2) = \frac{12}{100}.$$

We are **interested in finding $P(E_2/A)$.**

$$\therefore \quad P(E_2/A) = \frac{P(E_2)\, P(A/E_2)}{P(E_1)\, P(A/E_1) + P(E_2)\, P(A/E_2)}$$

$$= \frac{(1/2)\,(12/100)}{(1/2)\,(92/100) + (1/2)\,(12/100)} = \frac{12}{104} = \frac{3}{26}.$$

Example 50. *There are 4 boys and 2 girls in room **A** and 5 boys and 3 girls in room **B**. A girl from one of the two rooms laughed loudly. What is the probability that the girl who laughed was from room B.*

Solution: Let

E_1: the event that the girl laughed from room A.

E_2: the event that the girl laughed from room B.

$$P(E_1) = \frac{1}{2}; \quad P(E_2) = \frac{1}{2}.$$

Let *G* denote the event that the girl laughed.

$$P(G/E_1) = \text{Probability that the girl selected belongs to room } A$$

$$\therefore \quad P(G/E_1) = \frac{2}{6} = \frac{1}{3}.$$

Similarly, $$P(G/E_2) = \frac{3}{8}.$$

We are interested in finding the probability

$$P(E_2/G) = \frac{P(E_2)\,P(G/E_2)}{P(E_1)\,P(G/E_1) + P(E_2)\,P(G/E_2)}$$

$$= \frac{(1/2)\times(3/8)}{(1/2)\times(1/3)+(1/2)\times(3/8)} = \frac{(3/16)}{(17/48)} = \frac{9}{17}.$$

TREE DIAGRAM

We can also solve this problem by a **Tree Diagram of the following type:**

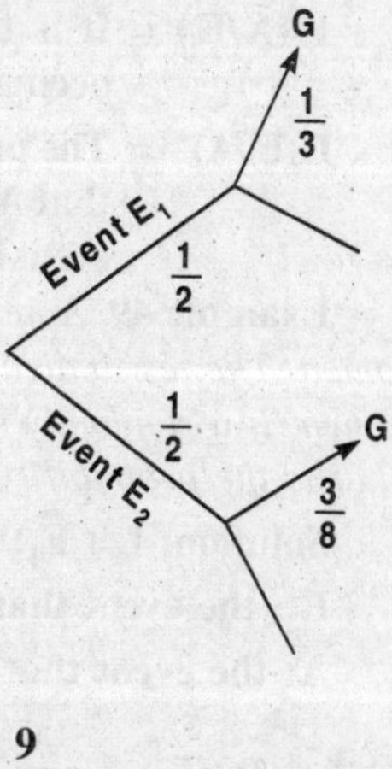

Event	**Probability**
$E_1 \cap G$	$P(E_1 \cap G) = \frac{1}{2}\times\frac{1}{3} = \frac{1}{6}.$
$E_2 \cap G$	$P(E_2 \cap G) = \frac{1}{2}\times\frac{3}{8} = \frac{3}{16}.$

Total $= P(E_1 \cap G) + P(E_2 \cap G)$

Also $$P(G) = P(E_1 \cap G) + P(E_2 \cap G)$$

$$= \frac{1}{6} + \frac{3}{16} = \frac{17}{48}.$$

$$P(G/E_2) = \frac{P(E_2 \cap G)}{P(G)} = \frac{(3/16)}{(17/48)} = \frac{9}{17}.$$

Hence, the probability that the girl who laughed was from room *B* is $\frac{9}{17}$.

Example 51. *The probabilities of X, Y and Z becoming managers are 4/9; 2/9 and 1/3 respectively. The probabilities that the Bonus scheme will be introduced if X, Y and Z becomes managers are 3/10, 1/2 and 4/5 respectively.*

(*a*) *What is the probability that the Bonus scheme will be introduced?*

(*b*) *If the Bonus scheme has been introduced, what is the probability that the manager appointed was X?*

Solution: Let $P(X)$ = **Probability that** X **becomes manager**

$P(Y)$ = **Probability that** Y **becomes manager**

$P(Z)$ = **Probability that** Z **becomes manager.**

Let $P(B/X)$ = Probability that Bonus scheme is introduced when X becomes manager. Similarly, we can define $P(B/Y)$ and $P(B/Z)$.

We are given: $P(X) = \frac{4}{9}$; $P(Y) = \frac{2}{9}$; $P(Z) = \frac{1}{3}$.

Also, $P(B/X) = \frac{3}{10}$; $P(B/Y) = \frac{1}{2}$; $P(B/Z) = \frac{4}{5}$.

Let B **= denote the event that the Bonus scheme is introduced.**

(*a*) **Required Probability:** $P(B) = P(B \cap X) + P(B \cap Y) + P(B \cap Z)$

$$= P(X)\, P(B/X) + P(Y)\, P(B/Y) + P(Z)\, P(B/Z)$$

$$= \frac{4}{9} \times \frac{3}{10} + \frac{2}{9} \times \frac{1}{2} + \frac{1}{3} \times \frac{4}{5} = \frac{12 + 10 + 24}{90} = \frac{46}{90} = \mathbf{\frac{23}{45}}.$$

(*b*) **Using Bayes' Theorem, the required probability is:**

$$P(X/B) = \frac{P(X \cap B)}{P(B)} = \frac{P(X) \cdot P(B/X)}{P(B)}$$

$$= \frac{4/9 \times 3/10}{23/45} = \frac{12}{90} \times \frac{45}{23} = \mathbf{\frac{6}{23}}.$$

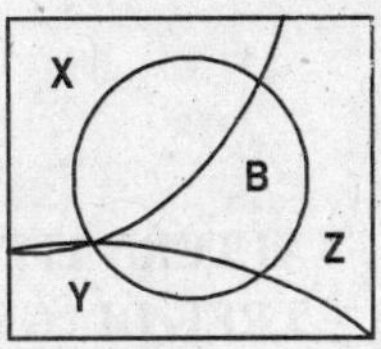

[**Note:**

$$B = X \cap B + Y \cap B + Z \cap B$$

$$\therefore \quad P(B) = P(X \cap B) + P(Y \cap B) + P(Z \cap B)$$

$$= P(X)\, P(B/X) + P(Y)\, P(B/Y) + P(Z)\, P(B/Z)]$$

Example 52. *A factory has two machines A and B. Past records show that machine A produces 30% of the total output and machine B the remaining 70%. Machine A produces 5% defective articles and machine B produces 1% defective items. An item is drawn at random and found to be defective. What is the probability that it was produced (a) by machine A (b) by machine B?*

Solution: Let D **denote the item drawn be defective.**

(***i***) Probability that the item came from machine A = $P(A_1)$: $P_1 = \frac{30}{100}$.

(***ii***) Probability that machine A produces a defective item:

$$P(D/A) = p_1 = \frac{5}{100}.$$

(***iii***) Joint probability of the two events: $P(A \cap D) = P(A)\, P(D/A)$

$$= P_1 p_1 = \frac{30}{100} \times \frac{5}{100} = \frac{150}{10{,}000}.$$

(***iv***) Probability that the item came from machine B = $P(B)$: $P_2 = \frac{70}{100}$

(***v***) Probability that machine B produces a defective item: $P(D/B) = p_2 = \frac{1}{100}$.

(*vi*) Joint probability of the two events: $P(B \cap D) = P(B)\, P(D/B)$

$$= P_2 p_2 = \frac{70}{100} \times \frac{1}{100} = \mathbf{\frac{70}{10000}}.$$

$$\therefore \; P(D) = P(A \cap D) + P(B \cap D) = \frac{150}{10000} + \frac{70}{10000} = \mathbf{\frac{220}{10000}}.$$

Now, **the probability that the defective item came from machine A:**

$$P(A/D) = \frac{P_1 p_1}{P_1 p_1 + P_2 p_2}$$

$$= \frac{(150/10{,}000)}{(150/10{,}000) + (70/10{,}000)} = \frac{150}{220} = \mathbf{0.682.}$$

The probability that the defective machine B:

$$P(B/D) = \frac{P(B \cap D)}{P(D)} = \frac{P_2 p_2}{P_1 p_1 + P_2 p_2}$$

$$= \frac{(70/10{,}000)}{(150/10{,}000) + (70/10{,}000)} = \frac{70}{220} = \mathbf{0.318.}$$

ALTERNATIVE METHOD
TREE DIAGRAM EVENT PROBABILITY

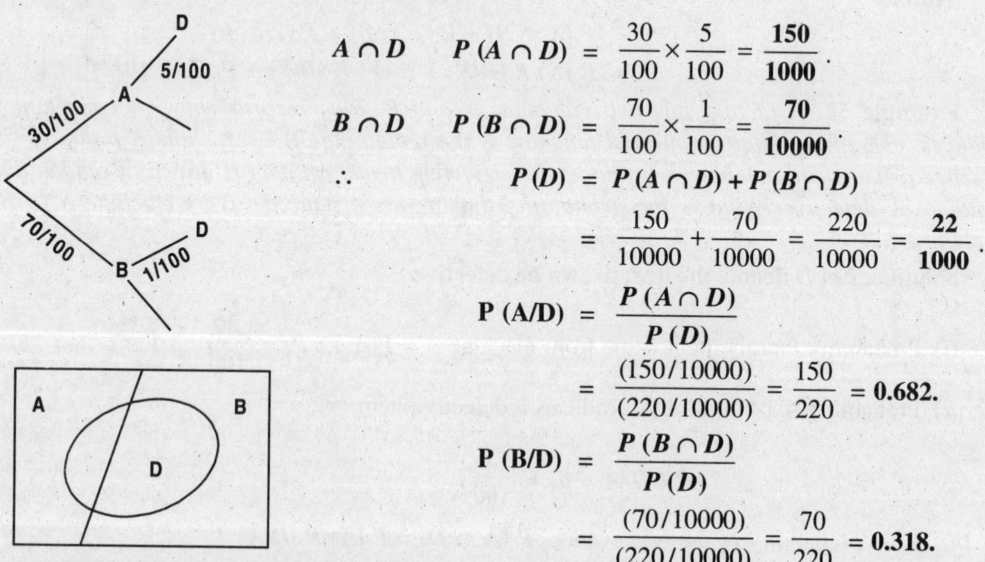

$A \cap D \quad P(A \cap D) = \frac{30}{100} \times \frac{5}{100} = \mathbf{\frac{150}{1000}}.$

$B \cap D \quad P(B \cap D) = \frac{70}{100} \times \frac{1}{100} = \mathbf{\frac{70}{10000}}.$

$$\therefore \quad P(D) = P(A \cap D) + P(B \cap D)$$

$$= \frac{150}{10000} + \frac{70}{10000} = \frac{220}{10000} = \mathbf{\frac{22}{1000}}.$$

$$\mathrm{P}(A/D) = \frac{P(A \cap D)}{P(D)}$$

$$= \frac{(150/10000)}{(220/10000)} = \frac{150}{220} = \mathbf{0.682.}$$

$$\mathrm{P}(B/D) = \frac{P(B \cap D)}{P(D)}$$

$$= \frac{(70/10000)}{(220/10000)} = \frac{70}{220} = \mathbf{0.318.}$$

Hence, (*i*) The probability that the defective item came from machine A = 0.682.

(*ii*) The probability that the defective item came from machine B = 0.318.

Example 53. *A company has two plants to manufacture scooters. Plant I manufactures 80% of the scooters, and Plant II manufacture, 20%. At plant 1; 85 out of 100 Scooters are rated standard quality or better. At Plant II only 65 out of 100 Scooters are rated standard quality or better.*

(*i*) *What is the probability that a Scooter selected at random came from Plant I if it is known that the Scooter is of standard quality?*

(*ii*) *What is the probability the Scooter came from Plant II if it is known that the scooter is of standard quality.*

Solution: Let P_1 = Probability that the scooter is manufactured by Plant I = **0.80.**

P_2 = Probability that the scooter is manufactured by Plant II = **0.20.**

Let the probabilities of standard quality scooter coming from plant I and II respectively be p_1 and p_2.

$\therefore$ p_1 = **0.85,** p_2 = **0.65.** **(Given)**

Let E = **denote the event that the scooter is of standard quality.**

The joint probabilities of standard quality scooter coming from Plant I and Plant II are

Plant I: $P_1p_1 = 0.80 \times 0.85 = \mathbf{0.68}.$

Plant II: $P_2p_2 = 0.20 \times 0.65 = \mathbf{0.13}.$

Probability of the event E: $P(E) = P_1p_1 + P_2p_2 = 0.68 + 0.13 = \mathbf{0.81}.$

(*i*) The probability that the standard quality scooter came from Plant I would be:

$$\text{P (Plant I/E)} = \frac{P_1p_1}{(P_1p_1)+(P_2p_2)} = \frac{P_1p_1}{P(E)} = \frac{0.68}{0.81} = \mathbf{0.84}.$$

(*ii*) The probability that the standard quality scooter came from Plant II would be:

$$\text{P (Plant II/E)} = \frac{P_2p_2}{(P_1p_1)+(P_2p_2)} = \frac{P_2p_2}{P(E)} = \frac{0.13}{0.81} = \mathbf{0.16}.$$

Example 54. *A factory produces a certain type of output by three types of machines. The respective daily production figures are Machine I: 3,000 units, Machine II: 2,500 units and Machine III: 4,500 units.*

Past experience shows that 1% of the output produced by Machine I is defective. The corresponding fraction of defectives for the other two machines are 1.2% and 2% respectively. An item is drawn at random from the day's production run and is found to be defective. What is the probability that it comes from the output of (a) Machine I, (b) Machine II and (c) Machine III?

Solution: Let

A = **the event that an item is produced by Machine I**

B = **the event an item is produced by Machine II**

C = **the event that an item is produced by Machine III.**

Then, $$P(A) = \frac{3000}{3000+2500+4500} = \frac{3000}{10000} = \frac{3}{10} = \mathbf{0.3}\ ;$$

$$P(B) = \frac{2500}{10000} = \mathbf{0.25}\ ;\ P(C) = \frac{4500}{10000} = \mathbf{0.45}.$$

Now, let D = **the event that an item drawn is defective.**

Then, the conditional probabilities are

$$P(D/A) = \frac{1}{100} = \mathbf{0.01}\ ;\ P(D/B) = \frac{1.2}{100} = \mathbf{0.012}\ ;\ P(D/C) = \frac{2}{100} = \mathbf{0.02}.$$

Again, the **joint probability that an item is produced by three machines and it is defective:**

$$P(A \cap D) = P(A).P(D/A) = 0.3 \times 0.01 = \mathbf{0.003}$$

$$P(B \cap D) = P(B).P(D/B) = 0.25 \times 0.012 = \mathbf{0.003}$$

$$P(C \cap D) = P(C).P(D/C) = 0.45 \times 0.02 = \mathbf{0.009}$$

Also, $P(D) = P(A \cap D) + P(B \cap D) + P(C \cap D) = 0.003 + 0.003 + 0.009 = \mathbf{0.015}.$

Now, the **revised or posterior probability is obtained by Bayes' Theorem.**

$$P(A/D) = \frac{P(A \cap D)}{P(D)} = \frac{P(A \cap D)}{P(A \cap D) + P(B \cap D) + P(C \cap D)} = \frac{0.003}{0.015} = \frac{3}{15} = \mathbf{0.2.}$$

$$P(B/D) = \frac{P(B \cap D)}{P(D)} = \frac{P(B \cap D)}{P(A \cap D) + P(B \cap D) + P(C \cap D)} = \frac{0.003}{0.015} = \frac{3}{15} = \mathbf{0.2.}$$

$$P(C/D) = \frac{P(C \cap D)}{P(D)} = \frac{P(C \cap D)}{P(A \cap D) + P(B \cap D) + P(C \cap D)} = \frac{0.009}{0.015} = \frac{9}{15} = \mathbf{0.6.}$$

Hence the Probability that the defective item comes from (a) Machine I = 0.2; from Machine II = 0.2; and from Machine III = 0.6.

ALTERNATIVE METHOD

TREE DIAGRAM **EVENT PROBABILITY**

Fig. 0.7

$A \cap D$	$P(A \cap D)$	$= 0.3 \times 0.01$	=	**0.003.**
$B \cap D$	$P(B \cap D)$	$= 0.25 \times 0.012$	=	**0.003.**
$C \cap D$	$P(C \cap D)$	$= 0.45 \times 0.02$	=	**0.009.**

$$P(D) = P(A \cap D) + P(B \cap D) + P(C \cap D)$$
$$= 0.003 + 0.003 + 0.009 = \mathbf{0.015.}$$

$$P(A/D) = \frac{P(A \cap D)}{P(D)} = \frac{0.003}{0.015} = \mathbf{0.2,}$$

$$P(B/D) = \frac{P(B \cap D)}{P(D)} = \frac{0.003}{0.015} = \mathbf{0.2,}$$

$$P(C/D) = \frac{P(C \cap D)}{P(D)} = \frac{0.009}{0.015} = \mathbf{0.6.}$$

Example 55. *Two sets of candidates are competing for the positions on the Board of Directors of a company. The probabilities that the first and the second sets will win are 0.6 and 0.4 respectively. If the first set wins, the probability of introducing a new product is 0.8 and corresponding probability if the second set wins is 0.3. What is the probability that the product will be introduced by the second set?*

Solution:

Let the probabilities of the winning of the first and second sets be P_1 and P_2.

$\therefore \quad P_1 = 0.6; \quad P_2 = 0.4.$

Let the probability of introducing the new product if set *A* wins be p_1; and if *B* wins be p_2.

$\therefore \quad p_1 = 0.8; \quad p_2 = 0.3.$

Joint probabilities P_1p_1 and P_2p_2 are:

$$P_1p_1 = 0.6 \times 0.8 = \mathbf{0.48.}$$

$$P_2p_2 = 0.4 \times 0.3 = \mathbf{0.12.}$$

Let E = **denote the event that the product will be introduced.**

$$\therefore \quad P(E) = P_1p_1 + P_2p_2 = 0.48 + 0.12 = \mathbf{0.60.}$$

$$\therefore \quad \text{P(Second set/E)} = \frac{P_2p_2}{P(E)} = \frac{P_2p_2}{P_1p_1 + P_2p_2} = \frac{0.12}{0.60} = \frac{1}{5} = \mathbf{0.2.}$$

Hence the probability that the product will be introduced by the second set is 0.2.

Example 56. *You note that your officer is happy on 60% of your calls, so you assign a probability of his being happy on your visit as 0.6. You have noticed also that if he is happy, he accedes to your request with a probability of 0.4 whereas if he is not happy, he accedes to the request with a probability of 0.1. You call one day, and he accedes to your request. What is the probability of his being happy?*

Solution: Let H be the hypothesis that the officer is happy and $\bar{H}$ be the hypothesis that the officer is not happy.

$$P(H) = \frac{6}{10}, \quad P(\bar{H}) = \frac{4}{10}$$

Let A be the event that he accedes to request.

$$(A/H) = 0.4 = \frac{4}{10},$$

Also $\quad P(A/\bar{H}) = 0.1 = \dfrac{1}{10}.$

$$\therefore \quad P(H/A) = \frac{P(H) \times P(A/H)}{P(H) \times P(A/H) + P(\bar{H} \times P(A/\bar{H})} \qquad \textbf{(Bayes' Theorem)}$$

$$= \frac{\frac{6}{10} \times \frac{4}{10}}{\frac{6}{10} \times \frac{4}{10} + \frac{4}{10} \times \frac{1}{10}} = \frac{24 \times 100}{100 \times 28} = \frac{24}{28} = \frac{6}{7} = \mathbf{0.857.}$$

Example 57. *A certain production process produces items that are 10 per cent defective. Each item is inspected before being supplied to customers but the inspector incorrectly classifies an item 10 per cent of the time. Only items classified as good are supplied. If 820 items in all have been supplied, how many of them are expected to be defective?*

Solution: Let $P(D)$ = Probability of defective item = $\dfrac{10}{100}$ = **0.1.**

$P(G)$ = Probability of good item = **1 – $P(D)$** = 1 – 0.2 = **0.9.**

$P(G/D)$ = P (classified as good when it is defective) = $\dfrac{10}{100}$ = **0.1.**

$P(G/G)$ = P (classified as good when it is good) = 1 – P (G/D) = 1 – 0.1 = **0.9.**

***P* (Defective/classified as good when actually it is defective)**

$$= \frac{P(D) \times P(G/D)}{P(D) \times P(G/D) + P(G) \times P(G/G)}$$

$$= \frac{0.1 \times 0.1}{0.1 \times 0.1 + 0.9 \times 0.9} = \frac{0.01}{0.82} = \frac{\mathbf{1}}{\mathbf{82}}.$$

∴ **Expected number of defective items out of 820 items = 820 × $\frac{1}{82}$ = 10.**

EXERCISE – 5.2

1. A coin is tossed. It turns up heads, two balls will be drawn from urn *A*; otherwise two balls will be drawn from urn *B*. Urn *A* contains three black and five white balls. Urn *B* contains seven black and one white ball. In both cases, selections are to be made with replacement. What is the probability that urn *A* is used given that both the balls drawn are black?
2. Assume that a factory has two machines. Past records show that machine *A* produces 30% of the items of output and machine *B* produces 70% of the items. Further 5% of the items produced by machine *A* were defective and only 10% produced by machine *B* were defective. If a defective item is drawn at random, what is the probability that it was produced by machine *B*?
3. A factory has three units, *A*, *B* and *C*. A produces 25% of its product, unit *B* produces 25% and unit *C* produces 50%. If the percentages of defective items produced by three units *A*, *B* and *C* are respectively 10%, 20% and 3% and an item selected randomly from the total products of the factory is found to be defective. What is the probability that it is produced by the unit *C*?
4. A box *A* contains 1 red and 2 white marbles and another box *B* contains 3 red and 2 white marbles. One marble is drawn at random from one of the boxes and it is found to be red. Find the probability that it was drawn from box *B*.
5. In a bulb factory, machines *A*, *B* and *C* manufacture 60%, 25% and 15% respectively. Of the total of their output 1%, 2% and 1% are defective bulbs. A bulb is drawn at random from the total production and found to be defective. From which machine, the defective bulb is expected to have been manufactured?
6. A company manufactures scooters at two plants *A* and *B*. Plant *A* produces 80% and plant *B* produces 20% of the total product, 85% of the scooters produced at plant *A* are standard in quality, while 65% of the scooters produced at plant *B* are standard in quality. A scooter produced by the company is selected at random and it is found to be standard in quality. What is the probability that it was manufactured at plant *A*?

   [**Hint.** Let E_1 and E_2 be the event of the scooter being manufactured at plants *A* and *B* respectively. Let *S* be the event of getting a scooter of standard quality.

   Then, $P(E_1) = \frac{4}{5}; P(E_2) = \frac{1}{5}; P(S/E_1) = \frac{17}{20}$ and $P(S/E_2) = \frac{13}{20}$.

   $$\therefore\ P(E_1/S) = \frac{P(S/E_1).P(E_1)}{P(S/E_1).P(E_1) + P(S/E_2).P(E_2)} = \frac{(4/5) \times (17/20)}{(4/5) \times (17/20) + (1/5) \times (13/20)} = \frac{\mathbf{68}}{\mathbf{81}}.]$$

7. There are 2000 scooter drivers, 4000 car drivers and 6000 truck drivers all insured. The probabilities of an accident involving a scooter, a car, a truck are 0.01, 0.03 and 0.15 respectively. One of the insured drivers met with an accident. What is the probability he is a scooter driver?

   [**Hint.** $P(S) = \frac{2000}{2000 + 4000 + 6000} = \frac{1}{6}; \ P(C) = \frac{4000}{12000} = \frac{1}{3}, \ P(T) = \frac{6000}{12000} = \frac{1}{2}.$

   $$P(A \cap S) = P(A)\,P(S/A) = \frac{1}{6} \times 0.01 = \frac{1}{600}$$

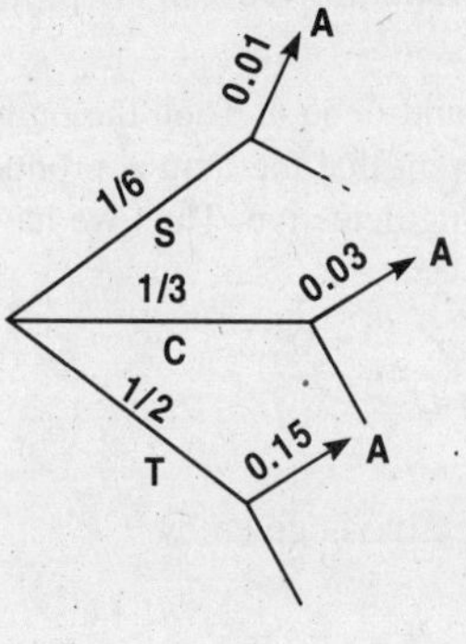

$$P(A \cap C) = P(A)\,P(C/A) = \frac{1}{3} \times 0.03 = \frac{1}{100}$$

$$P(A \cap T) = P(A)\,P(T/A) = \frac{1}{2} \times 0.15 = \frac{15}{200}$$

$$P(A) = P(A \cap S) + P(A \cap C) + P(A \cap T)$$

$$= \frac{1}{600} + \frac{1}{100} + \frac{15}{200} = \frac{52}{600}.$$

$$\boldsymbol{P(S/A)} = \frac{\boldsymbol{P(A \cap S)}}{\boldsymbol{P(A)}} = \frac{(1/600)}{(52/600)} = \frac{\mathbf{1}}{\mathbf{52}}.]$$

8. A letter is known to have come either from TATA NAGAR or CALCUTTA. On the envelop just two consecutive letters TA are visible. Show that the probability that the letter have come from CALCUTTA is 4/11.

[**Hint.** Let E_1**:** the event that letter came from TATA NAGAR.

E_2**:** the event that the letter came from CALCUTTA.

∴ A**:** the event that two consecutive visible letters on the envelop are TA.

$$\therefore\ P(E_1) = \frac{1}{2} = P(E_2); \quad P(A/E_1) = \frac{2}{8} = \frac{1}{4}; \quad P(A/E_2) = \frac{1}{7}.$$

$$P(E_2/A) = \frac{(1/2)\times(1/7)}{(1/2)\times(1/4)+(1/2)\times(1/7)} = \frac{\mathbf{4}}{\mathbf{11}}.]$$

9. By examining the chest *X*-ray probability that T.B. is detected when a person is actually suffering from T.B. is 0.99. The probability that the doctor digagnose incorrectly that a person has a T.B. on the basis of *X*-ray is 0.001. In a certain city, 1 in 1000 persons suffers from T.B. A person selected at random is diagnosed to have T.B. What is the chance that he actual has T.B.?

10. In a competition an examinee either guesses or copies or knows the answer to a multiple choice question with four choices. The probability that he makes a guess is 1/3 and the probability that he copies the answer is 1/6. The probability that his answer is correct, given that he copied it, is 1/8. Find the probabiliity that he knew the answer to the question, given that he correctly answered it.

[**Hint.** Let ***G, C, K*** respectively denote the event that he **guesses, Copies** and **Knows** the answer to question. **Let H denote the event that his answer is correct.**

$$\therefore \quad P(G) = 1/3, \quad P(C) = 1/6. \quad P(K) = 1 - \frac{1}{3} - \frac{1}{6} = \frac{1}{2}.$$

$P(H/G) = 1/4$ (∵ only one choice is correct out of four choices)

$P(H/C) = 1/8$, $P(H/K) = 1$.

$$\therefore \quad \boldsymbol{P(K/H)} = \frac{(1/2)\times 1}{(1/2)\times 1 + (1/3)\,(1/4) + (1/6)\,(1/8)} = \frac{\mathbf{24}}{\mathbf{29}}.]$$

11. T.V. compoenents are produced by two machines *A* and *B*. 50% of the components are produced by machine *A* with an estimate of 10% of them being defective. On machine *B*, 20% of the components produced are defective. If a component taken at random is found to be defective, what is the probability that the component was produced by machine *A*?

12. Suppose that a product is produced in three factories *A, B* and *C*. It is known that factory *A* produces twice as many items as factory *B*; and that factories *B* and *C* produce the same number of products. Assume that it is known that 2 per cent of the items produced by each of the factories *A* and *B* are defective while 4 per cent of those manufactured by factory *C* are defective. All the items produced in the three factories are stocked, and an item of product is selected at random. What is the probability that this product is defective?

[**Hint.** Let the number of items produced by each of the factories B and C be n. Then the number of items produced by the factory A is $2n$. Let A_1, A_2 and A_3 denote the event that the item is produced by factory A, B and C respectively and let E be the event of the item being detective. Then we have

$$\therefore \quad P(A_1) = \frac{2n}{2n+n+n} = \frac{2n}{4n} = 0.5,\ P(A_2) = \frac{n}{4n} = 0.25,\ \ P(A_2) = \frac{n}{4n} = 0.25.$$

$$P(E/A_1) = P(E/A_2) = 0.02 \text{ and } P(E/A_3) = 0.04 \text{ (Given)}$$

The probability that an item selected at random from the stock is defective is given by

$$P(E) = P(A_1 \cap E) + P(A_2 \cap E) + P(A_3 \cap E).$$

$$= P(A_1)\, P(E/A_1) + P(A_2)\, P(E/A_2) + P(A_3)\, P(E/A_3) = \mathbf{0.07}.]$$

13. A, B and C are bidding on a contract for the construction of a bridge. The probabilities that A, B, C will get the contract are 0.5, 0.3 and 0.2 respectively. If A gets he will select E as the sub-contractor with probability 0.8. If B or C gets it, E will be chosen with probabilites 0.4 and 0.1 respectively. E get the sub contract. What is the probability that E get the sub-contract whern A is selected?

14. An insurance company insured 1500 scooter drivers, 3,500 car drivers and 5,000 truck drivers. The probability of an accident is 0.05, 0.02 and 0.10 respectively in case of scooter, car and truck drivers. One of the insured persons meets an accident. What is the probability that he is a car driver?

15. A factory has a machine shop in which three machines A, B and C produce 100 cm. aluminium tubes. An inspector is equally likely to sample tubes from A and B, and three times as likely to select tubes from C as he is from B. The defective rates from the three machines are: A = 10%, B = 10% and C = 20%. What is the probability that a tube selectd by the inspector:

(*i*) is from machine A? (*ii*) is defective?

(*iii*) comes from machine A, given that it is defective?

16. Three machines A, B, C produce respectively 50%, 30% and 20% of the total number of items of a factory. The percentages of defective outputs of these machines are respectively 3%, 4% and 5%. If an item is selected at random and is found to be defective. What is the probability that this item is from machine A?

17. Suppose that one of three men, a politician, a businessman, and an educator will be appointed as the vice-chancellor of a university. The respective probabilities of their appointments are 0.50, 0.30, 0.20. Tlhe probabilities that research activities will be promoted by these people if they appointed are 0.30, 0.70 and 0.80 respectively. A research activity is promoted. What is the probability that research activity is promoted by the politician vice-chancellor?

18. There are three urns having the following compposition and white balls:

Urn I: 7 white and 3 black balls

Urn II: 4 white and 6 black balls

Urn III: 2 white and 8 black balls

One of the Urns is chosen at random with probabilities 0.2, 0.6 and 0.2 respectively. From the chosen Urn, two balls are drawn at random without replacement. Both the balls happen to the white. Calculate probability that the balls drawn were from Urn III.

19. A factory produces a certain type of output by three types of machines. The respective daily production figures are Machine I: 3000 units, Machine II: 2500 units, Machine III: 4500 units. Past experience shows that 3 per cent of the output produced by Machine I is defective. The corresponding fraction of defectives, for the other two machines are respectively, 1.2 per cent and 2 per cent respectively. An item is drawn at random from the day's production and is found to be defective. What is the probability that it comes from the output of (*a*) Machine I, (*b*) Machine II, and (*c*) Machine III?

20. The chance that a female worker in a chemical factory will contract an occupational disease is 0.04 and the chance for a male worker is 0.06. Out of 1,000 workers in a factory 200 are females. One worker is selected at random and the worker is found to have contracted the disease. What is the probability that the worker is a feamle?

21. Supoose that there are three Urns containing 2 white and 3 black balls; 3 white and 2 black balls; and 4 white and one black balls respectively. There is equal probability of each Urn being taken. One ball is drawn from an urn at random. We are told that a white ball has been drawn. Find the probability that it was drawn from the first urn.

22. A factory has two machines A and B. Past record shows that machine A produced 60% of the items of output and machine B produce 40% of the items of output. Further 20% of items produce by machine A were defective and 1% produced by machine B were defective. If a defective item is drawn at random . What is the probability that it was produced by machine A.

[**Hint.** Let D be defective events,

Here $P(A) = 0.6; P(B) = 0.4; P(D/A) = 0.02, P(D/B) = 0.01$

$$\therefore \quad P(A/D) = \frac{P(A)\,P(D/A)}{P(A)\,P(D/A) + P(B)\,P(D/B)} = \frac{0.6 \times 0.02}{0.6 \times 0.02 + 0.4 \times 0.01} = \mathbf{\frac{3}{4}}.]$$

23. A car manufacturing factory has two plants. Plant A manufacture 70% of cars and Plant B manufacture 30%. At plant A, 80% of cars are rated of standard quality and at plant B, 90% of cars are rated of standard quality. A car is pieced up at random and is found to be of standard quality. What is the probability that it has come from plant A?

[**Hint.** Let E be the event that cars is of standard quality.

$P(A) = 0.7; P(B) = 0.3; P(E/A) = 0.8, P(E/B) = 0.9$

$$\therefore \quad P(A/E) = \frac{P(A)\,P(E/A)}{P(A)\,P(E/A) + P(B)\,P(E/B)} = \frac{0.7 \times 0.8}{0.7 \times 0.8 + 0.3 \times 0.9} = \mathbf{\frac{56}{83}}.]$$

24. Suppose 5 men out of 100 and 25 women out of 1000 are good orators. An orator is chosen at random. Find the probability that a male person is selected. Assume that there are equal number of men and women.

[**Hint.** Let E be the event of being an orator.

$$P(M) = \frac{1}{2};\ P(W) = \frac{1}{2};\ P(E/M) = \frac{5}{100},\ P(E/W) = \frac{25}{1000}$$

$$\therefore \quad P(M/E) = \frac{P(M)\,P(E/M)}{P(M)P(E/M) + P(W)\,P(E/W)} = \frac{\frac{1}{2} \times \frac{5}{100}}{\frac{1}{2} \times \frac{5}{100} + \frac{1}{2} \times \frac{25}{1000}} = \mathbf{\frac{2}{3}}.]$$

25. A bag contains 7 white balls and 5 red balls, of which 3 white balls and 3 red balls marked A, and 4 white balls and 2 red balls marked B. A white ball is drawn randomly from the bag. Find the probability that the white ball drawn is (*i*) marked A and; (*ii*) also marked B.

26. Suppose that A is known to tell the truth in five cases out of six and he states that a white ball was drawn from a bag containing 9 black and one white ball. What is the probability that the white ball was really drawn?

[**Hint.** The probability that a white ball is drawn in any case is 1/10. Also the probability that the white ball was drawn and that A told the truth is $1/10 \times 5/6$.

Furthermore, the probability that a black ball was drawn and A told a lie about it is $9/10 \times 1/6$.

Hence the probability that a white ball was drawn is

$$= \frac{(1/10) \times (5/6)}{(1/10) \times (5/6) + (9/10) \times (1/6)} = \mathbf{\frac{5}{14}}.]$$

27. A bag contains 4 white balls and 2 red balls. Another bag contains 1 white and 3 red balls. One ball is transferred from the first bag to the second bag and then one ball is drawn from the second bag. If it is found to be a red ball, what is the probability that the ball transferred is white?

28. A company has two plants to manufactures secoters. Plant I manufacture 70% of the scooters and Plant II manufactures 30%. At Plant I, 80% of the scooters produced are of standard quality and at Plant II, 90% of the scooters produced are of standard quality. A scooter is picked at random and found to be of standard quality. What is the chance that it has come from Plant II.

29. A firm produces pipes in two Plants I and II with daily production of 500 and 1000 pipes respectively. Plant I produces 5% depective pipes and Plant II produces 8% defective pipes. A defective pipe is selected at random from the total production, what is the chance that the pipe is produced by (*i*) Plant I or (*ii*) Plant II?

30. A pack of playing cards was found to contain only 51 cards. If the first 13 cards which are examined are all red, what is the probability that the missing card is black?

[**Hint.** Let A_1, A_2 be the events that black and red card is lost respectively. Let B donote the occurrence of first 13 cards.

$$P(A_i) = \frac{1}{2},\ P(A_2) = \frac{1}{2},\ P(B/A_1) = \frac{{}^{26}C_{13}}{{}^{51}C_{13}};\ P(B/A_2) = \frac{{}^{25}C_{13}}{{}^{51}C_{13}}$$

$$P(A_1/B) = \frac{P(A_1)\,P(B/A_1)}{P(A_1)\,P(B/A_1) + P(A_2)\,P(B/A_2)} = \frac{(1/2)\ {}^{26}C_{13}}{(1/2)\left[{}^{26}C_{13} + {}^{25}C_{13}\right]} = \mathbf{\frac{2}{3}}.]$$

31. State and illustrate Bayes' theorem.

32. Distinguish between a priori and posterior probability.

33. Explain the concept of posterior probability.

34. Explain the significance of Bayes' Theorem.

35. A company has three plants to manufacture 8000 scooters in a month. Out of 8000 scooters, Plant I manufactures 4000 scooters, Plant II manufactures 3000 scooters and Plant III manufactures 1000 scooters. At Plant I, 85 out of 100 scooters are rated of standard quality or better; at Plant II, only 65 out of 100 scooters are rated of standard quality or better and at Plant III, 60 out of 100 scooters are rated of standard quality or better. What is the probability that the scooter selected at random came from (i) Plant I, (ii) Plant II, and (iii) Plant III, if it is known that the scooter is of standard quality.

[**Hint.** Let S be the event that the scooter selected is of standard quality.

$$\therefore \quad P(A_1) = \frac{1}{2},\ P(A_2) = \frac{3}{8};\ P(A_3) = \frac{1}{8}.$$

$$P(S/A_1) = \frac{85}{100} = 0.85,\ P(S/A_2) = \frac{65}{100} = 0.65,\ P(S/A_3) = 0.60.$$

$$P(S) = P(S \cap A_1) + P(S \cap A_2) + P(S \cap A_3) = \frac{1}{2}\times\frac{85}{100} + \frac{3}{8}\times\frac{65}{100} + \frac{1}{8}\times\frac{60}{100} = \frac{595}{800}.$$

(i) $$P(A_1/S) = \frac{P(S \cap A_1)}{P(S)} = \frac{(85/200)}{595/800)} = \frac{340}{595} = \frac{69}{119}.$$

(ii) $$\mathbf{P(A_2/S)} = \frac{P(S \cap A_2)}{P(S)} = \frac{(195/800)}{(595/800)} = \mathbf{\frac{39}{119}}.$$

(iii) $$\mathbf{P(A_3/S)} = \frac{P(S \cap A_3)}{P(S)} = \frac{(60/800)}{(595/800)} = \mathbf{\frac{12}{119}}.]$$

36. Police plan to enforce speed limits by using randar traps at 4 different locations within the city limits. The radar traps at each of the locations L_1, L_2, L_3 and L_4 are operated 40%, 30%, 20% and 10% of the time and if a person is speeding on his way to work has probabilities 0.2, 0.1, 0.5 and 0.2 respectively, of passing through these locations. What is the probability that he will receive speeding ticket?

[**Hint.** Let A_1, A_2, A_3 and A_4 be the event that the radar traps will be active at locations L_1, L_2, L_3 and L_4. Let S denote the event of speeding ticket.

$$P(A_1) = 0.4, \quad P(A_2) = 0.3, \quad P(A_3) = 0.2, \quad P(A_4) = 0.1$$

$$P(S/A_1) = 0.2, \quad P(S/A_2) = 0.1, \quad P(S/A_3) = 0.5, \quad P(S/A_4) = 0.2$$

$$\begin{aligned} P(S) &= P(S \cap A_1) + P(S \cap A_2) + P(S \cap A_3) + P(S \cap A_4) \\ &= P(A_1)\,P(S/A_1) + P(A_2)\,P(S/A_2)\,P(A_3)\,P(S/A_3) + P(A_4)\,P(S/A_4) \\ &= 0.4 \times 0.2 + 0.3 \times 0.1 + 0.2 \times 0.5 + 0.1 \times 0.2 = \mathbf{0.23}. \end{aligned}$$

Probability that a person will receive a speeding ticket = 0.23]

37. In a railway reservation office, two clerks are engaged in checking reservation forms. On an average, the first clerk checks 55% of the forms, while the second does the remaing. The first clerk has an error rate of 0.03 and second has an error rate of 0.02. A reservation form is selected at random from the total number of forms checked during a day, and is found to have an error. Find the probabilities that it was checked by first and second clerk respectively.

[**Hint.** Let E = the event that the selected from has an error.

Let A_1, A_2 denote the event that the form has been checked respectively by the first and second clerk.

$$P(A_1) = 0.55, \quad P(A_2) = 0.45; \quad P(E/A_1) = 0.03, \quad P(E/A_2) = 0.02$$

$$\mathbf{P(E)} = P(A_1 \cap E) + P(A_2 \cap E) = 0.55 \times 0.03 + 0.45 \times 0.02 = \mathbf{0255.}$$

(*i*) $$\mathbf{P(A_1/E)} = \frac{P(A_1 \cap E)}{P(E)} = \frac{0.55 \times 0.03}{0.0255} = \frac{\mathbf{11}}{\mathbf{17}}.$$

(*ii*) $$\mathbf{P(A_2/E)} = \frac{\boldsymbol{P(A_2 \cap E)}}{\boldsymbol{P(E)}} = \frac{0.45 \times 0.02}{0.0255} = \frac{\mathbf{6}}{\mathbf{17}}.]$$

38. A factory has three units A, B and C. A produces 25% of its product, unit B produces 25% and the unit C produces 50%. If the percentages of defective items produced by three units A, B and C are respectively 1%, 2% and 3% and an item selected randomly from the total production of the factory is found to be defective, what is the probability that it is produced by the unit C?

39. Suppose that a product is produced in three factories A, B and C. It is known that factory A produces thrice as many items as factory B, and that factories B and C produce the same number of product. Assume that it is known that 4 per cent of the items produced by each of the factories A and B are defective while 5 per cent of those manufacturd by factory C are defective. All the items produced in three factories are stocked, and an item of product is selected at random. What is the probability that this item is defective and is produced by the factory A.

40. In a population of workers, suppose 40% are school graduates, 50% are high school graduates, and 10% are college graduates. Among the school graduates, 10% are unemployed; among the high school graduates, 5% are unemployed, and among the college graduates 2% are unemployed. If a worker is chosen at random and found to be unemployed, what is the probability that he is a college graduate?

41. A company uses a 'selling aptitude test' in the selection of salesmen. Past experience has shown that only 70% of all persons applying for a sales position achieved a classification "dissatisfactory" in actual selling, whereas the remainder were classified as "satisfactory." 80% has scored a passing grade on the aptitude test. Only 25% of those classified unsatisfactory has passed the test on the basis of this information. What is the probability that a candidate would be a satisfactory salesman given that he passed the aptitutde test?

[**Hint.** Let S stand for a 'satisfactory' classification as a salesman and P stand for 'passing the test'. The probability that a candidate would be "satisfactory" salesman given that he passed the aptitude test is:

$$P(S/P) = \frac{(0.70)(0.85)}{(0.70)(0.85) + (0.30)(0.25)} = \frac{0.595}{0.595 + 0.075} = \mathbf{0.888.}$$

The result indicates that the tests are of value in screening candidates. Assuming no change in the type of candidates applying for the selling positions, the probability that a random applicant would be satisfactory is 70%. On the other hand, if the company only accepts an applicant if he passed the test, the probability increases to 0.888.]

42. A manufacturing firm produces units of a product in four plants. Define event A_i, : a unit is produced in plant i, i = 1, 2, 3, 4 and event B : a unit is defective. From the past records of the proportions of defectives produced at each plant the following conditional probabilities are set:

$$P(B/A_1) = 0.05, \ P(B/A_2) = 0.10, \ P(B/A_3) = 0.15, \ P(B/A_4) = 0.02.$$

The first plant produces 30 per cent of the units of the product, the second, plant 25 per cent, third plant 40 per cent and the fourth plant 5 per cent. A unit of the product made at one of these plants is tested and is found to be defective. What is the probability that the unit was produced in plant 3?

43. If a machine is correctly set up it will produce 90% acceptable items. If it is incorrectly set up it will produce 30% acceptable items. Past experiance shows that 80% of set ups are correctly done. If after a certain set up, first items produced is acceptable, what is the probability that the machine is correctly set up?

[**Hint.** $P(A_1) = 0.80$, $P(A_2) = 0.20$; $P(E/A_1) = 0.90$, $P(E/A_2) = 0.30$.

$$E = (A_1 \cap E) \cup (A_2 \cap E)$$

$$P(E) = P(A_1 \cap E) + P(A_2 \cap E) = 0.90 \times 0.80 + 0.30 \times 0.20.$$

$$P(A \cap E) = P(A_1)\,P(E/A_k)\ ;\ P(A_2 \cap E) = P(A_2)\,P(E/A_2)$$

$$\therefore \quad P(A_1/E) = \frac{P(A_1 \cap E)}{P(E)} = \frac{0.90 \times 0.80}{0.90 \times 0.80 + 0.30 \times 0.20} = \frac{\mathbf{12}}{\mathbf{13}}.]$$

44. A factory has three machines X, Y and Z producing 1000, 2000 and 3000 bolts per day respectively. The machines X produces 1% defective bolts, Y produces 1.5% and Z produces 2% defective bolts. At the end of a day, a bolt is drawn at random and is found defective. What is the probability that this defective bolt has been produced by the machine X?

45. If bag A conlans 2 white and 3 red balls and a bag B contain 4 while and 5 red balls. One ball is drawn at random from are of the bags and is found to the red. Find the probability it was drawn from the bag B.

ANSWERS

1. 0.125. **2.** $\frac{7}{22}$. **3.** $\frac{2}{3}$. **4.** $\frac{9}{14}$. **5.** Machine A. **6.** $\frac{68}{81}$.

7. $\frac{1}{52}$. **8.** $\frac{4}{11}$. **9.** $\frac{110}{221}$. **10.** $\frac{24}{29}$ **11.** $\frac{1}{3}$. **12.** 0.07.

13. 0.54. **15.** (*i*) 0.02; (*ii*) 0.16; (*iii*) 0.125. **16.** $\frac{3}{74}$. **17.** $\frac{15}{52}$. **18.** $\frac{1}{40}$.

19. (*a*) 0.43; (*b*) 0.14; (*c*) 0.43. **20.** $\frac{1}{7}$. **21.** $\frac{2}{9}$. **22.** $\frac{3}{4}$. **23.** $\frac{56}{83}$.

24. $\frac{2}{3}$. **25.** $\frac{3}{7}, \frac{4}{7}$. **26.** $\frac{5}{14}$. **27.** $\frac{3}{5}$. **28.** 0.325.

29. $P(A_1/E) = \frac{5}{21}$; $P(A_2/E) = \frac{16}{21}$. **30.** 0.656 **35.** (*i*) $\frac{69}{119}$, (*ii*) $\frac{39}{119}$, (*iii*) $\frac{12}{119}$. **36.** 0.23.

THEORETICAL DISTRIBUTIONS

(PROBABILITY DISTRIBUTION, BINOMIAL, POISSON, NORMAL, AND UNIFORM DISTRIBUTIONS)

RANDOM VARIABLE

6.1 RANDOM VARIABLE

A random variable x is a real valued function defined on a sample space of an experiment. In other words, to **each one of the outcomes of an experiment or a sample point E_i of the sample spaces, there corresponds a unique real number x_i, known as the value of the random variable X.**

Illustration 1 : Sample Space of rolling of a single dice.

When we roll a single dice, its possible outcomes are 6. It can be the number 0, 1, 2, 3, 4, 5, 6 appearing on the uppermost face. Thus in this case the random variable X **takes the value 1, 2, 3, 4, 5, 6.**

Illustration 2 : The sample space corresponding to tossing of two coins.

When two coins are tossed, its outcomes or sample points can be (Head, Head), (Head, Tail), (Tail, Head), (Tail, Tail). *i.e.*, $S = \{HH, HT, TH, TT\}$. After the performance of the experiment, we count the number of Tails and denote it by X. The first outcome HH has 0 Tail, so $X = 0$. Similarly $X = 1$, denote the outcomes HT or TH, and $X = 2$, represents the outcome TT.

Thus X takes the values 0, 1, 2, *i.e.*, $X = 0, 1, 2$.

A **random variable** is also called a **stochastic variable** or simply a **variable** or a **chance variable.** The name **random variable** or **variable** or **chance variable** ***is given to the variate X because it is defined on a sample space, the outcomes of which are uncertain and hence depend on chance.***

6.1.1 Types of Random Variables

A random variable is of two types.

1. Discrete Random Variable. **2. Continuous Random Variable.**

6.1.2 Discrete Random Variable

A random variable is said to be discrete if it takes only a finite or an infinite but countable number of values.

Tossing of a coin or throwing a dice are the examples of discrete random variables.

6.1.3 Probability Function of a Discrete Random Variable

If for a discrete random variable X, the real valued function p (x) is such that

$$P(X = x) = p(x),$$

then p (x) is called probability function or probability density function of a discrete random variable X. Probability function p (x) gives the measure of probability for different values of X.

6.1.4 Properties of a Probability Function

If $p(x)$ is a probability function of a random variable X, then it possesses the following properties:

1. ***p (x) is positive for all value of x, i.e., $p(x) \geq 0$ for all x.***

2. *Total of all values of p (x) for different x is always 1, i.e., Σ p(x) = 1, summation is taken over for all values of x.*
3. *p (x) measures the probability for any given value of x.*
4. *p (x) cannot be negative for any value of x.*

6.1.5 Probability Mass Function

If X is a discrete random variable, then its probability function p (x) is a discrete probability function. It is also called Probability Mass Function.

In other words "If $x_1, x_2, x_3, \ldots\ldots, x_n$ are n different values of a discrete random variable X and $p(x_1), p(x_2), \ldots\ldots, p(x_n)$ be their respective probabilities such that

(i) $p(x_i) \geq 0,$ $\quad i = 1, 2, 3, \ldots\ldots, n$

(ii) $\Sigma p(x_i) = 1,$ $\quad i = 1, 2, 3, \ldots\ldots, n;$

then p (x) is known as the Probability Mass Function of the variable X.

The values $p(x_i)$, i = 1, 2, 3, ..., n; are known as the probability distribution of the random variable X.

I. Probability Mass Function of a Binomial Distribution is:

$P(x) = {}^nC_x p^x q^{n-x}$, x = 0, 1, 2, ..., m; p + q = 1, p ≥ 0, Σ p = 1.

II. Probability Mass Function of a Poisson Distribution is:

$P(x) = \dfrac{e^{-m} m^x}{x!}$, x = 0, 1, 2, 3, ..., n; where m = mean of poisson distribution.

6.1.6 Continuous Random Variable

A continuous random variable is a random variable that can take on any value in an interval of two numbers.

6.1.7 Probability Density Function of a Continuous Random Variable

The probabilities associated with a continuous random variable X are determined by the probability density function f(x) of a random variable X.

Properties of Probability Density Function

The probability density function '**f(x)**' has the following properties.

1. *f(x) ≥ 0 for all values of x.*
2. *The probability that x will lie between two numbers a and b is equal to the area under the curve y = f(x) between x = a and x = b.*
3. *The total area under the entire curve y = f(x) is equal to one, i.e., 1.00.*

REMARKS :

1. *The probability of any single value is zero in a continuous sample space. Thus for a continuous random variable, the probability of occurrence of any given value is zero. The result is obvious from property 2, as the area under a curve between a point and itself is the area of the line which is zero.*
2. *For a continuous random variable, non-zero probabilities are associated only with interval of numbers.*

PROBABILITY DISTRIBUTION

? PROBABILITY DISTRIBUTION

A random variable X taking the values $x_1, x_2, x_3, ..., x_n$ with probabilities $p_1, p_2, p_3, ..., p_n$ respectively [where $p_r = P(X = x_r)$] such that

(i) every probability is a non-negative number not greater than 1

i.e.,$p_i \geq 0, i$ =1, 2, 3, ..., n ; $0 \leq p_i \leq 1$.

(ii) the sum of all the probabilities is always unity

i.e.,$p_1 + p_2 + ... + p_n$ = 1

is said to possess the probability distribution. It is represented in a tabular form as given below: Here $P(X = x_i) = p_i$.

X	x_1	x_2	x_3	...	x_r	...	x_n
$P(X)$	p_1	p_2	p_3	...	p_r	...	p_n

Also random variable *X* will have

Mean : $\mu = \sum_{i=1}^{n} p_i x_i$, **Variance :** $\sigma^2 = \sum_{i=1}^{n} p_i x_i^2 - \mu^2$.

Example 1. *Find the probability distribution of the number of heads in two tosses of a coin.*

Solution. When we toss a coin twice, we get the four outcomes, *viz*., ***HH, HT, TH, TT*** (*T* for tail, *H* for head). In the above case we notice **that the number of heads can be zero, one or two.**

∴ ***P* (zero head) = 1/4, P (one head) = 2/4, *P* (two heads) = 1/4.** Thus, ***X* takes the values 0, 1, 2 (zero head, one head, two heads) with probabilities 1/4, 1/2, 1/4, respectively.**

Hence, we have the probability distributions.

X	0	1	2
$P(X)$	1/4	1/2	1/4

Example 2. *Two cards are drawn successively, with replacement, from a well shuffled pack of 52 cards. Find the probability distribution of the number of kings.*

Solution. Let *X* denote the number of kings. Here *X* can take the values 0, 1, or 2.

$$P\text{ (king)} = \frac{4}{52} = \frac{1}{13}, \qquad P\text{ (not kings)} = 1 - \frac{1}{13} = \frac{12}{13}.$$

$$P(X = 0) = P\text{ (not king and not king)} = P\text{ (not king) } P\text{ (not king)} = \frac{12}{13} \times \frac{12}{13} = \frac{144}{169}.$$

$$P(X = 1) = P\text{ (king and not king)} + P\text{ (not king and king)} = \frac{1}{13} \times \frac{12}{13} + \frac{12}{13} \times \frac{1}{13} = \frac{24}{169}.$$

$$P(X = 2) = P\text{ (king and king)} = P\text{ (king) } P\text{ (king)} = \frac{1}{13} \times \frac{1}{13} = \frac{1}{169}.$$

∴ We have following **probability distribution:**

X :	0	1	2
$P(X)$:	$\frac{144}{169}$	$\frac{24}{169}$	$\frac{1}{169}$

Example 3. *Find the probability distribution of the number of sixes in three tosses of a dice and sketch its graph.*

Solution. Let X be the number of sixes. Then X takes the values 0, 1, 2, 3.

Now $\quad P(\text{six}) = \frac{1}{6}; \qquad P(\text{not six}) = 1 - \frac{1}{6} = \frac{5}{6};$

$\therefore \quad P(X = 0) = \frac{5}{6} \times \frac{5}{6} \times \frac{5}{6} = \frac{125}{216}; \qquad P(X = 1) = 3\left(\frac{1}{6} \times \frac{5}{6} \times \frac{5}{6}\right) = \frac{75}{216};$

$P(X = 2) = 3 \times \frac{5}{216} = \frac{15}{216}; \qquad P(X = 3) = \frac{1}{6} \times \frac{1}{6} \times \frac{1}{6} = \frac{1}{216}.$

We have the required **probability distribution** as

X	0	1	2	3
$P(X)$	$\frac{125}{216}$	$\frac{75}{216}$	$\frac{15}{216}$	$\frac{1}{216}$

and its graph is given by Fig. 9.1.

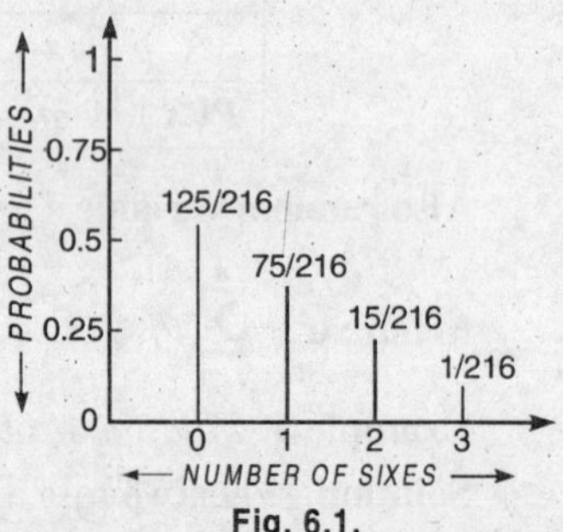

Fig. 6.1.

Example 4. *An urn contains 4 white and 3 red balls. Find the probability distribution of the number of red balls in three draws, with replacement, from the urn. Also find its mean and variance.*

Solution. Let X be the number of red balls in three draws. Then X takes the value 0, 1, 2 and 3.

$P(\text{red ball}) = 3/7, \qquad P(\text{not red ball}) = 1 - 3/7 = 4/7.$

Here $\quad P(X = 0) = \frac{64}{343}, \qquad P(X = 1) = \frac{144}{343},$

$P(X = 2) = \frac{108}{343}, \qquad P(X = 3) = \frac{27}{343}.$

$\therefore$ The required **probability distribution** is :

X	0	1	2	3
$P(X)$	$\frac{64}{343}$	$\frac{144}{343}$	$\frac{108}{343}$	$\frac{27}{343}$

Now, $\quad$ **Mean:** $\mu = \Sigma p_i x_i = 0 \times \frac{64}{343} + 1 \times \frac{144}{343} + 2 \times \frac{108}{343} + 3 \times \frac{27}{343} = \frac{441}{343} = \frac{9}{7}.$

Variance : $\sigma^2 = \Sigma p_i x_i^2 - (\mu)^2 = \left(0 + \frac{144}{343} + \frac{432}{343} + \frac{243}{343}\right) - \frac{81}{49} = \frac{252}{343} = \frac{36}{49}.$

Example 5. *Two cards are drawn with replacement from a well shuffled pack of 52 cards. Find μ and σ for the number of aces.*

Solution. The probability distribution of this question is that X takes the values 0, 1, 2 with probabilities $\frac{144}{169}, \frac{24}{169}$ and $\frac{1}{169}$ respectively. **[See Example 2 on page 9.2]**

$\therefore \qquad \mu = \Sigma p_i x_i = 0 \times \frac{144}{169} + 1 \times \frac{24}{169} + 2 \times \frac{1}{169} = \frac{26}{169} = \frac{2}{13}.$

$$\sigma^2 = \Sigma p_i x_i^2 - (\mu)^2 = 0 \times \frac{144}{169} + 1 \times \frac{24}{169} + 4 \times \frac{1}{169} - \frac{4}{169} = \frac{28}{169} - \frac{4}{169} = \mathbf{\frac{24}{169}}.$$

$$\therefore \qquad \sigma = \sqrt{\frac{24}{169}} = \frac{4.9}{13} = \mathbf{0.377}.$$

Example 6. *A dice is tossed twice. Getting a number greater than 4 is considered a success. Find the variance of the probability distribution of the number of successes.*

Solution. Let X be the number of successes, then X takes the values 0, 1 or 2.

The number greater than 4 is 5 or 6. Let S denote the success in getting 5 or 6.

$$\therefore \qquad P(S) = P(5 \text{ or } 6) = \frac{2}{6} = \mathbf{\frac{1}{3}}; \qquad \therefore P(\bar{S}) = 1 - \frac{1}{3} = \mathbf{\frac{2}{3}}.$$

$$\therefore \qquad P(X = 0) = P(\bar{S}, \bar{S}) = \frac{2}{3} \times \frac{2}{3} = \mathbf{\frac{4}{9}};$$

$$P(X = 1) = P(S, \bar{S}) + P(\bar{S}, S) = \frac{1}{3} \times \frac{2}{3} + \frac{2}{3} \times \frac{1}{3} = \mathbf{\frac{4}{9}};$$

$$P(X = 2) = P(S, S) = \frac{1}{3} \times \frac{1}{3} = \mathbf{\frac{1}{9}}.$$

Thus X takes the values 0, 1, 2, with probabilities $\frac{4}{9}, \frac{4}{9}$, and $\frac{1}{9}$ respectively.

Hence $$\mu = \Sigma p_i x_i = 0 \times \frac{4}{9} + 1 \times \frac{4}{9} + 2 \times \frac{1}{9} = \frac{6}{9} = \mathbf{\frac{2}{3}}.$$

Also $$\Sigma p_i x_i^2 = \frac{4}{9} \times 0 + \frac{4}{9} \times 1 + \frac{1}{9} \times 4 = \frac{8}{9}$$

$$\therefore \qquad \sigma^2 = \Sigma p_i x_i^2 - (\mu)^2 = \frac{8}{9} - \frac{4}{9} = \mathbf{\frac{4}{9}}.$$

Example 7. *A dice is tossed twice. A success is getting an odd number on a toss. Find the variance of the probability distribution of the number of successes.*

Solution. Let X be the random variable associated with getting an odd number in two tosses of a dice. Let $P(S)$ be the probability of getting a success, *i.e.*, getting an odd number (1 or 3 or 5) in a toss of a dice, and $P(\bar{S})$ be the probability of not a success.

$$P(S) = \frac{3}{6} = \mathbf{\frac{1}{2}}; \qquad P(\bar{S}) = 1 - \frac{1}{2} = \mathbf{\frac{1}{2}}.$$

$$\therefore \ P(X = 0) = P(\text{not success and not success}) = P(\bar{S})\ P(\bar{S}) = \frac{1}{2} \times \frac{1}{2} = \mathbf{\frac{1}{4}}.$$

$$P(X = 1) = P(S \text{ and } \bar{S}) + P(\bar{S} \text{ and } S) = \frac{1}{2} \times \frac{1}{2} + \frac{1}{2} \times \frac{1}{2} = \mathbf{\frac{2}{4}} = \mathbf{\frac{1}{2}}.$$

$$P(X = 2) = P(S \text{ and } S) = \frac{1}{2} \times \frac{1}{2} = \mathbf{\frac{1}{4}}.$$

Thus X takes the values 0, 1, 2, with respective probabilities of $\frac{1}{4}, \frac{1}{2}, \frac{1}{4}$.

Here $\Sigma p_i x_i = 0 + (1/2) + (1/2) = 1$; $\quad \Sigma p_i x_i^2 = 0 + (1/2) + 1 = \mathbf{3/2}$.

$$\therefore \ \mu = \Sigma p_i x_i = \mathbf{1}, \qquad \sigma^2 = \Sigma p_i x_i^2 - (\mu)^2 = \frac{3}{2} - 1 = \frac{1}{2} = \mathbf{0.5}.$$

Example 8. *Calculate the mean, variance and standard deviation of a number obtained as a result of throwing an unbiased dice.*

Solution. Here the random variable X takes the values 1, 2, 3, 4, 5, 6 with probabilities 1/6, 1/6, 1/6, 1/6, 1/6, 1/6 respectively.

$\therefore$ **Mean:** $$\mu = \sum_{i=1}^{6} p_i x_i = \frac{1+2+3+4+5+6}{6} = \frac{21}{6} = \frac{7}{2}.$$

Also $$\sum_{i=1}^{6} p_i x_i^2 = \frac{1+4+9+16+25+36}{6} = \frac{91}{6}.$$

$\therefore$ **Variance:** $$\sigma^2 = \sum_{i=1}^{6} p_i x_i^2 - (\mu)^2 = \frac{91}{6} - \frac{49}{4} = \frac{35}{12} = 2.92.$$

$\therefore$ **S.D.** : $$\sigma = \sqrt{2.92} = 1.709.$$

Example 9. *Two bad oranges are accidentally mixed with ten good ones. Three oranges are drawn at random, without replacement from this lot. Find mean and variance for the number of bad oranges.*

Solution. Let X be the random variable associated with the drawing of bad oranges, without replacement. Then obviously, X can take the values 0, 1, 2 (*i.e.*, no bad orange, one bad orange, two bad oranges).

Now $$P(X=0) = \frac{^{10}C_3}{^{12}C_3} = \frac{10 \times 9 \times 8}{12 \times 11 \times 10} = \frac{6}{11}.$$

$$P(X=1) = P(\text{2 good oranges and one bad orange}) = \frac{^{10}C_2 \times {^2C_1}}{^{12}C_3} = \frac{9}{22}.$$

$$P(X=2) = P(\text{one good orange and two bad oranges}) = \frac{^{10}C_1 \times {^2C_2}}{^{12}C_3} = \frac{1}{22}.$$

Thus X takes the values 0, 1, 2 with probability $\frac{6}{11}, \frac{9}{22}, \frac{1}{22}$ respectively. Now we have

$$\mu = \Sigma p_i x_i = 0 + \frac{9}{22} + \frac{2}{22} = \frac{1}{2}. \quad \text{Also } \Sigma p_i x_i^2 = 0 + \frac{9}{22} + \frac{4}{22} = \frac{13}{22}.$$

$$\sigma = \Sigma p_i x_i^2 - (\mu)^2 = \left(0 + \frac{9}{22} + \frac{4}{22}\right) - \left(\frac{1}{2}\right)^2 = \frac{13}{22} - \frac{1}{4} = \frac{15}{44}.$$

Example 10. *Find mean and standard deviation for the following distribution.*

X	*1*	*2*	*3*	*– 2*	*– 1*
P(X)	*0.10*	*0.35*	*0.20*	*0.30*	*0.05*

Solution.

Here $\Sigma p_i x_i = 1(0.10) + 2(0.35) + 3(0.20) - 2(0.30) - 1(0.05) = \mathbf{0.75}.$

and $\Sigma p_i x_i^2 = 1(0.10) + 4(0.35) + 9(0.20) + 4(0.30) + 1(0.05) = \mathbf{4.55}.$

Now **mean:** $$\mu = \Sigma p_i x_i = 0.75.$$

Variance: $$\sigma^2 = \Sigma p_i x_1^2 - (\mu)^2 = 4.55 - (0.75)^2 = 3.9875.$$

$\therefore$ **Standard deviation:** $\sigma = \sqrt{3.9875} = \mathbf{1.997}$ (approximately).

BINOMIAL DISTRIBUTION

6.3 BINOMIAL DISTRIBUTION

Binomial distribution is a discrete probability distribution which is obtained when the probabililty p of the happening of an event is same in all the trials, and there are only two events in each trial.

For example, the probability of getting a head, when a coin is tossed a number of times, must remains same in each toss, *i.e.*, $p = 1/2$.

Let an experiment consisting of n trials be performed. Let the occurrence of an event in any trial be called a success and its non-occurrence a failure. Let p be the probability of success and q be the probability of its failure in a single trial, where $q = 1 - p$, so that $p + q = 1$.

Let us assume that trials are independent and the probability of success is same in each trial. Let us claim that we have n trials, then the probability of happening of an event r times and failing (n – r) times in any specified order is $p^r q^{n-r}$ (by the theorem on multiplication of probability). But the total number of ways in which the event can happen r times exactly in n trials is C (n, r). These C (n, r) ways are equally likely, mutually exclusive and exhaustive.

Therefore, **the probability for r successes and $(n - r)$ failures in n trials in any order, whatsoever is = $^nC_r p^r q^{n-r}$.**

It can also be expressed in the form

$$P(X = r) = P(r) = {}^nC_r\, p^r q^{n-r}; \qquad r = 0, 1, 2, 3, ..., n;$$

where $P(x = r)$ or $P(r)$ is the probability distribution of a random variable X of r successes. Giving different values to r, *i.e.*, putting $r = 0, 1, 2, 3, ..., n$; we get the corresponding probabilities $^nC_0 q^n$, $^nC_1 q^{n-1}p$, $^nC_2 q^{n-2} p^2$, $^nC_3 q^{n-3} p^3$, ..., p^n, **which are the different terms in the Binomial expansion of $(q + p)^n$.**

As a result of it, the distribution $P(r) = C(n, r) p^r q^{n-r}$ **is called Binomial Probability Distribution.** The two independent constants, *viz.*, n and p in **the distribution are called the parameters of the distribution.**

Again if the experiment (each consisting of n trials) be repeated N times, the **Frequency Function of the Binomial Distribution** is given by

$$f(r) = N \times P(r) = N \times C(n, r)\, p^r q^{(n-r)}$$

The expected frequencies of 0, 1, 2, 3, ..., n successes in the above set of experiment are the successive terms in the Binomial expansion of $N(q + p)^n$; where $p + q = 1$, which is also called the Binomial Frequency Distribution.

6.4 MEAN AND VARIANCE OF BINOMIAL DISTRIBUTION

The probability distribution of Binomial distribution for r successes in n trials is given by

$$P(r) = {}^nC_r\, q^{n-r} p^r.$$

It has **MEAN = np** and **VARIANCE = npq.**

6.5 CONDITIONS FOR APPLICATION OF BINOMIAL DISTRIBUTION

1. *The variable should be discrete, i.e., the values of x could be 1, 2, 3, 4 or 5 etc. and never 1.5, 2.1 or 3.41 etc.*

2. *A dichotomy exists. In other words, the happening of events must have two alternatives. It must be either a success or a failure.*

3. *The number of trials 'n' should be finite and small.*

4. *The trials or events must be independent. The happening of one event must not affect the happening of other events. In other words, Statistical Independence must exist.*
5. *The trials or events must be repeated under identical conditions.*

6.6 EXPECTED OR THEORETICAL FREQUENCIES OF BINOMIAL DISTRIBUTION

Let the experiment of repeating n trials be performed N times. The probability of 'r' successes in n trials is

$$P(r) = {}^nC_r \, p^r \, q^{n-r}, \; r = 0, 1, 2, 3, ..., n.$$

is same for every experiment. Out of N performances of the experiment, the number of times we expect r successes to occur is

$$f(r) = N \times P(r) = N \times {}^nC_r \, p^r \, q^{(n-r)}.$$

This is called the **Expected or Theoretical frequency of a success.**

Obviously, $$\sum_{r=0}^{n} f(r) = N \sum_{r=0}^{n} P(r) = N. \qquad [\because \sum_{r=0}^{n} P(r) = 1]$$

The expected or theoretical frequencies $f(0), f(1), f(2), f(3), ...; f(n)$ of 0, 1, 2, 3, ..., n successes are respectively the first, second, third, ..., $(n + 1)$th term in the expansion of $N(q + p)^n$.

In practice, the **expected frequencies differ from observed frequencies due to chance factor.**

6.7 MODE OF BINOMIAL DISTRIBUTION

Mode *is that value of the variate X which maximises the probability function.* Thus if $X = r$ gives the mode of distribution, then

$$P(r) > P(r-1) \quad \textbf{and} \quad P(r) < P(r+1).$$

Working Rule to Find Mode of Binomial Distribution.

Let X be a Binomial variate with parameters n and p.

CASE I. *When $(n + 1)p$ is an integer = k, (say).*

In this case the distribution is bi-modal, the two modal values are: $X = k$ and $X = k - 1$.

CASE II. *When $(n + 1)p$ is not an integer.*

Let $(n + 1)p = k_1 + f$,

where k_1 is the integral part and f is the fractional part of $(n + 1)p$. In this case the distribution has a **unique mode at $X = k_1$, the integral part of $(n + 1)p$.**

***Illustration* 1.** If $n = 7$ and $p = 0.5$, then $(n + 1)p = 8 \times 0.5 = 4$, which is an **integer.**

Hence in this case the distribution is bi-modal, the two modal values being 4 and 4 – 1 = 3. Hence, the two modal values are 3 and 4.

***Illustration* 2.** If $n = 6$ and $p = 0.3$ then $(n + 1)p = 7 \times 0.3 = 2.1$, which is not an integer, then mode is the integral part *i.e.*, the integral part of 2.1, which is 2. **Hence the modal value is 2.**

Example 11. *Find the most probable number of times that an ace will appear if a die is tossed (i) 50 times; (ii) 53 times.*

Solution. (*i*) Here $n = 50$, p = probability of ace with a die = $\frac{1}{6}$.

$$\therefore \quad (n+1)p = 51 \times \frac{1}{6} = \frac{17}{2} = 8.5, \text{ which is not an integer}$$

∴ Most probable number of times an ace is obtained is the integral part of $(n + 1)\, p = 8.3$, which is 8.

Hence the mode is 8.

(*ii*) Here $\quad n = 53, \quad p = \dfrac{1}{6}$

$\therefore \quad (n + 1)\, p = 54 \times \dfrac{1}{6} = 9$, which is an integer.

∴ Most probable number of times an ace is obtained is x, where $x = (n + 1)\, p,\ (n + 1)\, p - 1$. *i.e.*, 9 and 8, both being equally probable.

Hence it has two modes, *viz.*, 9 and 8.

6.8 CHARACTERISTICS OF BINOMIAL DISTRIBUTION

1. *It is a discrete distribution which gives the theoretical probabilities.*
2. *It depends on the parameters p or q, the probability of success or failure and n (the number of trials). The parameter n is always a positive integer.*
3. *The distibution will be symmetrical if p = q. It is skew-symmetric or asymmetric if p ≠ q although with n tending to be large it is approximately so.*
4. *The statistics of the Binomial distribution are mean = np; variance = npq; and standard deviation =* $\sqrt{npq}$.
5. *The mode of the Binomial distribution is equal to that value of X which has the largest frequency.*
6. *It can be represented graphically, taking the X-axis to represent the number of successes and Y-axis to represent the probabilities or frequencies. Its graph will be vertical lines, with spaces in between them. Drawing a smooth curve by free hand is inadmissible as the variable X is a discrete one.*
7. *The shape and location of a Binomial distribution changes as p changes for a given n or n changes for a given p.*
8. *The Binomial coefficients are given by the Pascal's Triangle.*

6.9 RECURSION FORMULA OR RECURRENCE RELATION FOR BINOMIAL DISTRIBUTION

$P(r + 1) = \dfrac{p}{q} \cdot \dfrac{n-r}{r+1} \cdot P(r)$ **is the required recurrence relation for the Binomial distribution.**

Example 12. *In a Binomial distribution, the mean and standard deviations are 12 and 2 respectively. Find n and p.*

Solution. We are given that

$$\text{Mean} = np = 12, \quad \text{S.D.} = \sqrt{npq} = 2 \Rightarrow npq = 4.$$

Now $\quad \dfrac{npq}{np} = \dfrac{4}{12} = \dfrac{1}{3} \Rightarrow \mathbf{q = \dfrac{1}{3}}.$

$\therefore \quad \mathbf{p = 1 - q} = 1 - 1/3 = \mathbf{2/3.}$

Also $\quad \mathbf{np} = 12 \Rightarrow n \times (2/3) = 12 \Rightarrow \mathbf{n = 18.}$

Hence $\quad \mathbf{n = 18,\ p = 2/3.}$

Example 13. *Using the formula for Binomial distribution, find the probability of rolling at most 2 sixes in 5 rolls of a dice.*

Solution. Here $n = 5$

Probability of rolling a six on a dice $= p = \frac{1}{6}$, $\therefore\ q = \frac{5}{6}$:

$P(x \le 2) = P(x = 0) + P(x = 1) + P(x = 2)$

$$= {}^5C_0\left(\frac{1}{6}\right)^0\left(\frac{5}{6}\right)^5 + {}^5C_1\left(\frac{1}{6}\right)^1\left(\frac{5}{6}\right)^4 + {}^5C_2\left(\frac{1}{6}\right)^2\left(\frac{5}{6}\right)^3 = \left(\frac{5}{6}\right)^5 + 5\left(\frac{1}{6}\right)\left(\frac{5}{6}\right)^4 + \frac{5\times4}{2\times1}\left(\frac{1}{6}\right)^2\cdot\left(\frac{5}{6}\right)^3$$

$$= \frac{5^4}{6^5}[5+5+2] = 12\times\frac{5^4}{6^5} = \frac{625}{648}.$$

Example 14. *If 10 coins are tossed 100 times, how many times would you except 7 coins to fall head upward ?*

Solution. The probability of getting a head in a single toss of a coin is $p = 1/2$.

$\therefore$ $q = 1 - p = 1 - 1/2 = 1/2.$

Also we are given $n = 10, N = 100$, and $r = 7$.

$\therefore$ **The required frequency** $= N \times {}^nC_r\, p^r\, q^{n-r}$

$$= 100\times{}^{10}C_7\left(\frac{1}{2}\right)^7\left(\frac{1}{2}\right)^3 = 100\frac{10!}{7!\times3!}\left(\frac{1}{2}\right)^{10} = 100\times\frac{10\times9\times8}{3\times2\times1}\left(\frac{1}{2}\right)^{10} = \frac{375}{32} = 11.7 \simeq 12.$$

Example 15. *There are 64 beds in a garden and 3 seeds of a particular type of flower are sown in each bed. The probability of a flower being white is 1/4. Find the number of beds with 3, 2, 1 and 0 white flowers.*

Solution. The probability p of a white flower $= \frac{1}{4}$.

$\therefore\ q = 1 - p = 1 - \frac{1}{4} = \frac{3}{4}$. Here $n = 3, N = 64$. $f(r) = N \times C(3, r)\left(\frac{1}{4}\right)^r\left(\frac{3}{4}\right)^{3-r}$

Beds with zero white flower $= Nf(0) = 64\times C(3,0)\left(\frac{1}{4}\right)^0\left(\frac{3}{4}\right)^3 = 64\times\frac{27}{64} = 27.$

Beds with 1 white flower $= Nf(1) = 64f(1)$

$$= 64 \times C(3,1)\left(\frac{1}{4}\right)^1\left(\frac{3}{4}\right)^2 = 64\times3\times\frac{1}{4}\times\frac{9}{16} = \frac{27}{64}\times64 = 27.$$

Beds with 2 white flowers $= Nf(2) = 64\times C(3,2)\left(\frac{1}{4}\right)^2\left(\frac{3}{4}\right) = 64\times\frac{9}{64} = 9.$

Beds with 3 white flowers $= Nf(3) = 64\times C(3,3)\left(\frac{1}{4}\right)^3\left(\frac{3}{4}\right)^0 = 64\times\frac{1}{64} = 1.$

6.10 FITTING OF A BINOMIAL DISTRIBUTION

The **probability of 0, 1, 2, 3, ..., n successes would be obtained from the Binomial expansion of $(q + p)^n$.**

The **probability of r successes in a single throw is given by :** $P(r) = {}^nC_r\, p^r\, q^{n-r}$

Suppose this experiment is repeated for N times, then

the frequency of r success is $= N \times P(r) = N \times {}^nC_r\, q^{n-r}\, p^r$

Putting r = 0, 1, 2..... n, we get the expected or theoretical frequencies of the binomial distribution as follows:

Number of successes : (r)	*Expected or Theoretical frequency : N P(r)*
0	$N q^n$
1	$N \times {}^nC_1\, q^{n-1}\, p$
2	$N \times {}^nC_2\, q^{n-2}\, p^2$
r	$N \times {}^nC_r\, q^{n-r}\, p^r$
n	$N p^n$

Example 16. *As a result of a certain experiment, the data obtained was:*

x :	*0*	*1*	*2*	*3*	*4*
f:	*8*	*32*	*34*	*24*	*5*

Fit a Binomial distribution to the above data.

Solution. Mean : $\quad \boldsymbol{np} = \dfrac{0+32+68+72+20}{103} = \dfrac{192}{103} = \mathbf{1.864} \Rightarrow p = \dfrac{1.864}{n}\cdot$

$\therefore \quad \boldsymbol{p} = \dfrac{1.864}{4} = \mathbf{0.466}$; and $\boldsymbol{q} = \mathbf{1} - \boldsymbol{p} = 1 - 0.466 = \mathbf{0.534.}$

Expected frequencies are given by : $N\,(q + r)^n = 103\,(0.534 + 0.466)^4$.

Putting $r = 0, 1, 2, 3, 4$, we get:

$\mathbf{103(0.534)^4;\ 103(0.534)^3\,(0.466);\ 103(0.534)^2\,(0.466)^2;\ 103(0.534)\,(0.466)^3;\ 103(0.466)^4.}$

Example 17. *Find the prabability of success p for a Binomial distribution, if n = 6, and 4 × P (X = 4) = P (X = 2).*

Solution. We know that in **binomial distribution, the probability of *r* successes in *n* trials is given by**

$$\boldsymbol{P\,(X = r) = {}^nC_r\, p^r\, q^{n-r}.}$$

According to the given condition :

$\mathbf{4}\boldsymbol{P\,(X = 4) = P\,(X = 2)} \Rightarrow 4\,({}^6C_4\, p^4\, q^2) = {}^6C_2\, p^2\, q^4$

$\Rightarrow 4p^2 = q^2 \Rightarrow 4p^2 = (1 - p)^2 \quad [\because {}^6C_4 = {}^6C_2 \text{ and } q = 1 - p]$

$\Rightarrow 3p^2 + 2p - 1 = 0$ or $p = \dfrac{-2 \pm \sqrt{4+12}}{6} = -1$ or $\mathbf{\dfrac{1}{3}}\cdot$

Since *p* cannot be negative so $\boldsymbol{p} = \mathbf{\dfrac{1}{3}}\cdot$

Example 18. *The overall percentage of failures in a certain examination is 30. What is the probability that out of a group of 6 candidates at least 4 passed the examination?*

Solution. Probability of a student failing in the examination : $\boldsymbol{p} = \mathbf{0.30}$

Probability of a student passing the examination : $\boldsymbol{q} = 1 - p = \mathbf{0.70.}$

Probability that out of 6 candidates at least 4 passed the examination

$= {}^6C_4\,(0.7)^4\,(0.3)^{6-4} + {}^6C_5\,(0.7)^5\,(0.3)^{6-5} + {}^6C_6\,(0.7)^6\,(0.3)^{6-6}$

$= (0.7)^4\left[\dfrac{6!}{4!\times 2!}(0.3)^2 + \dfrac{6!}{5!\times 1!}(0.7)\,(0.3) + \dfrac{6!}{6!\times 0!}(0.7)^2\right]$

$= (0.7)^4\left[\dfrac{6\times 5}{2\times 1}\times 0.09 + 6\times 0.21 + 0.49\right] = 0.2401\,[1.35 + 1.26 + 0.49] = \mathbf{0.74431.}$

Example 19. *Find the binomial distribution if the mean is 12 and the standard deviation is 2.*

Solution. Here **mean :** $\bar{X} = 12 = np.$ ... (*i*)

Variance : $\sigma^2 = npq = 4.$ ... (*ii*)

Now (*i*) and (*ii*) $\Rightarrow \frac{npq}{np} = \frac{4}{12} \Rightarrow q = \frac{1}{3}. \therefore p = 1 - q = 1 - 1/3 = 2/3.$

Again $np = 12 \Rightarrow n \Rightarrow 12 \times \frac{3}{2} = 18$

Therefore, the distribution is $f(x) = {}^{18}C_x\,(2/3)^x\,(1/3)^{18-x}$, $x = 0, 1, \ldots\ldots, 18$

Example 20. *An experiment succeeds twice as often as it fails. What is the probability that in next five trials there will be*

(*i*) *three successes,* (*ii*) *at least three successes.*

Solution. Let p and q be the probabilities of successes and failure respectively.

$$\therefore \quad p = 2q \Rightarrow p = 2(1-p) \Rightarrow p = \frac{2}{3} \text{ and } q = \frac{1}{3}.$$

(*i*) **The probability of getting three successes** $= P(X = 3) = {}^5C_3\left(\frac{2}{3}\right)^3\left(\frac{1}{3}\right)^2 = \frac{80}{243}.$

(*ii*) **Probability of getting at least three successes** $= P(X = 3) + P(X = 4) + P(X = 5)$

$$= {}^5C_3\left(\frac{2}{3}\right)^3\left(\frac{1}{3}\right)^2 + {}^5C_4\left(\frac{2}{3}\right)^4\left(\frac{1}{3}\right) + {}^5C_5\left(\frac{2}{3}\right)^5 = \frac{80}{243} + \frac{80}{243} + \frac{32}{243} = \frac{192}{243}.$$

Example 21. *Find the Binomial distribution whose mean is 3 and variance is 2.*

Solution. Let $(q+p)^n$ be the Binomial distribution. We are given that its mean is 3 and variance is 2.

But **mean :** $np = 3$ ... (*i*) and **variance :** $npq = 2.$ ... (*ii*)

Dividing (*ii*) by (*i*), we get : $\frac{npq}{np} = \frac{2}{3} \Rightarrow q = \frac{2}{3}. \therefore p = 1 - q = 1 - \frac{2}{3} = \frac{1}{3}.$ ... (*iii*)

Now (*i*) and (*iii*) $\Rightarrow n = 9.$

Hence, **the distribution is** $\left(\frac{2}{3} + \frac{1}{3}\right)^9.$ **[Binomial Distribution = $(q+p)^n$]**

Example 22. *Bring out the fallacy, if any in the statement: "The mean of a Binomial distribution is 5 and its variance is 9".*

Solution. Let $(q+p)^n$, $q + p = 1$, be the Binomial distribution so that

Mean $= np = 5$ ... (*i*) **Variance** $= npq = 9.$ ... (*ii*)

Dividing (*ii*) by (*i*), we have : $\frac{npq}{np} = \frac{9}{5} \Rightarrow q = \frac{9}{5} > 1,$

which is wrong because we know that $0 \le q \le 1$. Hence the given statement is false.

Example 23. *If the probability of a defective bolt is 0.1, find the mean and standard deviation for the distribution of defective bolt in a total of 500.*

Solution. Let $(q+p)^n$, $q + p = 1$, be the Binomial distribution. Here we are given that

$$p = 0.1, n = 500$$

$\therefore$ **Mean** $= np = 0.1 \times 500 = 50.$

Now $p = 0.1 \Rightarrow q = 1 - p = 1 - 0.1 = 0.9.$

$\therefore$ **Variance** $= npq = 500 \times 0.1 \times 0.9 = 45.$

$\therefore$ **Standard deviation** $= \sqrt{45} = \mathbf{6.7}.$

Example 24. *Six dice are thrown 729 times. How many times do you expect at least three dice to show a five or a six?*

Solution. Let $N(q+p)^n$, $q+p=1$, be the Binomial function

Now p = the probability of getting 5 or 6 with one dice $= \frac{2}{6} = \frac{1}{3}$.

$\therefore$ $q = 1 - p = 1 - \frac{1}{3} = \frac{2}{3}$. Here $N = 729$, $n = 6$.

The Binomial distribution is $N(q+p)^n = 729\left(\frac{2}{3}+\frac{1}{3}\right)^6$.

The expected number of times at least three dice showing 5 or 6,

$$= 729\left[{}^6C_3\left(\frac{2}{3}\right)^3\left(\frac{1}{3}\right)^3 + {}^6C_4\left(\frac{2}{3}\right)^2\left(\frac{1}{3}\right)^4 + {}^6C_5\left(\frac{2}{3}\right)\left(\frac{1}{3}\right)^5 + {}^6C_6\left(\frac{1}{3}\right)^6\right]$$

$$= \frac{729}{3^6}[160+60+12+1] = \mathbf{233}.$$

Example 25. *Five coins are tossed 3,200 times. Find the frequencies of the distribution of heads and tails and tabulate the result. Calculate the mean number of successes and standard deviation.*

Solution. $$P(x) = C(5,x)\left(\frac{1}{2}\right)^x\left(\frac{1}{2}\right)^{5-x} = C(5,x)\left(\frac{1}{2}\right)^5$$

Frequency of x heads $= f(x) = N\,P(x) = \mathbf{3200 \times C(5,x)\left(\frac{1}{2}\right)^5; x = 0, 1, 2, 3, 4, 5;}$

Mean $= 5 \times \frac{1}{2} = 2.5$, **S.D.** $= \sqrt{npq} = \sqrt{5 \times \frac{1}{4}} = \mathbf{1.118}.$

Example 26. *Eight coins are thrown simultaneously. Find the probability of getting at least six heads.*

Solution. Let p denote the probability of getting a head; q be the probability of not getting a head, then $p = 1/2$, $q = 1/2$.

$\therefore$ **Probability of getting at least six heads when 8 coins are thrown simultaneously**
$= P(6) + P(7) + P(8)$

$$= {}^8C_6\left(\frac{1}{2}\right)^6\left(\frac{1}{2}\right)^2 + {}^8C_7\left(\frac{1}{2}\right)^7\left(\frac{1}{2}\right) + {}^8C_8\left(\frac{1}{2}\right)^8 = \left(\frac{1}{2}\right)^8 [{}^8C_6 + {}^8C_7 + {}^8C_8]$$

$$= \frac{1}{256}[28+8+1] = \frac{\mathbf{37}}{\mathbf{256}}.$$

Example 27. *Assuming that half the population is consumer of rice so that the chance of an individual being a rice consumer is 1/2 and assuming that 100 investigations each take 10 individuals to see whether they are rice consumers. How many investigations would you except to report that three people or less were consumers ?*

Solution. Here $p = \frac{1}{2}$, $q = \frac{1}{2}$, $n = 10$, $N = 100$.

∴ The probability that r persons out of 10 persons are consumers of rice is given by

$$P(r) = {}^{10}C_r \left(\frac{1}{2}\right)^r \left(\frac{1}{2}\right)^{10-r}$$

∴ **The expected number of investigations (*i.,e.*, expected frequencies) who would report that three or less people were consumers of rice = 100 [$P(0) + P(1) + P(2) + P(3)$]**

$$= 100\left[{}^{10}C_0\left(\frac{1}{2}\right)^{10} + {}^{10}C_1\left(\frac{1}{2}\right)^{10} + {}^{10}C_2\left(\frac{1}{2}\right)^{10} + {}^{10}C_3\left(\frac{1}{2}\right)^{10}\right]$$

$$= \frac{100}{2^{10}}[1+10+45+120] = \frac{17600}{1024} = \mathbf{17} \text{ approximately.}$$

Example 28. *A dice is tossed thrice. If getting a four is considered a success, find the mean and variance of the probability distribution of the number of successes.*

Solution. Let $(q + p)^n$, $q + p = 1$, be the Binomial distribution.

Here we are given that : $n = 3$, $p = \frac{1}{6}$, $q = 1 - \frac{1}{6} = \frac{5}{6}$.

$$\textbf{Mean : } np = 3 \times \frac{1}{6} = \frac{1}{2}. \qquad \textbf{Variance : } npq = 3 \times \frac{1}{6} \times \frac{5}{6} = \frac{5}{12}.$$

Example 29. *Determine the Binomial distribution whose mean is 9 and whose standard deviation is 3/2.*

Solution. Let $(q + p)^n$, $q + p = 1$, be the Binomial distribution.

Mean : $np = 9$. ... (1) **Variance = (Standard deviation)² = $npq = \frac{9}{4}$.** ... (2)

Dividing (2) by (1), we get :

$$\frac{npq}{np} = \frac{1}{4} \Rightarrow q = \frac{1}{4} \qquad \therefore\ p = 1 - q = 1 - \frac{1}{4} = \frac{3}{4}. \qquad \text{... (3)}$$

From (1) and (3) $\Rightarrow n = 12$.

Hence the Binomial distribution is $(q + p)^n = \left(\frac{1}{4} + \frac{3}{4}\right)^{12}$

Example 30. *If the sum of mean and variance of a Binomial distribution is 4.8 for five trials, find the distribution.*

Solution. In a Binomial distribution: **Mean = np ; Variance = npq.**

Mean + Variance = $np + npq$=4.8 $\Rightarrow$ $np(1 + q) = 4.8$.

$\therefore \quad 5p[1 + (1 - p)] = 4.8,$ $\quad [\because n = 5]$

or $\quad 5p(2 - p) = 4.8$, or $10p - 5p^2 = 4.8$.

or $\quad 5p^2 - 10p + 4.8 = 0$, or $p^2 - 2p + 0.96 = 0$

$$\Rightarrow \quad p = \frac{2 \pm \sqrt{4 - 4 \times 1 \times 0.96}}{2 \times 1} = \frac{2 \pm \sqrt{4 - 3.84}}{2} = \frac{2 \pm \sqrt{0.16}}{2} = \frac{2 \pm 0.4}{2} = \frac{2.4}{2}, \frac{1.6}{2}$$

$\Rightarrow \quad p = 1.2, 0.8.$ But $p \neq 1.2$ as $0 \leq p \leq 1$,

$$\therefore \quad p = 0.8 = \frac{8}{10} = \frac{4}{5}; \quad q = 1 - p = 1 - \frac{4}{5} = \frac{1}{5}.$$

Hence tne required distribution is $(q + p)^n = \left(\frac{1}{5} + \frac{4}{5}\right)^5$.

Example 31. *If the probability that a mcn aged 60 will live to be 70 is 0.65, what is the probability that out of 10 men now 60, at least 7 will up to 70 ?*

Solution. Probability of a man living up to 70 = 0.65 = $\frac{65}{100} = \frac{\mathbf{13}}{\mathbf{20}} = \boldsymbol{p}$ *(say)*

$\therefore$ Probability of dying = $1 - \frac{13}{20} = \frac{\mathbf{7}}{\mathbf{20}} = \boldsymbol{q}$ (say).

Total number of men = 10 and atleast **7** will live up to **70.** We have the following **four cases:**

Case I. The probability of 7 living and 3 dying = $\boldsymbol{{}^{10}C_7 p^7 q^3}$

$$= \frac{(10)!}{7!3!}\left(\frac{13}{20}\right)^7\left(\frac{7}{20}\right)^3 = \frac{10\times9\times8}{3\times2\times1}\times\left(\frac{13}{20}\right)^7\times\left(\frac{7}{20}\right)^3 = \boldsymbol{p_1}.$$

Case II. The probability of 8 living and 2 dying = $\boldsymbol{{}^{10}C_8 p^8 q^2} = \frac{10\times9}{2\times1}\left(\frac{13}{20}\right)^8\left(\frac{7}{20}\right)^2 = \boldsymbol{p_2}.$

Case III. The probability of 9 living and 1 dying = $\boldsymbol{{}^{10}C_9 \left(\frac{13}{20}\right)^9\left(\frac{7}{20}\right)} = 10\left(\frac{13}{20}\right)^9\left(\frac{7}{20}\right) = \boldsymbol{p_3}.$

Case IV. The probability of all the 10 living = $\boldsymbol{{}^{10}C_{10} p^{10}} = \left(\frac{13}{20}\right)^{10} = \boldsymbol{p_4}$

Required probability = $\boldsymbol{p_1 + p_2 + p_3 + p_4}$

$$= \frac{10\times9\times8}{3\times2\times1}\times\left(\frac{13}{20}\right)^7\times\left(\frac{7}{20}\right)^3 + \frac{10\times9}{2\times1}\left(\frac{13}{20}\right)^8\left(\frac{7}{20}\right)^2 + 10\left(\frac{13}{20}\right)^9\left(\frac{7}{20}\right) + \left(\frac{13}{20}\right)^{10}$$

$$= \left(\frac{13}{20}\right)^7\left[\frac{120\times343}{8000} + \frac{45\times13\times49}{8000} + \frac{10\times169\times7}{8000} + \frac{2197}{8000}\right]$$

$= (0.65)^7\ [5.145 + 3.583 + 1.479 + 0.275]$

$= (0.65)^7\ (10.482) = 0.04902 \times 10.482 =$ **0.5139.**

Hence the **required probability = 0.5139.**

Example 32. *The sum and the product of the mean and variance of a Binomial distribution are 24 and 128. Find the distribution.*

Solution.

Here $\qquad \boldsymbol{np + npq = 24},\quad$ *i.e.*, $np\,(1 + q) = 24 \qquad$ ... **(1)**

and $\qquad \boldsymbol{np \times npq = 128},\quad$ *i.e.*, $n^2p^2q = 128. \qquad$ ... **(2)**

From (1) $\qquad np = \frac{24}{1+q} \Rightarrow n^2p^2 = \left(\frac{24}{1+q}\right)^2.$

Putting this value of $n^2\,p^2$ in (2), we have :

$$\left(\frac{24}{1+q}\right)^2 q = 128 \Rightarrow \frac{576}{(1+q)^2}q = 128 \Rightarrow 9q = 2\,(1+q)^2$$

or $9q = 2\,(1 + 2q + q^2) \Rightarrow 2q^2 - 5q + 2 = 0 \Rightarrow (2q - 1)\,(q - 2) = 0$

$\Rightarrow \qquad \boldsymbol{q = \frac{1}{2}}\ :\ \boldsymbol{\therefore p = 1 - q = \frac{1}{2}}. \qquad [\because \boldsymbol{q \neq 2}]$

Putting $p = \frac{1}{2}$ in (1), we get :

$$n\times\frac{1}{2}\times\left(1+\frac{1}{2}\right) = 24 \Rightarrow n \times \frac{1}{2}\left(\frac{3}{2}\right) = 24 \Rightarrow n\times\frac{3}{4} = 24 \text{ or } \boldsymbol{n} = \frac{24\times 4}{3} = \mathbf{32.}$$

Hence **the Binomial Distribution is $(q + p)^n$ =** $\left(\frac{1}{2}+\frac{1}{2}\right)^{32}$

Example 33. *What the recursion formula for probabilities in the Binomial distribution with parameters n and p ? A dice is thrown 5 times. Getting a number 5 on the dice is considered a success. Calculate P (r) for r = 1, 2, 3, 4 from the recursion formula.*

Solution. Recursion formula : $P(r + 1) = \frac{n-r}{r+1} \cdot \frac{p}{q} \cdot P(r)$

Here $p = \frac{1}{6},\ q = \frac{5}{6},\ n = 5,\ P(0) = \left(\frac{5}{6}\right)^5 = \frac{3125}{7776}$

$$P(1) = P(0) = \frac{3125}{7176}; \quad P(2) = \frac{5-2}{2+1}\times\frac{1}{5}\times P(1) = \frac{625}{3888}.$$

Similarly, $P(3) = \frac{125}{3888}; \quad P(4) = \frac{25}{7776}$

Example 34. *Suppose A and B are two equally strong table tennis players. Which of the following two events is more probable :*

(a) A beats B exactly in 3 games out of 4 or

(b) A beats B exactly in 5 games out of 8 ?

Solution. Here $p = q = 1/2$

(*a*) Probability of *A* beating *B* in exactly 3 games out of 4

$$= {}^4C_3\left(\frac{1}{2}\right)^3\left(\frac{1}{2}\right) = 4\times\frac{1}{8}\times\frac{1}{2} = \frac{1}{4} = \mathbf{25\%}$$

(*b*) Probability of *A* beating *B* in exactly 5 games out of 8

$$= {}^8C_5\left(\frac{1}{2}\right)^5\left(\frac{1}{2}\right)^3 = 56\times\frac{1}{32}\times\frac{1}{8} = \frac{7}{32} = \mathbf{21.875\%}.$$

Clearly, the **first event is more probable.**

Example 35. *The probability that an evening college student will be graduate is 0.4. Determine the probability that out of 5 students (a) none, (b) one, and (c) at least one will be graduate.*

Solution. Here $p = 0.4$, $q = 0.6$. Applying the Binomial distribution : $P(r) = {}^nC_r\, q^{n-r}\, p^r$

(*a*) Probability that none will graduate is: $P(0) = {}^5C_0\,(0.6)^5\,(0.4)^0 = \mathbf{0.0777}$ or **0.08.**

(*b*) Probability that one will be graduate is:

$$P(1) = {}^5C_1\,(0.6)^4\,(0.4)^1 = \mathbf{0.26.}$$

(*c*) Probability that atleast one will be graduate

= 1 – (Probability that none will be graduate) = 1 – P (0) = 1 – 0.08 = **0.92.**

Example 36. *Is there any fallacy in the statement "the mean of a Binomial distribution is 20 and its standard deviation 7"*

Solution. **Mean** $= np = 20$

Standard Deviation: $\sigma = \sqrt{npq} = 7 \Rightarrow \sigma^2 = npq = 7^2 = 49$

Now $\dfrac{npq}{np} = \dfrac{49}{20} \Rightarrow q = 2.45.$

The value of q lies between 0 and 1. But here the value of q exceeds one.

So, the given statement is wrong.

Example 37. *Three per cent of a given lot of manufactured parts are defective. What is the probability that in a sample of four items none will be defective ?*

Solution. Probability of defective item : $p = 0.03$

$$q = 1 - p = 1 - 0.03 = 0.97. \text{ Also } n = 4.$$

Binomial Distribution is: $P(r) = {}^nC_r\, p^r\, q^{n-r}$.

Probability of getting 0 defective by binomial distribution is:

$$P(0) = {}^4C_0 (0.03)^0 (0.97)^4 = (0.97)^4 = \mathbf{0.885}.$$

Example 38. *The incidence of occupational disease in an industry is such that the workmen have a 25% chance of suffering from it. What is the probability that out of six workmen 4 or more will contact the disease ?*

Solution. Let p denotes chance of suffering and q chance of not suffering

$$\therefore \quad p = 25\% = \frac{1}{4} \text{ and } q = \frac{3}{4}$$

The Binomial distribution is : $P(x = r) = {}^nC_r\, p^r q^{n-r}$

The probability of 4 or more (*i.e.*, 4, 5 and 6) successes = $P(4) + P(5) + P(6)$

$$= 15q^2p^4 + 6qp^5 + p^6 = 15\left(\frac{1}{4}\right)^4\left(\frac{3}{4}\right)^2 + 6\left(\frac{1}{4}\right)^5\left(\frac{3}{4}\right) + \left(\frac{1}{4}\right)^6$$

$$= \frac{15 \times 9}{4096} + \frac{6 \times 3}{4096} + \frac{1}{4096} = \frac{135 + 18 + 1}{4096} = \frac{154}{4096} = \mathbf{0.0376}.$$

Example 39. *A student obtained the following answer to a certain problem given to him. Mean = 2.4; Variance = 3.2 for a binomial distribution. Comment on the result.*

Solution. The mean of binomial distribution is np and variance = npq.

We are given mean $= np = 2.4$; and Variance $= npq = 3.2$

$$\therefore \quad npq = 3.2 \Rightarrow 2.4\, q = 3.2 \Rightarrow q = \frac{3.2}{2.4} = \mathbf{1.333}. \qquad [\textbf{Note.} \because p + q = 1]$$

As the value of q is greater than one, so the given result is inconsistent.

Example 40. *If hens of a certain breed lay eggs on 5 days a week on an average, find how many days during a season of 100 days a poultry keeper with 5 hens of this breed, will expect to receive atleast 4 eggs ?*

Solution. In the usual notations we have ; $n = 5$, $N = 100$.

p = Probability that a hen of the given breed lays egg on any day = 5/7

$$\Rightarrow \quad q = 1 - p = 1 - 5/7 = 2/7$$

The probability that out of $n = 5$ hens, r hens lay eggs on any given day is given by:

$$P(r) = P(X = r) = {}^nC_r\, p^r\, q^{n-r} = {}^5C_r\, (5/7)^r\, (2/7)^{5-r} \qquad \text{... (1)}$$

Hence the probability that a poultry keeper with $n = 5$ hens of the given breed will receive at least 4 eggs on any day is given by :

$$P(X \geq 4) = P(X = 4) + P(X = 5) = P(4) + P(5) = {}^5C_4\,(5/7)^4\,(2/7) + {}^5C_5\,(5/7)^5 \qquad \text{[From (1)]}$$

$$= \frac{1}{7^5}\left[5 \times 5^4 \times 2 + 5^5\right] = \left(\frac{5}{7}\right)^5 [2+1]$$

$$= (0.7143)^5 \times 3 = 0.1859 \times 3$$

$$= \mathbf{0.5577.}$$

Hence the number of days during a season of 100 days, a poultry keeper of 5 hens will expect to receive at least 4 eggs is :

$$N \times P(X \geq 4) = 100 \times 0.5577 = 55.77 \simeq \mathbf{58.}$$

Example 41. *In a binomial distribution consisting of 5 independent trials, probabilities of 1 and 2 successes are 0.4096 and 0.2048. Find the parameter 'p' of the distribution.*

Solution. $n = 5,\ P(r = 1) = 0.4096,\ P(r = 2) = 0.2048$

$$P(r) = {}^nC_r\, q^{n-r}\, p^r;\ P(r = 1) = {}^5C_1\, q^4 p = 5q^4 p\ ;\ P(r = 2) = {}^5C_2\, q^3 p^2 = 10 q^3 p^2.$$

$$\frac{P(r=2)}{P(r=1)} = \frac{10q^3p^2}{5q^4p} = \frac{0.2048}{0.4096} \Rightarrow \frac{2p}{q} = \frac{1}{2}$$

$$\Rightarrow 4p = q,\ 4p = 1 - p \Rightarrow 5p = 1$$

$$\Rightarrow \boldsymbol{p = 1/5 = 0.2.}$$

Example 42. *Show that for the Binomial distribution with parameters n and p :*
(a) Variance is less than mean ; (b) Variance cannot exceed n/4.

Solution. (*a*) **Variance** $= npq = [np \times q] < np =$ **Mean** $[\because 0 < q < 1]$

(*b*) Variance $= npq = np(1 - p) = n(p - p^2) = \boldsymbol{f(p)}$, (say)

For maxima or minima of $f(p)$ for variations in 'p', we have

$$f'(p) = 0 \Rightarrow n(1 - 2p) = 0 \Rightarrow 1 - 2p = 0 \Rightarrow p = \frac{1}{2}$$

$$f''(p) = -2n < 0$$

$\Rightarrow$ **Variance $= f(p)$ is maximum at $p = 1/2$.**

$\therefore$ Maximum value of variance of Binomial distribution is: $(npq)_{\text{at } p = 1/2}$

$$(npq)_{\text{at } p = 1/2} = n \times \frac{1}{2} \times \frac{1}{2} = \frac{\boldsymbol{n}}{\mathbf{4}}.$$

Hence Variance = *n*/4.

Example 43. *Out of 320 families with 5 children each, what percentage would be expected to have (i) 2 boys and 3 girls, (ii) at least one boy? Assume equal probability for boys and girls.*

Solution. Here $n = 5,\ N = 320.$

Probability of a boy $= p = \dfrac{1}{2}$;

Probability of a girl $= q = \dfrac{1}{2}$.

Binomial Distribution is : $\boldsymbol{P(r) = {}^nC_r\, p^r\, q^{n-r}}$

(*i*) **Probability of 2 boys and 3 girls:** $P(r=2) = {}^5C_2\left(\frac{1}{2}\right)^3\left(\frac{1}{2}\right)^2 = \frac{\mathbf{10}}{\mathbf{32}}$.

Expected number of families with 2 boys and 3 girls: $N.P(r) = 320 \times \frac{10}{32} = \mathbf{100.}$

Percentage of families expected to have 2 boys and 3 girls $= \frac{10}{32} \times 100 = \mathbf{31.25\%}$

(*ii*) **Probability of at least one boy = $P(r \geq 1)$ = 1 – probability of no boy.**

$$P(r=0) = {}^5C_0\left(\frac{1}{2}\right)^5\left(\frac{1}{2}\right)^0 = \frac{\mathbf{1}}{\mathbf{32}}.$$

$$\therefore \quad P(r \geq 1) = 1 - P(r=0) = 1 - \frac{1}{32} = \frac{31}{32}.$$

Percentage of families expected to have at least one boy $= \frac{31}{32} \times 100 = \mathbf{96.875\%}.$

Expected number of families with at least one boy: $= 320 \times \frac{31}{32} = \mathbf{310.}$

Example 44. *A machine produces on an average 20% defective bolts. A batch is accepted if a sample of 5 bolts taken from that batch contains no defective and rejected if the sample contains 3 or more defectives. In other cases, a second sample is taken. What is the probability that the second sample is required ?*

Solution. Probability of defective bolt = p = 20% = 1/5

Probability of non-defective bolt = q = 4/5

Also $n = 5$. Binomial Distribution is : $P(x = r) = {}^nC_r\, p^r\, q^{n-r}$.

The second sample is taken only when number of defective bolts is either 1 or 2.

$\therefore$ **Probability of second sample is taken = $P(r=1) + P(r=2)$**

$$= {}^5C_1(4/5)^4\,(1/5) + {}^5C_2(4/5)^3\,(1/5)^2 = \frac{5 \times 256}{3125} + \frac{10 \times 64}{3125} = \frac{1280 + 640}{3125} = \frac{1920}{3125} = \mathbf{0.6144.}$$

Example 45. *What is the probability of guessing correctly at least six of the ten asnwers in a TRUE /FALSE objective Test?*

Solution. Probability 'p' of guessing an answer correctly is

$$p = 1/2 \quad \Rightarrow \quad q = 1 - p = 1/2$$

Probability of guessing correctly x answers in 10 questions

$$P(x) = {}^nC_x\, p^x\, q^{n-x} = {}^{10}C_x (1/2)^x\, (1/2)^{10-x},\ x = 0, 1, 2, 3, ..., 10$$

$$\textbf{Required Probability} = P(6) + P(7) + P(8) + P(9) + P(10)$$

$$= (1/2)^{10}\,[{}^{10}C_6 + {}^{10}C_7 + {}^{10}C_8 + {}^{10}C_9 + {}^{10}C_{10}]$$

$$= \frac{1}{1024}\left[\frac{10\times9\times8\times7}{4!} + \frac{10\times9\times8}{3!} + \frac{10\times9}{2!} + 10 + 1\right]$$

$$= \frac{1}{1024}[210 + 120 + 45 + 10 + 1] = \frac{386}{1024} = \frac{\mathbf{193}}{\mathbf{512}}.$$

EXERCISE – 6.1

1. Bring out the fallacy, if any in the following statement

(a) The mean of a Binomial distribution is 15 and its standard deviation is 5.

(b) The mean of a Binomial distribution is 5 and its standard deviation is 3. Is there any inconsistency in the given information?

[**Hint.** $np = 5$ and variance $= npq = 3^2 = 9 \Rightarrow q = 1.5 \Rightarrow$ the statement is inconsistent.]

(c) Is there any inconsistency in the statement 'the mean of Binomial distribution is 80 and standard deviation 8'? If no inconsistency is found, what shall be the value of p, q and n?

(e) [**Hint.** Mean $= np = 80$; Variance $= (\text{S.D.})^2 = npq = 8^2 = 64$.

$$\therefore \frac{npq}{np} = \frac{64}{80} \Rightarrow q = 0.8. \text{ Also } p = 1 - q = 0.2.$$

More over Mean $= np = 80 \Rightarrow n \times 0.2 = 80 \Rightarrow n = 400$]

(d) The mean of a binomial distribution is 15 and its standard deviation is 5.

[**Hint.** Mean $= np = 15$, Variance $= npq = 25$.

$$\therefore \frac{npq}{np} = \frac{25}{15} \Rightarrow q = 1.67,$$ which is not true as the value of q cannot exceed one.

Thus the statement is wrong]

(e) Comment: For a binomial distribution, mean = 7 and variance = 11.

2. Find the binomial distribution whose mean is 6 and variance 4.

[**Hint.** $np = 6$ and $npq = 4 \Rightarrow q = 1/3$. Thus $p = 1 - q = 2/3$.

Also $np = 6 \Rightarrow n \times \frac{1}{3} = 6 \Rightarrow n = 18$. $\therefore$ Binomial distribution is: $(q+p)^n = \left(\frac{1}{3} + \frac{2}{3}\right)^{18}$.]

3. How many tosses of a coin are needed so that the probability of getting atleast one head is 87.5%.

[**Hint.** Here $p = q = 1/2$. Also $P(x = r) = {}^nC_r\, p^r q^{n-r}$. $\therefore P(r = 0) = {}^nC_0\, p^0 q^n = (1/2)^n$.

Now $P(r \geq 1) = 1 - P(r = 0) \Rightarrow 0.875 = 1 - \left(\frac{1}{2}\right)^n \Rightarrow$ **$n = 3$**.]

4. If on an average 9 ships out of 10 arrive safely to ports, find the mean and standard deviation of ships returning safely out of a total of 500 ships.

[**Hint.** $p = 0.9$, $q = 0.1$, $n = 500$;

Mean $np = 500 \times 0.9 =$ **450**. S.D. $= \sqrt{npq} = \sqrt{500 \times 0.9 \times 0.1} =$ **6.71**.]

5. Three perfect coins are tossed together. What is the probability of getting atleast one head ?

[**Hint.** P (getting atleast one head) $= 1 - P(r = 0) = 1 - {}^3C_0\,(1/2)^0\,(1/2)^3 = 1/8$.]

6. The mean of a binomial distribution is 40 and standard deviation is 6. Calculate n, p and q.

7. A binomial random variable satisfies the relation $9\,P(X = 4) = P(X = 2)$ for $n = 6$. Find the value of the parameter p.

8. For a binomial distribution, the mean and variance are respectively 4 and 3. Calculate the probability of getting a non-zero value of this distribution.

9. If we take 1280 sets each of 10 tosses of a fair coins, in how many sets should we expect to get 7 heads and 3 tails ?

10. A production unit is made up of 20 indentical components and each component has a probability of 0.25 being defective. What is the average number of defective components in a unit? Further what is the probability in a unit (i) less than 3 components are defective? (ii) exactly 3 components are defective ?

11. It is known from the past experience that 80% of the students in a school do their home work. Find the probability that during a random check of 10 students : (i) all have done their home work; (ii) at the most two have not done their homework; (iii) at least one has not done the home work.

12. An oil exploration firm finds that 5% of the test wells it drills yield a deposit of natural gas. If it drills 6 wells, find the probability that at least one well will yield gas. Simplification is not necessary.

[**Hint.** $p = 0.05, q = 0.95, n = 6, P(x) = {}^nC_x (0.05)^x (0.95)^{n-x}$

$\therefore$ $P_{(0)} = {}^9C_0 (0.05)^0 (0.95)^6 = (0.95)^6$. Required Probability = $1 - P(8) = \mathbf{1 - (0.95)^6}$.]

13. An accountant is to audit 24 accounts of a firm. Sixteen of these are of highly-valued customers. If the accountant selects 4 of the accounts at random, what is the probability that he chooses at least one highly-valued account?

[**Hint.** $p = \frac{16}{24} = \frac{2}{3}$, $q = \frac{1}{3}$, $n = 4$. Required Probability = $1 - P_{(0)} = 1 - {}^4C_0 (2/3)^0 (1/3)^4$.]

14. The incidence of occupational disease in an industry is such that the workmen have a 20% chance of suffering from it. What is the probability that out of six workmen, 4 or more will contract the disease?

[**Hint.** Here $p = \frac{1}{5}$, $q = \frac{4}{5}$, $n = 6$. $P(x = r) = {}^nC_r p^r q^{n-r}$. Required Probability = $P(4) + P(5) + P(6)$]

15. An insurance salesmen sells policies to 5 men, all of identical age and good health. According to the actuarial tables the probability that a man of this particular age will be alive 30 years hence is 2/3. Find the probability that 30 years hence (*i*) at least 1 man will be alive, (*ii*) at least 3 men will be alive, (*iii*) all 5 men will be alive, (*iv*) at most three men will be alive.

[**Hint.** $p = \frac{2}{3}$, $q = \frac{1}{3}$, $n = 5$.

$P(r) = P$ (r out of n men will be alive 30 years hence = ${}^nC_r p^r q^{n-r} = {}^5C_r (2/3)^r (1/3)^{5-r}$.

(*i*) Probability $= 1 - P(0) = 1 - \left(\frac{1}{3}\right)^5 = \frac{242}{243}$.

(*ii*) Probability $= P(3) + P(4) + P(5) = {}^5C_3\left(\frac{2}{3}\right)^3\left(\frac{1}{3}\right)^2 + {}^5C_4\left(\frac{2}{3}\right)^4\left(\frac{1}{3}\right) + {}^5C_5\left(\frac{2}{3}\right)^5 = \frac{64}{81}$

(*iii*) $P(5) = \left(\frac{2}{3}\right)^5 = \frac{32}{243}$.]

16. Assuming that it is true that 2 in 10 industrial accidents are due to fatigue, find the probability that:

(*i*) exactly 2 of 8 industrial accidents will be due to fatigue.

(*ii*) at least 2 of 8 industrial accidents will be due to fatigue.

[**Hint.** $p = \frac{2}{10} = 0.2$, $q = 0.8$, $n = 8$.

(*i*) Required Probability = $P(2) = {}^8C_2 (0.2)^2 (0.8)^6 = 0.2936$.

(*ii*) Required Probability = $P(X \geq 2) = 1 - [(0.8)^8 + {}^8C_1 (0.2)(0.8)^7] = \mathbf{0.4967}$.

ANSWERS

1. (*a*) The statement is wrong; (*b*) The statement is consistent;
(*c*) $n = 400, p = 0.2, q = 0.8$; (*d*) The statement is wrong;
(*e*) The statement is wrong

2. $\left(\frac{2}{3} + \frac{1}{3}\right)^{18}$. **3.** 3. **4.** $\bar{X} = 450; \sigma = 6.71$. **5.** (1/8).

6. $n = 400, p = 0.1, q = 0.9$. **7.** $p = (1/8)$. **8.** $1 - (3/4)^{16}$.

9. 150 sets. **10.** 5; (*i*) 0.0912; (*ii*) 0.1339.

11. (*i*) 0.1074; (*ii*) 0.6778; (*iii*) 0.8926. **12.** $1 - (0.95)^6$. **13.** $1 - (1/3)^4$.

14. 0.01696. **15.** (*i*) 242/243; (*ii*) 64/8; (*iii*) 32/243. **16.** (*i*) 0.2936; (*ii*) 0.4967.

POISSON DISTRIBUTION

6.11 POISSON DISTRIBUTION

Poisson distribution was discovered by a French Mathematician **Simen Denis Poisson in 1837. It is a discrete distribution** and is very widely used. **Poisson distribution is a limiting form of the Binomial distribution in which *n*, the number of trials, becomes very large and *p*, the probability of the success of the event is very very small such that *np* is a finite quantity and not necessarily large.** In the Binomial distribution, the probability of x successes is ${}^nC_x\, p^x\, q^{n-x}$. We require the limiting value of the expression $f(x) = {}^nC_x\, p^x\, q^{n-x}$, where $p = m/n$; and n ultimately tends to ∞ and m is a constant.

Moreover, *in Binomial distribution there is a sample of definite size and one can count the number of times a certain event is observed but in Poisson distribution one cannot count the number of times an event occurs or does not occur* such as the accident in a factory, or the number of deaths in a city in one year by a rare disease. In such cases the Binomial distribution is not applicable because we don't know the value of n in the expression $(q + p)^n$. In order to deal such situations, Poisson distribution comes to our rescue.

Definition. The probability distribution of a random variable X is said to have a Poisson distribution if it takes only non-negative values and if its distribution is given by

$$P(X = x) = \frac{m^x e^{-m}}{x!},\ x = 0, 1, 2, \ldots..,$$

where m is the parameter and $e = 2.7183 = 1 + \frac{1}{1!} + \frac{1}{2!} + \frac{1}{3!} + \frac{1}{4!} + \ldots.$

Hence the probability of 0, 1, 2, 3, ..., x successes are given by

$$e^{-m}, \frac{e^{-m} m}{1!}, \frac{e^{-m} m^2}{2!}, \ldots., \frac{e^{-m} m^x}{x!}, \text{ respectively.}$$

Also the formula of Poisson distribution for computing the Theoretical or relative frequencies of a random variable x is given by :

Theoretical Frequency $= N\left(\frac{e^{-m} m^x}{x!}\right)$, **where N is number of trials, m = mean.**

The following are the **statistical measures of the Poisson distribution.**

1. **Mean = $np = m$**
2. **Variance = $np = m$**
3. **Standard Deviation = $\sqrt{np}$ or $\sqrt{m}$**
4. **Moment measure of skewness (γ_1) = $\frac{1}{\sqrt{m}}$**
5. **Moment measure of Kurtosis (γ_2) = $\frac{1}{\sqrt{m}}$.**

6.11.1 Some Examples of Poisson Distribution

1. *The number of cars passing through a certain street (say Janpath) in a time t.*
2. *The number of deaths in a city in one year by a rare disease.*
3. *The number of defective screws per box of 100 screws.*
4. *The number of suicides or deaths by heart attack in time t.*

5. *The number of printing mistakes in each page of the first proof of a book.*
6. *The emission of radioactive (alpha) particles.*
7. *The number of air accidents in India in one year.*
8. *The number of pieces of a certain merchandise sold by a Super Bazar in time t.*
9. *The number of telephone calls received at a particular telephone exchange in some unit of time.*
10. *The number of defective materials in a packing material manufactured by a goods concern.*
11. *The number of fragments received by a surface area A from a fragment Atom bomb.*
12. *Occurrence of a number of scratches on a sheet of glass.*
13. *Flashing of a number of lightings per second.*
14. *Scoring of number of goals in a game etc.*
15. *Scoring of runs in ODI Cricket match with Pakistan by Sachin Tendulkar.*

6.11.2 Conditions Under Which Poisson Distribution is Used

1. *The random variable X should be discrete.*
2. *A dichotomy exists, i.e., the happening of the event must be of two alternatives such as success and failure; occurrence and non-occurrence etc.*
3. *It is applicable in those cases where the number of trials n is very large and the probability of success p is very small but the mean $np = m$ is finite.*
4. *p should be very small (close to zero). If $p \to 0$, then the distribution is J-shaped and unimodal.*
5. *Statistical independence is assumed. In other words, it is applicable to those cases where the happening of one event does not affect the happening of the other events.*

6.11.3 Characteristics of Poisson Distribution

1. *Poisson distribution is a discrete distribution. It gives theoretical probabilities and theoretical frequencies of a discrete variable.*
2. *It depends mainly on the value of the mean m.*
3. *This distribution is positively skewed to the left. With the increase in the value of the mean m, the distribution shifts to the right and the skewness diminishes.*
4. *Its arithmetic mean in relative distribution is p and in absolute distribution is np.*
5. *If n is large and p is small, this distribution gives a close approximation to Binomial distribution. Since the arithmetic mean of Poisson is same as that of Binomial, so the Poisson distribution can be used instead of Binomial if n or p is not known.*
6. *Poisson distribution has only one parameter, viz., m, the arithmetic mean. Thus the entire distribution can be determined once the arithmetic mean is known.*
7. *Poisson distribution is based on the following assumptions:*
   - **(i)** *Statistical independence is assumed, i.e., the occurrence or non-occurrence of an event does not influence the other events.*
   - **(ii)** *The probability of happening of more than one event in a very small interval is negligible.*
   - **(iii)** *The probability of success for a small space or a short interval of time is proportional to the space or length of time interval as the case may be.*

6.12 BINOMIAL APPROXIMATION TO POISSON DISTRIBUTION

Poisson distribution can be derived from Binomial distribution under the following conditions

(*i*) *p*, the probability of the occurrence of the event is very small

(*ii*) *n* is very very large, where *n* is number of trials, *i.e.*, $n \to \infty$.

(*iii*) *np* is a finite quantity, say $np = m$, then *m* is called the parameter of the Poisson distribution. In Poisson distribution, the probability of *r* success is given by

$$P(r) \text{ or } P(X = r) = \frac{e^{-m} m^r}{r!}.$$

6.13 MEAN AND VARIANCE OF POISSON DISTRIBUTION

Poisson Distribution is : $P(r) = \frac{e^{-m} \times m^r}{r!}, r = 0, 1, 2, ...$

Mean: $\mu = \sum_{i=0}^{i=n} x_i p_i = m$

Also **Variance:** $\sigma^2 = \Sigma x_i^2 p_i - \mu^2 = m$

Standard deviation $= \sqrt{\text{Variance}} = \sqrt{m}$.

6.14 RECURRENCE RELATION

$$\Rightarrow \qquad P(r+1) = \frac{m}{r+1} P(r)$$

which is the required **recurrence relation.**

With this formula we can find *P* (1), *P* (2), *P* (3), *P* (4), ...; if *P* (0) is given.

Example 46. *Suppose a book of 585 pages contains 43 typographical errors. If these errors are ramdomly distributed throughout the book, what is the probability that 10 pages, selected at random, will be free from errors? (Use $e^{-0.735} = 0.4795$).*

Solution. Here $n = 10$, $p = \frac{43}{585} = 0.0735$, mean $= np = 10 \times 0.0735 = 0.735$.

Then **Poisson distribution is given by :**

$$P(r) = \frac{e^{-m} m^r}{r!} = \frac{e^{-0.735} \times (0.735)^r}{r!}.$$

$\therefore$ **Probability of zero error** $= P(0) = \frac{e^{-0.735} \times (0.735)^0}{0!} = e^{-0.735} = \mathbf{0.4795}$. **(given)**

Example 47. *The mortality rate for a certain disease is 7 in 1000. What is the probability for just 2 deaths on account of this disease in a group of 400 ? Given $e^{-2.8} = 0.06$.*

Solution. Here $p = \frac{7}{1000}$ and $n = 400$.

$$\therefore \qquad m = np = 400 \times \frac{7}{1000} = \mathbf{2.8}$$

Now the **required probability is given** by

$$P(2) = \frac{m^2}{2!} e^{-m} = \frac{(2.8)^2}{2!} e^{-2.8} = \frac{2.8 \times 2.8}{2} \times 0.06 = \mathbf{0.2352} \text{ or } \mathbf{23.52\%}.$$

Example 48. *It is given that 3% of the electric bulbs manufactured by a company are defective. Using Poisson distribution, find the probability that a sample of 100 bulbs will contain no defective bulb. Given that $e^{-3} = 0.05$.*

Solution. Let p be the probability of defective bulb. Then

$$p = \frac{3}{100}, \qquad n = 100. \text{ Also } \boldsymbol{\lambda = np} = 100 \times \frac{3}{100} = \mathbf{3}.$$

We know that : $$P(r) = \frac{e^{-\lambda}\lambda^r}{r!}.$$

Now probability that the sample will contain no defective bulb is

$$P(0) = \frac{e^{-3}(3)^0}{0!} = e^{-3} = \mathbf{0.05}.$$

Example 49. *If x is a Poisson variate such that*

$$P(x = 2) = 9P(x = 4) + 90P(x = 6),$$

find the mean of x.

Solution. The probability distribution of Poisson distribution with mean m is

$$P(x) = \frac{e^{-m}m^x}{x!}, x = 0, 1, 2, 3, \ldots\ldots, x, \ldots\ldots$$

Now $$P(x = 2) = 9P(x = 4) + 90P(x = 6)$$

$$\Rightarrow \frac{e^{-m}m^2}{2!} = 9.\frac{e^{-m}m^4}{4!} + 90\frac{e^{-m}m^6}{6!} \Rightarrow 1 = \frac{9m^2}{12} + \frac{90\,m^4}{6\times5\times4\times3} \Rightarrow 1 = \frac{9m^2}{12} + \frac{m^4}{4}$$

$$\Rightarrow 1 = \frac{3}{4}m^2 + \frac{m^4}{4} \Rightarrow m^4 + 3m^2 - 4 = 0 \Rightarrow (m^2 + 4)(m^2 - 1) = 0$$

$$\Rightarrow m^2 - 1 = 0 \Rightarrow m^2 = 1 \Rightarrow m = \pm 1 \Rightarrow \boldsymbol{m = 1}. \qquad (\because \boldsymbol{m^2 = -4} \textbf{ is rejected})$$

5.15 FITTING OF A POISSON DISTRIBUTION

The probability of 'r' sucesses in Poisson distribution is given by

$$P(r) = \frac{e^{-m} \times m^r}{r!}, r = 0, 1, 2, 3, \ldots, n.$$

and 'm' is mean of the distribution.

Suppose this experiment is repeated N times,

Then the frequency of 'r' successes is: $= N \times P(r) = N \times \dfrac{e^{-m} \times m^r}{r!}$

Putting $r = 0, 1, 2, \ldots. n$, we can get the expected or theoretical frequencies which are given below.

Number of successes (r):	**0**	**1**	**2**	**3**	**...**	e^n
Expected frequency	$N e^{-m}$	$N \times e^{-m} \times \frac{m}{1!}$	$N \times \frac{e^{-m} \times m^2}{2!}$	$N \times \frac{e^{-m} m^3}{3!}$	**...**	$N \times \frac{e^{-m} m^n}{n!}$

Example 50. *Assuming that the typing mistakes per page committed by a typist follows a Poisson distribution, find the expected frequencies for the following distribution of typing mistakes:*

No. of mistakes per Page:	0	1	2	3	4	5
No. of Pages :	40	30	20	15	10	5

(*Value of* $e^{-1.5} = 0.22313$)

Solution. Here $N = 120$

$$\textbf{Mean} = m = \frac{40\times0+30\times1+20\times2+3\times15+4\times10+5\times5}{120} = \frac{180}{120} = \mathbf{1.5.}$$

Frequencies are: ***P* (0), *P* (1), *P* (2), ..., *P* (5),** where

$P(0) = e^{-1.5} = \mathbf{0.22313}$; $P(1) = e^{-1.5} \times 1.5 = \mathbf{0.334695}$ $[\because e^{-1.5} = 0.22313]$

$P(2) = e^{-1.5} \times \frac{(1.5)^2}{2!} = \mathbf{0.25}$; $P(3) = e^{-1.5} \times \frac{(1.5)^3}{3!} = \mathbf{0.13}$;

$P(4) = e^{-1.5} \times \frac{(1.5)^4}{4!} = \mathbf{0.05}$; $P(5) = e^{-1.5} \times \frac{(1.5)^5}{5!} = \mathbf{0.01}$.

The expected frequencies are given by

$$N \times \frac{e^{-1.5}\ (1.5)}{r!};\ r = 0, 1, 2, 3, 4, 5, ...$$

No. of Mistakes	*No. of Pages*	*Expected Frequency* $Ne^{-m}\ m^x/x!$
0	40	120×0.22313 = **27**
1	30	120×0.334695 = **40**
2	20	120×0.25 = **30**
3	15	120×0.13 = **16**
4	10	120×0.05 = **6**
5	5	120×0.01 = **1**
	N = 120	**120**

Example 51. *The average number of customers, who appear at a counter of a certain bank per minute is two. Find the probability that during a given minute.*

(*i*) *No customer appears,*   (*ii*) *Three or more customers appear.* [*Given* $e^{-2} = 0.1353$.

Solution. Here $m = 2$, so that the **Poisson distribution is :**

$$P(x) = \frac{e^{-2}2^x}{x!}; x = 0, 1, 2, ...$$

(*i*) $P(0) = e^{-2} = \mathbf{0.1353}$.

(*ii*) $P(x \geq 3) = 1 - P(x \leq 2) = 1 - [P(x = 0) + P(x = 1) + P(x = 2)]$

$= 1 - (e^{-2} + 2e^{-2} + 2e^{-2}) = 1 - 5e^{-2} = 1 - 5\ (0.1353) = 1 - 0.6765 = \mathbf{0.3235}.$

Example 52. *Which probability distribution is appropriate to describe the situation where 100 misprints are distributed randomly throughout the 100 pages of a book ? For this distribution find the probability that a page selected at random will contain at least three misprints.*

(*Value of* $e^{-1} = 1/2.718$)

Solution. Since the number of trials is large and the probability of occurrence of printing mistake is very small, so **Poisson distribution** is appropriate.

$$m = np = 100 \cdot \frac{1}{100} = \mathbf{1.} \quad P(x = r) = \frac{e^{-m}\,m^r}{r!}$$

$$P(x \leq 3) = 1 - P(0) - P(1) - P(2)$$

$$= 1 - \left[\frac{e^{-1}}{0!} + \frac{e^{-1}}{1!} + \frac{e^{-1}}{2!}\right] = 1 - \left[\frac{1}{2.718} + \frac{1}{2.718} + \frac{1}{5.436}\right] = 1 - \frac{5}{5.436} = \mathbf{0.08}.$$

Example 53. *Suppose that the chance of an individual coal-miner being killed in a mine accident during a year is 1/1400. Use the Poisson distribution to calculate the probability that in the mine employing 350 miners, there will be at least one fatal accident in a year.* (*Use* $e^{-0.25} = 0.78$).

Solution. Here $p = \frac{1}{1400}$, $n = 350$.

$\therefore$ **Mean** $= m = np = \frac{350}{1400} = \mathbf{0.25}$.

The Poisson distribution is : $P(r) = \frac{e^{-0.25}(0.25)^r}{r!}$; $r = 0, 1, 2, 3, \ldots\ldots$

Probability of zero fatal accident $= P(0) = \frac{e^{-0.25}(0.25)^0}{0!}$ or $P(0) = e^{-0.25} = \mathbf{0.78}$.

P **(at least one fatal accident)** $= 1 - P(0) = 1 - 0.78 = \mathbf{0.22}$.

Example 54. *In a certain factory turning out razor blades, there is a small chance of 1/500 for any blade to be defective. The blades are supplied in a packets of 10. Use Poisson distribution to calculate the approximate number of packets containing no defective, one defective, two defective and three defective blades in a consignment of 10,000 packets.* [*Given* = $e^{-0.02}$ = 0.9802]

Solution. Here $N = 10{,}000$, $p = \frac{1}{500}$, $n = 10$.

$\therefore$ **Mean:** $m = np$

$$= 10 \times \frac{1}{500} = \frac{1}{50} = \mathbf{0.02}$$

Let x be the number of defective blades in a packet.

Let $P(x)$ be the number of packets containing x defective blades, then

$$P(x) = N \times \frac{e^{-m} m^x}{x!}; \quad x = 0, 1, 2, 3, \ldots\ldots$$

$P(0)$ = **Number of packets with no defective blades**

$$= 10{,}000 \times \left[0.9802 \times \left(\frac{0.02^0}{0!}\right)\right]$$

$$= 10{,}000\,[0.9802] = \mathbf{9802.}$$

$\therefore$ **Number of packets with no defective blade = 9808 packets.**

$P(1)$ = **Number of packets with 1 defective blade**

$$= 10{,}000\left[0.9802 \times \frac{0.02}{1!}\right] = 10{,}000\,[.9802 \times 0.02]$$

$$= 10{,}000\,[0.019604] = \mathbf{196.04}$$

$\therefore$ **Number of packets with one defective blade = 196.04 or 196 packets.**

$P(2)$ = **Number of packets with 2 defective blades**

$$= 10{,}000\left[0.9802 \times \frac{(0.02)^2}{2!}\right]$$

$$= 10{,}000\left[0.9802\times\frac{0.0004}{2}\right]$$

$$= 10{,}000\ [0.00019604]^2 = \mathbf{1.96 = 2\ packets.}$$

P (3) = **Number of packets with 3 defective blades**

$$= 10{,}000\left[0.9802\times\frac{(0.02)^3}{3!}\right] = \mathbf{0.0134 \simeq 0 \Rightarrow No\ packet.}$$

Example 55. *In a Poisson Distribution 3 P (x = 2) = P (x = 4). Find P (x = 3).*

Solution. Since $$P(x) = \frac{e^{-\lambda}\lambda^x}{x!};\quad x = 0, 1, 2, 3, \ldots\ldots$$

Now, $$3P(x = 2) = P(x = 4) \Rightarrow 3\left(\frac{\lambda^2 e^{-\lambda}}{2!}\right) = \frac{\lambda^4 e^{-\lambda}}{4!}$$

$$\Rightarrow \frac{3\lambda^2 e^{-\lambda}}{2\times 1} = \frac{\lambda^4 e^{-\lambda}}{4\times 3\times 2\times 1} \Rightarrow = \frac{3}{2} = \frac{1}{24}\lambda^2 \Rightarrow \lambda^2 = 36 \Rightarrow \lambda = \pm 6.$$

Since mean is always positive, therefore, $\lambda = 6$

Now, $$P(x) = 3 = \frac{\lambda^3 e^{-\lambda}}{3!} = \frac{(6)^3 e^{-6}}{6} = \mathbf{36e^{-6}}.$$

Example 56. *A car hire firm has two cars, which it hires out day by day. The number of demands for a car on each day is distributed as a Poisson distribution with mean 1.5. Calculate the proportion of days on which no car is used and the proportion of days on which some demand is refused.* [$e^{-1.5}$ = *0.2231*].

Solution. Since the number of demands for a car on each day is distributed as **Poisson Distribution** with mean 1.5, therefore, the probability of having r demands on a day is given by

$$P(r) = \frac{(1.5)^r e^{-1.5}}{r!};\quad r = 0, 1, 2, 3, \ldots\ldots$$

And the proportion of days on which no car is used

= **Probability of no demand = $P(0) = e^{-1.5}$ = 0.2231.**

And the proportion of days on which some demand is refused = Probability of more than two demands in a day = $1 - [P(0) + P(1) + P(2)] = 1 - \left[e^{-1.5} + 1.5e^{-1.5} + \frac{(1.5)^2}{2}e^{-1.5}\right]$

= 1 – 0.2231 × 3.625 = **0.1913.**

Example 57. *It is known from the past experience that in a certain plant there are on the average 4 industrial accidents per month. Find the probability that in a given year, there will be less than 4 accidents. Assume Poisson distribution.* [*Given* e^{-4} = *0.0183*]

Solution. Let the random variable X denote the number of accidents in the plant per month.

$$\therefore\quad P(X = r) = \frac{e^{-m}m^r}{r!}, r = 0, 1, 2, 3, \ldots.$$

Here $m = 4$

$$\therefore\quad P(X = r) = \frac{e^{-4}4^r}{r!}, r = 0, 1, 2, 3, \ldots..$$

Again $P(X < 4) = P(X = 0) + P(X = 1) + P(X = 2) + P(X = 3)$

$$= e^{-4}\left[1+4+\frac{4^2}{2!}+\frac{4^3}{3!}\right] = e^{-4}[1+4+8+10.67] = e^{-4} \times 23.67 = 0.0183 \times 23.67 = \mathbf{0.4332}.$$

Example 58. *Six coins are tossed 6,400 times. Using the Poisson distribution, what is approximate probability of getting six heads x times ?*

Solution. Probability of getting a head in a throw of a coin = $\frac{1}{2}$.

Probability of getting six heads in a throw of six coins = $\frac{1}{2^6} = \frac{1}{64}$.

$$\therefore \qquad \textbf{Mean} = m = np = 6400 \times \frac{1}{64} = \mathbf{100}.$$

Hence, **the probability of getting six heads x times according to Poisson distribution is**

$$P(x) = \frac{e^{-100}(100)^x}{x!}.$$

Example 59. *An office switch board receives telephone calls at the rate of 3 calls per minute on average. What is the probability of receiving no calls in one minute interval ?*

Solution. Let the random variable x denote the number of telephone calls per minute, then x follows Poisson distribution with parameter $m = 3$ and probability density function

$$P(x = r) = \frac{m^r e^{-m}}{r!} = \frac{3^r e^{-3}}{r!}, r = 0, 1, 2, 3, \ldots \qquad [\because \ m = 3]$$

Probability of no calls in one minute = $P(0)$, where

$$P(0) = \frac{3^0 e^{-3}}{0!} = \mathbf{0.04979} \qquad [e^{-3} = 0.4979 \text{ from Table}]$$

6.16 MODE OF POISSON DISTRIBUTION

Mode is that value of the Poisson variate X which maximises the probability function.

Thus if $X = r$ gives the mode of the distribution, then

$$P(r) > P(r-1) \quad \text{and} \quad P(r) < P(r+1) \text{ for all values of } r.$$

Working Rule:

Let X be a Poisson variate with parameter m.

Case I. ***When m is an integer.*** If m is an integer, equal to k, (say), then the Poisson distribution is bi-modal, the two modes being at the points $X = k$ and $X = k - 1$.

Case II. ***When m is not an integer.*** If m is not an integer then the distribution is unimodal, the unique modal value being the integral part of m. For example, if $m = 5.6$, then mode is 5, the integral part of 5.6.

Example 60. *Assuming one in 80 births is a case of twins, calculate the probability of 2 or more sets of twins on a day when 30 births occur.*

Solution. $\quad P(\text{twin birth}) = p = \frac{1}{80} = \mathbf{0.0125};\ n = 30$

$\therefore \ m = np = 30 \times 0.0125 = \mathbf{0.375}.$ Also $P(x) = \dfrac{e^{-m} \times m^x}{x!}$.

$$\therefore \qquad P(0) = \frac{e^{-0.375} \times (0.375)^0}{0!} = e^{-0.375} \qquad \ldots (1)\ [\because m = 0.375]$$

$$P(1) = \frac{e^{-0.375} \times (0.375)^1}{1!} = 0.375 \times e^{-0.375} \qquad \ldots (2)$$

Required probability $= \mathbf{1 - [P(0) + P(1)]} = 1 - e^{-0.375}[1 + 0.375]$

$= 1 - e^{-0.375} \times 1.375 = 1 - (0.6873 \times 1.375)$

$= 1 - 0.9450 = \mathbf{0.055.}$ $\quad$ **[$\because$ From tables $e^{-0.375} = 0.6873$]**

Example 61. *Six coins are tossed 6,400 times. Using the Poisson distribution, what is the approximate probability of getting six heads x times ? Assume that the coins are unbiased.*

Solution. In a random toss of 6 unbiased coins, the probability (p) of getting 6 heads is given by :

$$p = P(HHHHHH) = P(H)\,P(H)\,P(H)\,P(H)\,P(H)\,P(H)$$

$$= \frac{1}{2} \times \frac{1}{2} \times \frac{1}{2} \times \frac{1}{2} \times \frac{1}{2} \times \frac{1}{2} = \mathbf{\frac{1}{64}}.$$

Using Poisson approximation to the Binomial distribution with $n = 6400$ and $p = \frac{1}{64}$, we have:

$$m = np = 6400 \times \frac{1}{64} = \mathbf{100}.$$

$$\therefore \quad \mathbf{P\,[\text{Getting 6 heads } x \text{ times}]} = \frac{e^{-m} m^x}{x!} = \frac{e^{-100} \cdot 100^x}{x!}\ ; x = 0, 1, 2, \ldots$$

Example 62. *A manufacturer of pins knows that on an average 5% of his product is defective. He sells pins in boxes of 100, and guarantees that not more than 4 pins will be defective.*

(a) What is the probability that a box will meet the guaranteed quality ?

(b) What is the probability that a box will fail to meet the guaranteed quality ? $e^{-5} = 0.0067$).

Solution. p = Probability of a defective pin = 5% = 0.05.

Since the probability of a defective pin is small, so we may use Poisson distribution. We are given $n = 100$.

Hence $\quad m = np = 100 \times 0.05 = \mathbf{5}.$

If the random variable X denotes the number of defective pins in a box of 100, then

$$P(X = r) = \frac{e^{-m} \cdot m^r}{r!} = \frac{e^{-5} \cdot 5^r}{r!} \quad \ldots (1)$$

(*a*) A box will meet the guaranteed quality if the number of defective pins in it is less than or equal to 4. Hence the required probability is given by :

$$P(X \le 4) = \sum_{r=0}^{4} P(r) = P(0) + P(1) + P(2) + P(3) + P(4)$$

$$= e^{-5}\left[1 + 5 + \frac{25}{2!} + \frac{125}{3!} + \frac{625}{4!}\right] \quad \text{[From (1)]}$$

$$= 0.0067 \times [1 + 5 + 12.5 + 20.83 + 26.04] = 0.0067 \times 65.37 = \mathbf{0.438.}$$

(*b*) A box will fail to meet the guaranteed quality if the number of defectives in it is more than 4. Hence the required probability is :

$$P(X > 4) = 1 - P(X \le 4) = 1 - 0.438 = \mathbf{0.562.}$$

EXERCISE – 6.2

1. Between 2 and 4 P.M. the average number of phone calls per minute coming into the switchboard of a company is 2.5. Find the probability that during one particular minute there will be (*i*) no phone call at all, (*ii*) exactly 3 calls. (Given $e^{-2} = 0.13534$ and $e^{-0.5} = 0.60650$)

[**Hint.** $P(0) = \dfrac{e^{-2.5}(2.5)^0}{0!} = e^{-2.5} = e^{-2} \times e^{-0.5} = 0.13534 \times 0.60650 = \mathbf{0.0821}$.

$P(3) = \dfrac{e^{-2.5}(2.5)^3}{3!} = 0.214$.]

2. One fifth per cent of the blades produced by a blade manufacturing factory turn out to be defective. The blades are supplied in packets of 10. Use Poisson distribution to calculate the approximate number of packets containing no defective, one defective and two defective blades respectively in a consignment of 1,00,000 packets. Given $e^{-.02} = .9802$.

[**Hint.** $m = 10 \times \dfrac{1}{500} = 0.02.$ $P(x) = \dfrac{e^{-0.02}(0.02)^x}{x!}$; $f(x) = 1{,}00{,}000 \times P(x)$.]

3. The following table gives the number of days in a 50-day period during which automobile accidents occurred in a certain part of a city. Fit a Poisson distribution to the data.

No. of defects :	0	1	2	3	4
No. of units :	19	18	8	4	1

[**Hint.** Mean : $m = \dfrac{\Sigma fx}{\Sigma f} = \dfrac{50}{50} = 1$. Also $P(r) = \dfrac{e^{-m} \times m^r}{r!}$; $P(0) = \dfrac{e^{-1} \times (1)^0}{0!} = e^{-1} = 0.3679$.

Expected frequencies are : $NP(0)$, $NP(1)$, $NP(2)$, $NP(3)$ and $NP(4)$.

$NP(0) = 50 \times 0.3679 = 18.395$; $NP(1) = P(0) \times m = 18.395$,

$NP(2) \times \dfrac{m}{2} = 9.198$; $NP(3) = NP(2) \times \dfrac{m}{3} = 3.066$ and $NP(4) = NP(3) \times \dfrac{m}{4} = 0.767$.]

4. In a certain factory manufacturing razor blades, there is a small chance, 1/50 for any blade to be defective. The blades are placed in packets, each containing 10 blades. Using the Poisson distribution, calculate the approximate number of packets containing not more than 2 defective blades in a consignment of 10,000 packets.

[**Hint.** $m = np = 10 \times \dfrac{1}{50} = 0.2.$ $P(0) = e^{-m} = e^{-0.2} = 0.8187$.

$N \times P(0) = P(0) \times 10000 = 0.8187 \times 10000 = 8187$.

Similarly $N \times P(1) = m \times N \times P(0) = 1637.4$. $N \times P(2) = \dfrac{m}{2} \times N \times P(1) = 163.74$.

Required Number $= [8187 + 1637.4 + 163.74] = \mathbf{9988}$.]

5. The distribution of the number of road accidents per day in a city is Poisson with mean 4. Find the number of days out of 100 days when there will be (*i*) no accident, (*ii*) at least 2 accidents and (*iii*) at most 3 accidents.

[**Hint.** $m = 4$, $P(0) = e^{-m} = e^{-4} = 0.0183$ [From the Tables]

$NP(0) = 100 \times 0.0183 = 1.83$. $NP(1) = NP(0) \times \dfrac{m}{1} = 1.83 \times 4 = 7.32$;

$NP(2) = [NP(1)] \times \dfrac{m}{2} = 7.32 \times \dfrac{4}{2} = 14.64 \simeq 15$; $NP(3) = [NP(2)] \times \dfrac{m}{3} = 14.64 \times \dfrac{4}{3} = 19.52; \simeq 20$,

$NP(4) = [NP(3)] \times \dfrac{m}{4} = 19.52 \times \dfrac{4}{4} = 19.52 \simeq 20$

$NP(5) = [NP(4)] \times \dfrac{m}{5} = 19.52 \times \dfrac{4}{5} = 15.62 \simeq 16$.

(*i*) No accident $= NP(0) = 1.83 \simeq$ **2 days.**

(*ii*) Atleast 2 accident $= 100 - [NP(0) + NP(1)] = 100 - (2 + 7) =$ **91 days.**

(*iii*) At most 3 accidents = $N\,P(0) + N\,P(1) + N\,P(2) + N\,P(3) = 2 + 7 + 15 + 20 =$ **44 days.**]

6. The probability that an individual suffers from a bad reaction from an infection of a given serum is 0.001. Determine the probability that out of 2000 individuals : (*i*) exactly 3 and (*ii*) more than 2 individual will suffer from a bad reaction.

7. The distribution of typing mistakes committed by a typist is given below. Assuming a poisson model, find out the expected frequencies:

Mistakes per Page :	0	1	2	3	4	5
Number of Pages :	142	156	69	27	5	1

[**Hint.** Mean: $m = \dfrac{\Sigma f x}{\Sigma f} = \dfrac{400}{400} = 1$. Also $e^{-1} = 0.3679$ [From Tables] and $N = 400$.

Expected frequencies: $N\,P(0) = 147$, $N\,P(1) = 147$, $N\,P(2) = 74$, $N\,P(3) = 25$, $N\,P(4) = 6$ $N\,P(5) = 1$.]

8. If 2% of electric bulbs manufactured by a certain company are defective, find the probability that in a sample of 200 bulbs (*i*) less than 2 bulbs are defective; (*ii*) more than 3 bulbs are defective

[Given $e^{-4} = 0.0183$]

9. If a Poisson distribution has a double mode at $X = 1$ and $X = 2$, find $P(X = 1)$

10. The following mistakes per page were observed in a book:

No. of mistakes per page	0	1	2	3	4
No. of times the mistake occurred	211	90	19	5	0.

Fit a Poisson distribution to the data.

[**Hint.** $m = \dfrac{\Sigma fx}{\Sigma f} = \dfrac{211\times 0 + 1\times 90 + 2\times 19 + 3\times 5 + 4\times 0}{143} = \mathbf{0.44}.$

Expected frequencies are $N\ \ P(r) = N \times \dfrac{e^{-m}\times m^r}{r\,!}$, $r = 0, 1, 2, 3, 4$.

11. If X is a Poisson variate and $P(X = 0) = P(X = 1) = k$, show that $k = 1/e$.

[**Hint.** $P(X = 0) = P(X = 1) \Rightarrow e^{-m} = m\,e^{-m} \Rightarrow m = 1$.

$\therefore$ $P(X = 0) = k \Rightarrow e^{-m} = k \Rightarrow e^{-1} = k \Rightarrow k = 1/e$.]

12. An industrial area has power breakdown once in 15 days on the average. Assuming that the number of breakdowns follow a Poisson process, what is the probability of (*i*) no power breakdown in the next six days, (*ii*) more than one power breakdown in the next six days ?

13. It is known that 0.5% of ball pen refills produced by a factory are defective. These refills are dispatched in packagings of equal numbers. Using Poisson distribution determine the number of refills in a packing to be sure that at least 95% of them contain no defective refills.

14. Find the probability that at most 5 defective bolts will be found in a box of 200 bolts if it is known that 2 per cent of such bolts are expected to be defective. (You may take the distribution to be Poisson).

(Take $e^{-4} = 0.0183$)

ANSWERS

1. (*i*) 0.0821; (ii) 0.214 **2.** $N\,P(1) = 1960$; $N\,P(2) = 20$.

3.

0	1	2	3	4
18.395	18.395	9.198	3.066	0.767.

4. 9988. **5.** (*i*) 2; (*ii*) 91; (*iii*) 44.

6. (*i*) $\dfrac{4\,e^{-2}}{3}$; (*ii*) $1 - 6e^{-2}$.

7. 147, 147, 74, 25, 6, 1. **8.** (*i*) 0.0915 ; (*ii*) 0.5669. **9.** $2\,e^{-2}$

10. 209.30; 92.09, 20.26, 2.97; 0.34. **12.** (*i*) 0.6703; (*ii*) 0.0616. **13.** 10.

14. 0.7845.

NORMAL DISTRIBUTION

6.17 NORMAL DISTRIBUTION

Normal distribution *is a continuous probability distribution in which the relative frequencies of a continuous variable are distributed according to the normal probability law. In other words, it is a symmetrical distribution in which the frequencies are distributed evenly about the mean of the distribution.*

The normal distribution of a variable, when represented graphically, takes the shape of a symmetrical curve, known as the **Normal Curve.** *This curve is asymptotic to x-axis (base line) on its either sides. It is also known as* **Normal Curve of Error** *as the normal curve is extensively used to describe errors made in repeated measurements. It helps us to find the proportion of measurement that falls within a certain range above, or below or between selected values.*

Normal distribution is a limiting form of Binomial distribution under the following conditions.

(*i*) n, the number of trials is infinitely large, *i.e.*, $n \to \infty$

(*ii*) neither p (or q) is very small, *i.e.*, p and q are fairly near equal.

A random variable x is said to have a normal distribution with mean μ and standard deviation σ if its probability density function is given by

$$f(x) = \frac{1}{\sigma\sqrt{2\pi}} e^{-(x-\mu)^2/2\sigma^2}, \ -\infty < x < \infty; \text{ where } e = 2.7183, \ \sqrt{2\pi} = 2.5060.$$

The probability density function with mean zero and standard deviation σ is given by

$$f(x) = \frac{1}{\sigma\sqrt{2\pi}} e^{-x^2/2\sigma^2}, \ -\infty < x < \infty$$

Normal distribution was first discovered by British mathematician **De-Movire in 1733. Normal distribution is also known as Gaussian distribution named after Karl Friedrich Gauss,** who used this distribution to describe the theory of accidental errors of measurements involved in the calculation of orbits of heavenly bodies. Now-a-days normal probability model is one of the most important probability models in statistical analysis.

6.18 STANDARD NORMAL DISTRIBUTION

A random variable Z which has a normal distribution with mean $\mu = 0$ and a standard deviation $\sigma = 1$ is said to be a standard normal distribution. Its probability density function is given by :

$$f(z) = \frac{1}{\sqrt{2\pi}} e^{-z^2/2}, \ -\infty < z < \infty$$

It is denoted by N **(0, 1).** In short, **standard normal variate** is written as **S.N.V.** or **s.n.v.**

The **area under any normal curve** ***is found from the table of a standard normal probability distribution showing the area between the mean and any value of the normally distributed random variable. For a given value of μ and σ, and a specific value, X, of the random variable, the standardized normal variate Z is derived from the following formula:***

$$Z = \frac{X - \mu}{\sigma}, \text{ where } \mu \text{ is the mean; and } \sigma \text{ is the standard deviation.}$$

The purpose of standardization of normal distribution is to enable us to make use of the tables of the area of the standard curve: $f(z) = \frac{1}{\sqrt{2\pi}} e^{-z^2/2}$ for various points along the x-axis. **[*The standard normal tables are given at the end of the book.*]**

The standard normal distribution is also known as **Unit Normal Distribution** or **Z-Distribution.**

The **standard normal curve** helps us to find the areas within two assigned limits under the curve. The areas between the standard normal curve drawn at two assigned limits a and b will give the proportion of cases for which the values of Z lie between *a* and *b*. ***Thus the area between two assigned limits a and b under the standard normal curve will represent the probability that Z will lie between a and b. It is denoted by P (a ≤ Z ≤ b).***

6.19 PROPERTIES OF A NORMAL CURVE

The normal probability curve with **mean μ and standard deviation σ** has the following properties:

1. *The equation of the curve is :* $y = \dfrac{1}{\sigma\sqrt{2\pi}} e^{-(X-\mu)^2/2\sigma^2} \quad -\infty < x < \infty$
   and it is bell-shaped. The top of the bell is directly above the mean μ.
2. *The curve is symmetrical about the line* $x = \mu$ *and x ranges from* $-\infty$ *to* $+\infty$.
3. *Mean, mode and median coincide at* $x = \mu$ *as the distribution is symmetrical.*
4. *It can be shown that it has arithmetic mean* = μ *and variance* = σ^2
5. *X-axis is asymptote to the curve.*
6. *The points of inflexion of the curve are at* $x = \mu + \sigma$, $x = \mu - \sigma$ *and the curve changes from concave to convex at* $x = \mu + \sigma$ *to* $x = \mu - \sigma$.
7. *The total area under the normal curve is equal to unity and the percentage distribution of area under the normal curve is given below and is shown also in the figure.*
   (*i*) *About 68% of the area falls between* $\mu - \sigma$ *and* $\mu + \sigma$
   (*ii*) *About 95.5% of the area falls between* $\mu - 2\sigma$ *and* $\mu + 2\sigma$
   (*iii*) *About 99.7% of the area falls between* $\mu - 3\sigma$ *and* $\mu + 3\sigma$.
8. *In a normal distribution: Q.D : M.D : S.D : : 10 : 12 : 15;*
   where Q.D. = Quartile Deviation, M.D. = Mean Deviation and S.D. = Standard deviation.
9. *The mean deviation from the mean in normal distribution is equal to (4/5) of its standard deviation i.e. M.D. = 0.8σ.*
10. *The first and third quartiles are given by :* $Q_1 = \mu - 0.675\sigma$ *and* $Q_3 = \mu + 0.675\sigma$. *Also Q.D. = 0.675σ.*
11. *The maximum ordinate lies at the mean, i.e., at* $x = \mu$
12. *The curve of normal distribution has a single peak, i.e., it is a unimodal.*
13. *The two tails of the curve extend indefinitely and never touch the horizontal line.*
14. *The mathematical equation is completely determined if* μ *and* σ *are known.*
15. *Since mean = median = mode, the coordinate at* $x = \mu$ *(or Z = 0) divides the whole area into two equal parts. The area to the right of the ordinate as well as to the left of the ordinate at* $x = \mu$ *(or Z = 0) is 0.5.*
16. *No portion of the curve lies below the x-axis as f (x), being the probability function can never be negative.*
17. *If X and Y are independent normal variables with means and standard deviations as* μ_1 *and* μ_2 *and* σ_1, *and* σ_2 *respectively, then* $Z = X + Y$ *also*

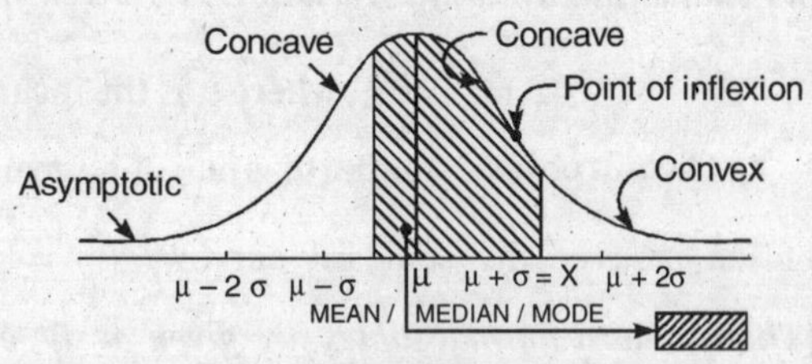

Fig. 6.2.

follows normal distribution with mean $(\mu_1 + \mu_2)$ *and SD* = $\sqrt{\sigma_1^2 + \sigma_2^2}$ *respectively.*

i.e. If $X \sim N(\mu_1, \sigma_1^2)$ *and* $Y - N(\mu_2, \sigma_2^2)$ *and x and y are independent, then*

$Z = X + Y \sim N(\mu_1 + \mu_2, \sigma_1^2 + \sigma_2^2)$.

6.20 USES OF NORMAL DISTRIBUTION

1. *The normal distribution can be used to approximate the Binomial and Poisson distributions.*
2. *It has extensive use in sampling theory. It helps us to estimate parameter from statistic and to find confidence limits of the parameter.*
3. *It has a wide use in testing Statistical Hypothesis and Tests of Significance in which it is always assumed that the population from which the samples have been drawn should have normal distribution.*
4. *It has significant applications in statistical quality control as the control chart in statistical quality control is closely related to normal distribution.*
5. *It can be used for smoothing and graduating a distribution which is not normal, simply by contracting a normal case.*
6. *It serves as a guiding instrument in the analysis and interpretation of statistical data.*

Example 63. *What is the probability that a standard normal variate Z will be (a) greater than 1.09; (b) less than – 1.65 (c) lying between – 1.00 and 1.96; (d) lying between 1.25 and 2.75?*

Solution. (*a*) The shaded area to the right of Z = 1.09 is the probability that Z will be greater than 1.09. In the table the area between 0 and 1.09 is 0.3621. The total area under the curve is equal to 1, so that the area to the right of zero must be 0.5000 (∵ curve is symmetrical at Z = 0)

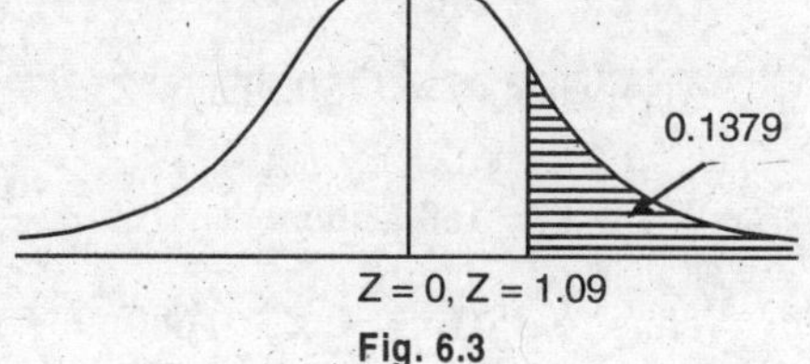

Fig. 6.3

$$P(Z > 1.09) = 0.5000 - 0.3621 = 0.1379.$$

(*b*) The shaded area to the left of Z = – 1.65 is the probability that Z will be less than – 1.65. Also the area between – 1.65 and 0 is the same as area between 0 and 1.65. In the table the area between zero and 1.65 is **0.4505**. But the area to the left of zero is 0.5.

$$\therefore P(Z \leq 1.65) = 0.5000 - 0.4505 = 0.0495.$$

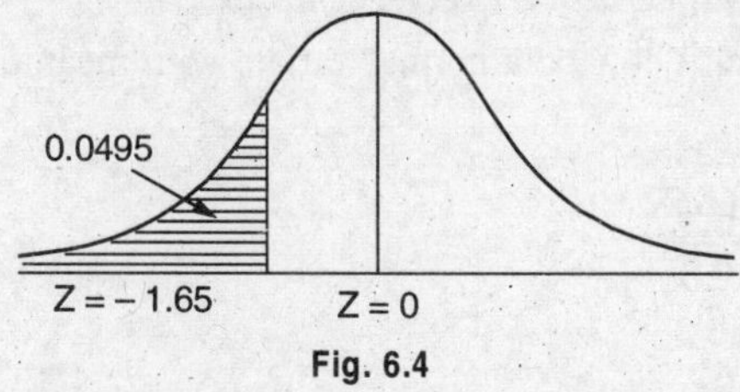

Fig. 6.4

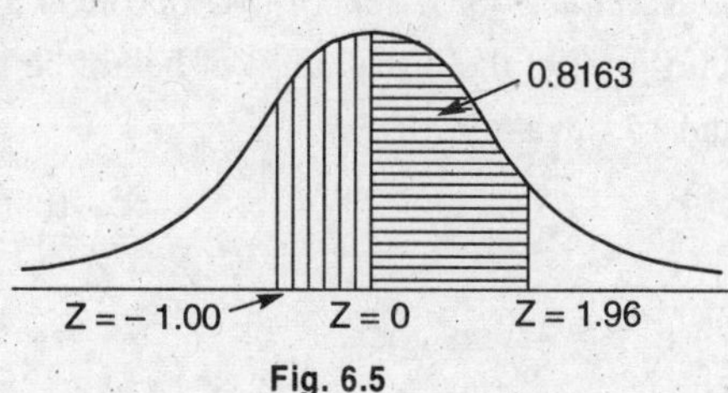

Fig. 6.5

(*c*) The probability that the random variable Z in between – 1.00 and 1.96 is the shaded area in the figure 9.7. It is found by adding the corresponding areas shown in the figure.

Also area between – 1.00 and 1.96 = Area between (– 1.00 and 0) + area between (0 and 1.96)

$$P(-1.00 < Z < 1.96) = P(-1.00 < Z < 0) + P(0 < Z < 1.96) = 0.3413 + 0.4750 = 0.8163.$$

(*d*) The shaded area between Z = 1.25 and 2.75 is the probability that Z will be between 1.25 and 2.75.

But **Area between Z = 1.25 and Z = 2.75 = (Area between Z = 0 and Z = 2.75) – (Area between Z = 0 and Z = 1.25)**

$$\therefore \quad P(1.25 < Z < 2.75) = P(0 < Z < 2.75) - P(0 < Z < 1.25)$$
$$= \mathbf{0.4970 - 0.3944 = 0.1026} \qquad \textbf{(From the table)}$$

Fig. 6.6

Fig. 6.7

Example 64. *Find the value of Z such that the probability of a larger value is 0.7881.*

Solution. The area to the right of Z is greater than 0.5000. We know that Z must be negative and that the area between **Z** and **0** must be 0.7881 – 0.5000 = 0.2881. In the table we search out the entry 0.2881 and find the corresponding Z is – 0.80. [See figure 9.9]

Hence $Z > -0.80$.

Example 65. *A certain type of wooden beam has a mean breaking strength of 1500 kgs and a standard deviation of 100 kgs. Find the relative frequency of all such beams whose breaking strengths lie between 1450 and 1600 kgs.*

Solution. Let X be the breaking strength. Then we are to find $\mathbf{P(1450 < X < 1600)}$.

Here X is a normal variate with mean 1500 and standard deviation 100.

$\therefore\ Z = \dfrac{X - 1500}{100}$ **is a standard normal variate $N(0, 1)$.**

Also when $X = 1450$, then $Z = \dfrac{1450-1500}{100} = -0.5$

When $X = 1600$, then $Z = \dfrac{1600-1500}{100} = 1$

Thus $P(1450 < X < 1600) = P(-0.5 < Z < 1) = P(-0.5 < Z < 0) + P(0 < Z < 1)$

$= 0.1915 + 0.3413 = \mathbf{0.5328}$ **[See tables]**

$\Rightarrow$ **53.28% of the beams have the breaking strenght between 1450 kgs and 1600 kgs.**

Example 66. *Assume the mean height of children to be 68.22 cm with a variance of 10.8 cm. How many children in a school of 1,000 would you expect to be over 72 cm tall ?*

Solution. Let the distribution of height be normal. Let X be a normal variate with mean 68.22 and standard deviation $\sqrt{10.8}$.

$$Z = \frac{X - \mu}{\sigma} = \frac{X - 68.22}{\sqrt{10.8}}.$$

When $X = 72$ cm, then

$$Z = \frac{72 - 68.22}{\sqrt{10.8}} = \frac{3.78}{3.286} = 1.15.$$

Now $P(X > 72) = P(Z > 1.15)$ = Area to the right of the cordinate at Z = 1.15 = (Total Area on the right of Z = 0) – (Area from Z = 0 to Z = 1.15) = (0.5000 – 0.3749) = **0.1251.**

Expected number of children to be above 72 cm out of 1000 = 0.1251 × 1000 = **125.1 or 125 children.**

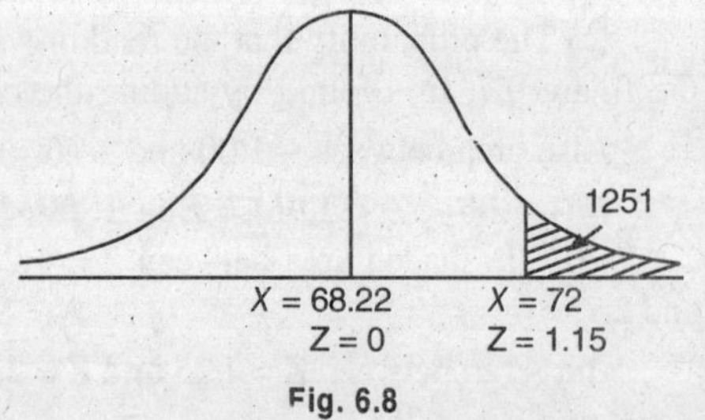

Fig. 6.8

Example 67. *The life time of a certain kind of battery has a mean of 300 hours and a standard deviation of 35 hours. Assuming that the distribution of life times, which are measured to the nearest hour is normal, find the percentage of batteries which have life time of more than 370 hours.*

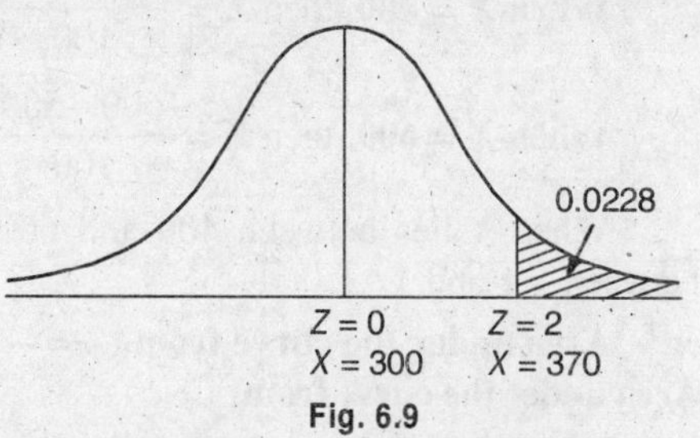

Fig. 6.9

Solution. Let X be a random normal variate measuring the life time of batteries

Here mean $\mu = 300, \sigma = 35$

$\therefore \quad Z = \dfrac{X - \mu}{\sigma} = \dfrac{X - 300}{35}$ **is a standard normal variate.**

When $\quad X = 370$, then $Z = \dfrac{370 - 300}{35} = 2.$

$P(X > 370) = P(Z > 2)$. **But [Area between Z = 0 and Z = 2] = 0.4772.**

Area of Z > 2 = 0.5000 – 0.4772 = **0.0228** (Shaded Region in figure 9.11)

$\therefore$ **Percentage of batteries having life more than 370 hours = 0.0228 × 100 = 2.28%.**

Example 68. *1,000 light bulbs with a mean life of 120 days are installed in a new factory. Their length of life is normally distributed with a standard deviation of 20 days.*

(*a*) *How many bulbs will expire in less than 90 days ?*

(*b*) *If it is decided to replace all bulbs together, what interval should be allowed between replacement if not more than 10% should expire before replacement ?*

Solution. Let X be the normal variate of life of light bulbs. Then its mean $\mu = 120$ and standard deviation $\sigma = 20$.

Now $\quad Z = \dfrac{X - \mu}{\sigma} = \dfrac{X - 120}{20}$ is an S.N.V.

(*a*) **When** $X = 90$, then $Z = \dfrac{90 - 120}{20} = -1.5$

$\therefore$ Number of bulbs expected to expire less than 90 days out of 1,000 bulbs = 1000 × 0.0668 ≈

6 6 8

or 67 bulbs.

(*b*) Since not more than 10% or 0.1 expire before replacement so the value of the Standard Normal Variate Z to an area 0.5 – 0.1 = **0.4** is **1.28.**

67 %
Z = – 1.5
X = 90
Z = 0
X = 120

Fig. 6.10

Thus the value of Z is less than – 1.28.

$\therefore \quad Z = \dfrac{X - 120}{20} = -1.28.$

$\Rightarrow \quad X = 120 - 25.6 = 94.4$ or **94 days**

Then **the bulbs may be replaced after 94 days.**

Example 69. *The scores made by a candidate in a certain test are normally distributed with mean 500 and standard deviation 100. What percentage of candidate receives the scores between 400 and 600 ?*

Solution. Let X the normal variate showing scores of candidates. Its mean $\mu = 500, \sigma = 100$.

Now $Z = \dfrac{X - \mu}{\sigma} = \dfrac{X - 500}{100}$ **is a standard normal variate** $N(0, 1)$.

When $X = 400$, then $\mathbf{Z} = \dfrac{400 - 500}{100} = \mathbf{-1.}$

When $X = 600$, then $\mathbf{Z} = \dfrac{600 - 500}{100} = \mathbf{1.}$

When X lies between 400 and 600, then Z lies between – 1 and 1.

68.27%

$Z = -1$, $X = 400$, $Z = 0$, $X = 500$, $Z = 1$, $X = 600$

Fig. 6.11

Area under the curve from (Z = – 1 to Z = 1) = Area under the curve from

(Z = – 1 to Z = 0) + Area under the curve from (Z = 0 to Z = 1).

= 2 × (area under the curve from Z = 0 to Z = 1) (Symmetric property)

$= 2 \times 0.34134 = \mathbf{0.68268} \simeq \mathbf{0.6827.}$

$\therefore$ **Percentage of candidates securing marks between 400 and 600**

$= 0.6827 \times 100 = \mathbf{68.27\%.}$

Example 70. *In a normal distribution 31% of the items are under 45 and 8% are over 64. Find the mean and standard deviation of the distribution.*

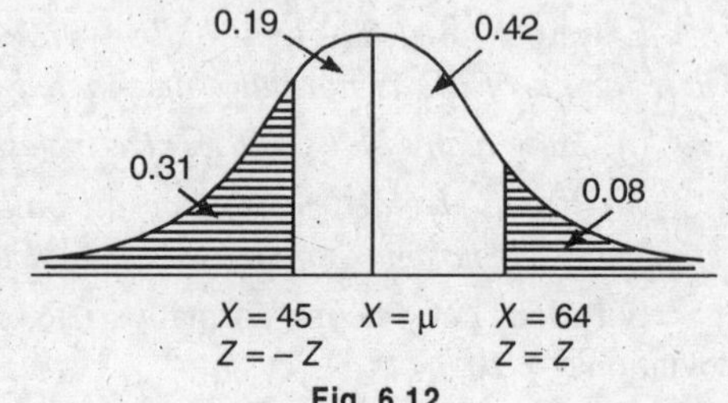

Fig. 6.12

Solution. 31% of items are under 45 ⇒ Area to the left is 0.31. But area right from this point (0.31) to the point 0.5 is. (0.50 – 0.31) = 0.19.

Let X be the random variate which is normally distributed with mean μ and standard deviation σ. Then,

$$\mathbf{Z} = \frac{X - \mu}{\sigma} \text{ is an S.N.V.}$$

The S.N.V. corresponding to $X = 45$ and $X = 64$ are as below:

When $X = 45$, then $\mathbf{Z} = \dfrac{45 - \mu}{\sigma} = \mathbf{Z_1},$ (Say) ... **(I)**

When $X = 64$, then $\mathbf{Z} = \dfrac{64 - \mu}{\sigma} = \mathbf{Z_2}.$ (Say) ... **(II)**

From the figure it is obvious that

$$P(0 < Z < Z_2) = 0.42 \Rightarrow Z_2 = 1.405 \quad \textbf{[From the Table]}$$

$$P(-Z_1 < Z < 0) = 0.19 \Rightarrow P(0 < Z < Z_1) = 0.19. \Rightarrow Z_1 = 0.496$$

[From Normal Table]

Substituting the values of Z_1 and Z_2 in (I) and (II), we get

$$\frac{45 - \mu}{\sigma} = -0.496 \Rightarrow 45 - \mu = -0.496\,\sigma \quad \textbf{... (III)}$$

$$\frac{64 - \mu}{\sigma} = 1.405 \Rightarrow 64 - \mu = 1.405\,\sigma \quad \textbf{... (IV)}$$

Solving (III) and (IV), we get $\boldsymbol{\sigma = 10, \mu = 49.96 \cong 50}$ **(approx.)**

Hence, **mean is 50** and **standard deviation is 10.**

Example 71. *An auditor has found that the credit rewards of a large mail order have approximately normally distributed and show an average account billing error of Rs. 10 and a standard deviation of Re. 1 (billing error may be positive or negative according to whether* *purchases were overcharged or undercharged). Suppose credit account is randomly selected from the files of the mail order house. Find the probability of a billing error between Rs. 10 and Rs. 11.50.*

Solution. Let X be the random normal variate of billing error. Then its mean $\mu = 10$ and $\sigma = 1$.

Let $Z = \dfrac{X-\mu}{\sigma} = \dfrac{X-10}{1}$ **be a S.N.V. ≃ N (0, 1).**

When $X = 10$, **then** $Z = 10 - 10 = 0$.

When $X = 11.50$, **then** $Z = 11.50 - 10 = 1.5$.

Billing error between Rs. 10 and Rs. 11.50

= Area under the curve from Z = 0 to Z = 1.5 = 0.4332 **[From Normal Table]**

Example 72. *A random variable X is normally distributed with mean $\mu = 12$ and s.d. $\sigma = 2$. find $P(9.6 < X < 13.8)$. Given that the area: A = 0.3159 for Z = 0.9 and A = 0.3849 for Z = 1.2, where Z is a standard normal variate.*

Solution. We know, standardised variate: $Z = \dfrac{X-\mu}{\sigma}$

Here $\mu = 12$; and $\sigma = 2$.

When $X = 9.6$, then $Z = \dfrac{9.6-12}{2} = -1.2$.

When $X = 13.8$, then $Z = \dfrac{13.8-12}{2} = 0.9$.

$$P(9.6 < X < 13.8) = P(-1.2 < Z < 0.9) = 0.3849 + 0.3159 = \mathbf{0.7008}.$$

Example 73. *The income of a group of 10,000 persons was found to be normally distributed with mean equal to Rs. 750 and standard deviation equal to Rs. 50. What was the lowest income among the richest 250 ?*

Solution. We have: $\mu = 750$ and $\sigma = 50$.

Let x_1 be the lowest income among the richest 250 people.

Then $P(X \geq x_1) = \dfrac{250}{10{,}000} = 0.025.$

$$P\left(\frac{X-\mu}{\sigma} \geq \frac{x_1-750}{50}\right) = 0.025 \Rightarrow P\left(Z \geq \frac{x_1-750}{50}\right) = 0.025$$

$$\therefore \quad \frac{x_1-750}{50} = 1.96 \quad \Rightarrow \quad x_1 = 750 + 50 \times 1.96 = 848.$$

Hence Rs. 848 is the lowest income among the richest 250 people.

Example 74. *The marks obtained in a certain examination follow normal distribution with mean 45 and standards deviation 10. If 1000 students appeared at the examination, calculate the number of students scoring (i) less than 40 marks and (ii) more than 60 marks.*

Solution. Given, $\bar{X} = 45$, $\sigma = 10$. Then, $Z = \dfrac{X-\bar{X}}{\sigma} = \dfrac{X-45}{10}$

(*i*) For $X = 40 \Rightarrow Z = \dfrac{40-45}{10} = -0.5$

$P(X < 40) = P(Z < -0.5)$

or $P(Z > 0.5) = 0.5 - P(0 < Z < 0.5) = 0.5 - 0.1915 = \mathbf{0.3085}$. **[From tables]**

Number of students = 1000 × 0.3085 = 309 (*approximately*)

(*ii*) For **$X = 60$** $\Rightarrow$ $Z = \dfrac{X - \overline{X}}{\sigma} = \dfrac{60-45}{10} = \mathbf{1.5}$

$$P(X > 60) = P(Z > 1.5) = 0.5 - P(0 < Z < 1.5) = 0.5 - 0.4332 = \mathbf{0.0668}$$

Number of students = 1000 × 0.0668 = **67 (*approximately*)**

Example 75. *Let X be a continuous variable and follows a normal distribution with mean 12 and standard deviation 2. What is the probability that the value of X selected at random lies in the interval [11, 14] ?*

Solution. Here $\overline{X} = 12, \sigma = 2$.

Converting the values from X scale to Z scale, we have.

For $X = 11$, $Z = \dfrac{X - \overline{X}}{\sigma} = \dfrac{11-12}{2} = \dfrac{-1}{2} = \mathbf{-0.50.}$

For $X = 14$, $Z = \dfrac{X - \overline{X}}{\sigma} = \dfrac{14-12}{2} = \mathbf{1.}$

We are, therefore, **interested in finding the probability that Z lies between, – 0.50 and 1.**

Thus $\mathbf{P(11 \le X \le 14) = P(-0.50 \le Z \le 1)} = P(0 \le Z \le 0.50) + P(0 \le Z \le 1)$

$$= 0.1915 + 0.3413 = \mathbf{0.5328.}$$

6.21 FITTING A NORMAL DISTRIBUTION or A NORMAL CURVE

Purpose. The purpose of fitting a normal curve is two-fold:

(*i*) ***To judge whether the normal curve is a good fit to the sample data.***

(*ii*) ***To use the smoothened normal curve instead of the regular curve representing the sample data to estimate the characteristics of population.***

There are two methods for fitting normal distribution:

1. **Ordinate method**
2. **Area Method**

Example 76. *The average test marks in a particular class is 79. The standard deviation is 5. If the marks are normally distributed, how many students in a class of 200 did not receive marks between 75 and 82 ?*

Solution. Here $\overline{X} = 79$ and $\sigma = 5$.

The value of Z when $X = 72$ is:

$Z = \dfrac{X - \overline{X}}{\sigma} \Rightarrow Z_1 = \dfrac{75-79}{5} = -\dfrac{4}{5} = \mathbf{-0.8}$

Area between Z = 0 and Z = – 0.8 is = 0.2881 ... (1)

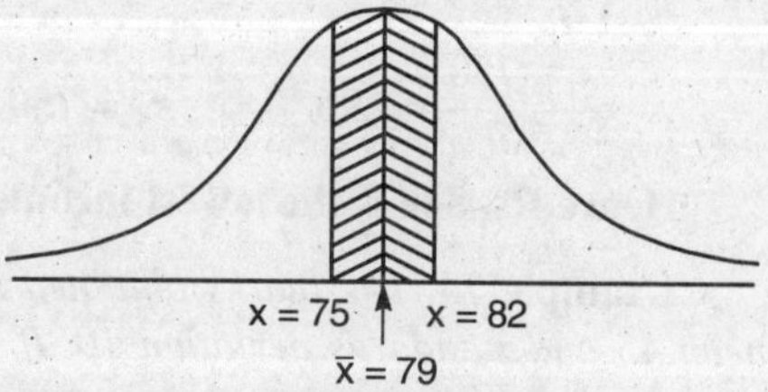

Fig. 6.13

The value of Z when $X = 82$ is:

$Z = \dfrac{X - \overline{X}}{\sigma} \Rightarrow Z = \dfrac{82-79}{5} = \dfrac{3}{5} = \mathbf{0.6.}$

Area between Z = 0 and Z = 0.6 is = 0.2257. ... (2)

Total area between ***X = 75*** and ***X* = 82** = 0.2881 + 0.2257 = **0.5138**

Number of students getting marks between 75 and 82

$= 200 \times 0.5138 = 102.76$ **or 103 students.**

Number of students did not getting marks between 75 and 82 = 200 – 103 = **97 students.**

Example 77. *The marks of the students in a certain examination are normally distributed with mean marks as 40% and standard deviation marks as 20%. On this basis, 60% of students failed. The result was moderated and 70% student passed. Find the pass marks before and after the moderation.*

Solution. Here $\bar{X} = 40\%$ and $\sigma = 20\%$.

Let X = Percentage of pass marks in the examination

X_1 = Pass marks before moderation

X_2 = Pass marks after moderation.

Fig. 6.14

Obviously, X_1 is located to the right of $X = \mu$ and consequently the corresponhding value of Z is positive.

The Standard normal variate corresponding to 0.1 (or 10%) is 0.25.

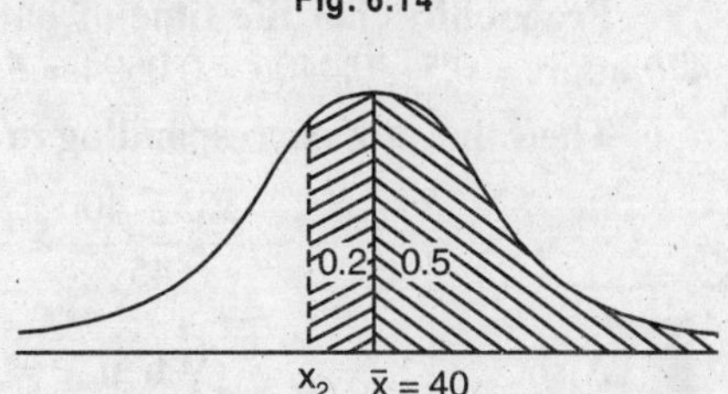

Fig. 6.15

Also
$$Z = \frac{X - \bar{X}}{\sigma}$$

$$\therefore \quad 0.25 = \frac{X_1 - 40}{20} \Rightarrow \mathbf{X_1 = 45.}$$

Hence the pass marks before moderation = 45.

After Moderation 70% of the students passed and pass marks X_2 %, so $P(X \geq X_2) = 0.70$. Obviously, X_2 is located to the left of $X = \mu$ and consequently, the corresponding value of Z will be negative.

Standard normal variate corresponding to 0.2 (or 20%) is = – 0.52.

$$\therefore \quad Z = \frac{X_2 - \bar{X}}{\sigma} \Rightarrow -0.52 = \frac{X_2 - 40}{20} \Rightarrow \mathbf{X_2 = 29.6.}$$

Hence the pass marks after moderation = 29.6 or 30 marks.

Example 78. *The monthly income distribution of workers in a certain factory was found to be normal with mean Rs. 500 and standard deviation equal to Rs. 50. There were 228 persons getting income above Rs. 600 per month. How many workers were there in all ?*

Extract of area under standard normal is given below :

Z :	*1*	*2*	*2.5*	*3*
Area :	*0.3413*	*0.4772*	*0.4938*	*0.4987*

Solution. Here $\bar{X} = 500;\ \sigma = 50;\ X = 600.$

Standard normal variate corresponding to $X = 600$ is:

$$\mathbf{Z} = \frac{X - \bar{X}}{\sigma} = \frac{600 - 500}{50} = \frac{100}{50} = 2.$$

Area between Z = 0 and Z = 2 is 0.4772

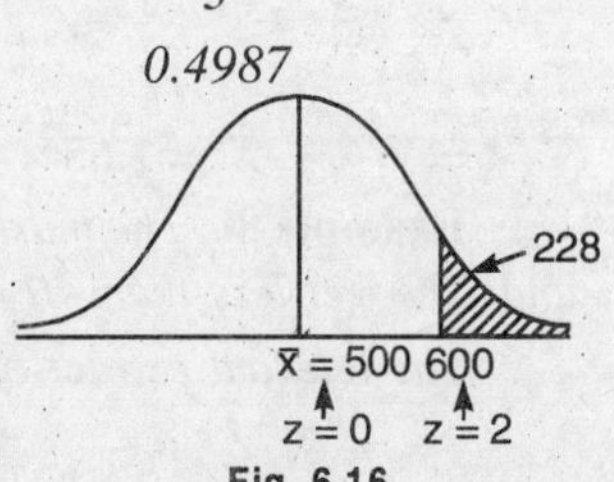

Fig. 6.16

[From the given extracts of the Table]

Area to the right of (Z = 2) = 0.5000 – 0.4772 = 0.0228 or **2.28%**

For the area 0.0228, the number of workers = 228

For the area 1.000, the number of workers $= \frac{228}{0.0228} \times 1.000$

= **10,000 workers.**

Example 79. *The life time of a certain kind of batteries has a mean life of 400 hours and standard deviation as 45 hours. Assuming the distribution of life time to be normal, find*

(*i*) *the percentage of batteries with a life time of atleast 470 hours.*

(*ii*) *the proportion of batteries with life time between 385 and 415 hours.*

(*iii*) *the minimum life of the best 5% of batteries*

Solution. Here Mean: $\bar{X} = 400$ hours; Standard deviation: $\sigma = 45$ hours.

(*i*) Standard Normal Variate *Z* for *X* = 470 is:

$$Z = \frac{X - \bar{X}}{\sigma} = \frac{470 - 400}{45} = \frac{70}{45} = \mathbf{1.56.}$$

$\therefore$ **Area between Z = 0 and Z = 1.56 = 0.4406 [See Table]**

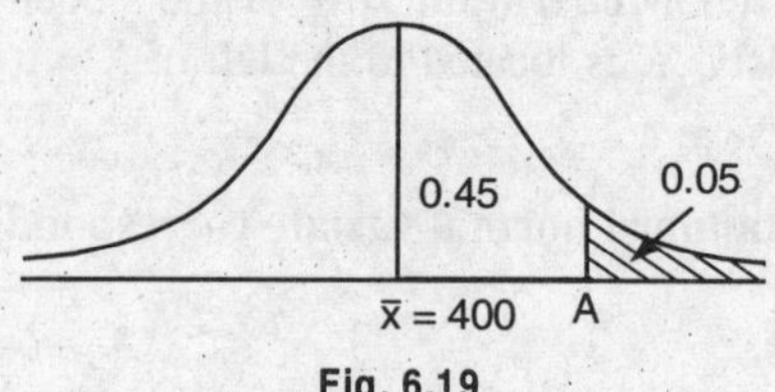

Fig. 6.17

Probability that life time of batteries is atleast 470 hours = 0.5 – 0.4406 = 0.0594 = **5.94% or 6%.**

The value of *Z* corresponding to *X* = 385 is

$$Z_1 = \frac{X - \bar{X}}{\sigma} = \frac{385 - 400}{45} = \frac{-15}{45} = \frac{-1}{3} = -0.33.$$

Also **When *X* = 415, then** $Z_2 = \frac{415 - 400}{45} = \frac{15}{45} = \frac{1}{3} = 0.33.$

Area between Z = 0 and Z = 0.33 is 0.1293 [From Table 3]

$\therefore$ **Area between *X* = 385 and *X* = 415 hours = (0.1293 + 0.1293) = 0.2586 or 26%.**

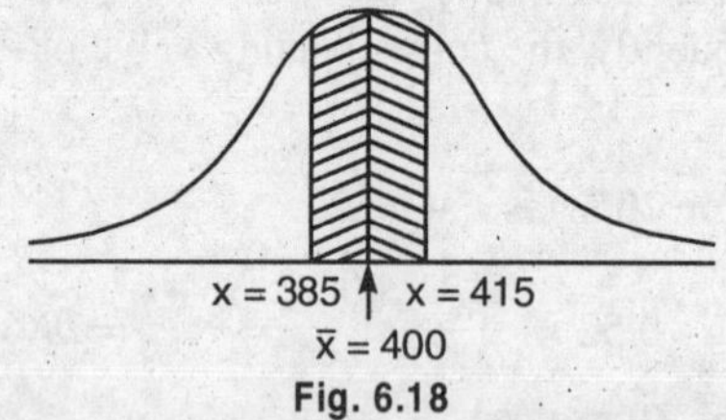

Fig. 6.18

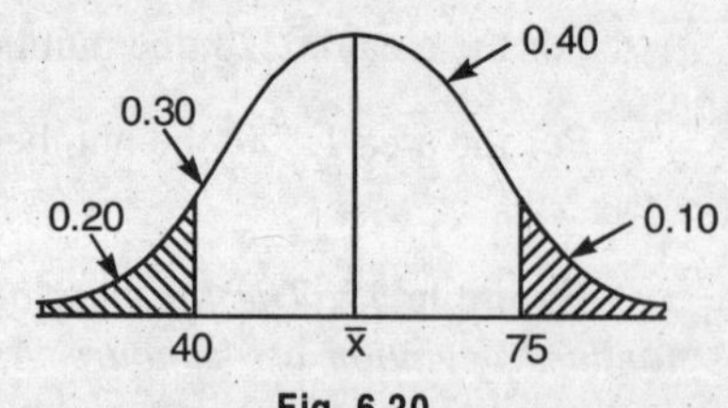

Fig. 6.19

(*iii*) Life of best 5% of batteries starts at *A* [See Figure 13.30].

The value of *X* at *A* gives the minimum life of best 5% batteries.

Value of *Z* corresponding to the area 0.45 to the right of $\bar{X} = 400$ = **1.65** **[From Table 3]**

$$\therefore \quad Z = \frac{X - \bar{X}}{\sigma} \Rightarrow 1.65 = \frac{X - 400}{45}$$

$\Rightarrow$ $\bar{X}$ = **1.65** × 45 + 400 = 74.25 + 400 = **474.25,** ***i.e.*****, 474 hours.**

Example 80. *The marks of the students are normally distributed. 10% get more than 75 marks and 20% get less than 40 marks. Find the mean and standard deviation of distribution.*

The relevant extract of Area Table (under the normal curve) is given below:

Z	0.84	1.28	2.0
Area	*0.2995*	*0.3997*	*0.4772*

Solution. Let $\bar{X}$ be the mean and σ be the standard deviation of the distribution.

The standard normal variate *Z* corresponding to

$30\% = \frac{30}{100} = 0.3$ is **0.84** [From the given table]

$$\therefore \quad \frac{40 - \bar{X}}{\sigma} = -0.84$$

$$\Rightarrow \quad \mathbf{0.84\,\sigma - \bar{X} = -40} \quad \text{... (1)}$$

Fig. 6.20

The standard normal variate corresponding to

$$40\% = \frac{40}{100} = 0.4 \text{ is } 1.28. \qquad \text{[From Tables]}$$

$$\therefore \quad \frac{75 - \bar{X}}{\sigma} = 1.28 \Rightarrow \mathbf{1.28\,\sigma + \bar{X} = 75.} \qquad \ldots (2)$$

Adding (1) and (2), we get: $2.12\,\sigma = 35 \Rightarrow \sigma = \frac{35}{2.12} = \mathbf{16.51}$

Putting $\sigma = 16.51$ in the equation (2), we get:

$$1.28 \times 16.51 + \bar{X} = 75 \Rightarrow \bar{X} = \mathbf{54.}$$

Hence the mean of the distribution is = 54 and its variance = 16.51.

Example 81. *The distribution of wages of a group of workers is known to be normal with mean Rs. 500 and SD Rs. 100. If the wages of 100 workers in the group are less than Rs. 430, what is the total number of workers in the group ?*

Solution. Let X denote the wage. It is given that X is normally distributed with mean as Rs. 500 and SD as Rs. 100 and $P(X < 430) = 100/N$, N being the total no. of workers in the group

$$\Rightarrow P\left(\frac{X-500}{100} < \frac{430-500}{100}\right) = \frac{100}{N} \Rightarrow P(Z < -0.70) = \frac{100}{N} \Rightarrow 1 - P(Z > 0.70) = \frac{100}{N}$$

$$\Rightarrow \qquad 1 - 0.758 = \frac{100}{N} \Rightarrow N = 4/3.$$

Example 82. *The mean of a normal distribution is 500 and 16 per cent of the values are greater than 600. What is the standard deviation of the distribution ?*

(Given that the area between $z = 0$ to $z = 1$ is 0.34)

Solution. Let σ denote the standard deviation of the distribution.

$$\therefore \qquad P(X > 600) = 0.16 \Rightarrow 1 - P(X \le 600) = 0.16 \Rightarrow P(X \le 600) = 0.84$$

$$\Rightarrow P\left(\frac{X-500}{\sigma} \le \frac{600-500}{\sigma}\right) = 0.84 \Rightarrow P\left(Z \le \frac{100}{\sigma}\right) = 0.84 \Rightarrow \frac{100}{\sigma} = 1 \text{ or } \boldsymbol{\sigma = 100.}$$

Example 83. *1,000 bulbs with a mean life 120 days are installed in a new factory. Their length of life is normally distributed with standard deviation 20 days. (i) How many bulbs will expire in less than 90 days ? (ii) If it is decided to replace all the bulbs together, what interval should be allowed between replacement if not more than 10% should expire before replacement ?*

Solution. Here $\bar{X} = 120$, and $\sigma = 20$.

(*i*) **When $X = 90$,** then $\quad Z = \frac{X - \bar{X}}{\sigma} \Rightarrow Z = \frac{90 - 120}{20} = \mathbf{-1.5.}$

Area between $Z = 0$ and $Z = -1.5 = 0.4332$ **[From Table]**

$\therefore$ Area to the left of $(Z = -1.5) = 0.5 - .4332$

$= 0.0668.$

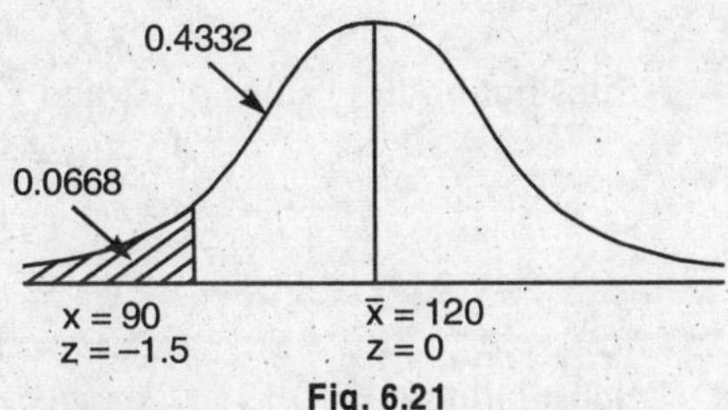

Fig. 6.21

Therefore, **number of bulbs expected to expire in less than 90 days** $= 0.0668 \times 1000 = \mathbf{66.8}$ or **67.**

(*ii*) The value of standard normal variate corresponding to area $(0.5 - 0.1) = 0.4$ is $= \mathbf{-1.28.}$ **[From Table]**

$$\therefore \quad \frac{X - 120}{20} = -1.28 \Rightarrow X = 120 - 1.28 \times 20 = \mathbf{94.4 \text{ or } 94.}$$

Hence the bulbs may be replaced after 94 days.

Example 84. *A multiple-choice quiz has 200 questions, each with 4 possible answers, of which only 1 is the correct answer. What is the probability (using normal approximation to binomal distribution) that sheer guesswork yields from 25 to 30 correct answers for 80 problems (out of the 200 problems) about which the student has no knowledge.*

[Area under the normal curve is 0.4996 from Z = 0 to Z = 2.71 and 0.3770 from Z = 0 to Z = 1.16]

Solution. Since each question has 4 possible answers and for $n = 80$ problems about which the student has no knowledge and uses sheer guess work, the probability of correct answer is given by:

$$p = 1/4. \qquad \therefore \quad q = 1 - p = 3/4.$$

If the random variable X denotes the number of correct answers (out of 80 problems done on sheer guess work) then using normal approximation to binomial distribution, $X \sim N(\mu, \sigma^2)$ where

$$\mu = \text{Mean} = np = 80 \times \frac{1}{4} = \mathbf{20.}$$

$$\sigma^2 = \text{Variance} = npq = 80 \times \frac{1}{4} \times \frac{3}{4} = 15 \Rightarrow \sigma = \sqrt{15} = 3.87.$$

The required probability that sheer guess work yields from 25 to 30 correct answers is:

$$P(25 \le X \le 30) = P\left(\frac{25-20}{3.87} \le Z \le \frac{30-20}{3.87}\right)$$

$$= P(1.29 \le Z \le 2.58) = P(0 \le Z \le 2.58) - P(0 \le Z \le 1.29)$$

$$= 0.4951 - 0.4015 = \mathbf{0.0936.}$$ **[From Normal Tables]**

Example 85. *At a certain examination, 10% of the students who appeared for the paper in Statistics got less than 30 marks and 97% of the students got less than 62 marks. Assuming the distribution to be normal, find the mean and the standard deviation of the distribution.*

Solution. Let X denote the marks in Statistics and let $X \sim N(\mu, \sigma^2)$. We are given:

$$P(X < 30) = 0.10 \text{ and } P(X < 62) = 0.97.$$

When $X = 30$, then $Z = \dfrac{30-\mu}{\sigma} = -Z_1$, (say). ... **(1)** [Note the negative sign]

When $X = 62$, then $Z = \dfrac{62-\mu}{\sigma} = Z_2$, (say). ... **(2)**

From the diagram, it can be seen that:

$$P(0 < Z < Z_2) = 0.47$$

$$\Rightarrow \quad Z_2 = \mathbf{1.88.}$$

Also, $$P(-Z_1 \le Z \le 0) = 0.40$$

$$\Rightarrow \quad P(0 \le Z \le Z_1) = 0.40 \text{ (By Symmetry)}$$

$$\Rightarrow \quad Z_1 = \mathbf{1.28.}$$

(From Normal Tables)

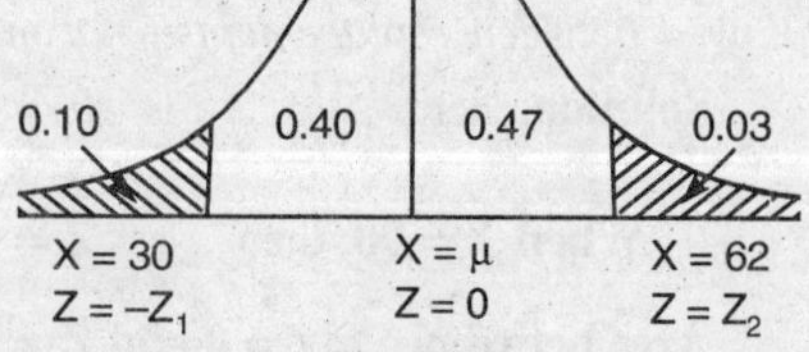

Fig. 6.22

Substituting the values of Z_1 and Z_2 in (1) and (2), we get:

$$30 - \mu = -1.28\,\sigma \qquad \text{... (3)}$$

$$62 - \mu = 1.88\,\sigma \qquad \text{... (4)}$$

Subtracting (3) from (4), we have:

$$32 = 3.16\,\sigma \Rightarrow \sigma = 32/3.16 = \mathbf{10.13}$$

Substituting $\sigma = 10.13$ in (3), we get :

$$\mu = 30 + 1.28 \times 10.13 = 30 + 12.97 = 42.97 \simeq \mathbf{43}$$

Hence mean is 43 and standard deviation is 10.13.

UNIFORM DISTRIBUTION

6.22 UNIFORM DISTRIBUTION

There are two types of Uniform distributions:

6.22.1 Uniform Distribution For Discrete Random Variable

A discrete random variable is said to follow a uniform distribution if it takes equal probabilities at its various discrete values.

If a discrete random variable X takes the values $x_1, x_2, x_3, ..., x_n$, each with probability '$\frac{1}{n}$', then the distribution of X is said to be uniform

In this case

$$\sum p(x_i) = p(x_1) + p(x_2) + ... + p(x_n) = \frac{1}{n} + \frac{1}{n} + ... + \frac{1}{n} = 1.$$

$$\textbf{MEAN}: E(X) = p(x_1)\,x_1 + p(x_2)\,x_2 + p(x_3)\,x_3 + ... + p(x_n)\,x_n$$

$$= \frac{1}{n}x_1 + \frac{1}{n}x_2 + ... + \frac{1}{n}x_n = \frac{x_1 + x_2 + ... + x_n}{n}.$$

6.22.2 Uniform Distribution For Continuous Random Variable

The **Uniform distribution of a continuous random** variable is the simplest continuous distribution. **Its probability density function $f(x)$ is given by :**

$$\frac{1}{(b-a)}, \qquad \text{for } a \le x \le b$$

$$0, \text{ for all other value of } x,$$

where a is the minimum possible value and b is the maximum possible value of x. It is also known as rectangular distribution.

6.22.3 GRAPH OF A UNIFORM DISTRIBUTION

The curve of $f(x)$ is a flat line. The area under it between any two points a and b will be the area of a rectangle with height $1/(b-a)$ and width $(b-a)$.

The total area under is curve is unity, *i.e.*

$$\int_a^b \frac{dx}{b-a} = \frac{1}{b-a} \left| x \right|_a^b = \frac{b-a}{b-a} = 1.$$

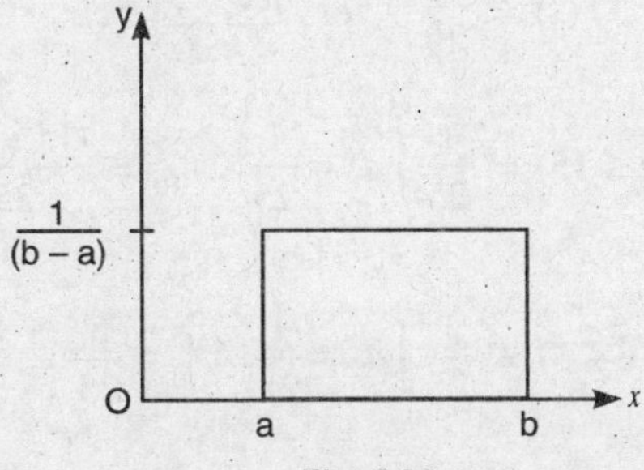

Fig. 6.23

MEAN: $$E(X) = \frac{1}{b-a}\int_a^b x\,dx = \frac{1}{b-a}\left|\frac{x^2}{2}\right|_a^b$$

$$= \frac{(b^2 - a^2)}{2(b - a)} = \frac{b+a}{2}.$$

Similarly, $$\textbf{Variance} = \frac{(b-a)^2}{12}.$$

6.22.4 PROPERTIES OF UNIFORM DISTRIBUTION

1. If X is uniformly distributed between a and b, we shall write $X \sim U(a, b)$.
2. $P(x_1 \le X \le x_2) = (x_2 - x_1)/(b - a)$
3. Its **mean** $= \dfrac{(a+b)}{2}$.
4. Its **variance** : $V(X) = \dfrac{(b-a)^2}{2}$.

6.22.5 Illustrations of a Uniform Distribution

1. A common instance of a uniform distribution is waiting time for a facility that goes in cycles such as elevator, which moves roughly in cycles with some cycle time.
2. Journey in Metro Rail or a tube in Delhi from Central Secretariat to Cannaught Place. If a user comes to a stop at random time and wait till the facility (tube) arrives, the waiting time will be uniformly distributed between a minimum of zero and maximum equal to cycle time (say 20 minutes).
3. If a shuttle bus has a cycle time of 30 minutes the waiting time would be uniformly distributed between 0 and 30 minutes.

Example 66. *The buses on a certain route run after every 25 minutes. If a person arrives at the bus stop at random, what is the probability that*

(*i*) *he has to wait between 10 to 15 minutes ?*

(*ii*) *he gets a bus within 5 minutes ?*

(*iii*) *he has to wait atleast 15 minutes ?*

Solution. Let x be the random variable denoting waiting time, which follows a uniform distribution with probability density function

$$f(x) = \frac{1}{25}; \quad \text{for} \quad 0 \le x \le 25.$$

(*i*) $$P(10 \le x \le 15) = \frac{1}{25}\int_{10}^{15} dx = \frac{1}{25}\left| x \right|_{10}^{15} = \frac{(15-10)}{25} = \frac{5}{25} = \frac{1}{5}.$$

(*ii*) $$P(0 \le x \le 5) = \frac{1}{25}\int_{0}^{5} dx = \frac{1}{25}\left| x \right|_{0}^{5} = \frac{5}{25} = \frac{1}{5}.$$

(*iii*) $$P(15 \le x \le 25) = \frac{1}{25}\int_{15}^{25} dx = \frac{1}{25}\left| x \right|_{15}^{25} = \frac{1}{25}(25-15) = \frac{10}{25} = \frac{2}{5}.$$

EXERCISE – 6.2

1. The I.Q.s of army volunteers in a given year are normally distributed with mean = 110 and standard deviation = 10. The army wants to give advance training to 20% of those recruits with the highest scores. What is the lowest I.Q. score acceptable for the advanced training ?

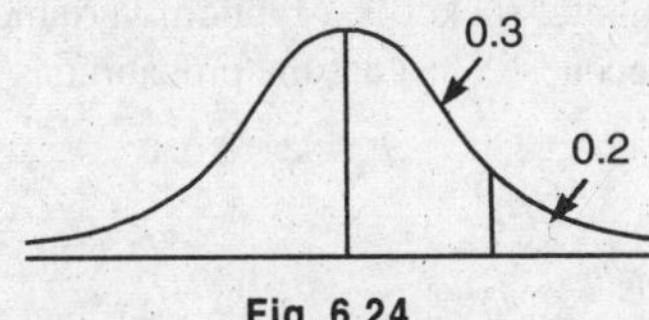

Fig. 6.24

[**Hint.** The value of Z for the area $(0.5 - 0.2) = 0.3$ is $= 0.84$.

$$\therefore \quad 0.84 = \frac{X - 110}{10} \Rightarrow \mathbf{X = 118.4.}]$$

2. A company, that sells 5000 batteries in a year, guarantees them for a life of 24 months. Life of the batteries is estimated to be approximately normal with mean equal to 34 months and standard diviation of 5 months. Find the number of batteries that will have to be replaced under the guarantee.

3. The weekly wages of 1,000 workers are normally distributed with a mean of Rs. 70 and a standard deviation of Rs. 5. Estimate the lowest weekly wages of the 100 highest paid workers.

4. Mean and standard deviation of chest measurements of 1,200 soldiers are 85 cms and 5 cms respectively. How many of them are expected to have their chest measurements exceeding 95 cms. assuming the measurements to follow the normal pattern ?

[**Hint.** $Z = \frac{95 - 85}{5} = 2.$ Area to the right of $Z = 2$ is $= 0.5000 - 0.4772 = 0.0228$.

$\therefore$ Number of soldiers $= 0.0228 \times 1200 = 27.$]

5. The mean life time of 60 watt light bulb produced by Bright light bulbs company is 200 hours. It is keeping the standard deviation 20 hours. Assuming that the life times of the bulbs are normally distributed, what are the probabilities that a single 60 watt light bulb extracted from the production lot will (*i*) burn out between 180 hours and 210 hours ; (*ii*) burn out at a time greater than 250 hours ?

6. The mean marks on a final examination was 72 and the standard deviation was 9. The top 10% of the students are to receive A's. What is the minimum marks a student must get in order to receive an A?

7. Two thousand electric bulbs with an average life of 1000 hours and a standard deviation of 200 hours are installed in a town. Assuming the lives of the bulbs to be normally distributed, answer the following

(*i*) What number of bulbs might be expected to fail in the first 700 burning hours?

(*ii*) What is the minimum burning life of top one quarter of bulbs?

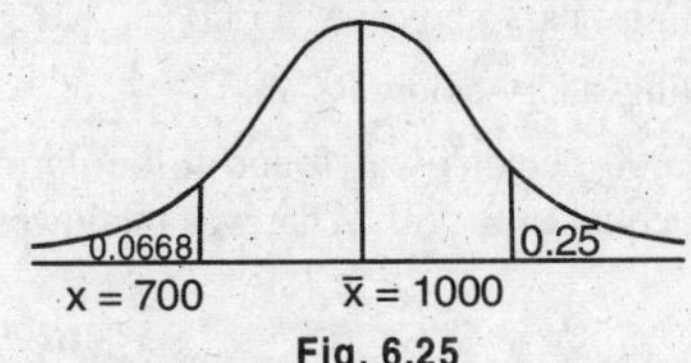

Fig. 6.25

[**Hint.** (*i*) $$Z = \frac{X - \bar{X}}{\sigma} = \frac{700 - 1000}{200} = -1.5.$$

$$P(Z < -1.5) = 0.5 - 0.4332 = 0.0668.$$

Expected number of bulbs to fail $= 2000 \times 0.0668 = 134$.

(*ii*) Probability of top one quarter bulbs = $\frac{1}{4}$ = 0.25.

$$\therefore \quad Z = \frac{X-1000}{200} \Rightarrow 0.67 = \frac{X-1000}{200} \Rightarrow \mathbf{X = 1134.\ hours.}]$$

8. The burning time of an experimental rocket is a random variable which has normal distribution with μ = 4.36 seconds and σ = 0.04 seconds. What are the probabilities that this kind of rocket will burn for

(*i*) less than 4.5 seconds ?

(*ii*) more than 4.40 seconds ?

(*iii*) between 4.30 to 4.42 seconds ?

9. Of a large group of men, 5 per cent are under 60 inches in height and 40 per cent are between 60 and 65 inches. Assuming a normal distribution, find the mean height and standard deviation.

10. In a normal distribution 30% of the items are under 50 and 10% are over 86. Find the mean and standard deviation of the distribution.

[**Hint.** $\frac{50-\bar{X}}{\sigma} = 0.52 \Rightarrow -\bar{X} + 0.52\sigma = -50$; ... (1) $\frac{86-\bar{X}}{\sigma} = 1.28 \Rightarrow -\bar{X} - 1.28\sigma = -86$ ... (2)

Solving (1) and (2), we get: **X = 60.4 and σ = 20.**]

11. 15,000 students appered in an examination. The mean marks were 49 and the standard deviation of marks was 6. Assuming the marks to be normally distributed:

(*i*) What proportion of students score more than 55 marks ?

(*ii*) If in the same examination, grade A is to be given to students scoring more than 70 marks, what proportion of students will receive grade A ?

[**Hint.** (*i*) $Z = \frac{55-49}{6} = 1$. The proportion of students scoring between $Z = 0$ and $Z = 1$ is 0.3413.

Proportion of students scoring more than 55 marks = 0.5 – 0.3413 = 0.1587 or 15.87%

(*ii*) $Z = \frac{70-49}{6} = 3.5$. Area corresponding to $Z = 3.5$ is 0.4998.

The proportion (0.5 – 0.4998) = 0.0002 or 0.02% would score more than 70 marks. Number of students getting 'A'-grade = 15000 × 0.0002 = **3.**]

12. In a manufacturing organisation the distribution of wages was perfectly normal and the number of workers employed in the organisation was 5000. The mean wages of the workers were calculated as Rs. 800 p.m. and the standard deviation was worked out to be Rs. 200. On the basis of the information, estimate:

(*i*) The number of workers getting salary between Rs. 700 and Rs. 900.

(*ii*) Percentage of workers getting salary above Rs. 1,000.

(*iii*) Percentage of workers getting salary below Rs. 600.

13. The income of a group of 10,000 persons was found to be normally distributed with mean equal to Rs. 750 and standard deviation equal to Rs. 50. What was the lowest income among the richest 250?

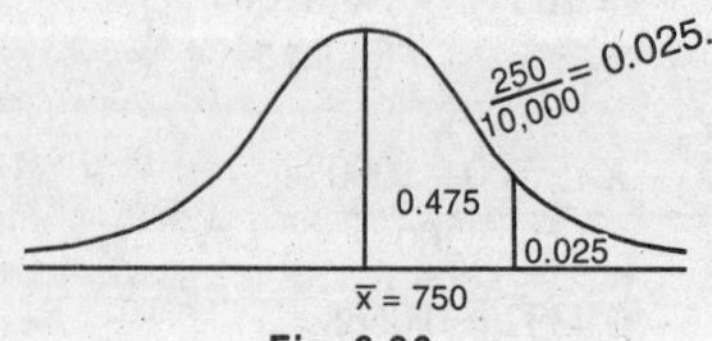

Fig. 6.26

[**Hint.** Area covered by richest 250 = $\frac{250}{10000} = 0.025$.

The value of Z corresponding to area 0.5 – 0.025 = 0.4750 is 1.96

$$\therefore \quad \frac{X - 750}{50} = 1.96 \Rightarrow \boldsymbol{X = \text{Rs. } 848.}]$$

14. For a normal distribution, the first moment about origin is 35 and the second moment about 35 is 10. Find the first four central moments.

15. The life time of a certain kind of battery has a mean of 300 hours and a standard deviation of 35 hours. Assuming that the distribution of life times, which are measured to the nearest hour is normal, find the percentage of batteries which have life time of more than 370 hours.

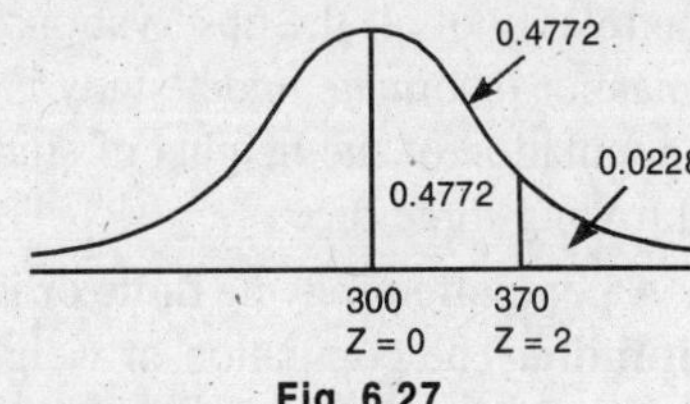

Fig. 6.27

[**Hint.** $Z = \frac{370 - 300}{35} = 2.$

The area between Z = 0 and Z = 2 is 0.4772.

Area of $Z > 2$ = 0.5000 – 0.4772 = 0.0228. Required Percentage = 0.0228 × 100 = **2.28%.**]

ANSWERS

1. 118.4. **2.** 114. **3.** 76.4. **4.** 27.

5. (*i*) 0.5328; (*ii*) 0.0062. **6.** 83.52.

7. (*i*) 134 Bulbs; (*ii*) 1134 hours. **8.** (*i*) 0.003; (*ii*) 0.1587; (*iii*) 0.8664.

9. $\mu = 65.42; \sigma = 3.29$. **10.** $\mu = 60.4, \sigma = 20$. **11.** (*i*) 15.87%; (*ii*) 3 students.

12. (*i*) 1915; (*ii*) 15.87%; (*iii*) 16%. **13.** Rs. 848.

14. 0; 10; 0; 300. **15.** 2.28%.

7

CHAPTER

SAMPLING DISTRIBUTION CONCEPTS

7.1 POPULATION AND SAMPLE

7.1.1 Population.

Population in statistics means the whole of the information which comes under the purview of statistical investigation. It is the totality of all the observations of a statistical enquiry. In other words, an aggregate of objects animate or inanimate under study is the population. It is also known as the **universe**. For example, the population of the heights of students in a school, the population of the sum of points obtained in throwing three dice.

Finite or Infinite Population. A population may be finite or infinite according as the number of individuals in it are finite or infinite. The population of weights of students of class XII in a government school is an example of a finite population. The population of pressures at different points in the atmosphere is an example of an infinite population.

7.1.2 Sample.

A part of the population selected for study is called a sample. In other words, ***the selection of a group of individuals or items from a population in such a way that this group represents the population, is called a sample.*** For example, a housewife tests a small quantity of rice to see that it has been cooked or not. This small quantity of rice is a sample and represents the entire quantity of rice cooked. A **sample is a selected portion of the population.** A sample drawn from a population provides a valuable information about its parent population. It gives a fairly accurate result and a reliable picture of the total observations under investigation. It is always used to measure and estimate the corresponding characteristics of its parent population. ***When the sample drawn is perfectly representative, it is identical with its parent population almost in every respect except that it is smaller than the population.***

Size of Sample. ***The number of individuals included in the finite sample is called the size of the sample.***

7.2 PARAMETER AND STATISTIC

There are various statistical measures in statistics such as mean, median, mode, standard deviation, coefficient of variation, variance etc. These statistical measures can be computed both from **population** (or universe) **data** and **Sample data.**

Parameter: ***Any statistical measure computed from population data is known as parameter.***

Statistic: ***Any statistical measure computed from sample data is known as statistic.***

Thus a **parameter** is a statistical measure which relates to the population and is based on **population,** whereas a **statistic** is a statistical measure which relates to the **sample** and is based on sample data. Thus a ***population mean, population median, population variance, population coefficient of variation etc., are all parameters.*** Statistic computed from a **Sample** drawn from the parent population plays an important role in (*i*) The theory of Estimation and, (*ii*) Testing of hypothesis.

The usual notation used for **parameter** (in the case of population data) and **Statistic** (in the case of sample data) are given below :

	Notations	
Statistical Mesure	*Population*	*Sample*
Mean	μ	$\overline{x}$
Standard deviation	σ	s
Population	P	p
Size	N	n

7.3 SAMPLING

It is the procedure or process of selecting a sample from the population. A sampling can also be defined as the process of drawing a sample from a population and of compiling a suitable **statistic** from such a sample in order to estimate the **parameter** drawn from the parent population and to test the significance of the statistic computed from such sample.

7.4 SAMPLING THEORY

The study of relationship existing between a population and the samples drawn from the population is called **Sampling. Sampling theory is based on sampling.** It deals with statistical inferences drawn from sampling results. **Statistical inference** made on the basis of sampling results are of the following three types:

(*i*) **Statistical Estimation.** ***It helps in estimating an unknown population parameter (such as population mean, median, mode, standard deviation, kurtosis etc.) on the basis of suitable statistic (such as sample mean, mode, media, variance etc.) computed from the sample drawn from such parent population.***

(*ii*) **Tests of Significance.** ***Sampling theory helps in testing of significance about the population characteristics on the basis of suitable statistic computed from a sample drawn from such parent population.*** In other words, it helps in determining whether observed differences between two samples are actually due to chance or whether they are really significant. Such questions help us in deciding whether one production is better than the other. The test of significance plays an important role in the theory of decisions.

(*iii*) **Statistical Inference.** ***Statistical inference means drawing conclusions about some matters on the basis of certain statistical results.*** These statistical results are obtained by drawing samples from the population and then by computing suitable statistic from these samples so as to make statistical inferences. These statistical inferences enable us to draw statistical conclusions about some measures of a population on the basis of such statistic.

7.5 OBJECTS OF SAMPLING THEORY

The main objects of the sampling theory are:

1. **To estimate population parameter on the basis of sample statistic.** Sampling aims at obtaining the information about the population on the basis of sample drawn from such population. This is done on the basis of estimation of unknown population **parameter** by using a suitable statistic computed from a sample drawn from such parent population.
2. **To set the limits of accuracy and degree of confidence of the estimates of the population parameter computed on the basis of sample statistic.** The estimates of the population **parameters** obtained on the basis of **sample** statistics may not give true results. Thus we set the limits of accuracy and degree of confidence on such estimates in order to determine how precise are the estimates. Thus the **precision of the estimate** is another main object of sampling theory.

3. **To test significance about the population characteristic on the basis of sample statistic.** Statistical inferences and statistical conclusions about population characteristic may be conveniently drawn on the basis of sample statistic. Thus the testing of statistical hypothesis and drawing of statistical conclusions are also the objective of the sampling theory.

7.6 METHODS OF SAMPLING

A sample can be selected from a population in various ways. Different situations call for different methods of sampling. There are two methods of Sampling:

1. **Random Sampling or Probability Sampling Method**
2. **Non-Random Sampling or Non-Probability Sampling Method.**

7.6.1. Random or Probability Sampling

Random or Probability sampling is the scientific technique of drawing samples from the population according to some laws of chance in which each unit in the universe or population has some definite pre-assigned probability of being selected in the sample. It is of two types.

(*a*) **Simple Random Sampling.**

It **is the method of selection of a sample in such a way that each and every member of population or universe has an equal chance or probability of being included in the sample.** Random sampling can be carried out in two ways.

(*i*) **Lottery Method.** It is the simplest, most common and important method of obtaining a random sample. Under this method all the members of the population or universe are serially numbered on small slips of a paper. They are put in a drum and thoroughly mixed by vibrating the drum. After mixing, the numbered slips are drawn out of the drum one by one according to the size of the sample. The number of slips so drawn constitute a **random sample.**

(*ii*) **Random Number Method.** In this method, sampling is conducted on the basis of random numbers which are available from the random number tables. The various random number tables available are

(*a*) Trippet's Random Number series;

(*b*) Fisher's and Yales Random Number series;

(*c*) Kendall and Badington Random Numbers series;

(*d*) Rand Corporation Random Number series;

One major disadvantage of random sampling is tthat all the members of the population must be known and be serially numbered. It will entail a lot of difficulties in case the population is of large size and will be impossible in case the population is of infinite size.

(*b*) **Restricted Random Sampling.**

It is of three types

(*i*) **Stratified Sampling**

(*ii*) **Systematic Sampling**

(*iii*) **Multi-stage Sampling**

Stratified Sampling. ***In stratified random sampling, the population is divided into strata (groups) before the sample is drawn.*** Strata are so designed that they do not overlap. Elementary units from each stratum is drawn at random and the units so drawn constitute a sample. Stratified sampling is suitable in those cases where the population is hetrogeneous but there is a homogeneity within each of the groups or strata.

Advantages

(*i*) ***It is a representative sample of the hetrogeneous population.***

(*ii*) ***It lessens the possibility of bias of one sidedness.***

Disadvantages

(*i*) ***It may be difficult to divide the population into hetrogeneous groups.***

(*ii*) ***There may be over-lapping of different strata of the population which will provide an unrepresentative sample.***

Systematic Sampling. ***In this method every elementary unit of the population is arranged in order and the sample units are distributed at equal and regular intervals.*** In other words, a sample of suitable size is obtained (from the orderly arranged population) by taking every unit say tenth unit of the population, one of the first units in this ordered arrangement is chosen at random and the sample is computed by selecting every tenth unit (say) from the rest of the lot. If the first unit selected is 4, then the other units constituting the sample will be 14, 24, 34, 44, and so on.

Advantages. ***It is most suitable where the population units are serially numbered or serially arranged.***

Disadvantages. ***It may not provide a desirable result due to large variation in the items selected.***

Multistage Sampling. ***In this sampling method, sample of elementary units is selected in stages. Firstly a sample of cluster is selected and from among them a sample of elementary units is selected.*** It is suitable in those cases where population size is very big and it contains a large number of units.

7.6.2 Non-Random or Non-Probability Sampling Method

A sample of elementary units that is being selected on the basis of personal judgement is called a **non-probability sampling.** It is of five types.

(*i*) ***Purposive Sampling;*** (*ii*) ***Cluster Sampling***

(*iii*) ***Quota Sampling;*** (*iv*) ***Convenience Sampling***

(*v*) ***Sequential Sampling***

Purposive Sampling. ***Purposive sampling is the method of sampling by which a sample is drawn from a population based entirely on the personal judgement of the investigator.*** It is also known as **Judgement Sampling** or **Deliberate Sampling.** A randomness finds no place in it and so the sample drawn under this method cannot be subjected to mathematical concepts used in computing sampling error.

Cluster Sampling. ***Cluster Sampling involves arranging elementary items in a population into hetrogeneous subgroups that are representative of the overall population.*** One such group constitutes a sample for study.

Quota Sampling. ***In quota sampling method, quotas are fixed according to the basic parameters of the population determined earlier and each field investigator is assigned with quotas of number of elementary units to be interviewed.***

Convenience Sampling. ***In convenience sampling, a sample is obtained by selecting convenient population elements from the population.***

Sequential Sampling. ***In sequential sampling a number of sample lots are drawn one after another from the population depending on the results of the earlier samples drawn from the same population.*** Sequential sampling is very useful in Statistical Quality Control. If the first sample is acceptable, then no further sample is drawn. On the other hand if the initial lot is completely unacceptable, it is rejected straightway. But if the initial lot is of doubtful and marginal character falling in the bond lying between the acceptance and rejection limits, a second sample is drawn and if need be a third sample of bigger size may be drawn in order to arrive at a decision on the final acceptance or rejection of the lot. Such sampling can be based on any of the random or non-random method of selection.

7.7 ADVANTAGES OF RANDOM OR PROBABILITY SAMPLING

1. ***Random sampling is objective and unbiased. As a result, it is defensible before the superiors or even before the court of law.***
2. ***The size of sample depends on demonstrable statistical method and therefore, it has a justification for the expenditure involved.***
3. ***Statistical measures, i.e., parameters based on the population can be estimated and evaluated by sample statistic in terms of certain degree of precision required.***
4. ***It provides a more accurate method of drawing conclusions about the characteristics of the population as parameters.***
5. ***It is used to draw the statistical inference.***
6. ***The samples may be combined and evaluated, even though accomplished by different individuals.***
7. ***The results obtained can be assessed in terms of probability, and the sample accepted or rejected on a consideration of the extent to which it can be considered representative.***

7.8 SAMPLING AND NON-SAMPLING ERRORS

A sample is a part of the whole population. A sample drawn from the population depends on chance and as such all the characteristics of the population may not be present in the sample drawn from the same population. **Any statistical measure say, mean of the sample, may not be equal to the corresponding statistical measure (mean) of the population from which the sample has been drawn. Thus there can be discrepancies in the statistical measure of population,** *i.e.*, **parameters** and the statistical measures of sample drawn from the same population, *i.e.*, **statistic**. These discrepancies are known as **Errors in Sampling.** Errors in sampling are of two types.

(*i*) **Sampling Erros**

(*ii*) **Non-sampling Errors or Bias.**

7.8.1 Sampling Errors

Sampling Errors is inherent in the method of sampling. Sampling depends on chance and due to the existence of chance in sampling, the sampling errors occur. Errors in sampling arise primarily due to the following reasons:

1. **Faulty selection of the sample.** This may be due to selection of defective sampling techniques which may introduce the element of bias, *e.g.*, purposive or judgement sampling, in which investigator deliberately selects a non-representative sample.
2. **Substitution.** Sometimes an investigator while collecting the information from a particular sampling unit, included in the random selection substitutes a convenient member of the population and this may lead to some bias as the characteristic possessed by the substituted unit may be different from those possessed by the original unit included in sampling.
3. **Faulty demarcation of sampling units.**
4. **Variability of the population.** Sampling error may also depend on the variability or hetrogenity of the population from which the samples are to be drawn.

7.8.2 Non-Sampling Errors or Bias

Non-**sampling errors** or **Bias** ***automatically creep in due to human factors which always varies from one investigator to another.*** Bias may arise in the following different ways.

(i) ***due to negligence and carelessness on the part of investigator;***

(ii) ***due to faulty planning of sampling;***

(iii) ***due to the faulty selection of sample units;***

(iv) ***due to incomplete investigation and sample survey;***

(v) ***due to framing of a wrong questionnaire;***

(vi) ***due to negligence and non-response on the part of the respondents;***

(vii) ***due to substitution of a selected unit by another;***

(viii) ***due to error in compilation;***

(ix) ***due to applying wrong statistical measure.***

7.9 SAMPLING FLUCTUATIONS

Any statistical measure or a **parameter** computed from a population is always **fixed and constant.** ***Thus there cannot be any fluctuation in the value of the parameter computed from the population. But this is not true in the case of statistical measure, i.e., statistic computed from different samples of the same size drawn from the same parent population, because various samples of the same size drawn from the same population may be different. Any statistical measure i.e., statistic computed from each of such samples will also vary from sample to sample.*** The **variation in the value of the statistic from one sample to another is called sampling fluctuation** or **sampling variation.** Thus a statistic has sampling fluctuations whereas a parameter has no sampling fluctuations.

7.10 SAMPLING DISTRIBUTION OF A STATISTIC

Sampling distribution of a statistic ***is the frequency distribution which is formed with various values of a statistic computed from different samples of the same size drawn from the same population.*** We can draw a large number of samples of same size from a population of fixed size, each sample containing different population members. Any **statistic** (statistical measure of sample) like mean, median, variance, standard, deviation etc. may be computed for each of the samples. As a result a series of **various values** of **that statistic** may be obtained. These various values can be arranged into a frequency distribution table, which is known as **the sampling distribution of the statistics.**

Sampling may be done with replacement or without replacement. **Sampling with replacement** ***means that the same unit of the population may be included in each sample more than once.*** **Sampling without replacement** ***means that the same unit of population may not be included in each sample more than once.*** In the case of sampling with replacement **the total number of possible samples each of size n drawn out of population of size N is N^n** . But if the sampling is without replacement the total number of possible samples will be **C (N, n) = m** (say): For each of these samples a value of satistic, say sample mean $\bar{x}$, is computed. As the samples are formed with different sample units so the value of each of the sample means will be different. ***The sample mean can be regarded as a random variable $\bar{x}$ and each sample mean then constitute as the observed value of this new random variable $\bar{x}$. Let these values be $\bar{x}_1, \bar{x}_2, \bar{x}_3, \ldots\ldots, \bar{x}_m$. These mean values $\bar{x}_1, \bar{x}_2, \bar{x}_3, \ldots\ldots, \bar{x}_m$ can be used to form a frequency distribution. Then this frequency distribution of the statistic $\bar{x}$ is known as the Sampling distribution of Sample mean.*** Similarly, the sampling distribution of standard deviation or coefficient of variation or variance may be constructed with the various values of standard deviation or coefficient of variation or variance respectively.

If the population **size is infinitely large** or **sampling is done with replacement**, then the total number of possible samples of the same size which may be drawn from the population cannot be determined. In such a case, ***a large number of repeated random samples from the population of fixed size can be drawn and the values of statistic for these samples may be computed. These values of the statistic can be used to form a frequency distribution. This frequency distribution of the statistic is known as the Sampling Distribution of Statistic.*** The main **characteristic** of the **Sampling distribution of a statistic** is that it approaches **normal distribution** ***even when the population distribution is not normal provided the sample size is sufficiently large*** (greater than 30). Another important feature of the sampling distribution of statistics is that the **mean and the standard deviation of the sampling distribution of sample mean bear a definite relation to the corresponding parameters,** *i.e.,* mean and standard deviation of parent population. These characteristics of the sampling distribution help us

(*i*) ***To estimate the unknown population parameter from the known statistic.***

(*ii*) ***To set the confidence limits of the parameter within which the parameter values are expected to lie.***

(*iii*) ***To test a hypothesis and to draw a statistical inference from it.***

7.11 BASIC STATISTICAL LAWS

The two popular and well defined statistical laws which indicate the utility of larger size of the sample for the purpose of reducing sampling errors are:

1. **Law of Statistical Regularity**
2. **Law of Inertia of Large Number.**

7.11.1 Law of Statistical Regularity

It states that a reasonably larger number of items selected at random from a large group of items, will on the average, represent the characteristics of the group. In the words of the statistician W.I. King, ***"The law of statistical regularity lays down that a moderately large number of items chosen at random from a large group, are almost sure (on the average) to possess the characteristics of the large group".***

In other words, this law explains that if a reasonably large sample is selected at random without bias (*i.e.,* probability sampling), it is almost certain that on an average, the sample so chosen, shall have the same characteristics as those of the parent population from where the units constituting the sample have been drawn. It is on the basis of this theory that the ***law of statistical theory tells us that a random selection is very likely to give a representative sample.***

7.11.2 Law of Inertia of Larger Number

It states that large groups or aggregate of data show high degree of stability because there is a greater possibility that the extremes on one side are compensated by the extremes on the other side.

The law of inertia of large number is a corollary to the law of statistical regularity. It emphasizes the fact that large numbers are relatively more stable and more reliable than small ones. In a larger number it is unlikely that the data would move in only one direction. Thus, *the greater the size of the sample, the greater will be the compensation or tendency to neutralise one another and consequently more stable would be the result*. For example, the birth rate, death rate etc. may vary from place to place, in India but India as a whole country, they will be found somewhat stable over a number of years.

The simplest method of increasing the accuracy of a sample is to increase its size. The larger the size of the sample, the more reliable is the result. The other things remaining unchanged, the sampling is inversely proportional to the root of the number of items in the sample.

The **Law of Statistical Regularity** and the **Law of Intertia of Large Numbers** *have great importance in the theory of sampling as the sampling error is reduced to minimum if these laws are correcttly applied.*

7.12 CENTRAL LIMIT THEOREM

If $x_1, x_2, x_3, ..., x_n$ is a random sample drawn from a population or universe having a mean μ and variance σ^2 then the distribution of sample means approaches a normal distribution with mean μ and variance σ^2/n as the sample size n increases, *i.e.*, **the sample mean is normally distributed with mean μ and variance σ^2/n.**

Central Limit Theorem is one of the most fundamental theorems in statistics. It states:

"If $x_1, x_2, ..., x_n$ is a random sample of size n drawn from any population, then the sample mean $(\bar{x})$ is normally distributed with mean μ and variance σ^2/n, provided n is sufficiently large, i.e., as $n \to \infty$, where μ and σ^2 respectively are population mean and variance."

Central limit assures us that the sampling distribution of the mean approaches normal with the increase in sample size. **Larger the sample size, better will be approximation.** From practical point of view, the approximation is farily good if the sample size $n > 30$. Thus if t is any statistic then by central limit theorem,

$$Z = \frac{t - E(t)}{\text{S.E.}(t)} \sim N(0,1) \text{ as } n \to \infty$$

The above result means **"The distribution of any statistic in its standardized form is asymptotically normal as $n \to \infty$."**

The above result is **extensively used in Large Sample Tests** and is also used in constructing **confidence limits** for the population parameters, when the samples are large.

7.13 STANDARD ERROR OF A STATISTIC

The statistical measure of **standard deviation** may be computed both from the observations of the **population** and also from the observations of the **Sampling distribution.** Also we know that the standard deviation is a measure of the average amount of the variability of all the observations of variable from their mean. *When the average amount of the variability of the observations of a population is computed, it is called the* **standard deviation.** *But when the average amount of the variability of the observations of a sampling distribution computed, it is known as* **Standard Error. Thus the standard deviation computed from the observations of a sampling distribution of a statistic is called the standard error of the statistic.** In other words, *the standard deviation used to measure the variability of the values of a statistic from sample to sample is called the stndard error of the statistic. Thus the standard deviation and standard error have the same meaning and same connection but are used in different cases and different circumstances. Both of them are used to measure the variability of observations.*

Standard error is used to measure the variability of the values of a statistic computed from the samples of the same size drawn from the population, whereas standard deviation is used to measure the variability of the observations of the population itself.

The following two standard errors are frequently used in statistics.

1. **Standard Error of Sample Mean.** ***It is the standard deviation of the sampling distribution of sample means. It is denoted by*** $\sigma_{\bar{x}}$ and is given by $\sigma_{\bar{x}} = \dfrac{\sigma}{\sqrt{n}}$, ***where σ is the standard deviation of population and n is the sample size.***

2. **Standard Error of Proportion.** It is computed from the proportions of all possible samples of the same size drawn from a population. It is denoted by s_p, and is given by

$$s_p = \sqrt{\frac{P(1-P)}{n}}, \text{ if } P \text{ is known,}$$

where P = population proportion, n = sample size.

7.14. UTILITY OF STANDARD ERROR OF A STATISTIC

The standard error of sample mean and the standard error of proportion are used in sampling in order to obtain the following facts:

1. ***Standard error is used to set up the confidence limits within which the population parameter may lie.***
2. ***Standard error is used to test the hypothesis and to draw a statistical conclusion from it.***
3. ***Standard error is used to measure the variability of the values of a statistic from its mean.***
4. ***It is generally, used for large samples and gives us the idea about the average amount of error which actually occurs in estimating the values of a parameter on the basis of a statistic.***

EXERCISE – 7.1

1. `Sampling is a necessity under certain conditions'. Explain this with illustrative examples.
2. Distinguish between a population and a sample. What is a random sample? Describe some methods of drawing random sample from a finite population.
3. What is a sampling? Give its objects and name the laws which form the basis of sampling.
4. Explain the terms `Parameter and statistics' as used in sampling.
5. What are the sampling and non-sampling errors.
6. What are the objectives of sampling?
7. Explain the various methods of sampling.
8. Explain the various types of restricted random sampling.
9. What are the advantages of random sampling?
10. What is a sampling distribution of a statistics?
11. Explain the concept of standard error. Discuss the role of standard error in large sample theory.
12. What is meant by statistic and its standard error? Give the expression for the standard error of the sample mean and sample proportion.
13. Distinguish between simple random sampling and purposive sampling.
14. Write short notes are the following:
    (*i*) Stratifed random Sampling;
    (*ii*) Multistage Sampling
15. Distinguish between parameter and statistic.
16. Distinguish between purposive and random sampling.
17. Distinguish between standard error and sampling error.
18. Write short notes an "Stratifud Random Sampling and Sequential Sampling."
19. Distinguish between judgement sampling and random sampling.
20. Name the methods of sampling.
21. What is Stratified Random Sampling? What are its advantages?

8

CHAPTER

THEORY OF ESTIMATION

8.1 THEORY OF ESTIMATION

It is possible to draw valid conclusion about the population parameters from sampling distribution. We have a sample from a population involving unknown parameter, such as the **sample mean.** The problem is **to construct a sample quantity that will serve to estimate the unknown parameter, *viz*, population mean μ. Such a sample quantity is called the estimator and the actual numerical value obtained by evaluating an estimator in a given instance is the estimate.** For example, the sample mean $\bar{x}$ is an estimator of the population mean μ. If for a specific sample, the sample mean is 7.85, then **7.85** is our **estimator** for the **population mean μ.** It is important to note that an estimator **must be a statistic** and **it must depend only on the sample and not on the parameter to be estimated.** In other words, **an estimator is a statistic which for all practical purposes, can be used in place of unknown parameter of the population.** Estimation of population paratmeters like mean, standard deviation, variance, correlation coefficient etc. from the corresponding sample statistic is one of the important problems of staistical inference.

There are many ways of estimating a parameter from the data of the sample and the knowledge of the form of sampling distribution. Some of these ways are better than other. The theory of estimation deals with these and related matters. Obviously, estimator are bound to differ from the true value of the population parameters. But the **tolerable divergence between the estimates and the true value of the population parameter may be specified before hand. The theory of estimation helps in determining the probablity of such divergence.**

8.2. QUALITIES OF A GOOD ESTIMATOR

A good estimator is one which is as close to the true value of the parameter as possible. A good estimator must possess the following characteristics:

(*i*) **Unbiasedness**
(*ii*) **Consistency**
(*iii*) **Efficiency**
(*iv*) **Sufficiency**

Unbiasedness. ***A statistic*** $t = t(x_1, x_2, x_3 \ldots\ldots, x_n)$ ***is said to be an unbiased estimate of the corresponding population parameter*** θ, ***if***

$$E(t) = \theta$$

⇒ ***the mean value of the sampling distribution of the statistic t is equal to the parameter of the population.*** For example, the sample mean $\bar{x}$ is an unbiased estimator of population parameter if

$$E(\bar{x}) = \mu.$$

Similarly, $\quad E(p) = P$, where

P = population proportion, p = sample proportion.

Let s^2 denote the sample variance, then s^2 is defined as

$$s^2 = \frac{1}{n}\sum_{i=1}^{n}(x_i - \bar{x})^2 .$$

But s^2 is not an unbiased estimator of population variance, then σ^2. But on the other hand $S^2 = \frac{1}{n-1}\sum (x_i - \bar{x})^2$ provides an unbiased estimate of the population variance σ^2. Thus

$$E(S)^2 = \sigma^2 \text{ but } E(s^2) \neq \sigma^2$$

It is important to note that

1. $ns^2 = (n-1)S^2 \Rightarrow S^2 = \left(\frac{n}{n-1}\right)s^2.$

2. When the sample size is large, theen $(n-1)$ can be approximated by n and in that case

$$S^2 = \frac{n}{n-1}s^2 \simeq s^2.$$

In other words for large samples, the sample variance s^2 gives an estimate of the population vaiance σ^2 *i.e.*, $E(s^2) = \sigma^2$ for large samples.

Consistency. A statistics $t = t_n = t(x_1, x_2, x_3, \ldots, x_n)$ based on the sample size n is said to be **consistent** estimator of the **parameter θ if** $t_n \to \theta$ as $n \to \infty$

Symballically,

$$\lim_{n \to \infty} P(t_n \to \theta) = 1$$

Efficiency. If t_1 and t_2 are two consistent estimators of a parameter θ such that **Var (t_1) < Var (t_2)** for all sample since n, then t_1 is said to be more efficient than t_2. In other words, ***an estimator with lesser variability is said to be more efficient and consquently more reliable than the other.***

Sufficiency. A statistic $t = t(x_1, x_2, x_3, \ldots, x_n)$ is **said to be a sufficient estimator of parameter θ if it contains all the information in the sample regarding the parameter.** In other words, if $t = t(x_1, x_2, x_3, \ldots, x_n)$ is a statistic based on a random sample $(x_1, x_2, \ldots, x_n)$ of size n from a population with probability density function $P.D.F = P(x, \theta)$, then it is a sufficient estimator of θ, if the conditional probability $P[x_1 \cap x_2 \cap x_3 \cap x_4 \ldots \cap x_n)/t = k]$ does not depend on θ, *i.e.*, $P(x, \theta) = P(x)$.

8.3 TYPES OF ESTIMATION

The theory of estimation is divided into two groups:

(*i*) Point Estimation

(*ii*) Interval Estimation

8.3.1 Point Estimation

In **point estimation** *a single statistic is used to provide an estimate of the population parameter.* In other words, **the estimate of a population parameter given by a single number is called the point estimation of the parameter.** In point estimation we find a statistic which may be used for to replace an unknown parameter of the population for all practical purposes.

8.3.2 Interval Estimation

There are situation where the point estimation is not desirable and we are interested in finding such limits within which with a known probability or to a known degree of reliance, the value of the population parameter is expected to lie. Such a process of estimation is called the interval estimation. In other words, an interval estimation is the range of values used in making estimation of a population parameter. Thus the interval estimation of the **population parameter is the estimation of the population parameter by an interval around the point. The inverval estimation of a population parameter θ is the estimation of the parameter θ with the help of the interval $[t - s, t + s]$, where t is the sample statistic *i.e.*, $t - s \leq \theta \leq t + s$.**

Both point and interval estimation are ways of drawing inductive inference and therefore, involve an element of uncertainty. But the interval estimation has advantage over the point estimation as it provides a measure of degree of uncertainty in terms of probability attached to the interval. In the interval estimation, the estimate for the parameter lies between two limits. The two limits within which the estimate for the parameter lies are known as **Confidence Limits or Fiducial Limits** *and the interval bounded by these two limits as* **Confidence Interval** or **Fiducial Interval.** The confidence interval depends upon the confidence that is required to set up. **The probability that we associate with an interval is called the confidence level.** It shows how confidently we can say that the interval estimate will include the population parameter. **The higher the probability, the more is the confidence.** Although any confidence level can be considered, but the most commonly used **confidence levels** are **90%, 95%, 98% and 99%.** In the light of above discussion the interval estimation is defined as:

$P[C_1 < \theta < C_2$, for given value of statistic $t] = 1 - \alpha$, where α is the level of significance, the interval $[C_1, C_2]$ within which the unknown parameter θ is expected to lie. It is known as Confidence Interval or Fiducial Interval and $1 - \alpha$ is called the confidence coefficient, depending upon the desired precision of the estimate [For example $\alpha = 0.01$ gives 99% confidence limits), C_1, C_2 are respectively known as lower limit and upper limit of the confidence Interval.

If 't' is the sample statistic used to estimate the corresponding population parameter θ, then $(1 - \alpha)$ % confidence limits for $\theta = t \pm S.E.(t) \times t_\alpha$, where S.E. (t) is the standard error of t, and t_α the significant or critical value at significance level α, $t + S.E.(t) \times t_\alpha$ is the upper limit and $t - S.E.(t) \times t_\alpha$ is the lower limit of confidence interval

$$[t - S.E.(t) \times t_\alpha,\ t + S.E.(t) \times t_\alpha]$$

8.4. METHOD TO COMPUTE CONFIDENCE LIMITS FOR A POPULATION PARAMETER θ

The following steps enable us to compute the confidence limits for the population **parameter** θ in terms of the sample **statistic** t.

1. Compute or **select** the appropriate sample **statistic** t.
2. Obtain **S.E.** (t), the standard error of the sampling distribution for the sample statistic t.
3. Select the confidence level and corresponding to that "**Specific level of confidence**" and note the **critical value** t_α.

8.5. INTERVAL ESTIMATION FOR LARGE SAMPLES

As per **central limit theorem, "For large samples, the underlying distribution of the standardized variate corresponding to the sampling distribution of the sample statistic t will be asymptotically normal,** ***i.e.*****,**

$$Z = \frac{t - E(t)}{S.E.(t)} \sim N(0,1)$$

is a standaridized normal variate with mean zero and variance one as $n \to \infty$.

8.6 CONFIDENCE INTERVAL OF THE MEAN

Let μ be the population mean and $\bar{x}$ be the sample mean of the sampling distributions of means. **It is also assumed that the sample mean is normal or the sample is large.** Then the interval estimate of population mean μ by the sample mean $\bar{x}$ of the sampling distribution of means is given by the following rule:

WORKING RULE

Step I. Computer $\bar{x}$ or take $\bar{x}$.

Step II. Select the confidence level and the corresponding to that specific level of confidence, note the value of confidence coefficient Z or t_α.

Step III. Compute S.E. ($\bar{x}$) with the help of following results:

Case I. **When σ, the standard deviation of the normal population σ is known.**

$$\text{S.E.}(\bar{x}) = \frac{\sigma}{\sqrt{n}}$$

where *n* is the sample size.

Case II. **When σ, the standard deviation of population is not known.**

$$\text{S.E.}(\bar{x}) = \frac{s}{\sqrt{n-1}},$$

where *n* = Sample size and *n* is large

***s* = standard deviation of the sampling distribution of sample.**

Step IV. Construct a confidence level as follows:

Case I. **When σ is known and population is normal or any population with large *n*.**

Confidence Interval = $[\bar{x} - \text{S.E.}(\bar{x}) \times Z_\alpha,\ \bar{x} + \text{S.E.}(\bar{x}) \times Z_\alpha]$

or $$\left[(\bar{x}) - \frac{\sigma}{\sqrt{n}} \times Z_\alpha \le \mu \le \bar{x} + \frac{\sigma}{\sqrt{n}} \times Z_\alpha\right] \qquad \left(\because \text{S.E.}(\bar{x}) = \frac{\sigma}{\sqrt{n}}\right)$$

Case II. **When σ is unknown and *n* is large.**

In this case the **Confidence Interval** = $[\bar{x} - S.E.(\bar{x}) \times Z_\alpha, \bar{x} + S.E.(\bar{x}) \times Z_\alpha]$

$$= \left[\bar{x} - \frac{s}{\sqrt{n-1}} \times Z_\alpha,\ \bar{x}\ \frac{s}{\sqrt{n-1}} \times Z_\alpha\right]$$

or $$\bar{x} - \frac{s}{\sqrt{n-1}} \times Z_\alpha \le \mu \le \bar{x} + \frac{s}{\sqrt{n-1}} \times Z_\alpha \qquad \left[\because\ \text{S.E.}(\bar{x}) = \frac{s}{\sqrt{n-1}}\right]$$

Step V. Select the value of Z_α from the following table

Table Value of Z_α

Confidence level (1 – α)	*Value of confidence coefficient* Z_α or t_α
90%	1.64
95%	1.96
98%	2.33
99%	2.58
Without any reference to the confidence level	3.00

Note: ***Where no reference to the confidence level is given, then always take t_α or $Z_\alpha = 3$.***

The above method is illustrated by the following examples.

Example 1. *The quality control manager of a tyre company has sample of 100 tyres and has found the mean life time to be 30,214 km. The population s.d. is 860. Construct a 95% confidence interval for the mean life time for this particular brand of tyres.*

Solution. Here $n = 100$, $\bar{x} = 30214$, $\sigma = 860$.

Also $(1 - \alpha)\% = 95\%$ so that $\mathbf{Z_\alpha = 1.96}$ **[From table]**

$$\therefore \quad S.E.(\bar{x}) = \frac{\sigma}{\sqrt{n}} = \frac{860}{\sqrt{100}} = \frac{860}{10} = \mathbf{86}$$

$$\therefore \quad \textbf{Confidence interval} = [\bar{x} - \frac{\sigma}{\sqrt{n}} \times Z_\alpha, \quad \bar{x} + \frac{\alpha}{\sqrt{n}} \times Z_\alpha]$$

$$= [30214 - 86 \times 1.96, 30214 + 86 \times 1.96]$$

$$= [30214 - 168.56, 30214 + 168.56] = \mathbf{[30045.44, \quad 30382.56]}$$

Example 2. *In a random selection of 64 of 600 road crossing in a town, the mean number of automobile accidents per year was found to 4.2 and the sample s.d was 0.8. Construct a 95% confidence interval for the mean number of automobile accidents per crossing per year.*

Solution. Here $n = 64$, $6m$ $s.d$ of sample $s = 0.8$, $\bar{x} = 4.2$.

$$\mathbf{(1 - \alpha)\ \% = 95\%, \text{ so the value of } Z_\alpha = 1.96}$$

$$\therefore \quad \textbf{Confidence Interval} = [\bar{x} - \text{S.E.}(\bar{x}) \times Z_\alpha, \quad \bar{x} + \text{S.E.}(\bar{x}) \times Z_\alpha]$$

Since we have a finite population of size 600 and $\frac{n}{N} = \frac{64}{600} = \mathbf{0.106}$, which is more than 0.05, so we use the following result to calculate the standard error of $\bar{x}$.

$$S.E.(\bar{x}) = \frac{s}{\sqrt{n-1}} \times \sqrt{\frac{N-n}{N-1}} = \frac{0.8}{\sqrt{63}} \times \sqrt{\frac{600-64}{600-1}}$$

$$= \frac{0.8}{7.93} \times \sqrt{\frac{536}{599}} = \frac{0.8}{7.93} \times 0.945 = \mathbf{0.0945.}$$

$$\textbf{Confidence Interval} = [4.2 - 0.0945 \times 1.96, 4.2 + 0.0945 \times 1.96]$$

$$= [4.2 - 0.185, 4.2 + 0.185] = \mathbf{[4.015, 4.385]}.$$

Example 3. *A random sample of size 100 has mean 15, the population variance being 25. Find the interval estimate of the population mean with a confidence level of (i) 99% and (ii) 95%.*

Solution. Here $n = 100$, $\sigma^2 = 25$, $\bar{x} = 15$.

$$S.E.(\bar{x}) = \frac{\sigma}{\sqrt{n}} = \frac{5}{\sqrt{100}} = \frac{5}{10} = \mathbf{0.5.}$$

(*i*) **99% confidence level** $\Rightarrow$ $\mathbf{Z_\alpha = 2.58}$

$$\therefore \quad \textbf{Confidence Interval} = \bar{x} \pm \text{S.E.}(\bar{x}) \times Z_\alpha$$

$$= 15 \pm 0.5 \times 2.58 = 15 \pm 1.29 \text{ i.e., } \mathbf{13.71} \text{ to } \mathbf{16.29.}$$

(*ii*) **95% Confidence level** $\Rightarrow$ $\mathbf{Z_\alpha = 1.96.}$

$$\textbf{Confidence Interval} = \bar{x} \pm \text{S.E.}(\bar{x}) \times Z_\alpha = 15 \pm (0.5) \times 1.96$$

$$= 15 \pm 0.98 \text{ i.e., } \mathbf{14.02} \text{ to } \mathbf{15.98}$$

Example 4. *A sample of 100 gave a mean of 7.4 kg and a standard deviation of 1.2 kg. Find 95% confidence limits for population mean.*

Solution. Standard error of mean S.E. $(\bar{x}) = \frac{\hat{\sigma}}{\sqrt{n}} = \frac{1.2}{\sqrt{100}} = \mathbf{0.1}$.

95% confidence limits for population mean are given by : $\bar{x} \pm \mathbf{1.96}\ \frac{\hat{\sigma}}{\sqrt{n}}$, *i.e.*, 74 ± 1.96 × 0.12.

The confidence limits are: 7.1648 and 7.6352.

Example 5. (*a*) *A random sample of size 65 was taken to estimate the mean annual income of 100 families and the mean and standard deviation were found to the Rs. 6300 and Rs. 9.5 respectively.*

Find a 95% confidence interval for the population.

Solution. (*a*) We have n = sample size = 65 **(large)**, N = 1000, $\bar{x}$ = 6300, s = 9.5.

Since the sample is large, so we can assume that the sampling distribution is Normal.

Since $\frac{n}{N} = \frac{65}{1000}$ = 0.065 > .05, therefore, the finite population correction factor is used. Thus we shall use the following result to calculate the standard Error of $\bar{x}$ *i.e.*,

$$\text{S.E. } (\bar{x}) = \frac{s}{\sqrt{n-1}} \times \sqrt{\frac{N-n}{n-1}} = \frac{9.5}{\sqrt{64}} \times \sqrt{\frac{1000-65}{1000-1}}$$

$$= \frac{9.5}{8} \times \sqrt{\frac{935}{999}} = \frac{9.5 \times \sqrt{.936}}{8} = \frac{9.5 \times .97}{8} = \frac{9.215}{8} = \mathbf{1.15.}$$

$$\textbf{Confidence interval} = \left[\bar{x} - S.E.(\bar{x})\, Z_\alpha,\ \bar{x} + S.E.(\bar{x}) Z_\alpha\right]$$

Here $\bar{x}$ = 6300, $SE\ (\bar{x})$ = 1.15; Z_α = 1.96

∴ **95% interval for the population mean are:**

[6300 – 1.15 × 1.96, 6300 + 1.15 × 1.96]

i.e., [6300 – 2.25. 6300 + 2.25] or **[6297.75, 6302.25].**

Example 6. *If sample mean is 20, population standard deviation is 3 and sample size is 64, find the interval estimate of the mean.*

Solution. Interval estimate of the sample mean is:

$$\bar{X} \pm \frac{\sigma}{\sqrt{n}} \times \mathbf{1.96} = \mathbf{20} \pm \frac{\mathbf{3}}{\mathbf{8}} \times \mathbf{1.96}$$

Hence the interval estimate of mean is: [19.265, 20.735].

8.7. CONFIDENCE INTERVAL ESTIMATE OF THE PROPORTION

Population proportion. The population proportion P is the ratio of the number of elements possessing a characteristic to the total number of elements in the population, *i.e.*,

$$P = \frac{\textbf{Number of elements possessing the characteristic}}{\textbf{Total number of elements in the Population}}$$

If we multiply the population by 100, then we get the **percentage** and we may make use of percentage for the proportion and *vice - versa.*

Sample Proportion. *The sample proportion p is the ratio of the number of elements possessing a characteristic to the total number of elements `n' in the sample.*

$$\textbf{Sample Proportion: } p = \frac{\text{Number of elements possessing a characteristic}}{n = \text{Total number of elements in the sample}}$$

It is important to note that **the mean of sampling distribution of p equals the population proportion,** *i.e.*, $E(p) = P$.

8.8. STANDARD ERROR (OR STANDARD DEVIATION) OF SAMPLE PORPORTION

Case I. When the population size is infinitely large or the sample is drawn with replacement.

(*i*) Standard Error of Sample Proportion is

$$\text{S.E.}(p) = \sqrt{\frac{PQ}{n}}, \text{ if } P \text{ is known,}$$

where $Q = 1 - P$, n = the sample size.

(*ii*) Standard error of Sample Proportion is

$$\text{S.E.}(p) = \sqrt{\frac{pq}{n}}, \text{ if } P \text{ is not known, where } q = 1 - p, n = \text{sample size.}$$

It is an estimated value.

Case II. When the population size is finite and the sample is drawn without replacement.

(*i*) Standard Error of Sample Proportion

$$\text{S.E.}(p) = \sqrt{\frac{PQ}{n}} \times \sqrt{\frac{N-n}{N-1}}, \text{ if } P \text{ is known, when } Q = 1 - P, N = \text{Size of Population,}$$

n = Sample size.

(*ii*) Standard Error of Sample Proportion

$$\text{S.E.}(p) = \sqrt{\frac{pq}{n}} \times \sqrt{\frac{N-n}{N-1}}, \text{ if } P \text{ is not known,}$$

where $q = 1 - p$, N = size of population, n = size of sample.

It is an estimated value.

8.9 CONFIDENCE INTERVAL ESTIMATE

Confidence Interval Estimate of the population proportion P is given by the following results:

Case I. When P is known. In this case the **confidence interval** of the sample proportion is given by $[p - Z_\alpha \times \text{S.E.}(p), p + Z_\alpha \times \text{S.E.}(p)]$,

where Z_α is the confidence coefficient, and S.E. (p) is the standard Error of p.

Case II. When P is not known. In this the confidence interval of the sample proportion is given by $[p - \text{S.E.}(p) \times Z_\alpha, p + \text{S.E.}(p) \times Z_\alpha]$,

where Z_α = confidence coefficient; Standard Error of p = S.E. (p).

Example 7. *A random sample of 800 units from a large consignment showed that 200 were damaged. Find 95% confidence limits for the population proportion of damaged units in the consignment.*

Solution. Here $n = 800$, Sample proportion $p = \frac{200}{800} = \mathbf{0.25}$, $q = 1 - 0.25 = \mathbf{0.75}$.

$$\text{S.E.}(p) = \sqrt{\frac{pq}{n}} = \sqrt{\frac{0.25 \times 0.75}{800}} = \mathbf{0.0153.}$$

Also $(1 - \alpha)\% = 95\%$ so that $Z_\alpha = 1.96$.

Confidence limits for the population proportion are: $p \pm Z_\alpha \times$ S.E. (p).

$$\textbf{Lower limit} = p - Z_\alpha \times \text{S.E.}(p) = 0.25 - 1.96 \times 0.0153$$
$$= 0.25 - 0.03 = \mathbf{0.22.}$$

$$\textbf{Upper limit} = p + Z_\alpha \times \text{S.E.}(p) = 0.25 + 1.96 \times 0.0153 = 0.25 + 0.03 = \mathbf{0.28.}$$

Hence, 95% confidence limits for the population proportion of damaged units in the consignment are 0.22 and 0.28.

Example 8. *Out of 300 households in a town 123 have T.V. sets. Find 95% confidence limits to the true value of the proportion of the households with T.V. sets in the whole town.*

Solution: Here $n = 300$, Sample proportion $p = \frac{123}{300} = \mathbf{0.41}$; $q = 1 - 0.41 = \mathbf{0.59}$

$$\text{S.E.}(p) = \sqrt{\frac{pq}{n}} = \sqrt{\frac{0.41 \times 0.59}{300}} = \mathbf{0.0283.}$$

Also $(1 - \alpha)\% = 95\%$, therefore, $Z_\alpha = 1.96$.

$$\textbf{Confidence limits} = p \pm Z_\alpha \times \text{S.E.}(p)$$
$$= 0.41 \pm 1.96 \times 0.0283 = 0.41 \pm 0.055$$
$$\textbf{Lower limit} = 0.41 - 0.055 = \mathbf{0.355}$$
$$\textbf{Upper limit} = 0.41 + 0.055 = \mathbf{0.465.}$$

Hence, 95% confidence limits to the true value of the proportion of the households with T.V. sets in the whole town are 0.355 and 0.465.

Example 9. *A factory is producting 50,000 pairs of shoes daily. From a sample of 500 pairs, 2% were found to be of substandard quality. Estimate the number of pairs that can be reasonably expected to be spoiled in the daily production and assign limits at 95% level of confidence.*

Solution. Here $n = 500$, Sample percentage $p = 2\%$

$$\text{S.E.}(p) = \sqrt{\frac{p(100 - p)}{n}} = \sqrt{\frac{2(100-2)}{500}} = \sqrt{\frac{2 \times 98}{500}} = \mathbf{0.63.}$$

Aslo $(1 - \alpha)\% = 95\% \Rightarrow Z_\alpha = 1.96$.

$\therefore$ **95% confidence limits for the percentage of substandard quality are:**

$$p \pm \text{S.E.}(p)\, Z_\alpha = 2 \pm (1.96 \times 0.63) = 2 \pm 1.23$$
$$= 2 - 1.23 \text{ and } 2 + 1.23 = \mathbf{0.77 \text{ and } 3.23.}$$

Number of pairs of shoes that can be expected to be spoiled in daily production lies between $\frac{0.77}{100} \times 50{,}000 = 385$ and $\frac{3.23}{100} \times 50{,}000 = 1615$.

Example 10. *Out of 20,000 customer's ledger accounts, a sample of 600 accounts was taken to test the accuracy of posting and balancing wherein 45 mistakes were found. Assign limits within which the number of defective cases can be expected at 5% level.*

[M.B.A; Anna University – December 2002]

Solution. Here $n = 600$, Sample proportion $p = \frac{45}{600}$ = **0.075.** $q = 1 - 0.075 =$ **.925.**

$$\text{S.E.}(p) = \sqrt{\frac{pq}{n}} = \sqrt{\frac{0.075 \times 0.925}{600}} = 0.011.$$

Also $\quad (1-\alpha)\% = 95\% \Rightarrow Z_\alpha = 1.96.$

95% confidence limits are $p \pm Z_\alpha \times$ S.E. (p)

$= 0.075 \pm 1.96 \times 0.011 = 0.075 \pm .022 =$ **0.053 to .097.**

Hence at 5% level of significance the expected number of mistakes would lie between 0.053 and 0.097. The number of defective cases will lie between 0.53 × 20,000 = 1060 and .097 × 20,000 = 1940.

Example 11. *A random sample of 500 pineapples was taken from large consignment and 65 of them were found to be bad. Show that the standard deviation of the population of bad one in a sample of this size is 0.015 and deduce that the percentage of bad pineapples in the consignment lie between 8.5 and 17.5.*

Solution. Here $n = 500$, Sample proportion $p = \frac{65}{500}$ = **0.13.** $q = 1 - 0.13 =$ **0.87.**

Standard deviation or standard error of population of bad pineapples

$$\text{S.E.}(p) = \sqrt{\frac{pq}{n}} = \sqrt{\frac{0.13 \times 0.87}{500}} = \mathbf{0.015}.$$

Here the level is not specific so we take Z = 3.

∴ 99.73% confidence limits for the proportion of bad pineapples

$= p \pm$ S.E. (p) $Z_\alpha = 0.13 \pm (3 \times 0.015) = 0.13 \pm 0.045$ *i.e.,* **0.085 to 0.175.**

Hence 99.73% confidence limits for percentage of bad pine apples lie between 0.085 × 100 and 0.175 × 100 *i.e.,* between 8.5% and 17.5%.

8.10 STUDENT'S *t*-DISTRIBUTION

In the last section, we studied that the sampling distribution of any statistic is based on the following asymptotic property.

The sampling distribution of any statistic in "the standardised form is *N* (0,1) asymptotically, *i.e.*, as $n \to \infty$". In other words the sampling distribution of any statistic t is normally distributed with mean 0 and variance 1 and is given by

$$Z = \frac{t - E(t)}{\text{S.E.}(t)} \sim N.(0.1) \text{ as } n \to \infty$$

The **theory of large sample ($n > 30$) is based on the above property of the sampling distribution.**

But if the **sample is small (*i.e.*, $n < 30$)**, then the distribution of the standarised variable Z of the statistic t will be far from normality and as a result the **normal test** cannot be applied. In order to deal with **small samples (when sample size $n < 30$)** new techniques and tests of significance known as **EXACT SAMPLE TECHNIQUES (TESTS)** have been developed.

It is important to note that "**Exact sample technique (test) can be applied even for large samples but the large sample theory cannot be applied for small samples.**

The theory of small (or Exact) sample was developed by Irish statistician William S. Cosset who used to write under pseudonym (pen name) of student in 1908. Cosset gave his statistic name as **Student's *t*-distribution or *t*-test.** The quantity t is defined as

$$t = \frac{\textbf{Difference of population parameter and the corresponding statistic}}{\textbf{Standard error of the statistic}}$$

with $(n-1)$ degrees of freedom if the samples size is n.

Note: The degree of freedom for t = (size of sample) – 1.

Student's distribution. If $x_1, x_2, ..., x_n$ is a random sample size of n drawn from a normal population with mean μ and variance σ^2 (not known) then the student's t statistic (mean) is defined as

$$t = \frac{\bar{x} - \mu}{(\text{S.E. of mean})} = \frac{\bar{x} - \mu}{\sqrt{s^2 / n-1}},$$

with $(n-1)$ degree of freedom, where s = standard deviation of the sample.

8.11 ASSUMPTION FOR *t*-test

The *t*-test is applied under the following assumptions:

1. ***Samples are drawn from normal population and are random.***
2. ***For testing the equality of two population means, the population variances are regarded as equal.***
3. ***The population standard deviation may not be known.***
4. ***In case of two samples some adjustments in degrees of freedom for t are made.***

8.12 PROPERTIES OF *t* – test

(i) t-distribution is asymptotic to x-axis (i.e., i.e.,) it extends to infinity on either side.

(ii) The shape of the curve or form of t-distribution varies with the degrees of freedom. The degree of freedom is defined as (size of sample - 1)

(iii) t-distribution is a symmetrical distribution with mean zero.

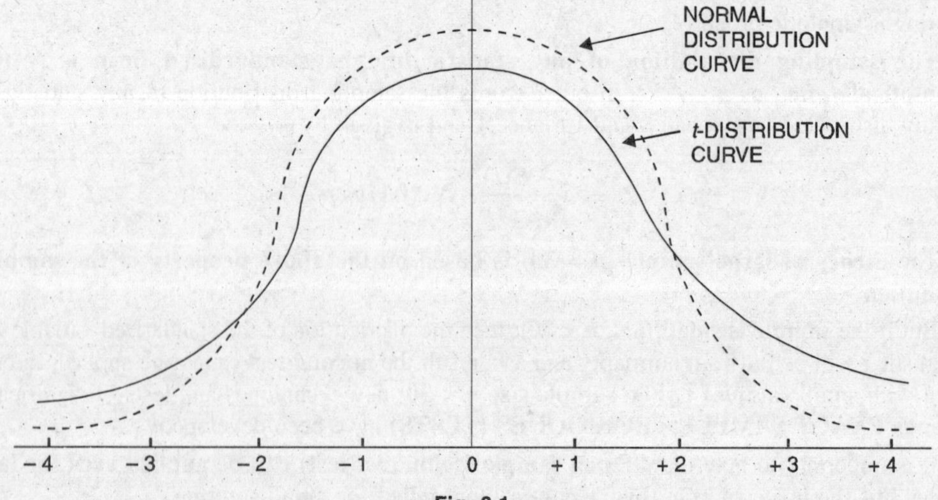

Fig. 8.1

(iv) ***Its graph is similar to that of normal distribution. There is more area in the tails of the t-distribution, and the standard normal curve is higher in the middle, i.e., t-distribution has a greater spread than normal distribution.***

(v) ***The larger the number of degrees of freedom the more closely t-distribution resembles standard normal distribution, i.e., t-distribution approaches normal distribution as the number of degrees of freedom approaches without limit.***

(vi) ***Sampling distribution of t does not depend on population parameter but it depends only on v (the degree of freedom) = n – 1 i.e., on the sample size.***

8.13. APPLICATION OF *t* – Distribution

The *t*-distribution has following three important applications in testing of hypotheses for small samples ($n < 30$).

1. **To test the significance of a sample mean, when the population variance σ is unknown**
2. **To test the significance of difference between two sample means the population variances being equal and unknown.**
3. **To test the significance of an observed sample correlation coefficient or difference between means of two samples (dependent samples or paired observations).**

8.14. INTERVAL ESTIMATE OF POPULATION MEAN

Let $\bar{x}$ be the sample mean, and n be the size of sample. Then the interval estimate of the population mean (μ) is given by

$$\bar{x} \pm t_{\alpha}\, S.E.(\bar{x})$$

where S.E. ($\bar{x}$) is standard error of mean and is defined as

$$\text{S.E.}\ (\bar{x}) = \frac{s}{\sqrt{n-1}} \text{ and } s^2 = \frac{\Sigma(x-\bar{x})^2}{n}.$$

The value of t at different level of $100(1-\alpha)\%$ are to be obtained from the t-table given at the end of the book. We shall see the column under 0.05, 0.02, 0.01 for confidence intervals of 95%, 98% and 99% respectively for the given degrees of freedom.

Example 12. *A sample of size 9 from a normal population gave* $\bar{x}$ = 15.8 *and* s_x^2 = *10.3. Find a 99% interval for population mean* **[C.A. (Intermediate), November 1985]**

Solution. We are given $\bar{x} = 15.8$; $s_x^2 = 10.3$ and $n = 9$.

∴ **Degree of freedom** = $n - 1 = 9 - 1 =$ **8.**

Also $t_{0.01}$ **for 8 *d.f.* = 3.36.** **[From *t* table]**

99% confidence limits for the population mean $\bar{x}$ are $= \bar{x} \pm t_{0.01} \dfrac{s}{\sqrt{n-1}}$

$$= 15.8 \pm 3.36 \times \sqrt{\frac{10.3}{8}}$$

$$= 15.8 \pm 3.36 \times 1.135 = 15.8 \pm 3.8136 = \mathbf{11.9864,\ 19.6136.}$$

Hence 99% confidence interval = [11.9864, 19.6136]

8.15. SAMPLING IN ACCOUNTING AND AUDITING

The application of sampling in accounting and auditing is very important from business point of view. It includes audit tests, estimation of inter-company transactions, estimation of values of inventories, control of clerical errors, costing operation etc. In a large business houses the number of accounts and entries therein are so numerous that a complete verification of all these is not possible. Test checking or sample checking is the usual method for verification of a large number of transactions on the basis of the selective test of a small set of items. Thus to have an adequacy of the reliability of check as also for proper generalisation of results, it is necessary for the auditor to apply scientific sampling techniques. He can specify and objectively state the risk he is willing to take for that as the sampling estimate will not be as accurate as desired. Sampling techniques can help the auditor in the following ways:

(*i*) ***To ascertain the appropriate size of the sample.***

(*ii*) ***To choose the method of selection.***

(*iii*) ***To evaluate the results.***

(*iv*) ***To determine the level of accuracy in accounts.***

Thus after specifying the accuracy and risk, a sample size can then be computed which will enable the auditor to achieve his objectives.

Accounting records which generally fulfil the criteria for sampling are often used as basis for different accounting or financial investigations inquiry, as may be required or as may be desired by the management. Auditing of accounts is one such area where the auditor generally resorts to sampling of data contained in accounting records for his audit purpose. The sample size may be determined on the basis of suitable statistical tables taking into account the confidence level, permissible error etc.

8.16 DETERMINATION OF SAMPLE SIZE

The determination of sample size for estimating a mean or proportion is a crucial question. By selecting a sample size lower than the correct size may affect reliability and a higher one will mean more cost and time. The determination of the size of a sample is the most important factor for the purposes of estimation of the value of the population parameters. We have following formula for it.

Sample size for Estimation a Mean

In order to determine the sample size for estimating a population mean, the following factors must be known:

(*i*) ***The desired confidence level.***

(*ii*) ***The permissible sampling error*** $E = \bar{x} - \mu$.

(*iii*) ***The standard deviation*** σ.

After having known the above mentioned three factors, **the size of sample mean *n* is given by**

$$n = \left(\frac{\sigma Z}{E}\right)^2$$

Sample size for Estimating a Proportion

In this case we must know the following three factors:

(*i*) ***the desired confidence level.***

(*ii*) ***the permissible sampling error E = difference between the estimate from the sample p and the parameter P to be estimated = P – p.***

(*iii*) ***the estimated true proportion of success.***

The sample size *n* is given by

$$n = \frac{Z^2 pq}{E^2}, \text{ where } q = 1 - p.$$

Example 13. *It is known that the population standard deviation in waiting time for L.P.G. gas cylinder in Delhi is 15 days. How large a sample should be chosen to be 95% confident, the waiting time is within 7 days of true average.*

Solution. The required sample size is

$$n = \left(\frac{\sigma Z}{E}\right)^2 = \left(\frac{15 \times 1.96}{7}\right)^2 = \mathbf{17.64.}$$

Hence, **the size of the sample is 18.**

Example 14. *A manufacturing concern wants to estimate the average amount of purchase of its product in a month by the customers whose standard error is Rs. 10. Find the sample size if the maximum error is not to exceed Rs. 3 with a probability of 0.99.*

Solution. It is given that

$$P[|\bar{x} - \mu| < 3] = 0.99. \quad \text{... (I)}$$

But $P\left[|\bar{x} - \mu| < 2.58 \frac{\sigma}{\sqrt{n}}\right] = 0.99$.... **(II)**

From (I) and (II), we have

$$2.58 \frac{\sigma}{\sqrt{n}} = 3 \Rightarrow \sqrt{n} = \frac{2.58 \times 10}{3}$$

or $$n = \left(\frac{2.58 \times 10}{3}\right)^2 = (8.6)2 = 73.96 \cong \mathbf{74.}$$

Hence, **the sample size should be 74.**

Example 15. *Mr. X wants to determine on the basis of sample study the mean time required to complete a certain job so that he may be 95% confident that the mean may remain within ± 2 days of the true mean. As per the available records the population variance is 64 days. How large should the sample be for his study ?*

Solution. Here $\sigma = \sqrt{64} = 8$. Also Z is $N(0,1)$.

It is given that

$$\boldsymbol{P[\,|\bar{x} - \mu\,| < 2] = 0.95} \quad \text{... (I)}$$

But $$\boldsymbol{P\left[(\bar{x} - \mu) \le 1.96 \frac{\sigma}{\sqrt{n}}\right] = 0.95.} \quad \text{... (II)}$$

From (I) and (II), we get

$$\frac{1.96 \times \sigma}{\sqrt{n}} = 2 \Rightarrow n = \left(\frac{1.96 \times 8}{2}\right)^2 = (7.84)2 = 61.42 \simeq \mathbf{62}$$

Hence, **the sample size should be 62.**

Example 16. *The business manager of a large company wants to check the inventory records against the physical inventories by a sample survey. He wants to almost assure that the maximum sampling error should not be more than 5% or below the true proportion of accurate records. The proportion of the accurate records is estimated at 35% from past experience. Determine the sample size.*

Solution. Here $P = 0.35$, $E = 5\% = 0.05$.

Confidence coefficient $1 - \alpha \cong 1$

We know that

$$P[|Z| \geq 3] = 0.9973 \cong 1$$ **[From standard normal Table]**

$\Rightarrow$ Z lies between –3 and +3 almost surely.

Here $Z = 3$.

Now, **sample size** $$n = \frac{PQZ^2}{E^2}$$

$$= \frac{0.35\,(1-0.35)\times 3^2}{(0.05)^2} = \frac{0.35\times 0.65\times 9}{0.0025} = \frac{2.0475}{0.0025} = \mathbf{819.}$$

Hence, **the sample size is 819.**

EXCERCISE - 8.1

1. A simple random sample of size 66 was drawn in the process of estimating the mean annual income of 950 families of a certain township. The mean and standard deviation of the samples were found to be Rs. 4730 and Rs. 7.65 respectively. Find a 95% confidence interval for the population mean (Table value of Z for confidence level at 95% is 1.96).

$$S.E.(\bar{x}) = \frac{\sigma}{\sqrt{n}} = \frac{7.65}{\sqrt{66}} = \mathbf{0.942.}$$

Confidence limits of population mean: $\bar{x} \pm 1.96 \times S.E.(\bar{x})$ = 4730 ± 1.96 (0.942)

= 4730 ± 1.846, [*i.e.*, **4728.154 to 4731.846**]

2. The mean height obtained from a sample of size 100 taken randomly from a population is 164 cm. If the standard deviation of the height distribution of the population is 3 cm, set up 95% confidence limits for the mean height of the population.

[**Hint:** Here $n = 100$, $\bar{x} = 164$, $\sigma = 3$, $Z = 1.96$. S.E. $= \frac{\sigma}{\sqrt{n}} = \frac{3}{\sqrt{100}} = 0.3$ cm.

Confidence limits = 164 ± 1.96 × (0.3) = 164 ± 0.588 *i.e.*, **163.412** to **164.588.]**

3. A random of 100 ball bearings selected from a shipment of 2000 ball bearing has an average diameter of 0.354 cm with a S.D. = 0.048 cm. Find 95% confidence interval for the average diameter of these 2000 ball bearings.

[**Hint:** S.E. $(\bar{x}) = \frac{s}{\sqrt{n}} \times \frac{N-n}{N-1} = \frac{0.048}{\sqrt{100}} \times \sqrt{\frac{2000-100}{2000-1}} = 0.00468.$

95% confidence limits = $\bar{x}$ ± 1.96 S.E. $(\bar{x})$

= 0.354 ± (1.96 × 0.0046) = 0.354 ± 0.0092 = **0.3448 and 0.3632.]**

4. A random sample of 100 items taken from a large batch of articles contains 85 defective items. Set up 96% confidence limits for the proportion of defective items in the batch (the value of Z for 96% confidence is 2.05]

5. 500 apples are taken at random from a large basket and 50 are found to be bad. Estimate the proportion of bad apples in the basket and assign limits within which the percentage most probably lie.

6. A random sample of 700 units from a large consignment showed that 200 were damaged. Find (*i*) 95% and (*ii*) 99% confidence limits for the proportion of damaged units in the consignment.

[**Hint:** Here $p = \frac{200}{700} = 0.286$, $q = 1 - 0.286 = 0.714$.

$\therefore$ **S.E. (p)** $= \sqrt{\frac{pq}{n}} = \sqrt{\frac{0.286 \times 0.714}{700}} = \mathbf{0.017.}$

(*i*) **95% confidence limits** $= 0.286 \pm 1.96 \times 0.017 = 0.286 \pm 0.033 =$ **(0.253,319).**

(*ii*) **99% confidence limits** $= 0.286 \pm 2.58 \times 0.017 =$ **(0.242, 0.330).**]

7. In a random sample of 400 oranges from a large consignment, 40 were considered bad. Estimate the percentage of defective oranges in the whole consignment and assign limits within which the percentage probably lies.

8. In a sample survey of 1000 house-wives in a city 23% prefer a particular brand of pressure cooker. Find 99% confidence limits for the percentage of all housewives in the city preferring that brand of cooker.

9. The foreman of a company has estimated the average quantity of extracted ore to be 36.8 tonnes per shift and the sample standard deviation to be 2.8 tonnes per shift, based on a random selection of 4 shifts. Construct 90% confidence interval around this estimate.

10. A random sample of 50 items drawn from a particular population has a mean 30 with a standard deviation 28. Construct a 98% confidence interval estimate of the population mean.

11. In a locality containing 18,000 families a sample of 840 families were selected at random. Out of these 840 families, 206 families were found to have a monthly income of Rs. 500 or less. It is desired to know how many out of 18,000 families have a monthly income of Rs. 500 or less.

12. For a given sample of 200 items drawn from a large population, the mean is 65 and the standard deviation is 8. Find the 95% confidence limits for the population mean.

[Hint: 95% confidence limits for population mean $= \bar{x} \pm \frac{s}{\sqrt{n}} Z = 65 \pm \frac{1.96 \times 8}{\sqrt{200}} = 65 \pm 1.109.$]

13. In a sample of 900 shareholders of companies, 400 stated that their major aim in holding stocks is capital application. What is the 90% confidence range within which lies the population of stock holders who hold stocks for appreciation.

[Area between Z = 0 and Z = 1.645 is 45%, where Z is the standard normal variate.]

[Hint: $p = \frac{400}{900} = \frac{4}{9} = 0.4444;\ q = 1 - \frac{4}{9} = \frac{5}{9}$

$$\therefore \quad \text{S.E. }(p) = \sqrt{\frac{pq}{n}} = \sqrt{\frac{4}{9} \times \frac{5}{9} \times \frac{1}{900}} = \mathbf{0.0166.}$$

90% confidence limits $= p \pm 1.645 \sqrt{\frac{pq}{n}}$

$= 0.444 \pm 1.645 \times 0.0166 =$ **0.4171, 0.4717.]**

14. A survey proposed to be conducted to know the annual earnings of old commerce graudates of Delhi University. How large should the sample size be taken in order to estimate the mean annual earnings within plus and minus Rs. 1000 at 95% confidence level ? The standard deviation of the annual earnings of the entire population is known to be Rs. 3000.

[Hints: $P[\,|\,\bar{x} - m\,| < 1000] = 0.95.$ **... (1)**

Also $P\left[\frac{\bar{x} - \mu}{(\sigma|\sqrt{n}} \le 1.96\right] = 0.95$ **... (2)**

(1) and (2) $\Rightarrow \frac{1.96\,\sigma}{\sqrt{n}} = 1000 \Rightarrow n = \left(\frac{1.96 \times 1000}{3000}\right)^2 = (5.88)^2 = 34.57.$

Hence, sample size should be **35.**]

15. A researcher wishes to estimate the mean of a population by using sufficiently large sample. The probability is 0.95 that the sample mean will not differ from the true mean by more than 25% of the standard deviation. How large sample should be taken ?

[Hint: $P[\,|\,\bar{x}-\mu\,|\ \leq\ 25\%\text{ of }s.d.] = 0.95 \ \Rightarrow\ P\left[|\,\bar{x}-\mu\,|\leq\frac{\sigma}{4}\right] = 0.95$ ... (1)

Also $P\left[|\bar{x}-\mu|\leq 1.96\frac{\sigma}{\sqrt{n}}\right] = 0.95$ ... (2)

From (1) and (2) $\Rightarrow 1.96\,\frac{\sigma}{\sqrt{n}} = \frac{\sigma}{4} \Rightarrow n = \mathbf{62.}$]

16. A firm wishes to estimate with an error of not more than 0.03 and a level of confidence of 98%, the proportion of consumers that prefers its brand of household detergent. Sales reports indicate that about 0.20 of all consumers prefer the firm's brand. What is the requisite sample size?

[Hint: $n = \frac{Z^2PQ}{E^2} = \frac{(2.33)^2 \times 0.2 \times 0.8}{(0.03)^2} = \mathbf{965.14.}$ [$\because$ $Z = 2.33$ from Table]

Hence, sample size should be $\boldsymbol{n} = \mathbf{965.}$]

17. It is known that population standard deviation in waiting time for new gas connection in a particular town is 20 days. How large sample should be chosen to the 95% confident that the average waiting time is within 3 days of true average?

[Hint: $n = \left(\frac{\sigma Z}{E}\right)^2 = \left(\frac{1.96 \times 20}{5}\right)^2 = \mathbf{61.46 \simeq 62.}$]

18. A study is made to estimate the average length call made from a telephone booth. The mean call length is to the estimated 0.3 minutes with a confidence level of 98%. Studies show the s.d. of call lengths to the approximetely 0.9 minutes. If call length are distributed normally, how large a sample should be selected?

ANSWERS

1. 4728.124 to 4731.846.
2. 63.412 to 64.588.
3. 0.3448 to 0.3632.
4. 0.005 to 0.095.
5. 6.1 to 13.9.
6. (i) 0.286 ± 0.033; (ii) 0.242 to 0.330.
7. 5.5% to 14.5%.
8. 19.57% to 26.43%.
9. 36.8 ± 3.802.
10. 20.68 to 39.32.
11. 5220 families.
12. 63.891 to 66.189.
13. 0.4171 to 0.4717.
14. 35.
15. 62.
16. 965.
17. 62.
18. $n = 49$; 8.0, 8185; 9.30, 124 ± 168.56.

9 CHAPTER

TESTS OF SIGNIFICANCE

9.1 INTRODUCTION

Sampling theory deals with two types of problems, *viz.*, **Estimation and Testing of hypothesis.** The theory of estimation was discussed in the last chapter. In this chapter we shall deal with the problem of **Testing of Hypothesis.** *Modern theory of probability plays an important role in decision making and the branch of Statistics which helps us in arriving at the criterion for such decisions is known as* **Testing of hypothesis,** *which was initiated by J. Neyman and E.S. Pearson. It employs statistical techniques to arrive at decisions in certain situations where there is an element of uncertainty on the basis of sample, whose size is fixed in advance.*

Hypothesis. A hypothesis is a statement about the population parameter. In other words, a hypothesis is a conclusion which is tentatively drawn on logical basis.

Statistical hypothesis. It is a tentative conclusion that specifies the properties of a distribution of a random variable. These properties, generally, refer to parameters of the population and are hypothetical values with which the values of statistic derived from a sample are compared in order to find the difference between statistic and corresponding parameter. In other words, **statistical hypothesis is some assumption or statement, which may or may not be true, about a population or about the probability distribution characterising the given population, which we want to test on the basis of the evidence from a random sample.**

Test of hypothesis. ***The testing of hypothesis is a procedure that helps us to ascertain the likelihood of hypothesised population parameter being correct by making use of the sample statistic.*** In other words, **it is a process of testing of significance which concerns with the testing of some hypothesis regarding a parameter of the population on the basis of statistic from the sample. In testing of hypothesis a statistic is computed from a sample drawn from the parent population and on the basis of this statistic, it is observed whether the sample so drawn has come from the population with certain specified characteristic.** The value of sample statistic may differ from the corresponding population parameter due to sampling fluctuations. The test of hypothesis discloses the fact whether the difference between sample statistic and the corresponding hypothetical population parameter is significant or not significant. Thus the test of hypothesis is also known as the **test of significance.**

9.2 PROCEDURE FOR TESTING A HYPOTHESIS

The following are the steps involved in hypothesis testing problems.

1. Setting up of Hypothesis. There are two types of hypothesis:

(*i*) **Null Hypothesis** and

(*ii*) **Alternative Hypothesis.**

Null Hypothesis. ***The statistical hypothesis that is set up for testing a hypothesis is known as*** **Null Hypothesis.** The null hypothesis is set up in testing a statistical hypothesis only to decide whether to accept or reject the null hypothesis. It asserts that **there is no difference between the sample statistic and population parameter** and whatever difference is there, is attributable to sampling errors. **Null hypothesis is usually denoted by H_0.**

Setting up Null Hypothesis.

The following **steps** must be taken into consideration while setting up a **null hypothesis.**

(*i*) ***In order to test the significance of the difference between a sample statistic and the population parameter or between the two different sample statistic, we set up the null hypothesis H_0 that the difference is not significant. There may be some difference but that is solely due to sampling fluctuations.***

(*ii*) ***To test any statement about the population, we hypothesise that it is true. for example: If we want to find the population mean has a specified value m_0, then the null hypothesis H_0 is set as follows:***

$$H_0: \mu = \mu_0.$$

Prof. R.A. Fisher remarked, **"Null hypothesis is the hypothesis which is to be tested for possible rejection under the assumption it is true".**

Alternative Hypothesis. *The negation of Null hypothesis is called the* **Alternative hypothesis.** In other words, any hypothesis **which is not a null Hypothesis is called an Alternative Hypothesis.** It is always denoted by H_1 or H_α. **It is set in such a way that the rejection of null hypothesis implies the acceptance of alternative hypothesis.** For example, if we want to find the null hypothesis that the average height of the students of a college is 165 cm, *i.e.*, μ_0 = 165 cms (say). Then the **null hypothesis** is

$$H_0: \mu = 165 (= \mu_0)$$

and the **alternative hypothesis** could be

(*i*) $H_1: \mu \neq \mu_0$ (*i.e.*, $\mu > \mu_0$ or μ_0)

(*ii*) $H_1: \mu > \mu_0$

(*iii*) $H_1: \mu < \mu_0$.

Thus **there can be more than one alternative hypothesis.**

2. Computation of Test Statistic. After setting up the null hypothesis and alternative hypothesis, we compute the **test statistic.** The **test statistic is a statistic based on appropriate probability distribution. It is used to test whether the null hypothesis set up should be accepted or rejected.** It is the main yard stick which helps us to decide whether to accept or reject the null hypothesis. Different probability distribution values are used in appropriate cases while testing the null hypothesis. The following are the most commonly used test statistics.

Z-Distribution. We use Z-distribution under normal curve for large sample (*i.e.*, **if the sample size $n \geq 30$.)** The Z-statistic is defined as

$$Z = \frac{t - E(T)}{S.E.(t)} \sim N(0, 1) \text{ as } n \to \infty$$

is a standard normal variable with mean zero and variance one.

Also *S.E.* (*t*) = Standard error of the statistic *t*.

***t*-Test. *t*-test is used for small sample *i.e.*, if the sample size $n < 30$. The student's *t*-statistic is defined as**

$$t = \frac{\bar{x} - \mu}{S.E.(\bar{x})} = \frac{\textbf{Difference of sample and population means}}{\textbf{Standard error of mean}}$$

with $n - 1$ degrees of freedom, where n is the size of sample.

The following table shows the conditions for using *Z*-test and *t*-test in testing null hypothesis about the means.

Table

Size of Sample	*Population s.d. σ is known*	*Population s.d. σ is not known*
$n \geq 30$	Z-test	Z-test
$n < 30$	*t*-test	*t*-test

3. Types of Errors in Hypothesis Testing. There is every chance that a decision regarding a null hypothesis may be correct or may not be correct. There are two types of errors.

(*a*) Type I Error. ***It is the error of rejecting null hypothesis H_0, when it is true.*** When a null hypothesis is true, but the difference (of means) is significant and the hypothesis is rejected, then a **Type I Error** is made. **The probability of making a** type I **error is denoted by α, the level of significance.** In order to control the type I error, the probability of type I error is fixed at a certain level of significance α. The probability of making a correct decision is then $(1 - \alpha)$.

(*b*) Type II Error. ***It is the error of accepting the null hypothesis H_0 when it is false.*** In other words when a null hypothesis is false, but the difference (of means) is insignificant and the hypothesis is accepted, a type II error is made. The probability of making a type II error is denoted by β.

4. Level of Significance. The next step is the fixation of level of significance. ***The level of significance is the maximum probability of making a type I error and it is denoted by α.*** *i.e.*, ***P[Rejecting H_0 when H_0 is true] = α. The probability of making a correct decision is then $(1 - \alpha)$.*** The best value for fixing the level of significance depends on the seriousness of the results of the types of error. The commonly used level of significance in practice are 5% (0.05) and 1% (0.01). If we use 5% level of significance ($\alpha = 0.05$), we shall mean that the probability of making type I error is 0.05 or 5% *i.e.*, **P [Rejecting H_0 when H_0 is true] = 0.05.**

This means that there is a probability of making 5 out of 100 (or making 1 out of 20) type I errors. This also means that we are 95% confident that a correct decision has been made. Similarly 1% level of significance ($\alpha = 0.01$) means that there is a probability of making 1 error out of 100. It is important to note that if no **level of significance is given, then we always take $\alpha = 0.05$.**

5. Critical Region or Rejection Region. ***The rejection region or critical region is the region of the standard normal curve corresponding to a predetermined level of significance α (which is fixed for knowing the probability of making a type I error rejecting the null hypothesis H_0 when it is true).*** The region under the normal curve which is not covered by the rejection region is known as **Acceptance Region.** Thus the statistic which leads to rejection of null hypothesis H_0 gives us a region known as **Rejection Region** or **Critical Region.** While those which lead to the acceptance of H_0 give us a region called as **Acceptance Region.** ***The value of the test statistic computed to test the null hypothesis H_0 is known as the*** **Critical Value. The critical value separates the rejection region from the acceptance region.**

6. Two Tailed Test And One Tailed Test. The probability curve of the sampling distribution of the test statistic is a normal curve. In any test, the critical region is represented by a portion of the area under this normal curve. This curve has two sides (or ends) known as **two tails.** ***The rejection region may be represented by a portion of area on each of the two sides or by only one side of the normal curve and correspondingly the test is known as*** **Two Tailed Test (or Two sided test) or one tailed Test (or one sided test).**

Two Tailed Test or **Two Sided Test.** ***When the test of hypothesis is made on the basis of rejection region represented by both sides of the standard normal curve,it is called a*** **two tailed test *or* two sided test**. In other words, a test of statistical hypothesis where the alternative hypothesis H_1 is two sided or two tailed such as

Null Hypothesis $H_0 : \mu = \mu_0$

Alternative Hypothesis : $H_1 : \mu \neq \mu_0$ ($\mu > \mu_0$ and $\mu < \mu_0$)

It is called as two tailed test or two sided test.

One Tailed or One Sided Test. ***A test of statistical hypothesis, where the alternative hypothesis is one sided is called as one tailed test or one sided test.*** There are two types of one tailed or one sided tests.

(1) **Right Tailed Test.** ***In the right tailed test the rejection region or critical region lies entirely on the right tail of the normal curve.*** (See figure 10.2)

(2) **Left Tailed Test.** ***In the left tailed test the critical region or rejection region lies entirely on the left tail of the normal curve.*** (See figure 10.3)

For example, in a test for testing the mean (μ) of the population

Null hypothesis $H_0 : \mu = \mu_0$

Alternative hypothesis $H_1 : \mu > \mu_0$ (Right Tailed) and $H_1 : \mu < \mu_0$ (Left Tailed)

Now, **the question arises which test is to be applied when *i.e.*, when a two tailed test will be applied and where to apply one tailed test?**

Application of Two Tailed Test : A two tailed test is applied in such cases when the difference between the sample mean and population mean is tending to reject the null hypothesis H_0, the difference may be positive or negative.

Application of One Tailed Test : On the other hand, **a one tailed test is applied in such cases when it is considered to see whether the population mean is at least as large as some specified value of the mean or atleast as small as some specified value of the mean. In the former case the right tailed tests and in the latter case a left tailed tests** are applied.

7. Critical Value. ***The value of the sample statistic that defines the regions of acceptance and rejection, is called the*** **critical value.** It is important to note that **"the critical value of Z for a single tailed test (left or right) at a level of significiant α is the same as the critical value of Z for a two tailed test at a level of significance 2α"**

REJECTION REGION (α/2)
ACCEPTANCE REGION
REJECTION REGION (α/2)
$Z = 0$

Fig. 9.1 Two Tailed Test at significance level

Further for **Right Tailed Test** :

$$P(Z > Z_\alpha) = \alpha_1$$

For **Left Tailed Test**: $P(Z < -Z_\alpha) = \alpha_1$.

Also for **a two tailed test**: $P(|Z| > Z_\alpha) = \alpha$.

$$\Rightarrow \quad P(Z > Z_\alpha) + P(Z < -Z_\alpha) = \alpha.$$

$$\Leftrightarrow \quad P(Z > Z_\alpha) + P(Z > Z_\alpha) = \alpha. \qquad \text{(By Symmetry)}$$

$$\Rightarrow \quad P(Z > Z_\alpha) = \frac{\alpha}{2} \Rightarrow \textbf{the area of each tail is } \frac{\alpha}{2}.$$

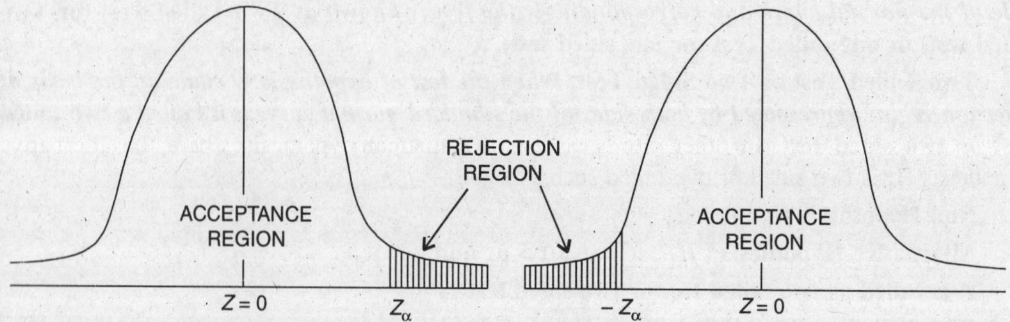

Fig. 9.2 Right Tailed Test at Significance level **Fig. 9.3** Left Tailed Test at Significance level

Two following table will give the critical value of Z *viz.*, Z_α at 1%, 5%, 10% level of significance.

Table for Critical Values Z_α of Z.

Critical value (Z_α)	*Level of Significance* α.		
	1 %	5 %	10 %
Two Tailed Test	$\lvert Z_\alpha \rvert = 2.58$	$\lvert Z_\alpha \rvert = 1.96$	$\lvert Z_\alpha \rvert = 1.645$
Right Tailed Test	$Z_\alpha = 2.33$	$Z_\alpha = 1.645$	$Z_\alpha = 1.28$
Left Tailed Test	$Z_\alpha = -2.33$	$Z_\alpha = -1.645$	$Z_\alpha = -1.28$

8. DECISION. The last step is the decision about the null hypothesis *i.e.*, whether to accept it or to reject it. In this regard we compare the computed value of Z (obtained in step 2) with the critical value or significant value or tabled value Z_α (given by table for critical values in step 7) at a given level of significance α and decide as under:

(*i*) **If $|Z| < Z_\alpha$, then we accept the null hypothesis H_0** $\Rightarrow$ *If the calculated value of Z is less than the tabled value Z_α of Z at a level of significance α, then the difference between $[t - E(t)]$ is not significant (and this difference may be due to fluctuation of sampling), so we accept the null hypothesis.* In this case the **test statistic falls in the region of acceptance.**

(*ii*) **If $|Z| > Z_\alpha$, then we reject the null hypothesis H_0 and accept the alternative hypothesis H_1.** In this case, *the computed value of Z is numerically greater than the critical value Z_α at a level of significance α, and therefore the computed value of test statistic falls in the rejection region. So we reject the null hypothesis and accept the alternative hypothesis at a level of significance α or confidence level $(1 - \alpha)$.*

9.3. TEST OF SIGNIFICANCE OF A SINGLE MEAN — LARGE SAMPLES

WORKING RULE

We have the following steps :

Step 1. Setting up of a Null hypothesis and Alternative Hypothesis: The **Null Hypothesis** is set up in the following form: "*There is no significance difference between the sample mean and population mean*" *or the sample has been drawn from the parent population*. Mathematically,

Null hypothesis :– H_0 : $\bar{x} = \mu$.

Alternative hypothesis :– H_1 : $\bar{x} \neq \mu$.

[Check the nature of test whether two tailed or one tailed.]

Step 2. Computation of test statistic Z : There are two ways of computing a test statistic Z.

(*a*) When the standard deviation σ of population is known.

In this case Standard Error of Mean = $\text{S.E.}(\bar{x}) = \dfrac{\sigma}{\sqrt{n}}$,

where n = sample size, σ = s.d. of population

$$\textbf{Test Statistic : } Z = \frac{\bar{x} - \mu}{\text{S.E.}(x)} = \frac{(\bar{x} - \mu)\sqrt{n}}{\sigma}$$, **where μ is the sample mean.**

(*b*) **When the standard deviation σ of population is not known.**

In this case **we take s, the standard deviation of sample to calculate the standard error of Mean** $\text{S.E.}(\bar{x}) = \dfrac{s}{\sqrt{n}}$

$$\textbf{Test Statistic : } Z = \frac{\bar{x} - \mu}{\text{S.E.}(\bar{x})} = \frac{(\bar{x} - \mu)\sqrt{n}}{s}$$

Step 3. Level of significance. Set the level of significance α.

Step 4. Critical value. Find the critical value Z_α of Z at the level of significance α **(of STEP 3)**, from the table **"Areas under the normal curve-Z_σ – values".**

Step 5. Decision. (*a*) ***If the modulus of the computed value of*** $Z = |Z| <$ ***critical value*** Z_α ***at a level of significance*** α, ***then*** **accept the null hypothesis.**

(b) If the modulus of computed value of $Z = |Z| >$ ***critical value*** Z_α ***at a level of significance*** α, ***then we*** **reject the null hypothesis** H_0 **and accept the alternative hypothesis** H_1.

Example 1. *A sample of size 400 was drawn and the sample mean was found to be 99. Test whether this sample could have come from a normal population with mean 100 and standard deviation 8 at 5% level of significance.*

Solution. Here $\bar{x} = 99, \ n = 400, \ \mu = 100, \ \sigma = 8.$

1. **Null hypothesis:** The sample have come from a normal population with mean 100 and s.d. 8, *i.e.*, $H_0 : \mu = 100$.

$\therefore$ **Alternative Hypothesis:** $H_1 : \mu \neq 100$.

$\Rightarrow$ It is a case of **two tailed test.**

2. **Test Statistic:**

Here $$\text{S.E.}(\bar{x}) = \frac{\sigma}{\sqrt{n}} = \frac{8}{\sqrt{400}} = \frac{8}{20} = \frac{2}{5}.$$

$$\therefore \quad \text{Test Statistic : } Z = \frac{\bar{x} - m}{\text{S.E.}(\bar{x})} = \frac{99 - 100}{(2/5)} = -\frac{5}{2} = -2.5.$$

$$\therefore \quad |Z| = 2.5.$$

3. **Level of significance:** $\alpha = 0.05$.

4. **Critical value:** The value of Z_α at 5% level of significance = 1.96 (from the table)

5. **Decision:** Since the calculated value of $|Z|$ is greater than the critical value as **2.5 > 1.96** $\Rightarrow |Z| > Z_\alpha \Rightarrow$ **Null hypothesis H_0 is rejected $\Rightarrow$ the sample has not been drawn from a normal population with mean 100 and s.d. 8.**

Example 2. *A stenographer claims that she can type at the rate of 120 words per minute. Can we reject her claim on the basis of 100 trials in which she demonstrates a mean of 116 words with a standard deviation of 15 words. Use 5% level of significance.*

Solution 1. Null Hypothesis : H_0 **:** Stenographer's claim is true, *i.e.*, $H_0 : \mu = 120$.

Alternative Hypothesis : $H_1 : \mu \neq 120$

$\Rightarrow$ It is a case of **two tailed test.**

2. **Calculation of Test Statistic.**

Here $n = 100, \ \bar{x} = 116, \ \mu = 120, \ s = 15.$

$$\text{Standard Error of Mean} = \text{S.E.}(\bar{x}) = \frac{\sigma}{\sqrt{n}} = \frac{15}{\sqrt{100}} = \frac{15}{10} = 1.5.$$

$$\therefore \quad \text{Test Statistic : } \quad Z = \frac{\bar{x} - m}{(s/\sqrt{n})} = \frac{116 - 120}{1.5} = 2.67.$$

3. **Level of significance:** $\alpha = 0.05$.

4. **Critical Value:** The value of Z_α at 5% level of significance is $Z_\alpha = 1.96$.

[From the Table]

5. **Decision.** Since $|Z| = 2.67$ **is greater than** $Z_\alpha = 1.96$ **at 5% level of significance, so the null hypothesis** H_0 **is rejected.** Thus **the** stenographer's claim of typing at the rate of 120 words per minute is not true.

Example 3. *A random sample of 900 items is taken from a normal population whose mean and the variance are 4. Can the sample with mean 4.5 be regarded as truly random one at 1% level of significance? (Table value at 1% level is 2.58).* **[C.A. (Found.); Nov. 1998]**

Solution.

1. **The Null hypothesis: $H_0 : m = 4$.**
2. **Alternative hypothesis: $H_1 : m \neq 4$.**
   ⇒ It is a case of two tailed test.
3. **Calculation of Test statistic:** $Z = \dfrac{\bar{x} - m}{\sigma / \sqrt{n}} = \dfrac{4.5 - 4}{2/30} = 7.5$.
4. **Decision: The modulus value of calculated Z is greater than 1% value of tabulated, which is given to be 2.58. Hence the null phypothesis is rejected and it is concluded that the sample cannot be regarded as a sample from a normal population with mean 4.**

Example 4. *A random sample of 400 items gives the mean 4.45 and the variance as 4. Can the sample be regarded as drawn from the normal population with mean 4? (At 5% level of significance).*

Solution.

1. **Null Hypothesis: $H_0 : \mu \neq 4$.**
2. **Alternative Hypothesis : $H_1 : \mu = 4$**
   ⇒ It is a case of one tail test.
3. **Test statistic :** $Z = \dfrac{\bar{x} - m}{S / \sqrt{n}} = \dfrac{4.45 - 4}{\sqrt{4} / \sqrt{400}} = \dfrac{0.45 \times \sqrt{400}}{2} = \mathbf{4.5}$
4. **Decision : The calculated value of Z is more than the tabulated value of $Z_\alpha = 1.96$ at 5% level of significance. Hence we reject Null Hypothesis. H_0 and conclude that sample is not from normal population with mean = 4.**

Example 5. *In a sample of 4000 parts manufactured by a company, the number of defective parts was found to be 30. The company, however, claimed that only 5% of their product is defective. Test at 5% level of significance whether the claim of the company is tenable.*

Solution.

Here $n = 400$; $P = \dfrac{30}{400} = 0.075$.

1. **Null Hypothesis :** Let us set $H_0 : P = 0.05$.
2. **Alternative Hypothesis :** $H_1 : P > 0.05$.
   ⇒ It is a case of one tailed test.
3. **Test statistic :** $Z = \dfrac{0.075 - 0.05}{\sqrt{\dfrac{0.05 \times (1 - 0.05)}{400}}} = \dfrac{0.025}{0.0108972} = \mathbf{2.29}$
4. **Critical Region :** At 5% level of significance the tabled value of Z is 1.64.

Decision: Since **the computed $|Z|$ > table value of Z_α, so H_0 is rejected and hence, claim of the supplier is not tenable.**

Example 6. *A sample of 400 male students of a college is found to have a mean height of 171.38 cm. Can it be regarded as a sample from a large population with mean height 171.17 cm and standard deviation 3.30 cm.*

Solution. Here $n = 400$ ⇒ It is a case of large sample.

Also $\mu = 171.17$ cms, $\bar{x} = 171.38$ cms, $\sigma = 3.30$ cms.

1. Null Hypothesis: H_0 : Sample has been drawn from a large population with mean height 171.17 cm, *i.e.*, $H_0 : \mu = 171.17$.

Alternative Hypothesis : $H_1 : \mu \neq 171.17$

It is a case of **Two failed test.**

Standard Error of mean: $\text{S.E.}(\bar{x}) = \frac{\sigma}{\sqrt{n}} = \frac{3.30}{\sqrt{400}} = \mathbf{0.165.}$

Test Statistic: $Z = \frac{\bar{x} - \mu}{\text{S.E.}(\bar{x})} = \frac{1.71.38 - 171.17}{0.165} = \frac{0.21}{0.165} = \mathbf{1.27.}$

$\therefore \quad |Z| = \mathbf{1.27.}$

3. Level of Significance : $\alpha = 5\%$.

4. Critical Value at $\alpha = 0.05$. The tabled value of Z at $\alpha = 0.05$ is $Z_\alpha = \mathbf{1.96}$ (from the table)

Decision. Here calculated value of Z is less than the tabled value or critical value of Z at $\alpha = 0.05$ as $1.27 < 1.96$, *i.e.*, $|Z| < Z_\alpha \Rightarrow$ the null hypothesis is accepted.

Hence the sample has been drawn from a population with mean height 171.17 cms and standard deviation 3.30 cms.

Example 7. *A sample of 900 members has a mean 3.4 cms and s.d. 2.61 cms. Can the sample be regarded as one drawn from a population with mean 3.25 cms. Using the level of significance as 0.05, is the claim acceptable ?*

Solution. Here $n = 900 \Rightarrow$ **it is a large sample.**

Also $\bar{x} = 3.4$ cm, s.d. of sample $s = 2.61$ cm, $\mu = 3.25$.

1. Null hypothesis: The sample has been drawn from a population with mean 3.25 cms, *i.e.* $H_0 : \mu = 3.25$.

Alternative hypothesis. $H_1 : \mu \neq 3.25$.

It is a case of **two tailed Test.**

2. Calculation of Test Statistic.

Standard Error of mean: $\text{S.E.}(\bar{x}) = \frac{s}{\sqrt{n}} = \frac{2.61}{\sqrt{900}} = \frac{2.61}{30} = \mathbf{0.087.}$

$\therefore$ **Test Statistic:** $Z = \frac{\bar{x} - m}{\text{S.E.}(\bar{x})} = \frac{3.4 - 3.25}{0.087} = \mathbf{1.72.}$

3. Level of Significance: Here $\alpha = 0.05$.

4. Critical Value: From the table $|Z|$ the value of Z at $\alpha = 0.05$ is $Z_\alpha = \mathbf{1.96.}$

5. Decision. Here calculated value of $|Z| = 1.72 <$ Critical value $Z_\alpha = 1.96$

$\Rightarrow$ Null hypothesis is accepted.

Hence, **the sample has been drawn from a population with mean 3.25 cms.**

Example 8. *The mean life time of a sample of 400 fluorescent light tube produced by a company is found to be 1570 hours with a standard deviation of 150 hours. Test the hypothesis that the mean life time of the bulbs produced by the company is 1600 hours against the alternative hypothesis that it is greater than 1600 hours at 1% level of significance.*

Solution. Here $n = 400 \Rightarrow$ it is a **large sample.**

Also $\bar{x} = 1570$, standard deviation of sample mean $s = 150$, $\mu = 1600$.

1. Null Hypothesis : The mean life time of the bulbs is 1600 hours, *i.e.*, $H_0 : \mu = 1600$.

Alternative Hypothesis : $H_1 : \mu > 1600 \Rightarrow$ It is a case of **right tailed test.**

2. Computation of Test Statistic Z :

$$\textbf{Standard error} = \text{S.E.}(\bar{x}) = \frac{s}{\sqrt{n}} = \frac{150}{\sqrt{400}} = \frac{150}{20} = \mathbf{7.5}$$

$$\textbf{Test Statistic} : Z = \frac{\bar{x}-\mu}{\text{S.E.}(\bar{x})} = \frac{1570-1600}{7.5} = \frac{30}{7.5} = -\mathbf{4.}$$

$$\therefore \quad |Z| = \mathbf{4.}$$

3. Level of significance : Here $\alpha = 0.01$.

4. Critical value : The critical value of Z at 1% level of significance for the Right tailed test (from the table) is $Z_\alpha = \mathbf{2.33}$.

5. Decision : Since the calculated value of $|Z|$ is 4 which is greater than the critical value Z_α (at $\alpha = 0.01$) = 2.33, so the null hypothesis is rejected and the alternative hypothesis is accepted.

Hence the **mean life time of bulbs is greater than 1600 hours.**

EXERCISE – 9.1

1. A random sample of 100 students gave a mean weight of 58 kilogram with s.d. of 4 kg. Test the hypothesis that the mean weight in the population is 60 kg.

   [**Hint.** $H_0 : \mu = 60$ kgs., $H_1 : \mu \neq 60$ kgs (Two tailed test)

   $$\text{Test Statistic} : Z = \frac{\bar{x}-\mu}{s/\sqrt{n}} = \frac{58-60}{4/\sqrt{100}} = 5 \Rightarrow |Z| = 5.$$

   Value of Z_α at $\alpha = 0.05$ is $1.96 < |Z| = 5 \Rightarrow H_0$ is rejected $\Rightarrow$ **the mean weight of the population is not 60 kgs.]**

2. A sample of 400 items is taken from a normal population whose mean is 4 and whose variance is also 4. If the sample mean is 4.45, can the sample be regarded as truly random sample?

   [**Hint.** Here the level of significance is not given so we take $\alpha = 0.05$. Also $\bar{x} = 4.45$, $\mu = 4$, $\sigma^2 = 4$. Let $H_0 : \mu = 4$.

   $$\therefore \quad \text{S.E.}(\bar{x}) = \frac{2}{\sqrt{400}} = \frac{2}{20} = 0.1.$$

   $$Z = \frac{\bar{x}-4}{S.E.(\bar{x})} = \frac{4.45-4}{0.1} = \mathbf{4.5}$$

   Also Z_α at $\alpha = 0.05$ is = 1.96.

   Since $|Z| = 4.5 > Z_\alpha = 1.96 \Rightarrow H_0$ is rejected at 5% level of significance

   $\Rightarrow$ **Sample cannot be regarded as having been drawn from the population with mean 4.]**

3. The mean I.Q. of a sample of 1600 children was 99. Is it likely that this was a rendom sample from a population with mean I.Q. 100 and standard deviation 15 ?

   [**Hint.** Here $n = 1600$, $\bar{x} = 99$, $\mu = 100$, $\sigma = 100$.

   Null hypothesis $H_0 : \mu = 100$, Alternative hypothesis $H_1 : \mu \neq 100 \Rightarrow$ Two tailed test.

   $$\textbf{S.E.}(\bar{x}) = \frac{\sigma}{\sqrt{n}} = \frac{15}{\sqrt{1600}} = \frac{15}{40} = \frac{3}{8}$$

   $$Z = \frac{\bar{x}-\mu}{\textbf{S.E.}(\bar{x})} = \frac{99-100}{3} \times 8 = -2.67 \Rightarrow |Z| = \mathbf{2.67.}$$

   Since a is not given so we take $\alpha = 0.05$.

   $\therefore \; Z_{0.05} = 1.96 < |Z| = 2.67 \Rightarrow H_0$ is rejected $\Rightarrow$ **sample has not been drawn from a population with mean 100 and σ = 15.]**

4. If a random sample of size 20 from a normal population with standard deviation 5.2 shows a mean of 16.9, test at 5% significance level that the population mean is 15.5. Also calculate 99% confidence limit for mean.

[**Hint.** Null hypothesis : $H_0 : \mu = 15.5$, Alternative hypothesis : $H_1 : \mu \neq 15.5$.

$Z = \frac{16.9 - 15.5}{(5.2/\sqrt{20})} = 1.20$. Also value of Z_α of $\alpha = 0.05 = 1.96$.

Again $|Z| = 1.20 < Z_\alpha = 1.96 \Rightarrow H_0$ is accepted $\Rightarrow$ population mean may be 15.5.

99% confidence limits are 16.9 $\pm \frac{5.2}{\sqrt{20}} \times 2.58 = 16.9 \pm 3 = 13.9, 19.9.$]

5. A sample of 100 students is taken from a large population. The mean height of these students is 64 cm and the standard devisation is 4 cm. Can it reasonable be regarded that in the population, the mean height is 60 cm ?

6. A new variety of potato grown in 250 plots gave rise to a mean yield of 82.7 quintals per hectare with a s.d. of 14.6 quintals per hectare. Is it reasonable to assert that the new variety is superior in yield to the standard variety with an established yield of 80.2 quintals per hectare ?

ANSWERS

1. The mean weight of the population is not 60 kgs.

2. Sample cannot be regarded as having been drawn from the population with mean 4.

3. Sample has not been drawn from the population with mean 100 and $\sigma = 15$.

4. 13.9 to 19.9 **5.** $Z = -5$, No. **6.** $Z = 2.71$, yes.

9.4 TEST OF SIGNIFICANCE FOR SINGLE PROPORTION – LARGE SAMPLES

WORKING RULE

1. Null Hypothesis: The sample has been drawn from a population with population proportion P, *i.e.*, $H_0 : P = P_0$, where P_0 is a particular value of P.

Alternative Hypothesis : $H_1 : P \neq P_0$ $\Rightarrow$ Two Tailed Test.

2. Selection of Test Statistic:

Standard Error of proportion = $\textbf{S.E.}(p) = \sqrt{\frac{P(1-P)}{n}}$

where n is the sample size and P is population proortion.

Or

Standard Error of Percentage = $\textbf{S.E.}(p) = \sqrt{\frac{P(100-P)}{n}}$

where P = population percentage and n is the sample size.

Test Statistic : $Z = \frac{p - P}{\text{S.E.}(p)}$

where p = sample proportion or percentage.

The other steps such as (3) level of significance; (4) Critical value of Z, (5) Decision are the same as those discussed in testing the significance of a sample means on page 368.

Example 9. *A dice was thrown 9000 times and of these 3220 yielded a 3 or 4. Is this consistent with the hypothesis that the dice was unbiased?*

Solution. Here the sample proportion : $p = \frac{3220}{9000} = \mathbf{0.358}$.

Also the Population proportion $P = \frac{2}{6} = \frac{1}{3}$.

1. **Null hypothesis :** The dice is unbiased, *i.e.*, $H_0 : P = \frac{1}{3} = 0.333$.

   Alternative hypothesis : $H_1 : P \neq \frac{1}{3}$ ⇒ **It leads to Two Tailed Test.**

2. **Selection of Test Statistic.**

   Standard Error of proportion: S.E.(*p*) $= \sqrt{\frac{PQ}{n}} = \sqrt{\frac{1}{3} \times \frac{2}{3} \times \frac{1}{9000}} = \sqrt{0.000025} = \mathbf{0.005}.$

   ∴ **Test Statistic:** $Z = \frac{p - P}{\text{S.E.}(p)} = \frac{0.358 - 0.333}{0.005} = \frac{0.025}{0.005} = \mathbf{5}.$

3. **Level of Significance.** Take $\alpha = 0.05$ as the level of significance as α is not given.
4. **Critical value.** The critical value of Z at $\alpha = 0.05$ is $Z_\alpha = \mathbf{1.96}$ (from Normal Table)
5. **Decision.** Now $|Z| > Z_\alpha$ as $5 > 1.96$ ⇒ **Null hypothesis H_0 is rejected**

   ⇒ **that the dice is a baised one.**

Example 10. *In a sample of 400 burners there were 12 whose internal diameters were not within tolerance. Is this sufficient evidence for concluding that the manufacturing process is turning out more than 2% defective burners. Take $\alpha = 0.05$.*

Solution. It is given that $P = \mathbf{0.02}$ so that $Q = 1 - P = 1 - 0.02 = \mathbf{0.98}$

Also $p = \frac{12}{400} = \mathbf{0.03}.$

1. **Null Hypothesis :** The process is in control, *i.e.*, $H_0 : P \leq 0.2$.

   Alternative Hypothesis : $H_1 : P > 0.2$ ⇒ it leads to **One Tailed Test.**

2. **Selection of Test Statistic :** The **test statistic is :**

$$Z = \frac{p - P}{\text{S.E of } (p)} = \frac{p - P}{\sqrt{PQ/n}} = \frac{0.03 - 0.02}{\sqrt{(0.02 \times 0.98)/400}}$$

$$= \frac{0.01}{\sqrt{.000049}} = \frac{0.01}{0.007} = \mathbf{1.429}.$$

Level of Significance. $\alpha = 0.05$ or $\frac{\alpha}{2} = 0.025.$

Critical Value. The critical value Z_α of Z for $\alpha = 0.05$ for One Tailed test is $Z_\alpha = \mathbf{1.645}.$

Decision. Since $|Z| < Z_\alpha$ **as 1.429 < 1.645 so the Null Hypothesis is accepted**

⇒ **the process is not out of control.**

Example 11. *In a random sample of 400 persons from a large population, 120 are females. Can it be said that males and females are in the ratio 5 : 3 in the population ? Use 10% level of significance.*

Solution. We are given the following data.

Population proportion of female $P = \frac{3}{5+3} = \frac{3}{8} = \mathbf{0.375}.$

∴ $Q = 1 - P = 1 - 0.375 = \mathbf{0.625}.$

Sample proportion of female $p = \frac{120}{400} = \mathbf{0.30}$

Also $n = 400$

1. **Null hypothesis : $H_0 : P = 0.375$**

   Alternative hypothesis : $H_1 : P \neq 0.375$. It leads to **Two Tailed Test.**

2. **Test Statistic.**

$$Z = \frac{p - E(p)}{\text{S.E.}(p)} = \frac{p - P}{\sqrt{PQ/n}}$$

$$= \frac{0.30 - 0.375}{\sqrt{0.375 \times 6.25/400}} = \frac{-0.075}{\sqrt{0.000586}} = \frac{-0.075}{0.024} = -3.125.$$

$$\therefore \qquad |Z| = 3.125.$$

3. **Level of significance :** Here $\alpha = 0.05$
4. **Critical value of Z at** $\alpha = 0.05$ is $Z_\alpha = 1.96$.
5. **Decision.** Here $|Z| > Z_\alpha$ as $3.125 > 1.96$, **so we reject the null hypothesis**

   ⇒ **the males and females in the population are not in the ratio 5 : 3.**

Example 12. *A manufacturer claimed that at least 95% of the equipment which he supplied to a factory conformed to specifications. An examination of a sample of 200 pieces of equipment revealed that 18 were faulty. Test his claim at a significance level of (i) 0.05; (ii) 0.01.*

Solution. We are given the following data :

$n = 200$. P = proportion of pieces conforming to specification in the lot is 95% = **0.95.**

$$\therefore \qquad Q = 1 - P = 1 - 0.95 = \mathbf{0.05}.$$

Also Sample proportion $\quad p = \dfrac{200-18}{200} = \dfrac{\mathbf{182}}{\mathbf{200}} = \mathbf{0.91}.$

1. **Null hypothesis :** Proportion of pieces conforming to specification is 95%, *i.e.*,

$$H_0 : P = 0.95.$$

   Alternative hypothesis : $H_1 : P < 0.95$ ⇒ It leads to **Left Tailed Test.**

2. **Calculation of Test Statistic :**

$$Z = \frac{p - E(p)}{\text{S.E. of } p} = \frac{p - P}{\sqrt{PQ/n}}$$

$$= \frac{0.91 - 0.95}{\sqrt{(0.95 \times 0.05)/200}} = \frac{-0.04}{\sqrt{0.00237}} = \frac{-0.04}{0.0154} = -2.6.$$

$$\therefore \qquad |Z| = 2.6.$$

(*i*) **Critical value of Z at $\alpha = 0.05$ for left tailed test is $Z_\alpha = 1.645$.**

3. **Decision : Since $|Z| > Z_\alpha$ as 2.6 > 1.6 so we reject the null hypothesis H_0 at 5% level of significance ⇒ the manufacturer's claim is rejected at 5% level of significance.**

(*ii*) **Critical Value** at 1% level of significance of Z is **$Z_\alpha = 2.33$ for one tailed test.**

Decision : Since $|Z| > Z_\alpha$ as 2.6 > 2.33 so we reject the null hypothesis at 1% level of significance ⇒ the manufacturer's claim is rejected at 1% level of significance.

9.5 TEST OF SIGNIFICANCE OF DIFFERENCE BETWEEN TWO SAMPLE PROPORTIONS FOR LARGE SAMPLES

WORKING RULE.

1. **Null hypothesis :** The two samples have been drawn from the same population *i.e.*

$$H_0 : P_1 = P_2.$$

Alternative hypothesis : $\quad H_1 : P_1 \neq P_2$. **It is a case of Two tailed test.**

2. Computation of Test Statistic : We have the following case :

Case I. When the population proportions P_1 and P_2 are known.

In this case $Q_1 = 1 - P_1$, $Q_2 = 1 - P_2$ and p_1 and p_2 are sample proportions.

$\therefore$ **Standard Error of Difference : S.E.$(p_1 - p_2)$** $= \sqrt{\dfrac{P_1Q_1}{n_1} + \dfrac{P_2Q_2}{n_2}}$

$\therefore$ **Test Statistic :** $Z = \dfrac{p_1 - p_2}{S.E.(p_1 - p_2)}$

Case II. When the population proportions P_1 and P_2 are not known but sample proportions p_1 and p_2 are known.

In this case we have two methods to estimate P_1 and P_2.

(*a*) **Method of substitution.** In this method p_1 and p_2 are subsituted for P_1 and P_2.

S.E. = Standard Error of Difference : S.E. $(p_1 - p_2) = \sqrt{\dfrac{p_1q_1}{n_1} + \dfrac{p_2q_2}{n_2}}$

Test Statistic : $Z = \dfrac{p_1 - p_2}{\text{S.E.}(p_1 - p_2)}$

The other steps such as (*i*) level of significance; (*ii*) Critical value; (*iii*) decision in respect of testing the significance of difference of two proportions are the same as those discussed in testing the significance of sample means.

(*b*) Method of Pooling. ***In this method the estimated value for the two population proportions is obtained by pooling the two sample proportions p_1 and p_2 into a single proportion p by the formula given below.***

Sample Proportion of Two Samples or Estimated p

$$p = \frac{n_1p_1 + n_2p_2}{n_1 + n_2}$$

so **that $q = 1 - p$.**

Standard Error of Difference : $\text{S.E.}(p_1 - p_2) = \sqrt{\dfrac{pq}{n_1} + \dfrac{pq}{n_2}}$,

where n_1 and n_2 are the sizes of the two samples

Test Statistic : $Z = \dfrac{p_1 - p_2}{S.E.(p_1 - p_2)}$

The other steps such as (*i*) level of significance; (*ii*) Critical value of Z at s, *i.e.*, Z_∞ and (*iii*) decision making in respect of testing the significance of the difference between two proportions is the same as those discussed in testing the difference of sample means.

Example 13. *A machine produced 20 defective articles in a batch of 400. After overnauling, it produced 10 defectives in a batch of 300. Has the machine improved ?*

Solution.

1. Null hypothesis : There is no significant difference in the improvement of the machine before and after overhaul *i.e.*, $H_0 : P_1 = P_2$.

Alternative hypothesis : $H_1 : P_2 > P_1 \Rightarrow$ the machine has imporved. $\Rightarrow$ It is a **One tailed test.**

2. Computation of Test Statistic :

p_1 = Sample proportion of defective articles before overhaul $= \dfrac{20}{400} = \mathbf{0.05}$.

$\therefore \quad q_1 = 1 - p_1 = 1 - 0.05 = \mathbf{0.95}$.

p_2 = Sample proportion of defective articles after overhaul $= \frac{10}{300} = \mathbf{0.033.}$

$\therefore \quad q_2 = 1 - p_2 = 1 - 0.033 = \mathbf{0.967.}$

Standard Error of Difference : $\text{S.E.}(p_1 - p_2) = \sqrt{\frac{p_1 q_1}{n_1} + \frac{p_2 q_2}{n_2}}$

$$= \sqrt{\frac{0.05 \times 0.95}{400} + \frac{0.033 \times 0.967}{300}} = \sqrt{0.00023} = \mathbf{0.015.}$$

Test Statistic : $Z = \frac{p_1 - p_2}{\text{S.E.}(p_1 - p_2)} = \frac{0.05 - 0.033}{0.015} = \frac{0.017}{0.015} = \mathbf{1.134.} \quad \therefore \ |Z| = \mathbf{1.134.}$

3. **Leval of significance :** Take $\alpha = 0.05$.
4. **Critical value** of Z at $\alpha = 0.05$ is $Z_\alpha = 1.645$ **for one tailed test.**
5. **Decision.** Since $|Z| = 0.134 < Z_\alpha = 1.645 \Rightarrow$ **Null hypothesis H_0 is accepted $\Rightarrow$ that the machine has not improved after overhauling.**

Example 14. *On a certain day, 74 trains were arriving on time at Delhi station during the rush hours and 83 were late. At New Delhi there were 65 on time and 107 late. Is there any difference in the proportions arriving on time at the two stations?*

Solution. p_1 = proportion of trains arriving on time at Delhi station $= \frac{74}{74+83} = \frac{74}{157} = \mathbf{0.471.}$

p_2 = proportion of trains arriving on time at New Delhi station $= \frac{65}{65+107} = \frac{65}{172} = \mathbf{0.378.}$

Mean proportion of trains arriving on time $= p = \frac{n_1 p_1 + n_2 p_2}{n_1 + n_2}$

$$= \frac{157 \times (74/157) + 172(65/172)}{157 + 172} = \frac{74 + 65}{329} = \frac{139}{329} = \mathbf{0.423.}$$

$\therefore \quad q = 1 - p = 1 - 0.423 = \mathbf{0.577.}$

1. **Null hypothesis :** There is no difference of train arriving 'on time' at the two stations, *i.e.*, $H : P_1 = P_2$.

   Alternative hypothesis : $H : P_1 \neq P_2 \Rightarrow$ It leads to **Two Tailed Test.**
2. **Calculation of Test Statistic.** Here $p = 0.423$, $q = 0.577$, $n_1 = 157$, $n_2 = 172$.

$\therefore$ **Standard Error of Difference :** $\text{S.E.}(p_1 - p_2) = \sqrt{\frac{pq}{n_1} + \frac{pq}{n_2}}$

$$= \sqrt{(0.423 \times 0.577)\left(\frac{1}{157} + \frac{1}{172}\right)}$$ **[Method of Pooling]**

$$= \sqrt{0.244 \times [0.00637 + 0.00581]} = \sqrt{0.244 \times 0.012} = \sqrt{0.00293} = \mathbf{0.054.}$$

$\therefore$ **Test Statistic :** $Z = \frac{p_1 - p_2}{\text{S.E.}(p_1 - p_2)} = \frac{0.471 - 0.378}{0.054} = \frac{0.093}{0.054} = \mathbf{1.722.}$

$\therefore \quad |Z| = \mathbf{1.722.}$

Level of significance : Take $\alpha = 0.05$.

Critical value. The critical value of Z at $\alpha = 0.054$ is $Z_{0.05} = \mathbf{1.96}$ (from the normal table).

Decision. Since $|Z| < Z_{0.05}$ as 1.722 < 1.96 $\Rightarrow$ Null Hypothesis H_0 is accepted $\Rightarrow$ that there is no significant difference of trains 'arriving' on time at the two stations.

Example 15. *In a sample of 600 men from a certain city, 450 men are found to be smokers. ı a sample of 900 from another city, 450 are found to be smokers. Do the data indicate that the vo cities are significantly different with respect to prevalence of smoking habits among men?*

Solution. Here we are given for one city; $n_1 = 600$, p_1 = proportion of smokers $= \frac{450}{600} = \mathbf{0.75}$.

Also for another city; $n_2 = 900$, p_2 = proportion of smokers $= \frac{450}{900} = \mathbf{0.5}$.

1. **Null hypothesis :** There is no significant difference in the smoking habits of two cities, *i.e.*, $H_0 : P_1 = P_2$.
   Alternative Hypothesis : $H_1 : P_1 \neq P_2 \Rightarrow$ a Two Tailed test.
2. **Computation of Test Statistic :**

   Sample mean proportion of the two sample proportions : $p = \dfrac{n_1 p_1 + n_2 p_2}{n_1 + n_2}$

$$= \frac{450+450}{600+900} = \frac{900}{1500} = \mathbf{0.6}. \qquad \textbf{[Method of Pooling]}$$

Standard Error of Difference : $\text{S.E.}(p_1 - p_2) = \sqrt{\dfrac{p(1-p)}{n_1} + \dfrac{p(1-p)}{n_2}}$

$$= \sqrt{\frac{0.6 \times 0.4}{600} + \frac{0.6 \times 0.4}{900}} = \sqrt{0.0004 + 0.000267} = \sqrt{0.000667} = \mathbf{0.0258}.$$

Test Statistic : $Z = \dfrac{p_1 - p_2}{\text{S.E.}(p_1 - p_2)} = \dfrac{0.75 - 0.5}{0.0258} = \mathbf{9.7}.$

3. **Level of Significance : $\alpha = 0.05$.**
4. **Critical Value :** The critical value of Z at $\alpha = 0.05$ is $\mathbf{Z_{0.05} = 1.96}$

[From the Normal table]

5. **Decision. Since $|Z|$ (= 9.7) > $Z_{0.05}$ (= 1.96) $\Rightarrow$ that the null hypothesis is rejected $\Rightarrow$ There is a significant difference in the smoking habits among men of two cities.**

Example 16. *In an infantile paralysis epidemic 500 persons contracted the disease. 300 received no serum treatment and of them 75 became paralysed. Of those who received serum treatment 65 became paralysed. Was the serum treatment effective?*

Solution. Let p_1, p_2 respectively be the proportions of persons who received no serum and who received serum treatment.

$$\therefore \quad p_1 = \frac{75}{300} = \mathbf{0.25}., \qquad p_2 = \frac{65}{500-300} = \frac{65}{200} = \mathbf{0.325}.$$

Sample mean proportion of the two sample proportions : $p = \dfrac{n_1 p_1 + n_1 n_2}{n_1 + n_2}$

$$= \frac{300 \times (75/300) + 200 \times (65/200)}{300 + 200} = \frac{75 + 65}{500} = \mathbf{0.28}. \qquad \textbf{[Method of Pooling]}$$

1. **Null hypothesis.** The serum treatment was not effective, *i.e.*, $H : P_1 = P_2$.
   Alternative hypothesis $H_1 : P_1 \neq P_2 \Rightarrow$ A Two Tailed test.
2. **Computation of Test Statistic :**

   Standard Error of Difference : $\text{S.E.}(p_1 - p_2) = \sqrt{\dfrac{p(1-p)}{n_1} + \dfrac{p(1-p)}{n_2}}$

$$= \sqrt{(0.28 \times 0.72)\left[\frac{1}{300} + \frac{1}{200}\right]} = \sqrt{\frac{1.008}{600}} = \sqrt{0.00168} = \mathbf{0.041.}$$

3. **Test Statistic :** $Z = \dfrac{p_1 - p_2}{\text{S.E.}(p_1 - p_2)} = \dfrac{0.25 - 0.325}{0.041} = -\dfrac{0.075}{0.041} = \mathbf{-1.83}.$ $\therefore |Z| = \mathbf{1.83.}$

4. **Level of significance.** Let us take $\alpha = 0.05$.

5. **Critical value :** The critical value of Z at $\alpha = 0.05$ is $\mathbf{Z_{0.05} = 1.96.}$

[From the normal Table values]

6. **Decision. Since the calculated value of $|Z|$ (= 1.83) < $Z_{0.05}$ (= 1.96) so the null hypothesis H_0 is accepted ⇒ the serum treatment was not effective.**

Example 17. *A machine put out 20 imperfect items in a sample of 500. After the machine was overhauled it put out 5 imperfect items in a batch of 150. Has the machine being improved after overhauling ?*

Solution. If p_1 and p_2 be the sample proportions of imperfect items put out by the machine before and after the overhauling and P_1 and P_2 be the corresponding population proportions

$$p_1 = \frac{20}{500} = \frac{\mathbf{1}}{\mathbf{25}}; \qquad p_2 = \frac{5}{150} = \frac{\mathbf{1}}{\mathbf{30}}.$$

We set, **Null hypothesis : $H_0 : P_1 = P_2$**

Alternative hypothesis : $H_1 : P_1 > P_2$

Now $$p = \frac{n_1p_1 + n_2p_2}{n_1 + n_2} = \frac{1}{26}; \qquad q = 1 - p = \frac{25}{26}$$

$$p_1 - p_2 = \frac{1}{25} - \frac{1}{30} = \frac{1}{150} = \mathbf{0.00666}$$

$$\text{S.E. } (p_1 - p_2) = \sqrt{pq\left(\frac{1}{n_1} + \frac{1}{n_2}\right)} = \sqrt{0.000296} = \mathbf{0.0179}$$

$$3 \times \text{S.E. } (p_1 - p_2) = 3 \times 0.0179 = 0.0516.$$

$$\text{Also } p_1 - p_2 = \mathbf{0.00666 < 3 \times S.E.}\ (p_1 - p_2)$$

So **the difference is not significant and we cannot reject the null hypothesis. Hence, the machine has not improved after overhauling.**

EXERCISE – 9.2

1. A manufacturer claims at least 95% of the items he produces are failure free. Examination of a random sample of 600 items showed 39 to be defective. Test the claim at a significance level of 0.05.
2. A bottle manufacturing process is 'under control' if no more than 1% of the bottles are defective. A random sample of 120 bottles showed 5 to be defective. Do these data indicate that the process is out of control? Use $\alpha = 0.10$.
3. In a sample of 400 parts manufactured by a factory, the number of defective parts was found to be 30. The company, however, claim that only 5% of their product is defective. Is the claim tenable?

[**Hint.** Here $n = 400$, $p = \dfrac{30}{400} = 0.075$. $P = 0.05$. $Q = 0.95$

Null hypothesis : $H_0 : P = 0.05$. $H_1 : P > 0.05$ (*Right Tailed Test*)

$$Z = \frac{p - E(p)}{\text{S.E.}(p)} = \frac{0.075 - 0.050}{\sqrt{(0.05 \times 0.95)/400}} = \mathbf{2.27.}$$

Value of Z_α at $\alpha = 0.05$ for one tailed test is 1.645.

Now $|Z| > Z_\alpha$ as $2.27 > 1.645$ so we reject the null hypothesis $H_0 \Rightarrow$ **Company's claim that only 5% of their product is defective is rejected.**]

4. A dice was thrown 400 times and 'six' resulted 80 times. Do the data justify the hypothesis of an unbiased dice?

[**Hint.** $H_0 : P = \frac{1}{6}$. $\quad Z = \frac{p - P}{\text{S.E. of } p} = \frac{(1/5) - (1/6)}{\sqrt{(1/6) \times (5/6) \times (1/400)}} = 1.79.$

Also Z_α at $\alpha = 0.05$ is 1.96.

Since $|Z| < Z_\alpha$ as $1.79 < 1.96$ so we accept the hypothesis that **the dice is an unbiased.**]

5. A sample of size 600 persons selected at random from a large city shows that the percentage of male in the sample is 53%. It is believed that male to the total population ratio in the city is 1/2. Test whether this belief is confirmed by the observation.

[**Hint.** Here $P = 50$, $Q = 100 - 50 = 50$, $p = 53$, $H_0 : P = 50\%$.

$$\text{S.E.}(p) = \sqrt{50(100-50)/600} = 2.04$$

$$\text{Test Statistic} : Z = \frac{p - P}{\text{S.E}(p)} = \frac{53 - 50}{2.04} = 1.5.$$

$$Z_\alpha \text{ at } \alpha = 0.05 \text{ in } 1.96.$$

Now $|Z| < Z_\alpha$ as $1.96 < 1.5 \Rightarrow$ Null hypothesis H_0 is accepted $\Rightarrow$ **male to the total population ratio in the city is $\frac{1}{2}$. *i.e.*, male population is 50%.**]

6. A machine puts out 16 imperfect articles in a sample of 500. After the machine is overhauled, it puts 3 imperfect articles in a batch of 100. Has the machine improved?

[**Hint.** Here $p_1 = \frac{16}{500} = 0.032 \Rightarrow q_1 = 0.968.$ $\quad p_2 = \frac{3}{100} = 0.030 \Rightarrow q_2 = 0.970.$

$H_0 : P_1 = P_2$. $\quad H_1 : P_1 > P_2$.

$$\text{S.E. of difference} = \textbf{S.E.}(\boldsymbol{p_1 - p_2}) = \sqrt{\frac{0.032 \times 0.968}{500} + \frac{0.030 \times 0.97}{100}} = \mathbf{0.0187.}$$

$$\textbf{Test Statistic : } Z = \frac{p_1 - p_2}{\text{S.E.}(p_1 - p_2)} = \frac{0.032 - 0.030}{0.0187} = \mathbf{0.107.}$$

Z_α at $\alpha = 0.05$ is 1.645. also $|Z| < Z_\alpha \Rightarrow H_0$ **is accepted.** $\Rightarrow$ **machine has not improved after over hauling.**]

7. In a random samples of 600 and 1000 men from two cities 400 and 600 men are found to be literate. Do the data indicate at 5% level of significance that the population are significantly different in the percentage of literacy?

[**Hint.** Here $p_1 = \frac{400}{600} = \frac{2}{3}$., $p_2 = \frac{600}{1000} = \frac{3}{5}$.∴ $\quad p = \frac{n_1 p_1 + n_2 p_2}{n_1 + n_2} = \frac{5}{8}.$

Null hypothesis : $H_0 : P_1 = P_2$. **Alternative hypothesis :** $H_1 : P_1 \neq P_2$.

Standard Error of Difference : $\textbf{S.E.}(\boldsymbol{p_1 - p_2}) = \sqrt{\frac{\boldsymbol{pq}}{\boldsymbol{n_1}} + \frac{\boldsymbol{pq}}{\boldsymbol{n_2}}} = \frac{\mathbf{1}}{\mathbf{40}}.$

Test statistic : $Z = \frac{p_1 - p_2}{S.E.(p_1 - p_2)} = \frac{(2/3) - (3/5)}{(1/40)} = \mathbf{2.67.}$

Critical value of Z at $\alpha = 0.05$ is $Z_{0.05} = 1.96$.

Also $|Z| > Z_{0.05}$ as $2.67 > 1.96 \Rightarrow H_0$ is rejected $\Rightarrow$ **population is significantly different in the percentage of literacy.**]

8. A company has the head office at Kolkata and a branch at Mumbai. The personnel director wanted to know if the workers at the two places would like the introduction of a new plan of work and a survey was conducted for the purpose. Out of a sample of 500 workers at Kolkata 62% favoured the new plan. At Mumbai out of a sample of 400 workers 41% were against the new plan. Is there any significant difference between the two groups in their attitude towards the new plan at 5% level?

[**Hint.** $H_0 : P_1 = P_2$, $H_1 : P_1 \neq P_2$.

$$p = \frac{500 \times 0.62 + 400 \times 0.59}{500 + 400} = \mathbf{0.607.}$$

$$Z = \frac{0.62 - 0.59}{\sqrt{0.067 \times 0.393 (1/500) + (1/400)}} = \mathbf{0.9155.}$$

$Z_{0.05} = 1.96$. Also $|Z| < Z_{0.05} \Rightarrow H_0$ is accepted $\Rightarrow$ **there is no significent difference in the two groups in their attitude towards the new plan.]**

9. Before an increase in excise duty on tea 400 people out of a sample of 500 persons were found to be tea drinkers. After an increase in the duty, 400 persons were known to the tea drinkers in a sample of 600 people. Do you think that there has been a significant decrease in the consumption of tea after the increase in the excise duty ?

[**Hint.** Here $p_1 = \frac{400}{500} = 0.80;$ $p_2 = \frac{400}{600} = 0.67.$

Null hypothesis : $H_0 : P_1 = P_2$.

Alternative hypothesis. $H_1 : P_2 < P_1 \Rightarrow$ **(One tailed test)**

Also $p = \frac{n_1 p_1 + n_2 p_2}{n_1 + n_2} = \frac{8}{11}, \quad q = 1 - \frac{8}{11} = \frac{3}{11}.$

Test Statistic : $Z = \frac{0.80 - 0.67}{\sqrt{(8/11) \times (3/11)[(1/500) + 1/600]}} = \mathbf{4.81.}$

Also $Z_{0.05} = 1.645$ for one tailed test.

Now $|Z| = 4.81 > Z_{0.05} = 1.645 \Rightarrow H_0$ is rejected $\Rightarrow$ there is a significant decrease in the consumption of tea after the increase in excise duty.]

10. Give the test for testing the proportion $H_0 : (\pi = \pi_0)$

ANSWERS

1. $Z = 1.69$; $Z_{0 \cdot 05} = 1.64$; Reject $H_0 : p = 0.05$.
2. $Z = 1.10$; $Z_{10} = 1.282$; No.
3. Company's claim that only 5% of their product is defective is rejected.
4. The die is an unbiased one.
5. Male population is 50%.
6. Machine has not improved after overhauling.
7. Population are significantly different in the percentage of literacy.
8. There is no significant difference in the two groups in their attitude towards the new plan.
9. There is a significant decrease in the consumption of tea after the increase in excise duty.

9.6 SMALL SAMPLE OR EXACT SAMPLE TEST

Small sample. A sample is said to be a small sample if, the size of the sample is less than 30, ***i.e.***, $n < 30$.

In this section we **shall discuss the tests of significance of a single mean for Small Samples ($n < 30$).** We know that the **mean of small samples conforms to the *t*-values contained in the *t*-distribution discussed in the last chapter.**

We shall use t-statistic. The t-values for different degrees of freedom and for different values of α, the level of significance are given in the table "Values of t for given probability levels." at the end of the book. The nature of the t-statistic used for the different tests of significance is as follows:

COMPUTATION OF TEST STATISTIC : *t* – values

To test the significance of a mean-Small Samples

Standard Error of Mean : $\text{S.E.}(\bar{x}) = \dfrac{s}{\sqrt{n-1}}$

where s **= sample standard deviation and** n **= size of sample.**

Test statistic : $t = \dfrac{\bar{x} - \mu}{\text{S.E.}(\bar{x})}$,

where $\bar{x}$ = *Sample mean,* μ = *population mean; S.E.* $(\bar{x})$ = *Standard Error of mean.*

The steps that are taken in testing the significance of single mean in the case of large samples should also be taken in testing the significance of single mean in the case of small samples ($n < 30$), except in respect of the application of test statistict.

9.7 TEST OF SIGNIFICANCE OF A SINGLE MEAN – SMALL SAMPLES

Example 18. *A soap manufacturing company was distributing a particular brand of soap through a large number of retail shops. Before a heavy advertisement compaign, the mean sales per shop was 140 dozens. After the campaign a sample of 26 shops was taken and the mean sales figure was found to be 147 dozens with standard deviation 16. Can you consider the advertisement effective at 5% level of significance.*

Solution. Here $n = 26 \Rightarrow$ **It is a case of small samples.**

Also $\bar{x} = 147$, $s = 16$. v = degree of freedom = 26 – 1 = 25.

1. **Null hypothesis:** There is no significant difference in between the sample means and population means ⇒ the advertisement is not effective, *i.e.*, $H_0 : \mu = 140$.

Alternative hypothesis: $H_1 : \mu > 140 \Rightarrow$ **Single Tailed Test ⇒** the advertisement is effective.

2. **Computation of Test Statistic :**

Standard Error of Mean: S.E. $(\bar{x}) = \dfrac{s}{\sqrt{n-1}} = \dfrac{16}{\sqrt{(26-1)}} = \dfrac{16}{5} = \mathbf{3.2.}$

Test Statistic : $t = \dfrac{\bar{x} - \mu}{S.E.(\bar{x})} = \dfrac{147 - 140}{3.2} = \dfrac{7}{3.2} = \mathbf{2.19.}$

3. **Level of significance.** Let us take $\alpha = \mathbf{0.05}$.
4. **Critical value.** The value of t for $\alpha = 0.05$ and for $v = 25$ is $t_{0.05,\,25} = \mathbf{1.708}$

[For single tailed test]

5. **Decision. The calculated value of** t **(= 2.19) >** $t_{0.05,\,25}$ (= 1.708) ⇒ **the null hypothesis** H_0 **is rejected ⇒ the advertisement is effective.**

Example 19. *A machine is designed to produce insulating washers for electrical devices of average thickness of 0.025 cm. A random sample of 10 washers was found to have an average thickness of 0.024 cm with a standard deviation of 0.002 cm. Test the significance of the deviation. Value of t for 9 degrees of freedom at 5% level is 2.262.*

Solution. Here we are given $n = 10 \Rightarrow$ **Small sample,** $\bar{x} = 0.024$ cm, $s = 0.002$ cm.

1. **Null Hypothesis:** $H_0 : \mu = \mathbf{0.025}$ **cm**, *i.e.,* there is no significant deviation between the sample mean and population mean.

Alternative Hypothesis: $H_1 : \mu \neq \mathbf{0.025}$ **cm**. (Two Tailed test)

2. **Computation of Test Statistic:**

Standard Error of Mean: $\text{S.E.}(\bar{x}) = \dfrac{s}{\sqrt{n-1}} = \dfrac{0.002}{\sqrt{9}} = \dfrac{\mathbf{0.002}}{\mathbf{3}}.$

$$\textbf{Test Statistic : } t = \frac{\bar{x}-\mu}{\text{S.E.}(\bar{x})} = \frac{0.024-0.025}{0.002/3} = \frac{-0.001\times 3}{0.002} = -\mathbf{1.5}.$$

$$\therefore \quad |t| = \mathbf{1.5}.$$

3. **Level of Significance:** Set $\alpha = 0.05$.
4. **Critical Value: Tabulated value of $t_{0.025}$ for 9 d.f = 2.262**
5. **Decision: Since calculated value $|t|$ (= 1.5) < $t_{0.05,\,9}$ (= 2.262) ⇒ Null hypothesis H_0 is accepted ⇒ there is no significant deviation between the sample mean and population mean.**

Example 20. *A random sample of size 16 has 53 as mean. The sum of squares of deviations from mean is 150. Can this sample be regarded as taken from the population having 56 as mean? Also obtain 95% confidence limits for the mean. (t_{15} = 2.131, t_{16} = 2.602)*

Solution. Null hypothesis: Let the sample be taken from population with mean as 56, *i.e.*,

$$H_0 : \mu = 56$$

Alternative Hypothesis: $\quad H_1 : \mu \neq 56$

$N = 16, \quad \bar{x} = 53\ ,\ \Sigma(x_1 - \bar{x})^2 = 150$

$$s^2 = \frac{1}{n-1}\sum (x_1 - \bar{x})^2 = \frac{150}{15} = 10 \quad \Rightarrow \quad s = \sqrt{10} = 3.162$$

$$\textbf{Test statistic : } \quad t = \frac{\bar{x}-\mu}{s/\sqrt{n}} = \frac{53-56}{3.162/\sqrt{16}} = \frac{-12}{3.162} = -3.795$$

Degree of freedom = n – 1 = 15

Tabulated value of t_{15} = 2.131. The value is significant at 5% level of significance.

Decision : Since the calculated value $|t|$ (= 3.795) > $t_{0.05,15}$ so the null hypothesis is rejected. Hence, **the sample cannot be regarded as taken from the population with mean 56.**

$$\textbf{Confidence interval is } = \bar{x} \pm t_{15}\frac{s}{\sqrt{n}} = 53 \pm \left(2.131 \times \frac{3.162}{16}\right) = \mathbf{51.316 < \mu < 54.684}$$

Example 21. *Ten objects are chosen at random from a large population and their weights are found to be in gms 63, 63,64, 65, 66, 69, 69, 70, 70, 71. In the light of the above data. Discuss the suggestion that the mean weight in the universe is 65 gms.*

Solution. Here n = 10 ⇒ a case of small samples, μ = 65, Let us calculate the sample mean and sample standard deviation from the following table.

Table for Sample Mean and S.D. Assuming A = 66

x	$d = x - 66$	d^2
63	– 3	9
63	– 3	9
64	– 2	4
65	– 1	1
66	0	0
69	3	9
69	3	9
70	4	16
70	4	16
71	5	25
	Σd = 10	Σd^2 = 98

$$\therefore \quad \bar{x} = A + \frac{\Sigma d}{n} = 66 + \frac{10}{10} = \mathbf{67.}$$

$$\therefore \quad \hat{\sigma}^2 = S^2 = \frac{1}{n-1}\left[\Sigma d^2 - \frac{(\Sigma d)^2}{n}\right] = \frac{1}{9}\left[98 - \frac{100}{10}\right] = \frac{1}{9}[98 - 10] = \frac{\mathbf{88}}{\mathbf{9}} = 9.78$$

$$\Rightarrow \mathbf{S = 3.127.}$$

1. **Null hypothesis:** $H_0 : \mu = 65$, *i.e.*, **there is no difference between the mean weight of Sample and Universe.**

   Alternative hypothesis: $H_1 : \mu \neq 65$ **[Two Tailed test]**

   Computation of Test Statistic:

   Test Statistic: $t = \frac{\bar{x} - \mu}{S / \sqrt{n}} = \frac{(67-65) \times \sqrt{10}}{3.127} = \frac{2 \times 3.162}{3.127} = \frac{6.33}{3.127} = \mathbf{2.024.}$

   Level of Significance: Take $\alpha = \mathbf{0.05.}$

   Critical Value: The tabulated value of t at $\alpha = 0.05$ for 9 d.f. is $t_{0.05,\,9} = \mathbf{2.262.}$

   Decision: Since $|t| < t_{0.05,\,9}$ **as 2.024 < 2.262** $\Rightarrow$ **so the Null hypothesis** H_0 **is accepted**

   $\Rightarrow$ **the mean weight of the universe is 65 gms.**

Example 22. *A salesman is expected to effect an average sales of Rs. 3500. A sample test revealed that a particular salesman had made the following sales: Rs. 3700; 3400; 2500, 5200, 3000, and 2000. Using 0.05 level of significance, conclude whether his work is below standard or not.*

Solution. Mean of the sample is

$$\bar{x} = \frac{3700 + 3400 + 2500 + 5200 + 3000 + 2000}{6} = \frac{19800}{6} = \mathbf{3{,}300.}$$

Table for Sample S.D.

Sale ('oo Rs.)	$d = x - \bar{x} = x - 33$	$d^2 = (x - \bar{x})^2$
37	4	16
34	1	1
25	– 8	64
52	19	361
30	– 3	9
20	– 13	169
	$\Sigma d = 0$	$\Sigma d^2 = 620$

The unbiased estimate of population standard deviation:

$$\hat{\sigma} = S = \sqrt{\frac{\Sigma (x - \bar{x})^2}{n-1}} = \sqrt{\frac{620}{5}} = \mathbf{11.13.}$$

1. **Null Hypothesis: $H_0 : \mu \geq 35$.**

   Alternative Hypothesis: $H_1 : \mu < 35$. (**One tailed test**).

2. **Computation of test statistic:**

   Test Statistic: $t = \dfrac{\bar{x} - \mu}{S/\sqrt{n}} = \dfrac{33 - 35}{(11.13)/\sqrt{6}} = \dfrac{-2 \times 2.45}{11.13} = -0.44 \Rightarrow |t| = \mathbf{0.44.}$

3. **Level of significance.** Take $\alpha = 0.05$.

4. **Critical value.** The critical or tabulated value of t at $\alpha = 0.05$ for 5 d.f. is $t_{0.05,5} = 2.015$ **for Left Tailed test.**

5. **Decision. Now calculated value of $|t| = 0.44 <$ Tabulated $t_{0.05,5} = 2.015$**

   $\Rightarrow H_0$ is accepted $\Rightarrow$ the salesman is up to standard.

Example 23. *The heights of ten children selected at random from a given locality had a mean 63.2 cms and variance 6.25 cms. Test at 5% level of significance the hypothesis that the children of the given locality are on the average less than 65 cms in all. Given for 9 degrees of freedom P(t > 1.83) = 0.05.*

Solution. Here $n = 10 \Rightarrow$ **Small Sample,** $\bar{x} = 63.2$ cms, $s^2 = 6.25$ cms $\Rightarrow s = 2.5$ cms; $\mu = 65$ cms.

1. **Null Hypothesis :** the average height of the children is 65 cms, *i.e.,* $H_0 : \mu = 65$

   Alternative Hypothesis: $H_1 : \mu < 65$ (One Tail test).

2. **Computation of Test Statistic:**

   Standard error of Mean: $\text{S.E.}(\bar{x}) = \dfrac{s}{\sqrt{n-1}} = \dfrac{2.5}{\sqrt{9}} = \dfrac{2.5}{3} = \mathbf{0.834.}$

   $\therefore$ **Test Statistic:** $t = \dfrac{\bar{x} - \mu}{\text{S.E.}(\bar{x})} = \dfrac{63.2 - 65}{0.834} = \dfrac{-1.8}{0.834} = \mathbf{-2.158}$

   $\therefore$ $|t| = 2.158$

3. **Level of significance.** $\alpha = 0.05$.

4. **Critical Value.** The tabulated or critical value of t for 9 d.f. at $\alpha = 0.05$ for one tail test is $t_{0.05,9} = \mathbf{1.83.}$

5. **Decision. Since $|t| = 2.158 > t_{0.05,9} = 1.83 \Rightarrow$ the null hypothesis H_0 is rejected $\Rightarrow$ Alternative hypothesis H_1 is accepted $\Rightarrow$ the mean height of the children is less than 65 cms.**

Example 24. *A random sample of size 16 has 53 as mean. Ths sum of the squares of the deviations taken from the mean is 150 Can this sample be regarded as taken from the population having 56 as mean? Obtain 95% and 99% confidence limits of the mean population. (For v = 15, $t_{0.01} = 2.95$, $t_{0.05} = 2.131$).*

Solution. We are given that $n = 16 \Rightarrow$ **Small Sample,**

Sample mean $\bar{x} = 53$; $\Sigma(x - \bar{x})^2 = 150$, $\mu = 56$.

$$\therefore \quad s^2 = \frac{\Sigma(x - \bar{x})^2}{n - 1} = \frac{150}{16 - 1} = \frac{150}{15} = 10 \Rightarrow s = \sqrt{10} = \mathbf{3.1623.}$$

1. **Null Hypothesis : $H_0 : \mu = 56$.**

   Alternative Hypothesis : $H_1 : \mu \neq 56$ (Two tailed test)

2. **Test Statistic :** $t = \dfrac{\bar{x} - m}{s/\sqrt{n}} = \dfrac{53 - 56}{3.162} \times \sqrt{16} = -\dfrac{12}{3.162} = \mathbf{-3.79}$

   $\therefore$ $|t| = \mathbf{3.79}$

3. Level of significance. (*i*) $\alpha = 0.05$, (*ii*) $\alpha = 0.01$.

4. Critical value. (*i*) The tabulated value of t for $\alpha = 0.05$ and 15 d.f. for two tailed test $= t_{0.05, 15} = 2.131$. (*ii*) Also $t_{0.01,15} = 2.95$.

5. Decision. (*i*) **Since $|t| = 3.79 > t_{0.05,15} = 2.131 \Rightarrow$ the Null hypothesis H_0 is rejected and we canclude that the population mean is not 56 at 5% level of significance.**

(*ii*) **Also $|t| = 3.79 > t_{0.01,15} = 2.95 \Rightarrow H_0$ is rejected at 1% level of significance $\Rightarrow$ Population mean is not 56 at 1% level of significance.**

95% confidence limits for μ(d.f = 15) are :

$$\bar{x} \pm t_{0.05} \times \frac{s}{\sqrt{n}} = 53 \pm 2.131 \times \frac{3.162}{4} = 53 \pm 1.684$$

$$\Rightarrow \qquad 50.316 < \mu < 54.684.$$

99% confidence limits for μ :

$$\bar{x} \pm t_{0.01} \times \frac{s}{\sqrt{n}} = 53 \pm \frac{3.16}{4} \times 2.95 = 53 \pm 2.33$$

$$\text{or} \qquad 50.67 < \mu < 55.33.$$

Example 25. *A random sample of size 7 from a normal population gave a mean of 977.51 and a standard deviation of 4.42. Find 95% confidence interval for the population mean.*

Solution. It is given to us that $n = 7$, $\bar{x} = 977.51$ and $s = 4.42$.

95% confidence limits for μ for 6 d.f. :

$$\bar{x} \pm t_{0.05,6} \times \text{S.E. of mean} = \bar{x} \pm t_{0.05,6} \frac{s}{\sqrt{n-1}}$$

$$= 977.51 \pm 2.447 \times \frac{4.42}{\sqrt{6}} = 977.51 \pm \frac{2.447 \times 4.42}{2.449} \qquad [\because \ t_{0.05,6} = 2.447]$$

$= 977.51 \pm 4.416$. *i.e.*, 973.094 and 981.926

$$\Rightarrow \quad 973.094 < \mu < 981.926.$$

9.8 PAIRED t–TEST FOR DIFFERENCE OF MEANS

(When the sample observations are not comletely independent.)

In this case we are given that

(*i*) *the sizes of the samples are equal, i.e.,* $n_1 = n_2 = n$ *(say).*

(*ii*) *the samples are not independent, and the sample observations* $(x_1, x_2, x_3, \ldots. x_n)$, $(y_1, y_2, y_3 \ldots. y_n)$ *are dependent in pairs. In other words the pairs of observation* (x_1, y_1), (x_2, y_2), (x_3, y_3), ..., (x_n, y_n) *correspond to 1st, 2nd, 3rd, ..., nth unit respectively. Although we may find two sets of sample values corresponding to two observations* $(x_1, x_2, \ldots, x_n)$ *and* $(y_1, y_2, y_3, \ldots, y_n)$ *for the sample elementary unit under different situations, the tests discussed earlier are not applicable, and we proceed as follows.*

9.9 WORKING RULE FOR ESTIMATED STANDARD ERROR OF DIFFERENCE

Step 1. Let $D = (x_i - y_i)$, $(i = 1, 2, \ldots. n)$ denote the difference in observations for the unit.

Step 2. Calculate $\overline{D} = \dfrac{\Sigma D}{n}$.

Step 3. Calculate $\Sigma(D - \overline{D})^2$ and $S = \sqrt{\dfrac{\Sigma (D - \overline{D})^2}{n-1}}$

or $$S = \sqrt{\frac{\Sigma D^2 - n(\overline{D})^2}{(n-1)}} = \sqrt{\frac{n\Sigma D^2 - (\Sigma D)^2}{n(n-1)}}$$

Step 4. Calculate Standard Error of Difference : S.E. $(\bar{x}_1 - \bar{x}_2) = \frac{S}{\sqrt{n}}$

9.10 WORKING RULE FOR PAIRED *t*-TEST

1. **Null Hypothesis:** $H_0 : \mu_1 = \mu_2$.

   Alternative Hypothesis: $H_1 : \mu_1 \neq \mu_2$ **(Two Tailed Test).**

2. **Calculation of Test Statistic:**

   Estimated Standard Error of Difference : S.E. $(\bar{x}_1 - \bar{x}_2) = \frac{S}{\sqrt{n}}$,

   where $S^2 = \frac{1}{n-1}\left[\Sigma D^2 - n(\overline{D})^2\right] = \frac{1}{n-1}\left[\Sigma D^2 - \frac{(\Sigma D^2)}{n}\right]$

   Test Statistic : $t = \frac{\overline{D}}{\text{S.E.}(\overline{D})}$

   where $\overline{D}$ **= difference between each pair of observations.**

3. **Level of significance :** Let $\alpha = 0.05$, or $\alpha = 0.01$ or $\alpha = 0.1$ as the case may be.

4. **Critical Value.** Find from the table the value of t at the level of significance α for $(n-1)$ degrees of freedom for the one tailed or two tailed test (for the case given by alternative hypothesis) *i.e.*, find $t_{d, n-1}$

5. **Decision.** *If the calculated value* $|t| < t_{\alpha, n-1}$, *then the null hypothesis is accepted and in case of calculated value of* $|t| > t_{\alpha, n-1}$, *then the null hypothesis is rejected and alternative hypothesis is accepted.*

Example 26. *An I.Q. Test was administered to 5 persons before and after they are trained. The results are given below.*

Candidates :	*I*	*II*	*III*	*IV*	*V*
I.Q. Before training :	*110*	*120*	*123*	*132*	*125*
I.Q. After training :	*120*	*118*	*125*	*136*	*121.*

Test whether there is any change in I.Q. after the training programme. It is given that $t_{0.01} = 4.6$ *for 4 d.f.*

Solution.

1. **Null Hypothesis:** $H_0 : \mu_1 = \mu_2$ or $\overline{D} = 0$ *i.e.*, there is no change in *I.Q.* after the training.

   Alternative Hypothesis : $H_1 : \mu_1 \neq \mu_2$.

2. **Computation of Test Statistic.** Let us compute mean and standard deviation from the following table.

Candidates	*I.Q. before training y*	*I.Q. after training x*	*Difference D = x – y*	D^2
I	110	120	10	100
II	120	118	– 2	4
III	123	125	2	4
IV	132	136	4	16
V	125	121	– 4	16
			$\Sigma D = 10$	$\Sigma D^2 = 140$

$$\textbf{Mean Difference: } \bar{D} = \frac{\Sigma D}{n} = \frac{10}{5} = \mathbf{2.}$$

$$\textbf{Estimated Variance: } S^2 = \frac{\Sigma D^2}{n-1} - \frac{(\Sigma D)^2}{n(n-1)}$$

$$= \frac{140}{5-1} - \frac{(10)^2}{5(5-1)} = \frac{140}{4} - \frac{100}{20} = 35 - 5 = \mathbf{30.}$$

$$\therefore \quad S = \sqrt{30}.$$

$$\textbf{Standard Error of Difference: } \text{S.E.}(\bar{D}) = \frac{S}{\sqrt{n}} = \frac{\sqrt{30}}{\sqrt{5}} = \sqrt{6}.$$

$$\textbf{Test Statistic: } t = \frac{\bar{D}}{S.E.(\bar{D})} = \frac{2}{\sqrt{6}} = \frac{2}{2.45} = \mathbf{0.816}$$

4. Critical Value at α = 0.01. The tabulated or critical value of t at $\alpha = 0.01$ for $5 - 1 = 4$ degrees of freedom is $t_{0.01,4}$ **= 4.6.**

5. Decision. The **calculated value of $|t|$ = 0.816 < $t_{0.01,4}$ = 4.6 ⇒ the Null hypothesis is accepted ⇒ There is no significant difference in I.Q. after the training programme.**

Example 27. *A drug was administered to 10 patients and the increments in their blood pressure were recorded to be 6, 3, – 2, 4, – 3, 4, 6, 0, 0, 2. Is it reasonable to believe that the drug has no effect on change of blood pressure? Use 5% significance level and assume that for 9 degrees of freedom, $t_{0.5,9}$ = 2.26.*

Solution. Let D denote the increment in blood pressure. We shall compute the mean difference and standard deviation from the following Table.

Table

Patient :	1	2	3	4	5	6	7	8	9	10	**Total**
D :	6	3	– 2	4	– 3	4	6	0	0	2	**ΣD = 20**
D^2 :	36	9	4	16	9	16	36	0	0	4	**ΣD² = 130**

$$\textbf{Mean Difference: } \bar{D} = \frac{\Sigma D}{n} = \frac{20}{10} = \mathbf{2.}$$

$$\textbf{Estimated variance of population: } S^2 = \frac{\Sigma D^2}{n-1} - \frac{(\Sigma D)^2}{n(n-1)}$$

$$= \frac{130}{9} - \frac{(20)^2}{10 \times 9} = \frac{130}{9} - \frac{40}{9} = \frac{90}{9} = \mathbf{10}$$

Now $\quad S^2 = 10 \Rightarrow S = \sqrt{10}.$

1. Null Hypothesis: $H_0 : \mu_1 = \mu_2$, *i.e.*, there is no chance in blood pressure due to the effect of drug.

Alternative Hypothesis: $H_1 : \mu_1 \neq \mu_2$. **(Two Tailed test)**

2. Computation of Test Statistic:

$$\textbf{Standard Error of Difference: } \text{S.E.}(\bar{D}) = \frac{S}{\sqrt{n}} = \frac{\sqrt{10}}{\sqrt{10}} = \mathbf{1.}$$

$$\textbf{Test Statistic: } t = \frac{\overline{D}}{S.E(\overline{D})} = \frac{2}{1} = 2.$$

$$\therefore \quad |t| = 2.$$

3. **Level of significance $\alpha = 0.05$.**
4. **Critical value.** The tabulated or critical value of t at $\alpha = 0.05$ for $10 - 1 = 9$ degrees of freedom for two tailed test is $t_{0.05,9} = 2.26$.
5. **Decision. Since the calculated value $|t| = 2 <$ tabulated value $t_{0.05,9} = 2.26$, so the Null hypothesis H_0 is accepted $\Rightarrow$ there is no change in blood pressure due to the effect of drug.**

Example 28. *Ten students were given intensive coaching for a month in statistic. The scores obtained in tests 1 and 5 are given below :*

S. No of Students	*1*	*2*	*3*	*4*	*5*	*6*	*7*	*8*	*9*	*10*
Marks in 1st test :	*50*	*52*	*53*	*60*	*65*	*67*	*48*	*69*	*72*	*80*
Marks in 5th test	*65*	*55*	*65*	*65*	*60*	*67*	*49*	*82*	*74*	*86*

Does the score from test 1 to test 5 show an improvement? Test at 5% level of significance. (The values of t for 9 degrees of freedom at 5% level for one tailed test is 1.833 and for two tailed test is 2.262).

Solution.

1. **Null Hypothesis $H_0 : \mu_1 = \mu_5$**

   Alternate Hypothesis : $H_1 : \mu_1 < \mu_5$ [One Tailed Test].
2. **Calculation of Test Statistic**

S.No.	*Marks in 1st Test = X_i*	*Marks in Vth Test = X_5*	*Difference $D = X_i - X_5$*	$D^2 = (X_i - X_5^2)$
1	50	65	– 15	225
2	52	55	– 3	9
3	53	65	–12	144
4	60	65	– 5	25
5	65	60	5	25
6	67	67	0	0
7	48	49	– 1	1
8	69	82	– 13	169
9	72	74	– 2	4
10	80	86	– 6	36
			$\Sigma D = - 52$	**$\Sigma D^2 = 638$**

$$\overline{D} = \frac{\Sigma D}{N} = \frac{-52}{10} = -5.2.$$

$$\therefore \quad S = \sqrt{\frac{\Sigma D^2 - n(\overline{D})^2}{n-1}} = \sqrt{\frac{638 - 10 \times 27.04}{9}} = 6.391. \qquad [\because n = 10]$$

$$\text{S.E.}(\overline{D}) = \frac{S}{\sqrt{n}} = \frac{6.391}{\sqrt{10}} = 2.021.$$

$$\textbf{Test Statistic: } t = \frac{\bar{D}}{S.E.(\bar{D})} = \frac{-5.2}{2.021} = -2.573.$$

$\therefore$ $|t| = 2.573.$

3. Level of Significance. $\alpha = 0.05$.

4. Critical Value. The critical value of t at $\alpha = \mathbf{0.05}$ for $10 - 1 = \mathbf{9}$ degrees of freedom for one tailed test is $t_{0.05,9} = \mathbf{1.833}$.

5. Decision. Since the calculated value of $|t| = 2.573 >$ tabulated $t_{0.05,9} = 1.833$ so, **the Null hypothesis is rejected $\Rightarrow$ Alternative hypothesis is accepted $\Rightarrow$ that the coaching has improved the standard.**

EXERCISE – 9.3

1. Ten cartons are taken at random from an automatic filling machine. The mean net weight of the 10 cartons is 11.8 kg and standard deviation is 0.15 kg. Does the sample mean differ significantly from the intented weight of 12 kg? You are given for $v = 9$, $t_{0.05} = 2.26$.

   [**Hint. Null hypothesis:** $H_0 : \mu = 12$.

   Alternative hypothesis: $H_1 : \mu \neq 12$

   $$\text{Also } t = \frac{\bar{x} - m}{s/\sqrt{n-1}} = \frac{11.8 - 12}{0.15/\sqrt{9}} = -4.0 \Rightarrow |t| = 4.$$

   Also $t_{0.05,9} = 2.26 < 4 =$ calculated $|t| \Rightarrow H_0$ is rejected $\Rightarrow$ **the sample mean differ significantly].**

2. The mean weekly sales of the chocolate bar in candy stores was 140.3 bars per store. After an advertisement campaign, the mean weekly sales in 22 stores for a typical week increased to 153.7 and showed a standard deviation of 17.2. Was the advertising campaign not successful?

3. Certain pesticide is packed into bags by a machine. A random sample of 10 bags is drawn and their contents are found to weigh (inkgs) as follows:

   50, 49, 52, 44, 45, 48, 46, 45, 49, 45.

   Test if the average packing can be taken to be 50 kg.

   [**Hint. Null hypothesis: $H_0 : \mu = 50$. Alternative hypothesis: $H_1 : \mu \neq 50$ (Two tailed test)**

   Here = 47.3, $S^2 = 7.12$.

   $$t = \frac{\bar{x} - \mu}{S/\sqrt{n}} = \frac{-2.7}{\sqrt{0.712}} = -3.2 \Rightarrow |t| = 3.2.$$

   $t_{0.05,\,9} = 2.262 < |t| = 3.2 \Rightarrow$ **H_0 is rejected $\Rightarrow$ Average packing cannot be taken as 50 kgs.]**

4. A fertiliser mixing machine is set to give 12 kg of nitrate for every quintal bag of fertiliser. Ten 100 kg bags are examined. The percentages of nitrate are as follows.

   11, 14, 13, 12, 13, 12, 13, 14, 11, 12.

   Is there reason to believe that the machine is defective? [Value of t for 9 degrees of freedom is 2.262].

   [**Hint.** Here $\bar{x} = \frac{125}{10} = 12.5$ kg. Also $S = \sqrt{\frac{\Sigma(x-\bar{x})^2}{n-1}} = \sqrt{\frac{10.5}{9}} = 1.08.$

   Null hypothesis: $H_0 : \mu = 12$. **Alternative hypothesis:** $H_1 : \mu \neq 12$ (Two tailed test)

   $$t = \frac{\bar{x} - m}{S/\sqrt{n}} = \frac{12.5 - 12}{(1.08)/\sqrt{10}} = \mathbf{1.464}.$$

   Now $|t| = 1.464 < t_{0.05,\,9} = 2.262 \Rightarrow$ **H_0 is accepted $\Rightarrow$ these is no reason to believe that machine is defective.]**

5. Ten students are selected at random from a college and their heights are found to be 100, 104, 108, 110, 118, 120, 122, 124, 126 and 128 cms. In the light of these data, discuss the suggestion that the mean height of the students of the college is 110 cms. Use 5% level of significance.

[**Hint.** $\bar{x} = \frac{1160}{10} = 116 \text{ cm.}$ Also $S = \sqrt{\frac{\Sigma(x-\bar{x})^2}{n-1}} = \sqrt{\frac{864}{9}} = 9.798.$

$$\textbf{Test Statistic: } t = \frac{\bar{x}-\mu}{S/\sqrt{n}} = \frac{(116-110)}{9.798} \times \sqrt{10} = \mathbf{1.937.}$$

Also $t_{0.05,\,9} = 2.262 > |t| = 1.937 \Rightarrow$ **H_0 is accepted $\Rightarrow$ the mean height of the students can be taken as 110 cms.]**

6. A random sample of 10 tins of oil filled in by an automatic machine gave the following weights, in kg?

2.05, 2.01, 2.04, 1.91, 1.96, 2.01, 1.97, 1.99, 2.04, 2.02

Can we accept at 5% level of significance, the claim that the average weight of the tin in 2 kg?

7. In the past a machine has produced washers having a mean thickness of 0.05 cm To determine whether the machine is in proper order, a sample of 10 washers is taken of which mean thickness is 0.053 cm and S.D. = 0.003. Test the hypothesis that machine is working in proper order.

[**Hint.** $H_0 : \mu \neq 0.05$, $H_1 : \mu \neq 0.05$ (Two tailed test.)

$$t = \frac{\bar{x}-\mu}{s/\sqrt{n-1}} = \frac{0.053-0.05}{0.003/\sqrt{9}} = \frac{.003}{.001} = \mathbf{-3.}$$

Also $t_{0.05,\,9}$ for two tail test is $2.262 < |t| = 3 \Rightarrow H_0$ is rejected $\Rightarrow$ **machine is not working in propei order.]**

8. Talcum powder is packed into tins by a machine. A random sample of 11 tins is drawn and their contents are found to weigh in kgs as follows:

0.44, 0.51, 0.49, 0.52, 0.45, 0.48, 0.46, 0.45, 0.47, 0.45 and 0.47

Test if the average packing can be taken to be 0.5 kgs.

9. A certain stimulus administered to each of 12 patients resulted in the following changes in blood pressure.

5, 2, 8, – 1, 3, 0 – 2, 1, 5, 0, 4, 6.

Can it be concluded that the stimulus will in general be accompanied by an increase in blood pressure?

[**Hint.** Here $\Sigma D = 31$, $\therefore \bar{D} = \frac{\Sigma D}{12} = \frac{31}{12} = \mathbf{2.58}$

Also $S^2 = \frac{\Sigma D^2}{n-1} - \frac{(\Sigma D)^2}{n(n-1)} = \frac{185}{11} - \frac{31\times 31}{12\times 11} = \mathbf{9.53} \Rightarrow \mathbf{S = 3.08.}$

Null hypothesis: $H_0 : \mu_1 = \mu_2.$

$$\textbf{Standard Error: S.E.}(\bar{D}) = \frac{S}{\sqrt{n}} = \frac{3.08}{\sqrt{12}} = 0.89.$$

$$\textbf{Test Statistic: } t = \frac{\bar{D}}{\text{S.E.}(\bar{D})} = \frac{2.58}{0.89} = \mathbf{2.89.}$$

Now $t_{0.05,\,11} = 2.2 <$ Calculated $|t| = 2.89 \Rightarrow$ Null hypothesis is rejected

$\Rightarrow$ **the stimulus will in general be accompanied by an increase in blood pressure.**

10. The sales data of an item in six shops before and after a special promotional compaign are as under

Shops	A	B	C	D	E	F
Before compaign:	53	28	31	48	50	42
After compaign:	58	29	30	55	56	45

Can the campaign be judged to be a success? Test at 5% level of significance.

11. Point out the various applications and properties of *t* distribution.

ANSWERS

1. The sample mean differs significantly.
2. $t = 1.9716$; the advertisement was successful at 5% level of significance.
3. $t = 3.2$; the average packing cannot be taken as 50 kgs.
4. There is no reason to believe that the machine is defective.
5. The mean height of the students can be taken as 110 cm.
6. $t = 0.707$; The claim that the average weight of the tin is 2 kg is accepted.
7. The machine is not in proper order.
8. $t = -3.43$; No.
9. The stimulus will in general be accompanied by an increase in blood pressure.
10. $t = -2.78$, the sales campaign has been a success.

9.11 TEST OF SIGNIFICANCE OF DIFFERENCE BETWEEN TWO MEANS — LARGE SAMPLES

Let *A* and *B* be two populations with means μ_1 and μ_2 and variances σ_1^2 and σ_2^2 respectively. Let us take two independent samples of size n_1 and n_2 from these two populations. Let $\bar{x}_1$ and $\bar{x}_2$ be the corresponding sample means. Then our problem is

(*i*) To test the equality of two population means, *i.e.*, to test whether $\mu_1 = \mu_2$.

(*ii*) To test the significance of the difference between two independent sample means *viz.*, $\bar{x}_1 - \bar{x}_2$.

The following steps should be taken for testing the significance difference between two means.

1. Null Hypothesis. The null hypothesis is set in any one of following forms. $H_0 : \mu_1 = \mu_2$, *i.e.*, the two samples have been drawn from different populations having the same means and equal standard deviation or, $H_0 : \bar{x}_1 = \bar{x}_2$. *i.e.*, the two samples have been drawn from the same parent population.

Alternative Hypothesis. $H : \mu_1 \neq \mu_2$ **(Two tailed test)**

or $H_1 : \mu_1 < \mu_2$ **(One tailed test)**

or $H_1 : \mu_1 > \mu_2$ **(One tailed test)**

2. Computation of Test Statistic: We have the following two cases :

Case I. When the population standard deviations σ_1 and σ_2 are known. In this case we have.

Standard Error of Difference of Means: $$\text{S.E.}(\bar{x}_1 - \bar{x}_1) = \sqrt{\frac{\sigma_1^2}{n_1} + \frac{\sigma_2^2}{n_2}}$$

Test statistic : $$Z = \frac{\bar{x}_1 - \bar{x}_2}{\text{S.E.}(\bar{x}_1 - \bar{x}_2)}.$$

Case II. When the population standard deviations σ_1 and σ_2 are not known. In this case we are given sample standard deviations s_1 and s_2 and we calculate the S.E. of difference of means by the following formula.

Standard Error of Difference of means: $S.E.(\bar{x}_1 - \bar{x}_2) = \sqrt{\dfrac{s_1^2}{n_1} + \dfrac{s_2^2}{n_2}}$

$\therefore$ **Test Statistic:** $Z = \dfrac{\bar{x}_1 - \bar{x}_2}{S.E.(\bar{x}_1 - \bar{x}_2)}$.

The other steps such as (*i*) Level of significence; (*ii*) Critical value Z_α; (*iii*) Decision for testing the significence of the difference between two means are the same as those given in testing the significance of a single mean.

Example 29. *A college conducts both day and night classes intended to be identical. A sample of 100 day students yields examination results as under.*

$$\bar{x}_1 = 72.4 \text{ and } \sigma_1 = 14.8.$$

A sample of 200 night students yields examination results as under.

$$\bar{x}_2 = 73.9 \text{ and } \sigma_2 = 17.9.$$

Are the two means statistically equal at 10% level ?

Solution. Here

$n_1 = 100$ $\quad n_2 = 200$

$\bar{x}_1 = 72.4$ $\quad \bar{x}_2 = 73.9$

$\sigma_1 = 14.8$ $\quad \sigma_2 = 17.9$

1. **Null Hypothesis:** $H_0 : \mu_1 = \mu_2$.

   Alternative Hypothesis: $H_1 : \mu_1 \neq \mu_2 \Rightarrow$ It is a case of **Two tailed test.**

2. **Calculation of Test Statistic:**

   S.E. of means: $S.E.(\bar{x}_1 - \bar{x}_2) = \sqrt{\dfrac{\sigma_1^2}{n_1} + \dfrac{\sigma_2^2}{n_2}} = \sqrt{\dfrac{(14.8)^2}{100} + \dfrac{(17.9)^2}{200}} = \mathbf{1.95.}$

   $\therefore$ **Test Statistic :** $Z = \dfrac{\bar{x}_1 - \bar{x}_2}{S.E.(\bar{x}_1 - \bar{x}_2)} = \dfrac{72.4 - 73.9}{1.95} = \mathbf{-0.769.}$

   $\therefore$ $\quad |Z| = \mathbf{0.769.}$

3. **Level of significance:** Here $\alpha = 0.10$.
4. **Critical Value at $\alpha = 0.10$ is $Z_\alpha = 1.645$.**
5. **Decision. Since the calculated value of *Z* (= 0.769) < critical value of Z_α (= 1.645) so H_0 is accepted.** Hence we conclude that the **two means are statistically equal.**

Example 30. *A random sample of 1000 workers from South India show that their mean wages are Rs. 47 per week with a standard deviation of Rs. 28. A random sample of 1500 workers from North India gives a mean wage of Rs. 49 per week with a standard deviation of Rs. 40. Is there any significant difference beween their mean level of wages ?*

Solution. We are given the following information

	South India	**North India**
Size of Sample:	$n_1 = 1000$	$n_2 = 1500$
Mean of Sample:	$\bar{x}_1 = 47$	$\bar{x}_2 = 49$
S.D. of Sample:	$s_1 = 28$	$s_2 = 40$
Population Mean:	μ_1 = Population mean	μ_2 = Population mean.

1. **Null Hypothesis:** There is no significant difference between two mean level of wages, *i.e.,* $H_0 : \mu_1 = \mu_2$.

   Alternative Hypothesis : $H_0 : \mu_1 \neq \mu_2 \Rightarrow$ **It represents a two tailed test.**

2. Calculation of Test Statistic:

$$\textbf{Standard Error: S.E. } (\bar{x}_1 - \bar{x}_2) = \sqrt{\frac{s_1^2}{n_1} + \frac{s_2^2}{n_2}} = \sqrt{\frac{(28)^2}{1000} + \frac{(40)^2}{1500}} = 1.36$$

$$\therefore \quad \textbf{Test Statistic : } Z = \frac{\bar{x}_1 - \bar{x}_2}{S.E.(\bar{x}_1 - \bar{x}_2)} = \frac{47 - 49}{1.36} = \frac{-2}{1.36} = -\mathbf{1.47.}$$

$$\therefore \quad |Z| = \mathbf{1.47}$$

3. Level of Significance. Since the level of significance is not given so we take $\alpha = 0.05$.

4. Critical Value. Value of Z_α at $\alpha = 0.05$ (from the table) is **1.96.**

5. Decision. The calculated value of $|Z| = 1.47 <$ Critical value of $Z_\alpha = 1.96 \Rightarrow$ the null hypothesis is accepted $\Rightarrow$ there is no significant difference between the two mean level of wages.

Example 31. *The research unit in an organisation wishes to determine whether scores on the scholastic aptitude test are different for male and female applicants. Random samples of applicant's file are taken and summarised below :*

Applicants

	Female	Male
$\bar{x}$	*502.1*	*510.5*
s	*86.2*	*90.4*
n	*399*	*204*

Using the above sample data, test the null hypothesis that the average score is same for the population male and female applicants. Use 5% significance level and assume that the scores are normalluy distributed in each case.

Solution. Let x_1 and x_2 be the random variables of female and male applicants. Then we have

$\bar{x}_1 = 502.1$ $\quad\quad$ $\bar{x}_2 = 510.5$

$s_1 = 86.2$ $\quad\quad$ $s_2 = 90.4$

$n_1 = 399$ $\quad\quad$ $n_2 = 204.$

$\mu_1 =$ population mean $\quad\quad$ $\mu_2 =$ population mean.

1. Null Hypothesis : The average score is same for the population of male and female, *i.e.*,

$$H_0 : \mu_1 = \mu_2$$

Alternative Hypothesis : $H_1 : \mu_1 \neq \mu_2 \Rightarrow$ It represents a **Two Tailed Test.**

2. Calculation of Test Statistic:

$$\textbf{S.E. of Difference of Means : S.E.}(\bar{x}_1 - \bar{x}_2) = \sqrt{\frac{s_1^2}{n_1} + \frac{s_2^2}{n_2}}$$

$$= \sqrt{\frac{(86.2)^2}{399} + \frac{(90.4)^2}{204}} = \sqrt{\frac{7430.44}{399} + \frac{8172.16}{204}}$$

$$= \sqrt{18.62 + 40.06} = \sqrt{58.68} = \mathbf{7.66.}$$

$$\textbf{Test Statistic : } Z = \frac{\bar{x}_1 - \bar{x}_2}{S.E.(\bar{x}_1 - \bar{x}_2)} = \frac{502.1 - 510.5}{7.66} = \frac{-8.4}{7.66} = -\mathbf{1.097.}$$

$$\therefore \quad |Z| = \mathbf{1.097.}$$

3. Level of significence. Here $\alpha = \mathbf{0.05.}$

4. **Critical Value.** The value of Z_α (at $\alpha = 0.05$ from the table) = **1.96.**
5. **Decision. Calculated value of $|Z| = 1.097$ < Critical value $Z_\alpha = 1.96 \Rightarrow$ Null hypothesis is accepted $\Rightarrow$ Average score is same for the population of male and female.**

Example 32. *The mean yield of wheat from a district A was 210 kgs. with standard deviation 10 kgs. per acre from a sample of 100 plots. In another district B, the mean yield was 220 kg. with standard deviation 12 kg. from a sample of 150 plots. Assuming that the standard deviation of the yield in the entire state was 11 kg, test whether there is any significant difference between the mean yield of crops in the two districts.*

Solution. We have the following data.

District A	*District B*
$n_1 = 100$	$n_2 = 150$
$\bar{x}_1 = 210$	$\bar{x}_2 = 220$
$s_1 = 10$	$s_2 = 12$
$\sigma_1 = 11$	$\sigma_2 = 11$
Population Mean μ_1	μ_2.

1. **Null hypothesis:** There is no significant difference between the mean yield of crops in the two district, *i.e.*, $H_0 : \mu_1 = \mu_2$

   Alternative Hypothesis: $H_1 : \mu_1 \neq \mu_2$. It represents a **Two Tailed Test.**
2. **Calculation of Test Statistic:**

Standard Error: $$\text{S.E.}(\bar{x}_1 - \bar{x}_2) = \sqrt{\frac{\sigma_1^2}{n_1} + \frac{\sigma_2^2}{n_2}} = \sqrt{\frac{(11)^2}{100} + \frac{(11)^2}{150}} = \sqrt{\frac{121}{100} + \frac{121}{150}}$$

$$= \sqrt{1.21 + 0.807} = \sqrt{2.017} = \mathbf{1.42.}$$

Test Statistic: $$Z = \frac{(\bar{x}_1 - \bar{x}_2)}{\text{S.E.}(\bar{x}_1 - \bar{x}_2)} = \frac{210 - 220}{1.42} = \frac{-10}{1.42} = \mathbf{-7.05.}$$

$$\therefore \qquad |Z| = \mathbf{7.05.}$$

3. **Level of significance.** We take $\alpha = 0.05$ as no level of significance is mentioned in the question.
4. **Critical Value at $\alpha = 0.05$.** The critical value of Z, *i.e.*,

$$Z_\alpha \text{ (at } \alpha = 0.05) = \mathbf{1.96.}$$

5. **Decision.** The calculated value of $|Z| = 7.05$ > Critical value value of $Z_\alpha = 1.96 \Rightarrow H_0$ **is rejected $\Rightarrow$ There is a significance difference between the mean yield of crops in the two districts.**

9.12 TEST OF SIGNIFICANCE FOR DIFFERENCE OF TWO STANDARD DEVIATIONS — LARGE SAMPLES.

Let A and B be two populations with standard deviations σ_1 and σ_2. Let n_1 and n_2 be the size of two samples with standard deviations s_1 and s_2 respectively drawn from the population A and B. Then we want to test the null hypothesis $H_0 : \sigma_1 = \sigma_2$. The sampling distribution of standard deviation s for large samples is normal with variance given by:

$$\text{Variance} = \frac{\sigma^2}{2n}$$

$\therefore$ **Standard Error of Standard deviation** $= \dfrac{\sigma}{\sqrt{2n}}$.

Case I. WHEN THE POPULATION STANDARD DEVIATIONS ARE KNOWN.

Here **Var** $(s_1 - s_2)$ **= Var**(s_1) **+ Var**(s_2) = $\frac{\sigma_1^2}{2n_1} + \frac{\sigma_2^2}{2n_2}$ $\Rightarrow$ **S.E.** = $\sqrt{\frac{\sigma_1^2}{2n} + \frac{\sigma_2^2}{2n_2}}$

Null hypothesis : H_0 : The sample standard deviation does not differ significantly. *i.e.*

$$H_0 : \sigma_1 = \sigma_2.$$

Alternative Hypothesis: $H_1 : \sigma_1 \neq \sigma_2 \Rightarrow$ **Two Tailed Test.**

Test Statistic: $Z = \frac{s_1 - s_2}{\text{S.E.}(s_1 - s_2)} = \frac{s_1 - s_2}{\sqrt{\frac{\sigma_1^2}{2n_1} + \frac{\sigma_2^2}{2n_2}}}$.

Also Z ~ N(0, 1).

The other steps **are the same as given in the test of significance of difference of two sample means.**

Case II. WHEN THE POPULATION STANDARD DEVIATIONS ARE NOT KNOWN.

Standard Error : $\text{S.E.}(s_1 - s_2) = \sqrt{\frac{s_1^2}{2n_1} + \frac{s_2^2}{2n_2}}$.

$\therefore$ **Test Statistic :** $Z = \frac{s_1 - s_2}{\text{S.E.}(s_1 - s_2)}$.

Example 33. *The standard deviation of the height of (Honours) students of a college is 4.0 cm. Two samples are taken. The standard deviation of 100 B. Com(Hons.) students is 3.5 cm and 50 B.A Eco (Hons.) students is 4.5 cm. Test the significance of the difference of standard deviations of the samples.*

Solution. We have the following data

B. Com(Hons.)	**B.A.(Hons.) Eco.**
$n_1 = 100$	$n_2 = 50.$
$s_1 = 3.5$ cm	$s_2 = 4.5$ cm
$\sigma = 4.00$ cm	

1. **Null Hypothesis:** There is no significance difference between the standard deviations of two sample, *i.e.*, $\boldsymbol{H_0 : \sigma_1 = \sigma_2}$.

   Alternate Hypothesis: $H_1 : \sigma_1 \neq \sigma_2 \Rightarrow$ It represents a **Two tailed test.**

2. **Calculation of Test Statistic:**

Standard Error : S.E. $(s_1 - s_2) = \sqrt{\frac{\sigma_1^2}{2n_1} + \frac{\sigma_2^2}{2n_2}} = \sqrt{\frac{(4)^2}{2 \times 100} + \frac{(4)^2}{2 \times 50}}$

$$= \sqrt{\frac{16}{200} + \frac{16}{100}} = \sqrt{0.24} = \mathbf{0.49}.$$

Test Statistic: $Z = \frac{s_1 - s_2}{\text{S.E.}(s_1 - s_2)} = \frac{3.5 - 4.5}{0.49} = \mathbf{-2.04}$.

$\therefore$ $|Z| = \mathbf{2.04}.$

3. **Level of Significance.** Since no level of significance is given so we take $\boldsymbol{\alpha = 0.05}$.
4. **Critical Value.** Critical value of Z at $\alpha = 0.05$ is $\boldsymbol{Z_\alpha = 1.96}$.

5. **Decision. The calculated value of $| Z | = 2.04 > Z_\alpha = 1.96 \Rightarrow$ the Null hypothesis H_0 is rejected $\Rightarrow$ there is a significant difference between the two standard deviations.**

Example 34. *You are given the following information relating to purchase of bulbs from two manufacturers A and B.*

Manufacturer	*No. of bulbs bought*	*Mean life*	*S.D.*
A	*50*	*1980 hours*	*80 hours*
B	*70*	*2010 hours*	*60 hours*

Is the difference between the standard deviation significant at 10% level of significance?

Solution. We are given the following data

A	*B*
$n_1 = 50$	$n_2 = 70.$
$\bar{x}_1 = 1980$ hours	$\bar{x}_2 = 2010$ hours
$s_1 = 80$ hours	$s_2 = 60$ hours.

1. **Null hypothesis:** The standard deviations of the two manufacturers do not differ significantly, *i.e.*, $H_0 : \sigma_1 = \sigma_2$.

   Alternative Hypothesis: $H_1 : \sigma_1 \neq \sigma_2$. It results in a **Two Tailed Test.**

2. **Selection of Test Statistic:**

$$\text{Here S.E of s.d. : S.E.}(s_1 - s_2) = \sqrt{\frac{s_1^2}{2n_1} + \frac{s_2^2}{2n_2}} = \sqrt{\frac{(80)^2}{2\times 50} + \frac{(60)^2}{2\times 70}}$$

$$= \sqrt{89.714} = \mathbf{9.47.}$$

$$\textbf{Test Statistic: } Z = \frac{s_1 - s_2}{\text{S.E.}(s_1 - s_2)} = \frac{80-60}{9.47} = \frac{20}{9.47} = \mathbf{2.11}$$

$$\therefore |Z| = \mathbf{2.11}$$

3. **Level of Significance.** Here it is given that $\alpha = 0.1$.
4. **Critical Value.** The value of Z_α at $\alpha = 0.1$ is $=$ **1.645** (from Table).
5. **Decision.** The calculated value of $| Z | > Z_\alpha$ as $2.11 > 1.645 \Rightarrow$ **Null hypothesis H_0 is rejected $\Rightarrow$ there is significent difference between the standard deviations of two manufacturers.**

EXERCISE – 9.4

1. In a certain factory there are two independent processes for manufacturing the same item. The average weight in a sample of 250 items produced from one process is found to be 120 gms with a standard deviation of 12 gms, while from other process are 124 and 14 in a sample of 400 items. Is the difference between the mean weights significant at 10% level of significance?

   [**Hint. Null hypothesis:** $H_0 : \mu_1 = \mu_2$. **Alternative hypothesis:** $H_1 : \mu_1 \neq \mu_2$.

$$\textbf{Standard error: S.E. } (\bar{x}_1 - \bar{x}_2) = \sqrt{\frac{s_1^2}{n_1} + \frac{s_2^2}{n_2}} = \sqrt{\frac{(12)^2}{250} + \frac{(14)^2}{400}} = \mathbf{1.0325.}$$

$$\textbf{Test Statistic: } Z = \frac{120-124}{1.0325} = \mathbf{-3.87.} \qquad \therefore |Z| = \mathbf{3.87.}$$

   Z_α at α of 10% = 1.645. Again $| Z | > Z_\alpha$ as $3.87 > 1.645 \Rightarrow H_0$ is rejected $\Rightarrow$ **there is a significant difference between two sample mean weights.]**

2. A random sample of 50 male employees is taken at the end of a year and the mean number of hours of absenteeism for the year is found to be 63 hours. A similar sample of 50 female employees has mean of 66 hours. Could these samples be drawn from a population with the same mean and s.d. 10 hours? State clearly the assumption you made.

[**Hint.** $H_0 : \mu_1 = \mu_2$. $\quad H_1 : \mu_1 \neq \mu_2$.

$$\textbf{S.E.}(\bar{x}_1 - \bar{x}_2) = \sqrt{\frac{\sigma^2}{n_1} + \frac{\sigma^2}{n_2}} = \sqrt{\frac{100}{50} + \frac{100}{50}} = \mathbf{2.}$$

$$\textbf{Test statistic: } Z = \frac{\bar{x}_1 - \bar{x}_2}{\textbf{S.E.}(\bar{x}_1 - \bar{x}_2)} = \frac{63 - 66}{2} = \mathbf{-1.5.} \qquad |Z| = \mathbf{1.5.}$$

Take $\alpha = 0.05$, then $Z_\alpha = 1.96$.

Now $|Z| < Z_\alpha$ as $1.5 < 1.96 \Rightarrow H_0$ is accepted $\Rightarrow$ **the two samples have been drawn from the same population.]**

3. A man buys 100 electric bulbs of each of two well known makes taken at random from stock for testing purpose. He finds that make *A* has a mean life of 1300 hours with s.d. of 82 hours and the make *B* has a mean life of 1248 hours with s.d of 93 hours. Discuss the significance of these results.

4. The mean yield of two sets of plots and their variability are as given below. Examine whether the difference in the variability in yields is significant

	Set of 40 plots	**Set of 60 plots**
Mean yield per plot	1258 kg	1243 kg
S.D. per plot	34	28.

[**Hint.** $\mathbf{H_0} : \sigma_1 = \sigma_2$, $\qquad \mathbf{H_1} : \sigma_1 \neq \sigma_2$.

$$\mathbf{Z} = \frac{s_1 - s_2}{\sqrt{(s_1^2/2n_1) + (s_2^2/2n_2)}} = \frac{34 - 28}{\sqrt{(34)^2/80 + (28)^2/120}} = \frac{6}{\sqrt{4.45 + 6.63}} = \mathbf{1.31.}$$

Take $\alpha = 0.05$, then $Z_\alpha = 1.96$.

Now $|Z| < Z_\alpha$ as $1.31 < 1.96 \Rightarrow$ $\boldsymbol{H_0}$ **is accepted** $\Rightarrow$ **there is no significant difference in the variability in yields.]**

5. Intelligence test of two groups of boys and girls gives the following results.

Girls : $\bar{x}_1 = 84, s.d.\ s_1 = 10, n_1 = 121$

Boys : $\bar{x}_2 = 81, s.d.\ s_2 = 12, n_2 = 81.$

(*a*) Is the difference in mean scores significant?

(*b*) Is the difference in s.d. significant?

[**Hint.** (*a*) Null Hypothesis $H_0 : \mu_1 = \mu_2$.

$$\text{Test Statistic } Z = \frac{\bar{x}_1 - \bar{x}_2}{\sqrt{s_1^2/n_1 + s_2^2/n_2}} = \frac{84 - 81}{\sqrt{100/121 + 144/81}} = 1.863.$$

Take $\alpha = 0.05$ so that value of $Z_\alpha = 1.96$

Now $|Z| < Z_\alpha$ as $1.863 < 1.96 \Rightarrow H_0$ is accepted $\Rightarrow$ Difference of means scores is not significant.

(*b*) Null hypothesis $\boldsymbol{H_0 : \sigma_1 = \sigma_2}$

$$\text{Test Statistic } Z = \frac{s_1 - s_2}{\sqrt{(s_1^2/2n_1) + (s_2^2/2n_2)}} = \frac{10 - 12}{\sqrt{(10)^2/(2 \times 121) + (12)^2/(2 \times 81)}} = \frac{-2}{\sqrt{1.302}} = -1.753$$

Take $\alpha = 0.05$ so that the value of Z_α at $\alpha = 0.05$ is 1.96. Again $|Z| < Z_\alpha$ as $1.753 < 1.96 \Rightarrow H_0$ is accepted $\Rightarrow$ the difference between the two standard deviations is significant.

6. Random samples drawn from two places gave the following data relating to the heights of children :

	Place A	Place B
Mean height (in cm)	68.50	68.58
Standard Deviation (in cm)	2.5	3.0
Number of items in sample	1200	1500

Test at 5% level that the mean height is the same for children at two places.

[**Hint.** $H_0 : \mu_1 = \mu_2$, $H_1 : \mu_1 \neq \mu_2$.

$$\text{S.E} = \sqrt{\frac{\sigma_1^2}{n_1} + \frac{\sigma_2^2}{n_2}} = \sqrt{\frac{(2.5)^2}{1200} + \frac{(3)^2}{1500}} = 0.1058$$

$$\text{Z} = \frac{\bar{x}_1 - \bar{x}_2}{\text{S.E.}} = \frac{68.50 - 68.58}{0.1058} = -\mathbf{0.756}. \quad \therefore \ |\text{Z}| = \mathbf{0.756}$$

But $Z_{0.05} = 1.96 >$ calculated value of $|Z| = 0.756 \Rightarrow$ the null hypothesis is accepted]

ANSWERS

1. There is a significant difference between the two sample mean weights.
2. The two samples have been drawn from the same population.
3. Z = 4.2, the difference is significant.
4. There is no significant difference in the variability in yields.
5. The difference between the two standard deviations is not significant.
6. The null hypothesis is accepted.

9.13 TEST OF SIGNIFICANCE OF DIFFERENCE BETWEEN TWO MEANS — SMALL SAMPLES

Example 35. *The average number of articles produced by two machines per day are 200 and 250 with standard deviations 20 and 25 respectively on the basis of records of 25 days production. Can you regard both the machines equally efficient at 1% level of significance?*

Solution. In the usual notations we are given

$$n_1 = 25,\ \bar{x}_1 = 200,\ s_1 = 20$$
$$n_2 = 25,\ \bar{x}_2 = 250,\ s_2 = 25.$$

Let S^2 **be the unbiased estimate of common population variance based on both the samples.** Then.

$$S^2 = \frac{n_1 s_1^2 + n_2 s_2^2}{n_1 + n_2 - 2} = \frac{25 \times (20)^2 + 25 \times (25)^2}{25 + 25 - 2} = \frac{10000 + 15625}{48} = \frac{25625}{48} = 533.85$$

$$\Rightarrow S = \sqrt{533.85} = \mathbf{23.1}$$

1. **Null Hypothesis :** $H_0 : \mu_1 = \mu_2$, *i.e.*, **both the machines are equally efficient.**
   Alternative Hypothesis: $H_1 : \mu_1 \neq \mu_2$ **(Two Tailed Test).**
2. **Computation of Test Statistic:**

Standard Error of Difference: S.E. $(\bar{x}_1 - \bar{x}_2) = S \times \sqrt{\left(\frac{1}{n_1} + \frac{1}{n_2}\right)}$

$$= 23.1 \times \sqrt{\frac{1}{25} + \frac{1}{25}} = 23.1 \times \sqrt{\frac{2}{25}} = 23.1 \times \sqrt{0.08} = 23.1 \times 0.28 = \mathbf{6.468.}$$

$\therefore$ **Test Statistic:** $t = \dfrac{\bar{x}_1 - \bar{x}_2}{\text{S.E.}(\bar{x}_1 - \bar{x}_2)} = \dfrac{200 - 250}{6.468} = \dfrac{-50}{6.468} = -\mathbf{7.7.}$

$\therefore \quad |t| = \mathbf{7.7.}$

3. **Level of significance.** Here $\alpha = 1\% = 0.01$
4. **Critical value.** The value of t at $\alpha = 0.01$ and $(25 + 25 - 2) = 48$ degrees of freedom for two tailed test is $t_{0.01,48} = 2.58$.
5. **Decision. The computed value of $|t| = 7.7 > t_{0.01,\,48} = 2.58 \Rightarrow$ Null hypothesis is rejected $\Rightarrow$ that the two machines are not equally efficient at 1% level of significance.**

Example 36. *Two salesman A and B are working in a certain district. From a sample survey conducted by the Head Office, the following results were obtained . State whether there is any significant difference in the average sales between the two salesmen?*

	A	*B*
No. of Sales	*20*	*18*
Average Sales (in Rs.)	*170*	*205*
Standard Deviation (in Rs.)	*20*	*25*

Solution. In the usual notations we are given

$$n_1 = 20, \quad \bar{x}_1 = 170, \quad s_1 = 20;$$
$$n_2 = 18, \quad \bar{x}_2 = 205, \quad s_2 = 25.$$

1. **Null hypothesis:** $H_0 : \mu_1 = \mu_2$, *i.e.*, **there is no significant difference in the average sales between the two salesmen.**

   Alternative hypothesis: $H_1 : \mu_1 \neq \mu_2$ **(Two Tailed Test)**
2. **Computation of Test Statistic:**

Estimated standard deviation: $S = \sqrt{\dfrac{n_1 s_1^2 + n_2 s_2}{n_1 + n_2 - 2}} = \sqrt{\dfrac{20(20)^2 + 18(25)^2}{20 + 18 - 2}}$

$$= \sqrt{\frac{8000 + 11250}{36}} = \sqrt{\frac{19250}{36}} = \frac{138.71}{6} = \mathbf{23.12.}$$

Standard Error of difference: $\text{S.E.}(\bar{x}_1 - \bar{x}_2) = S \times \sqrt{\dfrac{1}{n_1} + \dfrac{1}{n_2}} = 23.12 \times \left[\sqrt{\dfrac{1}{20} + \dfrac{1}{18}}\right]$

$$= 23.12 \times \sqrt{0.1060} = 23.12 \times 0.32 = \mathbf{7.4.}$$

Test Statistic $t = \dfrac{\bar{x}_1 - \bar{x}_2}{\text{S.E.}(\bar{x}_1 - \bar{x}_2)} = \dfrac{170 - 205}{7.4} = \dfrac{-35}{7.4} = -\mathbf{4.73.}$

$\therefore \quad |t| = \mathbf{4.73.}$

3. **Level of Significance :** $\alpha = 0.05$.
4. **Critical value : The tabulated or critical value of t at $\alpha = 0.05$ for $20 + 18 - 2 = 36$** degrees of freedom for two tailed test is $t_{0.05,\,36} = \mathbf{1.96.}$

5. **Decision. The calculated value of** $|t| = 4.73 >$ **tabulated value** $t_{0.05,\ 36} = \mathbf{1.96}$ **⇒ the Null hypothesis is rejected ⇒ there is a significant difference in the average sales between the two salesmen.**

Example 37. *Following table contains the data resulting from a sample of managers trained under different programs.*

Program Sampled	*Mean sensitivity of the program*	*Number of Managers observed*	*Estimated s.d. for sensitivity after the program*
Formal	*92%*	*12*	*15%*
Informal	*84%*	*15*	*19%*

Test at 0.05 level of significance whether the sensitivity achieved by the formal program is significantly higher than that achieved under the informal program.

Solution. In the usual notations, we are given

$$n_1 = 12, \quad \bar{x}_1 = 92\%, \quad s_1 = 15\%$$
$$n_2 = 15, \quad \bar{x}_2 = 84\%, \quad s_2 = 19\%$$

1. **Null Hypothesis:** $H_0 : \mu_1 = \mu_2$, *i.e.*, **there is no significant difference between the formal and Informal program.**

   Alternative Hypothesis: $H_0 : \mu_1 > \mu_2$. **(One Tailed Test)**

2. **Calculation of Test Statistic:**

Estimated common standard deviation: $S = \sqrt{\dfrac{n_1 s_1^2 + n_2 s_2^2}{n_1 + n_2 - 2}} = \sqrt{\dfrac{12 \times (15)^2 + 15 \times (19)^2}{12 + 15 - 2}}$

$$= \sqrt{\frac{2700 + 5415}{25}} = \sqrt{\frac{8115}{25}} = \sqrt{324.6} = \mathbf{18.02.}$$

Standard Error of Difference: $\text{S.E.}(\bar{x}_1 - \bar{x}_2) = S \times \sqrt{\dfrac{1}{n_1} + \dfrac{1}{n_2}} = 18.02 \times \left(\sqrt{\dfrac{1}{12} + \dfrac{1}{15}}\right)$

$$= 18.02\sqrt{\frac{27}{180}} = 18.02 \times \sqrt{0.15} = 18.02 \times 0.387 = \mathbf{6.97.}$$

Test Statistic: $t = \dfrac{\bar{x}_1 - \bar{x}_2}{\text{S.E.}(\bar{x}_1 - \bar{x}_2)} = \dfrac{92 - 84}{6.97} = \dfrac{8}{6.97} = 1.15$

$\therefore \quad |t| = \mathbf{1.15.}$

3. **Level of Significance.** Take $\alpha = 0.05$.
4. **Critical value.** The tabulated or critical value of t and $\alpha = 0.05$ for $12 + 15 - 2 = 25$ degrees of freedom for one tailed test is $t_{0.05,25} = \mathbf{1.708}$.
5. **Decision. The calculated value of** $|t| = 1.15 >$ **tabulated value of** $t_{0.05,\ 25} = \mathbf{1.708}$ **⇒ the null hypothesis** H_0 **is accepted ⇒ the sensitivity achieved by the Formal program is higher than that achieved under the Informal program.**

Example 38. *The mean life of a sample of 10 electric bulbs was found to be 1456 hours with a standard deviation of 423 hours. A second sample of 17 bulbs chosen from a different batch showed a mean life of 1280 hours with standard deviation 398 hours. Is there significant difference between the means of the two batches?*

Solution. In the usual notation, it is given that

$n_1 = 10, \quad \bar{x}_1 = 1456, \quad s_1 = 423$

$n_2 = 17, \quad \bar{x}_2 = 1280, \quad s_2 = 398$

1. **Null Hypothesis:** $H_0 : \mu_1 = \mu_2$, *i.e.*, **There is no significant difference between the means of two samples.**

   Alternative Hypothesis: $H_1 : \mu_1 \neq \mu_2$ **(Two Tailed test)**

2. **Computation of test Statistic:**

   Estimated standard deviation of population: $S = \sqrt{\dfrac{n_1 s_1^2 + n_2 s_2^2}{n_1 + n_2 - 2}}$

$$= \sqrt{\frac{10 \times (423)^2 + 17\,(398)^2}{10 + 17 - 2}} = \sqrt{\frac{1789290 + 2692869}{28}} = \frac{2117.11}{5} = \mathbf{423.42.}$$

Standard Error of Difference: $\text{S.E.}(\bar{x}_1 - \bar{x}_2) = S \times \sqrt{\dfrac{1}{n_1} + \dfrac{1}{n_2}} = 423.42 \times \sqrt{\left(\dfrac{1}{10} + \dfrac{1}{17}\right)}$

$$= 423.42(\sqrt{0.1588}) = 423.42 \times 0.398 = \mathbf{168.52.}$$

Test Statistic: $t = \dfrac{\bar{x}_1 - \bar{x}_2}{\text{S.E}(\bar{x}_1 - \bar{x}_2)} = \dfrac{1456 - 1280}{168.52} = \dfrac{176}{168.52} = \mathbf{1.04.}$

3. **Level of significance :** Take $\alpha = 0.05$.
4. **Critical value.** The tabulated or critical value of t at $\alpha = 0.05$ for $10 + 17 - 2 = 25$ degrees of freedom for two tailed test is $t_{0.05,\,25} = 2.06$.
5. **Decision. The calculated value of $|t| = 1.04 < t_{0.05,\,25} = 2.06 \Rightarrow$ the null hypothesis H_0 is accepted $\Rightarrow$ there is no significance difference between the means of two samples.**

Example 39. *Two types of batteries are tested for their lengths of life and the following data are obtained.*

	No. of Samples	*Mean life*	*Variance*
Type A	9	*600 hours*	121
Type B	8	*640 hours*	144

Is there significance difference in the two means? Value of t for 15 degrees of freedom at 5% level is 2.131.

Solution. In the usual notations we are given that

$n_1 = 9, \quad \bar{x}_1 = 600, \quad s_1^2 = 121.$

$n_2 = 8, \quad \bar{x}_2 = 640, \quad s_2^2 = 144.$

1. **Null Hypothesis:** $H_0 : \mu_1 = \mu_2$, *i.e.*, **there is no significance difference in the two means.**

   Alternative Hypothesis: $H_1 : \mu_1 \neq \mu_2$ **(Two Tailed test)**

2. **Computation of test Statistic:**

   Estimated standard deviation of Population: $S = \sqrt{\dfrac{n_1 s_1^2 + n_2 s_2^2}{n_1 + n_2 - 2}} = \sqrt{\dfrac{9 \times 121 + 8 \times 144}{9 + 8 - 2}}$

$$= \sqrt{\frac{2241}{15}} = \sqrt{149.4} = \mathbf{12.22.}$$

Standard Error of difference: $\text{S.E.}(\bar{x}_1 - \bar{x}_2) = S \times \sqrt{\left(\frac{1}{n_1} + \frac{1}{n_2}\right)}$

$$= 12.22 \times \sqrt{\frac{1}{9} + \frac{1}{8}} = 12.22 \times \sqrt{0.236} = 12.22 \times 0.486 = \mathbf{5.94.}$$

Test Statistic: $t = \frac{\bar{x}_1 - \bar{x}_2}{\text{S.E.}(\bar{x}_1 - \bar{x}_2)} = \frac{600 - 640}{5.94} = \frac{-40}{5.94} = -\mathbf{6.73.}$ $\quad \therefore \ |t| = 6.73$

3. **Level of significance.** $\alpha = 0.05$.
4. **Critical value.** Value of t at $\alpha = 0.05$ for $9 + 8 - 2 = 15$ degrees of freedom is $t_{0.05,\, 15} = 2.131$.
5. **Decision. The calculated value of $|t| = 6.73 >$ tabulated value of $t_{0.05,\, 15} = 2.131 \Rightarrow$ the null hypothesis H_0 is rejected $\Rightarrow$ there is a significant difference in the two means.**

Example 40. *A group of five patients treated with medicine 'A' weight 42, 39, 48, 60 and 41 kg. A second group of 7 patients from the same hospital treated with medicine 'B' weight 38, 42, 56, 64, 68, 69, and 62 kg. Do you agree with the claim that the medicine 'B' increases the weight significantly. (The value of 't' at 5% level of significance for 10 degrees of freedom is 2.2281).*

Solution. Let the variables x and y denote the weight (in kgs) of the patients treated with medicine A and B respectively.

1. **Null Hypothesis:** $H_0 : \mu_A = \mu_B$, *i.e.*, there is no significant difference between the medicines A and B as regards their effect on increase in weight.

   Alternative Hypothesis: $H_1 : \mu_A \neq \mu_B$ (Two Tailed Test)
2. **Computation of Test Statistic:** We shall first compute the sample means and standard deviation by means of the following table.

Table — Computation of Sample Mean and S.D.

Medicine A

x	$d_1 = x - \bar{x}$ $= x - 46$	d_1^2
42	– 4	16
39	– 7	49
48	2	4
60	14	196
41	– 5	25
$\Sigma x = 230$	$\Sigma d_1 = 0$	$\Sigma d_1^2 = 290$

Medicine B

y	$d_2 = y - \bar{y}$ $= y - 57$	d_2^2
38	– 19	361
42	– 15	225
56	– 1	1
64	7	49
68	11	121
69	12	144
62	5	25
$\Sigma y = 399$	$\Sigma d = 0$	$\Sigma d_2^2 = 926$

$$\bar{x} = \frac{\Sigma x}{n} = \frac{230}{5} = \mathbf{46.} \qquad \bar{y} = \frac{\Sigma x}{n} = \frac{399}{7} = \mathbf{57.}$$

Estimated standard deviation of the population: $S = \sqrt{\dfrac{\Sigma d_1^2 + \Sigma d_2^2}{n_1 + n_2 - 2}}$

$$= \sqrt{\frac{290+926}{5+7-2}} = \sqrt{\frac{1216}{10}} = \sqrt{121.6} = \mathbf{11.03.}$$

Standard Error of Difference: $\text{S.E.}(\bar{x} - \bar{y}) = S \times \sqrt{\dfrac{n_1 + n_2}{n_1 n_2}} = 11.03 \times \sqrt{\dfrac{5+7}{35}}$

$$= 11.03 \times \sqrt{0.34} = 11.03 \times 0.58 = \mathbf{6.4.}$$

Test Statistic: $t = \dfrac{\bar{x} - \bar{y}}{S.E.(\bar{x} - \bar{y})} = \dfrac{46-57}{6.4} = \dfrac{-11}{6.4} = \mathbf{-1.72.}$

$\therefore \quad |t| = \mathbf{1.72.}$

3. Level of significance : $\alpha = 0.05$.

4. Critical value. The tabulated or critical value of t at $\alpha = 0.05$ for $5 + 7 - 2 = 10$ degrees of freedom for two-tailed test is $t_{0.05,\, 10} = \mathbf{2.2281.}$

5. Decision. The calculated value of $|t| = 1.72 <$ tabulated value of $t_{0.05,\, 10} = 2.2281 \Rightarrow$ the null hypothesis H_0 is accepted $\Rightarrow$ There is no significant difference between the medicines A and B as regards the increase in the weight.

Example 41. *The following data shows the cost in hundred rupees per square metre of the floor area concerning randomly selected 7 schools and 5 office blocks from those completed during the period 1997 to 2002.*

Schools :	*28*	*31*	*26*	*27*	*23*	*38*	*37*
Office blocks :	*37*	*42*	*34*	*37*	*35*		

Do the data support the hypothesis that the cost per square metre for the office blocks was greater than that for the schools. Test at 5% level of significance.

Solution. Null Hypothesis: There is no difference in the costs, *i.e.*, $H_0 : \mu_1 = \mu_2$.

Alternative Hypothesis: $H_1 : \mu_1 < \mu_2$.

(μ_1 = average cost per school, μ_2 = average cost for office block.)

2. Computation of Test Statistics.

Computation Table

S.No.	X_1	$X_1 - \bar{X}_1 = X_1 - 30$	$(X_1 - \bar{X}_1)^2$	*S.No.*	X_2	$X_2 - \bar{X}_2 = X_2 - 37$	$(X_2 - \bar{X}_2)^2$
1	28	– 2	4	**1**	37	0	0
2	31	1	1	**2**	42	5	25
3	26	– 4	16	**3**	34	– 3	9
4	27	– 3	9	**4**	37	0	0
5	23	– 7	49	**5**	35	– 2	4
6	38	8	64				
7	37	7	49				
$\Sigma X_1 = 210$		$\Sigma(X_1 - \bar{X}_1)^2 = 192$			$\Sigma X_2 = 185$	$\Sigma(X_2 - \bar{X}_2)^2 = 38$	

$$\therefore\ X_1 = \frac{\Sigma X_1}{n} = \frac{210}{7} = 30, \qquad \overline{X}_2 = \frac{\Sigma X_2}{n_2} = \frac{185}{5} = \mathbf{37.}$$

$$\text{S.E.}(\overline{X}_1 - \overline{X}_2) = \sqrt{\frac{\Sigma(\overline{X}_1 - \overline{X}_2)^2 + \Sigma(\overline{X}_2 - \overline{X}_2)^2}{n_1 + n_2 - 2}} \times \sqrt{\left(\frac{1}{n_1} + \frac{1}{n_2}\right)}$$

$$= \sqrt{\frac{192+38}{7+5-2}} \times \sqrt{\frac{1}{7} + \frac{1}{5}} = 4.796 \times 0.586 = \mathbf{2.81}$$

$$\textbf{Test Statistic: } t = \frac{\overline{X}_1 - \overline{X}_2}{\text{S.E.}(\overline{X}_1 - \overline{X}_2)} = \frac{30-37}{2.81} = \mathbf{-2.491.}$$

$$\therefore \qquad |t| = \mathbf{2.491.}$$

3. **Level of Significance :** $\alpha = 0.05$.
4. **Critical Value:** The critical value of t at $\alpha = 0.05$ for 10 degrees of freedom for one tailed test is 1.812.
5. **Decision. The calculated $|t| = 2.491 >$ tabulated $t_{0.05,\,10} = 1.812 \Rightarrow$ the null hypothesis is rejected and so we accept Alternative Hypothesis H_1 which concludes that the cost per square metre for office block is greater than that for school block.**

Example 42. *Six guinea pigs injected with 0.5 mg of a medication took on the average 15.4 seconds to fall asleep with an unbiased standard deviation 2.2 seconds while six other guinea pigs injected with 1.5 mg of the medication took on the average 11.2 seconds to fall asleep with an unbiased standard deviation of 2.6 seconds. Use the 5% level of significance to test the null hypothesis that the difference in dosage has no effect.*

Solution.

1. **Null Hypothesis :** Difference in dosage has no effect, *i.e.*, $H_0 : \mu_1 = \mu_2$.
   Alternative Hypothesis : $H_1 : \mu_1 \neq \mu_2$ (Two Tailed test).
2. **Calculation of Test Statistic :** It is given that

$$2.2 = s_1\sqrt{\frac{n_1}{n_1 - 1}} = s_1\sqrt{6/5} = s_1 \times 1.095 \ \Rightarrow\ s_1 = 2.2/1.095 = \mathbf{2.009.}$$

Also

$$2.6 = s_2\sqrt{\frac{n_2}{n_2 - 1}} = s_2\sqrt{6/5} = s_2 \times 1.095 \ \Rightarrow\ s_2 = 2.6/1.095 = \mathbf{2.374.}$$

Standard Error of Difference :

$$\text{S.E.}\ (\bar{x}_1 - \bar{x}_2) = \sqrt{\frac{s_1^2 + s_2^2}{n}} = \sqrt{\frac{(2.009)^2 + (2.374)^2}{6}} = \mathbf{1.269.}$$

Test Statistic:

$$t = \frac{|\bar{x}_1 - \bar{x}_2|}{\text{S.E.}(\bar{x}_1 - \bar{x}_2)} = \frac{15.4 - 11.2}{1.269} = \mathbf{3.30.} \quad \therefore |t| = 3.30$$

3. **Critical Value.** The tabulated value of t at $\alpha = 0.05$ for $6 + 6 - 2 = 10$ degrees of freedom for a two tailed **test is 2.228.**
4. **Decision. Now $|t| = 3.30 > t_{0.05,\,10} = 2.28 \Rightarrow$ Null hypothesis H_0 is rejected $\Rightarrow$ Difference in dosage has significant effect.**

Example 43. *Samples of two types of electric bulbs were tested for length of life and the following data were obtained :*

	Type I	*Type II*
Number in the sample	*8*	*7*
Mean of the sample (in hours)	*1134*	*1024*
Standard deviation of the sample in hours	*12*	*25*

Solution. In the usual notations we are given that

$n_1 = 8 \quad \bar{x}_1 = 1134 \quad s_1 = 35$

$n_2 = 7 \quad \bar{x}_2 = 1024 \quad s_2 = 40.$

1. **Null Hypothesis :** $H_0 : \mu_1 = \mu_2$.
   Alternative Hypothesis : $H_1 : \mu_1 \neq \mu_2$ **(Two Tailed Test).**
2. **Calculation of Test Statistic :**
   The **unbiased estimate of the Common Population standard deviation is given by**

$$s_P = \sqrt{\frac{(n_1-1)s_1^2+(n_2-1)s_2^2}{n_1+n_2-2}} = \sqrt{\frac{7\times(35)^2+6\times(40)^2}{8+7-2}} = \sqrt{\frac{7\times1225+6\times1600}{13}}$$

$$= \sqrt{1398.08} = \mathbf{37.39}$$

Standard Error of Difference: $\text{S.E.}(\bar{x}_1 - \bar{x}_2) = s_P\sqrt{\frac{1}{n_1}+\frac{1}{n_2}}$

$$= 37.39 \times \sqrt{\left(\frac{1}{8}+\frac{1}{7}\right)} = \mathbf{19.368.}$$

Test Statistic: $t = \dfrac{\bar{x}_1 - \bar{x}_2}{\text{S.E.}(\bar{x}_1 - \bar{x}_2)} = \dfrac{1134-1024}{19.368} = \mathbf{5.679.}$

3. **Critical Value.** The tabulated or critical value of t at $\alpha = 0.05$ for $8 + 7 - 2 = 13$ degrees of freedom for a two tailed test is $t_{0.05,\,13}$ = **2.161.**
4. **Decision.** Since the calculated value $|t|$ = **5.679** > tabulated value $t_{0.05,13}$ = **2.161, so the null hypothesis is rejected ⇒ the two types of electric bulbs differ significantly in their mean values.**

Example 44. *Below are given the gain in weights in kgs of cows fed on two diets X and Y :*

Diet X:	25	32	30	32	24	14	32			
Diet Y:	24	34	22	30	42	31	40	30	32	35

Test at 5% level whether the two diets differ as regards their effect on mean increase in weight. (Tabulated value of 't' for 15 degrees of freedom at 5% = 2.131)

Solution.

1. **Null Hypothesis :** $H_0 : \mu_1 = \mu_2$
   Alternative Hypothesis : $H_1 : \mu_1 \neq \mu_2$ **(Two Tailed Test).**

Table

x	$d_1 = x - \bar{x}$ $= x - 27$	$d_1^2 = (x - \bar{x})^2$	y	$d_2 = y - \bar{y}$ $= y - 32$	$d_2^2 = (y - \bar{y})^2$
25	– 2	4	24	– 8	64
32	5	25	34	2	4
30	3	9	22	– 10	100
32	5	25	30	– 2	4
24	– 3	9	42	10	100
14	– 13	169	31	– 1	1
32	5	25	40	8	64
			30	– 2	4
			32	0	0
			35	3	9
$\Sigma x = 189$	$\Sigma d_1 = 0$	$\Sigma d_1^2 = 266$	$\Sigma y = 320$	$\Sigma d_2 = 0$	$\Sigma d_2^2 = 350$

Now $\bar{x} = \frac{\Sigma x}{n_1} = \frac{189}{7} = 27$; $\quad \bar{y} = \frac{\Sigma y}{n_2} = 32.$

$$\textbf{Sample variance for } x = s_1^2 = \frac{\Sigma(x-\bar{x})^2}{n_1} = \frac{266}{7} = 38.$$

$$\textbf{Sample variance for } y = s_2^2 = \frac{\Sigma(y-\bar{y})^2}{n_2} = \frac{350}{10} = 35.$$

The unbiased estimate of common population s.d. is given by

$$s_P = \sqrt{\frac{(n_1-1)s_1^2 + (n_2-1)s_2^2}{n_1+n_2-2}} = \sqrt{\frac{6\times 38 + 9\times 35}{7+10-2}} = \sqrt{\frac{543}{15}} = \sqrt{36.2} = \mathbf{6.017}.$$

Standard Error of difference : $\text{S.E.}(\bar{x}-\bar{y}) = s_P\sqrt{\left(\frac{1}{n_1}+\frac{1}{n_2}\right)} = 6.017 \times \sqrt{\left(\frac{1}{7}+\frac{1}{10}\right)}$

$$= 6.017 \times 0.493 = \mathbf{2.966}.$$

$$\textbf{Test Statistic: } t = \frac{\bar{x}-\bar{y}}{\text{S.E.}(\bar{x}-\bar{y})} = \frac{27-32}{2.966} = \frac{-5}{2.966} = -\mathbf{1.686}.$$

$\therefore \quad |t| = 1.686.$

Critical value. The tabulated value of t at $\alpha = 0.05$ for 15 degrees of freedom for a two tailed test is $t_{0.05,\,15} = 2.131$.

Decision. Since tabulated $|t| = 1.686 < t_{0.05,\,15} = 2.131$, **so the null hypothesis is accepted** ⇒ **The two diets do not differ significantly as regards their effect on mean increase in weights.**

9.14. MEASURING THE POWER OF A HYPOTHESIS TEST

In this section we shall descuss "*What a good hypothesis test should be*?" Recall that *a type I error occurs*, when we reject a null hypothesis that is true. The probability of making a type I error is α, where is α is the level of significance of the test. *A type II errors occurs* when we acept a null hypothesis that is false. The probability of type II error is β. **Ideally for a good hypolhesis, α and β should both be small.**

Since rejeting a null hypothesis when it is false in exactly what a good test out to do, a high value of $(1 - \alpha)$ (approximately equal to 1.0) means that the test is working quite well (i.e., it is rejecting the null hypothesis when it is false). A low value of $(1 - \beta)$ (nearly equal to 0.0) means that the test is the working very poorly, i.e., it is not rejecting the null hypothesis when it is false. Since the value of $(1 - \beta)$ is the measure of how well the test is working, it is known as the power of the test.

EXERCISE – 9.5

1. Samples of the types of electric bulbs were tested for the length of life and the following data were obtained

	Type I	**Type II**
Sample No.	8	7
Sample Mean	1234 hours	1036 hours
Sample S.D.	36 hours	40 hours

Is the difference in the means sufficient to warrant an inference that type I is superior to type II regarding the length of life?

2. Strength tests carried out on samples of two yarns spun to the same count gave the following results:

	Yarn A	**Yarn B**
No. of Sample	4	9
Sample Mean	50	42
Sample variance	42	56

The strengths are expressed in kgs. Is the difference in mean strengths significant of real difference in the mean strength of the sources from which the sample are drawn?

[Hint. Null hypothesis : $H_0 : \mu_1 = \mu_2$.

Alternative hypothesis : $H_1 : \mu_1 \neq \mu_2$.

$$S^2 = \frac{4 \times (42)^2 + 9 \times (56)^2}{4 + 9 - 2} = \frac{35280}{11} = \mathbf{3207.27.}$$

$$\therefore \textbf{ Test Statistic: } t = \frac{\bar{x}_1 - \bar{x}_2}{\sqrt{S^2[(1/n_1) + (1/n_2)]}} = \frac{50 - 42}{\sqrt{3207.27[(1/4) + (1/9)]}} = \frac{8}{34.03} = \mathbf{0.235.}$$

Also $t_{0.05,\,11} = 2.201$.

Now $|t| = 0.235 < t_{0.05,\,11} = 2.201 \Rightarrow H_0$ is accepted $\Rightarrow$ **the difference in the mean strength is insignificant of real difference in the mean strength of the sources.]**

3. Two samples of 6 and 5 gave the following data

	Sample I	**Sample II**
Mean	40	50
S.D.	8	10

Is the difference of the mean significant? The value of t for $v = 9$ at 5% of level of signifance is 2.26.

[Hint. $H_0 : \mu_1 = \mu_2$. $H_1; \mu_1 \neq \mu_2$ (Two tailed test). $t = 1.84$, $t_{0.05,\,9} = 2.26$

Since $|t| < t_{0.05,\,9} \Rightarrow H_0$ is accepted $\Rightarrow$ **the difference in means is not significant.]**

4. Two batches of the same product are tested for their mean life. Assuming that the lives of the product follow a normal distribution with an unknown variance, test the hypothesis that the mean life is the same for both branches, given the following information.

Batch	Sample size	Mean life in hrs.	S.D
I	10	750	12
II	8	820	14

Take $\alpha = 5\%$

[Hint. H_0; $\mu_1 = \mu_2$, $H_1 : \mu_2$. **(Two Tailed test)**

$$t = \frac{\bar{x}_1 - \bar{x}_2}{S} \times \sqrt{\frac{n_1 n_2}{n_1 + n_2}} = \mathbf{11.43.}$$

Also $t_{0.05,\,16} = 2.120$.

Since $|t| = 11.43 > t_{0.05,\,16} = 2.120 \Rightarrow H_0$ is rejected $\Rightarrow$ **the mean life differs significantly for both the batches.]**

5. Two kinds of manure were applied to sixteen one-hectare plots, other conditions remaining the same. The yields in quintals are given below.

Manure I	18	20	36	50	49	36	34	49	41
Manure II	29	28	26	35	30	44	46		

Is there any significant difference between the mean yield? Use 5% significance level?

6. The means of two random samples of size 9 and 7 are 196.42 and 198.82 respectively. The sum of the squares of the deviations from the mean are 26.94 and 18.73 respectively. Can the sample be considered to have been drawn from the same normal population?

[Hint. Null hypothesis: $H_0 : \mu_1 = \mu_2$. **Alternative hypothesis:** $H_1 : \mu_1 \neq \mu_2$.

$$\textbf{Estimated variance } S^2 = \frac{1}{n_1 + n_2 - 2}[\Sigma(x - \bar{x})^2 + \Sigma(y - \bar{y})^2] = \frac{26.94 + 18.73}{9 + 7 - 2} = \mathbf{3.26.}$$

$$\textbf{Test Statistic: } t = \frac{\bar{x} - \bar{y}}{\sqrt{S^2\left(\frac{1}{n_1} + \frac{1}{n_2}\right)}} = \frac{196.42 - 198.82}{\sqrt{3.26\left(\frac{1}{9} + \frac{1}{7}\right)}} = \mathbf{-2.638.}$$

$|t| = 2.638$. But $t_{0.05,\,14} = 2.15$.

Now $|t| = 2.638 > t_{0.05,\,14} = 2.15 \Rightarrow$ Null hypothesis H_0 is rejected $\Rightarrow$ **the samples cannot be considered to have come from the same normal population.]**

7. Two sets A and B of ten students selected at random from a college were taken : one was given the memory test as they were and the set was given a memory test after two weeks' training and the scores were given below :

Set A:	10	8	7	9	8	10	9	6	7	8
Set B:	12	8	8	10	8	11	9	8	9	9.

Do you think that there is any significant effect due to training ? (Given $t = 2.10$ at d.f = 18, $\alpha = 0.05$).

8. The incomes of a random sample of labourers in the industry I are : Rs. 630, 650, 680, 690, 710 and 720 per month. The incomes of a similar sample from industry II are Rs. 610, 620, 650, 680, 690, 700, 710, 720 are 730 per month. Discuss the validity of the suggestion that the industry I pays labourers much better than the industry II.

[Hint. $\bar{x} = \frac{4080}{6} = \mathbf{680},\quad \bar{y} = \frac{6780}{10} = \mathbf{678.}$

$$S^2 = \frac{1}{n_1 + n_2 - 2}[\Sigma(x - \bar{x})^2 + \Sigma(y - \bar{y})^2] = \frac{6000 + 15360}{6 + 10 - 2} = \mathbf{1525.71.}$$

Let $H_0 : \bar{x} = \bar{y}$, and $H_1 : \bar{x} \neq \bar{y}$.

Test statistic: $t = \dfrac{\bar{x} - \bar{y}}{\sqrt{S^2[(1/n_1) + (1/n_2)]}} = \dfrac{680 - 678}{\sqrt{1525.71[(1/6) + (1/10)]}} = \mathbf{0.1.}$

Also $t_{0.05,\,14} = 1.761 > |t| = 0.1 \Rightarrow H_0$ is accepted

⇒ there is no significant difference between the salaries of Industry I and Industry II.]

9. The following data present the yield in quintals of corn on ten subdivisions of equal area of two agricultural plots:

Plot 1:	6.2	5.7	6.5	6.0	6.3	5.8	5.7	6.0	6.0	5.8
Plot 2:	5.6	5.9	5.6	5.7	5.8	5.7	6.0	5.5	5.7	5.5.

Find out whether the difference between the mean yields of two plots is significant?

10. Of the two salesman, *X* claims that he has made larger sales than *Y*. For the accounts examined which are comparable for the two men, the results were:

	X	*Y*
Number of sales	10	17
Average size	Rs. 6200	Rs. 5600
Standard deviation	Rs. 690	Rs. 600.

Do these two average sizes of the sale figures differ significantly? Explain your result?

ANSWERS

1. $t = 9.35$. Type I bulb is superior to type II.

2. The difference in the mean strength is insignificant of the real difference in the mean strength of the sources.

3. The difference in means is not significant.

4. The mean life differs significantly for both the batches.

5. The mean yields do not differ significantly.

6. The sample cannot be considered to have come from the same normal population.

7. There is no significant difference due to training.

8. There is no significant difference between the salaries of Industry I and Industry II.

9. The mean yields of the two plots differ significantly; $t = 3.03$.

10. $t = 2.28$, *X's* claim is justified.

10

CHAPTER

CHI-SQUARE TEST

10.1 CHI – SQUARE TEST

The **chi-square test, written as ψ^2 - test, is a useful measure of comparing experimentally obtained results with those expected theoretically and based on the hypothesis.** It is used as a **test statistic in testing a hypothesis that provides a set of theoretical frequencies with which observed frequencies are compared.** In general **Chi-square** test is applied *to those problems in which we study whether the frequency with which a given event has occurred, is significantly different from the one as expected theoretically. The measure of* **Chi-square** *enables us to find out the degree of discrepancy between observed frequencies and theoretical frequencies and thus to determine whether the discrepancy so obtained between observed frequencies and theoretical frequencies is due to error of sampling or due to chance.*

The Chi-square is computed on the basis of frequencies in a sample and thus the value of Chi-square so obtained is a **statistic. Chi-square is not a parameter as its value is not derived from the observations in a population.** Hence **Chi-square test is a Non-Parametric test.** Chi-square test is not concerned with any population distribution and its observations.

The ψ^2-test was first used in testing statistical hypothesis by Karl Pearson in the year 1900. It is defined as

$$\psi^2 = \sum_{i=1}^{n} \frac{(O_i - E_i)^2}{E_i}$$

where O_i = observed frequency of ith event

E_i = Expected frequency of ith event.

We require the following steps to calculate ψ^2.

Step 1. Calculate all the expected frequencies, *i.e.*, E_i for all values of $i = 1, 2, ..., n$.

Step 2. Take the difference between each observed frequency O_i and the corresponding expected frequency E_i for each value of i, *i.e.*, **find $(O_i - E_i)$**

Step 3. Square the difference for each value of i, *i.e.*, **calculate $(O_i - E_i)^2$** for all values of $i = 1, 2, 3, ..., n$.

Step 4. Divide each square difference by the corresponding expected frequency, *i.e.*, **calculate** $\frac{(O_i - E_i)^2}{E_i}$ for all values of $i = 1, 2, 3, ... n$.

Step 5. Add all these quotients obtained in **STEP 4**, then

$$\psi^2 = \sum_{i=1}^{n} \left[\frac{(O_i - E_i)^2}{E_i} \right]$$ is the required value of **Chi-square**.

It should be noted that

(*a*) ***The value of ψ^2 is always positive as each pair is squared one.***

(*b*) ***ψ^2 will be zero if each pair is zero and it may assume any value extending to infinity, when the difference between the observed frequency and expected frequency in each pair is unequal. Thus ψ^2 lies between 0 and ∞.***

(c) ***The significance test on ψ^2 is always based on One Tailed Test of the right hand side of standard normal curve as ψ^2 is always non-negative.***

(d) ***As ψ^2 is a statistic and not a parameter, so it does not involve any assumption about the form of original distribution from which the observations have come.***

10.2 DEGREES OF FREEDOM

The number of data that are given in the form of a series of variables in a row or column or the number of frequencies that are put in cells in a contingency table, which can be calculated independently is called the **degrees of freedom and is denoted by ν.**

Case I If the data is given in the form of a series of variables in a row or column, then the **degrees of freedom = (number of items in the series) – 1,** *i.e.*, **$\nu = n - 1$,** where ***n*** is the number of variables in the series in a row or column.

Case II When the number of frequencies are put in cells in a contingency table, the degrees of freedom will be the product of **(number of rows less one)** and the **(number of columns less one)** *i.e.*, **$\nu = (R - 1)(C - 1)$**, where ***R* is the number of rows and *C* is the number of columns.**

10.3 CHI–SQUARE DISTRIBUTION

ψ^2-distribution is a continuous distribution whose probability density function is given by

$$P(\psi^2) = y_0(\psi^2)^{\frac{1}{2}(\nu-2)} e^{\frac{-1}{2}\psi^2}, \text{ where } \nu = \text{degrees of freedom} = n - 1$$

and $\quad y_0$ **= constant depending on the degrees of freedom.**

10.4 PROPERTIES OF ψ^2 – DISTRIBUTION

1. ***Chi-square curve is always positively skewed.***
2. ***The mean of ψ^2 distribution is the number of degrees of freedom.***
3. ***The standard deviation of ψ^2 – distribution = $\sqrt{2\nu}$, where ν is the degrees of freedom.***
4. ***Chi-square values increases with the increase in degrees of freedom.***
5. ***The value of ψ^2 lies between zero and infinity, i.e., $0 \le \psi^2 < \infty$.***
6. ***The sum of two ψ^2-distribution is again a ψ^2 distribution, i.e., if ψ_1^2 and ψ_2^2 are two independent and they have a ψ^2 distribution with n_1 and n_2 degrees of freedom respectively, then $(\psi_1^2 + \psi_1^2)$ is also a ψ^2-distribution with $(n_1 + n_2)$ degrees of freedom.***
7. ***For different degrees of freedom, the shape of the curve will be different. See figure 11.1.***

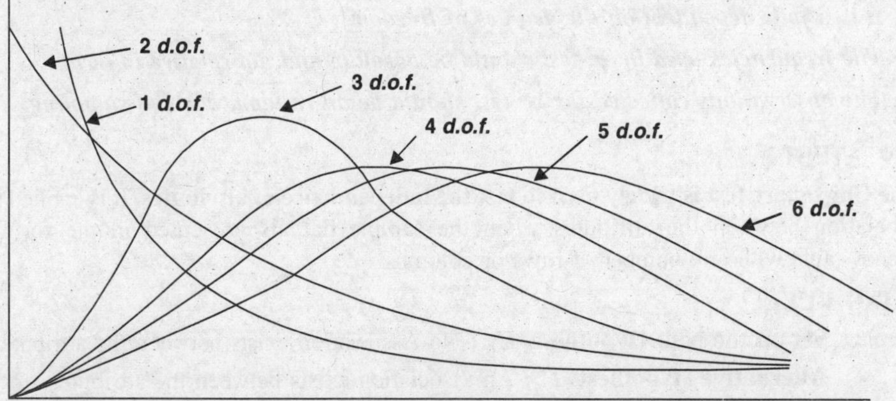

Fig. 10.1

8. ***Its shape depends on the degree of freedom but it is not a symmetrical distribution.***

10.5 USES OF ψ^2 – TEST

The ψ^2 test is a very powerful test for testing the hypothesis of a number of statistical problems. The important uses of ψ^2 test are:

1. Test of Goodness of Fit. If the two curves, *viz,* (*i*) **Observed frequency curve and** (*ii*) **the expected frequency curve, are drawn, then the Chi-square statistic may be used to determine whether the two curves so drawn are fitted good or not.** Thus, the term **goodness of fit is used to test the concordance of the fitness of these two curves.** Under this test there is only one variable, so the degrees of freedom is $v = n - 1$.

2. Test of Independence of Attributes. The Chi-square test is used to see that the principles of classification of attributes are independent. In this test the attributes are classified into a two way table or a contingency table as the case may be. The observed frequency in each cell (square) is known as **Cell frequency.** The total frequency in each row or column of the two way contingency table is known as **Marginal frequency.**

The degrees of freedom is $v = (R - 1)(C - 1)$,

where R = number of rows, C = number of columns in the two way contingency table. This test discloses whether there is any association or relationship between two or more attributes.

3. Test of Homogeneity or a Test for a Specified Standard deviation.

The **Chi-square test may be used to test the homogeneity of the attributes in respect of a particular characteristic or it may also be used to test the population variance.** In the case of a specified standard deviation the test statistic is given by $\psi^2 = (n-1)s^2 / \sigma_0^2$, where s^2 = sample variance and σ_0^2 is the hypothesized value of population variance.

10.6 CONDITIONS FOR APPLYING THE CHI–SQUARE TEST

1. ***Each of the observations making up the sample for this test should be independent of each other.***
2. ***The expected frequency of any item or cell should not be less than 5. If it is less than 5, then frequencies taking from the adjacent items or cells are pooled together in order to make it 5 or more than 5.***
3. ***The total number of observations used in this test must be large i.e., $n \geq 30$.***
4. ***This test is used only for drawing inferences by testing hypothesis. It cannot be used for estimation of parameter or any other value.***
5. ***It is wholly dependent on the degrees of freedom.***
6. ***The frequencies used in ψ^2-test should be absolute and not relative in terms.***
7. ***The observations collected for ψ^2-test should be on random basis of sampling.***

10.7 ψ^2 –TEST

The Chi-square test is widely used to test the independence of attributes. It is applied to test the association between the attributes when the sample data is presented in the form of a contingency table with any number of rows or columns.

WORKING RULE:

Step 1. Set up the Null Hypothesis H_0 : No Association exists between the attributes.

Alternative Hypothesis H_1: An association exists between the attributes.

Step 2. Calculate the expected frequency E corresponding to each cell by the formula

$$E_{ij} = \frac{R_i \times C_j}{n}$$

where R_i = Sum total of the row in which E_{ij} is lying

C_j = Sum total of the column in which E_{ij} is lying

n = Total sample size.

Step 3. Calculate ψ^2-statistic by the formula $\psi^2 = \sum \frac{(O-E)^2}{E}$.

The characteristics of this distribution are completely defined by the number of degrees of freedom v which is given by $v = (R-1)(C-1)$

where R = number of rows and C = number of columns in the contingency table.

Step 4. Find from the table the value of ψ^2 for a given value of the level of significance α and for the degrees of freedom v, **calculated in STEP 2**

If no value for α is mentioned, then take $\alpha = 0.05$.

Step 5. ***Compare the computed value of ψ^2, with the tabled value of ψ^2 found in step 4.***

(*a*) ***If calculated value of ψ^2 < tabulated value of ψ^2_α then accept null hypothesis H_0.***

(*b*) ***If calculated value of ψ^2 > tabulated value of ψ^2_α, then reject null hypotheses H_0 and accept the alternative hypothesis H_1.***

10.8 ψ^2–TEST FOR GOODNESS OF FIT

ψ^2-test is a measure of probabilities of association between the attributes. It gives us an idea about the divergence between the observed and expected frequencies. Thus the test is also described as the test of "**Goodness of Fit**". **If the curves of these two distributions, when superimposed do not coincide or appear to diverge much we say that the fit is poor. On the other hand if they don't diverge much, then the fit is less poor.**

This concept is illustrated by the following examples.

Example 1. *A survey of 320 families with 5 children each revealed the*

No. of boys:	*5*	*4*	*5*	*2*	*1*	*0*
No. of girls:	*0*	*1*	*2*	*3*	*4*	*5*
No. of families:	*14*	*56*	*110*	*88*	*40*	*12*

Is this result consistent with the hypothesis that the male and female births are equally probable ?

Solution. Null Hypothesis: H_0: Male and female births are equally probable.

Alternative Hypothesis: H_1: Male and female births are not equally probable.

2. Calculation of Test Statistic. On the basis of null hypothesis the expected frequencies are

given by $320\left(\frac{1}{2}+\frac{1}{2}\right)^5 = 320\left(\frac{1}{32},\frac{5}{32},\frac{10}{32},\frac{10}{32},\frac{5}{32},\frac{1}{32}\right)$ **= 10, 50, 100, 100, 50, 10.**

TABLE FOR ψ^2

O	E	$O-E$	$(O-E)^2$	$\frac{(O-E)^2}{E}$
14	10	4	16	1.60
56	50	6	36	1.72
110	100	10	100	1.00
88	100	– 12	144	1.44
40	50	10	100	2.00
12	10	2	4	0.40
			$\psi^2 = \Sigma \frac{(0-E)^2}{E} = 7.16$	

3. Level of Significance. Take $\alpha = 0.05$.

4. Critical Value. The table value of ψ^2 at $\alpha = 0.05$ for $(6-1) = 5$ degrees of freedom is $\psi^2_{\ 0.05} = $ **11.07.**

5. Decision. Since the calculated value of $\psi^2 = 7.16 <$ table value $\psi^2_{\ 0.05}$ for 5 d.f. = 11.07, so the null hypothesis is accepted $\Rightarrow$ the male and female births are equally probable.

Example 2. *The following table gives the number of aircraft accidents that occurred during the various days of the week. Find whether the accidents are uniformly distributed over the week.*

Days :	*sun.*	*Mon.*	*Tue.*	*Wed.*	*Thu.*	*Fri.*	*Sat.*
No. of Accidents :	*14*	*16*	*8*	*20*	*11*	*9*	*14*

[$\psi^2_{6,0.05} = 12.59$]

Solution. Null Hypothesis: H_0 : Accidents are uniformly distributed over the week.

Alterntive Hypothesis: H_1 : Accidents are not uniformly distributed oure the week.

Assuming uniformity, **expected frequency for each day** = $\frac{\textbf{total frequency}}{\textbf{No. of days}} = \frac{92}{7} = 13.14 = e_i$ for all value of i.

CALCULATION OF TEST STATISTIC

$\chi^2 = \Sigma \frac{(O_i - e_i)^2}{e_i}$, **where O_i = observed frequency and e_i = expected frequency.**

$$\chi^2 = \left[\frac{(14-13.14)^2}{13.14} + \frac{(16-13.14)^2}{13.14} + \frac{(8-13.14)^2}{13.14} + \frac{(20-13.14)^2}{13.14} + \frac{(11-13.14)^2}{13.14} + \frac{(9-13.14)^2}{13.14} + \frac{(14-13.14)^2}{13.14}\right]$$

$$= \frac{0.7396}{13.14} + \frac{8.1796}{13.14} + \frac{26.4196}{13.14} + \frac{47.0596}{13.14} + \frac{4.5796}{13.14} + \frac{17.1396}{13.14} + \frac{0.7396}{13.14} = \frac{104.8572}{13.14}$$

$$= \textbf{7.98.}$$

Decision: The calculated value of χ^2 (which is 7.98) is less than the given tabulated value (which 12.59), so we accept the assumption of uniformity, *i.e.*, number of accidents per day is uniform.

Example 3. *The following table gives the number of screws declared fit and also unfit by three inspectors X, Y and Z. Test the hypothesis that the proportion of screws declared unfit by the three inspectors are same:*

Inspector	*Number of Screws*			
	X	*Y*	*Z*	*Total*
Fit Screws	*50*	*47*	*56*	*153*
Unfit Screws	*5*	*14*	*8*	*27*
Total	*55*	*61*	*64*	*180*

(Table value $\chi^2_{1d.f.} = 3.84, \chi^2_{2d.f.} = 5.99$)

Solution. Let the Null hypothesis be that the **proportion of screws declared unfit by the three inspectors are same.**

Alternative hypothesis becomes that the proportion of screws declared unfit by the three inspectors are not same.

Expected Values = (Row Total × Column Total) ÷ Grand Total.

The expected frequencies are given in parenthesis.

Inspector	*Number of Screws*						
		X		*Y*		*Z*	*Total*
Fit Screws	50	**(46.75)**	47	**(51.85)**	56	**(54.40)**	153
Unfit Screws	5	**(8.25)**	14	**(9.15)**	8	**(9.60)**	27
Total	**55**		**61**		**64**		**180**

$$\chi^2 = \sum \frac{(O-E)^2}{E} = \frac{(50-46.75)^2}{46.75} + \frac{(47-51.85)^2}{51.85} + \frac{(56-54.40)^2}{54.4}$$

$$+ \frac{(5-8.25)^2}{8.25} + \frac{(14-9.15)^2}{9.15} + \frac{(8-9.60)^2}{9.60} = \mathbf{4.84},$$

which is less than the given value of $\chi^2 = 5.99$ for 2 degree of freedom, therefore, we accept the hypothesis and conclude that the proportions are statistically same for all inspectors.

Example 4. In *60 throws of a dice, face one turned up 6 times, face two or three 18 times, face four or five 24 times and face six 12 times. Test at 10% significance level, if the dice is honest, it being given that $P(\psi^2 > 6.25) = 0.1$ for 3 degrees of freedom.*

Solution 1. Null Hypathesis: H_0: The dice is honest.

Alternative Hypothesis: H_1: The dice is not honest.

2. Computation of Test Statistic. Under the assumption of null hypothesis that the dice is honest, the expected frequency for each face is $60 \times \frac{1}{6} = 10$ [$\because$ Probability of turning up any one of the numbers 1, 2, 3, 4, 5, 6, is $\frac{1}{6}$].

We prepare the following table.

COMPUTATION TABLE FOR ψ^2

Face of the die	*Observed frequency O*	*Epected frequency E*		*O – E*	$\frac{(O-E)^2}{E}$
1	6	10		– 4	1.6
2, 3	18	10, 10	20	– 2	0.2
4, 5	24	10, 10	20	4	0.8
6	12	10	10	2	0.4
Total	**60**	**60**		$\psi^2 = \Sigma \frac{(O-E)^2}{E}$	**= 3.0**

$\therefore \qquad \psi^2 = 3.0$

3. **Critical Value.** The table value of ψ^2 at $\alpha = 0.1$ for $4 - 1 = 3$ degrees of freedom is **6.25.**
4. **Decision. The calculated value of $\psi^2 = 3.0 < \psi^2_{0.1}$ for 3 d.f. = 6.25. Therefore, the Null hypothesis that the die is honest is accepted.**

Example 5. *In a sample survey of public opinion, answers to the questions*

(*i*) *Do you drink* ?

Question (1)

	Yes	*No*	*Total*
Yes	*56*	*31*	*87*
No	*18*	*6*	*24*
Total	*74*	*37*	*111*

(*ii*) *Are you in favour of local option on sale of liquor* ? *Are tabulated below:*

Can you infer whether or not the local option on the sale of liquor is dependent on individual drink.

Solution. Null Hypothesis H_0: The option on the sale of liquor is independent or not associated with individual drinking.

The theoretical frequencies are tabulated as below

Question (1)

	Yes	***No***
Yes	$\frac{87 \times 74}{111} = \mathbf{58}$	$\frac{37 \times 87}{111} = \mathbf{29}$
No	$\frac{74 \times 24}{111} = \mathbf{16}$	$\frac{24 \times 37}{111} = \mathbf{8}$
Total	**74**	**37**

Computation of Test Statistic

Applying the formula

$$\psi^2 = \Sigma \frac{(O-E)^2}{E}, \text{ we get}$$

$$\psi^2 = \frac{(56-58)^2}{58} + \frac{(18-16)^2}{16} + \frac{(31-29)^2}{29} + \frac{(6-8)^2}{8}$$

$$= \frac{4}{58} + \frac{4}{16} + \frac{4}{29} + \frac{4}{8} = \frac{111}{116} = \mathbf{0.957.}$$

3. **Degree of Freedom $v = (R-1)(C-1)$** $= (2-1)(2-1) =$ **1**.
4. **Decision.** The tabulated value of ψ^2 at $\alpha = 0.05$ and one degree of freedom is $\psi^2_{.05,1}$ = **3.411**

Since, **Calculated $\psi^2 = 0.957$ < Tabulated value $\psi_{0.05,1} = 3.411 \Rightarrow$ the Null Hypothesis H_0 is accepted $\Rightarrow$ Sale of liquor is independent or not associated with the individual drinking.**

Example 6. *The following figures show the distribution of digits in number chosen at random from a telephone directory:*

Digits:	*0*	*1*	*2*	*3*	*4*	*5*	*6*	*7*	*8*	*9*
Frequency:	*1026*	*1107*	*997*	*966*	*1075*	*933*	*1107*	*972*	*964*	*853*

Test at 5% level whether the digits may be taken to occur equally frequently in the directory. (The table value of ψ^2 for 9 degrees of freedom = 16.919).

Solution.

1. **Null Hypothesis:** H_0 : **The digits occur equally frequently in the directory.**
   Alternative Hypothesis: H_1 : **The digits do not occur equally frequently.**
2. **Calculation of Test Statistic.** Here the total number of frequencies n = 1026 + 1107 + 997 + 966 + 1075 + 933 + 1107 + 972 + 964 + 853 = **10,000.**

$\therefore$ Expected frequency for each of the digits 0, 1, 2, 3, ..., 9 is $E = \frac{10,000}{10}$ = **1000.** Thus the value of ψ^2 is

$$\psi^2 = \frac{(1026-1000)^2}{1000} + \frac{(1107-1000)^2}{1000} + \frac{(997-1000)^2}{1000} + \frac{(966-1000)^2}{1000} + \frac{(1075-1000)^2}{1000}$$

$$+ \frac{(933-1000)^2}{1000} + \frac{(1107-1000)^2}{1000} + \frac{(972-1000)^2}{1000} + \frac{(964-1000)^2}{1000} + \frac{(853-1000)^2}{1000}$$

$$= 0.676 + 11.449 + 0.009 + 1.156 + 5.625 + 4.489 + 11.449 + 0.784 + 1.296 + 21.609$$

= **58.542.**

3. **Critical Value.** The tabulated value or critical value of ψ^2 at $\alpha = 0.05$ for $10 - 1 = 9$ degrees of freedom is $\psi^2_{0.05,9}$ = **16.919**.

Since, **calculated value of $\psi^2 = 58.542$ is > Tabulated $\psi^2 = 16.919$, so the null hypothesis H_0 is rejected and the alternative hypothesis H_1 is accepted $\Rightarrow$ the digits do not occur equally frequently.**

Example 7. *A dice is tossed 120 times with the following results :*

Number turned up:	*1*	*2*	*3*	*4*	*5*	*6*	*Total*
Frequency:	*30*	*25*	*18*	*10*	*22*	*15*	*120*

Test the hypothesis that the dice is unbiased.

Solution. 1. Null Hypothesis H_0: The dice is unbiased one.

Alternative Hypothesis H_1: The dice is a biased one.

2. Calculation of Test Statistic. On the hypothesis that the dice is unbiased, the expected frequency is $120 \times \frac{1}{6} = 20$. We calculate ψ^2 from the following table.

0	*E*	*O – E*	*(O – E)²*	$\frac{(O-E)^2}{E}$
30	20	10	100	5.00
25	20	5	25	1.25
15	20	– 2	4	0.20
10	20	– 10	100	5.00
22	20	2	4	.20
15	20	– 5	25	1.25
$\psi^2 = \Sigma \frac{(O-E)^2}{E} = 12.90$				

3. **Critical Value.** The table value of ψ^2 at $\alpha = 0.05$ and for $6 - 1 = 5$ degrees of freedom is $\psi^2_{0.05,5}$ **= 11.070.**

4. **Decision.** Since the calculated value of $\psi^2 = 12.90$ is $> \psi^2_{0.05,5}$ **= 11.07, so the null hypothesis H_0 is rejected and the alternative hypothesis H_1 is accepted ⇒ the dice is a biased one.**

10.9. Ψ² –TEST AS A TEST OF INDEPENDENCE

ψ^2 -test can also be applied to test, ***the independence between various attributes when the sample data is presented in the form of a contingency table with any number of rows 'r' and columns 'c' the null hypothesis and alternative hypothesis are set as follows:***

Null Hypothesis H_0 : The attributes are independent.

Alternative Hypothesis H_1 : The attributes are not independent.

We then calculate ψ^2 as per the produce exparied larlier. If the calculated value of ψ^2 is less than the tabled value at a given level of significance, the hypothesis is accepted and vice-versa.

Example 8. *50 students selected at random from 500 students enrolled in a computer crash programme were classified according to age and grade points giving the following data.*

Grade Points	*Age (in years)*		
	20 and under	*21 – 30*	*Above 30*
Upto 5.0	*3*	*5*	*2*
5.1 to 7.5	*8*	*7*	*5*
7.6 to 10.0	*4*	*8*	*8*

Test at 5% level of significance the hypothesis that age and grade point are independent.

Solution. Null Hypothesis H_0: Age and Grade points are independent

Alternative Hypothesis H_1: Age and grade points are not independent.

On the basis of Null hypothesis H_0, the expected frequencies obtained are :

Grade Points	*20 and under*	*21 – 30*	*Above 30*	*Total*
Upto 5	3	4	3	10
5.1 to 7.5	6	8	6	20
7.6 to 10.0	6	8	6	20
Total	**15**	**20**	**15**	**50**

Since the values which are less than 5 are occuring in some cells of the expected frequencies, so we have to amalgamate these cells to its neighbours. After amalgamation, the new frequencies are :

Expected Frequencies (Age in years)

Grade Points	*20 and under*	*21 – 30*	*Above 30*	*Total*
Upto 7.5	9	12	9	30
7.6 to 10	6	8	6	20
Total	15	20	15	50

OBSERVED FREQUENCIES (Age in years)

Grade Points	*20 and under*	*21 – 30*	*Above 30*	*Total*
Upto 7.5	11	12	7	30
7.6 to 10	4	8	8	20
Total	15	20	15	50

Calculation of Test Statistic.

Here: $$\psi^2 = \frac{(11-9)^2}{9} + \frac{(12-12)^2}{12} + \frac{(7-9)^2}{9} + \frac{(4-6)^2}{6} + \frac{(8-8)^2}{8} + \frac{(8-6)^2}{8} = 2.222.$$

Decision. The table value of ψ^2 at a = 0.05 for 1 × 2 = 2 degrees of freedom is $\psi^2_{0.05} = 5.991$. Since the computed value of $\psi^2 = 2.222$ is less than the table value $\psi^2_{0.05}$ for 2 d.f. = 5.991, so we accept the Null Hypothesis $H_0 \Rightarrow$ that the grade points and age are independent of each other.

Example 9. *One thousand girls in a college were graded according to their I.Q. and the economic conditions of their homes. Use χ^2-test to find out whether there is any association between economic conditions at home and I.Q. of girls :*

	I.Q.		
Economic Condition	*High*	*Low*	*Total*
Rich	*100*	*300*	*400*
Poor	*350*	*250*	*600*
Total	*450*	*550*	*1,000*

Given for v = 1, $\chi^2_{0.05} = 3.84$

Solution. Null Hypothesis: H_0: **There is no association between Economic Condition at home and I.Q.**

Alternative Hypothesis: H_1: **The exists an association between economic conditions at home and I.Q.**

Calculation of test Statistic. Using the formula $E = \frac{\textbf{row total} \times \textbf{column total}}{\textbf{Grand total}}$, the expected frequencies are calculated in the table below :

Observed value O	*Expected value* E	$(O-E)^2$	$\frac{(O-E)^2}{E}$
100	**180**	6400	35.5
350	**270**	6400	23.7
300	**220**	6400	29.1
250	**330**	6400	19.4
			$\sum\left[\frac{(O-E)^2}{E}\right] = \mathbf{107.7}$

$$\chi^2 = \sum\left[\frac{(O-E)^2}{E}\right] = \mathbf{107.7}$$

Degrees of freedom: $\nu = (\mathbf{r} - \mathbf{1})\ (\mathbf{C} - \mathbf{1}) = (2-1)\ (2-1) = \mathbf{1}$

$\nu = 1,\ \chi^2_{0.05} = \mathbf{3.84}$

Decision: **The calculated value of x^2 is greater than the table value. The hypothesis is rejected. Hence there is association between Economic Condition at home and I.Q.**

Example 10. *In an industry, 200 workers, employed for a specific job, were classified according to their performance and training received/not received to test independence of a specific training and performance. The data is summarised as follows :*

	Performance		*Total*
	Good	*Not Good*	
Trained	*100*	*50*	*150*
Untrained	*20*	*30*	*50*
Total	*120*	*80*	*200*

Use χ^2 test of independence at 5% level of significance and write your conclusion.
[Data from χ^2-table; χ^2 (1 d.f. 5%) = 3.84]

Solution. We have the following table for expected values.

	Performance		***Total***
	Good	***Not Good***	
Trained	$\frac{150 \times 120}{200} = \mathbf{90}$	$\frac{150 \times 80}{200} = \mathbf{10}$	**150**
Untrained	$\frac{120 \times 50}{200} = \mathbf{30}$	$\frac{80 \times 50}{200} = \mathbf{20}$	**50**
Total	**120**	**80**	**200**

Null Hypotheses H_0: No association exist between performance and training of workers.

Alternate Hypothesis H_1: An association exist between performance and training of workers.

Now $$\chi^2 = \sum \frac{(O-E)^2}{E}$$

$$= \frac{(100-90)^2}{90} + \frac{(50-60)^2}{60} + \frac{(20-30)^2}{30} + \frac{(30-20)^2}{20}$$

$$= \frac{100}{90} + \frac{100}{60} + \frac{100}{30} + \frac{100}{20} = \frac{2000}{180} = \mathbf{11.11.}$$

Degree of freedom: $v = (r-1)(c-1) = (2-1)(2-1) = 1$.

Decision: Since computed value of χ^2 (= 11.11) at 1 d.f. is greater than critical value of χ^2 at 5% significance level and 1 d.f., *i.e.*, 3.84 so we reject null hypothesis and accept the alternative hypothesis. Hence an association exist between performance and training workers.

Example 11. *A sample of 400 students of undergraduate and 400 students of postgraduate classes were taken to know their opinion about autonomous colleges. 290 of the undergraduate and 310 of the postgraduate students favoured the autonomous status. Present these facts in the form of a table and test at 5% level, that the opinion regarding autonomous status of colleges are independent of the level of classes of students (table value of ψ^2 at 5% level is 3.84 for 1 d.f.).*

Solution. The contingency table of observet frequencies

	Favour	*Not in Favour*	*Total*
Undergraduate	290	110	400
Postgraduate	310	90	400
Total	**600**	**200**	**800**

Table of theoretical or expected frequencies

	Favour	*Not in Favour*
Undergraduate	$\frac{600 \times 400}{800} = \mathbf{300}$	$\frac{600 \times 400}{800} = \mathbf{300}$
Postgraduate	$\frac{200 \times 400}{800} = \mathbf{100}$	$\frac{200 \times 400}{800} = \mathbf{100}$

Null Hypothesis H_0: Option regarding an autonomous status are independent.

$$\therefore \quad \psi^2 = \sum \frac{(O-E)^2}{E} = \frac{(290-300)^2}{300} + \frac{(310-100)^2}{100}$$

$$+ \frac{(110-300)^2}{300} + \frac{(90-100)^2}{100} = \frac{100}{300} + \frac{210 \times 210}{100} + \frac{210 \times 210}{300} + \frac{100}{100}$$

$$= \frac{1}{3} + 441 + \frac{441}{3} + 1 = \mathbf{556.}$$

$$d.f. = (2-1) \cdot (2-1) = \mathbf{1.}$$

Table value of ψ^2 at $\alpha = 0.05$ for 1 d.f. is **3.84.**

Now **calculated $\psi^2 = 556 > \psi^2_{0.05} = 3.84 \Rightarrow$ Null Hypothesis H_0 is rejected.**

Example 12. *In a survey of 200 boys of which 75 were intelligent, 40 had educated fathers. While 85 of the unintelligent boys had uneducated fathers. Do these figures support the hypothesis that educated fathers have intelligent boys ?* (*Value of χ^2 for 1 df is 3.841*)

Solution. Let the **null hypothesis be that the education of fathers has no effect on the intellegence of boys.**

CONTINGENCY TABLE

	Intelligent boys	*Unintelligent boys*	*Total*
Educated fathers	40 (**30**)	40 (**50**)	80
Un-educated fathers	35 (**45**)	85 (**75**)	120
Total	**75**	**125**	**200**

Expected frequencies are given in brackets and are calculated by multiplying row total by column total and then dividing it by gross total

$$X^2 = \Sigma \frac{(O-E)^2}{E} = \frac{(40-30)^2}{30} + \frac{(40-50)^2}{50} + \frac{(35-45)^2}{45} + \frac{(85-75)^2}{75} = \mathbf{8.88.}$$

Since the calculated value is higher than the tabular value of 3.841, so we reject the null hypothesis $\Rightarrow$ that the education of fathers has an effect on the intelligence of the boys.

Example 13. *A sample analysis of examination results of 500 students was made. It was found that 180 students has failed, 170 had secured a third class, 110 were placed in second class and 40 got a first class. Are these figures commensurate with the general examination result which is in the raio of 4 : 3 : 2 : 1 for the various categories respectively ?* (*Answer at $\alpha = 0.05$*)*. Table value of chi-square at $\alpha = 0.05$ for 2d.f, 3d.f, and 4.d.f are 5.99, 7.81 and 9.49 respectively.*

Solution. Null Hypothesis, H_0 : That the given data is commensurate with the general examination result.

Calculations for Chi-Square

Category	*Observed Freq. (O)*	*Expected Freq. (E)*	*(O – E)$_2$*	$\frac{(O-E)^2}{E}$
Failed	180	$\frac{4}{10} \times 500 = \mathbf{200}$	400	2.000
Third class	170	$\frac{3}{10} \times 500 = \mathbf{150}$	400	2.6667
Second class	110	$\frac{2}{10} \times 500 = \mathbf{100}$	100	1.000
First class	40	$\frac{1}{10} \times 500 = \mathbf{50}$	100	2.000
Total	**500**			**7.6667**

$$\therefore \quad \psi^2 = \sum \frac{(O-E)^2}{E} = 7.6667$$

$$\text{d.f.} = 4 - 1 = 3. \ \chi^2_{0.05} \text{ for 3d.f.} = \mathbf{7.81}.$$

Since the calculated value of χ^2 is less than the tabulated value of χ^2 at 5% level of significance, so it is not significant. **We accept our hypothesis the given data is commensurate with the general examination result.**

Example 14. *From the table given below, whether the colour of son's eyes is associated with that of father's eyes.*

EYES COLOUR IN SONS

		Not light	*Light*
Eye colour	*Not light*	*230*	*148*
in fathers	*Light*	*151*	*471*

Solution. The observed frequencies are given by the following Table.

EYES COLOUR IN SONS

		Not light	*Light*	*Total*
Eye colour in fathers	Not light	230	148	378
	Light	151	471	622
	Total	381	619	1000

The theoretical frequencies or Expected frequencies are given by the table

EYES COLOUR IN SONS

		Not light	*Light*
Eye colour in	Not light	$\frac{381 \times 378}{1000} = 144$	$\frac{619 \times 378}{1000} = 234$
Fathers	Light	$\frac{381 \times 622}{1000} = 237$	$\frac{619 \times 622}{1000} = 385.$

1. **Null Hypothesis H_0: The colour of the son's eyes is not associated with the colour of father's eyes.**
2. **Calculation of test Statistic.**

$$\psi^2 = \frac{(230-144)^2}{144} + \frac{(151-237)^2}{237} + \frac{(148-234)^2}{234} + \frac{(471-385)^2}{385}$$

$$= \frac{7396}{144} + \frac{7396}{237} + \frac{7396}{234} + \frac{7396}{385} = 51.36 + 31.21 + 31.61 + 19.21 = \mathbf{133.39.}$$

3. **Degress of Freedom.** $v = (R-1)(C-1) = (2-1)(2-1) = \mathbf{1}$.
4. **Decision. The tabulated value of ψ^2 for $\alpha = 0.05$ and for 1 degree of freedom is $\psi^2_{0.05,1} = 3.84$.**

Now calculated $\psi^2 = 133.39 > \psi^2_{0.05,1} = 3.84 \Rightarrow$ the Null hypothesis is rejected $\Rightarrow$ there is an association between the colours of eyes of son's and colours of eyes of father's.

Example 15. *Calculate the expected frequencies for the following data presuming the two attributes, viz., conditions of home and condition of child as independent*

		Condition of hence	
		Clean	*Dirty*
Condition of child	*Clean*	*70*	*50*
	Fairly clean	*80*	*20*
	Dirty	*35*	*45*

Use chi-square test at 5% level of significance to state whether the two attributes are independent.

(*Table value of* ψ^2 *at 5% for d.o. f is 5.991 and for 3.d.o.f. is 7.815 and for 4.d.o.f. is 9.488*)

Solution. The expected frequency E, corresponding to each cell in the table is given by

$$E = \frac{R \times C}{n}$$

where R = *a* row total, C = *a* column total and n = the sample size.

We form the table of expected frequencies, with the help of the above rule and write the observed frequencies in each cell within brackets.

	Condition of Home		***Total***
	Clean	***Dirty***	
Clean	**(70)** $\frac{185 \times 120}{300} = \mathbf{74}$	**(50)** $\frac{115 \times 120}{300} = \mathbf{46}$	**120**
Fairly clean	**(80)** $\frac{185 \times 100}{300} = \mathbf{61.67}$	**(20)** $\frac{115 \times 100}{300} = \mathbf{38.33}$	**100**
Dirty	**(35)** $\frac{185 \times 80}{300} = \mathbf{49.33}$	**(45)** $\frac{115 \times 80}{300} = \mathbf{30.67}$	**80**
Total	**185**	**115**	**300**

1. Null Hypothesis H_0: There does not exist any association between the attributes.

2. Alternative Hypothesis H_1: An Association exists between the attributes.

3. Calculation of Test Statistics

$$\psi^2 = \frac{\Sigma(O-E)^2}{E} = \left[\frac{(70-74)^2}{74} + \frac{(50-46)^2}{46} + \frac{(80-61.67)^2}{61.67} + \frac{(20-38.33)^2}{38.33} + \frac{(35-49.33)^2}{49.33} + \frac{(45-30.67)^2}{30.67}\right]$$

$$= 0.2162 + 0.3478 + 5.4482 + 8.7657 + 4.1627 + 6.6954 = \mathbf{25.636.}$$

Degrees of Freedom: $v = (R-1)(C-1) = 2 \times 1 = 2.$

Decision. The table value of ψ^2 at $\alpha = 0.05$ for 2 degrees of freedom is $\psi^2_{0.05,2} = \mathbf{5.991}$

Also the calculated $\psi^2 = 25.636 > \psi^2_{0.05}$ for 2.*d.f.* = 5.991 (from the tables)

⇒ the **Null Hypothesis is rejected ⇒ Alternative Hypothesis H_1 is accepted from which we conclude that there exists an association between the attributes.**

10.10. YATES CORRECTION FOR CONTINUITY

Yates correction for continuity in a 2 × 2 table is applied when the number of degrees of freedom is 1 (*i.e.*, $v = 1$) and also any cell frequency in 2 × 2 table is less than 5. In such case ψ^2-test is applied by pooling the frequencies with the proceeding or succeeding so that the total frequency of the cell is greater than 5. This results in the loss of **one degree of freedom**. We know that for a 2 × 2 table degrees of **freedom = (2 – 1) × (2 – 1) = 1** (*i.e. d..f.* = 1). Since there is a loss of one degree of freedom in pooling, so the degrees of freedom left after adjusting for pooling are $v = 1 - 1 = \mathbf{0}$, which is absurd. [∵ ψ^2 must have at least 1 d.f.]. Under such circumstances, *i.e.*, when any cell frequency in a 2 × 2 contingency table is less than 5, we apply the correction due to **F. Yates,** which is popularly known as **Yates correction for continuity.** This consists in adding 0.5 to the cell frequency which is less than 5 and adjusting the remaining frequencies accordingly so that the sum total of each row and column in 2 × 2 contingency table remains the same. It is important to note that the value of ψ^2 is not affected much by **Yates Correction** if the grand total, N is sufficiently large.

Example 16. *In an experiment on immunization of pigs from Anthrax, the following results were obtained. Derive your inference on the efficacy of the vaccine.*

	Died of Anthrax	*Survived*	*Total*
Inoculated with vaccine:	*2*	*10*	*12*
Not inoculated	*6*	*6*	*12*
Total	*8*	*16*	

[*Table value ψ^2 at 5% for 1 d.f. is 3.841*]

Solution. The Expected frequency E, corresponding to each cell in the table is given by

$$E = \frac{R \times C}{n}, \text{ where R = a row total, c = column total and n = sample size.}$$

$$\therefore\ E(2) = \frac{8 \times 12}{24} = 4,\ E(10) = \frac{12 \times 16}{24} = 8,\ E(6) = \frac{8 \times 12}{24} = 4\ \ E(6) = \frac{16 \times 12}{24} = 8.$$

In the first cell, the cell frequency is 2, which is less than 5, so we apply **Yates correction for Continuity** for computing the value of ψ^2. It consists of adding 0.5 to the cell frequency which is less than 5 and adjusting the remaining frequencies as shown by the Table 2.

Table 1 : Expected Frequency (E)

4	8	12
4	8	12
8	16	24

Table 2 : Observed Frequencies after Yates correction

2.5	9.5	–12
5.5	6.5	12
8	16	24

Null Hypothesis H_0 : *'Inoculation with vacine' and 'survival from Anthrax' are independent.*

Computation of Test Statistics

$$\psi^2 = \frac{\Sigma(O-E)^2}{E} = \frac{(2.5-4)^2}{4} + \frac{(9.5-8)^2}{8} + \frac{(5.5-4)^2}{4} + \frac{(6.5-8)^2}{8}$$

$$= \frac{2.25}{4} + \frac{2.25}{8} + \frac{2.25}{4} + \frac{2.25}{8}$$

$$= 0.56250 + 0.28125 + 0.56250 + 0.28125$$

$$= \mathbf{1.68750}$$

Degrees of Freedom : $v = (R-1) \times (C-1) = 1 \times 1 = 1$.

Critical Value The table value of ψ^2 for 1 d.f. at $\alpha = 0.05$ is $\boldsymbol{\psi^2 = 3.841}$.

Decision : Since the calculated value of ψ^2 (= 1.68750) is less than the table value of $\Psi^2_{0.05}$ (=3.841), so *the null hypothesis is accepted.* ⇒ *'Inoculation with vacine'* and *'survival from Anthrax' are independent.*

10.11. TESTING A HYPOTHESIS ABOUT THE VARIANCE OF A NORMALLY DISTRIBUTED POPULATION

The decision rule for **One tailed test** as well as **Two tailed test** is stated below.

The one tailed test with null hypothesis $H_0 : \sigma^2 \le \sigma_0^2$ is similar to the other one tailed tests mentioned above with the exception that the one tailed rejection region appears in the lower tail at a critical value of $\psi^2_{(1-\alpha)}$. Also σ_0^2. = hypothesied value for population variance.

One Tailed Test		*Two Tailed Test*	
Hypothesis	**Decision Rule**	**Hypothesis**	**Decision Rule**
1. H_0: $\sigma^2 \le \sigma_0^2$	Accept H_0 if computed $\psi^2 <$ Table ψ^2_α	**1. H_0:** $\sigma^2 = \sigma_0^2$	Accept H_0 if $\psi^2_{(1-\alpha/2)} \le$ computed $\psi^2 \le \psi^2_{\alpha/2}$
2. H_i: $\sigma^2 < \sigma_0^2$	Reject H_0 if Computed $\psi^2 >$ Tabled ψ^2_α	**2. H_i:** $\sigma^2 \ne \sigma_0^2$	Reject H_i if computed $\psi^2 > \psi^2_{a/2}$ or computed $\psi^2 < \psi^2_{(1-\alpha/2)}$.

Example 17. *Heights in cms of 10 students are given below :*

60, 65, 62, 66, 68, 70, 65, 64, 68, 72

Can we say that variance of the distribution of heights of all students from which the above sample of 10 students was drawn is equal to 50.

Solution. Mean of the given sample is

$$\bar{x} = \frac{60+65+62+66+68+70+65+64+68+72}{10} = \frac{660}{10} = \mathbf{66.}$$

CALCULATION TABLE FOR s^2

x_i	$x_i - \bar{x} = (x_i - 66)$	$(x_i - \bar{x})^2 = (\bar{x}_i - 66)^2$
60	– 6	36
65	– 1	1
62	– 4	16
66	0	0
68	2	4
70	4	16
65	– 1	1
64	– 2	4
68	2	4
72	6	36
$\Sigma x_i = 660$		$\Sigma (x_i - \bar{x})^2 = 116$

$\therefore$ **Sample variance** $= s^2 = \dfrac{\Sigma(x_i - \bar{x})^2}{n-1} = \dfrac{116}{9}$

Null Hypothesis: $H_0 : \sigma^2 = \sigma_0^2 = 50.$

Alternative Hypothesis: $H_1 : \sigma^2 \neq 50$ **(Two Tailed test)**

Calculation of Test Statistic: The test statistic is given by $\psi^2 = \dfrac{(n-1)s^2}{\sigma_0^2}$

and the population height is assumed to be approximately normally distributed.

$$\therefore \quad \psi^2 = \frac{(n-1)s^2}{\sigma_0^2} = \frac{(n-1)}{\sigma_0^2} \times \frac{\Sigma(x_i - \bar{x})^2}{n-1} = \frac{\Sigma(x_i - \bar{x})^2}{\sigma_0^2} = \frac{116}{50} = \mathbf{2.32.}$$

Level of Significance. Let the level of significance be $\boldsymbol{\alpha} = \mathbf{0.05}$.

Statement of Decision Rule. At 5% level of significance and 10 – 1 = 9 d.f. $\psi^2_{0.025} = \mathbf{19.02}$ and $\psi^2_{0.957} = \mathbf{2.70}$.

Now reject H_0 if computed $\psi^2 > \psi^2_{\sigma/2}$ or computed $\psi^2 > \psi^2_{(1-\alpha/2)}$

Making a statistical decision. Here computed $\psi^2 = 2.32 < \psi^2_{0.975} = 2.70$, **so the null hypothesis is accepted and we conclude that the population variance may be 50.**

Example 18. *Specification for the manufacture of a particular type of ornament state that the variance in weight shall not exceed 0.015 mgm squared. A random sample of 15 such ornaments yields a variance of 0.027. Can we say that the specifications are not being met ? Use* $\alpha = 0.05$.

Solution. Null Hypothesis: H_0: $\sigma^2 = \sigma_0^2 \leq 0.015$.

Alternative Hypothesis: H_1: $\sigma^2 > 0.015$. **(One Tailed Test)**

Calculation of Test Statistic.

$$\textbf{Test statistic: } \psi^2 = \frac{(n-1)\,s^2}{\sigma_0^2} = \frac{(15-1)\,0.027}{0.015} = \mathbf{25.20.}$$

Level of significance. Let the level of significance be $\alpha = 0.05$.

Critical value. The table value of ψ^2 for 19 d.f. = 23.685 at $\alpha = 0.05$.

Decision. Since computed value $\psi^2 = 25.20 >$ Table value of $\psi^2_{0.05} = 23.685$ so the null hypothesis H_0 is rejected $\Rightarrow$ the data indicate that the specifications are not being met.

10.11 ψ^2 –DISTRIBUTION OF SAMPLE VARIANCE

Let σ^2, s^2 respectively be the population variance and sample variance. The sampling distribution $(n-1)\,s^2/\sigma^2$ has a **chi-square (ψ^2) distribution** with $n-1$ degrees of freedom. It is a very useful distribution in making inferences about the population variance σ^2 by using sample variance s^2. **It is used in making interval estimate of the population variance**, which is given by the result.

$$\frac{(n-1)\,s^2}{\psi^2_{\alpha/2}} \le \alpha < \frac{(n-1)s^2}{\psi^2_{(1-\alpha/2)}},$$

where α is the level of significance and ψ^2_α = value for the chi-square distribution giving an area to the right of stated ψ^2_α and $(1-\alpha)$ confidence coefficient.

It can also be used as hypothesis test for the value of the population variance.

Example 19. *A market analyst took a sample of 25 markets in Delhi in an attempt to determine how much variation is there in the butter prices. The 25 prices that were quoted to him for four samples of butter yielded the sample value $\bar{x} = 90$ and $s = 7$. The problem now is to find a 95% confidence interval for the standard deviation of all the markets.*

Solution. Here $n = 25$, $s = 7 \Rightarrow s^2 = 49$.

Now $\alpha = 0.05$, so $\psi^2_{\alpha/2}$ for $(n-1)$ d.f. is $\psi^2_{0.025}$ for $(25-1) = 24$ d.f. = 39.36 and

$\psi^2_{.975}$ for 24 d.f. = 12.40 (from ψ^2 – table)

$\therefore$ Required confidence interval is

$$\frac{(n-1)s^2}{\psi_{\alpha/2}} \le \sigma^2 \le \frac{(n-1)s^2}{\psi^2_{(1-\alpha/2)}} \Rightarrow \frac{24\times 49}{39.36} \le \sigma^2 \le \frac{24\times 49}{12.40}$$

$$\Rightarrow \frac{1176}{39.36} \le \sigma^2 \le \frac{1176}{12.40} \Rightarrow \mathbf{28.88 \le \sigma^2 \le 94.84.}$$

Example 20. *A sample of 30 light tubes yielded a standard deviation of 90 hours burning time whereas the long experience with the particular brand showed standard deviation of 105 hours. Using $\alpha = 0.05$, test if there is any difference in the standard deviation.*

Solution. Null Hypothesis: H_0: $\sigma_0 = 105$, *i.e.*, there is no difference in standard deviation.

Alternative Hypothesis: H_1: $\sigma_0 \ne 105$.

Degrees of freedom: $v = 30 - 1 =$ **29.**

$$\textbf{Test Statistic: } \psi^2 = \frac{(n-1)\,s^2}{\sigma_0^2} = \frac{29\times(90)^2}{(105)^2} = \frac{234900}{11025} = \mathbf{21.31.}$$

Critical Value. The table value of $\psi^2_{0.025}$ for 29 d.f. = 45.72 and $\psi^2_{0.975}$ for 29 d.f. = 16.05.

Decision. The calculated value of ψ^2 = 21.31 is less than tabled value $\psi^2_{0.025}$ for 29 d.f. = 45.72, **so the null hypothesis is accepted**.

EXERCISE – 10.1

1. A smaple fo 300 students of Undergraduate and 300 students of Postgraduate classes of a University were asked to give their opinion toward the autonomous colleges. 190 of the Under-Graduate and 210 of the Postgraduate students favoured the autonomous status.

Present the above data in the form of a frequency table and test at 5% levels, the opinions of Under-Graduate and Postgraduate students on autonomous status of colleges are independent. (Table value of ψ^2 at 5% level for 1 d.f. is 3.84)

[**Hint.** The contingency table of observed frequencies in which expected frequencies are shown in braces is

Undergraduate:	Favour	Not in Favour	Total
	190 (**200**)	110 (**100**)	**300**
Postgraduate:	210 (**200**)	90 (**100**)	**300**
	400	**200**	**600**

Null Hypothesis H_0: Opinion on autonomous status and level of Graduation are independent.

$$\psi^2 = \frac{100}{200} + \frac{100}{200} + \frac{100}{100} + \frac{100}{100} = 3.$$

d.f. = (2 – 1) (2 – 1) = **1.**

Table value of ψ^2 at α = 0.05 and for 1 d.f. is $\psi^2_{0.05}$ = **3.84.**

Now calculated ψ^2 = 3 < $\psi^2_{0.05}$ = 3.84 ⇒ **Null hypothesies is accepted ⇒ Opinion on autonomous status and level of Graduation are independent.**]

2. A certain drug is claimed to be effective in curing colds. In an experiment on 164 people with cold, half of them were given the drug and half of them given sugar pills. The patient's reactions to the treatment are recorded in the following table. Test the hypothesis that the drug is not better than sugar pills for curing colds

	Helped	*Harmed*	*No effect*
Drug:	52	10	20
Sugar Pills	44	12	26

3. In a survey of 200 boys, of which 75 were intelligent, 40 had skilled fathers, while 85 of the unintelligent boys had unskilled fathers. Do these support the hypothesis that skilled fathers have intelligent boys. Use ψ^2 test. Values of ψ^2 for 1-degree of freedom at 5% level is 3.84.

4. Four dice were thrown 112 times and the number of times 1, 3 or 5 were as under :

Number of dice showing 1, 3 or 5:	0	1	2	3	4
Frequency:	10	25	40	30	7

[**Hint.** Let the probability fo showing 1, 3, or 5 be the success = $\frac{3}{6} = \frac{1}{2}$. Probability fo x successes on a throw of 4 dice 112 times = $112\ {}^4C_x \left(\frac{1}{2}\right)^x \left(\frac{1}{2}\right)^{4-x} = 112\ {}^4C_x \left(\frac{1}{2}\right)^4$; $x = 0, 1, 2, 3, 4.$

Calculated ψ^2 = 1.845, $\psi^2_{0.05}$ for 5 – 1 = 4 d.f. = **9.488**.

Now calculated $\psi^2 < \psi^2_{0.05}$ ⇒ **All the four dice are fair.**]

	Frequency			
Number of days	Observed (O)	Expected (E)	$(O-E)^2$	$\frac{(O-E)^2}{E}$
0	10	7	9	1.286
1	25	28	9	0.391
2	40	42	4	0.025
3	30	28	4	0.143
4	7	7	0	0
Total	**112**	**112**		$\Sigma\frac{(O-E)^2}{E} = \mathbf{1.845}$

5. To test the efficiency of a new drug a controlled experiment was conducted where in 300 patients were administered the new drug and 200 other patients were not given the drug. The patients were monitored and results were obtained as follows

	Cured	*Condition worsens*	*No effect*	*Total*
Given the drug	200	40	60	300
Not given the drug	120	30	50	200
Total	320	70	110	500.

Use ψ^2 test for finding the effect of the drug.

6. The following table gives the classification of 100 workers according to sex and the nature of work. Test whether the nature of work is independent of the sex of the work.

	Skilled	*Unskilled*
Male	40	20
Female	10	30

7. Two groups of 100 people each were taken for testing the use of a vaccine. 15 persons contracted the disease out of the inoculated persons, while 25 contracted the disease in the other group. Test the efficiency of vaccine using ψ^2 – test.

8. There is a general belief that high income families send their children to public schools and low income families send their children to Government schools. For this 1000 families were selected in a city and the following results were obtained.

Income	*Public Schools*	*Govt. Schools*	*Total*
Low	100	200	300
High	500	200	700
Total	600	400	1000

Use ψ^2 test to determine whether income level and the type of schooling were associated.

9. A market anlayst took a sample of 20 markets of Bombay to determine how much variation is there in bread prices. The 20 prices that were quoted to him for the four samples of the bread yielded the sample values $\bar{x} = 60$ and $s = 9$. Find the 95 % confidence interval for the standard deviation of the entire market.

10. A sample of 101 light bulbs yielded a standard deviation of 80 hours burning time, whereas long experience with the particular brand showed a standard deviation of 90 hours. Using $\alpha = 0.05$ test if there is any difference in the standard deviation.

11. Weights in kg of 10 students are given as 38, 40, 45, 53, 47, 43, 55, 48, 52, 49. Can we say that variance of distribution of weights of all students from which the above sample of 10 students was drawn, is equal to 20 kg?

12. From the adult male population of seven large cities random sample giving 2 × 7 contingency table of married and unmarried men, as given below were taken. Can it be said that there is a significant variation among the cities in the tendency of men to marry?

City	*A*	*B*	*C*	*D*	*E*	*F*	*G*	*Total*
Married	133	164	155	106	153	123	146	980
Unmarried	36	57	40	37	55	33	36	294
Total	169	221	195	143	208	156	182	1274

(At (2 – 1) (7 – 1) d.f. Take $\psi^2_{0.05,6} = 12.6$)

[**Hint. Null Hypothesis:** H_0: There is no significant variation among the cities in the tendency of men to marry.

Alternative Hypothesis: H_1: There is a significant variation among the cities in the tendency of men to marry.

Married	130	170	150	110	160	120	140
Unmarried	39	51	45	33	48	36	42

$\psi^2 = \Sigma \dfrac{(O-E)^2}{E} = 5.337.$ < tabled value of $\psi^2_{0.05}(= 12.6)$ $\Rightarrow$ H_0 **is accepted.**]

13. 1000 extensive sets of trials for an event of small probability, frequencies O_i of the number x_i of successes proved to be.

x_i :	0	1	2	3	4	5	6	7
O_i :	305	365	210	80	28	9	2	1.

Find the mean and variance of successes. Discuss the possibility of variable following the Poisson distribution. Also test the goodness of fit by chi-square test.

[**Hint.** Since mean is approximately equal to the variance, so there are reason to belive that success conform to poisson distribution = H_0. The expected frequencies are given by $f(x) = N \times \dfrac{e^{-m} \times m^x}{x!}$ and they are 301.1, 361.4, 216.8, 86.7, 26.0, 6.2, 1.2, 0.2. Then $\psi^2 = 3.5 < \psi_{0.05}$, to $\Rightarrow$ H_0 **is accepted.**

14. A printer wishes to determine the proportion of lines typeset with error is affected by the complexity of the text. The following sample data have been obtained. At the 5% significant level, should the printer conclude that the error rate depends on the complexity of text ?

	Complexity		
Low	*Medium*	*High*	
Number of lines	52	10	20
Number of lines correct	44	12	26

ANSWERS

1. Opinion on autonomous status and level of graduations are independent.
2. Computed $\psi^2 = 1.622$, $\psi^2_{0.05} = 5.991$, Drug is not better than sugar pills.
3. $\psi^2 = 0.88$, $\psi^2_{0.05}$ for $1.d.f = 3.84$; skilled fathers have intelligent boys.
4. All the four dice are fair.
5. $\psi^2 = 2{,}434$, $\psi^2_{0.05} = 5.99$; Drug is not effective.
6. $\psi^2 = 16.666$, $\psi^2_{0.05} = 3.84$. The nature of work does not seem to be independent of the sex of the worker.
7. $\psi^2 = 3.125$, $\psi^2_{0.05}$ 3.84; Inoculation is effective in contracting the disease.
8. The level and type of schooling are associated.
9. $6.84 \leq \sigma^2 < 13.4$,
10. There is no difference in standard deviation.
11. The population variance is not 20.
12. There is no significant variation among the cites in the tendency of men to marry.
13. Mean = 1.20, variance = 1.28. H_0: success conform to Poisson distribution.
14. The error rate does not depend on the complexity of text.

CHAPTER 11

F-DISTRIBUTION AND ANOVA TABLE

11.1 F–TEST OR FISHER'S F–TEST

The F-test was first originated by the statistician R.A. Fisher. This test is also known as **Fisher's F-test or simply F-test. It is based on the F-distribution, which is defined as the ratio of two independent chi-square variates** which is derived by dividing each variable by its corresponding degrees of freedom, *i.e.*,

$$F = \frac{\psi^2}{v_1} \div \frac{\psi^2}{v_2}$$

The **F-test refers to a test of hypothesis concerning two variances derived from two samples.** The various tests of hypothesis discussed previously are not suitable for the test of hypothesis concerning two or more sample variances.

11.2 F–STATISTIC

The F–Statistic is the ratio of independent estimates of population variances and expressed as

$F = \frac{\hat{\sigma}_1^2}{\hat{\sigma}_2^2}$ with $n_1 - 1$ degree of freedom for the numerator and $n_2 - 1$ degree of freedom for the denominator, where $\hat{\sigma}_1^2$, $\hat{\sigma}_2^2$ are the unbaised estimate on the basis of sample variance s_1^2 and s_2^2 given by the population variances based on sample sizes n_1 and n_2, where,

$$\hat{\sigma}_1^2 = \frac{n_1 s_1^2}{n_1 - 1}, \qquad \hat{\sigma}_2^2 = \frac{n_2 s_2^2}{n_2 - 1}$$

Generally $\hat{\sigma}_1^2$ is greater than $\hat{\sigma}_2^2$, but if $\hat{\sigma}_2^2$ is greater than $\hat{\sigma}_1^2$, then in such cases the two variances should be interchanged so that the value of F is always greater than 1.

11.3 ASSUMPTIONS IN F-TEST

The F-Test is based on the following assumptions:

1. **Normality.** The values in each group should be normally distributed.
2. **Independence of Error.** The variation of each value around its own group mean, *i.e.*, error should be independent of each value
3. **Homogeneity.** The variances within each group should be equal for all groups,

$$i.e., \ \sigma_1^2 = \sigma_2^2 = \sigma_3^2 = \ldots = \sigma_n^2$$

If however, the sample sizes are large enough, we donot need the assumption of normality.

11.4 TESTS OF HYPOTHESIS ABOUT THE VARIANCE OF TWO POPULATIONS

We have the following steps:

1. **Null Hypothesis :– H_0: $\hat{\sigma}_1^2 = \hat{\sigma}_2^2$.**

   Alternate Hypothesis :– H_1 : $\hat{\sigma}_1^2 \neq \hat{\sigma}_2^2$ (Two Tailed test).

2. **Calculation of Test statistic.** Calculate the F-statistic as given below:

$$F = \frac{\hat{\sigma}_1^2}{\hat{\sigma}_2^2} \text{ if } \hat{\sigma}_1^2 \geq \hat{\sigma}_2^2 \text{ so that } F \geq 1$$

or

$$F = \frac{\hat{\sigma}_2^2}{\hat{\sigma}_1^2} \text{ if } \hat{\sigma}_2^2 \geq \hat{\sigma}_1^2 \text{ so that } F \geq 1$$

Level of Significance. Take the level of significance $\alpha = 0.05$ if α is not known.

Decision: Accept H_0 if computed $F \leq$ tabled F_α

Reject H_0 if computed $F >$ tabled F_α.

STEPS FOR ONE TAILED TEST

1. **Set the Null Hypothesis :– $H_0 : \hat{\sigma}_1^2 \geq \hat{\sigma}_2^2$**

   Alternative Hypothesis :– $H_1 : \hat{\sigma}_1^2 \geq \hat{\sigma}_2^2$ (One Tailed Test)

Note. ***For one tailed test H_0 is set in such a way that the rejection region appears in the upper tail. This is done by judiciously numbering the population variances, so that the alternative hypothesis takes the form: $H_1 : \hat{\sigma}_2^2 \geq \hat{\sigma}_1^2$. If H_0 is of the form $\hat{\sigma}_1^2 \geq \hat{\sigma}_2^2$, then we compute $F = \frac{\hat{\sigma}_2^2}{\hat{\sigma}_1^2}$ which has F distribution, but with $n_2 - 1$ d.f. in the numerator and $n_1 - 1$ d.f. in the denominator.***

2. **Computation of Test Statistic.** Calculate the F-statistic $F = \frac{\hat{\sigma}_1^2}{\hat{\sigma}_2^2}$, if $\hat{\sigma}_1^2 > \hat{\sigma}_2^2$
3. **Level of significance.** Set the level of significance $\alpha = 0.05$ if no value of α is given.
4. **Decision.** If computed $F <$ tabled $F_{\alpha/2}$, then accept the null hypothesis H_0. or Reject H_0 if **Computed** $F >$ tabled $F_{\alpha/2}$

A sumarry of procedure is given below:

Null hypothesis: H_0	*Alternative hypothesis: H_1*	*Type of Test*	*Rejection Region*	*Decision Rule*
$H_0 : \hat{\sigma}_1^2 = \hat{\sigma}_2^2$	$H_1 : \hat{\sigma}_1^2 \neq \hat{\sigma}_2^2$	**Two tailed**	**Both tails**	**Accept H_0 if $F_{cal} \geq F_{\alpha/2}$ otherwise accept H_1**
$H_0 : \hat{\sigma}_1^2 \geq \hat{\sigma}_2^2$	$H_1 : \hat{\sigma}_1^2 < \hat{\sigma}_2^2$	**One tailed**	**Right tails**	**Accept H_0 if $F_{cal} \geq F_\alpha$**

For one tailed test H_0 is set in such a way that the rejection region appears in the upper tail. This is done by a judiciously numbering of populations variances, so that alternative hypothesis takes the form $H_1 : \hat{\sigma}_1^2 < \hat{\sigma}_2^2$. If H_1 is of the form $\hat{\sigma}_1^2 < \hat{\sigma}_2^2$, we compute the ratio $(\hat{\sigma}_2^2 / \hat{\sigma}_1^2)$ (in stead of $\hat{\sigma}_1^2 / \hat{\sigma}_2^2$, which also has a F distribution, but with $n_2 - 1$ d.o.f for the numerator and $n_1 - 1$ d.o.f for the denominator. The critical value in the left tail of F distribution is found from the table of critical values in the right tail of F distribution.

Example 1. *The standard deviations calculated from two random samples of sizes 9 and 13 are 2.1 and 1.8 respectively. May the samples be regarded as drawn from the normal populations with the same standard deviation. The value of F from the table with degrees of freedom 8 and 12 is 2.85.*

Solution. With usual notations, we have

$n_1 = 9, \quad s_1 = 2.1, \quad n_2 = 13, \quad s_2 = 1.8.$

$$\therefore \quad \hat{\sigma}_1^2 = \frac{n_1 s_1^2}{n_1 - 1} = \frac{9 \times (2.1)^2}{8} = \mathbf{4.96.}$$

$$\hat{\sigma}_2^2 = \frac{n_2 s_2^2}{n_2 - 1} = \frac{13 \times (1.8)^2}{12} = \mathbf{3.51.}$$

Null Hypothesis :– H_0 : $\hat{\sigma}_1^2 = \hat{\sigma}_2^2$.

Alternate Hypothesis :– H_1 : $\hat{\sigma}_1^2 \neq \hat{\sigma}_2^2$ (Two Tailed Test)

$$\therefore \quad F = \frac{\hat{\sigma}_1^2}{\hat{\sigma}_2^2} = \frac{4.96}{3.51} = \mathbf{1.41.}$$

Decision. Since calculated $F = 1.41 <$ tabled F for 8 and 12 d.f. = 2.85, so we axcept the null hypothesis $H_0 \Rightarrow$ the samples may be regarded as drawn from the normal population with same standard deviation.

Example 2. *Two independent samples of sizes 9 and 8 gave the sum of squares of deviations from their respective means as 160 and 91 respectively. Can the samples be regarded as drawn from the normal populations with equal variance? Given: $F_{0.05}$ (8, 7) = 3.73 ; $F_{0.05}$ (7, 8) = 3.50.*

Solution. We set H_0 : $(\sigma_1^2 = \sigma_2^2)$.

$H_1 : (\sigma_1^2 \neq \sigma_2^2)$

$$\text{Also} \quad s_1^2 = \frac{\sum (x - \bar{x})^2}{n_1 - 1} = \frac{\mathbf{160}}{\mathbf{8}} = \mathbf{20}; \quad s_2^2 = \frac{\sum (y - \bar{y})^2}{n_2 - 1} = \frac{\mathbf{91}}{\mathbf{7}} = \mathbf{13}.$$

$$\therefore \quad F_{cal} = \frac{\sigma_1^2}{\sigma_2^2}.$$

$$\text{Now} \quad \hat{\sigma}_1^2 = \frac{n_1}{n - 1} s_1^2 = \frac{9}{9 - 1} \times 20 = \mathbf{45}$$

$$\hat{\sigma}_2^2 = \frac{n_2}{n_2 - 1} s_2^2 = \frac{8}{8 - 7} \times 13 = \frac{\mathbf{104}}{\mathbf{7}}$$

$$\therefore \quad F = \frac{\hat{\sigma}_1^2}{\hat{\sigma}_2^2} = \frac{45}{(104/7)} = \frac{45 \times 7}{104} = \frac{315}{104} = \mathbf{3.029}$$

Here $F_{cal} = 3.029 < F_{0.05}$ (8.7) = 3.73 (given)

$\Rightarrow H_0$ is accepted $\Rightarrow$ that the samples be regarded as drawn from the normal population with equal variance.

Example 3. *Random samples are drawn from two populations and the following results were obtained:*

Sample X:	16	17	18	19	20	21	22	24	26	27		
Sample Y:	19	22	23	25	26	28	29	30	31	32	35	36.

Find the variance of two populations and test whether the two samples have same variance.

Solution. Null hypothesis :– H_0: $\hat{\sigma}_x^2 = \hat{\sigma}_y^2$, *i.e.*, the two samples have the same variance.

Alternative Hypothesis :– H_1 : $\hat{\sigma}_x^2 \neq \hat{\sigma}_y^2$ (Two Tailed Test).

Calculation of Test Statistic. We have the following table to evaluate $\hat{\sigma}_x^2$ and $\hat{\sigma}_y^2$

Computation Table

x	$x - \bar{x}$ $= x - 21$	$(x - \bar{x})^2$ $= (x - 21)^2$	y	$y - \bar{y}$ $= y - 28$	$(y - \bar{y})^2$ $= (y - 28)^2$
16	– 5	25	19	– 9	81
17	– 4	16	22	– 6	36
18	– 3	9	23	– 5	25
19	– 2	4	25	– 3	9
20	– 1	1	26	– 2	4
21	0	0	28	0	0
22	1	1	29	1	1
24	3	9	30	2	4
26	5	25	31	3	9
27	6	36	32	4	16
			35	7	49
			36	8	64
$\Sigma x = 210$		$\Sigma (x - \bar{x})^2 = 126$	$\Sigma y = 336$		$\Sigma (y - \bar{y})^2 = 298$

Now $\bar{x} = \dfrac{\Sigma x}{n} = \dfrac{210}{10} = 21$, $\qquad \bar{y} = \dfrac{\Sigma y}{n} = \dfrac{336}{12} = 28.$

$$\therefore \qquad \hat{\sigma}_1^2 = \frac{\Sigma(x - \bar{x})^2}{n_1 - 1} = \frac{126}{9} = 14$$

$$\hat{\sigma}_2^2 = \frac{\Sigma(y - \bar{y})^2}{n_2 - 1} = \frac{298}{11} = 27.09.$$

Test Statistic $\qquad F = \dfrac{\hat{\sigma}_2^2}{\hat{\sigma}_1^2} = \dfrac{27.09}{14} = 1.94.$

Critical Value. The tabled value of F at $\alpha = 0.05$ for 11 and 9 degrees of freedom is

$$F_{0.05} = 2.6458.$$

Decision. The computed value of $F = 1.94 <$ Table value $F_{0.05} = 2.6458$

$\Rightarrow$ the null hypothesis H_0 is accepted $\Rightarrow$ The two samples have the same variance.

Example 4. *In a laboratory experiment, two samples gave the following results:*

Sample	*Size*	*Sample mean*	*Sum of squares of deviation from the mean*
1	*10*	*15*	*90*
2	*12*	*14*	*108*

Test the equality of sample variances at 5% level of significance.

Solution. We set the Null hypothesis :– H_0: $\sigma_1^2 = \sigma_2^2$.

Alternate Hypothesis :– H_1 : $\sigma_1^2 \neq \sigma_2^2$. **(Two Tailed Test)**

An unbiased estimate of the variance for sample I

$$\hat{\sigma}_1^2 = \frac{\Sigma(x_i - \bar{x})^2}{n-1} = \frac{90}{9} = 10$$

An unbiased estimate of the variance for sample II

$$\hat{\sigma}_2^2 = \frac{\Sigma(y_i - \bar{y})^2}{n-1} = \frac{108}{11} = 9.82$$

$\therefore$ **Test Statistic** $$F = \frac{\hat{\sigma}_1^2}{\hat{\sigma}_2^2} = \frac{10}{9.82} = 1.018.$$

Critical value: The table value of F at $\alpha = 0.05$ for 9 and 11 degrees of freedom is $F_{0.05} = 2.90$.

Decision. Since the computed value of $F = 1.018 <$ Tabled value of $F_{0.05} = 2.90$, so the null hypothesis H_0 is accepted $\Rightarrow$ the two populations have the same variance.

Example 5. *In a test given to two groups of students drawn from two normal populations, the marks obtained were as follows:*

Group A:	*18*	*20*	*36*	*50*	*49*	*36*	*34*	*49*	*41*
Group B:	*29*	*28*	*26*	*35*	*30*	*44*	*46*		

Examine at 5% level, whether the two populations have the same variance.

Solution. Null Hypothesis :– H_0: $\sigma_A^2 = \sigma_B^2$.

Alternate Hypothesis :– H_1: $\sigma_A^2 \neq \sigma_B^2$ **(Two Tailed Test).**

Calculation of Test Statistic. We have the following table to find the sample variances.

$\therefore$ $\bar{x} = \frac{\Sigma x}{n} = \frac{333}{9} = 37$; $\bar{y} = \frac{\Sigma y}{n} = \frac{238}{7} = 34$.

Variance of group A $= \hat{\sigma}_A^2 = \frac{\Sigma(x-\bar{x})^2}{n-1} = \frac{1134}{8} = 141.75.$

Variance of group B $= \hat{\sigma}_B^2 = \frac{\Sigma(y-\bar{y})^2}{n-1} = \frac{386}{6} = 64.33.$

Group A			*Group B*		
x	$(x-\bar{x})$ $= x - 37$	$(x-\bar{x})^2$	y	$(y-\bar{y})$ $= y - 34$	$(y-\bar{y})^2$
18	– 19	361	29	5	25
20	– 7	289	28	6	36
36	– 1	1	26	8	64
50	13	169	35	1	1
49	12	144	30	4	16
36	– 1	1	44	10	100
34	– 3	9	46	12	144
49	12	144			
41	4	16			
$n_1 = 9$; $\Sigma (x - \bar{x})^2 = 1134$; $\Sigma x = 333$			$n_2 = 7$; $\Sigma (y - \bar{y})^2 = 386$; $\Sigma y = 238$		

$\therefore$ Test Statistic: $F = \dfrac{\hat{\sigma}_A^2}{\hat{\sigma}_B^2} = \dfrac{141.75}{64.33} = \mathbf{2.203.}$

Critical Value. The table value of F at 5% level for 8 and 6 degrees of freedom is $F_{0.05} = \mathbf{4.15}$.

Decision. The **computed value of $F = 2.203 <$ Tabled value of $F_{0.05} = 4.15 \Rightarrow$ the null hypothesis H_0 is accepted $\Rightarrow$ The populations from where the samples have been taken have the same variances.**

EXERCISE – 11.1

1. The standard deviations calculated from two random samples of size 9 and 13 are 2 and 1.9 respectively. May the sample be regarded as drawn from the normal populations with the same standard deviation.

   [Hint: H_0: $\sigma_1^2 = \sigma_2^2$, $\quad H_1: \sigma_1^2 \neq \sigma_2^2$.

   $$\hat{\sigma}_1^2 = \frac{n_i s_1^2}{n_1 - 1} = \frac{9}{8} \times 4 = \mathbf{4.5}, \qquad \hat{\sigma}_2^2 = \frac{13}{12} \times 3.61 = \mathbf{3.91}.$$

   $$\therefore \quad F = \frac{\hat{\sigma}_1^2}{\hat{\sigma}_2^2} = \frac{4.5}{3.91} = 1.15 < F_{0.05} = \mathbf{3.51.}$$

   $\Rightarrow$ the null hypothesis is accepted $\Rightarrow$ the samples can be regarded as drawn from the populations with the same standard deviation.]

2. Following results were obtained from two samples, each drawn from two different populations A and B.

Population	A	B
Sample	**I**	**II**
Sample Size	$n_1 = 25$	$n_2 = 17$
Sample s.d.	$s_1 = 3$	$s_2 = 2$

   Test the hypothesis that the variance of brand A is more than that of B.

   [Hint. $H_0 = \sigma_1^2 \le \sigma_2^2$, $\quad H_1; \sigma_1^2 > \sigma_2^2$

   $$\hat{\sigma}_1^2 = \frac{n_1}{n_1 - 1} s_1^2 = \frac{25 \times 9}{24} = \mathbf{9.375}, \qquad \hat{\sigma}_2^2 = \frac{17 \times 4}{16} = \mathbf{4.25.}$$

   $$F = \frac{\hat{\sigma}_1^2}{\hat{\sigma}_2^2} = \frac{9.375}{4.25} = \mathbf{2.205.}$$

   Also from table $F_{0.05}$ for 24 and 16 d.f. = 2.24.

   Now **Computed $F = 2.205 < F_{0.05} = 2.24 \Rightarrow$ the null hypothesis is accepted $\Rightarrow$ variance of brand A is not more than the variance of brand B].**

3. Two sample are drawn from two normal populations. From the following data test whether the two samples have the same variance at 5% level.

Sample I :	60	65	71	74	76	82	85	87		
Sample II :	61	66	67	85	78	63	85	86	88	91

   [Hint: Here $\Sigma x = 600, \bar{x} = \dfrac{600}{8} = 75, \qquad \Sigma(x - \bar{x})^2 = 636,$

   $\Sigma y = 770, \bar{y} = \dfrac{770}{10} = 77. \qquad \Sigma(y - \bar{y})^2 = 1200,$

   $$\hat{\sigma}_1^2 = \frac{636}{8-1} = \mathbf{90.857}, \qquad \hat{\sigma}_2^2 = \frac{1200}{10-1} = \mathbf{133.333.}$$

Let H_0 : $\sigma_1^2 = \sigma_2^2$. $F = \frac{\hat{\sigma}_1^2}{\hat{\sigma}_2^2} = \frac{13333}{90.857} = \mathbf{1.468}.$

F_{cal} **= 1.468 <** $F_{0.05}$ **= 3.68 for 9 and 7** ***d.f.*** **⇒** H_0 **is accepted ⇒ the samples have the same variance].**

4. The time taken by workers in performing a job by method I and method II is given below.

Method I :	20	16	26	27	22		
Method II:	27	33	42	35	32	34	38.

Do the data show that the variances of time distribution in a population from which these samples are drawn do not differ significantly?

5. Two samples were drawn from two normal populations and their values are:

A :	66	67	75	76	82	84	88	90	92		
B :	64	66	74	78	82	85	87	92	93	95	97.

Test whether the two populations have the same variance at 5% level of significance.

6. Two sources of raw materials are under consideration of a company. Both sources seem to have similar characteristics, but the company is not sure about their respective uniformity. A sample of ten lots from source *A* yields a variance of 225, and a sample of eleven lots from source *B* yields a variance of 200. It is likely that the variance of source *A* is significantly greater than the variance of source *B*. Take $\alpha = 0.01$.

[Hint: H_0**:** variance of source *A* is not significantly greater than the variance of source *B*., *i.e.*,

$H_0 : \sigma_1^2 = \sigma_2^2$. $H_1 : \sigma_1^2 > \sigma_2^2$. **(One tailed Test)**

$\hat{\sigma}_1^2 = 250, \hat{\sigma}_2^2 = 220, F = \frac{250}{220} = 1.17$. Critical value of *F* at 10% level of significance for $v_1 = 9, v_2 = 10$ degrees of freedom is $F_{(0.01)}$ (9, 10) = 4.93 > computed value of *F* = 1.17 ⇒ H_0 is accepted ⇒ **that variance of source *A* is not significantly greater than the variance of source *B*].**

7. It is known that the mean diameters of rivets produced by two firms *A* and *B* are practically the same but their standard deviations differ. For 22 rivets produced by firm *A*, the standard deviation is 2.9 mm, while for 16 rivets manufactured by firm *B*, the standard deviation is 3.8 mm. Compute the statistic you would use to test whether the product of firm *A* has the same variability as those of firm *B*. Also state how you would proceed further.

8. The following data relate to a random sample of government employees in two States of the Indian Union:

	State I	**State II**
Sample size	16	25
Mean monthly income of the sample employees (in Rs.)	440	460
Sample variance	40	42

First, cary out a test of hypothesis that the variance of the two populations are equal.

In the light of the result of the above test, carry out a test of the hypothesis that the means of the two populations are equal.

ANSWERS

1. The sample can be regarded as drawn from the population with the same standard derivation.
2. The variance of brand *A* is not more than the variance of brand *B*.
3. The samples have the same variance.
4. *F* = 1.37, the variance of time distribution in a population from which the samples are drawn do not differ significantly.
5. *F* = 1.415, the two populations have the same variance.
6. The variance of source *A* is not significantly greater than the variance of source *B*.
7. *F* = 2.14; It is not significant at 5% level of significance.
8. *F* = 1.33; Variances are equal; No significant difference in means, *t* = 1.44.

11.5 ANALYSIS OF VARIANCE – A TEST FOR HOMOGENEIETY OF MEAN

The analysis of variance is a method which separates the variation ascribable to one set of causes from the variation ascribable to other set. In other words, analysis of variance is a method of splitting the total variation of a data into constituent parts which measures different sources of variations. **The total variation is split up into the following two components:**

(*a*) Variation within the subgroups of samples.

(*b*) Variation between the subgroups of the samples.

After obtaining the above two variations, viz., (a) and (b), these two variations are tested for their significance by **F-test which is also known as Variance Ratio Test.**

11.6 OBJECTS OF ANALYSIS OF VARIANCE

The **first object of the analysis of variance is to obtain a measure of the total variation within the series** and the **second object is to find a measure of variation between or among the components. Then the test of significance of difference between the variations in two series or more may be measured.** In other words, with the help of the teachnique of analysis of variance **we can test the hypothesis that the means of all the components constituting a population are equal to the mean of the population or that the samples have come from the same population.**

11.7 ANOVA TABLE

The technique of analysis of variance is referred to as ANOVA. A table showing the **source of variation, the sum of squares, degrees of freedom, mean square (variance) and the formula for the *F*-ratio is known as ANOVA TABLE.**

This technique was developed by R.A. Fisher.

11.8 COMPUTATION OF TEST STATISTIC

The actual analysis of variance is carried out on the **basis of ratio between the variances. The variance ratio is obtained by dividing the variance between the samples by the variance within the samples.** This ratio forms the **test statistic known** as **F-Statistic, *i.e.*,**

$$\textbf{F-Statistic} = \frac{\textbf{Variance between the samples}}{\textbf{Variance within the samples}}.$$

11.9 ASSUMPTIONS

The underlying assumptions for the study of analysis of variance are:

1. ***Each of the samples is a simple random sample.***
2. ***Populations from which the samples are selected is normally distributed (If, however, the sample sizes are large enough, this assumption of normality is not required).***
3. ***Each one of the samples is independent of the other samples.***
4. ***Each one of the populations has the same variance*** $(\sigma_1^2 = \sigma_2^2 = ... = \sigma_n^2)$ ***and indentical means*** $(\mu_1 = \mu_2 = \mu_3 = ... = \mu_n)$.
5. ***The effect of various components are additive.***

11.10 USES

It is used to test whether the means of a number of populations (more than two) are equal. (We know that t-statistic is used for testing whether two population means are equal. Thus the analysis of variance of test may be taken as an extension of t-test for the case of more than two population means).

11.11 CLASSIFICATION OF ANALYSIS OF VARIANCE

The analysis of variance is mainly carried on under the following two classifications:

(*i*) One way Analysis of variance or One way Classification.

(*ii*) Two way analysis of variance or Two way Classified Data or Manifold Classification.

11.12 ONE WAY CLASSIFICATION

Under one way classification the influence of only one attribute or factor is considered.

METHODS FOR ONE WAY ANALYSIS OF VARIANCE–ANOVA TABLE

11.12.1 Direct Method

The following steps are required under the direct method for one way classification.

1. **Null Hypothesis :** H_0: $\mu_1 = \mu_2 = ... = \mu_p$, *i.e.*, the means of the populations from which the samples are drawn are equal to one another.

   Alternate Hypothesis : H_i **:** At least two of the means of the populations are unequal, or **All the μ_i's are not equal.**

2. **Computation of Test Statistic.** We define certain terms before obtaining the test statistic known as *F*-value.

   Meaning of Sample Observation x_{ij}. A sample observation x_{ij} has two subscripts. The subscript ***i*** **denotes the row *i*th or *i*th sample observation,** while the subscript ***j*** **denotes the *j*th column or *j*th population** from which the observation came. **Thus x_{ij} is the *i*th observation from the *j*th population.** The observation can be regarded as the element of the following matrix *A* of order $m \times n$.

Population

$$A = \begin{bmatrix} x_{11} & x_{12} & x_{13} & --- & x_{1j} & --- & x_{1n} \\ x_{21} & x_{22} & x_{23} & --- & x_{2j} & --- & x_{2n} \\ x_{31} & x_{32} & x_{33} & --- & x_{3j} & --- & x_{3n} \\ --- & --- & --- & & --- & & --- \\ --- & --- & --- & & --- & & --- \\ x_{i1} & x_{i2} & x_{i3} & --- & x_{ij} & --- & x_{in} \\ --- & --- & --- & & --- & & --- \\ --- & --- & --- & & --- & & --- \\ x_{m1} & x_{m2} & x_{m3} & --- & x_{mj} & --- & x_{mn} \end{bmatrix} \begin{matrix} \\ \text{S} \\ \text{A} \\ \text{M} \\ \text{P} \\ \text{L} \\ \text{E} \\ \text{S} \\ \\ \end{matrix}$$

where x_{ij} is the element lying in the *i*th row and *j*th column.

The above matrix *A* can be regarded as a matrix of **(*m* samples × *n* populations.)**

The total observations in above matrix **A is $m \times n$.**

The various **Sums of squares involved in the computation of *F*-statistic are :**

(*i*) Sum of the Squares of Variations Amongst the Columns : SSC = It is the sum of the squares of deviation between the columns or group means and the grand mean, *i.e.*,

$$\text{SSC} = r\,\Sigma(\bar{x}_j - \bar{x})^2,$$

where $\bar{x}_j$ = **mean of the *j*th sample**

$\bar{x}$ = **mean of the sample (Column) means**

r = **number of rows or size of each sample.**

Variance amongst columns : MSC = $\dfrac{\text{SSC}}{c-1}$**, where *c* is the number of columns.**

The variance amongst the column MSC indicates the degree of Explained Variance due to sampling variations. In other words, it indicates some thing more than a chance variation which is probability due to unlike population.

(*ii*) Sum of the squares of variations within Columns : SSE = It is the sum of the squares of variations between individual items and the column means, *i.e.*,

$$\text{SSE} = \Sigma_i \Sigma_i (x_{ij} - \bar{x}_j)^2,$$

where x_{ij} = **ith observation in jth column;**

$\bar{x}_j$ = **mean of jth column.**

Mean of the square of Column Errors : MSE = $\dfrac{\text{SSE}}{c\,(r-1)}$,

where c = number of columns; r = number of rows.

This is also called **Un-explained Variance because it indicates only the chance variation which cannot be explained in terms of variation in population.**

(*iii*) Total Sum of Squares of Variation : SST = The total sum of squares is given by SST which is defined as

$$\text{SST} = \sum_j \sum_i x_{ij}^2 - C,$$

where C = **Correction Factor =** $\dfrac{T^2}{rc}$,

T = **grand total of the values in all the samples;**

r = **number of rows;**

c = **number of columns.**

It is the sum of the square of observations between the individual values and the grand mean $\bar{x}$.

Also $\text{SST} = \text{SSC} + \text{SSE}$.

Total Variance. ***The total variance comprise of both the explained and the unexplained variance and is defined as***

$$\textbf{Total variance} = \frac{\text{SST}}{n-1}$$

where $n = r \times c$ = total number of observations in all the samples and $(n - 1)$ is the degrees of freedom.

ALTERNATIVE APPROACH FOR CALCULATION OF SSC, SST AND SSE.

Correction factor : $C = \dfrac{T^2}{rc}$. **where**

T = **grand total of values in all the samples and** $T = \sum_i \sum_j x_{ij}$

r = **number of rows**

c = **number of columns**

$$\text{SSC} = \frac{\sum_j T_j^2}{r} - C, \text{ where}$$

T_j = **total of *j*th column**

SST = $\sum_i \sum_j x_{ij}^2 - C$, **where *C* = correction factor**

2. Test Statistic. The test statistic is the *F*-value or *F* **Statistics** and is defined as

$$F\text{-statistic or } F\text{-value} = \frac{\text{Explained variance}}{\text{Unexplained variance}} = \frac{\text{MSC}}{\text{MSE}}.$$

i.e., $F = \frac{\text{MSC}}{\text{MSE}}$, **with (*c* – 1) and *c* (*r* – 1) degrees of freedom.**

ANOVA Table for one-way classification for equal sample size.

Source of variation	*Sum of squares*	*Degree of freedom.*	*Mean Square*	*F*
Between samples (Column means)	**SSC**	$d.o.f_1 = c - 1$	$MSC = \frac{SSC}{c-1}$	$F = \frac{MSC}{MSE}$
Within Samples (Errors)	**SSE**	$d.o.f_2 = c(r-1)$	$MSE = \frac{SSE}{c(r-1)}$	
Total	**SST**	$cr - 1$		

If the sample sizes are unequal ***c* (*r* – 1)** is to be replaced by ***n* – *c*** and ***cr* – 1** by ***n* – 1, *n*** being the total number of observations in all the samples. **The rationale for this test is simple, if the variance amongst the column of groups MSC is equal to the variance within the group MSE, *i.e., if MSC = MSE, then the F-value will be equal to one. It shows that no extra variation in volumes than what is there within the columns, and therefore, no variance worth noticing. If however, the variance amongst columns is more than that within the columns, the computed F value has to be compared with the tabled F value at certain level of significance α, generally α = 0.05.***

3. Level of Significance. It is denoted by α and the value taken, generally is α = **0.05**

4. Decision. (*i*) If the computed value of *F* < Tabled value of *F* at α level of significance, then the null hypothesis H_0 is accepted ⇒ the means of the population from which samples are drawn are equal.

(*ii*) If the computed value of *F* > Table value of F at α level of significance, then the null hypothesis is rejected and the alternate hypothesis H_1 is accepted ⇒ at least two of the means of the population are unequal.

The actual application is illustrated by the following examples.

Example 6. *The three samples below have been obtained from normal populations with equal fariances. Test the hypothesis at 5% level that the population means are equal:*

8	7	12
10	5	9
7	10	13
14	9	12
11	9	14

(The table value of F at 5% level of significance for $v_1 = 2$ *and* $v_2 = 12$ *is 3.88).*

Solution. Null Hypothesis :– H_0: There is no significant difference in means of the three samples.

Using analysis of variance technique we have the following table.

	X_1	X_2	X_3
	8	7	12
	10	5	9
	7	10	13
	14	9	12
	11	9	14
Total	50	40	60
	$\bar{X}_1 = \frac{50}{5} = 10$	$\bar{X}_2 = \frac{40}{5} = 8$	$\bar{X}_3 = \frac{60}{5} = 12$

Grand Mean: $\bar{X} = \frac{10+8+12}{3} = 10.$

Variance Between Samples

$(\bar{X}_1 - \bar{X})^2$	$(\bar{X}_2 - \bar{X})^2$	$(\bar{X}_3 - \bar{X})^2$
0	4	4
0	4	4
0	4	4
0	4	4
0	4	4
0	**20**	**20**

SSC = Sum of squares between samples = 0 + 20 + 20 = 40.

Variance Within Samples

X_1	$(X_1 - \bar{X}_1)^2$	X_2	$(X_2 - \bar{X}_2)^2$	X_3	$(X_3 - \bar{X}_3)^2$
8	4	7	1	12	0
10	0	5	9	9	9
7	9	10	4	13	1
14	16	9	1	12	0
11	1	9	1	14	4
	30		**16**		**14**

SSE = Sum of squares within samples = 30 + 16 + 14 = 60.

ANOVA Table

Source of variation	*Sum of squares*	*v*	*Mean squares*
SSC = Between samples	40	2	$\frac{40}{2} = 20$
SSE = Within samples	60	12	$\frac{60}{12} = 5$
Total	**SST = 100**	**14**	

Test Statistic : $F = \frac{20}{5} = 4.$

For $v_1 = 2$, $v_2 = 12$ and for $\alpha = 0.05$, the table value of F is $F_{0.05} = 3.80$.

Since the calculated value of $F = 4$ is greater than the tabled value $F_{0.05} = 3.80$, so that the null hypothesis is rejected. Hence there is a significant difference in the means of the three given samples.

Example 7. *The following table gives the yield on 15 sample fields under three varieties of seeds (viz., A. B. C) :*

A	*B*	*C*
20	18	25
21	20	28
23	17	22
16	25	28
20	*15*	*32*

Test at 5% level of significance whether the average yields of land under different varieties of seed show significance differences.

(Table value of F at 5% level for $v_1 = 2$ and $v_2 = 12$ is 3.88)

Solution. Null hypothesis be H_0 **: that the average yields of land under different varieties do not show significant differences.**

							T_i	$\sum_j x_{ij}^2$
Varieties	*A*	20	21	23	16	20	100	**2026**
	B	18	20	17	25	15	95	**1863**
	C	25	28	22	28	32	123	**3701**
	T_j	63	69	62	69	67	330	
	$\sum_j x_{ij}^2$	**1349**	**1625**	**1302**	**1665**	**1649**		**7590**

$$\text{T.S.S.} = \sum_i \sum_j x_{ij}^2 - \frac{(\Sigma\Sigma x_{ij})^2}{n} = 7590 - \frac{(330)^2}{15} \qquad [n = 5 \times 3 = 15]$$

$$= 7590 - 7260 = \mathbf{330.}$$

$$\text{S.S. between varieties} = \sum_j \frac{T_i^2}{n_j} - \frac{(\Sigma\Sigma x_{ij})}{n}$$

$$= \frac{(100)^2}{5} + \frac{(95)^2}{5} + \frac{(135)^2}{5} - \frac{(300)^2}{15} = 7450 - 7260 = \mathbf{190.}$$

ANOVA Table

Source of Variation	*Sum of Squares SSC*	*Degree of freedom*	*Mean sum of of squares*	*Variance Ratio*
Between varieties	SSC = 190	d.o.f. = 2	MSC = $\frac{SSC}{d.o.f.}$ = **95.00**	
Within Varieties	SSE = 140	d.o.f. = 12	MSE = $\frac{SSE}{d.o.f.}$ = **11.67**	$F = \frac{MSC}{MSE} = 8.14$
Total	**SST = 330**	**cr = 14**		

Since the calculated value of F is larger than the tabulated value of F at a 5% level of significance, **so we reject the null hypothesis.** Hence, **the average yields of land under different varieties show significant differences.**

11.13 ANALYSIS OF VARIANCE IN TWO WAY OR IN MANIFOLD CLASSIFICATIONS:

In **the manifold classifications we consider two or more characteristics or altributes.** The table dealing with data on such grouping has a number of columns and rows. The analysis in that case will get extended to include the sum of squares between rows. The sample of **ANOVA table for manifold classification is given below.**

ANOVA Table for Manifold Classification

Source of variation	*Sum of squares*	*d.o.f.*	*Variance Ratio*	
Between columns	**SSC**	$c-1$	$\frac{SSC}{c-1} = MSC$	$F_0 = \frac{MSC}{MSE}$
Between Rows	**SSR**	$r-1$	$\frac{SSR}{r-1} = MSR$	
Error (Residual)	**SSE**	$(c-1)(r-1)$	$\frac{SSE}{(r-1)(c-1)} = \text{MSE}$	$F_R = \frac{MSR}{MSE}$

6.13.1. Short Cut Method

The following are the various steps under Short Cut Method.

1. ***Set the null hypothesis H_0 and alternate hypothesis H_1.***
2. ***Computation of Test statistic. The test statistic is obtianed by the following steps.***

Step I. ***Find the sum of the values of all the items of all the samples and denote it by T, i.e., obtain*** $T = \sum_i \sum_j x_{ij}$.

Step II. ***Compute the correction factor C, where***

$C = \frac{T^2}{N}$, ***N is the total number of all the observations of all the samples and T is obtained in STEP 1.***

Step III. ***Find the sum of the squares of all items of all the samples i.e., find*** $\sum_i \sum_j x_{ij}^2$.

Step IV. ***Find the total sum of the squares SST, where***

$$SST = \sum_i \sum_j x_{ij}^2 - C, \text{ or } SST = \sum_i \sum_j x_{ij}^2 - \frac{T^2}{N}.$$

where C is the correction factor obtained in STEP 2.

Step V. ***Find the sum of the squares between the samples, i.e., SSC by the following steps***

(A). ***Square the totals and divide it by the number of items in each sample.***

$$P = \frac{(\Sigma X_i)^2}{n_i} + \frac{(\Sigma X_2)^2}{n_2} + \frac{(\Sigma X_3)^2}{n_3}$$

(B). ***Subtract the correction factor*** $C = \frac{T^2}{N}$ ***from P, then the resulting figure would be the sum of the squares between the samples SSC, i.e.,***

$$SSC = P - \frac{T^2}{N}$$

Step VI. ***Find the sum of the squares with in the smaples sse from the result SSE = SST – SSC.*** ***[From Step 4 and 5]***

Step VII. ***Set up the ANOVA table and calculate F, which is the test statistic.***

3. Level of Significance. Let the level of significance be $\alpha = 0.05$.

4. Decision (a). If the computed value of F is less than the tabled value of F at $\alpha = 0.05$, then accept the null hypothesis H_0.

(b). If the computed vlaue of F is greater than the tabled value of F at $\alpha = 0.05$, then reject the null hypothesis H_0 and accept the alternate hypothesis H_1.

The method will become clear from the following examples.

Example 8. *A farmer applied three types of fertilizers on 4 separate plots. The figure on yield per acre are tabulated below:*

Fertilizers Plots	*Yield*				
	A	*B*	*C*	*D*	*Total*
Nitrogen	*6*	*4*	*8*	*6*	*24*
Potash	*7*	*6*	*6*	*9*	*28*
Phosphates	*8*	*5*	*10*	*9*	*32*
Total	*21*	*15*	*24*	*24*	*84*

Find out if the plots are materially different in fertility, as also, if the three fertilizers make any material differnece in yields.

Solution. Correction Factor: $C = \frac{T^2}{N} = \frac{(84)^2}{12} = \mathbf{588}$

SST = Total sum of square = $\Sigma\Sigma\, x_{ij}^2 - C$

$= [(6^2 + 7^2 + 8^2) + (4^2 + 6^2 + 5^2) + (8^2 + 6^2 + 10^2) + (6^2 + 9^2 + 9^2)] - \mathbf{588.}$

$= 624 - 588 = \mathbf{36.}$

SSC = Sum of square between plots = $\frac{\Sigma\left(\sum_j x_i\right)^2}{n_j} - C$

$$= \left[\frac{(21)^2}{3} + \frac{(15)^2}{3} + \frac{(24)^2}{3} + \frac{(24)^2}{3}\right] - \mathbf{588} = \frac{1818}{3} - \mathbf{588} = 606 - 588 = \mathbf{18.}$$

Sum of squares between Fertilizers = $\left(\sum_{i=1}^{r} T_i^2 / c\right) - C$

$$= \frac{(24)^2}{4} + \frac{(28)^2}{4} + \frac{(32)^2}{4} - \mathbf{588} = \frac{2384}{4} - \mathbf{588} = 596 - 588 = \mathbf{8.}$$

Error sum of squares = 36 – 18 – 8 = **10.**

Null Hypothesis: ***(i)*** **Plots are equally fertile**

(ii) **Fertilizer are equally effective.**

ANOVA Table

Sources of variation	*d.o.f.*	*S.S.*	$MSS = \dfrac{S.S.}{d.o.f}$	*FC*	F_{Tab} (5%)
Between Plots	3	18	$MSC = \dfrac{18}{3} = 6$	$\dfrac{MSC}{MSE} = 3.6$	**4.76**
Between Fertilizer	2	8	$MSR = \dfrac{8}{2} = 4$	$\dfrac{MSR}{MSE} = 2.4$	**5.14**
Error	6	10	$MSE = \dfrac{10}{6} = 1.667$		
Total	$C_{r-1} = 11$	**36**			

Conclusion: We accept the hypothesis at (*i*) and (*ii*), that is, the plots are equally effective and the fertilizer have the same effect.

Example 9. (*a*) *A sample of 400 items is taken from a normal population whose mean as well as variance is 4.* (*b*) *Set up a two-way ANOVA table for the following per hectare yield for 4 varieties of y old wheat on 3 plots.*

	Yield			
Plot of Land	*A*	*B*	*C*	*D*
I	*3*	*4*	*6*	*6*
II	*6*	*4*	*5*	*3*
III	*6*	*6*	*4*	*7*

Solution. (*i*) Let us take the null hypothesis that there is no significant difference in the yield of the four varieties of wheat and

(*ii*) there is no significant difference in the plots of land with regard to yield.

			Yield		
Plot of Land	***A***	***B***	***C***	***D***	***Total***
I	3	4	6	6	19
II	6	4	5	3	18
III	6	6	4	7	23
Total	**15**	**14**	**15**	**16**	**60**
$\overline{X}$	**5**	**4.67**	**5**	**5.33**	

[∵ N = 3 × 4 = 12]

Correction Factor: $C = \dfrac{T^2}{N} = \dfrac{(60)^2}{12} = 300.$

Total sum of squares $= \sum_i \sum_j x_{ij}^2 - C$

$= (3)^2 + (6)^2 + (4)^2 + (4)^2 + (6)^2 + (5)^2 + (4)^2 + (6)^2 + (3)^2 + (7)^2 - \boldsymbol{C}.$

$= [9 + 36 + 36 + 16 + 16 + 36 + 36 + 25 + 16 + 36 + 9 + 49] - 300 + 320 - \mathbf{300} = \mathbf{20.}$

Sum of squares between columns (yield)

$$= \frac{1}{3}[(15)^2 + (14)^2 + (15)^2 + (16)^2] - C = \frac{1}{3}[225 + 196 + 225 + 256] - \mathbf{300}$$

$$= \frac{1}{3}[902] - 300 = 300.67 - 300 = \mathbf{0.67.}$$

Sum of squares between rows (Plots of land) $= \frac{1}{4}[(19)^2 + (18)^2 + (23)^2] - C$

$$= \frac{1}{4}[361 + 324 + 529] - \mathbf{300} \quad = \frac{1}{4}[1214] - 300 = 303.5 - 300 = \mathbf{3.5}$$

Residual or Error = Total sum of squares – (sum of squares between columns + sum of squares between rows)

= 20 – 0.67 – 3.5 = **15.83**

Analysis of Variance Table

Source of variation	*S.S.*	*v*	*M.S.*	*F*
Between yields :	0.67	3	**0.223**	**0.076**
Between plots :	3.50	2	**1.750**	**0.597**
Residual :	15.83	6	**2.930**	
Total	**20.0**	**11**		

For $v_1 = 3$ and $v_2 = 6$, $F_{0.05} = 4.76$. The calculated value of F is much less than the tabled value. Hence, **there is no significant difference in the yield of four varieties of wheat.**

For $v_1 = 2$ and **$v_2 = 6$, $F_{0.05} = 5.14$. The calculated value of F is less than the tabled value.** Hence **there is no significant difference in the plots of land with regard to yield.**

Example 10. *The three samples below have been obtained from the normal population with equal variances. Test the hypothesis at 5% level that the population means are equal.*

8	7	12
10	5	9
7	10	13
14	9	12
11	9	14.

Solution. Null Hypothesis H_0: *All the population means are equal.*

Alternate Hypothesis H_1. *All the population means are not equal.*

2. Computation of Test Statistic. We first of all form the **following table.**

X_1	X_1^2	X_2	X_2^2	X_3	X_3^2	*Total: T =* $X_1 + X_2 + X_3$
8	64	7	49 –	12	144	$27 = T_1$
10	100	5	25	9	81	$24 = T_2$
7	49	10	100	13	169	$30 = T_3$
14	196	9	81	12	144	$35 = T_4$
11	121	9	81	14	196	$34 = T_5$
$\Sigma X_1 = 50$	$\Sigma X_1^2 = 530$	$\Sigma X_2 = 40$	$\Sigma X_2^2 = 336$	$\Sigma X_3 = 60$	$\Sigma X_3^2 = 734$	$T = 150 = \Sigma x_{ij}$

Here $N = 3 \times 5 = \mathbf{15}.$

Correction Factor: $C = \frac{T^2}{N} = \frac{(150)^2}{15} = \mathbf{1500}.$

$$SSC = \frac{(\Sigma X_1)^2}{n_1} + \frac{(\Sigma X_2)^2}{n_2} + \frac{(\Sigma X_3)^2}{n_3} - C$$

$$= \frac{(50)^2}{5} + \frac{(40)^2}{5} + \frac{(60)^2}{5} - 1500 = 1540 - 1500 = \mathbf{40}.$$

$$SST = \sum_i \sum_j X_{ij}^2 - C = \Sigma X_1^2 + \Sigma X_2^2 + \Sigma X_3^2 - C$$

$$= 530 + 336 + 734 - \mathbf{1500} = 1600 - 1500 = \mathbf{100}$$

$\therefore$ **SSE = SST – SSC =** 100 – 40 = **60.**

The various sums of squares along with their degree of freedom are given by the following analysis of variance table (**ANOVA Table**).

ANOVA TABLE

Source of variation	*Sum of squares*	*Degree of freedom*	*Mean Square*	*F*
Between samples	*SSC* = 40	2	$MSC = \frac{40}{2} = \mathbf{20}$	
Within samples (Errors)	*SSE* = 60	12	$MSE = \frac{60}{12} = 5$	$F = \frac{MSC}{MSC} = \frac{20}{5} = \mathbf{4}$
Total	***SST* = 100**	***d.f.* = 15 – 1 = 14**		

Critical value. The critical value of *F* for 2 and 12 degrees of freedom at $\alpha = 0.05$ is $F_{0.05\,(2,\,12)} = \mathbf{3.88}.$

Decision. The computed value of $F = 4$ > Tabled value of $F_{0.05}$ (2, 12) = 3.88 ⇒ the Null hypothesis H_0 is rejected and the Alternate hypothesis H_1 is accepted ⇒ All the population means are not equal.

Example 11. *To test the significance of the variations of the retail prices of the commodity in three principal cities; Bombay, Calcutta and Delhi, the four shops were chosen at random in each city and prices observed in rupees were as follows:*

Bombay:	*16*	*8*	*12*	*14*
Calcutta:	*14*	*10*	*10*	*6*
Delhi :	*4*	*10*	*8*	*8*

Do the data indicate that the prices in the three cities are significantly different?

Solution. We test the significance by coding of data method.

1. Null Hypothesis :– H_0: $\mu_1 = \mu_2 = \mu_3$, *i.e.*, **the mean prices in the three cities are same.**

Alternate Hypothesis :– H_1: **At least two of the three population means are different.**

2. Computation of Test statistic: First of all we shall set up **ANOVA Table.** The observations each reduced by 10 are given by the following table.

CITY						
	Bombay		*Calcutta*		*Delhi*	
	x_1	x_1^2	x_2	x_2^2	x_3	x_3^2
	6	36	4	16	– 6	36
	– 2	4	0	0	0	0
	2	4	0	0	– 2	4
	4	16	– 4	16	– 2	4
Total	**10**	**60**	**0**	**32**	**– 10**	**44**

Here $\Sigma x_1 = 10, \Sigma x_2 = 0, \Sigma x_3 = -10, \quad T = \Sigma x_1 + \Sigma x_2 + \Sigma x_3 = 0.$

$\Sigma x_1^2 = 60, \qquad \Sigma x_2^2 = 32, \qquad \Sigma x_3^2 = 44,$

$n_1 = 4\ n_2 = 4\ n_3 = 4; N = n_1 + n_2 + n_3 = 12.$

$\Sigma x_{ij}^2 = \Sigma x_1^2 + \Sigma x_2^2 + \Sigma x_3^2 = 60 + 32 + 44 = 136.$

Total sum of squares: $\text{SST} = \Sigma x_{ij}^2 - \frac{T^2}{N} = 136 - \frac{0}{12} = 136.$

Sum of the squares between the samples: $\text{SSC} = \frac{(\Sigma x_1)^2}{n_1} + \frac{(\Sigma x_2)}{n_2} + \frac{(\Sigma x_3)^2}{n_3} - \frac{T^2}{N}$

$= \frac{(10)^2}{4} + \frac{0}{4} + \frac{(-10)^2}{4} - \frac{0}{12} = \frac{100}{4} + 0 + \frac{100}{4} - 0 = 50.$

Sum of the squares within the samples: SSE = SST – SSC = 136 – 50 = 86.

The various **Sum of Squares** along with their degrees are given in the following **ANOVA table:**

ANOVA TABLE

Source of variation	*Sum of squares*	*Degree of freedom.*	*Mean square*	*F – value.*
Between samples (city)	***SSC* = 50**	3 – 1 = 2	$\frac{50}{2} = 25 = MSC.$	$F = \frac{MSC}{MSE}$
Within samples City (Errors)	***SSE* = 86**	9	$\frac{86}{9} = 9.56 = MSE$	$= \frac{25}{9.56} = 2.62$
Total	**136**	**12 – 1 = 11**		

Critical value. The table value of F at $\alpha = 0.05$ for 2 and 9 degrees of freedom is $F_{0.05}\ (2, 9) = 4.26.$

Decision. The computed value of $F = 2.62 <$ Tabled value of $F_{0.05}\ (2, 9) = 4.26$

$\Rightarrow$ the null hypothesis is accepted $\Rightarrow$ the population means are all equal.

Example 12. *Set up* ***ANOVA*** *for the following per hectare yield for three varieties of wheat each grown on four plots.*

Plot of land	*Variety of wheat*			
	A_1	A_2	A_3	*Total*
1	*6*	*5*	*5*	*16*
2	*7*	*5*	*4*	*16*
3	*3*	*3*	*3*	*9*
4	*8*	*7*	*4*	*19*
Total $X^2{}_j \rightarrow$	*24*	*20*	*16*	*60*

Solution. Correction factor : $C = \dfrac{T^2}{N = rc} = \dfrac{(60)^2}{4 \times 3} = 5 \times 60 = \mathbf{300}.$

Sum of the squares between the columns (SSC) = $\dfrac{(\Sigma x_j)^2}{n_j} - C$

$$= \frac{(24)^2}{4} + \frac{(20)^2}{4} + \frac{(16)^2}{4} - \mathbf{300} \;=\; 144 + 100 + 64 - 300 = \mathbf{8}.$$

Sum of the squares between the rows (SSR) = $\dfrac{T_i^2}{c} - C$

$$= \frac{(16)^2}{3} + \frac{(16)^2}{3} + \frac{9^2}{3} + \frac{19^2}{3} - \mathbf{300} = 318 - 300 = \mathbf{18}.$$

Total sum of the squares (SST) = $\sum_i \sum_j x_{ij}^2 - C$

$= [6^2 + 7^2 + 3^2 + 8^2 + 5^2 + 5^2 + 3^2 + 7^2 + 5^2 + 4^2 + 3^2 + 4^2] - \mathbf{300}$
$= [36 + 49 + 9 + 64 + 25 + 25 + 9 + 49 + 25 + 16 + 9 + 16] - 300 = 332 - 300 = \mathbf{32}.$

Error = Sum of squares (SSE) = SST – (SSR + SSC) = 32 – (8 + 18) = 6.

ANOVA TABLE

Sources of variation	*Sum of squares*	*d.o.f.*	*Mean square*	*F-Value*
Between columns (SSC)	*SSC* = 8	3 – 1 = 2	$\dfrac{SSC}{d.o.f.} = \dfrac{8}{2} = \mathbf{4}$	$F_c = \dfrac{(SSC/v)}{(SSE/v)} = \dfrac{4}{1} = \mathbf{4}$
Between (rows) *SSR*	*SSR* = 18	4 – 1 = 3	$\dfrac{SSR}{d.o.f.} = \dfrac{18}{3} = \mathbf{6}$	
Error (SSE)	*SSE* = 6	6	$\dfrac{SSE}{d.o.f.} = \dfrac{6}{6} = \mathbf{1}$	$F_R = \dfrac{(SSR/v)}{(SSE/v)} = \dfrac{6}{1} = \mathbf{6}$
Total	**32**	C_{r-1} **= 11**	**11**	

Example 13. *The following data represent the number of units of a product produced by 3 different workers using 3 different types of machines.*

Workers	*Machines*		
	A	*B*	*C*
X	*8*	*32*	*20*
Y	*28*	*36*	*38*
Z	*6*	*28*	*14*

Test (i) whether the mean productivity is the same for the different machine types, and (ii) whether the three workers differ with respect to mean productivity.

Solution 1. Null hypothesis :

H_0: (a) The mean productivity is same for the three different machines.

(b) The workers do not differ with respect to mean productivity.

Alternate Hypothesis :

H_1: (a) The mean productivity is not same for the three different machines.

(b) The workers differ with regard to mean productivity.

2. Computation of Test Statistic. We have the following table to calculate different **sum of squares.**

	Machine			
Workers	***A***	***B***	***C***	***Total***
X	8	32	20	60
Y	28	36	38	102
Z	6	28	14	48
Total	**42**	**96**	**72**	
Mean $\bar{X}$	$\frac{42}{3} = \mathbf{14}$	$\frac{96}{3} = \mathbf{32}$	$\frac{72}{3} = \mathbf{24}$	**T = 210**

Here $N = 9$.

Correction Factor $C = \frac{T^2}{N} = \frac{(210)^2}{9} = \mathbf{4900}.$

SST = Total Sum of Squares = $\Sigma\Sigma X_{ij}^2 - C$

$= [(8)^2 + (28)^2 + (6)^2 + (32)^2 + (36)^2 + (28)^2 + (20)^2 + (38)^2 + (14)^2] - C$

$= 64 + 784 + 36 + 1024 + 1296 + 784 + 400 + 1444 + 196 - \mathbf{4900} = 6028 - 4900 = \mathbf{1128}.$

SSC = Sum of squares between Columns (Machines) = $\frac{(42)^2}{3} + \frac{(96)^2}{3} + \frac{(72)^2}{3} - C$

$= \frac{1}{3}[1764 + 9216 + 5184] - \mathbf{4900} = [5388 - 4900] = \mathbf{488}.$

SSR = Sum of squares between rows (workers) = $\frac{1}{3}[(60)^2 + (102)^2 + (48)^2] - C$

$= \frac{1}{3}[3600 + 10404 + 2304] - \mathbf{4900} = \mathbf{536}.$

Residual or Error SSE = SST – (SSC + SSR) = 1128 – (488 + 536) = **104.**

The various sums of squares are put in the following Two - way **ANOVA** Table.

ANOVA TABLE

Source of variation	*Sum of squares*	*Degree of freedom (v)*	*Mean square* $= \frac{\text{Sum of squares}}{d.o.f.}$	*F – Ratio*
Between machines	488 = *SSC*	3 – 1 = 2	$\frac{SSC}{d.o.f.} = \frac{488}{2} = \mathbf{244}$	$F_C = \frac{244}{26} = \mathbf{9.38}$
Between workers	536 = *SSR*	3 – 1 = 2	$\frac{SSR}{d.o.f.} = \frac{536}{2} = \mathbf{268}$	$F_R = \frac{268}{26} = \mathbf{10.31}$
Residual (Error)	104 = *SSE*	(3 – 1) (3 – 1) = 4	$\frac{SSE}{d.o.f.} = \frac{104}{4} = 26$	
Total	**1128**	**8**		

Decision. **(*a*) The Computed value of F_c = 9.38 > the table value of F at α = 0.05 and for 2 and 4 degrees of freedom = $F_{0.05}$ (2, 4) = 6.95 ⇒ the null hypothesis is rejected ⇒ H_1 is accepted ⇒ the mean productivity is not same for the three different machines.**

(*b*) Since F_R = 10.31 > table value of F at α = 0.05 and for 2 and 4 degrees of freedom = $F_{0.05}$ (2, 4) = 6.95 ⇒ H_0 is rejected and H_1 is accepted ⇒ the three workers differ with regard to mean productivity.

Example 14. *In a certain factory, production can be accomplished by four different workers on five different types of machines. A sample study was made with two-fold objectives of examining whether the four workers differ with respect to mean productivity and whether the mean productivity is the same for the five different machines. The researcher involved in this study, reports while analysing the gathered data as under.*

(*i*) *Sum of squares for variance between machines = 35.2.*

(*ii*) *Sum of squares for variance between workmen = 53.8.*

(*iii*) *Sum of squares for variance = 174.2.*

Set up Anova table for the given information and draw the inference about variances at 5% level of significance.

Solution. Null Hypothesis H_0:

(a) The four workers do not differ with respect to their mean productivity,

(b) The mean productivity is the same for the five different machines.

$$\textbf{F-statistic or F-value} = \frac{\textbf{Explained variance}}{\textbf{Unexplained variance}} = \frac{\textbf{MSC}}{\textbf{MSE}}$$

Analysis of Variance Table

Source of variation	*Sum of squares*	*d.o.f (x)*	*Mean squares* $= \dfrac{\text{Sum of Squares}}{d.o.f.}$	*F-Ratio*
Between machines	$35.2 = SSC$	$5 - 1 = 4$	$\dfrac{SSC}{d.o.f.} = 8.8$	$8.8/7.1 = 1.24 = F_C$
Between workmen	$53.8 = SSE$	$4 - 1 = 3$	$\dfrac{SSE}{d.o.f.} = 17.93$	$17.93/7.1 = 2.53 = F_R$
Residual (Error)	$174.2 - 35.2 - 53.8 = 85.2$	$19 - 4 - 3 = 12.$	7.1	

Critical Value. The value of F for 4 and 12 *d.f.* at $\alpha = 0.05$ is 3.49 and for 3 and 12 *d.f.* is 3.4.

Decision. (*a*) The calculated value of $F_c = 1.24 <$ Table value of $F_{0.05} = 3.49$

$\Rightarrow$ accept the null hypothesis.

(*b*) The Calculated value of $F_R = 2.53 <$ Table value of $F_{0.05}$ for 3 and 12 *d.f.* = 3.4

$\Rightarrow$ accept the null Hypothesis H_0. Hence **there is no significant difference in the productivity of different types of machines as well as between mean productivity of workers.**

Example 15. *For the following data representing the number of units of production per day turned out by five workers using four machines, set-up the Anova table (assumed the origin at 20):*

Worker	*Machine Type*			
	A	*B*	*C*	*D*
I	*4*	*– 2*	*7*	*– 4*
II	*6*	*0*	*12*	*3*
III	*– 6*	*– 4*	*4*	*– 8*
IV	*3*	*– 2*	*6*	*– 7*
V	*– 2*	*2*	*9*	*– 1*

Solution.

Worker	***Machine Type***				***Total***
	A	***B***	***C***	***D***	
I	4	– 2	7	– 4	5
II	6	0	12	3	21
III	– 6	– 4	4	– 8	– 14
IV	3	– 2	6	– 7	0
V	– 2	2	9	– 1	8
Total	**5**	**– 6**	**38**	**– 17**	**20**

$$\sum_i \sum_j x_{ij}^2 = 16+4+49+16+36+0+144+9+36+16+16+64+9+4+36+49+4+4+81+1$$

$$= \mathbf{594}$$

$$\text{TSS} = \sum_i \sum_j x_{ij}^2 - \frac{G^2}{N} = 594 - \frac{(20)^2}{20} = \mathbf{574} \qquad [\because G = 20]$$

$$SSC\ (SS\ \text{Machines}) = \sum_{j=1}^{c} \frac{C^2}{r} - \frac{G^2}{N} = \frac{5^2 + (-6)^2 + 38^2 + (-17)^2}{5} - \frac{(20)^2}{20} = \mathbf{338.8}$$

$$SSR\ (SS\ \text{Workmen}) = \sum_{i=1}^{c} \frac{R^2}{c} - \frac{G^2}{N} = \frac{5^2 + 21^2 + (-14)^2 + 8^2}{4} - \frac{(20)^2}{20} = \mathbf{161.5}$$

$$SSE = TSS - SSC - SSR = 574 - 338.8 - 161.5 = \mathbf{73.7.}$$

ANOVA TABLE

Sources of Variations	*d.o.f.*	*Sum of Squares*	*Mean Squares* $\left(= \frac{\text{Sum of Squares}}{d.o.f.}\right)$	*Variance Ratio*
Machine	3 $(c-1)$	338.8	$112.93 = MSC$	**18.39** $\left(= \frac{MSC}{MSE}\right)$
Workmen	4 $(r-1)$	161.5	$40.38 = MSR$	**6.58** $\left(= \frac{MSR}{MSE}\right)$
Eror	12 $(c-1)(r-1)$	73.7	$6.14 = MSE$	
Total	**19** **$(cr-1)$**	**574.0**		

Example 16. *Following are the weekly sale records (in Rs.) of three salesmen A, B and C of a company during 13 sale-calls:*

A :	*300*	*400*	*300*	*500*	
B :	*600*	*300*	*300*	*400*	
C :	*700*	*300*	*400*	*600*	*500*

Test whether the sales of three salesmen are different.

$[F_{2,10,0.05} = 4.10 \text{ and } F_{2,13,0.05} = 3.81]$

Solution. $H_0 : \mu_1 = \mu_2 = \mu_3$

For making computation easier, let us take $x_{ij} = y_{ij} / 100$, where y_{ij} is the original observational value and x_{ij} is the new observational value when the original one has been divided by 100.

	A	*B*	*C*	
	3	6	7	
	4	3	3	
	3	3	4	
	5	4	6	
			5	
Total	**15**	**16**	**25**	**56**

Correction factor: $C = \frac{56 \times 56}{13} = \mathbf{241.23}$

$$\text{SSC} = \frac{225}{4} + \frac{256}{4} + \frac{625}{5} - \mathbf{241.23}$$
$$= 56.25 + 64 + 125 - 241.23 = 245.25 - 241.23 = \mathbf{4.02}$$
$$\text{SST} = 9 + 16 + 9 + 25 + 36 + 9 + 9 + 16 + 49 + 9 + 16 + 36 + 25 - \mathbf{241.23}$$
$$= 264 - 241.23 = \mathbf{22.77}$$
$$\text{SSE} = SST - SSC = 22.77 - 4.02 = \mathbf{18.75}$$

ANOVA TABLE

Source of variation	*Sum of squares*	*d.o.f.*	*Mean squares*	*F*
Column Means	4.02	2	2.01	
Error	18.75	10	1.875	**1.072**

Thus we see that the calculated value of F is less than the tabulated value of F, i.e., the differences are insignificant and we accept the hypothesis H_o. Therefore, we can say that the sales of the three persons do not differ significantly.

11.14 ANALYSIS OF VARIANCE BY CODING METHOD

The coding of data method or coding process further simplifies the calculations. **Coding here, denotes addition, subtraction, multiplication or division of data by a constant number.** ***The test static F, which a ratio, remains unaffected if the calculations for the significance test are made by putting the data in a coded form.*** The following step is required for coding – **Subtract 10 from each item and divide the result by 2, *i.e.*,** use

$$\frac{x - 10}{2}.$$

The rest of steps for calculating SSC, SSE and SST and Test statistic F and its decision are the same as for short cut method discussed in section --------------

Example 17. *Set up an analysis of variance of the following data:*

A	*B*	*C*
20	*20*	*12*
10	*14*	*18*
18	*16*	*20*
8	*14*	*22*

Solution. We shall set the analysis of variance by **coding process.**

			Code Data		
A	*B*	*C*	*A*	*B*	*C*
20	20	12	5	5	1
10	14	18	0	2	4
18	16	20	4	3	5
8	14	22	– 1	2	6
		Total	**8**	**12**	**16**
		$\bar{x}$	**2**	**3**	**4**

Grand Total: $T = 8 + 12 + 16 = \mathbf{36}$.

Correction Factor: $C = \dfrac{T^2}{N} = \dfrac{36 \times 36}{12} = \mathbf{108}.$

Total Sum of Squares (SST) $= \Sigma x_{ij}^2 - C$

$= (5^2 + 0^2 + 4^2 + (-1)^2 + 5^2 + 2^2 + 3^2 + 2^2 + 1^2 + 4^2 + 5^2 + 6^2) - \mathbf{108}.$

$= 25 + 0 + 16 + 1 + 25 + 4 + 9 + 4 + 1 + 16 + 25 + 36 - \mathbf{108}$

$= \mathbf{162 - 108 = 54}.$

Also **degrees of freedom** = 12 – 1 = **11.**

Sum of the squares between the samples (SSC) $= \dfrac{\Sigma X_1^2}{n_1} + \dfrac{\Sigma X_2^2}{n_2} + \dfrac{\Sigma X_3^2}{n_3} - C$

$= \dfrac{8^2}{4} + \dfrac{12^2}{4} + \dfrac{16^2}{4} - \mathbf{108 = 8}.$

Now degrees of freedom between the samples = 3 – 1 = 2.

Sum of the squares within the samples $SSE = SST - SSC = 54 - 8 = 46.$

Then the table of analysis of variance is set up as follows:

ANOVA TABLE

	Sum of squares	*d.o.f.*	*Variance*
Between Samples	8	2	$\frac{8}{2} = \mathbf{4}$
Within samples	46	9	$\frac{46}{9} = \mathbf{5.11}$
	54	**11**	

Test Statistic : F $= \dfrac{\textbf{Greater Variance}}{\textbf{Smaller Variance}} = \dfrac{5.11}{4} = \mathbf{1.28}.$

Example 18. *Set up a two way ANOVA table for the data given below:*

Pieces of field	*Treatment*			
	A	*B*	*C*	*D*
P	45	40	38	37
Q	43	41	45	38
R	39	39	41	41

(*Use coding method subtracting 40 from the given numbers*)

Solution. Substracting 40 from the given numbers, the Coding table can be written as

Coding Table

Pieces of field	*Treatment* A	B	C	D	*Total* (T_i)	T_i^2
P	5	0	– 2	– 3	0	0
Q	3	1	5	– 2	7	49
R	– 1	– 1	1	1	0	0
Total x_j	**7**	**0**	**4**	**– 4**	**$T_i = 7$**	**$T^2 = 49$**

Correction Factor : $C = \frac{T^2}{N} = \frac{0^2 + 7^2 + 0^2}{12} = \frac{49}{12}.$

Sum of the square between Treatments (Columns): $SSC = \frac{\Sigma x_j^2}{r} - C$

$$= \frac{7^2}{3} + \frac{0^2}{3} + \frac{4^2}{3} + \frac{(-4)^2}{3} - C = \frac{49}{3} + 0 + \frac{16}{3} + \frac{16}{3} - \frac{49}{12} = \mathbf{22.912.}$$

Sum of the squares between Fields $(SSR) = \Sigma \frac{T_i^2}{c} - C = \frac{0^2}{4} + \frac{49}{4} + \frac{0}{4} - \frac{49}{12} = \mathbf{8.167.}$

Total Sum $(SST) = 25 + 9 + 1 + 0 + 1 + 1 + 4 + 25 + 1 + 9 + 4 + 1 - \frac{49}{12} = \mathbf{76.917.}$

$[\because SST = \Sigma\Sigma x_{ij}^2 - C]$

$\therefore$ $SSE = SST - (SSC + SSR)$=76.912 – 22.912 – 8.167 = **45.033**

TWO WAY ANOVA TABLE

Sources of variation	*Sum of squares*	*d.o.f.*	*Mean square* $= \frac{Sum\ of\ Squares}{d.o.f.}$	*F = ratio*
Between treatments **SSC**	22.917	3	$\frac{SSC}{d.o.f.} = \mathbf{7.639}$	$F_C = \frac{7.639}{7.639} = \mathbf{1.0}$
Between fields **(SSR)**	8.167	2	$\frac{SSR}{d.o.f.} = \mathbf{4.083}$	$F_R = \frac{4.083}{7.639}$
Error (SSE)	45.833	6	$\frac{SSE}{d.o.f.} = \mathbf{7.639}$	$= \mathbf{0.534.}$
Total	**76.917**	**11**		

Example 19. *There are three main brands of a certain powder. A set of its 120 sales is examined and found to be allocated among four groups (A, B, C, and D) and brands (I, II and III) as shown here under*

Brands	*Groups*			
	A	*B*	*C*	*D*
I	*0*	*4*	*8*	*15*
II	*5*	*8*	*13*	*6*
III	*18*	*19*	*11*	*13*

Is there any significant difference in brands preference? Answer at 5 percent level using one-way Anova. (Take 10 as the code value to subtract it from all given values in your working.)

Solution. We shall prepare the ANOVA table using the coding method by subtracting 10 from all the items.

CODING TABLE

		I	*II*	*III*
Group :	A	– 10	– 5	8
	B	– 6	– 2	9
	C	– 2	3	1
	D	5	– 4	3
	Total	**– 13**	**– 8**	**21**
			Grand Total T = 0	

We compute the following terms for setting up an ANOVA TABLE:

C = Correction factor $= \dfrac{T^2}{N} = \dfrac{0 \times 0}{12} = \mathbf{0}.$

Sum of squares for total variance: $SST = \Sigma x_{ij}^2 - C$

$= (100 + 36 + 4 + 25 + 25 + 4 + 9 + 16 + 64 + 81 + 1 + 9) - 0 = \mathbf{374}.$

Sum of squares for variance amongst brands: $SSC = \dfrac{\Sigma x_j^2}{r} - C$

$$= \frac{(-13)^2}{4} + \frac{(-8)^2}{4} + \frac{(21)^2}{4} - 0 = \frac{169}{4} + \frac{64}{4} + \frac{441}{4} - 0 = \frac{169 + 64 + 441}{4} = \frac{674}{4} = \mathbf{168.5}.$$

Now $SSR = SST - SSC = 374 - 168.5 = \mathbf{205.5}.$

One-Way ANOVA Table

Source of variance	*S.S.*	*d.o.f.*	*Mean* squares $= \dfrac{\textit{Sum of Squares}}{d.o.f.}$	*F-ratio*
SSC = between brands	168.5	2	84.25	$\dfrac{84.25}{22.83}$
SSR = Within brands	205.5	9	22.83	= **3.69**
Total	**374.0**	**11**		

Null Hypothesis H_0: There is no significant difference in brand preference.

Decision. Since the computed value of $F = 3.69 <$ Table value of F at $\alpha = 0.05 = 4.26$ so the null hypothesis H_0 is accepted ⇒ there is no significant difference in brands preference.

EXERCISE – 11.2

1. Four salesmen were posted in different areas by a company. The number of units of commodity X sold by them are as follows:

A	20	23	28	29
B	25	32	30	21
C	23	28	35	18
D	15	21	19	25

On the basis of his information, can it be concluded that there is a significant difference in the performance of the four salesmen? (Given for $v_1 = 3$ and $v_2 = 12$, $F_{0.05} = 3.24$).

2. A manufacturing company has purchased three new machines of different makes and wishes to determine whether one of them is faster than the other in producing a certain output. Five hourly producing figures are observed at random from each machine and the results are given below:

Observations	A_1	A_2	A_3
1	25	31	24
2	30	39	30
3	36	38	28
4	38	42	25
5	31	35	28

Use analysis of variance and determine whether the machines are significantly different in their mean speed. (Given at 5% level F_2, 12 = 3.89).

[**Hint.**

Analysis of Variance Table

Sources of variation	*Sum of squares*	*Degree of freedom*	*Mean squares.*
Between samples	250	2	125
Within samples	200	12	16.667
Total	450	14	

$$F = \frac{\textbf{Variance between samples}}{\textbf{Variance within samples}} = \frac{\textbf{125}}{\textbf{16.667}} = \textbf{7.5.}$$

Given at 5% $F_{0.05}$ (2, 12) = 3.89 < computed $F \Rightarrow$ The hypothesis is rejected $\Rightarrow$ the machines are significantly different in their mean speed.]

3. To study the performance of three detergents and three different water temperatures, the following 'Whiteness' readings were obtained with specially designed equipment.

Water Temp.	*Detergent A*	*Detergent B*	*Detergent C*
Cold water	57	55	67
Warm water	49	52	60
Hot water	54	46	58

Perform a two-way analysis of variance, using 5% level of significance.

4. The following table gives the yield on 15 sample plots under three varieties of seed:

A	*B*	*C*
20	18	25
21	20	28
23	17	22
16	15	28
20	25	32

Find out whether the average yield of land under different varieties of seed shows significant differences.

5. The three samples given below have been obtianed from normal populations with equal variances. Test the hypothesis at 5% level that the population means are equal.

8	7	12
10	5	9
7	10	13
14	9	12
11	9	14

(The table value of F at 5% level of significance for $v_1 = 2$, and $v_2 = 12$ is 3.88).

6. The Amit Merchandising company wishes to test whether its three salsmen *A*, *B* and *C* tend to make sales of the same size or whether they differ in their selling ability as measured by the average size of their sales. During the last week there have been 14 sale calls – *A* made 5 calls, *B* made 4 calls and *C* made 5 calls. Following are the weekly sales record of the three salesmen:

A	*B*	*C*
Rs.	Rs.	Rs.
300	600	700
400	300	300
300	300	400
500	400	600
0		500

Perform the analysis of variance and draw your conclusions.

7. Four experiments determine the moisture contents of samples of a powder, each man taking a sample from each of six consignments. Their assessments are given below:

	Consignments					
Observers	1	2	3	4	5	6
1	9	10	9	10	11	11
2	12	11	9	11	10	10
3	11	10	10	12	11	10
4	12	13	11	14	12	10

Analyse the data and discuss whether there is any significant difference between consignments or between observers.

8. A tea company appoints four salesmen, *A*, *B*, *C* and *D* and observes their sales in three seasons–summer, winter and monsoon. The figures (in lakhs) are given in the following table:

Seasons	*Salesmen*				*Season's Total*
	A	*B*	*C*	*D*	
Summar	36	36	21	35	128
Winter	28	29	31	32	120
Monsoon	26	28	29	29	112
Salesmen's Total	90	93	81	96	360

Carry out an analysis of variance and test whether there is any significant difference in the salesman and in the seasons, so far as sales are concerned.

[**Hint.** To carry out an analysis of variance and test whether there is a significant difference in the salesmen and in the seasons as far as sales are concerned, we have the following table:

ANOVA Table

Sources of variations	Sum of squares	d.f.	mean square	F
Between columns (salesmen)	42	3	14	$F_C = \frac{14}{22.67} = 0.618$
Between rows (seasons)	32	2	16	
Residual	136	6	22.67	$F_R = \frac{16}{22.67} = 0.796$
	210	11		

Here $F_0 = 0.618 < F_{0.05}$ (3, 6) = 4.76 ⇒ Sales of different salesman do not differ significantly. Also $F_R = 0.706 < F_{0.05}$ (3, 6) = 5.14 ⇒ there is no significant difference in seasons as far as sales are concerned.]

ANSWERS

ANOVA Table

1.

Sources of variation	Sum of squares	Degree of freedom	Mean squares
Between	116	3	38.67
Within	338	12	28.17
Total	454	12	

2. The machines are significantly different in their mean speed.
3. $F = 9.85$, there is a significant difference in the three varieties of detergents.

ANOVA Table

Sources of variation	Sum of squares	d.o.f.	mean squares
Between Columns	304.22	2 (Detergents)	152.110
Between Rows (Temp.)	73.55	2	36.775
Residual	61.79	4	15.445
	439.56	8	

4. The average yields of land under different varieties of seed show difference.
5. Population means are not equal.
6. The three salesmen do not differ in their selling activity as measured by the average size of their sales.
7.

Sources of Variation	Sum of Squares	Degree of Freedom	Mean sum of Squares	Variance Ratio
Between columns (consignments)	9.71	3	3.237	$F_C = \frac{3.237}{0.875} = \mathbf{3.70}$
Between rows (observers)	13.13	5	2.626	$F_R = \frac{2.626}{0.875} = \mathbf{3.0}$
Residual	13.12	15	0.875	
Total	35.96	23		

There is a significant difference in the sales of different salesmen.

8. (*i*) There is no significant difference in the sales of different salesmen.
(*ii*) There is no significant difference in the seasons as far as their sales are concerned.

12
CHAPTER

CORRELATION ANALYSIS

12.1 INTRODUCTION

We know that the area A of a circle of radius r is given by $A = \pi r^2$. This gives us that a circle with larger radius will always have a larger area than a circle with a smaller radius. Thus there exists a functional relationship between A and r. Now consider the two variables h and w the height and weight of a given group of people. We know that in most of the cases the taller people will have higher weights as compared to persons of shorter height. But we also come across, at the same time, a short fat person who may have a higher weight as compared to a lean tall person. Thus we cannot determine the weight of person from his height though we could try to make a reasonable guess about his weight, *i.e.*, we cannot put the statement. "*A tall person is more likely to be heavier in weight than a short person*", as an exact rule to be true in all cases.

In this chapter we shall study the relationship of the type: *relation between height and weight* or *height and age or price and demand or wage and price index*, etc. Such a relationship is called **statistical relationship.** Special methods have been developed to discover the existence of **statistical relationship** *between the two variables from the bivariate data. When both variables in the bivariate data are quantitative, we use the term* **Correlation Analysis** *to describe the methods designed to find out if the statistical relationship between the two variables exists or not.* But in a bivariate data having both the variables as quantitative variables, *the methods of estimating the likely value of one variable from the known value of the other variable form a part of* **Regression Analysis.** We shall also study some methods of correlation and regression analysis when both the variables are quantitative.

Croxton and Cowden defined the **Correlation** as

"The relationship of quantitative nature. The appropriate statistical tool for discovering and measuring the relationship and expressing it in brief formula is known as correlation".

According to the statistician **A.M. Tuttle,**

"Correlation is an analysis of the covariation between two or more variables."

12.2 CONCEPT OF CORRELATION

Correlation is a statistical tool which studies the relationship between two variables, and Correlation Analysis involves various methods and techniques used for studying and measuring the extent of the relationship between two variables.

The correlation expresses rates between the groups of items but not between the individual items. The relationship between the two variables is **not functional.**

In other words, *correlation analysis is a statistical procedure by which we can determine the degree of association or relationship between two or more variables.* In this chapter we shall restrict ourselves to the case of two variables only. The amount of correlation in a sample (of data) is measured by the sample **coefficient of correlation,** which is, generally, denoted by ***r* or by ρ.**

12.3 COEFFICIENT OF CORRELATION

Correlation *may be defined as a tendency towards interrelation variation and the* **Coefficient of correlation** *is a measure of such a tendency, i.e., the degree to which the two variables are interrelated is measured by a coefficient which is called the* **Coefficient of correlation.** *It gives the degree of correlation.*

Definition. *The relationship between two variables such that a change in one variable results in a positive or negative change in the other variable and also a greater change in one variable results in corresponding greater or smaller change in the other variable is known as* **correlation.**

The coefficient of correlation between the two variables x, y is generally denoted by r or r_{xy} or $\rho\ (x, y)$ or ρ.

12.3.1 Properties of co-efficient of correlation

1. ***It is a measure of the closeness of a fit in a relative sense.***
2. ***Correlation coefficient lies between – 1 and + 1, i.e., $-1 \le r \le 1$.***
3. ***The correlation is perfect and positive if r = 1 and it is perfect and negative if r = – 1.***
4. ***If r = 0, then there is no correlation between the two variables and thus the variables are said to be independent.***
5. ***The correlation coefficient is a pure number and is not affected by a change of origin and scale.***
6. ***It is a relative measure of association between two or more variables.***

12.4 BIVARIATE DISTRIBUTION

There are two types of distributions.

(*i*) **Univariate Distribution.** These are the distributions in which there is only **one variable** such as the heights of the students of a class or marks obtained by the students of a class.

(*ii*) **Bivariate Distribution.** Distribution involving **two discrete variables** is called a **Bivariate distribution.** For example,

1. The heights and weights of the students of a class in a school.
2. The daily petrol used by a scooter owner and the mileage covered.

Frequency. Let (x_i, y_j), $i = 1, 2, 3, ..., m$; $j = 1, 2... n$, be a bivariate distribution. If the pair (x_i, y_j) occurs f_{ij} times, then f_{ij} is called the **frequency** of the pair (x_i, y_j) and $\sum_{i-1}^{m} \sum_{j=1}^{n} f_{ij} = N =$ the total frequency.

In this section, we shall be dealing with a bivariate population. In a bivariate population we are interested to know whether there exists some sort of functional relationship between the two variables involved. In other words, we wish to know if the change in one variable affects a change in the other variable. The knowledge of such a relationship enables us to analyse the bivariate population and the variables are said to be correlated (*i.e.;* to predict and control the events). **The relationship between the two variables can be established and measured by means of correlation.**

12.5 COVARIANCE

Before we study the correlation analysis we introduce the concept of **covariance between two quantitative variables X and Y.** Let the corresponding values of the two variables X and Y on the given set of n units of observations be given by the ordered pairs

$$(x_1, y_1), (x_2, y_2), (x_3, y_3), ..., (x_n, y_n).$$

Then **the covariance between X and Y is denoted by Cov. (X, Y).** It is defined as

$$\text{Cov. (X, Y)} = \frac{(x_1 - \bar{x})(y_1 - \bar{y}) + (x_2 - \bar{x})(y_2 - \bar{y}) + ...+ (x_n - \bar{x})(y_n - \bar{y})}{n}$$

$$= \frac{1}{n} \sum_{i=1}^{n} (x_i - \bar{x})(y_i - \bar{y}) \qquad ...(1)$$

where $\bar{x}$ and $\bar{y}$ are the means of X and Y respectively.

$$i.e.,\ \bar{x} = \frac{x_1 + x_2 ... + x_n}{n};\ \bar{y} = \frac{y_1 + y_2 + ... + y_n}{n}$$

Covariance is an "absolute measure".

12.5.1 Calculation of Covariance

The above formula for the calculation of covariance is complicated and may have more chances of the occurrence of an error. We now give below a formula which makes the calculations easier to carry out and which also reduces the chances of error.

The formula is

$$\text{Cov. (X, Y)} = \frac{1}{n}\left[\sum_{i=1}^{n} x_i y_i - \frac{1}{n}\left(\sum_{i=1}^{n} x_i\right)\left(\sum_{i=1}^{n} y_i\right)\right] \quad ...\ (2)$$

$$\text{Cov. (XY)} = \text{E (XY)} - \text{E(X) E(Y)}, \quad ...\ (3)$$

where **E (X), E (Y), E (XY)** are the expectations of **X, Y,** and **XY** respectively.

WORKING RULE

STEP I. Calculate the sums $\sum_{i=1}^{n} x_i, \sum_{i=1}^{n} y_i$.

STEP II. Calculate the sum $\sum_{i=1}^{n} x_i\ y_i$ of the products $x_i\ y_i$.

STEP III. Obtain the difference $\sum_{i=1}^{n} x_i\ y_i - \frac{1}{n}\left(\sum_{i=1}^{n} x_i\right)\left(\sum_{i=1}^{n} y_i\right)$

STEP IV. Divide the difference obtained in **STEP III by n to get the Cov. (X, Y)**

Example 1. *Calculate the covariance of the following pairs of observations of the two variates X and Y.*

(1, 6), (2, 9), (3, 6), (4, 7), (5, 8), (6, 5), (7, 12), (8, 3), (9, 17), (10, 1).

Solution. Here $\Sigma x_i = 1 + 2 + 3 + 4 + 5 + 6 + 7 + 8 + 9 + 10 = \mathbf{55}.$

$\Sigma y_i = 6 + 9 + 6 + 7 + 8 + 5 + 12 + 3 + 17 + 1 = \mathbf{74}.$

$\Sigma\ x_i\ y_i = 1 \times 6 + 2 \times 9 + 3 \times 6 + 4 \times 7 + 5 \times 8 + 6 \times 5 + 7 \times 12 + 8 \times 3 + 9 \times 17 + 10 \times 1$
$= 6 + 18 + 18 + 28 + 40 + 30 + 84 + 24 + 153 + 10 = \mathbf{411}.$

$$\therefore \quad \text{Cov. (X, Y)} = \frac{1}{n}(\Sigma x_i\ y_i - \frac{1}{n}\Sigma x_i \Sigma y_i)$$

$$= \frac{1}{10}\left(411 - \frac{1}{10} \times 55 \times 74\right) = \frac{1}{10}(411 - 407) = \mathbf{0.4}.$$

Example 2. *Calculate the covariance of the following pairs of observations of the variables X and Y.*

(15, 44), (20, 43), (25, 45), (30, 37), (40, 34), (50, 37).

Solution. Here $\bar{x} = \frac{15 + 20 + 25 + 30 + 40 + 50}{6} = \frac{180}{6} = \mathbf{30};$

$$\bar{y} = \frac{44 + 43 + 45 + 37 + 34 + 37}{6} = \frac{240}{6} = \mathbf{40}.$$

Table

x_i	$x_i - 30$	y_i	$y_i - 40$	$(x_i - 30)(y_i - 40)$
15	– 15	44	4	– 60
20	– 10	43	3	– 30
25	– 5	45	5	– 25
30	0	37	– 3	0
40	10	34	– 6	– 60
50	20	37	– 3	– 60

Here $\Sigma (x_i - \bar{x})(y_i - \bar{y}) = -60 - 30 - 25 + 0 - 60 - 60 = -235$. Also $n = 6$.

$$\therefore \text{Cov.}(X, Y) = \frac{\Sigma(x_i - \bar{x})(y_i - \bar{y})}{n} = \frac{235}{6} = -39.17 \text{ nearly.}$$

12.6 TYPES OF CORRELATION

12.6.1 Positive Correlation

If the values of the two variables deviate in the same direction, i.e., an increase in the value of one variable results, on an average, in a corresponding increase in the values of the other variable or if a decrease in the values of one variable say X results on an average a corresponding decrease in the other variable say Y, the correlation is said to be **positive** or **direct.** The height and weight of a growing child is an example of positive correlation.

12.6.2 Negative or Inverse Correlation

The correlation is said to be **negative** *or* **inverse** *if the two variables X and Y deviate in the opposite direction, i.e., if the increase or decrease in the values of the variable X results, on the average, in a corresponding decrease (or increase) in the values of other variable Y.*

The price and demand of a commodity or the consumption of electricity and its bill etc. are examples of negative or inverse correlation.

12.6.3 Linear Correlation

If the plotted points (x_i, y_i) are approximately on or near about a straight line, than the correlation between the variables is said to be linear (see Fig. 12.1). For example, the correlation between the savings and earnings of a man is a linear correlation.

12.6.4 Perfectly Linear Correlation

When all the plotted points lie exactly on a straight line, then the correlation is said to be perfectly linear. (See Fig. 12.2)

12.6.5 Perfect Correlation

If the deviation in one variable is followed by a corresponding and proportional deviation in the other, then the correlation is said to be **perfect correlation.**

Perfect Positive Correlation

Fig. 12.1

12.6.6 Direct or Perfect Positive Correlation

If the correlation is perfectly linear and the line runs from the ***lower left hand corner to the upper right-hand corner,*** *then it is called the direct or positive perfect correlation.*

In other words, *if the equal proportional changes in the two variates x any y are in same direction, then it is a perfect positive correlation* (see Fig. 12.1).

In this case $r = 1$ or $\rho(x, y) = 1$.

12.6.7 Inverse or Perfect Negative Correlation.

It is a perfectly linear correlation in which the line runs ***from upper left-hand corner to the lower right-hand corner*** (see Fig. 12.2).

In other words, if the equal proportional changes in the two variates x and y are in opposite direction, then the relation between x and y is *inverse or perfect negative correlation.*

In this case $r = -1$, *i.e.*, $\rho(x, y) = -1$.

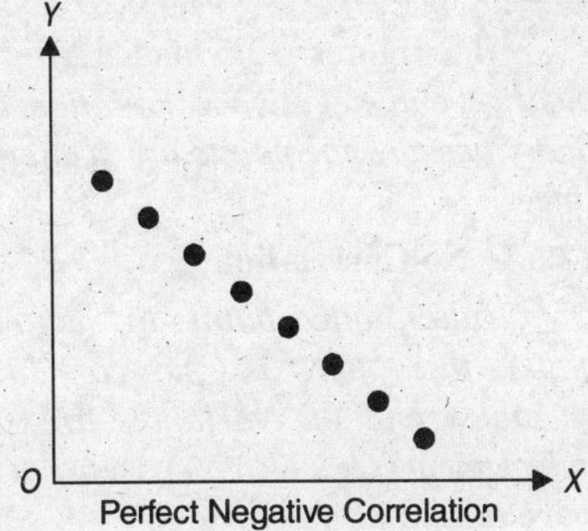

Perfect Negative Correlation

Fig. 12.2

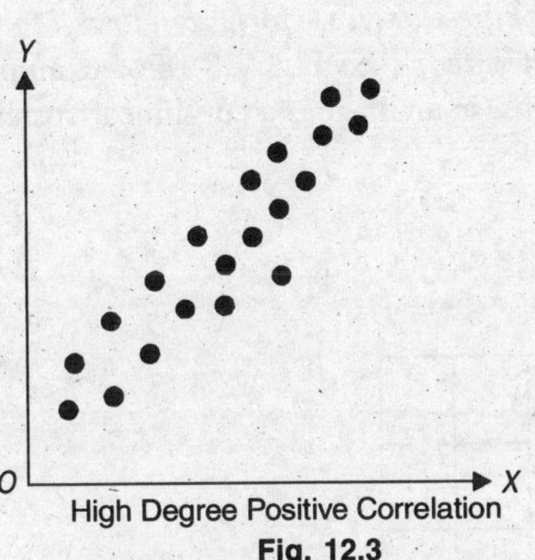

High Degree Positive Correlation

Fig. 12.3

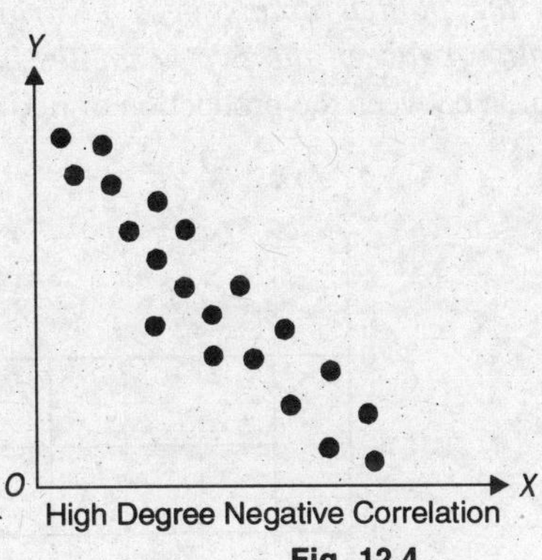

High Degree Negative Correlation

Fig. 12.4

7.6.8 High Degree Positive Correlation

If the plotted points (x_i, y_i), $i = 1, 2, 3, \ldots, n$ *fall in a narrow band and the points are rising from* ***lower left hand corner to the upper right hand corner,*** *then there will be a high degree of positive correlation between x and y* (see Fig. 12.3).

7.6.9 High Degree Negative Correlation

If the plotted points (x_i, y_i), $i = 1, 2, \ldots, n$ *fall in a narrow band from* ***upper left hand corner to the lower right hand corner***, *then there will be a high degree of negative correlation between x and y* (see Fig. 12.4).

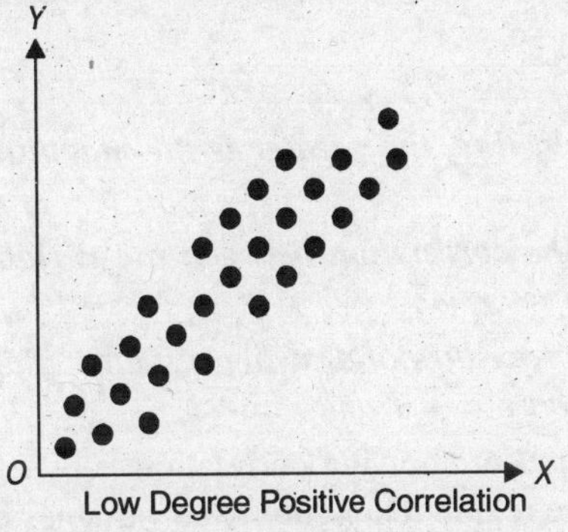

Low Degree Positive Correlation

Fig. 12.5

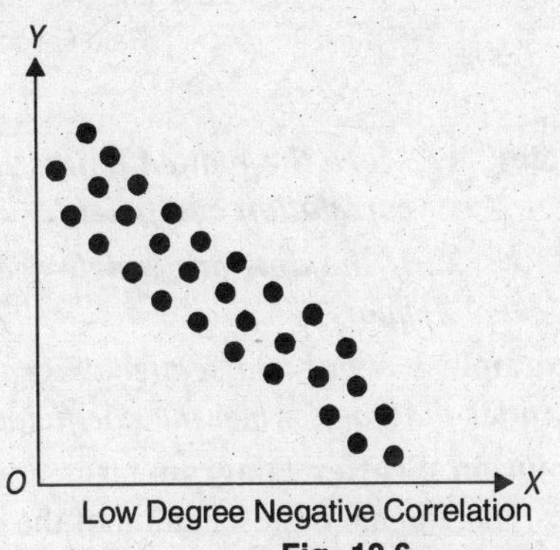

Low Degree Negative Correlation

Fig. 12.6

12.6.10 Low Degree Positive or Negative Correlation

The figures 12.5 and 12.6 *show a low degree of positive and negative correlation between the variates x and y because the points are scattered much away from the line.*

12.6.11 No Correlation

If the plotted points lie scattered all over the graph paper, then there is no correlation between the two variables and the variables are said to be statistically independent (see Fig. 12.7). Here ***r* = 0. The variables *x* and *y* are said to be independent.**

No Correlation

Fig. 12.7

12.6.12 Curvilinear Correlation

If the plotted points in a bivariate distribution more or less form a curve, then the correlation between the two variables is said to be curvilinear (see Fig. 7.8). For example the correlation between the production of rice and rainfall in cms. in a village is a curvilinear correlation.

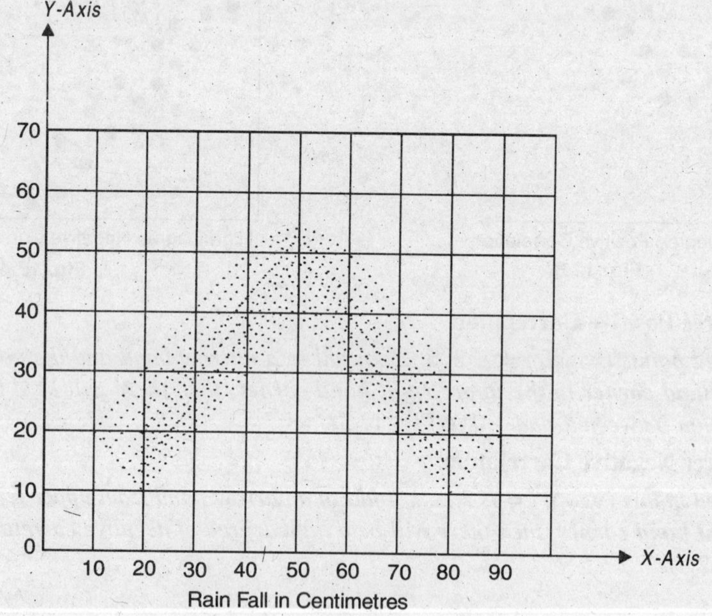

Fig. 12.8.

Notes. 1. ***More the plotted points lie on a straight line, the greater is the magnitude of correlation coefficient.***

2. ***If the dots are widely scattered, then the correlation between the variables is poor.***

Example 3. *What is a scatter diagram? Give the procedure of drawing a scatter diagram. Draw scatter diagrams when the coefficient of correlation r = + 1 and r = – 1.*

Solution. Scatter Diagram. It is a graphical method of showing correlation between two variables.The graphic representation of the data in a square region representing the variables in two axes, X, Y by way of points is called a scatter diagram.

On a paper draw two axes at right angles one axis corresponding to X and the other to Y. For each of the data, there will correspond a pair of values X, Y which in turn will correspond to a point whose abscissa on the diagram is X and the ordinate is Y. The data represented in this way will give a swarm of points on the diagram. We can interpret the way scatter of cluster as the properties of relationship between the two variables.

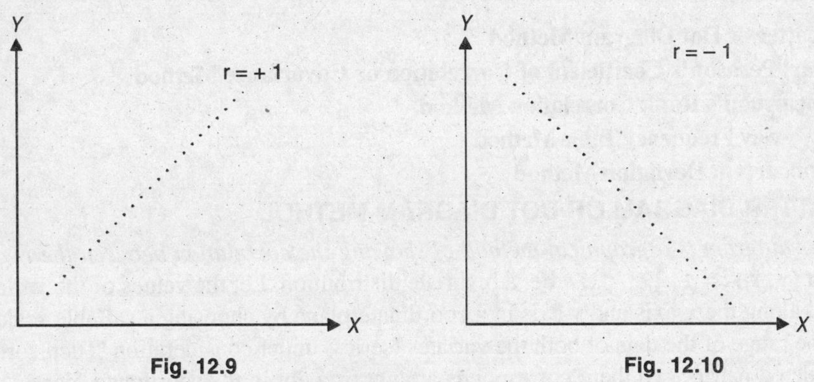

Fig. 12.9

Fig. 12.10

EXERCISE – 12.1

1. Define the term correlation. Explain the concept of positive and negative correlation with examples.
2. State the nature of the following correlation (positive, negative or no correlation)
   (*i*) The amount of rainfall and the yield of crops.
   (*ii*) The colour of a saree and the intelligence of the girl who wears it.
   (*iii*) The sale of woollen garments and the day temperature.
   (*iv*) Age of applicants for life insurance and the premium of insurance.
   (*v*) Demand for goods and their prices under normal time.
   (*vi*) Production of pig iron and soot contents in Durgapur.
   (*vii*) Unemployment index and the purchasing power of the common man.
3. While drawing a scatter diagram if all points appear to form a straight line going downward from left to right, then it is inferred that there is......
   (*a*) Perfect positive correlation, (*b*) Simple positive correlation
   (*c*) Perfect negative correction, (*d*) No correction
4. Write a short note on scatter diagram. Draw sketches of scatter diagram to show the following correlation between two variables x and y.
   (*i*) Linear, (*ii*) Non-linear (*iii*) linear and perfect (*iv*) x and y uncorrelated
5. Distinguish clearly between
   (*i*) Positive and Negative correlation (*ii*) linear and Non-linear correlation
6. Define scatter diagram and draw four different types of scatter diagrams for different degrees of correlation between two correlated variables.
7. Define scatter diagram and depict the scatter diagrams for perfect correlation, perfect negative correlation, non-linear correlation and no correlation.
8. What is correlation ? Explain the implications of positive and negative correlation. Show by means of scatter diagram, the presence of perfect positive and perfect negative correlation.

ANSWERS

2. (*i*) Positive, (*ii*) No correlation, (*iii*) positive (*iv*) positive
   (*v*) Negative, (*vi*) positive, (*vii*) negative.
3. Perfect negative correlation.

12.7 METHODS TO STUDY CORRELATION

The intensity of correlation between two variates x and y can be measured by the following methods.

(i) Scatter or Dot Diagram Method
(ii) Karl Perason's Coefficient of Correlation or Covariance Method
(iii) Spearman's Rank Correlation Method.
(iv) Two-way Frequency Table Method
(v) Concurrent Deviation Method

12.8 SCATTER DIAGRAM OR DOT DIAGRAM METHOD

Scatter diagram is a graphical method of showing the correlation between the two variables x and y. Let (x_i, y_i) $i = 1, 2, 3, ..., n$ be a bivariate distribution. Let the values of the variables x and y be plotted along the x-axis and y-axis in a coordinate plane by choosing a suitable scale, so that it measures the range of the data of both the variates (series) under consideration. Then corresponding to every ordered pair (x_i, y_i) there corresponds a point or a dot in the coordinate plane.

The diagram of dots or points so obtained is called a **scatter diagram** *or a* **dot diagram.**

The scatter diagram may indicate both degree and the type of correlation. From scatter diagram we can form a fairly good, though rough idea about the relationship between the two variables. The diferent types of correlation are depicted by means of scattered diagrams as shown by figures **12.1, 12.2, 12.3, 12.4, 12.5, 12.6. and 12.7.** **on pages 364 to 366.**

12.8.1 Advantages and Disadvantages of Scatter Diagram

Advantages

1. *It is readily comprehensible and enables us to form a rough idea of the nature of relationship between the two variables x and y.*
2. *It is not affected by extreme observations.*
3. *It is not influenced by the size of extreme items.*

Disadvantages

1. *It is not a suitable method if the number of observations is very large.*
2. *It enables us to obtain an approximate estimating line or line of best fit by this method.*
3. *It is only a rough measure of correlation where the exact magnitude cannot be known.*

12.9 KARL PEARSON'S COEFFICIENT OF CORRELATION

Karl Pearson (1857 – 1936) was a great statistician. He gave the following mathematical formula for measuring the magnitude of linear correlation coefficient between two variables. If X and Y are two variables, then the correlation coefficient $\rho\,(X, Y)$ between them is given by

FIRST FORM

$$\rho\,(X, Y) = \frac{\textbf{Cov.}\,(X, Y)}{\sqrt{\textbf{Var.}\,(X) \times \textbf{Var.}\,(Y)}} \quad ...\ (1)$$

SECOND FORM

$$\rho\,(X, Y) = \frac{\sum_{i=1}^{n}(x_i - \bar{x})(y_i - \bar{y})}{\sqrt{\left\{\sum_{i=1}^{n}(x_i - \bar{x})^2\right\}\left\{\sum_{i=1}^{n}(y_i - \bar{y})^2\right\}}} \quad ...\ (2)$$

where $\bar{x}, \bar{y}$ are means of X and Y series respectively.

$\rho\,(X, Y)$ is also written as r_{xy} or r or simply ρ.

THIRD FORM

We know that

$$\Sigma(x_i - \bar{x})^2 = \Sigma x_i^2 - \frac{(\Sigma x_i)^2}{n}, \Sigma(y_i - \bar{y}) = \Sigma y_i^2 - \frac{(\Sigma y_i)^2}{n}$$

and $$\Sigma(x_i - \bar{x})(y_i - \bar{y}) = \Sigma x_i y_i - \frac{(\Sigma x_i)(\Sigma y_i)}{n}$$

Substituting these results in the form (II) of ρ (X, Y) we get

$$\rho(X, Y) = \frac{\Sigma x_i y_i - \dfrac{(\Sigma x_i)(\Sigma y_i)}{n}}{\sqrt{\left\{\Sigma x_i^2 - \dfrac{(\Sigma x_i)^2}{n}\right\}\left\{\Sigma y_i^2 - \dfrac{(\Sigma y_i)^2}{n}\right\}}} \quad \text{... (3)}$$

This formula saves a lot of computational labour. It reduces the error due to computation and rounding off.

Example 4. *Calculate the coefficient of correlation for the following data:*

(1, 2), (2, 4), (3, 8), (4, 7), (5, 10), (6, 5), (7, 14), (8, 16), (9, 2), (10, 20).

Solution. Here $n = 10$.

$\Sigma x_i = 1 + 2 + 3 + 4 + 5 + 6 + 7 + 8 + 9 + 10 = \mathbf{55}$

$\Sigma y_i = 2 + 4 + 8 + 7 + 10 + 5 + 14 + 16 + 2 + 20 = \mathbf{88}$

$\Sigma x_i^2 = 1 + 4 + 9 + 16 + 25 + 36 + 49 + 64 + 81 + 10 = \mathbf{385}$

$\Sigma y_i^2 = 4 + 16 + 64 + 49 + 100 + 25 + 196 + 256 + 4 + 400 = \mathbf{1114}$

$\Sigma x_i y_i = 2 + 8 + 24 + 28 + 50 + 30 + 98 + 128 + 18 + 200 = \mathbf{586.}$

Now

$$\rho(X, Y) = \frac{\Sigma x_i y_i - \dfrac{(\Sigma x_i)(\Sigma y_i)}{n}}{\sqrt{\left\{\Sigma x_i^2 - \dfrac{(\Sigma x_i)^2}{n}\right\}\left\{\Sigma y_i^2 - \dfrac{(\Sigma y_i)^2}{n}\right\}}} = \frac{586 - \dfrac{55 \times 88}{10}}{\sqrt{\left\{385 - \dfrac{55 \times 55}{10}\right\}\left\{1114 - \dfrac{88 \times 88}{10}\right\}}}$$

$$= \frac{586 - 484}{\sqrt{(385 - 302.5)(1114 - 774.5)}} = \frac{102}{\sqrt{(82.5)(339.5)}} = \frac{102}{\sqrt{2808.75}} = \frac{102}{167.35} = \mathbf{0.61 \text{ nearly.}}$$

The above example shows that the computation is not simple. We give below another method to make the computation simpler.

7.9.1 Direct Method

If X and Y are two variates having their means $\bar{x}$ and $\bar{y}$ respectively, then

$$\rho(X, Y) = \frac{\Sigma dx \times dy}{\sqrt{\Sigma dx^2 \times \Sigma dy^2}},$$

where $dx = x_i - \bar{x}$, $dy = y_i - \bar{y}$, $dx^2 = (x_i - \bar{x})^2$ and $dy^2 = (y_i - \bar{y})^2$.

It can also be written as

$$r_{xy} = \frac{\Sigma dx \times dy}{n\sigma_x \times \sigma_y}$$

where n is the number of observations in X or Y series, σ_x, σ_y are standard deviation of X and Y respectively.

The following formula can also be deduced from equation (3) of THIRD FORM

$$\rho (X, Y) = \frac{\Sigma dxdy - \frac{(\Sigma dx)(\Sigma dy)}{n}}{\sqrt{\Sigma dx^2 - \frac{(\Sigma dx)^2}{n}} \times \sqrt{\Sigma dy^2 - \frac{(\Sigma dy)^2}{n}}},$$

where n is the number of observations.

WORKING RULE

The coefficient of correlation is calculated by the following steps :

Step I. Denote one series by x and the other series by y.

Step II. Calculate $\bar{x}$ and $\bar{y}$ of the x and y series respectively.

Step III. Take the deviations of the observations in x-series from $\bar{x}$ and write it under the column headed by $dx = x - \bar{x}$. Take the deviations of the observations in y-series from $\bar{y}$ and write it in a column headed by $dy = y - \bar{y}$.

Step IV. Square these deviations and write them under the columns headed by dx^2 and dy^2.

Step V. Multiply the respective dx and dy and write it under the column headed by $dxdy$.

Step VI. Apply the following formula to calculate r or r_{xy}, the coefficient of correlation

$$r = \frac{\Sigma dxdy}{\sqrt{\Sigma dx^2 \times \Sigma dy^2}} \quad \text{or } r = \frac{\Sigma dxdy}{n\sigma_x \times \sigma_y},$$

where n is the number of observations in x or y series, σ_x and σ_y are the standard deviations of x and y.

The above method is illustrated by the following example.

Example 5. *Find the coefficient of correlation between the heights of brothers and sisters from the following data:*

Heights of brothers (in cm) (x)	*65*	*66*	*67*	*68*	*69*	*70*	*71*
Heights of sisters (in cm (y)	*67*	*68*	*66*	*69*	*72*	*72*	*69*

Solution. Let the heights of brothers be denoted by x and that of sisters by y, then

$$\bar{x} = \frac{65+66+67+68+69+70+71}{7} = \mathbf{68}, \text{ and}$$

$$\bar{y} = \frac{67+68+66+69+72+72+69}{7} = \mathbf{69}.$$

Let us prepare the following Table:

x	$dx = (x-68)$	$dx^2 = (x-68)^2$	y	$dy = (y-69)$	$dy^2 = (y-69)^2$	$dxdy$
65	– 3	9	67	– 2	4	6
66	– 2	4	68	– 1	1	2
67	– 1	1	66	– 3	9	3
68	0	0	69	0	0	0
69	1	1	72	3	9	3
70	2	4	72	3	9	6
71	3	9	69	0	0	0
	$\mathbf{\Sigma\ dx = 0}$	$\mathbf{\Sigma\ dx^2 = 28}$		$\mathbf{\Sigma\ dy = 0}$	$\mathbf{\Sigma\ dy^2 = 32\Sigma}$	$\mathbf{\Sigma dxdy = 20}$

$$\text{Now } r = \frac{\Sigma dx\ dy}{\sqrt{\Sigma\ (dx^2)\ (\Sigma dy^2)}} = \frac{20}{\sqrt{28 \times 32}} = \frac{5}{7.5} = \mathbf{0.67} \text{ (approx)}$$

Example 6. *Calculate the coefficient of correlation between X and Y for the following data.*

X :	*1*	*2*	*3*	*4*	*5*	*6*	*7*	*8*	*9*
Y :	*9*	*8*	*10*	*12*	*11*	*13*	*14*	*16*	*15*

Solution. Here $\bar{X} = \frac{\Sigma X}{n} = \frac{\mathbf{45}}{\mathbf{9}} = 5$; $\bar{Y} = \frac{\Sigma Y}{n} = \frac{108}{9} = \mathbf{12.}$

x	$dx = (X-5)$	$dx^2 = (X-5)^2$	*Y*	$dy = (Y-12)$	$dy^2 = (Y-12)^2$	*dxdy*
1	– 4	16	9	– 3	9	12
2	– 3	9	8	– 4	16	12
3	– 2	4	10	– 2	4	4
4	– 1	1	12	0	0	0
5	0	0	11	– 1	1	0
6	1	1	13	1	1	1
7	2	4	14	2	4	4
8	3	9	16	4	16	12
9	4	16	15	3	9	12
	$\Sigma dx = 0$	$\Sigma dx^2 = 60$		$\Sigma dy = 0$	$\Sigma dy^2 = 60$	$\Sigma dxdy = 57$

Now $r = \frac{\Sigma dx\, dy}{\sqrt{\Sigma dx^2 \times \Sigma dy^2}} = \frac{57}{\sqrt{60 \times 60}} = \frac{57}{60} = \mathbf{.95.}$

Example 7. *Calculate the correlation coefficient between the following data:*

X	*5*	*9*	*13*	*17*	*21*
Y	*12*	*20*	*25*	*33*	35

Solution. Here $\bar{X} = \frac{65}{5} = \mathbf{13,}$ $\bar{Y} = \frac{125}{5} = \mathbf{25.}$

x	$dx = (X-13)$	$dx^2 = (X-13)^2$	*y*	$dy = (Y-25)$	$dy^2 = (Y-25)^2$	*dxdy*
5	– 8	64	12	– 13	169	104
9	– 4	16	20	– 5	25	20
13	0	0	25	0	0	0
17	4	16	33	8	64	32
21	8	64	35	10	100	80
	$\Sigma dx = 0$	$\Sigma dx^2 = 160$		$\Sigma dy = 0$	$\Sigma dy^2 = 358$	$\Sigma dxdy = 236$

$$r = \frac{\Sigma dX\, dY}{\sqrt{\Sigma dX^2 \times \Sigma dY^2}} = \frac{236}{\sqrt{160 \times 358}} = \frac{236}{239.33} = \mathbf{0.986.}$$

Example 8. *Calculate the correlation coefficient from the following data:*

X:	*12*	*9*	*8*	*10*	*11*	*13*	*7*
Y:	*14*	*8*	*6*	*9*	*11*	*12*	*3*

Let now each value of X is multiplied by 2 and then 6 is added to it. Similarly, each value of Y is multiplied by 3 and 2 is subtracted from it. What will be the correlation coefficient between the new series of X and Y ?

Solution.

Table

X	Y	$dx = X - \overline{X}$	$dy = Y - \overline{Y}$	dx^2	dy^2	$dxdy$
12	14	2	5	4	25	10
9	8	−1	−1	1	1	1
8	6	−2	−3	4	9	6
10	9	0	0	0	0	0
11	11	1	2	1	4	2
13	12	3	3	9	9	9
7	3	−3	−6	9	36	18
70	**63**	$\Sigma dx = 0$	$\Sigma dy = 0$	$\Sigma dx^2 = 28$	$\Sigma dy^2 = 84$	$\Sigma dxdy = 46$

Here $\overline{X} = \frac{70}{7} = \mathbf{10},$ $\overline{Y} = \frac{63}{7} = \mathbf{9}.$

$\therefore$ **Coefficient of correlation:** $r = \frac{\Sigma dxdy}{\sqrt{\Sigma dx^2 \times \Sigma dy^2}} = \frac{46}{\sqrt{28 \times 84}} = \mathbf{0.949.}$

Example 9. *Calculate Karl Pearson's coefficient of correlation for the data given below:*

Independent Variable X :	*3*	*7*	*5*	*4*	*6*	*8*	*2*	*7*
Dependent Variable Y :	*7*	*12*	*8*	*8*	*10*	*13*	*5*	*10*

Solution. Let the assumed mean for x and y series be 5 and 9 respectively

Table for Correlation Coefficent

X	Y	$dx = x - 5$	$dy = y - 5$	dx^2	dy^2	$dxdy$
3	7	−2	−2	4	4	4
7	12	2	3	4	9	6
5	8	0	−1	0	1	0
4	8	−1	−1	1	1	1
6	10	1	1	1	1	1
8	13	3	4	9	16	12
2	5	−3	−4	9	16	12
7	10	2	1	4	1	2
Total		**2**	**1**	**32**	**49**	**38**

Now
$$r = \frac{\Sigma\, dxdy - \frac{(\Sigma\, dx)(\Sigma\, dy)}{n}}{\sqrt{\left\{\Sigma\, dx^2 - \frac{(\Sigma dx)^2}{n}\right\}\left\{\Sigma\, dy^2 - \frac{\Sigma\,(dy)^2}{n}\right\}}}$$

$$= \frac{38 - \frac{2\times 1}{8}}{\sqrt{\left\{32 - \frac{(2)^2}{8}\right\}\left\{49 - \frac{(1)^2}{8}\right\}}} = \frac{37.75}{39.24} = \mathbf{0.96.}$$

7.9.2 Short-cut Method

The above direct method for calculating r is not convenient when (*i*) the terms of the series x and y are big and the calculation of $\bar{x}$ and $\bar{y}$ becomes difficult or (*ii*) the means $\bar{x}$ or $\bar{y}$ are not integers. In these cases we apply the following formula of **assumed mean**

$$r_{xy} = \frac{\Sigma\,(dxdy) - \frac{(\Sigma dx)(\Sigma dy)}{n}}{\sqrt{\Sigma dx^2 - \frac{(\Sigma dx)^2}{n}}\sqrt{\Sigma dy^2 - \frac{(\Sigma dy)^2}{n}}}$$

where $dx = x - a$, a is the assumed mean of x-series, $dy = y - b$, b is the assumed mean of y-series and n is number of observations of x or y.

WORKING RULE

STEP I. Take any term a (preferably the middle one) of x series as assumed mean and any term b (preferably middle one) as assumed mean for y series.

STEP II. Take deviations of the observations in x series from a, *i.e*, $dx = x - a$. Take deviations of the observations in y series from b, *i.e.*, $dy = y - b$ and write it under columns dx and dy.

STEP III. Find dx^2 and dy^2 and write it under columns dx^2 and dy^2.

STEP IV. Find $dxdy$ and write it under the column $dxdy$.

STEP V. Apply the formula

$$r_{xy} = \frac{\Sigma\,(dxdy) - \frac{(\Sigma dx)(\Sigma dy)}{n}}{\left(\sqrt{\Sigma dx^2 - \frac{(\Sigma dx)^2}{n}}\sqrt{\Sigma dy^2 - \frac{(\Sigma dy)^2}{n}}\right)}$$

The method is illustrated by the following example.

Example 10. *Calculate the coefficient of correlation between x and y for the following data.*

x :	– 3	– 2	– 1	0	1	2	3
y :	9	4	1	0	1	4	9

Solution. Let **2 be assumed mean for *x*-series** and **4 be the assumed mean for *y*-series.** Then we have the following table:

x	$dx = x - 2$	$dx^2 = (x-2)^2$	y	$dy = (y-4)$	$dy^2 = (y-4)^2$	$dxdy$
– 3	– 5	25	9	5	25	– 25
– 2	– 4	16	4	0	0	0
– 1	– 3	9	1	– 3	9	9
0	– 2	4	0	– 4	16	8
1	– 1	1	1	– 3	9	3
2	0	0	4	0	0	0
3	1	1	9	5	25	5
	$\Sigma dx = -14$	**$\Sigma dx^2 = 56$**		**$\Sigma dy = 0$**	**$\Sigma dy^2 = 84$**	**$\Sigma dxdy = 0$**

Now $$r = \frac{\Sigma(dxdy) - \frac{(\Sigma dx)(\Sigma dy)}{N}}{\sqrt{\Sigma dx^2 - \frac{(\Sigma dx)^2}{N}} \times \sqrt{\Sigma dy^2 - \frac{(\Sigma dy)^2}{N}}} = \frac{0-0}{\sqrt{56 - \frac{256}{7}} \times \sqrt{84-0}} = 0.$$

$\therefore$ $r = 0 \Rightarrow$ **x and y are uncorrelated.**

Example 11. *Calculate the coefficient of correlation between x and y for the following data:*

x :	1	2	4	5	7	8	10
y :	2	6	8	10	14	16	20.

Solution. Let **5 be the assumed mean for the values of x and 14 be the assumed mean for the values of y.**

Let $dx = x - 5$, $dx^2 = (x-5)^2$, $dy = (y-14)$, $dy^2 = (y-14)^2$. We have the following table.

x	y	dx	dy	dx^2	dy^2	$dxdy$
1	2	– 4	– 12	16	144	48
3	6	– 2	– 8	4	64	16
4	8	– 1	– 6	1	36	6
5	10	0	– 4	0	16	0
7	14	2	0	4	0	0
8	16	3	2	9	4	6
10	20	5	6	25	36	30
$n = 7$	**$n = 7$**	**$\Sigma dx = 3$**	**$\Sigma dy = -22$**	**$\Sigma dx^2 = 59$**	**$\Sigma dy^2 = 300$**	**$\Sigma dxdy = 106$**

Since we have taken the deviations from assumed means, so we shall apply the following formula of correlation for assumed mean

$$r_{xy} = \frac{\Sigma dxdy - \frac{(\Sigma dx)(\Sigma dy)}{n}}{\left(\sqrt{\Sigma dx^2 - \frac{(\Sigma dx)^2}{n}}\right)\left(\sqrt{\Sigma dy^2 - \frac{(\Sigma dy)^2}{n}}\right)} = \frac{106 + \frac{66}{7}}{\sqrt{\left(59 - \frac{9}{7}\right)}\sqrt{\left(300 + \frac{484}{7}\right)}}$$

$$= \frac{808}{\sqrt{404} \times \sqrt{1616}} = \frac{808}{\sqrt{652864}} = \frac{\mathbf{808}}{\mathbf{808}} = \mathbf{1}.$$

Thus the coefficient of correlation between x and y is positive and perfect.

Example 12. *Calculate the coefficient of correlation for the following pairs of values of x and y:*

x :	*17*	*19*	*21*	*26*	*20*	*28*	*26*	*27*
y :	*23*	*27*	*25*	*26*	*27*	*25*	*30*	*33.*

Solution. Let the assumed means for x and y be 23 and 27 respectively. $dx = x - 23$, $dy = (y - 27)$, we have the following table.

x	y	$dx = (x - 23)$	$dy = (y - 27)$	dx^2	dy^2	$dxdy$
17	23	– 6	– 4	36	16	24
19	27	– 4	0	16	0	0
21	25	– 2	– 2	4	4	4
26	26.	3	– 1	9	1	– 3
20	27	– 3	0	9	0	0
28	25	5	– 2	25	4	– 10
26	30	3	3	9	9	9
27	33	4	6	16	36	24
$n = 8$	**$n = 8$**	**$\Sigma\, dx = 0$**	**$\Sigma\, dy = 0$**	**$\Sigma\, dx^2 = 124$**	**$\Sigma\, dy^2 = 70$**	**$\Sigma\, dxdy = 48$**

Now
$$r = \frac{\left[\Sigma dxdy - \dfrac{(\Sigma dx)(\Sigma dy)}{n}\right]}{\sqrt{\Sigma dx^2 - \dfrac{(\Sigma dx)^2}{n}} \times \sqrt{\Sigma dy^2 - \dfrac{(\Sigma dy)^2}{n}}} = \frac{[48 - 0]}{\sqrt{124-0}\sqrt{70-0}} = \frac{48}{\sqrt{124 \times 70}}$$

$$= \frac{48}{93.166} = \mathbf{0.515.}$$

Hence r = **0.515.**

Example 13. *From the following data compute the coefficient of correlation between x and y.*

(*i*) *Arithmetic Mean of x series is 25 and that of y is 18.*

(*ii*) *Sum of the products of deviations of x and y series from their respective means = 122.*

(*iii*) *Sum of the squares of deviation from their respective means are 136, 138 respectively for x series and y series.*

(*iv*) *Number of pairs of values = 15.*

Solution. We are given the following information

$$n = 15 \qquad \bar{x} = 25 \qquad \bar{y} = 18.$$

$$\Sigma (x - \bar{x})^2 = 136, \qquad \Sigma (y - y)^2 = 138, \qquad \Sigma (x - \bar{x})(y - \bar{y}) = 122.$$

$$\text{Now } r = \frac{\Sigma (x-\bar{x})(y-\bar{y})}{\sqrt{\Sigma (x-\bar{x})^2 \Sigma (y-\bar{y})^2}} = \frac{122}{\sqrt{136 \times 138}} = \frac{122}{136.996} = \mathbf{0.89.}$$

Hence the coefficient of correlation is 0.89.

Example 14. *From the data given below, find the number of items n.*

$$r = 0.5, \Sigma xy = 120, \sigma_y = 8, \Sigma x^2 = 90$$

where x and y are deviations from arithmetic mean.

Solution. Here x and y are deviations from arithmetic mean, *i.e.*, $x = X - \bar{X}$

Now $$\sigma_y = \sqrt{\frac{1}{n}\Sigma (Y-\bar{Y})^2} = 8 \Rightarrow \Sigma (Y-\bar{Y})^2 = 64n.$$

Also $$\Sigma x^2 = \Sigma (X-\bar{X})^2 = 90.$$

Again $$r = \frac{\Sigma (X-\bar{X})(Y-\bar{Y})}{\sqrt{\Sigma(X-\bar{X})^2 \Sigma(Y-\bar{Y})^2}} = 0.5 \Rightarrow \frac{120}{\sqrt{90\times 64n}}$$

$$\Rightarrow \quad 0.25 = \frac{120\times 120}{90\times 64n} \Rightarrow 64n = \frac{120\times 120}{900\times 0.25}$$

or $$n = \frac{120\times 120}{64\times 90\times 0.25} = \mathbf{10.}$$

Example 15. *Calculate Karl Pearson's correlation coefficient between the marks in English and Hindi obtained by 10 students.*

Marks in English :	*10*	*25*	*13*	*25*	*22*	*11*	*12*	*25*	*21*	*20*
Marks in Hindi :	*12*	*22*	*16*	*15*	*18*	*18*	*17*	*23*	*24*	*17.*

Solution. Let the **assumed means for the values of x and y be 18 and 18 respectively.** We have the following table.

x	$dx = x - 18$	$dx^2 = (x-18)^2$	y	$dy = (y-18)$	$dy^2 = (y-18)^2$	$dxdy$
10	−8	64	12	−6	36	48
25	7	49	22	4	16	28
13	−5	25	16	−2	4	10
25	7	49	15	−3	9	−21
22	4	16	18	0	0	0
11	−7	49	18	0	0	0
12	−6	36	17	−1	1	6
25	7	49	23	5	25	35
21	3	9	24	6	36	18
20	2	4	17	−1	1	−2
	$\mathbf{\Sigma dx = 4}$	$\mathbf{\Sigma dx^2 = 350}$		$\mathbf{\Sigma dy = 2}$	$\mathbf{\Sigma dy^2 = 128}$	$\mathbf{\Sigma dxdy = 122}$

Here $$r = \frac{\Sigma dxdy - (\Sigma dx\, \Sigma dy)/N}{\sqrt{\left[\Sigma dx^2 - \frac{(\Sigma dx)^2}{N}\right]\left[\Sigma dy^2 - \frac{(\Sigma dy)^2}{N}\right]}} = \frac{122 - \frac{4\times 2}{10}}{\sqrt{\left(350 - \frac{16}{10}\right)}\sqrt{\left(128 - \frac{4}{10}\right)}}$$

$$= \frac{122 - 0.8}{\sqrt{348.4}\sqrt{127.6}} = \frac{121.2}{210.834} = \mathbf{0.575.}$$

Example 16. *Calculate the correlation coefficient between X and Y for the following data:*

X :	3	4	5	8	7	9	6	2	1
Y :	5	3	4	7	8	7	6	9	2

Solution. Let the assumed mean for the values of X and Y series be 5 and 6 respectively.

X	$dX = (X-5)$	$dX^2 = (X-5)^2$	Y	$dY = (Y-6)$	$dY^2 = (Y-6)^2$	$dXdY$
3	−2	4	5	−1	1	2
4	−1	1	3	−3	9	3
5	0	0	4	−2	4	0
8	3	9	7	1	1	3
7	2	4	8	2	4	4
9	4	16	7	1	1	4
6	1	1	6	0	0	0
2	−3	9	9	3	9	−9
1	−4	16	2	−4	16	16
	$\Sigma dX = 0$	**$\Sigma dX^2 = 60$**		**$\Sigma dY = -3$**	**$\Sigma dY^2 = 45$**	**$\Sigma dXdY = 23$**

$$\therefore \quad r = \frac{\Sigma dXdY - (\Sigma dX\, \Sigma dY)/N}{\sqrt{\left[\Sigma dX^2 - \frac{(\Sigma dX)^2}{N}\right]\left[\Sigma dY^2 - \frac{(\Sigma dY)^2}{N}\right]}}$$

$$= \frac{23-0}{\sqrt{(60-0)\,(45-(-3)^2/9)}} = \frac{23}{\sqrt{60\times 44}} = \frac{23}{51.38} = \mathbf{0.448.}$$

Example 17. *A computer operator while calculating the coefficient between two variates x and y for 25 pairs of observations obtained the following constants:*

$$n = 25,\ \Sigma x = 125,\ \Sigma x^2 = 650,\ \Sigma y = 100,\ \Sigma y^2 = 460,\ \Sigma xy = 508.$$

It was, however later discovered at the time of checking that he had copied down two pairs as (6, 14) and (8, 6) while the correct pairs were (8, 12) and (6, 8). Obtain the correct value of the correlation coefficient.

Solution. Here **Corrected Σx** = Incorrect Σx – (6 + 8) + (8 + 6) = **125**

Corrected Σy = Incorrect Σy – (14 + 6) + (12 + 8) = **100**

Corrected Σx^2 = $650 - (6^2 + 8^2) + (8^2 + 6^2)$ = **650**

Corrected Σy^2 = $460 - (14^2 + 6^2) + (12^2 + 8^2)$ = **436**

Corrected Σxy = 508 – (84 + 48) + (96 + 48) = **520**

Correct value of correlation is

$$r_{xy} = \frac{520 - \frac{125\times 100}{25}}{\sqrt{\left(650 - \frac{(125)^2}{25}\right)\left(436 - \frac{(100)^2}{25}\right)}}$$

$$= \frac{520-500}{\sqrt{(650-625)\,(436-400)}} = \frac{20}{\sqrt{25\times 36}} = \frac{2}{3} = \mathbf{0.67.}$$

Example 18. *In two sets of variables X and Y with 50 observations each, the following data were observed*

$$\bar{x} = 10, \text{ S.D. of } x = 3, \ \bar{y} = 6, \text{ S.D of } x = 2$$

Correlation coefficient between X and Y is 0.3. However, on subsequent verification it was found that one value of X (= 10) and one value of Y (= 6) were inaccurate and hence weeded out. With remaining 49 pairs of values, how is the original value of correlation coefficient affected.

Solution. We know that: $r = \dfrac{\Sigma (x-\bar{x})(y-\bar{y})/n}{\sigma_x \sigma_y}$

$$\Rightarrow \frac{1}{n} r \sigma_x \sigma_y = \Sigma (x-\bar{x})(y-\bar{y})$$

$$\Rightarrow \Sigma xy - \bar{x}\Sigma y - \bar{y}\Sigma x + \Sigma \bar{x}\,\bar{y} = 50 \times 0.3 \times 3 \times 2 = 90$$

$$\Rightarrow \Sigma xy - \bar{x}(n\bar{y}) - \bar{y}(n\bar{x}) + n\bar{x}\bar{y} = 90 \qquad [\because \Sigma x = n\bar{x}, \Sigma y = n\bar{y}, \Sigma \bar{x}\,\bar{y} = n\bar{x}\,\bar{y}]$$

$$\Rightarrow \Sigma xy = 90 + n\bar{x}\,\bar{y} \Rightarrow \Sigma xy = 90 + 50 \times 10 \times 6$$

$$\Rightarrow \Sigma xy = 3090.$$

Corrected Σxy = 3090 – 10×6 = **3030**

$\mathbf{\Sigma x} = n\bar{x} - 10 = 50 \times 10 - 10 =$ **490**

$\mathbf{\Sigma y} = n\bar{y} - 6 = 50 \times 6 - 6 =$ **294**

Again $\sigma_x^2 = \dfrac{\Sigma x^2}{n} - \left(\dfrac{\Sigma x}{n}\right)^2 = \dfrac{1}{n}\Sigma x^2 - \bar{x}^2$

$$\Rightarrow \Sigma x^2 = n(\sigma_x^2 + \bar{x}^2) = 50\,(9 + 100) = \mathbf{545.0}$$

Similarly, $\Sigma y^2 = n\,(\sigma_y^2 + \bar{y}^2) = 50\,(4 + 36) =$ **2000**

Corrected $\Sigma x^2 = 5450 - (10)^2 =$ **5350**

Corrected $\Sigma y^2 = 2000 - (6)^2 =$ **1964**

Corrected value of correlation coefficient when $N = 49$ is

$$r = \frac{N\Sigma xy - (\Sigma x)(\Sigma y)}{\sqrt{[N\Sigma x^2 - (\Sigma x)^2][N(\Sigma y^2) - (\Sigma y)^2]}} = \frac{49 \times 3030 - 490 \times 294}{\sqrt{[49 \times 5350 - (490)^2][49 \times 1964 - (294)^2]}}$$

$$= \frac{49\,(3030 - 2940)}{\sqrt{49\,(5350 - 4900)\,49\,(1964 - 1764)}} = \frac{90}{\sqrt{450 \times 200}} = \frac{90}{\sqrt{90000}} = \frac{9}{30} = 0.3$$

Thus the original value of correlation coefficient remains unaffected.

EXERCISE – 12.2

1. Calculate the correlation coefficient between the height of sisters and height of brothers from the given data.

Height of sisters (in cm) :	64	65	66	67	68	69	70
Height of brothers (in cm) :	66	67	65	68	70	68	72.

[**Hint.** Here $\bar{x} = 67, \bar{y} = 68$, $\Sigma x = 0$, $\Sigma y = 0$, $\Sigma x^2 = 28$, $\Sigma y^2 = 34$. Also $\Sigma xy = 25$.

Use $r = \dfrac{\Sigma xy}{\sqrt{\Sigma x^2 \times \Sigma y^2}} = \dfrac{25}{\sqrt{28 \times 34}} = 0.81$.]

2. Calculate the coefficient of correlation between x and y series from the following data :

	X series	Y series
Number of pairs of observation	15	15
Arithmetic mean	25	18
Standard deviation	3.01	3.03
Sum of the squares of deviations from mean	136	138

Sum of the product of the deviations of x and y series from their respective means = 122.

[**Hint** $r = \frac{\Sigma xy}{\sqrt{\Sigma x^2 \times \Sigma y^2}} = \frac{122}{\sqrt{136 \times 138}} = \mathbf{0.89}$]

3. Given $r = 0.8$, $\Sigma\, xy = 60$, $\Sigma\, y = 2.5$ and $\Sigma\, x^2 = 90$. Find the number of items (x and y are deviations from their respective A.M).

[**Hint.** $\sigma_x^2 = \frac{1}{n}\Sigma x^2 = \frac{90}{n}$. Also $r = \frac{\Sigma\,(x-\bar{x})(y-\bar{y})}{n\sigma_x\sigma_y} = \frac{\Sigma\, xy}{n\sigma_x\sigma_y}$

$$\Rightarrow \quad r^2 = \frac{(\Sigma\, xy)^2}{n^2\sigma_x^2\,\sigma_y^2} \Rightarrow 0.64 = \frac{(60)^2}{n^2\,(90/n) \times 6.25} \Rightarrow \mathbf{n = 10.}]$$

4. With the following data in 6 cities, calculate the coefficient of correlation by Pearson's method between the density of population and the death rate.

Cities	*Area in Sq. metres*	*Population in '000*	*No. of deaths*
A	150	30	300
B	180	90	1440
C	100	40	560
D	60	42	840
E	120	72	1224
F	80	24	312

[**Hint. Using Density** = $\frac{\textbf{Population}}{\textbf{Area}}$; **Death rate** = $\frac{\textbf{No. of deaths}}{\textbf{Population}}$, we are to find r of the following data:

Cities :	*A*	*B*	*C*	*D*	*E*	*F*
Density :	200	500	400	700	600	300
Death rate :	10	16	14	20	17	13.

Here $\bar{x} = \frac{2700}{6} = 450$, $\bar{y} = \frac{90}{6} = 15$, then $\Sigma\, x^2 = 175000$, $\Sigma\, y^2 = 60$, $\Sigma\, xy = 3200$

$$r = \frac{\Sigma\, xy}{\sqrt{\Sigma\, x^2 \times \Sigma\, y^2}} = \frac{3200}{\sqrt{1{,}75{,}000 \times 60}} = \mathbf{0.9875.}]$$

5. If $N = 50$, $\Sigma\, x = 75$, $\Sigma\, y = 80$, $\Sigma\, x^2 = 150$, $\Sigma\, y^2 = 140$, $\Sigma\, xy = 120$, find the value of r.

[**Hint.** $r = \frac{\Sigma xy - \frac{(\Sigma x) \times (\Sigma y)}{N}}{\sqrt{\Sigma x^2 - \frac{(\Sigma x)^2}{N}}\sqrt{\Sigma y^2 - \frac{(\Sigma y)^2}{N}}} = \frac{120 - \frac{75 \times 80}{50}}{\sqrt{130 - \frac{(75)^2}{5}}\sqrt{140 - \frac{(80)^2}{50}}} = \mathbf{0.}$]

6. Calculate Karl Pearson coefficient of correlation from the following data, using 20 as working mean for price and 70 as working mean for demand.

Price :	14	16	17	18	19	20	21	22	23
Demand :	84	78	70	75	66	67	62	58	60.

7. From the following information relating to the Stock Exchange Quotations for two shares A and B, ascertain by using Pearson's coefficient of correlation how shares A and B are correlated, in their prices?

Price share (A) Rs:	160	164	172	182	166	170	178
Price share (B) Rs:	292	280	260	234	266	254	230.

8. Find the correlation coefficient between the income and expenditure of a wage earner and comment on the result.

Month :	Jan.	Feb.	March	April	May	June	July
Income :	46	54	56	56	58	60	62
Expenditure :	36	40	44	54	42	58	54.

9. Calculate correlation coefficient from the following data

$N = 10, \Sigma X = 140, \Sigma Y = 150, \Sigma (X-10)^2 = 180, \Sigma (Y-15)^2 = 215, \Sigma (X-10)(Y-15) = 60$

[**Hint.** $\Sigma u = \Sigma (x-10) = 40, \Sigma v = \Sigma (y-15) = 0,$

$\Sigma u^2 = \Sigma (x-10)^2 = 180 ; \Sigma v^2 = \Sigma (y-15)^2 = 215$

$\Sigma uv = \Sigma (x-10)(y-15) = 60$

$$\therefore \quad r = \frac{n \Sigma uv - (\Sigma u)(\Sigma v)}{\sqrt{\{n \Sigma u^2 - (\Sigma u)^2\}\{n \Sigma x^2 - (\Sigma v)^2\}}}$$

$$= \frac{10 \times 60 - 40 \times 0}{\sqrt{\{10 \times 180 - (40^2)\}\{10 \times 215 - (0)^2\}}} = \mathbf{0.915}]$$

10. Calculate the correlation coefficient between the following pairs of values

X :	100	110	115	116	120	120	125	130	135
Y :	18	18	17	16	16	16	15	13	10.

11. From the following data, examine whether input of oil and output of electricity can be said to be correlated

Input of Oil :	6.9	8.2	7.8	4.8	9.6	8.0	7.7
Output of Electricity :	1.9	3.5	6.5	1.3	5.5	3.5	2.2.

12. If $N = 50, \Sigma X = 75, \Sigma Y = 80, \Sigma X^2 = 130, \Sigma Y^2 = 140, \Sigma XY = 120$, find the values of r.

13. A computer while calculating the correlating coefficient between the variables X and Y obtained the following constants, $N = 30, \Sigma X = 120, \Sigma X^2 = 600, \Sigma Y = 90, \Sigma Y^2 = 250, \Sigma XY = 356$.

It was, however, later discovered at the time of checking that it had copied down two pairs of observations as

x	y
8	10
12	7

while the corrected values were

x	y
8	12
10	8

Obtain the correct value of correlation coefficient between X and Y.

14. Find the Karl Pearson's co-efficient of correlation between the age and the playing habit of people from the following information :

Age group (years)	*No. of People*	*No. of Players*
15 and less than 20	200	150
20 and less than 25	270	162
25 and less than 30	340	170
30 and less than 35	360	180
35 and less than 40	400	180
40 and less than 45	300	120

15. Find the correlation between sales and advertising expenditure from the following data:

Sales (Rs lakh) :	65	66	67	67	68	69	70	22
Adv. Expenditure :	67	68	65	68	72	72	69	71.

16. Calculate coefficient of correlation between X and Y

X:	78	89	97	69	59	79	68	61
Y:	125	137	156	112	107	136	123	108.

[**Hint.** Take $dx = x - 69$, $dy = y - 112$. The $\Sigma\, dx = 48$, $\Sigma\, dy = 108$, $\Sigma\, dx = 1530$, $\Sigma\, dy^2 = 3468$, $\Sigma\, dxdy = 2160$.

$$\therefore \quad r = \frac{8 \times 2160 - 48 \times 108}{\sqrt{\{8 \times 1530 - (48)^2\} \times \{8 \times 3468 - (108)^2\}}} = \mathbf{0.957}.]$$

ANSWERS

1. 0.81. **2.** 0.89. **3.** $n = 10$. **4.** $r = 0.9875$.
5. $r = 0$. **6.** $r = -0.954$. **7.** $r = -0.96$ **8.** $r = 0.769$.
9. $r = 0.915$. **10.** $r = -0.915$. **11.** $n = 0.696$. **13.** $r = 0.05$.
14. $r = 0.939$. **15.** 0.603. **16.** 0.957.

12.10 SPEARMAN'S RANK CORRELATION COEFFICIENT

The **coefficient of rank correlation** is based on the various values of the variates and is denoted by R. *It is applied in the problems in which data cannot be measured quantitatively but qualitative assessment is possible such as beauty, honesty etc*. In this case the best individual is given rank number 1, next rank 2 and so on. The **coefficient of rank correlation is given by the formula**

$$\mathbf{R = 1 - \frac{6\,\Sigma\, D^2}{n\,(n^2 - 1)}},$$

where D^2 is the square of the difference of corresponding ranks, and n is the number of pairs of observations.

TYPE I. When the ranks are given.

WORKING RULE

Step I. Calculate the difference of ranks of x from the ranks of y and write it under the column headed by D.

Step II. Square the difference D and write it under the column headed by D^2.

Step III. Apply the formula

$$\mathbf{R = 1 - \frac{6\,\Sigma\, D^2}{n\,(n^2 - 1)}}$$

where n is the total number of pairs of observations.

The method is illustrated by the following examples.

Example 19. *Two judges in a beauty competition rank the 12 entries as fallows.*

X:	1	2	3	4	5	6	7	8	9	10	11	12
Y:	12	9	6	10	3	5	4	7	8	2	11	1

Calculate the rank correlation coefficient between X and Y.

Solution. Here $n = 12$

Table for Computation of Rank Correlation

Rank X	*Rank Y*	*D*	D^2
1	12	– 11	121
2	9	– 7	49
3	6	– 3	9
4	10	– 6	36
5	3	2	4
6	5	1	1
7	4	3	9
8	7	1	1
9	8	1	1
10	2	8	64
11	11	0	0
12	1	11	121
			$\Sigma D^2 = 416$

Now $$R = 1 - \frac{6\Sigma D^2}{n(n^2-1)} = 1 - \frac{6 \times 416}{12(12^2-1)} = 1 - \frac{6 \times 416}{1728-12} = 1 - \frac{2496}{1716}$$

$$= 1 - 1.4545 = -\mathbf{0.4545}.$$

TYPE II. When the ranks are not given.

In this case we are given only the data. We assign the ranks to both the series *X* and *Y* by giving the rank 1 to highest values in both the series (or to the lowest values in both the series) and so on.

WORKING RULE

STEP I. Assign ranks to each item of both the series, if they are not given.

STEP II. Calculate the difference of ranks of *X* from the ranks of *Y* and write it under the column headed by *D*.

STEP III. Square the difference *D* and write it under the column D^2.

STEP IV. Apply the formula

$$R = 1 - \frac{6\Sigma D^2}{n(n^2-1)},$$

where *n* is the total number of pairs of observations.

The above method is explained with the help of the following example.

Example 20. *For the following data, calclate the coefficient of Rank Correlation:*

x :	*80*	*91*	*99*	*71*	*61*	*81*	*70*	*59*
y :	*123*	*135*	*154*	*110*	*105*	*134*	*121*	*106*

Solution.

x	*Rank x*	*y*	*Rank y*	*Rank difference D*	*Squared rank difference* D^2
80	4	123	4	0	0
91	2	135	2	0	0
99	1	154	1	0	0
71	5	110	5	– 1	1
61	7	105	8	– 1	1
81	3	134	3	0	0
70	6	121	5	1	1
59	8	106	7	1	1
$n = 8$					$\Sigma(D^2 = 4)$

Coefficient of rank correlation is given by

$$R = 1 - \frac{6\Sigma D^2}{n(n^2-1)} = 1 - \frac{6\times 4}{8(64-1)} = 1 - \frac{3}{63} = \frac{60}{63} = \frac{20}{21} = 0.952. \qquad \textbf{[Here } n = 8\textbf{]}$$

Example 21. *From the data given below, calculate the coefficient of rank correlation between X and Y:*

X :	*78*	*89*	*97*	*69*	*59*	*79*	*68*	*57*
Y :	*125*	*137*	*156*	*112*	*107*	*136*	*123*	*108*

Solution. Here $n = 8$.

Computation Table:

X	*Y*	*Rank in X* (R_1)	*Rank in Y* (R_2)	*Rank difference* $d = R_1 - R_2$	d^2
78	125	4	4	0	0
89	137	2	2	0	0
97	156	1	1	0	0
69	112	5	6	– 1	1
59	107	7	8	– 1	1
79	136	3	3	0	0
68	123	6	5	1	1
57	108	8	7	1	1
				$\Sigma d = 0$	$\Sigma d^2 = 4$

Coefficient of rank correlation: $R = 1 - \dfrac{6\Sigma d^2}{n(n^2-1)} = 1 - \dfrac{6\times 4}{8(64-1)} = 1 - \dfrac{3}{63} = 0.952.$

Example 22. *Calculate the rank correlation coefficient from the data given below:*

X :	*75*	*88*	*95*	*70*	*60*	*80*	*81*	*50*
Y :	*120*	*134*	*150*	*115*	*110*	*140*	*142*	*100.*

Solution. Here $n = 8$.

Table for Rank Correlation Coefficient

X	*Rank X*	*Y*	*Rank Y*	*D*	D^2
75	5	120	5	0	0
88	2	134	4	– 2	4
95	1	150	1	0	0
70	6	115	6	0	0
60	7	110	7	0	0
80	4	140	3	1	1
81	3	142	2	1	1
50	8	100	8	0	0
		Total			$\Sigma D^2 = 6$

Rank Correlation Coefficient: $R = 1 - \dfrac{6\Sigma D^2}{n(n^2-1)} = 1-\dfrac{6\times 6}{8(64-1)} = 1-\dfrac{36}{8\times 63}$

$$= 1 - 0.07143 = \mathbf{0.929}.$$

Example 22. *Calculate rank correlation coefficient between the two series X and Y, give below*

X :	*70*	*65*	*71*	*62*	*58*	*69*	*78*	*64*
Y :	*91*	*76*	*65*	*83*	*90*	*64*	*55*	*48*

Also comment on the result.

Solution. We prepere the following Table for finding rank correlation. Here $n = 8$.

Table

Rank X	*Rank Y*	$d = X - Y$	d^2
3	1	2	4
5	4	1	1
2	5	– 3	9
7	3	4	16
8	2	6	36
4	6	– 2	4
1	7	– 6	36
6	8	2	4
$n = 8$			$\Sigma d^2 = 110$

Rank correlation coefficient: $R = 1 - \dfrac{6\,\Sigma d^2}{n(n^2-1)}$

$$= 1-\frac{6\times 110}{8\times 63} = \mathbf{-0.3095}$$

$\Rightarrow$ **The correlation between X and Y is negative.**

Example 24. *The coefficient of rank correlation of marks obtained by 10 students, in English and Economics was found to be 0.5. It was later discovered that the difference in ranks in the two subjects obtained by one student was wrongly taken as 3 instead of 7. Find the correct coefficient of rank correlation.*

Solution. Here $R' = 0.5, N = 10.$

Also $$\boldsymbol{R} = 1 - \frac{6\,\Sigma\,D^2}{N(N^2-1)} \Rightarrow 0.5 = 1 - \frac{6\Sigma D^2}{10(10^2-1)} \Rightarrow \frac{6\Sigma D^2}{990} = 1 - 0.5$$

or $$\boldsymbol{6\,\Sigma\,D^2} = 0.5 \times 990 \Rightarrow \Sigma\,D^2 = \frac{0.5\times 990}{6} = \mathbf{82.5.}$$

Corrected value of $\Sigma\,D^2$ $= 82.5 - 3^2 + 7^2 = 122.5.$

$$\text{Corrected } R = 1 - \frac{6\times 122.5}{990} = 1 - \frac{735}{990}$$

$$= \mathbf{1 - 0.7424 = 0.2576}$$

Hence **Corrected $R \simeq$ 0.26.**

Example 25. *Ten students got the following percentage of marks in Mathematics and Physics.*

Mathematics (X) :	8	36	98	25	75	82	92	62	65	35
Physics (Y) :	84	51	91	60	68	62	86	58	35	49.

Find the coefficient of rank correlation.

Solution. We first assign ranks to Y and Y series giving rank 1 to the highest value in both the series.

X	Y	$x_i =$ *Rank in X*	$y_i =$ *Rank in Y*	$D = x_i - y_i$	D^2
8	84	10	3	7	49
36	51	7	8	– 1	1
98	91	1	1	0	0
25	60	9	6	3	9
75	68	4	4	0	0
82	62	3	5	– 2	4
92	86	2	2	0	0
62	58	6	7	– 1	1
65	35	5	10	– 5	25
35	49	8	9	– 1	1
				$\Sigma\,D = 0$;	**$\Sigma\,D^2 = 90$**

$$\therefore\ \mathbf{R = 1 - \frac{6\,\Sigma\,D^2}{n\,(n^2-1)}} = 1 - \frac{6\times 90}{10(100-1)} = 1 - 0.545 = \mathbf{0.455.}$$

Example 26. *The coefficient of rank correlation of marks obtained by 10 students in English and Economics was found to be 0.5. It was later discovered that the difference in ranks in the two subjects obtained by one of the students was wrongly taken as 3 instead of 7. Find the correct coefficient of rank correlation.*

Solution.

$$R = 1 - \frac{6 \Sigma D^2}{N(N^2 - 1)}$$

Substituting the values in the above formula, we get

$$0.5 = 1 - \frac{6 \Sigma D^2}{10(100-1)}$$

$$\Rightarrow \frac{6 \Sigma D^2}{990} = 1 - 0.5 = 0.5 \Rightarrow 6 \Sigma D^2 = 82.5$$

Corrected value of ΣD^2 = In corrected value of $\Sigma D^2 - 32 + 72$

$$= 82.5 - 9 + 49 = 122.5$$

$$\textbf{Corrected } R = 1 - \frac{6 \times 122.5}{10 \times (900-1)} = 1 - \frac{735}{990} = \frac{255}{990} = \mathbf{0.2576.}$$

Example 27. *The marks obtained by the students in Physics and Mathematics are as follows:*

Marks in Physics :	*35*	*23*	*47*	*17*	*10*	*43*	*9*	*6*	*28*
Marks in Mathematics :	*30*	*33*	*45*	*23*	*8*	*49*	*12*	*4*	*31*

Compute their ranks in the two subjects and the coefficient of correlation of ranks.

Solution. Assign rank 1 to the highest value in both the series. We have, the following table.

X	Y	x_i = Rank in X	y_i = Rank in Y	$D = x_i - y_i$	D^2
35	30	3	5	– 2	4
23	33	5	3	2	4
47	45	1	2	– 1	1
17	23	6	6	0	0
10	8	7	8	– 1	1
43	49	2	1	1	1
9	12	8	7	1	1
6	4	9	9	0	0
28	31	4	4	0	0
					$\Sigma D^2 = 12$

Now, **Rank Correlation:** $R = 1 - \frac{6 \Sigma D^2}{n(n^2-1)} = 1 - \frac{6 \times 12}{9(81-1)}$

$$= 1 - \frac{72}{9 \times 80} = 1 - \frac{1}{10} = \frac{9}{10} = \mathbf{0.9.}$$

Example 28. *Quotation of index numbers of equity share prices of certain joint stock company and prices of preference share are given below:*

Years :	*1961*	*1962*	*1963*	*1964*	*1965*	*1966*	*1967*
Equity :	*97.5*	*99.4*	*90.6*	*96.2*	*95.1*	*98.4*	*97.1*
Preference :	*75.1*	*75.9*	*77.1*	*78.2*	*79.0*	*74.8*	*70.1*

Use the method of rank correlation to determine the relationship between equity share and preference share prices.

Solution. Here first we assign the ranks in the descending orders.

Computation Table

X	Y	Rank of X x	Rank of Y y	d = x − y	d^2
97.5	75.1	4	5	− 1	1
99.4	75.9	1	4	− 3	9
98.6	77.1	2	3	− 1	1
96.2	78.2	6	2	4	16
95.1	79.0	7	1	6	36
98.4	74.0	3	6	− 3	9
97.1	70.2	5	7	− 2	4
				Σ d = 0	**$\Sigma d^2 = 76$**

Here $n = 7$

$$R = 1-\frac{6\Sigma d^2}{n(n^2-1)} = 1-\frac{6\times 76}{7(49-1)} = 1-\frac{456}{336}$$

$$= 1 - 1.357 = -\mathbf{0.357}$$

Example 29. *Ten competitors in a beauty contest are ranked by three judges in the following orders:*

1st Judge :	*1*	*6*	*5*	*10*	*3*	*2*	*4*	*9*	*7*	*8*
2nd Judge :	*3*	*5*	*8*	*4*	*7*	*10*	*2*	*1*	*6*	*9*
3rd Judge :	*6*	*4*	*9*	*8*	*1*	*2*	*3*	*10*	*5*	*7*

Use the correlation coefficient to determine which pair of judges has the nearest approach to common taste in beauty.

Solution. Let R_1, R_2, R_3, respectively be the ranks given by first, second and third judge. Let r_{ij} be the rank correlation coefficient between the rank given by *i*th and *j*th judges, $i \neq j$, $i = 1, 2, 3, j = 1, 2, 3$. Let $D_{ij} = R_i - R_j$ be the difference of ranks of an individual given by *i*th and *j*th judge.

Table of Calculations

R_1	R_2	R_3	$D_{12} = R_1 - R_2$	$D_{13} = R_1 - R_3$	$D_{23} = R_2 - R_3$	D_{12}^2	D_{13}^2	D_{23}^2
1	3	6	− 2	− 5	− 3	4	25	9
6	5	4	1	2	1	1	4	1
5	8	9	− 3	− 4	− 1	9	16	1
10	4	8	6	2	− 4	36	4	16
3	7	1	− 4	2	6	16	4	36
2	10	2	− 8	0	8	64	0	64
4	2	3	2	1	− 1	4	1	1
9	1	10	8	− 1	9	64	1	81
7	6	5	1	2	1	1	4	1
8	9	7	− 1	1	2	1	1	4
			$\Sigma D_{12} = 0$	$\Sigma D_{13} = 0$	$\Sigma D_{23} = 0$	$\Sigma D_{12}^2 = 200$	$\Sigma D_{13}^2 = 60$	$\Sigma D_{23}^2 = 214$

Here $n = 10$

$$\therefore \quad R_{12} = 1-\frac{6\Sigma D_{12}^2}{n(n^2-1)} = 1-\frac{6\times 200}{10\times 99} = -\frac{7}{33} = \mathbf{0.2121.}$$

$$R_{13} = 1-\frac{6\Sigma D_{13}^2}{n(n^2-1)} = 1-\frac{6\times 60}{10\times 99} = \frac{7}{11} = \mathbf{0.6363.}$$

$$R_{23} = 1-\frac{6\Sigma D_{23}^2}{n(n^2-1)} = 1-\frac{6\times 214}{10\times 99} = \frac{-49}{165} = \mathbf{-0.2970.}$$

Since R_{13} **is maximum, so the pair of first and third judge has the nearest approach to the common taste of beauty.**

TYPE III. When Equal Ranks are given to more than two attributes. Or TIE CASE

If two or more individuals are placed together in any classification with respect to an attribute, i.e., if in case of variable data, there are more than one item with the same rank in either or both the series (*i.e.*, **Tie Rank**), **then the Spearman's Rank Correlation coefficient formula given above does not give the correlation coefficient for Tie Rank.** The problem is solved by assigning average rank to each of these individuals who are put in tie. For example, suppose an item is repeated at rank 5, (i.e., the 5th and 6th items are having same values), then the **common rank assigned to 5 and 6th is (5 + 6)/2 = 5.5,** which is the average of 5 and 6, the ranks which these items would have been assigned if they were different. **The next rank assigned will be 7.** But if an item is repeated thrice at rank 2, then the common rank assigned to each value will be (2 + 3 + 4)/3 = 3, which is the arithmetic mean of 2, 3 and 4. The **next ranks to be assigned would be 5.** In order to find the Rank Correlation coefficient of **Repeated Ranks** or **Tie Ranks**, an **Adjustment** or **Correction Factor** is added to the **Spearman's Rank Correlation formula,** *i.e.,* to

$$R = 1-\frac{6\Sigma D^2}{n(n^2-1)} \qquad \textbf{... (A)}$$

Correction Factor. "In the formula (A), add the factor $\frac{m(m^2-1)}{12}$ **to** ΣD^2, **where m is the number of times an item is repeated. This correction factor is to be added for each repeated value in both the series."**

The modified formula for Tie Rank Correlation Coefficient is given by

$$R = 1-\frac{6\left[\Sigma D^2 + \frac{1}{12}(m_1^3 - m_1) + \frac{1}{12}(m_2^3 - m_2) + ...\right]}{n(n^2-1)}$$

where $m_1, m_2, ...$ **are the number of times a value is repeated.**

The method is illustrated by the following example.

Example 30. *From the following table, calculate the rank correlation coefficient:*

X :	*48*	*33*	*40*	*9*	*16*	*16*	*65*	*24*	*16*	*57*
Y :	*13*	*13*	*24*	*6*	*15*	*4*	*20*	*9*	*6*	*19.*

Solution. We prepare the following table.

Table for Calculations

X	*Y*	*Rank X* R_1	*Rank Y* R_2	$D = R_1 - R_2$	D^2
48	13	8	5.5	2.5	6.25
33	13	6	5.5	0.5	0.25
40	24	7	10	– 3	9
9	6	1	2.5	–1.5	2.25
16	15	3	7	– 4	16
16	4	3	1	2	4
65	20	10	9	1	1
24	9	5	4	1	1
16	6	3	2.5	0.5	0.25
57	19	9	8	1	1
					$\Sigma D^2 = 41$

Remarks

In the *X*-series, we notice that the value **16 is repeated thrice.** The common rank given to these values is **3**, which is the average of 2, 3 and 4 (*i.e.*, (2 + 3 + 4)/3 = 3), ranks which these values would have assumed if they were different. The next rank given to the next value is **5**. Here $m_1 = 3$. The **correction factor for X-series is** $\frac{m_1(m_1^2-1)}{12} = \frac{3(3^2-1)}{12} = 2$. In the Y series the value 6 and 13 are repeated twice. The value 6 occurs twice and its average rank is 2.5 = (2 + 3)/2. The next value 9 is assigned the rank 4. Here $m_2 = 2$. Again the value 13 is repeated twice at different places. The average rank is 5.5 = $\frac{5+6}{2}$. The next value 15 is assigned the rank 7. Here $m_3 = 2$. The total correction for the Y series is

$$\frac{m_2(m_2^2-1)}{12} + \frac{m_3(m_3^2-1)}{12} = \frac{2(2^2-1)}{12} + \frac{2(2^2-1)}{12} = 0.5 + 0.5 = \mathbf{1}.$$

Thus the rank correlation is

$$\mathbf{R} = 1 - \frac{6\left[\Sigma D^2 + \frac{m_1(m_1^2-1)}{12} + \frac{m_2(m_2^2-1)}{12} + \frac{m_3(m_3^2-1)}{12}\right]}{n(n^2-1)}$$

$$= 1 - \frac{6[41+2+0.5+0.5]}{10(10^2-1)} = 1 - \frac{264}{990} = 1 - 0.267 = \mathbf{0.733}.$$

12.10.1 Merits and Limitations of Rank Correlation Coefficient

Merits

1. ***It is simpler to understand and easy to calculate as compared to Karl Pearson's method.***
2. ***It is useful for qualitative data such as beauty, honesty, efficiency etc.***
3. ***It is a useful method when the actual data is not given but only ranks are given.***

Limitations

1. ***It cannot be used for grouped frequency distribution.***
2. ***It is not as accurate as Karl Pearson's coefficient of correlation.***
3. ***It cannot be used in a continuous series.***
4. ***When the number of items is more than 30 and if the ranks are not known, this method consumes more time and therefore cannot conveniently be used.***

EXERCISE – 12.3

1. The ranking of ten students in Statistics and Accountancy are as follows:

Statistics :	3	5	8	4	7	10	2	1	6	9
Accountancy :	6	4	9	8	1	2	3	10	5	7.

What is the coefficient of rank correlation.

2. Calculate the coefficient of correlation from the following data by the method of rank correlation:

X :	75	88	95	70	60	80	81	50
Y :	120	134	150	115	110	140	142	100.

3. Compute Rank correlation from the following table of Index Number of Supply (X) and price (Y):

X :	115	134	120	130	124	128
Y :	130	132	128	130	127	125.

4. Ten competitors in a beauty contest are ranked by three judges in the following order :

First judge	1	5	4	8	9	6	10	7	3	2
Second judge	4	8	7	6	5	9	10	3	2	1
Third judge	6	7	8	1	5	10	9	2	3	4.

Use the rank correlation to gauge which pair of judges have the nearest approach to common taste in beauty.

5. Following are the scores of ten students in a class and their I.Q.

Student :	1	2	3	4	5	6	7	8	9	10
Score :	35	40	25	55	85	90	65	55	45	50
I.Q. :	100	100	110	140	150	130	100	120	140	110.

Calculate the rank correlation coefficient between the score and I.Q.

[**Hint.** Use correction factor method. It is a case of repeated ranks)

6. Calculate Spearman's rank correlation coefficient between advertisement cost and sales from the following data:

Adv. Cost (Rs. 1000)	39	65	62	90	82	75	25	98	36	78
Sales (Rs. 100,000)	47	53	58	86	62	68	60	91	51	84.

7. Obtain the Rank correlation from the following data:

X :	68	64	75	50	64	80	75	40	55	64
Y :	62	58	68	45	81	60	68	48	50	70.

8. The following table gives the score obtained by 11 students in English and Tamil Translation. Find the rank correlation coefficient.

Scores in English :	40	46	54	60	70	80	82	85	85	90	95
Scores in Tamil :	45	45	50	43	40	75	55	72	65	42	70.

9. Calculate Spearman's coefficient of rank correlation for the following the data:

X :	53	98	95	81	75	71	59	55
Y :	47	25	32	37	30	40	39	45.

10. Calculate the rank correlation coefficient for the following table of marks of students in two subjects :

First subject :	80	64	54	49	48	35	32	29	20	18	15	10
Second subject :	36	38	39	41	27	43	45	52	51	42	40	52.

11. The coefficient of rank correlation of marks obtained by 10 students in English and Economics was found to be 0.5. It was later discovered that the difference in ranks in the two subjects obtained by one of the students was wrongly taken as 3 instead of 7. Find the correct coefficient of rank correlation.

12. Quotations of index number of security prices of a certain joint Stock Company and of prices of preferences shares and debentures are given below:

Preference shares Price (X)	73.2	85.8	78.9	75.8	77.2	81.2	83.8
Debenture Price (Y)	97.8	99.2	98.8	98.3	98.3	96.7	97.1.

Find the rank coefficient of correlation between preference share prices and debenture prices.

ANSWERS

1. – 0.3. **2.** 0.9. **3.** 0.33.

4. Second and third judges are in nearest approach to the sense of beauty.

5. 0.47. **6.** 0.82. **7.** 0.545. **8.** 0.36.

9. – 0.905. **10.** – 0.685 **11.** 0.26. **12.** 0.125.

12.11 TWO WAY FREQUENCY TABLE

A **bivariate distribution** refers to data concerning two variables x and y. In this case the two variables are paired and then their coefficient of correlation is calculated. If in a bivariate distribution data, the number of pairs of observation is large, then it becomes very difficult to calculate the coefficient of correlation. In order to facilitate the calculation of coefficient of correlation in such a large pairs of observations, the data are summarised in the form of a **Two way Frequency Table.** It is also called **Bivariate Frequency Table or Contingency Table or Correlation Table,** or simply **Two way Table.**

In a two way frequency table of two variables x and y, the class intervals of x series are put in the first row on the top of the table and the class intervals of second series y are put in the first column on the left of the table.

TWO WAY FREQUENCY TABLE

<table>
<tr><td colspan="2" rowspan="3">X-SERIES →
Y-SERIES
↓</td><td colspan="6">CLASS INTERVALS</td><td rowspan="3">Total
Frequency
of Y</td></tr>
<tr><td>$x_1 - x_2$</td><td>$x_2 - x_3$</td><td>$x_3 - x_4$</td><td>...</td><td>...</td><td>$x_m - x_{m-1}$</td></tr>
<tr><td colspan="6">MID-POINTS</td></tr>
<tr><td></td><td></td><td>x_{12}</td><td>x_{23}</td><td>x_{34}</td><td></td><td></td><td>$x_{(m-1)m}$</td><td></td></tr>
<tr><td>CLASS INTERVALS
$y_1 - y_2$</td><td>MID POINTS
y_{12}</td><td>$f(x_{12}, y_{12})$</td><td>$f(x_{23}, y_{12})$</td><td></td><td></td><td></td><td></td><td rowspan="5">f_y</td></tr>
<tr><td>$y_2 - y_3$</td><td>y_{23}</td><td>$f(x_{12}, y_{23})$</td><td>$f(x_{23}, y_{23})$</td><td>CELLS</td><td></td><td></td><td></td></tr>
<tr><td>$y_3 - y_4$</td><td>y_{34}</td><td>$f(x_{12}, y_{34})$</td><td>$f(x_{23}, y_{34})$</td><td></td><td></td><td></td><td></td></tr>
<tr><td></td><td></td><td colspan="6">CELLS</td></tr>
<tr><td></td><td></td><td colspan="6"></td></tr>
<tr><td>$y_{(n-1)} - y_n$</td><td>$y_{(n-1)n}$</td><td>$f(x_{12}, y_{(n-1)n})$</td><td></td><td></td><td></td><td></td><td></td><td rowspan="2">Total frequency
$\Sigma f_x =$
$\Sigma f_y = N$,
the number of pairs</td></tr>
<tr><td>Total of Frequency of X
$= f(x_{12}, y_{12})$
$+ f(x_{12}, y_{23})$
$+ \ldots$</td><td></td><td>Frequency of x_{12}</td><td></td><td>f_x</td><td></td><td></td><td></td></tr>
</table>

It there are *m* classes for *x* variable series and *n* classes for *y* variable series, then there will be $m \times n$ cells or squares in the body of a two way table. By going through the different pairs of the values of (x, y) and using tally marks just as in case of univariate distribution, one can find the frequency $f(x, y)$ of each cell. The frequencies which are put in the cells are called **cell frequencies.** The nature of concentration of frequencies in various cells (squares) reveals the type of correlation-positive or negative. **A univariate frequency distribution table obtained from a bivariate frequency table is called the marginal distribution. Marginal frequencies are the total of cell frequencies in the last row and the last column.** A model of two way frequency table is given on the previous page:

Working Rule

It is important to note that

1. **x_{ij} is the mid point of interval $x_i - x_j$, *i.e.*, $x_{ij} = \dfrac{x_i + x_j}{2}$.**
2. **y_{pl} is the mid-point of the interval *i.e.*, $y_p - y_t = \dfrac{y_p + y_l}{2}$.**
3. **$f(x_{ij}; y_{pl})$ is the frequency of x lying in the interval $(x_i - x_j)$ and y lying in the interval $y_p - y_l$ and so on.**
4. **Frequency of $x_{12} = f(x_{12}, y_{12}) + f(x_{12}; y_{23}) + ... + f(x_{12}, y)_{(n-1)n}$ and so on.**

Example 31. *The marks scored by 22 students in Mathematics and Accountancy are given below.*

(7, 3), (15, 13), (0, 1), (1, 2), (3, 7), (16, 8), (2, 9), (18, 12), (10, 10), (5, 9), (4, 17), (17, 16), (6, 7), (19, 18), (14, 11), (13, 11), (11, 14), (9, 3) (8, 5), (9, 15), (5, 19) and *(13, 4).*

The first figures in brackets indicate the marks in mathematics, and the second marks in accountancy. Prepare a two-way table taking the magnitude of each class interval as 4 marks, the first being equal to 0 and less than 4.

Solution.

Two way Table

Marks in Mathematics → / *Marks ↓ in Accountancy*	0 – 4	4 – 8	8 – 12	**12 – 16**	16 – 20	*Total frequencies of y*
0 – 4	| (2)	| (1)	| (1)			2+1+1 = **4**
4 – 8	| (1)	| (2)	| (1)	| (1)		1+2+1+1 = **5**
8 – 12	| (1)	| (1)	| (1)	| (2)	| (1)	1+1+1+2 +1 = **6**
12 – 16			| (2)	| (1)	| (1)	2+1+1= **4**
16 – 20		| (1)			| (2)	1+2 = **3**
Total frequencies of *x*	2+1+1 = **4**	1+2+1+ 1 = **5**	1+1+1+ 2 = **5**	1+2+1 = **4**	1+1+2 = **4**	**22**

Example 32. *Twenty five values of two variables X and Y are given below. Form a two-way frequency table showing the relationship between the two. Take class intervals of X as 10 – 20, 20 – 30 etc. and Y as 100 – 200, 200 – 300 etc.*

X:	12	24	33	22	44	37	26	36	55
Y:	140	256	360	470	470	380	480	315	420
X:	48	27	57	21	51	27	42	43	52
Y:	390	440	390	590	250	550	360	570	290
X:	57	44	48	48	52	41	69		
Y:	416	380	492	370	312	330	590		

Solution.

Two Way Frequency Table

X series → / Y series ↓	10 – 20	20 – 30	30 – 40	40 – 50	50 – 60	60 – 70	**Total**
100–200	| (1)						**1**
200–300		| (1)			| (2)		1+2=**3**
300–400			||| (3)	~~||||~~ (5)	| (2)		3+5+2 = **10**
400–500		||| (3)		| (2)	| (2)		3+2+2 = **7**
500–600		| (2)		| (1)		| (1)	2+1+1= = **4**
Total	**1**	**6**	**3**	**8**	**6**	**1**	**25**

12.12 CORRELATION COEFFICIENT BY TWO WAY TABLE

WORKING RULE

STEP I. Find the mid-points of the x intervals and denote it by x's. Also find the mid point of the y intervals and denote it by y's

STEP II. Take **A** and the assumed mean for ***x*-series** (of mid points) and **B** as the assumed mean for ***y*-series** (of mid points)

STEP III. Take the deviations of ***x*-series** from the assumed mean **A**. Take out the common factor from it and denote it by ***dx***.

STEP IV. Take the deviations of ***y*-series** from the assumed mean **B**. Take out the common factor from it and denote it by *dy*.

STEP V. Multiply ***dx***, ***dy*** and the respective frequency ***f*** of each cell to get ***fdxdy*** and write down the figure so obtained in a box **or** in a corner. Also find the sum $\Sigma f\, dx\, dy$

STEP VI. Multiply the deviation dx of $\boldsymbol{x}$ by its respective frequency f_x and get the total $\Sigma f_x\, dx$

STEP VII. Find the sum $\Sigma f_x dx^2$

STEP VIII. Multiply the deviation ***dy*** of y by its respective frequency f_y and get the sum $\Sigma f_y dy$

STEP IX. Find the sum $\Sigma f_y dy^2$

STEP X. Substitute the values of Σ ***fdxdy***, $\Sigma f_x dx$, $\Sigma f_x dx^2$, $\Sigma f_y dy$, $\Sigma f_y dy^2$ in the following formula to get the value of coefficient of correlation.

$$r = \frac{\Sigma fdxdy - \frac{(\Sigma f_x dx)(\Sigma f_y dy)}{N}}{\left(\sqrt{\Sigma f_x dx^2 - \frac{(\Sigma f_x dx)^2}{N}}\right)\sqrt{\Sigma f_y dy^2 - \frac{(\Sigma f_y d_y)^2}{N}}}$$

Example 33. *From the following table of a bivariate Frequency distribution, calculate the coefficient of correlation between the heights and weights of children.*

Height (in inches)

Weight (in pounds)	*40–44*	*44–48*	*48–52*	*52–56*	*56–60*	*60–64*	*Total*
35–55	*4*	*40*	*60*	—	—	—	*104*
55–75	—	—	*24*	*88*	*12*	—	*124*
75–95	—	—	—	*8*	*32*	*8*	*48*
95–115	—	—	—	—	*4*	*8*	*12*
115–135	—	—	—	*4*	—	—	*4*
135–155	—	—	—	—	*4*	*4*	*8*
Total	*4*	*40*	*84*	*100*	*52*	*20*	*300*

Solution. Let **A = 50** so that $dx = \frac{X - A}{4}$, and **B = 85** so that $dy = \frac{Y - 85}{20}$

			Height in (inches)									
			40-44	44-48	48-52	52-56	56-60	60-64				
			42	46	50	54	58	62				
y	*m*	*dy*	– 2	– 1	0	1	2	3	*fy*	*fydy*	$fydy^2$	*fdxdy*
35-55	45	– 2	16 4	80 40	0 **60**	0	**– 0**	**0**	**104**	– 208	416	96
55-75	65	– 1	**0**	**0**	0 24	–88 88	–24 12		**124**	– 124	124	–112
75-95	85	0	**0**	**0**	**0**	**0** 8	**0** 32	0 8	**48**	0	0	0
95-115	105	1	**0**	**0**	**0**	**0**	**0** 4	24 8	12	**12**	12	32
115-135	125	2	**0**	**0**	**0**	**8** 4	**0**	0 4	**4**	8	16	8
135-155	145	3	**0**	**0**	**0**	**0**	**24** 4	**36** 4	**8**	24	72	60
		$f_x \rightarrow$	**4**	**40**	84	100	52	20	**N = 300**	**$\Sigma fydy$ = –288**	**$\Sigma f_y dy^2$ = 640**	**$\Sigma fdxdy$ = 84**
		$f_x d_x \rightarrow$	**–8**	–40	0	100	104	60	**$\Sigma fxdx$ = 216**			
		$f_x dx^2 \rightarrow$	**16**	40	0	100	208	180	**$\Sigma fxdx^2$ = 544**			
		$\Sigma fdxdy \rightarrow$	**16**	80	0	–80	–8	60	**$\Sigma fdxdy$ = 84**			

Now
$$r = \frac{\Sigma\, fdxdy \times N - (\Sigma\, f_x dx)(\Sigma\, f_y dy)}{\sqrt{\Sigma\, f_x dx^2 \times N - (\Sigma\, f_x dx)^2}\ \sqrt{\Sigma\, f_y dy^2 \times N - (\Sigma\, f_y dy)^2}}$$

Here N = 300 $\Sigma fdxdy = 84$ $\Sigma f_x dx = 216$, $\Sigma f_x dx^2 = 544$

$\Sigma f_y dy = -288$, $\Sigma f_y dy^2 = 640$.

$$r = \frac{84 \times 300 - (216)(-288)}{\sqrt{544 \times 300 - (216)^2} \times \sqrt{300 \times 640 - (-288)^2}}$$

$$= \frac{25200 + 62208}{\sqrt{163200 - 46656} \times \sqrt{192000 - 82944}}$$

$$= \frac{87408}{\sqrt{116544} \times \sqrt{109056}} = \frac{87408}{(341.39) \times (330.23)}$$

$$= \frac{87408}{11273722} = \mathbf{0.78.}$$

Example 34. *The following table gives the number of students having different heights and weights:*

Height in → *Centimetres ↓*	*Weight in Kilograms* 55–60	*60–65*	*65–70*	*70–75*	*75–80*	*Total*
150 – 155	*1*	*3*	*7*	*5*	*2*	*18*
155 – 160	*2*	*4*	*10*	*7*	*4*	*27*
160 – 165	*1*	*5*	*12*	*10*	*7*	*35*
165 – 170	—	*3*	*8*	*6*	*3*	*20*
Total	*4*	*15*	*37*	*28*	*16*	*100*

Find the coefficient between the heights and weights of the students.

Solution. Let us take the assumed mean for series as A = 67.5 so that the deviation $dx = \mathbf{\frac{x - 67.5}{10}}$. Let the assumed mean for y series be

B = 157.5 so that $dy = \frac{y - 157.5}{5}$

Coefficient of Correlation *r* is:

$$r = \frac{\Sigma\, fdxdy \times N - (\Sigma\, f_x dx)(\Sigma f_y dy)}{\sqrt{\Sigma f_x dy^2 \times N - (\Sigma f_x dx)^2}\ \sqrt{\Sigma\, f_y dy^2 \times N - (\Sigma f_y dy)^2}}$$

$\Sigma fdxdy = 31$; N = 100 ; $\Sigma f_x dx = 37$; $\Sigma f_y dy = 57$; $\Sigma f_x dx^2 = 123$; $\Sigma f_y dy^2 = 133$.

$$r = \frac{31 \times 100 - (37 \times 57)}{\sqrt{123 \times 100 - (37)^2}\ \sqrt{133 \times 100 - (57)^2}}$$

$$= \frac{3100 - 2109}{\sqrt{12300 - 1369}\ \sqrt{13300 - 3249}} = \frac{991}{\sqrt{10931} \times \sqrt{10051}}$$

Computation Table for Coefficient of Correlation

$$= \text{Anti log [log 991} - \frac{1}{2} \text{(log 10931 + log 10051]}$$

$$= \text{Anti log} \left[2.996 - \frac{1}{2}(4.04 + 4.0002) \right]$$

$$= \text{Antilog } [2.996 - 4.020] = \text{Antilog } [-1.024] = \mathbf{0.0946.}$$

Series →		x	55-60	60-65	65-70	70-75	75-80				
			Weight in kilograms								
Series ↓		m	57.5	62.5	67.5	70.75	77.5	mid points			
		dx	– 2	– 1	0	1	2				
y	m	dy						f_y	$f_y dy$	$f_y dy^2$	$fdxdy$
150–155	152.5	– 1	**2** 1	**3** 3	**0** 7	– 5 5	– 4 2	18	– 18	18	– 4
155–160	157.5	0	**0** 2	**0** 4	**0** 10	**0** 7	**0** 4	27	0	0	0
160–165	162.5	1	–2 1	**–5** 5	0 12	**10** 10	14 7	35	35	35	17
165–170	167.5	2		**–6** 3	**0** 8	**12** 6	**12** 3	20	40	80	18
		f_x	4	15	37	28	16	N = 100	$\Sigma f_y dy$ = 57	$\Sigma f_y dy^2$ = 133	$\Sigma fdxdy$ = 31
		$f_x dx$	– 8	– 15	0	28	32	$\Sigma f_x dx$ = 37			
		$f_x dx^2$	16	15	0	28	64	$\Sigma f_x dx^2$ = 123			
		$f_x dxdy$	0	– 8	0	17	22	$\Sigma f_x dxdy$ = 31			

EXERCISE – 12.4

1. The marks obtained by 25 students in Statistics and Economics are given below. The first figure in bracket indicates the marks in Statistics and the second indicates the marks in Economics:

   (14, 12), (0, 2), (1, 5), (7, 3), (15, 9), (2, 8), (12, 18), (9, 11), (5, 3), (17, 13), (19, 18), (11, 7), (10, 13), (13, 16), (16, 14), (16, 10), (4, 1), (11, 14), (18, 3), (9, 15), (13, 11), (14, 17), (10, 10), (11, 7), (15, 15).

   Prepare a two way table taking the magnitude of each class interval as 4 marks, the first being equal to 0 and less than 4.

2. The bivariate frequency distribution based on monthly salary and age of 100 clerks working in some large scale commercial organisation is as under:

Age (years)	*Monthly salary (in Rs.)*			
	600–700	*700–800*	*800–900*	*900–1000*
20 and less than 30	16	16	—	—
30 and less than 40	4	10	4	—
40 and less than 50	—	4	18	12
50 and less than 60	—	—	10	12

   Compute Karl Pearson's coefficient of correlation between age and monthly salary of clerks and comment on it.

3. Determine Karl Pearson's correlation coefficient from the following bivariate frequency distribution.

Age of husbands	*Age of wives (years)*				
	16–23	*23–30*	*30–37*	*37–44*	*44–51*
18–25	9	3	—	—	—
25–32	—	20	10	4	—
32–39	—	—	12	5	3
39–46	—	—	8	7	5
46–53	—	—	—	10	4

4. Calculate Karl Pearson's coefficient of correlation from the data given below:

Marks	*Age in years*				
	18	19	20	21	22
20–25	5	2	—	—	—
10–15	—	5	4	—	—
10–20	—	—	7	10	—
5–10	—	—	—	3	2
0–5	—	—	—	3	1

5. Calculate from the data reproduced below pertaining to 66 selected villages in Meerut District, the value of r between total cultivable area and the area under wheat.

Area under wheat (in hectares)	*Total cultivable area (in hectares)*					
	0 – 500	500 – 1000	1000 – 1500	1500 – 2000	2000 – 2500	Total
0 – 200	12	6	—	—	—	18
200 – 400	2	18	4	2	1	27
400 – 600	—	4	7	3	—	14
600 – 800	—	1	—	2	1	4
800 – 1000	—	—	—	1	2	3
Total	14	29	11	8	4	66

6. From the table given below, calculate the coefficient of correlation between the ages of husbands and wives

Age of wives Y-series	*Age of Husband X-series*					
	20–30	*30–40*	*40–50*	*50–60*	*60–70*	*Total*
15–25	5	9	3	—	—	17
25–35	—	10	25	2	—	37
35–45	—	1	12	2	—	15
45–55	—	—	4	16	5	25
55–65	—	—	—	4	2	6
Total	5	20	44	24	7	100

ANSWERS

2. 0.763. **2.** – 0.7768. **4.** – 0.8373. **5.** 0.746. **6.** 0.80

12.13. COEFFICIENT OF CONCURRENT DEVIATION

Coefficient of Concurrent Deviation. ***The coefficient of concurrent deviation is of use, generally, in indicating relationship between short term oscillations.*** The **principle** underlying in the coefficient of concurrent deviations is as follows:

"If the short-term fluctuations of two series are correlated positively, their deviations would be concurrent and their curves would move in the same direction indicating positive correlation between the series".

The **coefficient of concurrent Deviation** is denoted by

$$r_c = \pm\sqrt{\pm\frac{(2C-n)}{n}}$$

where C = **number of pairs of concurrent deviations,**

n = **number of pairs of deviations, which is one less than the actual numbers N, *i.e.*, $n = (N-1)$.**

In both the series, each of the values is compared with one immediately preceding it (starting from the second which is to be compared with first) and the plus (+) or minus (–) or zero (0) deviation is recorded depending whether there has been an increase or decrease or no change in the magnitude of the variable. **If $\left(\frac{2C-n}{n}\right)$ is a negative quantity, its square root cannot be calculated. In such case, a minus sign is placed before $\left(\frac{2C-n}{n}\right)$ as well as outside the square root. The coefficient of Concurrent Deviation** r_c would be **negative** if $\frac{2C-n}{n}$ **is negative** and it would be **positive if** $\frac{2C-n}{n}$ **is positive.**

WORKING RULE

Step I. Find the direction of change of x variable *i.e.*, "Take the first two values into consideration and find out the direction of x variable. Put + or – sign if the difference '**Second value – first value**' is positive or negative. If the two values are identical, then put **0.**

Similarly, as compared with the second value, find out whether the third value is increasing or decreasing or constant, *i.e.*, (**third value – second value**) is positive or negative or zero. Continue this process for all other values and denote this column by C_x.

Step II. Repeat the process for Y series in the same manner as done in **STEP I** for x – series, *i.e.*, find the direction of change of y variable and denote it by C_y.

Step III. Multiply C_x with C_y and determine the number of positive signs with the rule (+) (+) = +, (+) (–) = (–), (–) (–) = +. Denote it under the column $\mathbf{C_xC_y}$

Step IV. Find the total number of positive signs under the column C_xC_y and denote it by C = sum of number of positive signs of $\mathbf{C_xC_y}$.

Step V. Find $n = N - 1$, where N is the total number of observation of x series or y series.

Step VI. Substitute the values of C, n in the formula

$$r_c = \pm\sqrt{\pm\frac{(2c-n)}{n}}$$

and attach positive sign with r_c if $\frac{2c-n}{n}$ is positive and negative sign with r_c if $\frac{2c-n}{n}$ is negative.

Example 35. *Calculate the coefficient of concurrent deviations for the following data :*

(5 marks)

Supply :	*65*	*40*	*35*	*75*	*63*	*80*	*35*	*20*	*80*	*60*	*50*
Demand :	*60*	*55*	*50*	*56*	*30*	*70*	*40*	*35*	*80*	*75*	*80*

Solution.

Supply (*X*)	C_x	*Demand* (*Y*)	C_y	C_xC_y
65		60		
40	–	55	–	+
35	–	50	–	+
75	+	56	+	+
63	–	30	–	+
80	+	70	+	+
35	–	40	–	+
20	–	35	–	+
80	+	80	+	+
60	–	75	–	+
50	–	80	+	–
				C = 9

Here $\boldsymbol{n} = \text{N} - 1 = 11 - 1 = \mathbf{10}.$

$$r = \pm\sqrt{\pm\frac{(2c-n)}{n}} = +1\sqrt{\pm\frac{2\times 9-10}{10}} = \sqrt{\pm\frac{18-10}{10}} = \pm\sqrt{\pm 0.8} = \mathbf{0.89.}$$

Example 36. *Calculate the coefficient of correlation using the method of concurrent deviations between supply and demand and an item for a 10 year period as given below:*

Year :	*1981*	*1982*	*1983*	*1984*	*1985*	*1986*	*1987*	*1988*	*1989*	*1990*
Supply :	*125*	*160*	*164*	*174*	*155*	*170*	*165*	*162*	*172*	*175*
Demand :	*115*	*125*	*192*	*190*	*165*	*174*	*124*	*127*	*152*	*169*

Solution.

TABLE: Coefficient of Correlation by Concurrent Deviation Method

Year	*Supply X*	*Direction of change in X D_x*	*Demand Y*	*Direction of change in Y d_y*	$d_x d_y$
1981	125		115		
1982	160	+	125	+	+
1983	164	+	192	+	+
1984	174	+	190	–	–
1985	155	–	165	–	+
1986	170	+	174	+	+
1987	165	–	124	–	+
1988	162	–	127	+	–
1989	172	+	152	+	+
1990	175	+	169	+	+
					C = 7

Also $$r_C = \pm\sqrt{\frac{\pm 2C - n}{n}}.$$

Here $C = 7$, $n = 10 - 1 = 9$

$$\therefore \quad r_C = \pm\sqrt{\frac{2\times 7-9}{9}} = \mathbf{0.745.}$$

Example 37. *Calculate the coefficient of correlation using the method of concurrent deviations between supply and demand of an item for a period of ten years, as given below:*

Year:	*1987*	*1988*	*1989*	*1990*	*1991*	*1992*	*1993*	*1994*	*1995*	*1996*
Supply:	*125*	*160*	*164*	*174*	*155*	*170*	*165*	*162*	*172*	*175*
Demand:	*115*	*125*	*192*	*190*	*165*	*174*	*124*	*127*	*152*	*169*

Solution.

TABLE: Concurrent Deviations

Year	*Supply*	C_x	*Demand*	C_y	$C_x C_y$
1987	125	+	115	+	+
1988	160	+	125	+	+
1989	164	+	192	+	+
1990	174	+	190	–	–
1991	155	–	165	–	+
1992	170	+	174	+	+
1993	165	–	124	–	+
1994	162	–	127	+	–
1995	172	+	152	+	+
1996	175	+	169	+	+

Here $n = 10 - 1 = 9$.

Coefficient of Correlation $= \pm\sqrt{\pm\frac{2c-n}{n}} = \pm\sqrt{\pm\frac{2\times7-9}{9}}$ $[\because c = 7]$

$= \mathbf{0.7454.}$

Example 38. *Find the coefficient of concurrent deviations for the following data :*

Year:	*1984*	*1985*	*1986*	*1987*	*1988*	*1989*	*1990*
Supply:	*150*	*154*	*160*	*172*	*160*	*165*	*180*
Price:	*200*	*180*	*170*	*160*	*190*	*180*	*172*

Solution. We have the following table of calculations.

Table of Computation

Year	*Supply X*	*Direction change of x* C_y	*Price Y*	*Direction change of* C_y	*Concurrent Deviation* C_xC_y
1984	150		200		
1985	154	+	180	(–)	(–)
1986	160	+	170	(–)	(–)
1987	172	+	160	(–)	(–)
1988	160	(–)	190	+	(–)
1989	165	+	180	(–)	(–)
1990	180	+	172	(–)	(–)
					$C = \Sigma C_xC_y = 0$

Here N = Total number of observation years = 7

$$n = N - 1 = 7 - 1 = 6$$

$$C = \Sigma C_xC_y = 0$$

$$\therefore r_c = \pm\sqrt{\pm\frac{2C-n}{n}} = \pm\sqrt{\pm\frac{(2\times0-6)}{6}} = -\sqrt{-(-1)} = -\sqrt{+1} = -1.$$

Note. *Here* $\frac{2C-n}{n} = -1$, ***which is negative, we shall multiply it by (– 1) to get*** $\sqrt{+1} = 1$. ***But ultimately, the value of*** r_c ***will be negative as*** $\frac{2C-n}{n}$ ***is negative.***

12.14. MERITS AND LIMITATIONS OF CONCURRENT DEVIATION METHOD

Merits

1. ***It is simple to compute and easy to understand.***
2. ***It is simplest of all the methods.***
3. ***It is suitable for measuring the extent of correlation if the variable includes item-values having short-term fluctuations/oscillations.***

Limitations

1. ***It is not useful if long-term changes are to be considered because it ignores the general trend of variation.***
2. ***It does not differentiate between small and big variations.***
3. ***The results of this method are rough indicators of the presence or absence of correlation.***

EXERCISE – 12.5

1. Calculate the correlation coefficient by using concurrent deviation method

X :	150	135	90	140	100
Y :	60	50	100	80	70.

2. Calculate the coefficient of correlation by the method of concurrent deviation.

Year :	1981	1982	1983	1984	1985	1986	1987	1988	1989
Supply :	80	82	86	91	83	85	89	96	93
Price :	146	140	130	117	133	127	115	94	100.

3. Calculate the coefficient of correlation using the method of current deviation.

Year :	1963	1964	1965	1966	1967	1968	1969	1970	1971
Supply :	160	164	172	182	166	170	178	192	186
Price :	292	280	260	234	266	254	230	190	200.

4. Compute the coefficient of correlation by concurrent deviation method for the following data in X and Y.

Months	X	Y	Months	X	Y
Jan.	8.5	60	July	5.1	130
Feb.	9.2	65	Aug.	6.6	106
March	9.3	61	Sept.	7.7	58
Apr.	8.5	74	Oct.	7.6	80
May	7.2	92	Nov.	8.2	52
June	5.9	157	Dec.	9.2	45

5. Calculate the correlation coefficient by concurrent deviation method from the following data.

Test Number :	1	2	3	4	5	6	7	8	9	10	11
Marks in Mathematics :	65	40	35	75	63	80	35	20	80	60	50
Marks in Accountancy :	60	55	50	56	30	70	40	35	80	75	80.

6. Define coefficient of concurrent Deviations and comment on its usefulness.

7. Calculate r_c from the following :

X:	368	384	385	361	347	384	395	403	400	385
Y:	22	21	24	20	22	26	26	29	28	27.

8. Compute the coefficient of the concurrent deviation for the following data in X and Y:

Months :	J	F	M	A	M	J	J	A	S	O	N	D
X:	8.5	9.2	9.3	8.5	7.2	5.9	5.1	6.6	7.7	7.6	8.2	9.2
Y:	60	65	61	74	92	157	130	106	58	80	52	45.

9. Find the coefficient of concurrent deviation for the following data:

Year :	1974	1975	1976	1977	1978	1979	1980
Supply :	150	154	160	172	160	165	180
Price :	200	180	170	160	190	180	172.

ANSWERS

1. 0. **2.** – 1. **3.** – 1. **4.** 0.798.

7. + 0.58. **8.** 0.798. **9.** – 1.

12.15 PROBABLE ERROR IN CORRELATION

The probable error of the coefficient of correlation is obtained by the following formula:

Probable Error of the co-efficient Correlation or

$$\text{P.E.} = 0.6745 \times \frac{(1-r^2)}{\sqrt{n}} = \frac{2(1-r^2)}{3\sqrt{n}}$$

where n = **number of pairs of observations,**

r = **Coefficient of correlation.**

The probable error of the coefficient of correlation helps in interpreting its value. Since the coefficients of correlation are, generally, computed from samples, they, like other statistical quantities, are subject to errors of sampling, so from interpretation point of view **probable error** of the coefficient of correlation is very useful.

12.15.1 Properties of Probable Error

It is used for interpreting the coefficient of correlation r, whether it is significant or not

(i) If r < 6 × P.E. then it is not significant. Perhaps there is no evidence of correlation.

(ii) If r ≥ 6 × P.E., then it is siginificant and the correlation exists.

(iii) By adding and subtracting the value of Probable Error from r, we get respectively the upper and lower limits within which the coefficient of correlation in the population can be expected to. It is given as:

correlation of the population = r ± P.E.

Thus P.E is used for testing the reliability of the value of r.

Standard Error. The standard error is difined as:

$$\textit{Standard Error or S.E.} = \frac{1-r^2}{\sqrt{n}}$$

where ***r = coefficient of correlation***

n = number of pairs of observations.

12.15.2 Conditions for the use of P.E. The concept of P.E. can be applied if the following conditions exist:

1. ***The samples must have been taken out in an unbiased manner and the individual items must be independent.***
2. ***The whole data is symmetrical and gives a normal frequency curve (bell shaped curve)***
3. ***The statistical measure for which the P.E. is computed must have been calculated from the sample.***
4. ***The items in two series should not be independent of each other.***

12.16 COEFFICIENT OF DETERMINATION

It is the square of the coefficient of correlation *i.e.*, r^2, where *r* is the coefficient of correlation.

If $r = 0.4$, then the coefficient of determination is $r^2 = (0.4)^2 = 0.16$.

The coefficient of determination is preferred to the coefficient of correlation because it explains the process of variation in the dependent variable which is explained by a change in the independent variable. It is a highly useful measure.

Example 39. *A student calculates the value of r as 0.7 when the value of n is 5 and concludes that r is highly significant. Is he corect?*

Solution. We know that if the value of *r* is more than six times the P.E., then it is considered to be significant.

$$\therefore \quad \textbf{P.E.} = \frac{\mathbf{2\,(1-r^2)}}{\mathbf{3\sqrt{n}}} = \frac{2\,[1-(0.7)^2]}{3\sqrt{5}} = \frac{2}{3}\cdot\frac{(.51)}{2.24}$$

$$= \frac{.34}{2.24} = 0.15$$

$$\therefore \quad \frac{r}{\text{P.E.}} = \frac{.7}{.15} = 4.67 \Rightarrow R = 4.67 \times \text{P.E.}$$

Since r is less than six times the P.E., so it is not significant.

Example 40. *In a correlation analysis, the value of the Karl Pearson's coefficient of correlation and its probable error were found to be 0.90 and 0.04 respectively. Find the value of n.*

Solution. Here $r = 0.9$, P.E. = 0.04

$$\text{Also} \quad \textbf{P.E.} = \mathbf{0.6745} \times \frac{\mathbf{(1-r^2)}}{\sqrt{\boldsymbol{n}}}$$

$$\Rightarrow \quad 0.04 = 0.6745 \times \frac{1-(0.9)^2}{\sqrt{n}} \Rightarrow 0.04\sqrt{n} = 0.6745 \times 0.19$$

$$\therefore \quad \sqrt{n} = \frac{0.6745 \times 0.19}{0.04} = 3.204.$$

Squaring, we get $n = (3.204)^2 = 10.287$ or **10**.

Example 41. *Find the coefficient of correlation r, when its probable error is 0.2 and the number of pairs of items is 9.*

Solution. Here P.E. = 0.2, n = 9.

$$\text{Also} \quad \text{P.E.} = 0.6745 \times \frac{(1-r^2)}{\sqrt{n}}$$

$$\Rightarrow \quad 0.2 = 0.6745 \times \frac{(1-r^2)}{\sqrt{9}} = \frac{0.6745 - 0.6745\,r^2}{3}$$

$$\Rightarrow \quad 0.6745r^2 = 0.6745 - 0.2 \times 3 = 0.6745 - 0.6$$

$$\Rightarrow \quad r^2 = \frac{0.0745}{0.6745} \Rightarrow r = \sqrt{\frac{0.0745}{0.6745}}$$

$$\Rightarrow \quad r = \sqrt{0.1105} = \mathbf{0.332.}$$

Example 42. *Test the significance of correlation for the following values based on the number of observations*

(i) n = 10, r = 0.4 and r = 0.9

(ii) n = 100, r = 0.4 and r = 0.9

Solution. We know that the value of r is significant if

$$\boldsymbol{r \leq 6 \times \text{P.E.}} \Rightarrow \frac{\boldsymbol{r}}{\textbf{P.E.}} \geq \mathbf{6.}$$

We Have the Following Table

No. of observations	*r*	*P.E.*	$\frac{r}{P.E.}$	*Test of significance*
10	0.4	$.6745\times\frac{(1-(0.4)^2)}{\sqrt{10}}=\mathbf{0.18}$	$\frac{0.4}{0.18}=\mathbf{2.22}$	**Not significant**
100	0.4	$.6745\times\frac{(1-(0.4)^2)}{\sqrt{100}}=\mathbf{0.06}$	$\frac{0.4}{0.06}=\mathbf{6}$	**Significant**
10	0.9	$.6745\times\frac{(1-(.9)^2)}{\sqrt{10}}=\mathbf{0.04}$	$\frac{0.9}{0.04}=\mathbf{22.5}$	**Significant**
100	0.9	$.6745\times\frac{1-(0.9)^2}{\sqrt{100}}=\mathbf{0.128}$	$\frac{0.9}{0.13}=\mathbf{7}$	**Significant**

Example 43. *The coefficient of correlation (r) between consumption expenditure (C) and disposable income (Y) in a study was found to + 0.8. What percentage of variation in C are explained by variation in Y.*

Solution. Here $r = + 0.8 \Rightarrow r^2 > 0.64$. It means that 0.64 or 64% of the variations in consumption expenditure are explained by variations income.

EXERCISE – 12.6

1. Tick the correct answers

   (a) The coefficient of determination is defined as (*i*) r^2 ; (*ii*) r^3 ; (*iii*) $1 - r^2$, (*iv*) None of those.

   (b) The limits of population correlation are given by:

   (*i*) $r \pm$ P.E., (*ii*) $r \pm 3.5$ P.E. (*iii*) $r \pm 3$P.E; (*iv*) None of these.

   (c) If $r = 0.8$, then the coefficient of correlation shall be:

   (*i*) 0.64 ; (*ii*) 0.4; (*iii*) 0.6, (*iv*) 0.8, (*v*) None of these

   (d) The probable error is: (*i*) 0.6745 × S.E; (*ii*) 0.6457 S.E; (*iii*) 0.6753 S.E. (*iv*) None of these.

   (e) the standard error is:

   (*i*) $\frac{1-r^2}{n}$, (*ii*) $\frac{1+r^2}{\sqrt{n}}$ (*iii*) $\frac{1-r^2}{\sqrt{n}}$ (*iv*) $\frac{1-r^2}{n}$.

   (f) The coefficient of correlation r is significant if:

   (*i*) $r > 5$ P.E.; (*ii*) $r < 6$ P.E.; (*iii*) $r > 6$ P.E. (*iv*) $r = 6$ P.E.

2. If $r = 0.6$ and N = 64, find out the P.E. of the coefficient of correlation and determine the limits for the measure of population.

   [**Hint.** P.E. = $0.6745\times\frac{1-r^2}{\sqrt{N}}=0.6745\times\frac{1-(0.0)^2}{8}=\mathbf{0.054}.$

   Limits of Population are 0.6 ± 0.054, *i.e.*, **0.546** to **0.654**.

3. If $r = 0.7885$ for a pair of 12 observations, find probable error of r and determine the limits for the measure of limitation.

4. If $r = 0.9643$ for a pair of 7 observations, find the probable error.

5. Calculate the probable error in each of the following cases and comment.

   (*i*) $n = 20, r = 0$; (*ii*) $n = 10, r = 1.0$;

   (*iii*) $n = 30, r = 0.8$; (*iv*) $n = 16, r = -0.7$.

6. If $n = 25$ and probable error = 0.072 find the values of

   (*i*) r and (*ii*) standard error of r.

7. Examine the significance of r in the following situations:
   (*i*) $n = 8$ and $r = 0.2$
   (*ii*) $n = 30$, $r = -0.4$.
   (*iii*) P.E. = 0.024282, $n = 100$.
8. Which of the two has more significant correlation coefficient.
   (A) $r = 0.7$ and P.E. = 0.07;
   (B) $r = -0.6$ and P.E. = 0.04.
9. "A correlation coefficient of – 0.5 does not mean that 50% of the data are explained." Comment.
10. "A correlation coefficient of 0.6 indicates a relationship twice as close as when $r = 0.3$" Comment.

ANSWERS

1. (*a*) (*i*); (*b*) (*i*); (*c*) (*iii*) ; (*d*) (*i*); (*e*) (*iii*) ; (*f*) (*iii*).
2. P.E. = 0.054; limits 0.546 to 0.654.
3. 0.0737.
4. 0.0179
5. (*i*) 0.1509, no evidence of correlation, (*ii*) PE = 0, perfect positive correlation; (*iii*) 16.46, significance correlation; (*iv*) 8.14 ; Evidence of correlation
6. (*i*) $r = 0.6828$, (*ii*) 0.1068.
7. (*i*) Not significant, (*i*) Not significant, (*iii*) Significant.
8. (A) 0.9.

13

CHAPTER

REGARESSION ANALYSIS

13.1 INTRODUCTION

We know that the correlation studies the relationship between two variables x and y. In this chapter we shall consider the related problem of **prediction** or **estimation of the value of a variable from a known value of other variable to which it is related.** This would be discussed by means of **regression lines.**

Regression *means to return or to go back.* So *it implies the act of returning to or going back to.* Now the question arises what returns and where it returns. Natural penomena generally have a tendency to return to normal. In statistics the term **'regression'** ***is used to denote backward tendency which means going back to average or normal.***

The term regression was first used by **Sir Francis Galton** in the study of heredity. He found that though "tall fathers have tall sons". If the average height of sons of tall fathers is x above the general height, the average height of their sons is (2/3) x above the general height. The recession in the average height was described by Galton as **Regression to Mediocrity.**

Though **Sir Galton** used the term **'regression'** in studying hereditary, the concept of regression is now extended to other spheres of phenomena which have a tendency to regress or set back to the normal or the general average. The concept of regression is very helpful in the study of correlation.

Regression shows a relationship between the average values of two variables. Thus regression is very helpful in estimating and predicting the average value of one variable for a given value of the other variable. The estimate or prediction may be made with the help of a regression line which shows the average value of one variable x for a given value of the other variable y. The best average value of one variable associated with the given value of the other variable may also be estimated or predicted by means of an equation and the equation is known as Regression equation.

In the words of **M.M. Blair,**

"Regression analysis is a mathematical measure of the average relationship between two or more variables in terms of the original units of the data".

According to **Taro Yamane,**

"One of the most frequently used techniques in Economics and Business research, to find a relation between two or more variables that are related usually is Regression Analysis."

Ya-Lun-Chou remarked,

"Regression Analysis attempts to establish the NATURE OF RELATIONSHIP between the variables-that is to study the functional relationship between the variates *x* and *y*, and thereby, provides a mechanism for prediction or forecasting".

There are two types of variables in regression analysis – **Independent variable** and **Dependent variable.** ***The variable whose value is influenced or is to be predicted is called*** **Dependent Variable,** ***whereas the variable which influences the value or is used for prediction is called*** **Independent Variable.** ***The independent variable is also known as*** **Regressor** or **Predictor** or **Explanator** ***while the dependent variable is called as*** **Regressed** or **Explained variable.**

13.2 TYPES OF REGRESSIONS

Simple Regression. ***The regression analysis confined to the study of only two variables at a time is called the* simple regression.**

Multiple Regression. ***The regression analysis for studying more than two variables at a time is known as multiple regression.***

This chapter is confined only to the study of simple regression.

Linear Regression. If the regression curve is a straight line, then there is a linear regression between the variables under study. In other words, in **linear regression the relationship between the two variables *x* and *y* is linear.** In order to estimate the best average values of the two variables, two regression equations are required and they are used separately. ***One equation is used for estimating the value of x variable for a given value of y variable and the second equation is used for estimating the value of y variable for a given value of x variable. In both cases the assumption is that one is an independent variable and the other is a dependent variable and vice versa.*** Again in linear regression, the dependent variable varies at a constant rate with a given change in the independent variable. The constant rate of change can be in absolute terms or in terms of rate or percentage.

Non-linear Regression. *If the curve or regression is not a straight line, i.e., a first degree equation in the variables x and y, then it is called a* **non-linear** *or* **curvilinear regression.** *In this case the regression equation will be a functional relation between the variables x and y involving terms in x and y of the degree higher than one, i.e., involving terms of the type* x^2, y^2, x^3, y^3, xy, *etc.*

13.3 LINES OF REGRESSIONS

A line of regression is the line which gives the best estimate of one variable X for any given value of the other variable Y.

I. Line of regression of X on Y. ***It is the line which gives the best estimate for the values of X for a specified value of Y.***

It is given by $$X - \overline{X} = r\frac{\sigma_x}{\sigma_y}(Y - \overline{Y}),$$

where $\overline{X}, \overline{Y}$ are means of *X* series and *Y* series respectively σ_x, σ_y are S.D. of *X* and *Y* series respectively and *r* is the correlation coefficient between *X* and *Y*.

It can also be put in the form

$$\text{X} = a + b\text{Y},$$

where *a* is intercept of the line and *b* is the slope of the line *X* on *Y*.

II. Line of Regression of Y on X. ***It is the line which gives the best estimate for the values of Y for any specified values of X.***

Regression equation fo *Y* on *X* is given by

$$Y - \overline{Y} = r\frac{\sigma_y}{\sigma_x}(X - \overline{X}).$$

It can also be put in the form

$$\text{Y} = a + b\text{X}$$

where *a* is the intercept of the line and *b* is the slope of the line *Y* on *X*.

13.4 REGRESSION COEFFICIENTS

The regression coefficient of y on x is $b_{yx} = r\,\dfrac{\sigma_y}{\sigma_x}$ and that of x on y is

$$b_{xy} = r\,\frac{\sigma_x}{\sigma_y}.$$

Proof. Regression coefficient of y on x.
We know that the coefficient of correlation

$$r = \frac{n\,\Sigma xy - \Sigma x\,\Sigma y}{\sqrt{n\,\Sigma x^2 - (\Sigma x)^2}\,\sqrt{n\,\Sigma y^2 - (\Sigma y)^2}} = \frac{\Sigma xy - \dfrac{\Sigma x\,\Sigma y}{n}}{\sqrt{\Sigma x^2 - \dfrac{(\Sigma x)^2}{n}}\sqrt{\Sigma y^2 - \dfrac{(\Sigma y)^2}{n}}}$$

Also we know that $\sigma_x = \sqrt{\Sigma x^2 - \dfrac{(\Sigma x)^2}{n}},\ \sigma_y = \sqrt{\Sigma y^2 - \dfrac{(\Sigma y)^2}{n}}$

$$\therefore \quad r = \frac{\Sigma xy - \dfrac{\Sigma x\,\Sigma y}{n}}{\sigma_x \sigma_y} \Rightarrow \Sigma xy - \frac{\Sigma x\,\Sigma y}{n} = r\,\sigma_x\sigma_y$$

Now,

$$b_{yx} = \frac{\Sigma xy - \dfrac{\Sigma x\,\Sigma y}{n}}{\Sigma x^2 - \dfrac{(\Sigma x)^2}{n}} = \frac{\Sigma xy - \dfrac{\Sigma x\,\Sigma y}{n}}{\sigma_x^2}$$

or

$$b_{yx} = r\frac{\sigma_x\,\sigma_y}{\sigma_x^2} \qquad \left[\because\ \Sigma xy - \frac{\Sigma x\,\Sigma y}{n} = r\,\sigma_x\sigma_y\right]$$

$$\therefore \quad b_{yx} = r\frac{\sigma_y}{\sigma_x}.$$

Similarly, it may be proved that $b_{xy} = r\dfrac{\sigma_x}{\sigma_y}$.

13.5 PROPERTIES OF REGRESSION COEFFICIENTS

1. The coefficient of correlation is the geometric mean of the coefficients of regression *i.e.*, $r = \sqrt{b_{yx} \times b_{xy}}$.

Proof. We have, $b_{yx} = r\dfrac{\sigma_y}{\sigma_x}$ and $b_{xy} = r\dfrac{\sigma_x}{\sigma_y}$

$$\therefore \quad b_{yx} \times b_{xy} = r\frac{\sigma_y}{\sigma_x} \times r\frac{\sigma_x}{\sigma_x} = r^2$$

$$\Rightarrow \quad r^2 = b_{yx} \times b_{xy} \Rightarrow r = \sqrt{b_{yx} \times b_{xy}}.$$

2. If one of the regression coefficients is greater than unity, then the other is less than unity.

Proof. Let $b_{yx} > 1$. Also we know that

$$r^2 \leq 1 \text{ and } r^2 = b_{yx} \times b_{xy}$$

$$\Rightarrow \quad b_{yx} \times b_{xy} \leq 1$$

$$\Rightarrow \quad b_{xy} \leq \frac{1}{b_{yx}} < 1.$$

3. **Arithmetic mean of the regression coefficient is greater than the correlation coefficient.**
4. **Regression coefficients are independent of change of origin but not of scale.**
5. **Both regression coefficients will have the same sign, *i.e.*, either both are positive or both are negative.**
6. **The sign of correlation is same as that of regression coefficients, *i.e.*, $r > 0$ if $b_{xy} > 0$ and $b_{yx} > 0$; and $r < 0$ if regression coefficients are negative.**

Example 1. *Find r, if $b_{xy} = 0.8$ and $b_{yx} = 0.46$.*

Solution. We know that $\quad r = \sqrt{b_{xy} \times b_{yx}} = \sqrt{0.8 \times 0.46} = \sqrt{0.368} = \mathbf{0.606}.$

Hence $\quad r = \mathbf{0.606}.$

13.6 PROPERTIES OF LINEAR REGRESSIONS

1. *The two regression lines x on y and y on x always intersect at the means $(\bar{x}, \bar{y})$.*
2. *Since $r^2 = b_{yx} \times b_{xy} \Rightarrow r = \sqrt{b_{yx} \times b_{xy}}$, then r, b_{xy}, b_{yx} all have the same sign.*
3. *If r = 0, then the regression coefficients are zero.*
4. *The regression lines become identical if $r = \pm 1$. It follows from the regression equations that $X = \bar{X}$ and $Y = \bar{Y}$ if $r = 0$, and these lines are perpendicular to each other.*

Example 2. *For the following observations, find the regression coefficients b_{yx} and b_{xy} and hence find correlation coefficient between x and y.*

$\{(x, y)\} = \{(4, 2), (2, 3), (3, 2), (4, 4), (2, 4)\}$.

Solution. Hence $\Sigma x = 4 + 2 + 3 + 4 + 2 = \mathbf{15}$;

$\Sigma y = 2 + 3 + 2 + 4 + 4 = 15$; $\Sigma x^2 = 16 + 4 + 9 + 16 + 4 = \mathbf{49}$;

$\Sigma y^2 = 4 + 9 + 4 + 16 + 16 = 49$; $\Sigma xy = 8 + 6 + 6 + 16 + 8 = \mathbf{44}$.

Now
$$b_{yx} = \frac{\Sigma x_i y_i - \dfrac{(\Sigma x_i)(\Sigma y_i)}{n}}{\Sigma x_i^2 - \dfrac{(\Sigma x_i)^2}{n}} = \frac{44 - \dfrac{15 \times 15}{5}}{49 - \dfrac{225}{5}} = \frac{44 - 45}{49 - 45} = \frac{-1}{4}.$$

Also
$$b_{xy} = \frac{\Sigma x_i y_i - \dfrac{(\Sigma x_i)(\Sigma y_i)}{n}}{\Sigma y_i^2 - \dfrac{(\Sigma y_i)^2}{n}} = \frac{44 - \dfrac{15 \times 15}{5}}{49 - \dfrac{225}{5}} = \frac{44 - 45}{49 - 45} = -\frac{1}{4}.$$

Now
$$\rho^2 (x, y) = b_{xy} \times b_{yx} = \left(-\frac{1}{4}\right) \times \left(-\frac{1}{4}\right) = \frac{1}{16}$$

$$\Rightarrow \quad \rho (x, y) = \frac{1}{4} = 0.25.$$

13.7 METHOD OF LEAST SQUARES

The Method of Least Squares *is a mathematical device. It is used for obtaining the equation of a curve which fits best to a given set of observations.*

It is based on the **assumption that the sum of the squares of differences between the estimated values and the actual observed values of the observations is minimum. The least squares method is used to obtain the best fitting straight line to the given data and consists in selecting the best fitting straight line to the values of independent variable for dependent variable.**

Mathematically the **best fitting straight line can be obtained by means of an equation which shows the relation between the two variables. When the relationship between them is linear, the method of least squares is used for obtaining the best values of the constants in an appropriate equation.** For obtaining the best values of the constants in the estimating equation, a set of equations, depending on the number of constant in the estimating equation, are required to be solved. ***The system of equations required to be solved for obtaining the values of constants in the estimating equation is known as Normal Equations.***

The method of least squares is used in obtaining the **equation for fitting a regression line.**

Direct Method

Equation of *a* straight line is $y = a + bx$

Normal Equations for obtaining the values of the constants a and b are:

(*i*) $\Sigma y = Na + b\Sigma x$ and

(*ii*) $\Sigma xy = a\Sigma x + b\Sigma x^2$.

where x = **Independent variable,** a = **Constant,**
y = **Dependent variable,** b = **Constant,**
Σ = **Sign of summation,** N = **Number of pairs of observations.**

The equation of the best fit is the straight line : $y = a + bx$,

where the values of a **and** b **are obtained by solving the normal equations.**

The method is illustrated by the following example.

Example 3. *Fit a straight line of Y on X from the following data*

X :	0	1	2	3	4	5	6
Y :	2	1	3	2	4	3	5

Solution. We have the table shown on next page.

The **regression line of** Y **on** X **is**

$$Y = a + bX. \quad \text{... (A)}$$

The two normal equations which evaluate a and b are :

$$\Sigma Y = aN + b\Sigma X$$
$$\Sigma XY = a\Sigma X + b\Sigma X^2$$

Table for Computation of Fitting a Straight Line

X	Y	X^2	XY
0	2	0	0
1	1	1	1
2	3	4	6
3	2	9	6
4	4	16	16
5	3	25	15
6	5	36	30
$\Sigma X = 21$,	$\Sigma Y = 20$,	$\Sigma X^2 = 91$	$\Sigma XY = 74$.

Substituting the values of ΣX, ΣX^2, ΣXY and N = 7, we get

$$20 = 7a + 21b \quad \text{... (1)}$$

$$74 = 21a + 91b. \quad \text{... (2)}$$

Multiplying (1) by 3, we get: $60 = 21a + 63b$. ... **(3)**

Subtracting (3) from (2), we have: $14 = 28b \Rightarrow$ ***b* = 0.5.**

Putting $b = 0.5$ in (1), we get: a = **1.357**

Substituting the values of ***a* = 1.357, and *b* = 0.5** in the equation (A), we get

Y = 1.357 + 0.5X, which is the line of best fit by least squares method.

Example 4. *Fit a straight line to the following data by the method of least squares.*

X :	*0*	*1*	*2*	*3*	*4*
Y :	*1*	*1.8*	*3.3*	*4.5*	*6.3*

Solution. We prepare the following table.

Computation Table for Fitting a Straight Line

x	*y*	x^2	*xy*
0	1.0	0	0
1	1.8	1	1.8
2	3.3	4	6.6
3	4.5	9	13.5
4	6.3	16	25.2
$\Sigma x = 10$	$\Sigma y = 16.9$	$\Sigma x^2 = 30$	$\Sigma xy = 47.1$

The equation of the line of fit is: $y = a + bx$, ... **(A)**

where *a* and *b* are constants to be evaluated by the normal equations

$$\Sigma y = Na + b\,\Sigma x$$

$$\Sigma xy = a\,\Sigma x + b\,\Sigma x^2$$

i.e., $16.9 = 5a + 10b$ ... **(1)**

and $47.1 = 10a + 16b$. ... **(2)**

Multiplying (1) by 2, we get: $33.8 = 10a + 20b$. ... **(3)**

Subtracting (2) from (3), we get: $-13.3 = 3.1b \Rightarrow$ ***b* = – 4.29.**

Substituting $b = -4.29$ in (1), we get:

$$16.9 = 5a - 42.9 \text{ or } 5a = 59.8 \Rightarrow \boldsymbol{a = 11.96.}$$

Substituting the values of $a = 11.96$, $b = -4.29$, in **(A)** we get the equation of line as

$$y = 11.96 - 4.29\,x.$$

Hence, **the line of best fit is: *y* = – 4.29*x* + 11.96.**

13.8 DERIVATIONS OF THE LINES OF REGRESSIONS DIRECTLY FROM DATA

To find the equations of regression of (*i*) *y* on *x* and (*ii*) *x* on *y*.

EQUATION OF LINE OF REGRESSION OF *y* ON *x*.

Let $(x_1, y_1), (x_2, y_2), \ldots, (x_n, y_n)$ be a set of n pairs of observations and a straight line be fitted in the form of $y = a + bx$ to these data. The normal equations for obtaining the values of the constants a and b are

$\Sigma y = na + b \Sigma x$ ... (*i*) and $\Sigma xy = a \Sigma x + b \Sigma x^2$. ... (*ii*)

Here, $\Sigma y = na + b \Sigma x$ ... (*i*)

or $\dfrac{\Sigma y}{n} = \dfrac{na}{n} + \dfrac{b \Sigma x}{n}$ **[Dividing both sides by *n*]**

or $\dfrac{\Sigma y}{n} = a + \dfrac{b \Sigma x}{n}$

or $\bar{y} = a + b\bar{x}$. $\therefore a = \bar{y} - b\bar{x}$.

We have $\mathbf{y = a + bx}$

or $y = (\bar{y} - b\bar{x}) + bx$. $[\because a = \bar{y} - b\bar{x}]$

$\therefore$ $\mathbf{y - \bar{y} = b(x - \bar{x})}$. ... (*iii*)

Again, $\Sigma y = na + b \Sigma x$ or $\Sigma x \Sigma y = na \Sigma x + b \Sigma x . \Sigma x$. ... (*i*)

[Multiplying both sides by Σ *x*]

$\therefore$ $\Sigma x \Sigma y = na \Sigma x + b (\Sigma x)^2$ ... (*iv*)

and $\Sigma xy = a \Sigma x + b \Sigma x^2$. ... (*ii*)

$\therefore$ $n \Sigma xy = na \Sigma x + nb \Sigma x^2$. ... (*v*)

[Multiplying both sides by *n*]

Now, $n \Sigma xy = na \Sigma x + nb \Sigma x^2$ ... (*v*)

$\Sigma x \Sigma y = na \Sigma x + b (\Sigma x)^2$. ... (*iv*)

Substracting (*iv*) from (*v*), we get

$$n \Sigma xy - \Sigma x \Sigma y = nb \Sigma x^2 - b (\Sigma x)^2$$

$$\Rightarrow \quad n \Sigma xy - \Sigma x \Sigma y = b [n \Sigma x^2 - (\Sigma x)^2]$$

$$\Rightarrow \quad \mathbf{b} = \frac{n \Sigma xy - \Sigma x \Sigma y}{n \Sigma x^2 - (\Sigma x)^2} \Rightarrow \mathbf{b} = \frac{\dfrac{n \Sigma xy}{n} - \dfrac{\Sigma x \Sigma y}{n}}{\dfrac{n \Sigma x^2}{n} - \dfrac{(\Sigma x)^2}{n}}$$

[Dividing both numerator and denominator by *n*]

$$\therefore \quad \mathbf{b} = \frac{\Sigma xy - (\Sigma x \Sigma y)/n}{\Sigma x^2 - ((\Sigma x)^2)/n} = \frac{\Sigma xy - n \bar{x} \bar{y}}{\Sigma x^2 - n(\bar{x})^2}$$

It is important to note that ***b*** **is** known as **the regression coefficient of *y* on *x* and is denoted by b_{yx}.**

$$\therefore \quad b_{yx} = \frac{\Sigma xy - (\Sigma x \Sigma y)/n}{\Sigma x^2 - ((\Sigma x)^2)/n} = \frac{r \sigma_x \sigma_y}{\sigma_x^2} = \frac{r \sigma_y}{\sigma_x} = \frac{\Sigma xy - n \bar{x} \bar{y}}{\Sigma x^2 - n(\bar{x})^2}.$$

Now, $y - \bar{y} = b\,(x - \bar{x})$

$$\left.\begin{aligned} y - \bar{y} &= b_{yx}(x - \bar{x}) \\ \therefore \quad y - \bar{y} &= r\frac{\sigma_y}{\sigma_x}(x - \bar{x}). \end{aligned}\right\} \quad \text{... (A)}$$

The equation of regression of *y* on *x* is: $\mathbf{y - \bar{y} = b_{yx}\,(x - \bar{x})}$

or $$y-\bar{y} = r\frac{\sigma_y}{\sigma_x}(x-\bar{x}).$$

Similarly, $$b_{xy} = \frac{r\sigma_x}{\sigma_y} = \frac{\Sigma xy - n\bar{x}\,\bar{y}}{\Sigma y^2 - n(\bar{y})^2}.$$

The equation of regression of x on y is :

$$\left.\begin{aligned} x-\bar{x} &= b_{xy}(y-\bar{y}) \\ \text{or} \quad x-\bar{x} &= r\frac{\sigma_x}{\sigma_y}(y-\bar{y}). \end{aligned}\right\} \quad \text{... (B)}$$

Example 5. *Find both the regression equations from the following:*

$\Sigma X = 60$ $\quad \Sigma Y = 40$ $\quad \Sigma XY = 1150$

$\Sigma X^2 = 4160$ $\quad \Sigma Y^2 = 1720$ $\quad N = 10$

Solution. The regression coefficients b_{XY} and b_{YX} given by:

$$b_{XY} = \frac{\Sigma XY - \dfrac{\Sigma X \times \Sigma Y}{N}}{\Sigma Y^2 - \dfrac{(\Sigma Y)^2}{N}} = \frac{1150 - \dfrac{60\times 40}{10}}{1720 - \dfrac{(40)^2}{10}} = \frac{910}{1560} = \mathbf{0.58.}$$

and $$b_{YX} = \frac{\Sigma XY - \dfrac{(\Sigma X \times \Sigma Y)}{N}}{\Sigma X^2 - \dfrac{(\Sigma X)^2}{N}} = \frac{910}{4160 - \dfrac{(60)^2}{10}} = \mathbf{0.24.}$$

$$\bar{X} = \mathbf{60/10 = 6;}\ \bar{Y} = \mathbf{40/10 = 4}$$

Now, the **regression equation of Y on X is: $Y - \bar{Y} = b_{YX}(X - \bar{X})$**

$\Rightarrow Y - 4 = 0.24\,(X - 6)$ or $Y = 0.24X + 2.56$.

The regression of X on Y is: $X - \bar{X} = b_{XY}(Y - \bar{Y})$

$\Rightarrow X - 6 = 0.58\,(Y - 4)$ or $X = 0.58Y + 3.68$.

Example 6. *State any two properties of linear regression. Calculate the regression coefficients for the following information:*

$$\Sigma X = 50,\ \Sigma Y = 30,\ \Sigma XY = 1000,\ \Sigma X^2 = 3000,\ \Sigma Y^2 = 180,\ N = 10.$$

Solution. Regression Coefficient of Y on X is given by:

$$b_{yx} = r\frac{\sigma_y}{\sigma_x} = \frac{\Sigma XY - N\bar{X}\bar{Y}}{\Sigma X^2 - N(\bar{X})^2} \quad \text{... (1)}$$

Here $$\bar{X} = \frac{\Sigma X}{N} = \frac{50}{10} = 5, \quad \bar{Y} = \frac{\Sigma Y}{N} = \frac{30}{10} = 3.$$

Substituting these values in (1), we get

$$b_{yx} = r\frac{\sigma_y}{\sigma_2} = \frac{1000 - 10\,(5)\,(3)}{3000 - 10\,(5)^2} = \frac{1000 - 150}{3000 - 250} = \mathbf{0.309.}$$

Regression coefficient of X on Y is, given by:

$$b_{xy} = r\frac{\sigma_x}{\sigma_y} = \frac{\Sigma XY - Y\,\overline{X}\,\overline{Y}}{\Sigma Y^2 - N(\overline{Y})^2}$$

$$= \frac{1000 - 10\,(5)\,(3)}{1800 - 10\,(3)^2} = \frac{1000 - 150}{1800 - 90} = \mathbf{0.497.}$$

Example 7. *Find the regression equation of X on Y and the coefficient of correlation from the following data*

$$\Sigma X = 60,\ \ \Sigma X^2 = 4160,\ \ \Sigma Y = 40,\ \ \Sigma Y^2 = 1720,\ \ \Sigma XY = 1150,\ \ N = 10.$$

Solution. Here

$$\overline{X} = \frac{\Sigma X}{N} = \frac{60}{10} = \mathbf{6,}$$

$$\overline{Y} = \frac{\Sigma Y}{N} = \frac{40}{10} = \mathbf{4.}$$

Also

$$b_{xy} = \frac{\Sigma XY - N\,\overline{X}\,\overline{Y}}{\Sigma Y^2 - N(\overline{Y})^2} = \frac{1150 - 10 \times 6 \times 4}{1720 - 10\,(4)^2}$$

$$= \frac{1150 - 240}{1720 - 160} = \frac{910}{1560} = \mathbf{0.583.}$$

The **regression equation of *X* on *Y* is given by :**

$$X - \overline{X} = b_{xy}\,(Y - \overline{Y})$$

$$\Rightarrow \quad X - 6 = 0.583\,(Y - 4)$$

or

$$X = 0.583Y - 2.332 + 6 \text{ Þ } \overline{X} = \mathbf{0.583Y + 3.668.}$$

Also.

$$b_{yx} = \frac{\Sigma \overline{X}\,\overline{Y} - N\,\overline{X}\,\overline{Y}}{\Sigma X^2 - N(\overline{X})^2} = \frac{1150 - 240}{4160 - 10\,(6)^2} = \frac{910}{3800} = \mathbf{0.239.}$$

Now

$$r = \sqrt{b_{xy} \times b_{yx}} = \sqrt{0.583 \times 0.239} = \mathbf{0.373.}$$

Example 8. *In a partially destroyed record the following data are available.*

Variance of x = 25.

Regression equation of x on y: 5x – y = 22.

Regression equation of y on x: 64x – 45y = 24.

Find (a) mean values of x and y; (b) coefficient of correlation between x and y. (c) Standard deviation of y.

Solution. (*a*) The Mean values of *x* and *y* lie on the regression lines and are obtained by solving the given regression equations.

$$5\bar{x} - \bar{y} = 22 \qquad \text{... (1)}$$

$$64\,\bar{x} - 45\,\bar{y} = 24. \qquad \text{... (2)}$$

Multiplying the equation (1) by 45, we get

$$225\bar{x} - 45\,\bar{y} = 990. \qquad \text{... (3)}$$

Subtracting (2) from (3), we get

$$161\bar{x} = 966 \Rightarrow \bar{x} = 6.$$

Putting $\bar{x} = 6$ in (1), we get

$$30 - \bar{y} = 22 \Rightarrow \bar{y} = 8.$$

Hence $\bar{x} = 6$ **and** $\bar{y} = 8$.

(b) The regression equation of y on x is

$$64x - 45y = 24 \text{ or } y = \frac{64}{45}x - \frac{24}{45}$$

$$\Rightarrow \qquad y = -\frac{8}{15} + \frac{64}{45}x.$$

$$\therefore \qquad b_{yx} = \frac{64}{45}.$$

Again regression equation of x on y is

$$5x - y = 22 \quad \text{or} \quad x = \frac{22}{5} + \frac{1}{5}y$$

$$\therefore \qquad b_{xy} = \frac{1}{5}.$$

But
$$r = \sqrt{b_{xy} \times b_{yx}} = \sqrt{\frac{64}{45} \times \frac{1}{5}} = \frac{8}{15}.$$

(+ve sign with r is taken as both the regression coefficients b_{xy} and b_{yx} are positive.)

Hence, **the coefficient of correlation** $r = \frac{8}{15}$.

(*c*) Now it is given that variance of $x = \sigma_x^2 = 25 \Rightarrow \sigma_x = 5$.

Also
$$r = \frac{8}{15}, \qquad b_{yx} = \frac{64}{45}.$$

But we know that
$$b_{yx} = r\frac{\sigma_y}{\sigma_x}$$

$$\Rightarrow \quad \therefore \quad \frac{64}{45} = \frac{8}{15} \times \frac{\sigma_y}{5} \Rightarrow \sigma_y = \frac{40}{3} = 13.33$$

Hence, **the standard deviation of y = 13.33.**

Example 9. *From the following results, obtain the two regression equations and estimate the yield, when the rainfall is 29 cms. and the rainfall, when the yield is 600 kg:*

	Yield in kg.	*Rainfall in cms.*
Mean	*508.4*	*26.7*
S.D.	*36.8*	*4.6*

Coefficient of correlation between yield and rainfall is + 0.52.

Solution. Let x represent rainfall and y represent yield.

Regression of y on x is given by: $y - \bar{y} = r\,\frac{\sigma_y}{\sigma_x}(x - \bar{x})$

or $y - 508.4 = 0.52 \times \dfrac{36.8}{4.6}(x - 26.7) \Rightarrow y - 508.4 = 0.52 \times 8\ (x - 26.7)$

$\Rightarrow y - 508.4 = 4.16x - 111.072 \Rightarrow y = 4.16x + 397.328$

When $x = 29$, yield $y = 517.968$ kg.

Regression of x on y is given by: $x - \bar{x} = r\dfrac{\sigma_x}{\sigma_y}(y - \bar{y})$

or $x - 26.7 = 0.52 \times \dfrac{4.6}{36.8}(y - 508.4) = 0.065\ (y - 508.4) = 0.065\ y - 33.046$

or $\qquad \mathbf{x = 0.065y - 6.346}$

When $y = 600$, rainfall $x = 32.654$ cms.

Example 10. *Given the regression lines as $3x + 2y = 26$ and $6x + y = 31$.*

Find their point of intersection and interpret it. Also find the Correlation Coefficient between x and y.

Solution. $3x + 2y = 26 \qquad$ ... (*i*)

$6x + y = 31 \qquad$ ... (*ii*)

From (*ii*), $y = 31 - 6x \qquad$ ... (*iii*)

Substituting (*iii*) in (*i*), we get: $3x + 2\ (31 - 6x) = 26$

$\Rightarrow \qquad 3x - 12x = 26 - 62 - 9x = -36$, *i.e.*, $\mathbf{x = 4}$

Substituting $x = 4$ in (*iii*), $\mathbf{y = 31 - 24 = 7}$.

Thus, **the point of intersection of (*i*) and (*ii*) is given by (4, 7), which corresponds to the variable means.**

Let **the regression line of y on x be: $3x + 2y = 26$** *i.e.*, $y = 13 - \dfrac{3}{2}x$

and **that of x on y be: $6x + y = 31$** *i.e.*, $x = \dfrac{31}{6} - \dfrac{1}{6}y$

Thus, $b_{yx} = -\dfrac{3}{2}$ **and** $b_{xy} = -\dfrac{1}{6}$

Required correlation coefficient is given by $r = \sqrt{b_{yx}b_{xy}} = \sqrt{\left(-\dfrac{3}{2}\right)\left(-\dfrac{1}{6}\right)} = -\dfrac{1}{2}$.

Example 11. *The following data are given regarding expenditure on advertising and sales of a particular firm:*

	Adversement Expenditure (X)	*Sales (Y)* *(Rs. lakhs)*
Mean	*10*	*90*
Standard Deviation	*3*	*12*

Correlation coefficient rxy = 0.8

(*i*) *Calculate the regression equation of Y on X.*

(*ii*) *Estimate the advertisement expenditure required to attain a sales target of Rs. 120 lakhs.*

Solution. ***(i)*** **The Regression Equation of *Y* on *X* is given by:**

$$(Y - \bar{Y}) = r\frac{\sigma_y}{\sigma_x}(X - \bar{X})$$

Here, $\bar{X} = 10$, $\bar{Y} = 90$, $\sigma_x = 3$, $\sigma_y = 12$, $r = 0.8$

Substituting these values in the above equation, we get

$$(Y - 90) = 0.8\left(\frac{12}{3}\right)(X - 10)$$

or $(Y - 90) = 0.8\,(4)\,(X - 10)$ or $(Y - 90) = 3.2\,(X - 10)$

or $(Y - 90) = 3.2\,X - 32$ or $\mathbf{Y = 3.2X + 58.}$

(ii) **For determining advertisement expenditure for a given sales target, we have to find the regression equation of *X* on *Y***

i.e., $$(X - \bar{X}) = r\frac{\sigma_x}{\sigma_y}(Y - \bar{Y})$$

or $$(X - 10) = 0.8 \times \frac{3}{12}(Y - 90)$$

or $(X - 10) = 0.2\,(Y - 90)$ or $\mathbf{X = 0.2\,Y - 8}$

Putting value of $Y = 120$ in this equation, we get

$X = 0.2\,(120) - 8 \Rightarrow X = 24 - 8 = \mathbf{16.}$

Hence, **the required advertisement expenditure is Rs. 16 lakhs.**

Example 12. *For 100 students of a class, the regression equation of marks in Statistics (X) on the marks in Commerce (Y) is 3Y – 5X + 180 = 0. The mean marks in Commerce is 50 and variance of marks in Statistics is 4/9th of the variance of marks in Commerce. Find the mean marks in Statistics and the coefficient of correlation between marks in the two subjects.*

Solution. Given $3y - 5x + 180 = 0$ or $\mathbf{3y + 180 = 5x}$

Let '*x*' *represents marks in Statistics and* '*y*' *represent marks in commerce.*

When $y = 50$, x will be given by $5x = 3\,(50) + 180$

$$\Rightarrow \quad 5x = 150 + 180 \quad \text{or} \quad \mathbf{x = \frac{330}{5} = 66}$$

Hence, **the mean marks in Statistics are 66.**

Calculation of Coefficient of Correlation:

$$b_{xy} = r\frac{\sigma_x}{\sigma_y}$$

Regression coefficient of *x* on *y* from the given equation is

$$5x = 3y + 180 \Rightarrow \mathbf{x = 0.6y + 36}$$

$$\therefore\ b_{xy} = \mathbf{0.6};\ r\frac{\sigma_x}{\sigma_y} = r\frac{\sqrt{4}}{\sqrt{9}}\ \textbf{(given)}\ 0.6 = r\left(\frac{\sqrt{4}}{\sqrt{9}}\right)$$

$$\Rightarrow 0.6 = r\left(\frac{2}{3}\right) \Rightarrow 2r = 1.8 \text{ or } \mathbf{r = +\,0.9}$$

Example 13. *The equations of the regression lines between two variables are expressed as 2X – 3Y = 0 and 4Y – 5X – 8 = 0. Find* $\bar{X}$ *and* $\bar{Y}$*, the regression coefficients and the correlation coefficient between X and Y.*

Solution. The two regression equations are given as

$$2X - 3Y = 0. \quad \text{... (i)}$$

and
$$4Y - 5X - 8 = 0. \quad \text{... (ii)}$$

The point $(\bar{X}, \bar{Y})$ lies at the intersection of both the lines of regression equations Solving the two regression equations.

$(i) \times 4 \Rightarrow 8X - 12Y = 0$ ... (iii)

$(ii) \times 3 \Rightarrow -15X + 12Y - 24 = 0$ ... (iv)

Adding (*iii*) and (*iv*), we get:

$$-7X - 24 = 0.$$

$$\Rightarrow \quad X = -\frac{24}{7} = -3.429$$

Also
$$2X - 3Y = 0$$

$$2 \times \left(-\frac{24}{7}\right) - 3Y = 0 \Rightarrow Y = -\frac{16}{7} = -2.286$$

Thus $\bar{X} = -3.429$ and $\bar{Y} = -2.286$

From equation (*i*), $Y = \frac{2}{3}X \Rightarrow b_{yx} = \frac{2}{3}$

From equation (*ii*), $X = \frac{4}{5}Y - \frac{8}{5} \Rightarrow b_{xy} = \frac{4}{5}$.

$\therefore r^2 = b_{yx} \times b_{xy} = \frac{2}{3} \times \frac{4}{5} = \frac{8}{15}$.

$\Rightarrow r = 0.73$ (r is positive since b_{yx} and b_{xy} are positive)

Example 14. *The two regression lines obtained from certain data were Y = X + 5 and 16X = 9Y – 94. Find the variance of X, if the variance of Y is 16. Also find the covariance between X and Y.*

Solution. The two regression lines are given by

$$Y = X + 5 \quad \text{... (i)}$$

and
$$16X = 6Y - 94 \Rightarrow X = \frac{9}{16}Y - \frac{47}{8} \quad \text{... (ii)}$$

$\therefore \quad b_{yx} = 1 \qquad \therefore \quad b_{xy} = \frac{9}{16}$

Also $\quad r^2 = b_{yx} \times b_{xy} \qquad r = \sqrt{1 \times 9/16} = 3/4$

But $b_{xy} = r\dfrac{\sigma_x}{\sigma_y} \Rightarrow 16 = \dfrac{6}{4}\dfrac{\sigma_x}{\sqrt{16}} \Rightarrow \sigma_x = 3$

Also $\quad \dfrac{\text{Cov}(X,Y)}{\sigma_x \sigma_y} = r \Rightarrow \text{Cov}(X, Y) = \dfrac{3}{4} \times 3 \times 4 = 9$

Example 15. *Find the mean value of two random variables x and y and the correlation coefficient between them when the two lines of regression are given by:*

$$5x + 7y - 22 = 0 \text{ and } 6x + 2y - 20 = 0.$$

If the variance of y is 15, find the standards deviation of x.

Solution. Solving the given equations, we get $x = 3$ and $y = 1$.

Hence $\bar{x} = 3$ and $\bar{y} = 1$.

$5x + 7y = 22 \Rightarrow y = \frac{22}{7} - \frac{5}{7}x$; Therefore, $\boldsymbol{b_{yx} = \frac{-5}{7}}$.

Also $6x + 2y = 2y = 20 \Rightarrow x = -\frac{1}{3}y + \frac{10}{3}$; therefore, $\boldsymbol{b_{xy}} = -\frac{\mathbf{2}}{\mathbf{6}}$.

Now $\boldsymbol{r} = -\sqrt{\boldsymbol{b_{yx} \times b_{xy}}} = -\sqrt{\frac{10}{42}} = -\sqrt{\frac{\mathbf{5}}{\mathbf{21}}}$.

Again $\sigma_y = \sqrt{15}$, $b_{xy} = r\frac{\sigma_x}{\sigma_y} \Rightarrow \frac{-1}{3} = \sqrt{\frac{5}{21}} \times \frac{\sigma_x}{\sigma_y}$

$\Rightarrow \boldsymbol{\sigma_x} = \frac{1}{3}\frac{21}{5} \times \sqrt{15} = \sqrt{7}$.

Hence the standard deviation of *x* is $\sqrt{7}$.

Example 16. *Given that the variance of x = 9 and the regression equations are 8x – 10y + 66 = 0, 40x – 18y = 214. Find (a) mean values of x and y; (b) coefficient of correlation between x and y. (c) standard deviation of y.*

Solution. (*a*) The mean values of *x* and *y*, *i.e.*, $(\bar{x}, \bar{y})$ **lie on both the lines.**

$$\therefore \quad 8\bar{x} - 10\bar{y} = -66. \quad \text{... (1)}$$

$$40\bar{x} - 18\bar{y} = 214. \quad \text{... (2)}$$

Multiplying (1) by 5 and subtracting it from (2), we get

$$32\bar{y} = 544 \text{ or } \boldsymbol{\bar{y} = 17}.$$

Substituting $\bar{y} = 17$ in (1), we get: $\boldsymbol{\bar{x} = 13}$.

Hence, **the mean values of *x* and *y* are $\bar{x} = 13$, $\bar{y} = 17$.**

(***b***) Let us assume that the **line 8*x* – 10*y* + 66 = 0 be the regression equation of *y* on *x*.**

$$\therefore \quad 10y = 8x + 66 \Rightarrow \boldsymbol{y = \frac{8}{10}x + \frac{66}{10}}$$

$$\Rightarrow \quad \boldsymbol{b_{yx} = \frac{4}{5}}.$$

Also **the regression equation of *x* on *y* is**

$$\boldsymbol{40x = 18y + 214 \Rightarrow x = \frac{18}{40}y + \frac{214}{48}}$$

$$\Rightarrow \quad \boldsymbol{b_{xy} = \frac{9}{20}}.$$

But we know that $\boldsymbol{r} = \sqrt{\boldsymbol{b_{xy} \times b_{yx}}} = \sqrt{\frac{4}{5} \times \frac{9}{20}} = \sqrt{\frac{36}{100}} = +\frac{\mathbf{6}}{\mathbf{10}}$.

(Positive sign is taken because both b_{xy} and b_{yx} are positive.)

$$\sigma_x^2 = 9 \Rightarrow \sigma_x = 3.$$

Also $$b_{yx} = r\frac{\sigma_y}{\sigma_x} \Rightarrow \frac{4}{5} = \frac{6}{10} \times \frac{\sigma_y}{3} \Rightarrow \sigma_y = \mathbf{4.}$$

Hence the standard deviation of y is **4.**

Example 17. *For a bivariate data the mean value of X is 20 and the mean value of Y is 45. The regression coefficient of Y on X is 4 and that of X on Y is (1/9). Find*

(*i*) *the coefficient of correlation*

(*ii*) *the standard deviation of X if the standard deviation of Y is 12.*

(*iii*) *Also write down the equations of regression lines.*

Solution.

(*i*) Here $\bar{X} = 20,\ \bar{Y} = 45,\ b_{yx} = 4,\ b_{xy} = \frac{1}{9}$.

Also $$r = \sqrt{b_{xy} \times b_{yx}} \Rightarrow \sqrt{4 \times \frac{1}{9}} \Rightarrow r = \frac{2}{3} = \mathbf{0.667.}$$

(+ve sign with r is taken because b_{xy} and b_{yx} both are positive.)

Hence $$r = \mathbf{0.667.}$$

(*ii*) $\therefore$ $$b_{yx} = r\frac{\sigma_y}{\sigma_x}.$$

$\therefore$ $$4 = \frac{2}{3} \times \frac{12}{\sigma_x} \Rightarrow \sigma_x = \mathbf{2.}$$

(*iii*) Regression equation of *X* on *Y* : $X - \bar{X} = b_{xy}(Y - \bar{Y})$

$\Rightarrow$ $$X - 20 = \frac{1}{9}(Y - 45), \text{ or } X = \frac{1}{9}Y + 15.$$

which is the required **regression line *X* on *Y*.**

Again **Regression equation of *Y* on *X x* :** $Y - \bar{Y} = b_{yx}(X - \bar{X})$

or $$Y - 45 = 4(X - 20) \Rightarrow Y = 4X - 35,$$

which is the required **regression equation of *Y* on *X*.**

Example 18. *Two variables gave the following data:*

$$\bar{X} = 20,\ \bar{Y} = 15,\ \sigma_x = 4,\ \sigma_y = 3,\ r = 0.7.$$

Obtain the regression equations and find the most likely value of Y, when X = 24.

Solution. The regression coefficients are

$$b_{yx} = r\frac{\sigma_y}{\sigma_x} = 0.7 \times \frac{3}{4} = \frac{2.1}{4} = \mathbf{0.525} \text{ and } b_{xy} = r\frac{\sigma_y}{\sigma_x} = 0.7 \times \frac{4}{3} = \mathbf{0.933.}$$

$\therefore$ **The regression equation of *X* on *Y* :** $X - \bar{X} = b_{xy}(Y - \bar{Y})$

$\Rightarrow$ $$X - 20 = 0.933(Y - 15)$$

or $$X = 0.933Y - 0.933 \times 15 + 20 = \mathbf{0.933Y + 6.}$$

Hence the regression equation *X* on *Y* is *X* = 0.933*Y* + 6.

The **regression equation of *Y* on *X*:** $Y - \bar{Y} = b_{yx}(X - \bar{X})$.

$\Rightarrow$ $$Y - 15 = 0.525(X - 20)$$

or $$Y = 0.525X + 15 - 20 \times 0.525$$

or $$Y = \mathbf{0.525X + 4.5.}$$

When $X = 24$, the most likely value of Y is

$Y = 0.525 \times 24 + 4.5 = \mathbf{17.1}$.

Example 19. *In a laboratory experiment on correlation research study the equations to the two regression lines were found to be $2x - y + 1 = 0$ and $3x - 2y + 7 = 0$. Find the mean of x and y. Also work out the values of regression coefficients and coefficient of correlation between the two variables x and y.*

Solution. The mean values of x and y, *i.e.*, $(\bar{x}, \bar{y})$ **lies on both the given lines;**

$$\therefore \quad 2\bar{x} - \bar{y} = -1 \qquad \text{... (1)}$$

$$3\bar{x} - 2\bar{y} = -7. \qquad \text{... (2)}$$

Multiplying (1) by 2 and subtracting it from (2), we get

$$-\bar{x} = -5 \Rightarrow \bar{x} = \mathbf{5}.$$

Putting $\bar{x} = 5$ in (1), we get $\bar{y} = \mathbf{11}$.

Hence the mean of x is 5 and that of y is 11.

To find the regression coefficient.

From the equation $2x - y + 1 = 0$, we have : $x = \frac{1}{2}y - \frac{1}{2}$.

$$\therefore \quad b_{xy} = \frac{\mathbf{1}}{\mathbf{2}}.$$

Similarly $3x - 2y + 7 = 0 \Rightarrow y = \frac{3}{2}x + \frac{7}{2}$

$$\therefore \quad b_{yx} = \frac{\mathbf{3}}{\mathbf{2}}.$$

To find correlation coefficient.

The **correlation coefficient r is given by : $r = \sqrt{b_{xy} \times b_{yx}}$** $= \sqrt{\frac{1}{2} \times \frac{3}{2}} = \frac{\sqrt{3}}{2} = \mathbf{0.866}$.

Hence $r = \mathbf{0.866}$.

(Positive sign with r is taken as both the regression coefficients are positive).

Example 20. *Using the following information you are required to*

(i) Obtain the linear regression of Y on X.

(ii) Estimate the level of defective parts delivered when inspection expenditure amounts to Rs. 28,000.

$\Sigma X = 424$, $\Sigma XY = 12815$, $\Sigma X^2 = 21926$, $\Sigma Y = 363$, $\Sigma Y^2 = 15123$, $N = 10$.

X means expenditure on inspection ('000 rupees)

Y means defective parts delivered ('000 numbers)

Solution. Here $\bar{X} = \frac{\Sigma X}{N} = \frac{424}{10} = \mathbf{42.4}$ and $\bar{Y} = \frac{\Sigma Y}{N} = \frac{363}{10} = \mathbf{36.3}$.

$$b_{yx} = r\frac{\sigma_y}{\sigma_x} = \frac{\Sigma XY - N\bar{X}\bar{Y}}{\Sigma X^2 - N(\bar{X})^2} = \frac{12815 - 10\,(42.4)\,(36.3)}{21926 - 10\,(42.4)^2}$$

$$= \frac{12815 - 15391.2}{21926 - 17977.6} = \frac{-2576.2}{3948.4} = \mathbf{-0.6525}.$$

$\therefore$ **Regression equation of Y on X : $Y - \bar{Y} = B_{XY}(X - \bar{X})$**

$$\Rightarrow \quad y - 36.3 = -0.6525\,(X - 42.4)$$

$$\Rightarrow \quad \mathbf{Y = -0.6525X + 63.967.}$$

When the inspection expenditure is Rs. 28,000, the estimated value of defective parts would be

$Y = -0.6525 \times 28 + 63.967 = -18.27 + 63.967 = 45.697$ or **Rs. 45697 approximately.**

Example 21. *Equations of the two lines of regression are*

$2x + 3y = 7$ and $5x + 4y = 9$. Find

(a) mean of x and mean of y.

(b) regression coefficients b_{yx} and b_{xy}.

(c) correlation coefficient between x and y.

Solution. (*a*) The two lines of regression pass through $(\bar{x}, \bar{y})$

$$\therefore \quad 2\bar{x} + 3\bar{y} = 7 \qquad \text{...(1)}$$

$$\text{and} \quad 5\bar{x} + 4\bar{y} = 9. \qquad \text{...(2)}$$

Solving (1) and (2), we get

$$\bar{x} = \frac{17}{7}, \bar{y} - \frac{5}{7}.$$

Hence, **mean of x is $\frac{17}{7}$ and mean of y is $\frac{5}{7}$.**

(*b*) Let us take the line **$2x + 3y = 7$ as the regression line of y on x.**

$$\therefore \quad 3y = -2x + 7 \text{ or } y = -\frac{2}{3}x + \frac{7}{3}, \quad \therefore \quad b_{yx} = -\frac{2}{3}.$$

The line $5x + 4y = 9 \Rightarrow 5x = -4y + 9$

$$\text{or} \quad x = -\frac{4}{5}y + \frac{9}{5}. \quad \therefore \quad b_{xy} = -\frac{4}{5}.$$

Hence, **the regression coefficients are $b_{yx} = -\frac{2}{3}, b_{yx} = -\frac{4}{5}$.**

(*c*) We know that $r = \sqrt{b_{xy} \times b_{yx}}$

$$\therefore \quad r = (-)\sqrt{\left(\frac{-2}{3}\right) \times \left(\frac{-4}{5}\right)} = -\sqrt{0.5333} = \mathbf{-0.730.}$$

Hence, **correlation coefficient between x and y is – 0.730.**

Example 22. *From the data given below estimate the most likely height of a the brother whose sister's height is 70 cm.*

Brother's : mean height is 67 cm with a s.d. of 3.5 cm

Sister's: Mean height is 65 cm with a s.d. of 2.5 cm

The coefficient of correlation between the heights of brothers and sisters is + 0.8.

Solution. Let x, y *be the variables corresponding to the heights of brothers and sisters.*

∴ Then $\bar{x} = 65$, $\bar{y} = 67$, $\sigma_x = 2.5$, $s_y = 3.5$, $r = 0.8$.

Now $$b_{yx} = \frac{r\sigma_y}{\sigma_x} = \frac{(0.8)(3.5)}{2.5} = 1.12.$$

∴ The equation of line of regression of y on x is

$$y - \bar{y} = b_{yx}(x - \bar{x}) \Rightarrow y - 67 = 1.12(x - 65)$$

$$\Rightarrow \quad y = 1.12x - 5.8. \quad \text{... (1)}$$

Height of the brother whose sister's height is 70 cm is = Estimate of y for $x = 70$

Putting $x = 70$ in (1), we get

$y = (1.12) \times 70 - 5.8 = 78.4 - 5.8 =$ **72.6 cm.**

Example 23. *Given the following results of the height and weight of 1,000 students:*

$\bar{Y} = 170$ cm, $\bar{X} = 75$ kg., $r = 0.60$, $\sigma_y = 6.0$ cm, $\sigma_x = 6$ kg. Amit weighs 50 kg. Sumeet is 1.5 m tall. Estimate the height of Amit from his weight and the weight of Sumeet from his height.

Solution. Here

Height = Y **Weight = X**

$\bar{Y} = 170$ cm $\bar{X} = 75$ kg. ; $r = 0.60$

$\sigma_Y = 6$ cm $\sigma_X = 6$ kg.

(*a*) The regression equation of Y on X is

$$Y - \bar{Y} = r\frac{\sigma_Y}{\sigma_X}(X - \bar{X}) \Rightarrow Y - 170 = 0.60 \times \frac{6}{6}(X - 75)$$

$$\Rightarrow \quad Y - 170 = 0.60(X - 75) \Rightarrow Y - 170 = 0.6X - 102$$

$$\Rightarrow \quad Y = 0.6X - 102 + 170$$

$$\therefore \quad \mathbf{Y = 0.6X + 68}$$

When Amit's weight $X = 50$ kg, his height $Y = 0.6 \times 50 + 68 = 30 + 68 = 98$ cm

∴ **Required height of Amit = 98 cm**

(*b*) The regression equation of X on Y is : $X - \bar{X} = r\dfrac{\sigma_X}{\sigma_Y}(Y - \bar{Y})$

$$\Rightarrow \quad X - 75 = 0.60 \times \frac{6}{6}(Y - 170)$$

$\Rightarrow X - 75 = 0.6(Y - 170) \Rightarrow X - 75 = 0.6Y - 102 \Rightarrow X = 0.6Y - 102 + 75$

$$\therefore \quad \mathbf{X = 0.6Y - 27}$$

When Sumeet's height $Y = 1.5$ m $= 150$ cm his **weight**

$X = 0.6 \times 150 - 27 = 90 - 27 = 63$ kg

∴ **Required weight of Sumeet = 63 kg.**

Example 24. *Find the regression of x on y from the following data:*

$\Sigma x = 24$, $\Sigma y = 44$, $\Sigma xy = 306$,

$\Sigma x^2 = 164$, $\Sigma y^2 = 574$, $N = 4$.

Find the value of x, when $y = 6$.

Solution. Here $\bar{x} = \frac{\Sigma x}{N} = \frac{24}{4} = 6.;$ $\quad \bar{y} = \frac{\Sigma y}{N} = \frac{44}{4} = 11.$

$$\textbf{Regression coefficient} = r\frac{\sigma_x}{\sigma_y} = \frac{N\Sigma xy - \Sigma x \Sigma y}{N\Sigma y^2 - (\Sigma y)^2}$$

$$= \frac{4\times 306 - 24\times 44}{4\times 574 - (44)^2} = \frac{1224 - 1056}{2296 - 1936} = \frac{168}{360} = \mathbf{0.47.}$$

The regression equation of x on y is

$$x - \bar{x} = r\frac{\sigma_x}{\sigma_y}(y - \bar{y})$$

$\Rightarrow x - 6 = 0.47\ (y - 11) \Rightarrow x - 6 = 0.47\ y - 5.17 \Rightarrow x = 0.47y - 5.17 + 6.$

$\therefore$ $\quad \mathbf{x = 0.47\ y + 0.83.}$

When $y = 6$, x $= 0.47 \times 6 + 0.83 = 2.82 + 0.83 =$ **3.65.**

$\therefore$ **Required value of x = 3.65.**

Example 25. *There are two series of index numbers, P for price index and S for stock of a commodity. The mean and standard deviations of P are 100 and 8 respectively and that of S are 103 and 4. The correlation coefficient between the two series is 0.4. With these data work out a linear equation to read off the values of P for various values of S. Can the same equation be used to read off the values of S for different values of P? If not, give the appropriate equation.*

Solution.

P = Price index $\qquad S$ = Stock of Commodity

$\bar{P} = 100 \qquad \bar{S} = 103$

$\sigma_P = 8 \qquad \sigma_s = 4.$

$r = 0.4.$

(*a*) The regression equation of *P* on *S* is

$$P - \bar{P} = r\frac{\sigma_P}{\sigma_S}(S - \bar{S}),$$

$$\Rightarrow \quad P - 100 = 0.4\times\frac{8}{4}(S - 103)$$

$\Rightarrow P - 100 = 0.8\ (S - 103) \Rightarrow P - 100 = 0.8S - 82.4 \Rightarrow P = 0.8\ S - 82.4 + 100.$

$\therefore$ $\quad \mathbf{P = 0.8\ S + 17.6.}$

(*b*) *The above equation cannot be used to read off the values of S for the different values of P. The reason is that P is a dependent variable and values depend upon the various values of S which is an independent variable.*

The appropriate regression equation in this case would be S on P:

$$S - \bar{S} = r\frac{\sigma_S}{\sigma_P}(P - \bar{P}),$$

$$\Rightarrow \quad S - 103 = 0.4\times\frac{4}{8}(P - 100)$$

$\Rightarrow S - 103 = 0.2\ (P - 100) \Rightarrow S - 103 = 0.2P - 20 \Rightarrow S = 0.2P - 20 + 103.$

$\therefore$ $\quad \mathbf{S = 0.2P + 83.}$

Example 26. *Equations of two lines of regression are*
$4x + 3y + 7 = 0$ and $3x + 4y + 8 = 0$. Find
(i) mean of x and mean of y
(ii) regression coefficient b_{yx} and b_{xy}, and
(iii) correlation coefficient between x and y.

Solution. (*i*) The mean values of *x* and *y*, *i.e.*, $(\bar{x}, \bar{y})$ **lie on both the lines.**

$$\therefore \quad 4\bar{x} + 3\bar{y} + 7 = 0 \quad \text{...(1)}$$

$$3\bar{x} + 4y + 8 = 0 \quad \text{...(2)}$$

Solving (1) and (2), we get

$$\bar{x} = -4/7, Y = -11/7.$$

(*ii*) Let us take the equation $4x + 3y + 7 = 0$ as the line of the regression of *x* on *y* and the equation $3x + 4y + 8 = 0$ as the line of regression of *y* on *x*.

Now $\quad 4x + 3y + 7 = 0 \Rightarrow x = -\frac{3}{4}y - \frac{7}{4}.$

$$\therefore \quad b_{xy} = -\frac{3}{4}. \quad \text{...(3)}$$

Also $\quad 3x + 4y + 8 = 0 \Rightarrow y = -\frac{3}{4}x - 2.$

$$\therefore \quad b_{yx} = -\frac{3}{4}. \quad \text{...(4)}$$

(*iii*) Now $\quad \rho^2 = b_{xy} \times b_{yx} = \left(\frac{-3}{4}\right) \times \left(\frac{-3}{4}\right) = \frac{9}{16}$

$$\Rightarrow \quad \rho = -\frac{3}{4}. \quad [\because \text{ Using (3) and (4)}]$$

Example 27. *For 10 observations on price (x) and supply (y), the following data were obtained (in appropriate units):*

$\Sigma x = 130$, $\Sigma y = 220$, $\Sigma x^2 = 2288$, $\Sigma y^2 = 5506$, $\Sigma xy = 3467$.

Obtain the line of regression of y on x and estimate the supply when the price is 16 units.

Solution. The regression line *y* on *x* is $y = a + bx$, where *a* and *b* are given by the normal equation $\quad 220 = 10a + 130b \quad$ ... (1), $\quad 3467 = 130a + 2288b \quad$ ... (2)

Solving (1) and (2), we get:

$$a = 8.8 \text{ and } b = 1.015$$

The line is $\quad y = \mathbf{8.8 + 1.015\,x.}$

Put $\quad x = \mathbf{16 \text{ to get } y = 25.04.}$

Example 28. *Given the following data, estimate the value of y when x = 12.*

Series	*x*	*y*	
Average	*7.6*	*14.8*	
Standard Deviation	*3.6*	*2.5*	*r = 0.99.*

Solution. Here

$$\bar{x} = 7.6, \quad \bar{y} = 14.8,$$

$$\sigma_x = 3.6, \quad \sigma_y = 2.5, \quad r = 0.99.$$

The regression equation of y on x is : $\boldsymbol{y - \bar{y} = r \frac{\sigma_y}{\sigma_x}(x - \bar{x})}$

$$\begin{aligned} \text{or} \quad y - 14.8 &= 0.99 \times \frac{2.5}{3.6}(x - 7.6) \\ \text{or} \quad y - 14.8 &= 0.687\,(x - 7.6) \\ \text{or} \quad y - 14.8 &= 0.687x - 5.22 \\ \Rightarrow \quad y &= 0.687x - 5.22 + 14.8. \\ y &= \mathbf{0.687x + 9.58} \end{aligned}$$

When $x = 12$, $y = 0.687 \times 12 + 9.58 = 8.24 + 9.58 = 17.82$.

13.9 EQUATIONS OF REGRESSION LINES

When the Deviations are taken from Actual Arithmetic of X and Y.

We know that the equation of line of regresion Y on X derived from the actual data is given by

$$Y = a + bX, \qquad \text{... (1)}$$

where a and b are constants to be derived from the normal equations.

$$\Sigma Y = Na + b\,\Sigma X \qquad \text{... (2)}$$

$$\Sigma XY = a\,\Sigma X + b\,\Sigma X^2. \qquad \text{... (3)}$$

If we take the deviations of X and Y variables from their respective means, then the calculations become simple. In this case the equation (1) reduces to

$$Y - \bar{Y} = b\,(X - \bar{X}) \qquad \text{... (4)}$$

Let $\quad y = Y - \bar{Y},\ x = X - \bar{X},$

so that equations (4) becomes regression equations

$$y = a + bx \qquad \text{... (5)}$$

Transforming equations (2) and (3) in term of x, y we get

$$\Sigma y = Na + b\,\Sigma x \qquad \text{... (6)}$$

$$\Sigma xy = a\,\Sigma x + b\,\Sigma x^2 \qquad \text{... (7)}$$

Since the deviations are taken from the actual means, so $\Sigma x = 0$, $\Sigma y = 0$. Substituting these values in equations (6) and (7), we get

$$Na = 0 \Rightarrow a = 0 \text{ as } N \neq 0.$$

$$\Sigma xy = b\,\Sigma x^2 \Rightarrow b = \frac{\Sigma xy}{\Sigma x^2}. \qquad \text{... (8)}$$

Substituting this value of b from (8) in (5) we get the regression line **y on x as**

$$y = \frac{\Sigma xy}{\Sigma x^2}x \quad \text{or} \quad Y - \bar{Y} = \frac{\Sigma xy}{\Sigma x^2}(X - \bar{X})$$

Similarly; the regression equation $X = a + bY$ reduces to

$$x = by, \text{ where } b = \frac{\Sigma xy}{\Sigma y^2}$$

$$\text{or} \quad X - \bar{X} = \frac{\Sigma xy}{\Sigma y^2}(Y - \bar{Y}).$$

The method is illustrated by the following examples.

Example 29. *Compute the two regression coefficients from the data given below and find the value of 'r' (the correlation coefficient) using the same.*

x:	7	4	8	6	5
y:	6	5	9	8	2

Solution. Here

$\Sigma x = 30$; $\Sigma y = 30$; $\Sigma xy = 42 + 20 + 72 + 40 + 10 = \mathbf{192}$;

$\Sigma x^2 = 49 + 16 + 64 + 36 + 25 = \mathbf{190}$;

$\Sigma y^2 = 36 + 25 + 81 + 64 + 6 = \mathbf{210}$.

$$\therefore \quad b_{yx} = \frac{\Sigma xy - (\Sigma x \Sigma y)/n}{\Sigma x^2 - (\Sigma x)^2/n} = \frac{192-(30\times 30/5)}{190-(30)^2/5} = \frac{\mathbf{6}}{\mathbf{5}}.$$

$$b_{xy} = \frac{\Sigma xy - (\Sigma x)(\Sigma y)/n}{\Sigma y^2 - (\Sigma y)^2/n} = \frac{192-\dfrac{30\times 30}{5}}{210-\dfrac{900}{5}} = \frac{\mathbf{2}}{\mathbf{5}}.$$

$$\therefore \quad = \sqrt{b_{xy}\times b_{yx}} = \sqrt{12/25} = \mathbf{0.693}.$$

Example 30. *Compute the two regression coefficients using the values of actual means of X and Y from the data given below and then work out the value of the correlation coefficient:*

X:	7	4	8	6	5
Y:	6	5	9	8	2

Solution.

Table for the Regression Coefficients

X	x = X – 6	Y	y = Y – 6	x^2	y^2	xy
7	1	6	0	1	0	0
4	– 2	5	– 1	4	1	2
8	2	9	3	4	9	6
6	0	8	2	0	4	0
5	– 1	2	– 4	1	16	4
30		**30**		**10**	**30**	**12**

Here $\overline{X} = \dfrac{30}{5} = \mathbf{6}$ and $\overline{Y} = \dfrac{30}{5} = \mathbf{6}$

Regression Coefficient of *X* on *Y*: $b_{xy} = \dfrac{\Sigma xy}{\Sigma y^2} = \dfrac{12}{30} = \mathbf{0.4}$

Regression Coefficient of *Y* on *X*: $b_{yx} = \dfrac{\Sigma xy}{\Sigma x^2} = \dfrac{12}{10} = \mathbf{1.2}.$

Correlation Coefficient: $r = \sqrt{b_{xy}\times b_{yx}} = \sqrt{(0.4)\times(1.2)} = \sqrt{\mathbf{0.48}} = \mathbf{0.693}.$

Example 31. *Calculate the correlation and find the two lines of regression from the following data:*

X:	57	58	59	59	60	61	62	64
Y:	67	68	65	68	72	72	69	71

Find the estimate of Y when X = 66.

Solution. We prepare the following table for the calculations of b_{yx} and b_{xy} by using deviations from actual means.

Table for Calculations

X	$x = X - \overline{X}$ $= X - 60$	x^2	Y	$y = Y - \overline{Y}$ $= Y - 69$	y^2	xy
57	– 3	9	67	– 2	4	6
58	– 2	4	68	– 1	1	2
59	– 1	1	65	– 4	16	4
59	– 1	1	68	– 1	1	1
60	0	0	72	3	9	0
61	1	1	72	3	9	3
62	2	4	69	0	0	0
64	4	16	71	2	4	8
$\Sigma X = 480$	$\Sigma x = 0$	$\Sigma x^2 = 36$m	$\Sigma Y = 552$	$\Sigma y = 0$	$\Sigma y^2 = 44$	$\Sigma xy = 24$

Here $N = 8$

Mean $\overline{X} = \frac{\Sigma X}{N} = \frac{480}{8} = 60;$

Mean $\overline{Y} = \frac{\Sigma Y}{N} = \frac{552}{8} = \mathbf{69.}$

Also we know that $b_{xy} = \frac{\Sigma xy}{\Sigma x^2} = \frac{24}{36} = \mathbf{0.67.}$

$$b_{yx} = \frac{\Sigma xy}{\Sigma y^2} = \frac{24}{44} = \mathbf{0.545.}$$

Regression Equation of Y on X : $Y - \overline{Y} = b_{yx}\ (X - \overline{X})$

$\Rightarrow Y - 69 = 0.545\ (X - 60)$ or $Y = 0.545X + 69 - 0.545 \times 60$

or $\quad Y = \mathbf{0.545\ X + 27.3.}$

When $X = 66$, $Y = 0.545 \times 66 + 27.3 = 35.97 + 27.3 = \mathbf{63.27}$.

Hence $Y = 63.27$ **when $X = 66$**

Regression Equation of X on Y : $X - \overline{X} = b_{xy}\ (Y - \overline{Y})$

$\Rightarrow X - 60 = 0.67\ (Y - 69)$ or $\Rightarrow$ $X = \mathbf{0.67Y + 13.77}$

13.10 EQUATIONS OF REGRESSION LINES WHEN THE DEVIATIONS ARE TAKEN FROM ASSUMED MEANS

Let A and B respectively be the **assumed mean for X and Y series.** Let $dx = X - A$ and $dy = X - B$**.** Then we have

$$b_{xy} = r\frac{\sigma_x}{\sigma_y} = \frac{(\Sigma dx\, dy) - (\Sigma dx)(\Sigma dy)/N}{\Sigma dy^2 - (\Sigma dy)^2/N} \quad \text{... (1)}$$

and

$$b_{yx} = r\frac{\sigma_y}{\sigma_x} = \frac{(\Sigma dx\, dy) - (\Sigma dx)(\Sigma dy)/N}{\Sigma dx^2 - (\Sigma dx)^2/N} \quad \text{... (2)}$$

The equations of **Line of Regressions are:**

$$X \text{ on } Y : X - \bar{X} = b_{xy}\ (Y - \bar{Y})$$

$$Y \text{ on } X : Y - \bar{Y} = b_{yx}\ (X - \bar{X}),$$

where b_{xy} and b_{yx} are given by (1) and (2).

The method is illustrated by the following example.

Example 32. *Find the two lines of regressions from the following data:*

X:	*158*	*160*	*163*	*165*	*167*	*170*	*172*	*175*	*177*	*181*
Y:	*163*	*158*	*167*	*170*	*160*	*180*	*170*	*175*	*172*	*175.*

Estimate Y, when X = 164.

Solution. Let $a = 170$, $b = 175$ be the assumed means for X and Y series so that $dx = X - 170$, $dy = Y - 175$.

We have the following table of calculation:

Table For Calculations

X	*dx = X – 170*	*dx^2*	*Y*	*dy = Y – 175*	*dy^2*	*dxdy*
158	– 12	144	163	– 12	144	144
160	– 10	100	158	– 17	289	170
163	– 7	49	167	– 8	64	56
165	– 5	25	170	– 5	25	25
167	– 3	9	160	– 15	225	45
170	0	0	180	5	25	0
172	2	4	170	– 5	25	– 10
175	5	25	175	0	0	0
177	7	49	172	– 3	9	– 21
181	11	121	175	0	0	0
Σ X = 1688	**Σ *dx* = – 12**	**Σ *dx^2* = 526**	**Σ *Y* = 1690**	**Σ *dy* = – 60**	**Σ *dy^2* = 806**	**Σ *dxdy* = 409**

Here $N = 10$, $\bar{X} = \frac{\Sigma X}{N} = \frac{1688}{10} = \mathbf{168.8}$

$$\bar{Y} = \frac{\Sigma Y}{N} = \frac{1690}{10} = \mathbf{169.}$$

$$b_{xy} = \frac{\Sigma dxdy - (\Sigma dx)(\Sigma dy)/N}{\Sigma dy^2 - (\Sigma dy)^2/N} = \frac{409 - (-12)\times(-60)/10}{806 - (-60)^2/10}$$

$$= \frac{409 - 72}{806 - 360} = \frac{337}{446} = \mathbf{0.756}$$

$$b_{yx} = \frac{\Sigma\, dxdy - (\Sigma\, dx)(\Sigma\, dy)/N}{\Sigma\, dx^2 - (\Sigma\, dx)^2/N} = \frac{409-72}{526-(-12)^2/10}$$

$$= \frac{337}{526-14.4} = \frac{337}{511.6} = \mathbf{0.659.}$$

Regression Equation of X on Y : $X - \bar{X} = b_{xy}\ (Y - \bar{Y})$

$$X - 168.8 = 0.756\,(Y - 169)$$

or $X = 0.756Y + 168 - (0.756)\,(169) = 0.756Y + 168 - 127.764$

$$X = 0.756\,Y + 40.236.$$

Regression Equation of Y on X : $Y - \bar{Y} = b_{yx}\ (X - \bar{X})$

$$Y - 169 = 0.659\ (X - 168.8)$$

or $\quad Y = 0.659X + 169 - (0.659)\ (168.8)$

or $\quad \mathbf{Y = 0.659X + 57.761.}$

When $X = 164$, $Y = 0.659 \times 164 + 57.761 = 108.076 + 57.761$

or $\quad \mathbf{Y = 165.837.}$

Example 33. *A panel of judges P and Q graded seven debators and independently awarded the following marks*

Debators	*1*	*2*	*3*	*4*	*5*	*6*	*7*
Marks by judge P :	*40*	*34*	*28*	*30*	*44*	*38*	*31*
Marks by judge Q :	*32*	*39*	*26*	*30*	*38*	*34*	*28.*

An eighth debator was awarded 36 marks by judge P while the judge Q was not present. If judge Q were also present, how many marks would you expect him to award to the eighth debator assuming the same degree of relationship exists between their judgement?

Solution. Let X, Y respectively denote the marks given by judge P and Q. Let $A = 38$ and $B = 30$ be the assumed means for X and Y series respectively so that $\mathbf{dx = X - 38, dy = Y - 30.}$ Here we shall find the regression equation of Y on X, and estimate Y, when $X = 36$.

Table For Calculations

X	$dx = X - 38$	dx^2	Y	$dy = Y - 30$	dy^2	$dxdy$
40	2	4	32	2	4	4
34	– 4	16	39	9	81	– 36
28	– 10	100	26	– 4	16	40
30	– 8	64	30	0	0	0
44	6	36	38	8	64	48
38	0	0	34	4	16	0
31	– 7	49	28	– 2	4	14
$\Sigma X = 245$	$\Sigma dx = -21$	$\Sigma dx^2 = 269$	$\Sigma Y = 227$	$\Sigma dy = 17$	$\Sigma dy^2 = 185$	$\Sigma dxdy = 70$

Here $\quad N = 7;\ \therefore\ \bar{X} = \dfrac{\Sigma X}{N} = \dfrac{245}{7} = \mathbf{35.}$

$$\bar{Y} = \frac{\Sigma Y}{N} = \frac{227}{7} = \mathbf{32.43.}$$

Also $$b_{yx} = \frac{\Sigma dxdy - (\Sigma dx)(\Sigma dy)/N}{\Sigma dx^2 - (\Sigma dx)^2/N} = \frac{70-(-21)(17)/7}{269-(-21)^2/7}$$

$$= \frac{70+51}{269-63} = \frac{121}{206} = \mathbf{0.587.}$$

∴ **Regression Equation Y on X : $Y - \bar{Y} = b_{yx}(X - \bar{X})$**

$\Rightarrow Y - 32.43 = 0.587(X - 35) = 0.587X - 0.587 \times 35$

or $Y = 0.587X + 32.43 - 20.545 = \mathbf{0.587X + 11.885}$

Hence $\qquad Y = \mathbf{0.587X + 11.885.}$

When $X = 36$, Y $= 0.587 \times 36 + 11.885 = 21.132 + 11.885 = 33.017 \simeq$ **33.**

Hence when $X = 36$, then $Y = 33$.

Example 34. *The following table gives the age of cars of a certain make and actual maintenance costs. Obtain the regression equation for costs related to age. Also estimate the maintenance cost for a ten years old car.*

Age of car (years)	*2*	*4*	*6*	*8*
Maintenance cost (Rs. hundred)	*10*	*20*	*25*	*30*

Solution. Let X denote the age of car and Y the maintenance cost. Also we know that the maintenance cost depends on the age of the car, so we are to find the regression equation of Y on X. We prepare the following table. Let 5 and 20 respectively be the assumed means for X and Y series.

Table For Calculations

X	$x = X - \bar{X}$ $= X - 5$	x^2	Y	$y = Y - \bar{Y}$ $= Y - 20$	y^2	xy
2	– 3	9	10	– 10	100	30
4	– 1	1	20	0	0	0
6	1	1	25	5	25	5
8	3	9	30	10	100	30
$\Sigma X = 20$	**$\Sigma x = 0$**	**$\Sigma x^2 = 20$**	**$\Sigma Y = 85$**	**$\Sigma y = 5$**	**$\Sigma y^2 = 225$**	**$\Sigma xy = 65$**

Here $\bar{X} = \frac{20}{4} = 5, \bar{Y} = \frac{85}{4} = \mathbf{21.25.}$

Also $b_{yx} = \frac{\Sigma xy - (\Sigma x)(\Sigma y)/N}{\Sigma x^2 - (\Sigma x)^2/N} = \frac{65-(0\times5)/4}{20-0/4} = \mathbf{3.25.}$

∴ **The regression line Y on X :** $\qquad Y - \bar{Y} = b_{yx}(X - \bar{X})$

$\Rightarrow \qquad Y - 21.25 = 3.25(X - 5)$

$\Rightarrow \qquad Y = 3.25X - 5(3.25) + 21.25$

or $\qquad Y = \mathbf{5 + 3.25\,X.}$ ... (1)

The maintenance cost for a ten years old car:

Putting $X = 10$ in the equation (I), we get **$Y = 37.5$ or Rs. 37.50.**

Example 35. *Find out the regression coefficient of Y on X from the following data:*

X:	1	2	3	4	5
Y:	160	180	140	180	200.

Solution. We prepare the following table to find b_{yx}.

Table For Regression Coefficient b_{yx}

X	$dx = X - \overline{X}$ $= X - 3$	dx^2	Y	$dy = Y - \overline{Y}$ $= Y - 172$	$dxdy$
1	– 2	4	160	– 12	24
2	– 1	1	180	8	– 8
3	0	0	140	– 32	0
4	1	1	180	8	8
5	2	4	200	28	56
$\Sigma X = 15$	$\Sigma dx = 0$	$\Sigma dx^2 = 10$	$\Sigma Y = 860$	$\Sigma dy = 0$	$\Sigma dxdy = 80$

Also $N = 5$, Here $\overline{X} = \dfrac{\Sigma X}{N} = \dfrac{15}{5} = \mathbf{3},\qquad \overline{Y} = \dfrac{\Sigma Y}{N} = \dfrac{860}{8} = \mathbf{172.}$

Regression coefficient Y **on** $X = b_{yx} = \dfrac{\Sigma\, dx\, dy}{\Sigma\, dx^2} = \dfrac{80}{10} = \mathbf{8.}$

Hence $\qquad b_{yx} = \mathbf{8.}$

Example 36. *The heights of a sample of 10 fathers and their eldest sons are given below (to the nearest cm):*

Height of Father (X) :	170	167	162	163	167	166	169	171	164	165
Height of Son (Y) :	168	167	166	166	168	165	168	170	165	168

(i) *Compute the correlation coefficient Y.*

(ii) *Find the regression of Y on X.*

(iii) *Compute the coefficient of determination and give your comments.*

Solution.

Table for Calculations

X	Y	$x = X–165$	$y = Y–165$	x^2	y^2	xy
170	168	5	3	25	9	15
167	167	2	2	4	4	4
162	166	– 3	1	9	1	– 3
163	166	– 2	1	4	1	– 2
167	168	2	3	4	9	6
166	165	1	0	1	0	0
169	168	4	3	16	9	12
171	170	6	5	36	25	30
164	165	– 1	0	1	0	0
165	168	0	3	0	9	0
Total		**14**	**21**	**100**	**67**	**62**

$\overline{X} = 165 + \dfrac{14}{10} = \mathbf{166.4}\,;\qquad \overline{Y} = 165 + \dfrac{21}{10} = \mathbf{167.1.}$

$$r = \frac{n\Sigma xy - \Sigma x \Sigma y}{\sqrt{[n\Sigma x^2 - (\Sigma x)^2]\times[n\Sigma y^2 - (\Sigma y^2)]}} = \frac{10\times 62 - 14\times 21}{\sqrt{[10\times 100 - 14^2][10\times 67 - 21^2]}}$$

$$= \frac{326}{\sqrt{804\times 229}} = \frac{326}{429.087} = 0.75975 \cong \mathbf{0.76}$$

The regression equation of Y on X is given by: $Y - \overline{Y} = b_{yx}\ (X - \overline{X})$,

where $b_{yx} = \frac{n\Sigma xy - \Sigma x \Sigma y}{n\Sigma x^2 - (\Sigma x)^2} = \frac{10\times 62 - 14\times 21}{10\times 100 - 14^2} = \frac{326}{804} = \mathbf{0.405}$

The required regression equation is $Y - 167.1 = 0.405\ (X - 166.4)$

or $$Y = 0.405\ X + 99.708$$

Coefficient of determination $= r^2 = (0.76)^2 = \mathbf{0.58}$

Thus, **the regression line accounts for 58% of total variation.**

Example 37. *From the following data, obtain he two regression equations:*

Sales:	*91*	*97*	*108*	*121*	*67*	*124*	*51*	*73*	*111*	*57*
Purchases:	*71*	*75*	*69*	*97*	*70*	*91*	*39*	*61*	*80*	*47.*

Solution. Let the variable x and y respectively denote the sales and purchases. We prepare the following table:

Table for Regression Equations

x	y	$dx = x - \bar{x}$ $= x - 90$	$dy = y - \bar{y}$ $= y - 70$	dx^2	dy^2	$dxdy$
91	71	1	1	1	1	1
97	75	7	5	49	25	35
108	69	18	– 1	324	1	– 18
121	97	31	27	961	729	837
67	70	– 23	0	529	0	0
124	91	34	21	1156	441	714
51	39	– 39	– 31	1521	961	1209
73	61	– 17	– 9	289	81	153
111	80	21	10	441	100	210
57	47	– 33	– 23	1089	529	759
$\Sigma x =$ **900**	$\Sigma y =$ **700**	$\Sigma dx = 0$	$\Sigma dy = 0$	$\Sigma dx^2 =$ **6360**	$\Sigma dy^2 =$ **2868**	$\Sigma dxdy =$ **3900**

$\therefore\ \bar{x} = \frac{\Sigma x}{n} = \frac{900}{10} = \mathbf{90};$ $\qquad \bar{y} = \frac{\Sigma y}{n} = \frac{700}{10} = \mathbf{70.}$

$$b_{yx} = \frac{\Sigma (x-\bar{x})(y-\bar{y})}{\Sigma (x-\bar{x})^2} = \frac{\Sigma dxdy}{\Sigma dx^2} = \frac{3900}{6360} = \mathbf{0.6132.}$$

$$b_{xy} = \frac{\Sigma dxdy}{\Sigma dy^2} = \frac{3900}{2868} = \mathbf{1.36.}$$

Regression line y on x : $y - \bar{y} = b_{yx}\ (x - \bar{x})$

$\Rightarrow y - 70 = 0.6132\ (x - 90) \quad \Rightarrow \quad \mathbf{y = 0.6132x + 14.812.}$

Regression line x on y : $(x - \bar{x}) = b_{xy}\ (y - \bar{y})$

$\Rightarrow x - 90 = 1.36\ (y - 70) \quad \Rightarrow \quad \mathbf{x = 1.36\ y - 5.2.}$

Example 38. *Find the two lines of regression from the following data:*

Age of husband:	25	22	28	26	35	20	22	40	20	18
Age of wife:	18	15	20	17	22	14	16	21	15	14.

Hence estimate: (i) *the age of husband when the age of wife is 19 and* (ii) *the age of wife when the age of husband is 30.*

Solution. We have the following table by taking assumed mean $A = 26$ for X series and assumed mean $B = 17$ for Y series.

Table for Calculations

Age of husband X	$x = (X-26)$	x^2	*Age of wife Y*	$y = (Y-17)$	y^2	xy
25	– 1	1	18	1	1	– 1
22	– 4	16	15	– 2	4	8
28	2	4	20	3	9	6
26	0	0	17	0	0	0
35	9	81	22	5	25	45
20	– 6	36	14	– 3	9	18
22	– 4	16	16	– 1	1	4
40	14	196	21	4	16	56
20	– 6	36	15	– 2	4	12
18	– 8	64	14	– 3	9	24
$\mathbf{\Sigma X = 256}$	$\mathbf{\Sigma x = -4}$	$\mathbf{\Sigma x^2 = 450}$	$\mathbf{\Sigma Y = 172}$	$\mathbf{\Sigma y = 2}$	$\mathbf{\Sigma y^2 = 78}$	$\mathbf{\Sigma xy = 172}$

$$\bar{X} = \frac{256}{10} = \mathbf{25.6}\ ; \qquad \bar{Y} = \frac{172}{10} = \mathbf{17.2.}$$

Regression coefficient $\quad b_{xy} = r\frac{\sigma_x}{\sigma_y} = \frac{N\,\Sigma xy - \Sigma x\,\Sigma y}{N\,\Sigma y^2 - (\Sigma y)^2}$

$$= \frac{10\times 172-(-4)\times 2}{10\times 78-(+2)^2} = \frac{1720+8}{780-4} = \frac{1728}{776} = \mathbf{2.23.}$$

Regression coefficient $\quad b_{yx} = r\frac{\sigma_Y}{\sigma_X} = \frac{N\,\Sigma xy - \Sigma x\,\Sigma y}{N\,\Sigma x^2 - (\Sigma x)^2}$

$$= \frac{10\times 172-(-4)^2}{10\times 450-(-4)^2} = \frac{1720+8}{4500-16} = \frac{1728}{4484} = \mathbf{0.385.}$$

The regression line of X on Y is

$$X - \bar{X} = r\frac{\sigma_X}{\sigma_Y}(Y - \bar{Y})$$

$$\Rightarrow \quad X - 25.6 = 2.23\ (Y - 17.2) = 2.23Y - 38.36$$

$$\Rightarrow \quad X = 2.23\ Y - 38.36 + 25.6$$

or $$\mathbf{X = 2.23\ Y - 12.76.}$$

The regression line of Y on X is

$$Y - \bar{Y} = r\frac{\sigma_Y}{\sigma_X}(X - \bar{X})$$

$$\Rightarrow \quad Y - 17.2 = 0.385\ (X - 25.6) = 0.385\ X - 9.86$$

$$\Rightarrow \quad Y = 0.385\ X - 9.86 + 17.2$$

or $$\mathbf{Y = 0.385X + 7.34.}$$

(*i*) When $Y = 19$, then $\quad \mathbf{X = 2.23 \times 19 - 12.76}$

$$= \mathbf{42.37 - 12.76 = 29.61 \text{ or } 30 \text{ nearly.}}$$

(*ii*) When **$X = 30$, then $Y = 0.385 \times 30 + 7.34$**

$$= 11.55 + 7.34 = \mathbf{18.89 \text{ or } 19 \text{ nearly.}}$$

Hence (*i*) **Required estimated age of the husband = 30 years.**

(*ii*) **Required estimated age of the wife = 19 years.**

Example 39. *The following data represents rainfall* (x) *and yield of paddy per hectare* (y) *in a particular area. Find the linear regression of x on y.*

x:	*113*	*102*	*95*	*120*	*140*	*130*	*125*
y:	*1.8*	*1.5*	*1.3*	*1.9*	*1.1*	*2.0*	*1.7*

Solution. We prepare the following table of calculation to compute b_{xy}. Let $A = 120$ be the assumed mean for X-series and $B = 1.8$ be the assumed mean for Y series.

Table for Calculations

x	*dx = x – 120*	*y*	*dy = y – 1.8*	*dy*²	*dxdy*
113	– 7	1.8	0	0	0
102	– 18	1.5	– 0.3	0.09	5.4
95	– 25	1.3	– 0.5	0.25	12.5
120	0	1.9	0.1	0.01	0
140	20	1.1	– 0.7	0.49	– 1.4
130	10	2.0	0.2	0.04	2.0
125	5	1.7	– 0.1	0.01	– 0.5
Σ x = 825	**Σ dx = – 15**	**Σ y = 11.3**	**Σ dy = – 0.13**	**Σ dy² = 0.89**	**Σ dxdy = 18**

Here, $n = 7$.

$$\therefore\ b_{xy} = \frac{\Sigma dxdy - \frac{(\Sigma dx)(\Sigma dy)}{n}}{\Sigma dy^2 - \frac{(\Sigma dy)^2}{n}} = \frac{18 - \frac{(-15)(-0.13)}{7}}{89 - \frac{(-0.13)^2}{7}} = \frac{18 - 0.279}{.89 - 00241} = \frac{17.721}{0.8876} = \mathbf{19.965.}$$

Also $$\bar{x} = \frac{825}{7} = \mathbf{117.86}, \quad \bar{y} = \frac{11.3}{7} = \mathbf{1.614}.$$

Equation of **regression line x on y is** $= x - \bar{x} = b_{xy}\ (y - \bar{y})$

or $$x - 117.86 = 19.965\ (y - 1.614)$$

$$x = 19.965y + 117.86 - 19.965 \times 1.614$$

or $$x = \mathbf{19.965\ y + 85.637} \text{ is required regression equation.}$$

Example 40. *For a certain bivariate data, the equations of line of regression of Y on X is 2.5 Y = X + 35 and putting X = 15, given Y = 20. The equation of the line of regression X on Y is 10X = Y + 70 and putting Y = 20 in this equation gives X = 9 and not the original value of X = 15. Explain why this is so. For what values of X and Y does this process give consistent results?*

Solution. The regression equation of Y on X can be used to obtain the best estimate of Y for a given value of X. It cannot be used to obtain the estimate of X for any given value of Y, for which we have to use the regression equation of X on Y.

The value $X = 9$ given by the regression equation of X on Y does not agree with the original value of $X = 15$, the reason being that the two equations of regression cannot be used interchangeably as the basis and assumptions for deriving them are quite different. The two equations will give consistent result at their point of intersection $\bar{X}, \bar{Y}$.

Solving the equations

$$2.5\,\bar{Y} = \bar{X} + 35 \qquad \text{... } (i)$$

$$\bar{Y} = -\,10\,\bar{X} + 70 \qquad \text{... } (ii)$$

we get $$\bar{X} = \frac{35}{4} \text{ and } \bar{Y} = \frac{35}{2}.$$

Hence, **the equations (*i*) and (*ii*), will give the consistent results for** $\bar{X} = \mathbf{\frac{35}{4}}, \bar{Y} = \mathbf{\frac{35}{2}}$.

13.11 VARIABILITY IN REGRESSION ANALYSIS

Let X_C and Y_C denote the estimated (or computed) values of X and Y obtained from the regression equations. Then we know that the **sum of the deviations of *Y*'s and *X*'s from their regression lines are zero.**

Therefore $$\Sigma\ (Y + Y_C) = 0 \text{ and } \Sigma\ (X - X_C) = 0$$

Variability: Variability is the measure of spread or scatter of actual values around the computed values Y_C. It is given by

$$\textbf{Variability} = \frac{(Y - Y_C)^2}{n} \quad \text{or} \quad \textbf{Variability} = \frac{(X - X_C)^2}{n}.$$

13.12 STANDARD ERROR OF ESTIMATE

The standard error of estimate is similar to standard deviation. The standard error of estimate is a measure of the variation or scatteredness about the line of regression, whereas standard deviation measures the variation or scatteredness about the arithmetic mean.

Standard error of Estimate indicates how precise the prediction of Y is based on X or conversely. There are two types of standard error of estimates.

(*a*) **Standard Error of Estimate *Y* on *X*.** It is denoted by S_{YX}. It is derived from the formula:

$$S_{YX} = \sqrt{\frac{\Sigma (Y - Y_C)^2}{n-2}} \quad \text{or} \quad S_{YX} = \sqrt{\frac{\Sigma Y^2 - a\Sigma Y - b\Sigma XY}{(n-2)}}$$

where *a* and *b* are to be obtained from the normal equation and *a* = intercept, *b* = slope of line.

Also $$S_{YX} = \sigma_y \sqrt{1-r^2},$$

where *r* is the coefficient of correlation *X* and *Y*. S_{YX} predicts the values of *Y* based on *X*.

(*b*) **Standard Error of Estimate of X on Y. It** is denoted by S_{XY}. It predicts the values of *X* based on *Y*.

$$S_{XY} = \sqrt{\frac{\Sigma(X - X_C)^2}{n-2}} \quad \text{or} \quad S_{XY} = \sqrt{\frac{\Sigma X^2 - a\Sigma X - b\Sigma XY}{(n-2)}}.$$

where *a* and *b* to be the evaluated from the normal equation and *a* = intercept, *b* = slope of line.

Also $$S_{XY} = \sigma_x \sqrt{1-r^2},$$

where *r* is the coefficient of correlation between *X* and *Y*.

13.13 EXPLAINED AND UNEXPLAINED VARIATIONS

Explained Variation. It is the variation which is explained by the variable *X*.

It is given by $\Sigma (Y_C - \overline{Y})^2$.

Unexplained Variation. It is the variation which is unexplained by the variable *X*. This variation is due to some other factors (variables) affecting the total variation in *Y*–values.

It is given by $\Sigma (Y - Y_C)^2$.

Now **Total Variation:** $\Sigma (Y - \overline{Y})^2 = \Sigma [(Y - Y_C) + (Y_C - \overline{Y})]^2$

$$= \Sigma (Y - Y_C)^2 + \Sigma (Y_C - \overline{Y})^2$$
$$= \Sigma (Y - Y_C)^2 + \Sigma (Y_C - \overline{Y})^2 + 2\Sigma (Y - Y_C)(Y_C - \overline{Y})$$

$$(\because \Sigma (Y_C - \overline{Y}) = 0.)$$

= Un-Explained Variation + Explained Variation.

Similarly, **Explained Variance** $= \dfrac{\Sigma (Y_C - \overline{Y})^2}{n}$

Unexplained Variance $= \dfrac{\Sigma (Y - Y_C)^2}{n}$

Example 41. *State some important properties of linear correlation and regression coefficients.*

Solution. Properties of correlation coefficient.

(i) It lies between – 1 and + 1.

(ii) It is a pure number and is independent of the units of measurement.

(iii) It is independent of the choice of both origin and the scale of observation.

Properties or regression coefficient.

(i) There are two regression lines (namely X on Y and Y on X) and they always interesect at the mean $(X, \overline{Y})$

(ii) If one regression coefficient is greater than 1, then the other has to be less than 1.

(iii) Since $b_{yx} \times b_{xy} = r^2 \Rightarrow r = \sqrt{b_{yx} \times b_{xy}}$*, and r,* b_{yx}*,* b_{xy} *all have same sign. The regression cofficients are zero if r = 0.*

(iv) Although regression equations are usually different, they become identical if r = ± 1.

(v) If r = 0 then regression lines are perpendicular to each other.

13.14 COEFFICIENT OF DETERMINATION

The **coefficient of determination** is the square of the coefficient of correlation. It is equal to r^2 or ρ^2. The maximum value of r^2 is unity and in that case all the variation in *Y* is explained by the variation in *X*. It is also defined as

$$\textbf{Coefficient of Determination} = \frac{\textbf{Explained Variance}}{\textbf{Total Variance}}.$$

$$\textbf{Coefficient of Non-Determination} = \frac{\textbf{Unexplained Variance}}{\textbf{Total Variance}}$$

It is **denoted by** k^2**.Also** $k^2 = 1 - r^2$

Coefficient of Allienation. It is the square root of coefficient of nondetermination. It is **denoted by** ***k*****.**

13.15 RELATION BETWEEN REGRESSION ANALYSIS AND CORRELATION ANALYSIS

Correlation Analysis	Regression analysis
1. It is a relationship between two or more variables.	**1. Regression means stepping back or returning to average value.**
2. Correlation coefficient *r* between *x* and *y* is a measure of direction and degree of linear relationship between *x* and *y*.	**2.** b_{xy} **and** b_{yx} **are mathematical measures expressing the average relationship between the two variables.**
3. It is symmetric in *x* and *y i.e.*, $r_{xy} = r_{yx}$**.**	**3. The regression coefficient** b_{xy} **and** b_{yx} **are not symmetric in *x* and *y i.e.*,** $b_{xy} \neq b_{yx}$**.**
4. The correlation coefficient does not reflect upon the nature of variable (independent or dependent variable)	**4. Regression coefficients reflect on the nature of variable *i.e.*, which is dependent variable and which is independent variable. In other words, it estimates the value of dependent variable for any given value of independent variable.**
5. It does not imply cause and effect relationship between the variables under study.	**5. It indicates the cause and effect relationship between the variables. The variable corresponding to cause is taken as independent variable, whereas corresponding to effect is taken as dependent variable.**
6. It is a relative measure and is independent of the units of measurement.	**6. Regression coefficents are absolute measure of finding out the relationship between two or more variables.**
7. It indicates the degree of association.	**7. It is used to forecast the nature of dependent variable when the value of independent variable is known.**
8. It has limited application as it is confined to the study of linear relationship between two variables.	**8. It has wider applications as it also studies non-linear relationship between the variables.**

13.16 UTILITY OF REGRESSION ANALYSIS

1. ***The cause and effect relations are indicated from the study of regression analysis.***
2. ***It establishes the rate of change in one variable in terms of the changes in another variable.***
3. ***It is useful in economic analysis as regression equation can determine an increase in the cost of living index for a particular increase in general price level.***
4. ***It helps in prediction and thus it can estimate the values of unknown quantities.***
5. ***It helps in determining the coefficient of correlation as***

$$r = \sqrt{b_{yx} \times b_{xy}}.$$

6. ***It enable us to study the nature of relationship between the variables.***
7. ***It can be useful to all natural, social and physical sciences, where the data are in functional relationship.***

13.17. REGRESSION EQUATION IN A BIVARIATE GROUPED FREQUENCY DISTRIBUTION WORKING RULE

Step I. Prepare a Two Way frequency Table as discussed in the Chapter 7 on correlation analysis [see § 7.12 on page 185]

Step I. Calculate σ_x, σ_y and the correlation coefficient r.

Step III. Calculate the regression coefficient b_{xy} and b_{yx} from the following formula.

$$b_{yx} = r\frac{\sigma_y}{\sigma_x} \quad \text{and} \quad b_{xy} = r\frac{\sigma_x}{\sigma_y}$$

$$b_{xy} = r.\frac{\sigma_x}{\sigma_y} = \frac{\sum fdxdy - \frac{\sum fdx \times \sum fdy}{N}}{\sum fdy^2 - \frac{(\sum fdy)^2}{N}} \times \frac{i_x}{i_y}$$

$$b_{yx} = r\frac{\sigma_y}{\sigma_x} = \frac{\sum fdxdy - \frac{\sum fdx \times \sum fdy}{N}}{\sum fdx^2 - \frac{(\sum fdx)^2}{N}} \times \frac{i_y}{i_x}$$

i_x = Width of the class interval of *x* variable

l_y = Width of the class interval of y variable.

Step IV. Find the repression equations

Regression Equation X and Y : $X - \overline{X} = b_{xy}(Y - \overline{Y})$

Degression Equation Y and X : $Y - \overline{Y} = b_{yx}(X - \overline{X})$

EXERCISE – 13.1

1. In the estimation of regression equations of two variables X and Y, the following results were obtained $\overline{X} = 90$ $\overline{Y} = 70$, $N = 10$, $\Sigma x^2 = 6360$, $\Sigma y^2 = 2860$, $\Sigma xy = 3900$, where *x* and *y* are deviations from their respective means. Obtain the two lines of regression.
2. By using the following data, find and the two lines of regressions and from them compute the Karl Pearson's coefficient of correlation.

$\Sigma X = 250, \Sigma Y = 300, \Sigma XY = 7900, \Sigma X^2 = 6500, \Sigma Y^2 = 10{,}000$ and $N = 10$.

[Hint. $\bar{X} = \frac{250}{10} = 25, \bar{Y} = \frac{300}{10} = 30.$

$$b_{yx} = \frac{N\Sigma XY - (\Sigma X)(\Sigma Y)}{N(\Sigma X^2) - (\Sigma X)^2} = \frac{10 \times 7900 - 250 \times 300}{10 \times 6500 - (250)^2} = \mathbf{1.6.}$$

Similarly $b_{xy} = \frac{10 \times 79000 - 250 \times 300}{10 \times 10000 - (300)^2} = \mathbf{0.4.}$ $\quad r = \sqrt{b_{xy} \times b_{yx}} = \sqrt{1.6 \times 0.4} = \mathbf{+0.8.}$**]**

3. Find the mean values of the variables X and Y and correlation coefficient between them from the following regression lines:

$$2y - x - 50 = 0, 3y - 2x - 10 = 0$$

[Hint. Solving the lines, we get $x = 130$, $y = 90$ as their point of interection, which is the means $(\bar{x}, \bar{y})$. $\quad \therefore \bar{x} = \mathbf{130}, \bar{y} = \mathbf{90}.$ Also $2y = x + 50 \Rightarrow y = \frac{1}{2}x + 25$

$\therefore b_{yx} = \frac{1}{2}$. Similarly, $x = \frac{3}{2}y - 5 \Rightarrow b_{xy} = \frac{3}{2}$.

$$\therefore \quad r = \sqrt{b_{xy} \times b_{yx}} = \sqrt{\frac{1}{2} \times \frac{3}{2}} = \frac{\sqrt{3}}{2} = \mathbf{0.866.}]$$

4. Find out the likely production corresponding to a rainfall of 40 cms. from the following data.

	Rainfall (in cms)	*Output (in quintals)*
Average	30	50
S.D.	5	10
r = 0.8.		

[Hint. Regression of Y on X: $Y - \bar{Y} = r\frac{\sigma_y}{\sigma_x}(X - \bar{X})$

$\Rightarrow y - 50 = 1.6\,(X - 30) \Rightarrow Y = 1.6\,X + 2.$

When $X = 40$ cm, then $Y = 1.6\,(40) + 2 =$ **66 cm.]**

5. The equations of two variables X and Y are as follows:

$$3X + 2Y - 26 = 0; 6X + Y - 31 = 0.$$

Find (*i*) the means; (*ii*) the regression coefficients; (*iii*) the coefficient of correlation between X and Y.

[Hint. Solve the given lines to get $\bar{X} = \mathbf{4}, \bar{Y} = 7.$

Assume $3X - 2Y - 26 = 0$ as the regression line of Y on X.

$$\therefore \quad \Rightarrow Y = -\frac{3}{2}X + 13 \Rightarrow b_{yx} = -\frac{3}{2}.$$

Similarly, from the second line, we have $b_{xy} = -\frac{1}{6}$.

$$\therefore \quad r = \sqrt{b_{yx} \times b_{xy}} = \sqrt{\frac{-3}{2} \times \frac{-1}{6}} = \mathbf{-0.5}$$

$\therefore$ Since $b_{xy} < 0$, $b_{yx} < 0$ so the sign of r must be negative.]

6. You are given the following data for the variables x and y.

$\bar{x} = 36, \bar{y} = 85, \sigma_x = 11, \sigma_y = 8$ and $r = 0.66$

(*i*) Find the two regression lines; (*ii*) Estimate the value of x when $y = 75$.

7. The following data give the correlation coefficient, means and standard deviation of rainfall and yield of paddy in a certain tract:

	Yield per hectare in Kgs.	Annual rainfall in cm.
Mean	973.5	18.3
S.D.	38.4	2.0

Coefficient of correlation = 0.58.

Estimate the most likely yield of paddy when the annual rainfall is 22 cm, other factors being assumed to remain the same.

[**Hint.** Regression of y on x is: $y - 973.5 = 11.136\ (x - 18.3)$
Put $x = 22$ to estimate $y =$ **1014.7.**]

8. The equation of two regression lines obtained in a correlation analysis are
$3x + 12y = 9$ and $3y + 9x = 46$. Find
(*i*) mean values of x and y
(*ii*) The correlation coefficient between x and y.

9. The lines of regression of y on x and x on y are given to be
$y = 5 + 0.2x$ and $x = 3 + y$.
Find the means for the two variables x and y.
[**Hint.** $(\bar{x}, \bar{y})$ is the point of intersection of the two given regression lines].

10. In a regression analysis problem, the following data are given:
The regression lines are: $x + 2y - 5 = 0$, $2x + 3y = 8$ and $\sigma_x^2 = 12$.
Find $\bar{x}, \bar{y}$, $r(x, y)$, and σ_y^2.

11. Find the lines of regression X on Y and Y on X for the following data:

X:	3	5	6	6	9
Y:	2	3	4	6	5.

12. For 10 observations on price (x) and supply (y), the following data were obtained (in appropriate units):
$\Sigma x = 130$, $\Sigma y = 220$, $\Sigma x^2 = 2288$, $\Sigma y^2 = 5506$, $\Sigma xy = 3467$.
Obtain the line of regression of y on x and estimate the supply when the price is 16 units.
[**Hint.** The line y on x is $y = a + bx$, where a and b are given by the normal equation $220 = 10a + 130b$, and $3467 = 130a + 2288b$
$\Rightarrow$ $a = 8.8$ and $b = 1.015$.
The line is $y = 8.8 + 1.015x$.
Put $x = 16$ to get $y =$ **25.04**].

13. From the following data find the regression equations and estimate the likely value of Y when $X = 100$

X:	72	90	76	81	56	76	92	88	49
Y:	124	131	117	132	96	120	136	97	85.

[**Hint.** Take $dx = x - 81$, $dy = Y - 120$, then

$$\Sigma dx = -41,\ \Sigma dy = -42,\ \Sigma dxdy = 1901,\ \Sigma dx^2 = 2239,\ \Sigma dy^2 = 2876,$$

$$\bar{x} = \frac{688}{9} = 76.44;\quad \bar{y} = \frac{1038}{9} = 115.33$$

$$b_{xy} = \frac{1901 - [(-41)\times(-4x)3/9}{2876 - \frac{(-42)^2}{9}} = \mathbf{0.64}$$

Similarly $byx = 0.83$.

Regression equation x on y: $x - 76.444 = 0.64\,(y - 1/5.33) \Rightarrow$ $\mathbf{x = 0.64\,y + 2.63}$

regression equation y on x: $y - 115.33 = 0.83\,(x - 76.44) \Rightarrow$ $\mathbf{y = 0.83\,x + 51.88.}$]

14. In a correlation study of two variables x and y, the following values are obtained

$\bar{x} = 65, \bar{y} = 67, \sigma_x = 2.5, \sigma_y = 2.5, \sigma_y = 3.5, r = 0.8.$

Find the two regression equations that are associated with the above values.

[**Hint.** $b_{yx} = r\dfrac{\sigma_y}{\sigma_x} = 0.8 \times \dfrac{3.5}{2.5} = 1.12;\quad b_{xy} = r\dfrac{\sigma_{xy}}{\sigma_y} = 0.8 \times \dfrac{2.5}{3.5} = 0.571$

$\therefore$ **Line x on y** is $x - 65 = 0.571\,(y - 67) \Rightarrow x =$ $\mathbf{26.72 + 0.5914\,y}$

Similarly, **y on x is** $y - 67 = 1.12\,(x - 65) \Rightarrow y =$ $\mathbf{1.12x - 5.8.}$]

15. For some bivariate data, the following results were obtained for the two variables x and y.

$\bar{x} = 53.2,\quad \bar{y} = 27.9,\quad b_{yx} = -1.5, b_{xy} = -0.2.$

Find the most probable value of y when $x = 60$.

[**Hint.** The regression equation of y on x is $y - \bar{y} = b_{yx}(x - \bar{x})$

$\Rightarrow\ y - 27.9 = (-1.5)\,(x - 53.2)$ or $y = 107.7 - 1.5x$.

When $x = 60$, then $y = 107.7 - 1.5 \times 60 =$ **17.7**]

16. Given is the following information

	X	Y
Arithmetic mean	6	8
Standard deviation	5	40/3.

Coefficient of correlation between X and $Y = 8/15$.

Find (*i*) the regression coefficient of Y on X.

(*ii*) the regression equation of X on Y.

(*iii*) the most likely value Y where $X = 100$.

[**Hint.** Here $b_{yx} = r\dfrac{\sigma_y}{\sigma_x} = \dfrac{8}{15} \times \dfrac{40}{3 \times 5} = 1.422.$

$\therefore$ Regression line Y on X is $Y - 8 = 1.422\,(X - 6)$

$\Rightarrow$ $\mathbf{Y = 1.422, X - 0.532}$

When $X = 100$, $Y =$ **141.668.**

Also $b_{xy} = \dfrac{\sigma_y}{\sigma_x} = \dfrac{8}{15} \times \dfrac{5 \times 3}{40} = .02$

Regression line X on Y is : $X - 6 = 0.2\,(Y - 8) \Rightarrow$ $\mathbf{X = 4.4 + 0.2Y}$]

17. The following calculations have been made for prices of 12 stocks (X) on Bombay Stock Exchange on a certain day along with the volume of sales in thousands of share (Y). From these calculations, find the regression equation of prices of stocks, on the volume of the sales of shares.

$\Sigma X = 580, \Sigma X^2 = 41658, \Sigma Y = 370, \Sigma Y^2 = 17206, \Sigma XY = 11494.$

[**Hint.** $b_{xy} = \dfrac{\Sigma XY - N\,\bar{X}\,\bar{Y}}{\Sigma Y^2 - N(\bar{Y})^2} = \dfrac{11494 - 12\,(48.333 \times 30.833)}{17236 - 12\,(30.833)^2} = $ $\mathbf{-1.102.}$

$\therefore$ Line X on Y: $X - 48.333 = (-1.102)\,(Y - 30.833) \Rightarrow$ $\mathbf{X = -1.102\,Y + 82.331.}$]

18. From the following table showing age of cars of a certain make and annual maintenance costs, obtain the regression equation for costs related to age.

Age of cars (years)	2	4	6	7	8	10	12.
Maintenance cost (Rs. in 100)	16	15	18	19	17	21	20.

19. The following figures relate to the years of service and income in hundreds of rupees of the employees of an organisation. Find the initial start for a person applying for a job after having served in another factory for a period of 12 years in a similar capacity.

Service X: (in years)	11	7	9	5	8	6	10
Income (in 100 rupees) Y:	7	5	3	2	6	4	8.

[**Hint.** Here $\bar{X} = \frac{56}{7} = \mathbf{8.}$ $\bar{Y} = \frac{35}{7} = \mathbf{5.}$

Also then $b_{yx} = \Sigma \frac{(X-\bar{X})(Y-\bar{Y})}{\Sigma(X-\bar{X})^2} = \frac{21}{28} = 0.75.$

Regression line of Y on X : $Y - \bar{Y} = b_{yx}(X - \bar{X})$

$\Rightarrow Y - 5 = 0.75(X-8) \Rightarrow Y = 0.75X - 1.$

When $X = 12$, then $Y = 0.75 \times 12 - 1 = 8.$ or **Rs. 800.**]

20. Given the bivariate data

X:	1	5	3	2	1	1	7	3
Y:	6	1	0	0	1	2	1	5.

(*a*) Fit a regression line of Y on X and hence predict X if $Y = 2$.

(*b*) Fit a regression line of X on Y and predict Y if $X = 5$.

(*c*) Calculate the Karl Pearson's correlation coefficient.

[**Hint.** Here $\bar{X} = \frac{23}{8} = \mathbf{2.875}$; $\bar{Y} = \frac{16}{8} = \mathbf{2.}$

Take the **assumed mean** $A = 3$ for X, and $B = 2$ for Y, so that $dx = X - 3$, $dy = Y - 2$.

Then $\Sigma dx = -1$, $\Sigma dx^2 = 33$, $\Sigma dy = 0$, $\Sigma dy^2 = 36$, $\Sigma dxdy = -10$.

$$\therefore \quad b_{xy} = \frac{\Sigma dxdy - (\Sigma dx)(\Sigma dy)/N}{(\Sigma dy^2) - (\Sigma dy)^2/N} = \frac{-10-(-1)(0)/8}{36-0} = -\mathbf{0.278.}$$

$$\text{Similarly,} \quad b_{yx} = \frac{-10-0}{33-(-1)^2/8} = -\mathbf{0.3.4.}$$

Regression equation Y on x : $Y - 2 = -0.304(X - 2.875)$

$\Rightarrow \quad Y = -0.304X + 2.874,$

When $\quad$ **$X = 5$,** then $Y = -0.304 \times 5 + 2.874 = \mathbf{1.354}$

Regression equation X on Y : $X - 2.875 = -0.27(Y-2) = -0.2784 + 0.556$

$\Rightarrow \quad X = \mathbf{-0.278Y + 3.431}$

When $\quad$ **$Y = 2.5$,** then $X = -0.278(2.5) + 3.431 = \mathbf{2.736.}$

Also $\quad r = \sqrt{b_{xy} \times b_{yx}} = \sqrt{(-0.278)\times(-0.304)} = -\mathbf{0.291.}$

21. Assuming that there is a linear relationship between height (X) and weight (Y) of residents of a certain locality, estimate the height of an individual whose weight is 75 kg from the following data

Mean height = 165 cm, Mean weight = 80 kg, Standard deviation of height = 8 cm, Standard deviation of weight = 10 kg.

Correlation coefficient between height and weight = 0.6.

Also find the most likely weight of a person whose height is 170 cm.

[**Hint. Regression equation of X on Y :** $X - \bar{X} = r\frac{\sigma_x}{\sigma_y}(Y - \bar{Y})$

$\Rightarrow X - 165 = 0.6 \times \frac{8}{10}(Y - 80) \Rightarrow X = \mathbf{0.48\,Y + 126.6}$

When $Y = 75$, then $X = 0.48 \times 75 + 126.6 =$ **162.6 cm**

Regression equation of Y on X : $Y - \overline{Y} = r\frac{\sigma_y}{\sigma_x}(X - \overline{X})$

$\Rightarrow Y - 80 = 0.6 \times \frac{10}{8}(X - 165) \Rightarrow$ $\mathbf{Y = 0.75X - 43.75}$

When $X = 170$ cm, then $Y =$ **83.75 kg.**

22. Calculate the two regression equations of X on Y and Y on X from the data given below taking deviations from actual means of X and Y.

Price (Rs.):	10	12	13	12	16	15
Amount demanded :	40	38	43	45	37	43.

Estimate the likely demand when the price is Rs. 20.

23. Calculate correlation and regression equations from the following data.

X:	2	4	6	8	10	12	14
Y:	4	2	5	10	4	11	12.

Find the estimate of Y when $X = 13$.

24. Given regression coefficient of x on y and y on x as 0.85 and 0.89, find the value of coefficient of correlation.

25. A student obtained the two regression lines as $2x - 5y - 7 = 0$; and $3x + 2y - 8 = 0$. Do you agree with him?

[**Hint.** No. here $b_{yx} = \frac{2}{5}, b_{xy} = -\frac{2}{3}$. It is impossible because both the regression coefficients must have the same sign.]

26. Given $x = 4y + 5$ and $y = kx + y$ are the lines of regression of x on y and y on x respectively. If k is positive, prove that it cannot exceed $\frac{1}{4}$. If $k = \frac{1}{6}$, find the means of two variables and the coefficient of correlation between them.

[**Hint.** $x = 4y + 5 \Rightarrow b_{xy} = 4$ and $y = kx + y \Rightarrow b_{yx} = k$.

Now $r^2 = b_{xy} \times b_{yx} = 4k$. Also $0 \le r^2 \le 1 \Rightarrow 0 \le 4k \le 1$ or $k \le = \frac{1}{4}$.

If $k = \frac{1}{16}$, then the result $r^2 = 4k \Rightarrow r^2 = \frac{1}{4}$ or $r = \pm\frac{1}{2} = \pm 0.5$.

As both b_{xy} and b_{yx} are positive so r must be positive, $r = 0.5$.

Now solving $x = 4y + 5$ and $y = \frac{1}{10}x + 4$, we get

$x = 28$, $y = 5.75$. Hence $\mathbf{\bar{x} = 28}$, $\mathbf{\bar{y} = 5.75}$].

27. Mark each of the following statements **True** or **False**.

(*a*) The regression lines are drawn by applying the principal of least squares.

(*b*) The correlation $r = b_{xy} \times b_{yx}$.

(*c*) b_{xy} and b_{yx} must be both either positive or negative.

(*d*) The sign of r is the same as that of b_{xy} or b_{yx}.

(*e*) The two regression lines x on y and y on x always intersect at $(\bar{x}, \bar{y})$.

(*f*) The values of '*a*' and '*b*' in a regression equation are determined by solving the two normal equations simultaneously.

(*g*) Both regression coefficients must be less than 1.

28. State the properties of linear regressions

29. Distinguish between correlation and regression analysis and indicate the utility of regression analysis in economic activities.

30. Using the following data, obtain the two regression equations

X:	14	19	24	21	26	22	15	20	19
Y:	31	36	48	37	50	45	33	41	39.

31. You are given the following data:

$\overline{X} = 36,\ \sigma_x = 11; \overline{Y} = 85, \sigma_y = 8.$ and correlation coefficient between *X* and *Y* = 0.66. Find (*i*) two regression equations; (*ii*) Estimate the value of *X* when *Y* = 75

Regression equation of *Y* on *X* : $Y - \overline{Y} = r\dfrac{\sigma_y}{\sigma_x}(X - \overline{X})$

$\Rightarrow Y - 85 = 0.66 \times \dfrac{8}{11}(x - 36) \ \Rightarrow\ \mathbf{Y = 0.48X + 67.72}$

Regression equation *X* on *Y* : $X - \overline{X} = \gamma\dfrac{\sigma_x}{\sigma_y}(Y - \overline{Y})$

$\Rightarrow X - 36 = 0.66 \times \dfrac{11}{8}(Y - 85) \ \Rightarrow\ X = 0.9075Y - 77.1375$

Put *Y* = 75 in (1) to get ***X* = 26.93.**]

32. Compare the meaning of *r* and r^2.

33. Is the relation $r^2 = 1 - \dfrac{500}{400}$ possible? If not, why?

ANSWERS

1. *X* on *Y* : *X* = 1.361 *Y* – 5.27; *Y* on *X* : *Y* = 0.6132*X* + 14.812.

2. *r* = + 0.8, **3.** $\bar{x} = 130,\ \bar{y} = 90;\ r = 0.866.$

4. 66 quintals. **5.** $\overline{X} = 4, \overline{Y} = 7, b_{yx} = -3/2,\ b_{xy} = -1/6,\ r = -0.5.$

6. Regression line *X* on *Y* is *X* = 0.9075*X* + 41.1375 and that of *Y* on *X* is *Y* = 0.48*X* + 67.72; When *Y* = 75, *X* = 26.93.

7. 1014.7. **8.** $\bar{x} = 5,\ \bar{y} = 0.333;\ r = -1/2\sqrt{3}$ **9.** $\bar{x} = 10,\ \bar{y} = 7.$

10. $\bar{x} = 1,\ \bar{y} = 2,\ \sigma_y^2 = 4,\ r = -0.866.$ **11.** *Y* = 0.6*X* + 0.4; *X* = 1.2*Y* + 1.20.

12. *y* = 8.8 + 1.015*x*; 25.04.

13. *X* = 0.64*Y* + 2.63; *Y* = 0.83*X* + 51.88, When *X* = 100, *Y* = 134.88. 51.88.

14. *x* = 26.72 + 0.5714*y* ; *y* = 1.12*x* **15.** 17.7. – 5.8.

16. *Y* = 1.422*X* – 0.532, 141.668; **17.** *X* = – 1.102*Y* + 82.331.

18. *Y* = .5286*X* + 14.299 or *Y* = 52.86 + 1429.98.

19. Regression line *Y* on *X* : *Y* = 0.75*X* – 1, *X* = 12 then *Y* = Rs. 800.

20. *X* = 2.736, *Y* = 1.354, *r* = 0.291 **21.** 170 cm, 83.75 kg.

22. *X* = – 0.12*Y* + 17.92, *Y* = – 0.25*X* + 44.25; When *X* = 20, *Y* = 49.25.

23. *Y* = 0.73 *X* + 1.02; When *X* = 13, *Y* = 10.51, $b_{xy} = 0.85,\ b_{yx} = 0.73,\ r = 0.79.$

24. *r* = 0.8698. **25.** No. **26.** $\bar{x} = 28,\ \bar{y} = 5.75,\ r = 0.5.$

27. (*a*). True, (*b*) False, (*c*) True; (*d*) True; (*e*) True; (*f*) True, (*g*) False.

30. *X* = 0.557*Y* – 2.28 ; *Y* = 1.608 *X* + 7.84.

31. (*i*) *X* = 0.9075*Y* – 41.1375; *Y* = 0.48*X* + 67.72; (*ii*) *X* = 26.93.

33. $r^2 = 1 - \dfrac{500}{400} = -160$, which is negative. So it is not possible as r^2 is always positive, whatever may be the values of *r* (positive or negative).

CHAPTER 14

TIME SERIES ANALYSIS

14.1 INTRODUCTION

Frequency distribution measure variation around the mean. Time series refers to a series in which one variable is time. ***A time series is an arrangement of statistical data in accordance to the time of occurrence or in a chronological order.***

It reflects the dynamic pace of the movements of a phenomenon over a period of time, *e.g.*, the figures of national income, production or cost or prices spread over a period of time. Most of the series relating to Economics, Business and Commerce are all time series spread over a period of time.

Time series analysis is done primarily for the purpose of making forecasts for future and also for the purpose of evaluating past performances. An economist or a businessman is interested in estimating the figures of national income, population, prices and wages etc. in future by analysing the past behaviour of the variable under study. The analysis of time series plays an important role in the study of all economic, business and natural and social sciences. Thus, we can conclude that ***the analysis of time series is helpful in studying any phenomenon whose values are or can be arranged chronologically over successive period of time.***

14.2 DEFINITION OF A TIME SERIES

Any sequence of measurement taken on a response, that is, variable over time is called a time series. It can be written as (t, u(t)) for t = 1, 2, 3, ...

The time series is usually represented by the mathematical equation taking the values of the response as a function of time or equivatently, as a curve on a graph whose vertical coordinates gives the value of the random response plotted against the time on the horizontal axis.

A few definitions of time series are given below to further clarify the concept:-

1. **A time series is a set of observations taken at specified times, usually at "equal intervals". Mathematically, a time series is defined by the values Y_1, Y_2 ...... of a variable Y (temperature, closing price of a share, etc.) at times t_1, t_2, ... Thus Y is a function of t, symbolised by $Y = F(t)$.** – ***Spiegel***
2. **A time series may be defined as a collection of magnitudes belonging to differentt time periods, of some variable or composite of variables, such as producttion of steel, per capita income, gross national product, price of tobacco, or index of industrial production.** – ***Ya-Wn-Chou***
3. **A time series consists of data arranged chronologically.** – ***Croxton and Cowden***
4. **A time series consists of statisical data which are collected, recorded or observed over successive increments.** – ***Patterson***
5. **A set of data depending on the time is called time series.** – ***Kenny***
6. **A time series is a set of statistical observations arranged in chronological order.** – ***Morris Hamburg***
7. **When quantitative data are arranged in the order of their occurrence the resulting series is called the time series.** – ***Wessel & Wallet***

14.3 UTILITY OF A TIME SERIES

The following points indicate the utility of time series analysis:

1. Analysis. It helps in the analysis of past behaviour of a variable. Analysis of past data discloses the effect of various factors on the variable under study. With the help of such analysis the future behaviour of the variable under study can be predicted.

2. Forecasting. It helps in fore casting. The analysis of past conditions is the basis of forecasting the future behaviour of the variable under study. This helps in making future plans of action. Various five-year plans of our country are based on the analysis of past data.

3. Evaluation. It helps in the evaluation of current achievements. The review and evaluation of progress made on the basis of a plan are done on the basis of time series data. The progress of plans is judged by the yearly rates of growth in GNP (Gross National Product).

4. Comparison. It helps in making comparative studies. Once the data is arranged chronologically, the comparison between one time period and another is facilitated. It provides a scientific basis for making comparisons by studying and isolating the effects of various components of a time series.

5. Approximation. It gives approximate indicators.

14.4 COMPONENTS OF A TIME SERIES

There are four types of components of a time series, which are as follows:

(1) **Secular Trend or Trend (T)**

(2) **Seasonal Variations (S)**

(3) **Cyclical Variations (C)**

(4) **Irregular Variations. (I)**

I. SECULAR TREND (T)

Secular Trend is the general tendency of the time series data to increase or decrease or stagnate during a long period of time. An upward tendency is usually observed in time series relating to population, production and sales, prices, income money in circulation while a downward tendency is noticed in data of birth's and death's and epidemics as a result of advancement in medical sciences, illiteracy, etc. Thus **trend is either upward or downward. It should be clearly understood that trend is the general, smooth, long-term average tendency.** It is not necessary that the increase or decrease should be in the same direction throughout the given period. With the help of a long-term trend, it is possible to determine and present the direction of change.

II. SEASONAL VARIATIONS (S)

Seasonal variation refer to such movements in a time series which are due to forces which are rhythmic in nature and which repeat themselves periodically in every season. These variations repeat themselves in less than one year time. Seasonal variations are usually measured in an interval. The **seasonal variation may be attributed to those resulting from natural forces and social customs and traditions. Seasonal variations are the results of such factors which uniformly and regularly rise and fall in the magnitude.**

III. CYCLICAL VARIATIONS (C)

Cyclic variations are the oscillatory movements in a time series with period of oscillation greater than one year. **These variations in a time series are due to ups and downs recurring after a period greater than one year.** These are not necessarily uniformly periodic, *i.e.*, they may or may not follow exactly similar patterns after equal intervals of time. One cyclic period normally lasts from 7 to 9 years.

IV. IRREGULAR VARIATIONS (I)

Random or Irregular variations do not exhibit any definite pattern and there is no regular period or time of their occurrence. **These are accidental changes which are purely random, unforeseen and unpredictable. They turn the series first in one way and then in the other way purely by chance with no regularity of occurrence, *e.g.*, earthquakes, wars, floods etc. Normally**

they are short-term variations but sometimes their effect is so intense that they may give rise to new cyclical or other movements.

14.5 ANALYSIS OF A TIME SERIES

The time series analysis consists of

(*i*) ***Identifying or determining the various forces or influences whose interaction produces the variations in the time series.***

(*ii*) ***Isolating, studying, analysing and measuring them independently, i.e., by holding other things constant.***

The time series analysis is of great importance not only to businessman or to an economist but also to people working in various disciplines in natural, social and physoical sciences.

14.6 USES OF A TIME SERIES

The time series analysis has the following uses:

(*i*) ***It enables us to study the past behaviour of the phenomenon under consideration.***

(*ii*) ***It helps to study the components which are of paramount importance to a businessman in the planning of future operations and in the formulation of executive and policy decisions.***

(*iii*) ***It helps to compare the actual current performance or accomplishments with the expected ones and analyse the causes of such variations.***

(*iv*) ***It helps us to compare the changes in the values of different phenomena at different places etc.***

14.7 MATHEMATICAL MODELS FOR A TIME SERIES

The following are the two models commonly used for the decomposition of a time series into its components.

1. Additive model. This **model assumes that the observed value is the sum of four components of time series,** ***i.e.,***

$Y = T + S + C + I$, **where** Y **= Original data,** T **= Trend value,** S **= Seasonal component,** C **= Cyclical component,** I **= Irregular component.**

The additive model for decomposition of ***time series assumes that all the four components of the time series operate independently of one another. It also assumes that the behaviour of components is of an additive character. It is to be noted that only absolute values are added or deducted from the trend value to arrive at the observed value.***

2. Multiplicative model. This model assumes that the observed value is obtained by multiplying the trend (T) by the rates of three other components as under, *i.e.*,

$Y = T \times S \times C \times I$, **where** Y **= Original data,** T **= Trend values,** S **= Seasonal component,** C **= Cyclical component,** I **= Irregular component.**

The multiplicative model assumes that the compoenents although due to different causes are not necessarily independent and they can affect one another. It also assumes that the behaviour of components is of a multiplicative character. ***It may be noted that except the value of trend all the other values on the right hand side are rates or index numbers.***

Most of the time series relating to economic and business phenomena conform to the multiplication model. In practice, additive model is rarely used.

14.8 EDITING OF A TIME SERIES

It is necessary to make certain adjustments in the available data. **Some important adjustments are:**

1. Time Variation. When data are available on monthly basis, the effect of time variation needs to be adjusted because all months of the year do not have the same number of days. This adjustment of time variation is done by dividing each monthly total by the number of days in the months. The daily average is then multiplied by $\frac{365}{12}$ which is the average number of days in a month.

2. Population Changes. Adjustment for population change becomes necessary when a variable is affected by change in population. If we are studying national income figures such adjustment is necessary. In this case, adjustment is to divide the income by the number of persons concerned. Then we can have per capita income figures.

3. Price Changes. Adjustment for price changes becomes necessary wherever we have a current value series and we are interested in real value changes. Current values are to be deflated by the ratio of current prices to base year prices.

4. Comparability. In order to have valid conclusion the data which are being analysed should be comparable. When we are dealing with the analysis of time series it involves the data relating to past which must be homogeneous and comparable. Therefore, efforts should be there to make the data as homogeneous and comparable as possible.

14.9 MEASUREMENT OF TREND

Following are the methods which are used to measure the trend.

1. FREEHAND OR GRAPHIC METHOD

2. METHOD OF SEMI-AVERAGES

3. METHOD OF MOVING AVERAGES

4. METHOD OF LEAST SQUARES.

We shall discuss each of these methods in the following paragraphs.

14.10 FREEHAND OR GRAPHIC METHOD

This is the simplest and the most flexible method of estimating the secular trend and consists in first obtaining a histogram by plotting the time series values on a graph paper and then drawing a freehand smooth curve through these points so that it accurately reflects the long-term tendency of the data. For **obtaining proper trend line or curve, the following points may be considered:**

(*i*) It should be smooth.

(*ii*) The number of points above the trend curve should be more or less equal to the number of points below it.

(*iii*) The sum of the vertical deviations of the given points above the trend line should be aproximately equal to the sum of vertical deviations of the points below the trend line so that the total positive deviations are more or less balanced against total negative deviations.

(*iv*) The sum of the squares of the vertical deviations of the given points from the trend line should be minimum possible.

MERITS AND DEMERITS

Merits: (*i*) ***It is very simple and time saving method and does not require any mathematical calculations.***

(*ii*) ***It is very flexible method in the sense that it can be used to describe all types of trend.***

Demerits: (*i*) ***The strongest objection to this method is that it is highly subjective in nature. The trend line so obtained will very much depend on the personal bias and judgement of the investigator handling the data and consequently different persons will obtain different trend curves for the same set of data. Thus this method requires great skill and expertise on the part of investigator and this restricts the popularity and utility of this method.***

(*ii*) ***Because of the subjective nature of the free hand trend line, it is not suitable for forecasting and for making decisions.***

Example 1. *From the following data of the sales figures determine the trend line by freehand curve method.*

Years:	*1978*	*1979*	*1980*	*1981*	*1982*	*1983*	*1984*	*1985*	*1986*	*1987*	*1988*
Sales`000:	*60*	*80*	*70*	*100*	*80*	*120*	*110*	*140*	*130*	*160*	*150*

Solution.

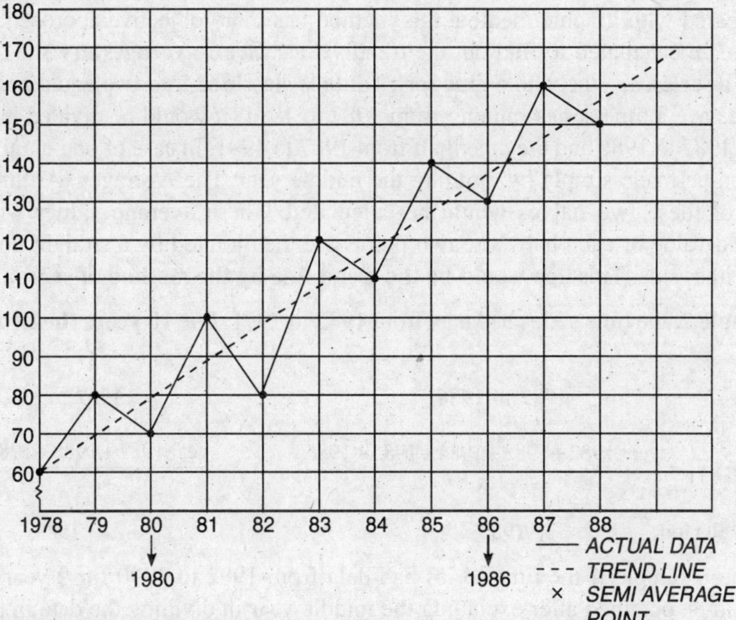

Example 2. *With the help of graph paper fit a smooth freehand curve for the following data:*

Year:	*1982*	*1983*	*1984*	*1985*	*1986*	*1987*	*1988*	*1989*	*1990*	*1991*
Values:	*128*	*164*	*194*	*142*	*156*	*224*	*230*	*262*	*176*	*260*

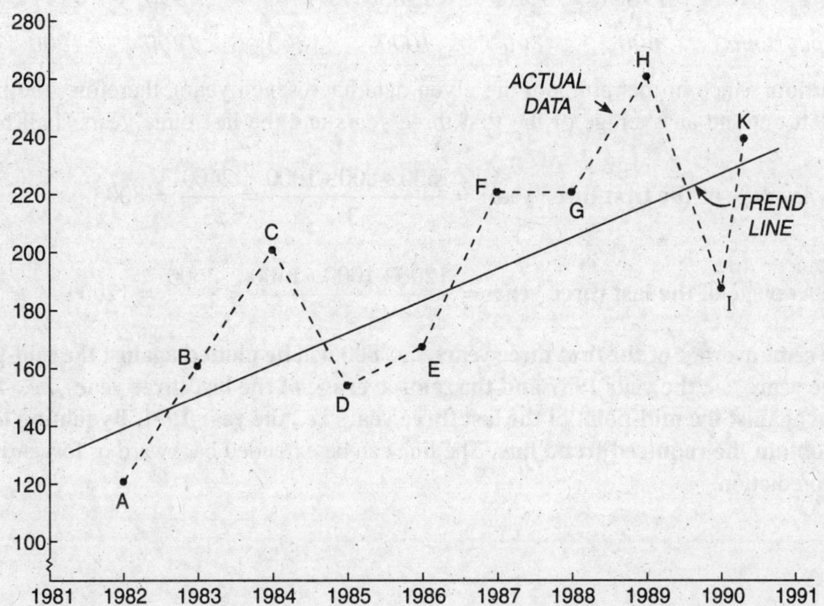

Plot the points *A* (1982, 128), *B* (1983, 164), *C* (1984, 194), *D* (1985, 142), *E* (1986, 156), *F* (1987, 224), *G* (1988, 230), *H* (1989, 262), *I* (1990, 176) and *K* (1991, 260). Join these points by a freehand curve to get the time graph as given by dotted lines.

14.11 METHOD OF SEMI-AVERAGES

As compared with graphic method, this method has more objective approach. In this method semi-averages are calculated to find out the trend values. **Semi-averages are the averages of the two halves of a series,** *i.e.*, the **whole time series data is classified into two equal parts with respect to time.** If we have a time series running from 1982 to 1991, it would be divided into two halves – one half from 1982 to 1986 and the other half from 1987 to 1991. In case of odd number of years, two equal parts can be made simply by omitting the middle year. The Averages (Arithmetic Averages) of the value of these two halves would be calculated. These average values would be plotted against the mid value of each half. The two points are then jointed by a straight line which can be extended on either side. This line **would be the trend line by the method of semi-averages.**

In example 2, the time series had data from 1982 to 1991. **For 10 years the two semi-averages would be**

Period: **1982 to 1986** **1987 to 1991**

Semi-Average: $\bar{x}_1 = \dfrac{\text{year } 1982+1983+1984+1985+1986}{5}$; $\bar{x}_2 = \dfrac{\text{year } 1987+1988+1989+1990+1991}{5}$

Middle of the Period: **1984** **1989**

In another example if the time series had data from 1982 to 1990 for 9 years the two semi-averages would be obtained after excluding the middle year in dividing the data in two halves. The middle year in this example is 1986.

Example 3. *Fit a trend line to the following data by the method of semi-averages:*

Year	*1986*	*1987*	*1988*	*1989*	*1990*	*1991*	*1992*
Output (units)	*600*	*800*	*1000*	*800*	*1200*	*1000*	*1400*

Solution. Since in this problem the given data have seven years, therefore, the middle year shall be left out and an average of the first three years and the last three years shall be obtained.

The **Average of the first three years** $= \dfrac{600+800+1000}{3} = \dfrac{2400}{3} = 800$

The **Average of the last three years** $= \dfrac{1200+1000+1400}{3} = \dfrac{3600}{3} = 1200$

The semi-average of the first three years, *i.e.*, 800 will be plotted against the mid-point of the first three years *i.e.*, the year 1987 and the semi-average of the last three years, *i.e.*, 1200 would be plotted against the mid-point of the last three years *i.e.*, the year 1991. By joining these points we shall obtain the required trend line. The line can be extended backward or forward and can be used for prediction.

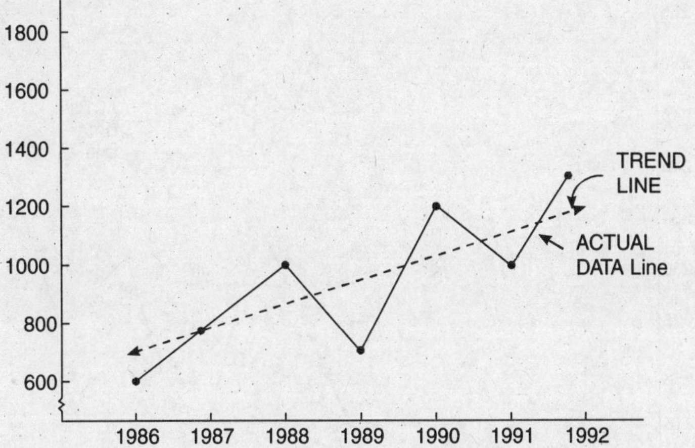

Thus if the semi-average are plotted against the mid-point and then joined together, it will give the trend line by the method of semi-averages.

MERITS AND LIMITATIONS OF THE METHOD OF SEMI – AVERAGES

MERITS :

(i) It is a simple method.

(ii) The trend figures are objective in the sense that any two persons will get the same trend line from a set of figures.

(iii) The line can be extended to obtain future or past estimates.

LIMITATIONS:

(i) This method assumes the presence of linear trend which would not be true in many cases.

(ii) In this method, trend is affected appreciably by higher or extreme values. The use of arithmetic mean for obtaining semi-averages is also questioned.

(iii) The trend values obtained by this method and the predicted values for future are not precise and reliable.

Example 4. *Draw a trend line by the method of semi-averages.*

Year	*1982*	*1983*	*1984*	*1985*	*1986*	*1987*	*1988*	*1989*	*1990*	*1991*
Sales (000)	*210*	*200*	*215*	*205*	*220*	*235*	*210*	*235*	*225*	*245*

Solution. There are 10 years figures available to us, so we have two halves of 5 years each 1982-1986 and 1987-1991.

$$\textbf{The average of the first half} = \frac{210+200+215+205+220}{5} = \frac{1050}{5} = \mathbf{210}$$

$$\textbf{The average of the second half} = \frac{235+210+235+225+245}{5} = \frac{1150}{5} = \mathbf{230}$$

The semi-average of the first half, *i.e.*, 210 will be plotted against the mid-point of the first half *i.e.*, the year 1984 and the semi-average of the second half *i.e.*, 230 would be plotted against the mid-point of the second half *i.e.*, the year 1989. These points would be joined and the line extended both backward and forward. This would be the trend line by the method of semi-average of the above problem. Both the original data and the trend line are plotted in the graph given below:

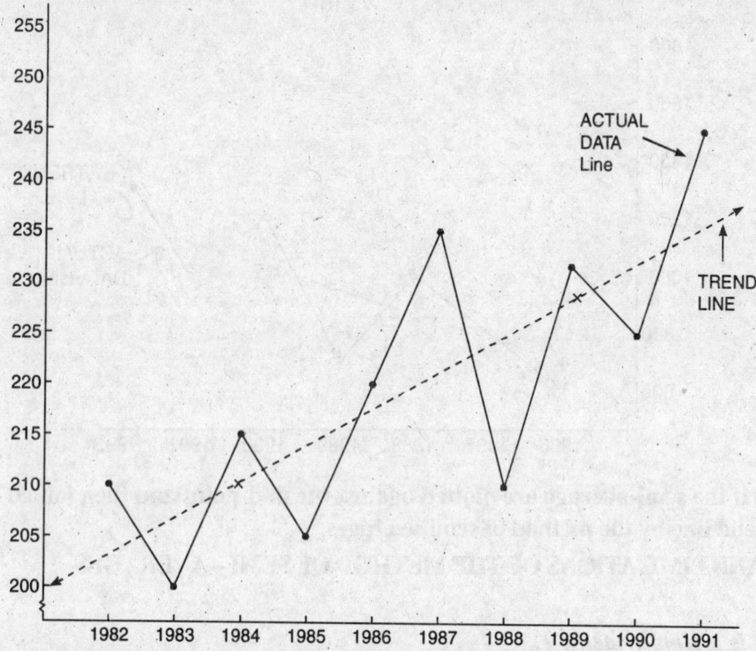

Example 5. *Draw a trend line by the method of semi-averages.*

Year:	*1991*	*1992*	*1993*	*1994*	*1995*	*1996*
Sales: ('000)	*60*	*75*	*81*	*110*	*106*	*120*

Solution. Step 1. Divide the 6 year in two equal parts from 1991 to 1993 and 1994 to 1996.

Step 2. Average of the 1st part = $\frac{60+75+81}{3}$ **= 72.**

Average of the 2nd part = $\frac{110+106+120}{3}$ **= 112.**

Step 3. Plot these values, *i.e.*, 72 and 112 against the mid-years of the two parts *i.e.*, against **1992 and 1995** respectively.

Step 4. Join the two points. The straight line so obtained, shown as a **thick line is the trend line.**

The **dotted curve** represents the graph of the **actual data.**

Year	*Sales ('000)*
1991	60
1992	75
1993	81
1994	110
1995	106
1996	120

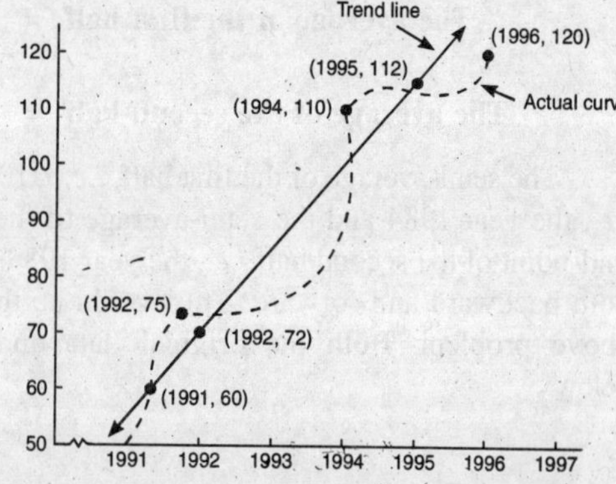

Example 6. *Draw a trend line by the method of semi-averages from the following data:*

Year:	*1988*	*1989*	*1990*	*1991*	*1992*	*1993*	*1994*	*1995*
Sales (Thousand Units):	*100*	*105*	*109*	*96*	*102*	*108*	*112*	*114*

Also predict the sales for the year 1998 from the graph.

Solution. There are 8 year sales available to us. So, we have two halves of 4 years each, *i.e.*, 1988 – 1991 and 1992–1995

The average of first half years sale $= \frac{100+105+109+96}{4} = \frac{410}{4} =$ **10.25.**

The average of second half years sale $= \frac{102+108+112+114}{4} = \frac{436}{4} =$ **109.**

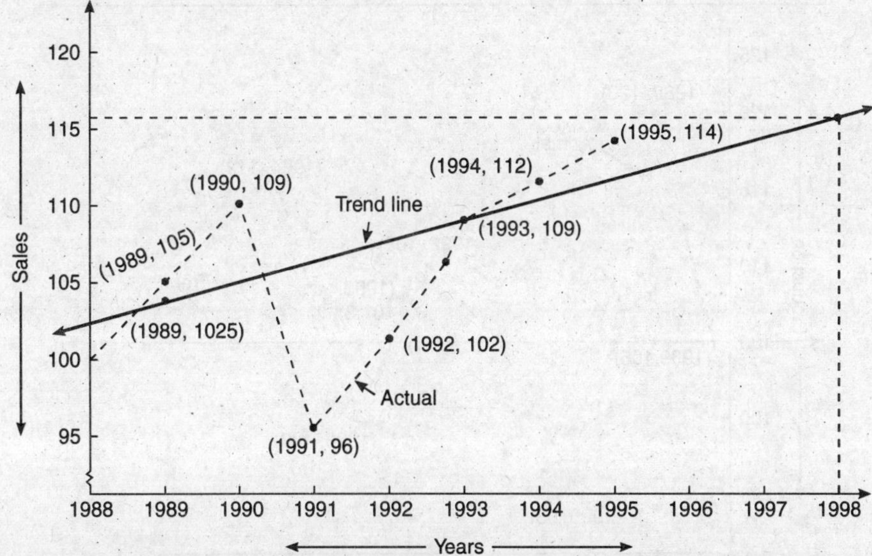

Plot the points 102.5 and 109 against the middle period *i.e.*, Ist July 1989 and 1st July 1993. Join them to get the **trend line shown as thick line.** The **dotted curve is the actual curve.**

In order to predict the value for 1998, we will draw perpendicular from 1998 on the trend line. From the point where it meets the line draw a horizontal line on the Y-axis parallel to the X-axis. The point where it meets the Y-axis will give the required value for 1998.

From the graph it is clear that **the likely sales for 1998 are 116 thousand units.**

Example 7. *Plot the following data on a graph paper and ascertain trend by the method of semi-averages.*

Year	*Production (Million tonnes)*	*Year*	*Production (Million Tonnes)*
1988	*100*	*1992*	*108*
1989	*120*	*1993*	*102*
1990	*95*	*1994*	*112*
1991	*105*		

Solution. Since the number of years is odd (seven), we will leave the middle year's item, *i.e.*, the item 105 of 1991 and obtain an average of the first three years and last three years.

Year	*Production (Million Tonnes)*	*Average*
1988	100	
1989	120	$\frac{100+120+95}{3}=\mathbf{105}$
1990	95	
1991	**105 ← left out**	
1992	108	$\frac{108+102+112}{3}$
1993	102	
1994	112	= **107.33**

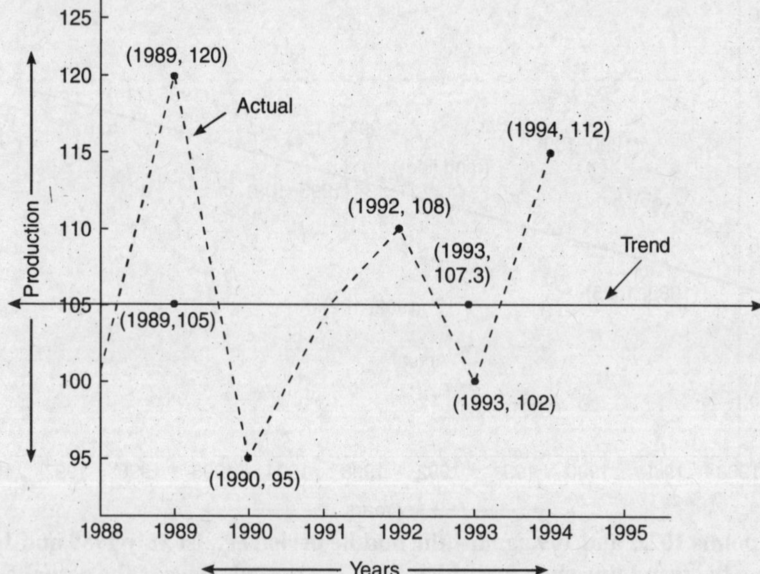

Now plot the points 105, 107.33 against the mid-years 1989 and 1993 respectively. Join these two points to get the required trend line. It is shown as a thick line. The **dotted curve** is the **actual curve** which is obtained by plotting the points (1988, 100), (1989, 120), (1990, 95), (1991, 105), (1992, 108), (1993, 102) and (1994, 112).

14.12 METHOD OF MOVING AVERAGES

Moving average method is a simple device of reducing fluctuations and obtaining trend values with a fair degree of accuracy. This method consists of taking arithmetic mean of the values for a certain time span and placing it at the centre of the time span. In this method the average value of a number of years (months, weeks or days) is taken as the trend value for the middle point of the period of moving averages. The process of averaging smoothens the curve and reduces the fluctuations. For applying the method of moving averages the period of the moving averages has to be first selected. This period can be 3-yearly moving averages, 5-yearly moving averages, 7-yearly moving average, 4-yearly moving averages, etc. The period of moving averages is to be decided in the light of the length of the series. After deciding the period of moving averages the figures are to be added by dropping the first year figure of the values and adding the figure of the next which had not been added previously. **For example,** 3-yearly moving average can be calculated from the data given by

$$\frac{a+b+c}{3}, \frac{b+c+d}{3}, \frac{c+d+e}{3}, \frac{d+e+f}{3}, \ldots\ldots\ldots$$

and these total to be written besides the year *b*, *c*, *d*, *e* and so on.

If the moving average is an *odd number, e.g.,* 3 years, there is no problem of centering it. because the moving total for 3 year average will be centered besides the 2nd year and for 5 year average be centered besides 3rd year.

But if the moving average is an *even number, e.g.,* 4 years moving average, then the average of first four figures will be placed between 2nd and 3rd year. Likewise, the average of the second group of 4 years will be placed between 3rd and 4th year. These two moving averages will then be averaged and this average would be written against the 3rd year. **This process is called centering of the averages. In case of even period moving averages, the trend values are obtained after centering the averages.**

MERITS AND LIMITATIONS OF METHOD OF MOVING AVERAGES.

Merits:

(i) This method is simple.

(ii) This method is flexible enough to add more figures to the data because the entire calculations are not changed.

(iii) This method is objective in the sense that any body working on a problem with this method will get the same results.

(iv) If the period of moving averages coincides with the period of cyclic fluctuations in the data, such fluctuations are automatically eliminated.

(v) This method is used for determining seasonal, cyclic and irregular variations besides the trend values.

Limitations:

(i) There are no trend values for some years in the beginning and some in the end. **For example, for 7-yearly moving average there will be no trend values for the first three years and the last three years.**

(ii) There is no functional relationship between the values and the time. Thus, this method is not helpful in forecasting and predicting the values on the basis of time.

(iii) The selection of the period of moving average is a difficult task. Therefore great care has to be taken in selecting the period, particularly, when there is no business cycle in the time series.

(iv) In case of non-linear trend the values obtained by this method are biased in one or the other direction.

14.13 CALCULATION OF MOVING AVERAGE WHEN THE PERIOD IS ODD

Example 8. *Calculate the 3-yearly moving averages of the data given below:*

Years:	*1980*	*1981*	*1982*	*1983*	*1984*	*1985*	*1986*	*1987*	*1988*	*1989*
Sales (millions of rupees):	*3*	*4*	*8*	*6*	*7*	*11*	*9*	*10*	*14*	*12*

Draw a graph to represent the moving averages. Also predict the sale for 1993.

Solution.

WORKING RULE:

Step 1. Add the values of the first 3 years (namely 1980, 1981 and 1982, *i.e.*, 3 + 4 + 8 = 15) and place the total against the middle year 1981.

Step 2. Leave the first year's value and add up the values of the next 3 years (*i.e.*, 1981, 1982 and 1983), *viz.*, 4 + 8 + 6 = 18 and place the total against the middle year, *i.e.*, year 1982.

Step 3. Continue this process until the last year's value is taken for calculating moving average.

Step 4. On dividing the 3-yearly totals by 3, you will obtains the 3-yearly moving average.

Step 5. Plot the points (1981, 5), (1982, 6), (1983, 7), (1984, 8), (1985, 9), (1986, 10), (1987, 11) and (1988, 12) and join them to get the **trend line shown as thick straight line.**

Table: Calculation of 3-year Moving Averages

Year	*Sales, (Millions of rupees)*	*3-yearly totals*	*3-yearly moving average (trends)*
1980	3		
1981	4	15	**5** = (15 ÷ 3)
1982	8	18	**6** = (18 ÷ 3)
1983	6	21	**7** = (21 ÷ 3)
1984	7	24	**8** = (24 ÷ 3)
1985	11	27	**9** = (27 ÷ 3)
1986	9	30	**10** = (30 ÷ 3)
1987	10	33	**11** = (33 ÷ 3)
1988	14	36	**12** = (36 ÷ 3)
1989	12		

When the 3-year moving averages are plotted on the graph as shown in the **figure, they fall on a straight line, and the cyclical fluctuations have been smoothed out. The straight line is the required trend line.** The hypothetical data have been adjusted so that the trend line is a straight line; however, in general, we shall have a smooth curve.

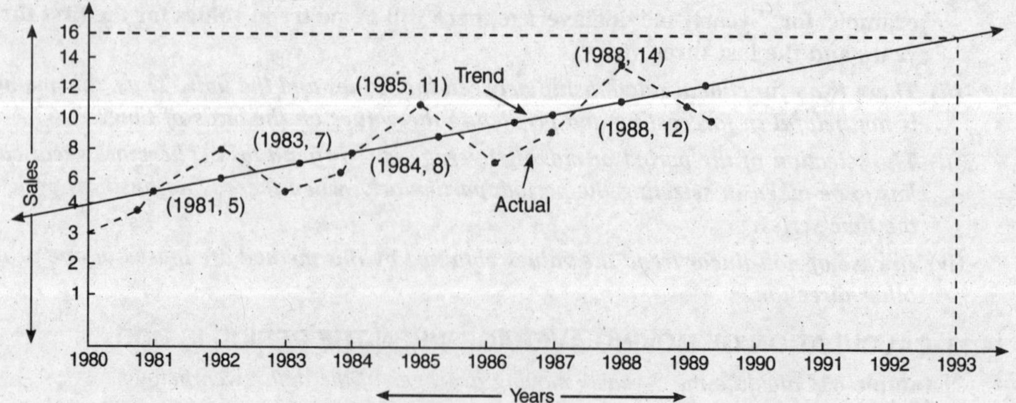

Sale for 1983. In order to find the sale for 1993, we draw a perpendicular from 1993 on the trend line meeting it at *P*. From *P* draw a line parallel to *X*-axis meeting *Y*-axis. The point where it **meets the *Y*-axis is the required sale, *i.e.*, 16 million units.**

Example 9. *Assuming 5-yearly moving averages, calculate trend values from the data given below and plot the results on a graph paper.*

Year:	*1971*	*1972*	*1973*	*1974*	*1975*	*1976*	*1977*	*1978*	*1979*	*1980*	*1981*	*1982*	*1983*
Production: (in thousand tonnes)	*105*	*107*	*109*	*112*	*114*	*116*	*118*	*121*	*123*	*124*	*125*	*127*	*129*

Solution.

Table: Calculation of 5-yearly Moving Averages

Year (t) (1)	*Values (y)* (2)	*5-yearly moving totals* (3)	*5-yearly moving averages (trend value) (column 3) ÷ 5* (4)	*Trend values rounded off (column 3) ÷ 5* (5)
1971	105	—	—	—
1972	107	—	—	—
1973	109	547	109.4	**109**
1974	112	558	111.6	**112**
1975	114	569	113.8	**114**
1976	116	581	116.2	**116**
1977	118	592	118.4	**118**
1978	121	602	120.4	**120**
1979	123	611	122.2	**122**
1980	124	620	124.0	**124**
1981	125	628	125.6	**126**
1982	127	—	—	—
1983	129	—	—	—

The trend values are given by the column (4), viz., 109.4, 111.6, 113.8, 116.2, 118.4, 120.4, 122.2, 124, 125.6.

Graph. In order to draw the graph, we rounded off the trend values as given by the **column (5).** Now plot the points (1973, 109), (1974, 112), (1975, 114), (1976, 116), (1977, 118),

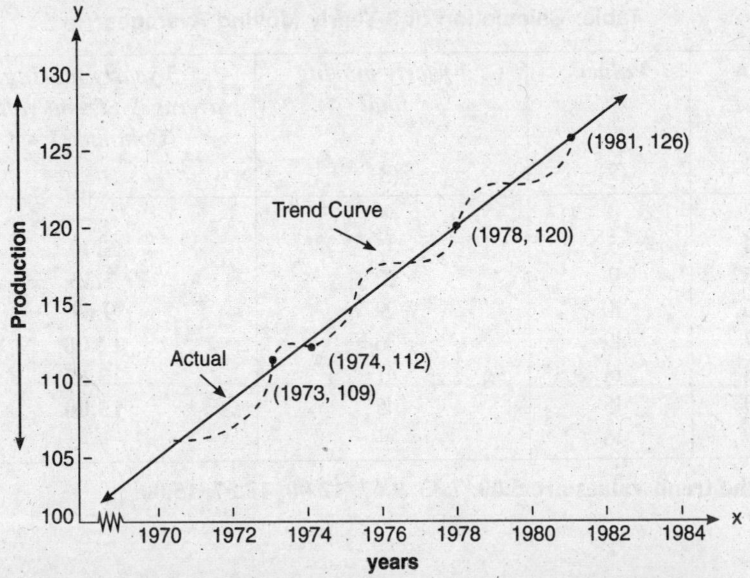

(1978, 120), (1979, 122), (1980, 124) and (1981, 126) and join them by free hand, get the **trend curve shown as dotted curve.** The graph of **the observed values is shown as thick line.**

Example 10. *Calculate five-yearly moving averages of number of students studying in a college from the following figures:*

Year	*1981*	*1982*	*1983*	*1984*	*1985*
No. of students	*332*	*317*	*357*	*392*	*402*
Year	*1985*	*1986*	*1987*	*1988*	*1990*
No. of students	*405*	*410*	*427*	*405*	*438*

Solution. Computation of Five Yearly Moving Averages

Year (*1*)	***No. of Students*** (2)	***5-yearly Moving Totals*** (*3*)	***5-yearly Moving Averages*** (*Col. 3*) ÷ *3*
1981	332	—	—
1982	317	—	—
1983	357	1800	**360.0**
1984	392	1873	**374.6**
1985	402	1966	**393.2**
1986	405	2036	**407.2**
1987	410	2049	**409.8**
1988	427	2085	**417.0**
1989	405	—	—
1990	438	—	—

Example 11. *Calculate 3-yearly moving averages or trend values for the following data:*

Year (t):	*1953*	*1954*	*1955*	*1956*	*1957*	*1958*	*1959*	*1960*
Values (Y):	*3*	*5*	*7*	*10*	*12*	*14*	*15*	*16*

Solution.

Table: Calculation of 3-Yearly Moving Averages

Year (1)	***Values*** (2)	***3-yearly moving total*** (3)	***3-yearly moving averages (Trend values) (Column 3) ÷ 3***
1953	3	—	—
1954	5	15	**5.00**
1955	7	22	**7.33**
1956	10	29	**9.67**
1957	12	36	**12.00**
1958	14	41	**13.67**
1959	15	45	**15.00**
1969	16	—	—

Hence, the trend values are 5.00. 7.33, 9.67, 12.00, 13.67, 15.00.

Example 12. *Calculate Trend values by taken 5-yearly moving average from the data given below:*

Year	*1977*	*1978*	*1979*	*1980*	*1981*	*1982*	*1983*	*1984*
Output ('000)	*4*	*5*	*6*	*7*	*9*	*6*	*5*	*7*
Year	*1985*	*1986*	*1987*	*1988*	*1989*	*1990*	*1991*	*1992*
Output ('000)	*8*	*7*	*6*	*8*	*9*	*10*	*7*	*9*

Solution.

Table for Computation of 5-Yearly Moving Averages

Year	*Output ('000)*	*5-Yearly Moving Totals*	*5-Yearly Moving Averages*
1977	4	—	—
1978	5	—	—
1979	6	31	(31 ÷ 5) = **6.2**
1980	7	33	(33 ÷ 5) = **6.6**
1981	9	33	(33 ÷ 5) = **6.6**
1982	6	34	(34 ÷ 5) = **6.8**
1983	5	35	(35 ÷ 5) = **7.0**
1984	7	33	(33 ÷ 5) = **6.6**
1985	8	33	(33 ÷ 5) = **6.6**
1986	7	36	(36 ÷ 5) = **7.2**
1987	6	38	(38 ÷ 5) = **7.6**
1988	8	40	(40 ÷ 5) = **8.0**
1989	9	40	(40 ÷ 5) = **8.0**
1990	10	43	(43 ÷ 5) = **8.6**
1991	7	—	—
1992	9	—	—

14.14 CALCULATION OF MOVING AVERAGES WHEN THE PERIOD IS EVEN

If the period of the moving average is even, (say 4 years) centre point of the group will lie between two years. It is, therefore, necessary to adjust or shift (technically known as *centre*) these averages so that they coincide with the years. The 4-yearly moving total and the 4-yearly moving average may be obtained by the methods already outlined for the odd period average. In order to *centre* the values, *a* 2-yearly moving average is taken in the case of the even period moving average.

The resulting average is located between the two 4-yearly moving average values and, therefore, coincides with the years. The end results (*i.e.*, *a* 2-yearly moving average of *a* 4-yearly moving average) is known as the **4-yearly moving average centred.**

We shall follow the steps given below in calculating the moving average when the order is even, say 4.

Illustration. *Calculate the trend by 4-yearly moving average of the data:*

Year:	y_1	y_2	y_3	y_4	y_5	y_6	y_7
Value:	*a*	*b*	*c*	*d*	*e*	*f*	*g*

WORKING RULE:

Step 1. Add the values of first four years, and place the total T_1, where $T_1 = a + b + c + d$ in between the 2nd and 3rd year (*i.e.*, in between y_2 and y_3).

Step 2. Leave the first year value and then add the values (b, c, d, e) of the next four years (y_2, y_3, y_3, y_4) and place the total T_2, ($T_2 = b + c + d + e$) in between the 3rd and 4th year (*i.e.*, in between y_3 and y_4). **Continue this process until the last year is taken into account.**

Step 3. Divide 4 yearly moving totals T_1, T_2, T_3,..... by 4. It will give **4-yearly moving averages** A_1, A_2, A_3,....

Step 4. Add first two moving Averages (A_1 and A_2) and divide it by 2 to get the **moving average centred** C_1. Place it against 3rd year.

Now, add next two moving average A_2 and A_3 and divide by 2 to get the next moving *average* centred C_2. Place it against the 4th year. Continue this process till the last moving average is included.

OR

ALTERNATIVELY:

Step 4A. Take the total of 4-yearly moving averages taking two terms at a time starting from the first and place the sum at the middle of these two terms. The same procedure is repeated for other averages.

Step 4B. Finally, take the two-period averages of the above moving averages by dividing each by 2. These are the required trend values. **This process is called centering of moving averages.**

Step 5. **Graph.** Plot the points (y_3, C_1), (y_4, C_2), (y_5, C_3) and join them to get the trend line.

Year	*Value*	*4-yearly Moving Totals (T)*	*4-yearly Moving Average (A)*	*4-yearly Moving Average Centred (C)*
y_1	a			
y_2	b			
		$\leftarrow a + b + c + d = T_1$	$\leftarrow \frac{T_1}{4} = A_1$	
y_3	c	$\leftarrow$	$\leftarrow$	$\leftarrow \frac{1}{2}(A_1 + A_2) = C_1$
		$\leftarrow b + c + d + e = T_2$	$\leftarrow \frac{T_2}{4} = A_2$	
y_4	d	$\leftarrow$	$\leftarrow$	$\leftarrow \frac{1}{2}(A_2 + A_3) = C_2$
		$\leftarrow c + d + e + f = T_3$	$\leftarrow \frac{T_3}{4} = A_3$	
y_5	e	$\leftarrow$	$\leftarrow$	$\leftarrow \frac{1}{2}(A_3 + A_4) = C_4$
		$\leftarrow d + e + f + g = T_4$	$\leftarrow \frac{T_4}{4} = A_4$	
y_6	f			
y_7	g			

Example 13. *Compute the 4-yearly moving averages from the following data:*

Year:	*1991*	*1992*	*1993*	*1994*	*1995*	*1996*	*1997*	*1998*
Annual Sales: (Rs. in crores)	*36*	*43*	*43*	*34*	*44*	*54*	*34*	*24*

Solution.

Calculation of 4-yearly Moving Averages

Year (1)	*Annual sales (Rs. in crores)* (2)	*4-yearly moving totals (T)* (3)	*4-yearly moving average (A) (column 3 ÷ 4)* (4)	*4-yearly centred moving average Trend values)* $\left(\frac{A}{4}\right)$ (5)
1991	36	—	—	—
1992	43	—	—	—
	—	← 156	← 39	
1993	43	—	—	$\frac{39+41}{2}=\frac{80}{2}=\mathbf{40}$
	—	← 164	← 41	
1994	34	—	—	$\frac{41+43.75}{2}=\frac{84.75}{2}=\mathbf{42.375}$
	—	← 175	← 43.75	
1995	44	—	—	$\frac{43.75+41.50}{2}=\frac{85.25}{2}=\mathbf{42.625}$
	—	← 166	← 41.50	
1996	54	—	—	$\frac{41.50+39}{2}=\frac{80.50}{2}=\mathbf{40.25}$
	—	← 156	← 39	
1997	34	—	—	—
1998	24	—	—	—

CALCULATIONS:

Here 36 + 43 + 43 + 34 = **156**; 43 + 43 + 34 + 44 = **164**;
43 + 34 + 44 + 54 = **165**; 34 + 44 + 54 + 34 = **166**;
44 + 54 + 34 + 24 = **156.**

Example 14. *The following table shows the average monthly production of coal in millions of tonnes for the year 1987-1996.*

Year	*Average monthly production of coal (In million tonnes)*	*Year*	*Average monthly production of coal (In million tonnes)*
1987	*50.0*	*1992*	*38.1*
1988	*36.5*	*1993*	*32.6*
1989	*43.0*	*1994*	*41.7*
1990	*44.5*	*1995*	*41.1*
1991	*38.9*	*1996*	*33.8*

Determine: (i) 4-yearly moving average figures; (ii) 4-yearly centered moving average. Also write the trend values.

Solution.

Table: Calculation of 4-yearly Moving Average and 4-yearly Centered Averages.

Year (*1*)	*Average Monthly Production of Coal (in Million Tonnes)* (2)	*4-yearly Moving Totals* (*3*)	*4-yearly Moving Average* (*4*)	*4-yearly Moving Average Centered* (*5*)	*Rounded off Average Centered* (*6*)
1987	50.0	—	—	—	—
1988	36.5	—	—	—	—
	—	← 174.0	← **43.5**		
1989	43.0	←	←	← **42.10**	**42**
	—	← 162.9	**40.7**		
1990	44.5	←	←	← **40.90**	**41**
	—	← 164.5	**41.1**		
1991	38.9	←	←	← **39.80**	**40**
	—	154.1	**38.5**		
1992	38.1	←	←	← **38.15**	**38**
	—	← 151.3	**37.8**		
1993	32.6	←	←	← **38.14**	**38**
	—	← 153.5	**38.4**		
1994	41.7	←	←	← **37.85**	**38**
	—	← 149.2	**37.3**		
1995	41.1	—	—	—	—
1996	33.8	—	—	—	—

The trend values are given by column (6), *viz.*, 42, 41, 40, 38, 38, 38.

Example 15. *Construct by two methods the four year centred moving averages of the following data. Show that the two methods give the same 4 year centred moving averages.*

Year:	*1987*	*1988*	*1989*	*1990*	*1991*	*1992*	*1993*	*1994*
Annual Sales (Rs. '0000):	*3.6*	*4.3*	*4.3*	*3.4*	*4.4*	*5.4*	*3.4*	*2.4*

Solution. FIRST METHOD

Table 1: 4-year Centered Moving Averages

Year (*1*)	*Data Annual sales (in Rs. '0000)* (2)	*4-year moving totals* (3)	*4-year moving Averages* (4)	*2-year moving totals of col. 4 (centered)* (5)	*4-year centered moving Averages (Col. 5) + 2*
1987	3.6	—	—	—	—
1988	4.3	—	—	—	—
	—	← 15.6	3.9		
1989	4.3			← 8.0	**4.0**
	—	← 16.4	4.1		
1980	3.4			← 8.5	**4.2**
	—	← 17.5	4.4		
1991	4.4			← 8.6	**4.3**
	—	← 16.6	4.2	—	**—**
1992	5.4	—	—	8.1	4.0
	—	← 15.6	3.9	—	—
1993	3.4	—	—	—	—
1994	—	—	—	—	—

SECOND METHOD

Table 2: 4-year Centered Moving Averages

Year *(1)*	*Annual Sales (Rs. '0000)* *(2)*	*4-year moving totals* *(3)*	*2-point moving total of col. 3 (Centered)* *(4)*	*4-year centered moving averages (Rs. '0000) (Col 4) ÷ 8*
1987	3.6	—	—	—
1988	4.3	—	—	—
	—	← 15.6	—	—
1989	4.3	—	← 32.0	**4.00**
	—	← 16.4	—	—
1990	3.4	—	← 33.9	**4.24**
	—	← 17.5	—	—
1991	4.4	—	← 34.1	**4.26**
	—	← 16.6	—	—
1992	5.4	—	← 32.2	**4.03**
	—	← 15.6	—	—
1993	3.4	—	—	—
1994	2.4	—	—	—

Explanation. In the above table 4-year moving totals are shown against the mid-points of the time intervals in **col. 3.** As the moving totals do not correspond to the given years, 2-point moving total of **col. 3** are found in **col. 4**, for centering them (*i.e.*, for syncorinising them with the original data).

The last column of table 1 and table 2 are identical. Hence the two methods give the same 4-yearly centered moving averages.

Example 16. *From the following data, find 3-yearly moving averages.*

Year :	*1977*	*1978*	*1979*	*1980*	*1981*	*1982*	*1983*
Values Rs. (in Lakhs):	*20.0*	*22.8*	*24.3*	*18.8*	*28.3*	*30.4*	*35.8*
Year :	*1984*	*1985*	*1986*	*1987*	*1988*	*1989*	*1990*
Values Rs. (in Lakhs):	*31.6*	*35.7*	*47.2*	*55.0*	*63.2*	*66.3*	*57.7.*

Solution. **Table for Computation of 3-yearly Moving Averages.**

Year	*Value Rs. (in lakhs)*	*3-yearly moving totals*	*3-yearly moving averages*
1977	20.0	—	—
1978	22.8	67.1	22.37
1979	24.3	65.9	21.97
1980	18.8	71.4	23.80
1981	28.3	77.5	25.83
1982	30.4	94.5	31.50
1983	35.8	97.8	32.60
1984	31.6	103.1	34.37
1985	35.7	114.5	38.17
1986	47.2	137.9	45.97
1987	55.0	165.4	55.13
1988	63.2	184.5	61.50
1989	66.3	187.2	62.40
1990	57.7	—	—

Example 17. *For the data given above in* **Example 16,** *calculate (a) 4 yearly moving average, (b) 4 yearly moving average centered*

Solution. **Calculation Table**

Year	*Value*	*4-yearly moving totals*	*4-yearly moving averages*	*4-yearly moving average total centered*	*4-yearly moving average centered*
1977	20.0	—	—	—	—
1978	22.8	—	—	—	—
—	—	**85.9**	**21.48**	—	—
1979	24.8	—	—	← 45.03	**22.52**
—	—	**94.2**	**23.55**	—	—
1980	18.3	—	—	← 49.00	**24.50**
—	—	**101.8**	**25.45**	—	—
1981	28.3	—	—	← 53.78	**26.89**
—	—	**113.3**	**28.33**	—	—
1982	30.4	—	—	← 59.86	**29.93**
—	—	**126.1**	**31.53**	—	—
1983	35.8	—	—	← 64.91	**32.46**
—	—	**133.5**	**33.38**	—	—
1984	31.6	—	—	← 70.96	**35.48**
—	—	**150.3**	**37.58**	—	—
1985	35.7	—	—	← 79.96	**34.98**
—	—	**169.5**	**42.38**	—	—
1986	47.2	—	—	← 92.66	**46.33**
—	—	**201.1**	**50.28**	—	—
1987	55.0	—	—	← 108.21	**54.11**
—	—	**231.7**	**57.93**	—	—
1988	63.2	—	—	← 118.48	**59.24**
—	—	**242.2**	**60.55**	—	—
1989	66.3	—	—	—	—
1990	57.7	—	—	—	—

Example 18. *From the following data calculate the 4-yearly moving average and determine he trend values.*

Year	*Value*	*Year*	*Value*
1968	*50.0*	*1973*	*38.1*
1969	*36.5*	*1974*	*32.6*
1970	*43.0*	*1975*	*41.7*
1971	*44.5*	*1976*	*41.1*
1972	*38.9*	*1977*	*33.8*

Solution.

Table for Computation of 4-yearly Moving Averages

Year (*1*)	*Value* (2)	*4-yearly Moving Totals* (3)	*4-yearly Moving Average (column 3 ÷ 4)* (4)	*4-yearly Moving Average Total Centered* (5)	*4-yearly Moving Average Centered*
1968	50.0	—	—	—	—
1969	36.5	—	—	—	—
		174.0	43.50 (174 ÷ 4)		
1970	43.0			**84.23**	$42.1 \left[= \frac{84.23}{2}\right]$
		162.9	40.73		
1971	44.5			**81.86**	**40.9**
		164.5	41.13		
1972	38.9			**79.66**	**39.8**
		154.1	38.53		
1973	38.1			**76.36**	**38.2**
		151.3	37.83		
1974	32.6			**76.21**	**38.1**
		153.5	38.38		
1975	41.7			**75.68**	**37.8**
		149.2	37.30		
1976	41.1	—	—	—	—
1977	33.8	—	—	—	—

.15 WEIGHTED MOVING AVERAGE

The weighted moving average is obtained on dividing the weighted moving totals by the sum weights. Let $x_1, x_2, x_3, ..., x_n$ occur with weights $w_1, w_2, w_3, ..., w_n$, then

$$\text{Weighted Moving Average} = \frac{\Sigma w_i x_i}{\Sigma w_i}.$$

A weighted moving average with appropriate weight is, generally, used when the moving erages are strongly affected by extreme values.

The method is illustrated by the following examples:

Example 19. *Find the trend for the following series using a three-year weighted moving average with weights 1, 2, 1:*

Year:	*1991*	*1992*	*1993*	*1994*	*1995*	*1996*	*1997*
Values:	2	4	5	7	8	10	13

Solution. Sum of weights = 1 + 2 + 1 = 4

Table: 3-year Weighted Moving Averages

Year (*1*)	*Values* (*2*)	*3-year weighted moving total* (*3*)	*3-year weighted moving average* (*Col. 3*) ÷ *Total weight* (*4*)
1991	2	—	—
1992	4	2 × 1 + 4 × 2 + 5 × 1 = **15**	15/4 = **3.75**
1993	5	4 × 1 + 5 × 2 + 7 × 1 = **21**	21/4 = **5.25**
1994	7	5 × 1 + 7 × 2 + 8 × 1 = **27**	27/4 = **6.75**
1995	8	7 × 1 + 8 × 2 + 10 × 1 = **33**	33/4 = **8.25**
1996	10	8 × 1 + 10 × 2 + 13 × 1 = **41**	41/4 = **10.25**
1997	13	—	—

Explanation. Col. 4 = Col. 3 ÷ total weight, where total weight = 4.

Example 20. *Given the numbers 2, 6, 1, 5, 3, 7, 2; Write down the weighted moving average of period 3, the weights being 1, 4, 1.*

Solution.

Sum of Weights = 1 + 4 + 1 = 6

Table: 3-year Weighted Moving Averages

Values (*X*) (*1*)	*Weighted moving totals* (*2*)	*Weighted moving average Col.* (*2*) ÷ *6*
2	—	—
6	1 × 2 + 6 × 4 + 1 × 1 = **27**	27 ÷ 6 = **4.5**
1	1 × 6 + 4 × 1 + 1 × 5 = **15**	15 ÷ 6 = **2.5**
5	1 × 1 + 4 × 5 + 1 × 3 = **24**	24 ÷ 6 = **4**
3	1 × 5 + 4 × 3 + 1 × 7 = **24**	24 ÷ 6 = **4**
7	1 × 3 + 4 × 7 + 1 × 2 = **33**	33 ÷ 6 = **5.5**
2	—	—

Example 21. *Calculate 5-year weighted moving averages for the following data, using the weights 1, 1, 3, 2, 1 respectively:*

Year :	*1*	*2*	*3*	*4*	*5*
Coded sales :	*40*	*33*	*72*	*81*	*76*

Solution. Sum of weights = 1 + 1 + 3 + 2 + 1 = 8

Table: 5-year Weighted Moving Averages

Year (*1*)	*Sales* (2)	*5-year Weighted Moving Totals* (*3*)	*5-year weighted moving averages* [*column* (*3*) ÷ (*total weights*)]
1	40		
2	33		
3	72	(40 × 1 + 33 × 1 + 72 × 3 + 81 × 2 + 76 × 1) = **527**	527 ÷ 8 = **65.88**
4	81	(33 × 1 + 72 × 1 + 81 × 3 + 76 × 2 + 68 × 1) = **568**	568 ÷ 8 = **71.00**
5	76	(72 × 1 + 81 × 1 + 76 × 3 + 68 × 2 + 91 × 1) = **630**	630 ÷ 8 = **78.75**
6	68	(81 × 1 + 76 × 1 + 68 × 3 + 91 × 2 + 87 × 1) = **630**	630 ÷ 8 = **78.75**
7	91	(76 × 1 + 68 × 1 + 91 × 3 + 87 × 2 + 91 × 1) = **689**	689 ÷ 8 = **86.125**
8	87	(68 × 1 + 91 + 1 + 87 × 3 + 98 × 2 + 97 × 1) = **713**	713 ÷ 8 = **89.125**
9	98		
10	97		

Example 22. *For the following data, verify that the 5 year weighted moving average with weights 1, 2, 3, 3, 1 respectively is equivalent to 4 year centred moving average.*

Year	*1979*	*1980*	*1981*	*1982*	*1983*	*1984*	*1985*	*1986*	*1987*	*1988*	*1989*
Sales Rs. in lakhs	*5*	*3*	*7*	*6*	*4*	*8*	*9*	*10*	*8*	*9*	*9*

Solution. Total weights = **1 + 2 + 3 + 3 + 1 = 8.**

Table 1: 5-Year Weighted Moving Averages

Year (*1*)	*Annual sale* (2)	*5-year weighted moving totals* (*3*)	*5-year weighted moving average* [*Column* (*3*) ÷ *8*]
1979	5		
1980	3		
1981	7	5 × 1 + 3 × 2 + 7 × 2 + 6 × 2 + 4 × 1 = **41**	**5.125**
1982	6	3 × 1 + 7 × 2 + 6 × 2 + 4 × 2 + 8 × 1 = **45**	**5.625**
1983	4	7 × 1 + 6 × 2 + 4 × 2 + 8 × 2 + 9 × 1 = **52**	**6.5**
1984	8	6 × 1 + 4 × 2 + 8 × 2 + 9 × 2 + 10 × 1 = **58**	**7.25**
1985	9	4 × 1 + 8 × 2 + 9 × 2 + 10 × 2 + 8 × 1 = **66**	**8.25**
1986	10	8 × 1 + 9 × 2 + 10 × 2 + 8 × 2 + 9 × 1 = **71**	**8.875**
1987	8	9 × 1 + 10 × 2 + 8 × 2 + 9 × 2 + 9 × 1 = **72**	**9.00**
1988	9		
1989	9		

Table 2: 4 Year Centred Moving Averages

Year	*Annual Sale*	*4 year moving total*	*2-year moving total of col.*	*4 year centred moving average [col. 4 + 8]*
(1)	(2)	(3)	(4)	(5)
1979	5			
1980	3			
		21		
1981	7		**41**	**5.125**
		20		
1982	6		**45**	**5.625**
		25		
1983	4		**52**	**6.5**
		27		
1984	8		**58**	**7.25**
		31		
1985	9		**66**	**8.25**
		35		
1986	10		**71**	**8.875**
		36		
1987	8		**72**	**9.00**
		36		
1988	9			
1989	9			

From the table 1 and 2, we notice that the column 5 in both the tables is same.

Hence 5 year weighted average = 4 year Centred Moving Average.

Example 23. *For the following series of observations, verify that the 4-year centered moving average is equivalent to a 5-year weighted moving average with weights 1, 2, 2, 2, 1 respectively:*

Year:	*1984*	*1985*	*1986*	*1987*	*1988*	*1989*	*1990*	*1991*	*1992*	*1993*	*1994*
Sales (Rs. '00000):	*2*	*6*	*1*	*5*	*3*	*7*	*2*	*6*	*4*	*8*	*3*

Solution. Sum of the weights = 1 + 2 + 2 + 2 + 1 = **8**

TABLE 1 : 5-year Weighted Moving Averages

Year	*Annual Sale (Rs '00000)*	*5-year weighted moving total*	*5-year weighted moving average (Col. 3 + total weight)*
(1)	(2)	(3)	
1984	2	—	—
1985	6	—	—
1986	1	2 × **1** + 6 × **2** + 1 × **2** + 5 × **2** + 3 × **1** = 29	**29/8 = 3.625**
1987	5	6 × **1** + 1 × **2** + 5 × **2** + 3 × **2** + 7 × **1** = 31	**31/8 = 3.875**
1988	3	1 × **1** + 5 × **2** + 3 × **2** + 7 × **2** + 2 × **1** = 33	**33/8 = 4.125**
1989	7	5 × **1** + 3 × **2** + 7 × **2** + 2 × **2** + 6 × **1** = 35	**35/8 = 4.375**
1990	2	3 × **1** + 7 × **2** + 2 × **2** + 6 × **2** + 4 × **1** = 37	**37/8 = 4.625**
1991	6	7 × **1** + 2 × **2** + 6 × **2** + 4 × **2** + 8 × **1** = 39	**39/8 = 4.875**
1992	4	2 × **1** + 6 × **2** + 4 × **2** + 8 × **2** + 3 × **1** = 41	**41/8 = 5.125**
1993	8	—	—
1994	3	—	—

TABLE 2 : 4-year Centered Moving Averages

Year (*1*)	*Annual Sales (Rs. '00000)* (*2*)	*4-year moving total* (*3*)	*2-year moving total of Col. 3* (*4*)	*4-year centered moving average (Col. 4 ÷ 8)* (*5*)
1984	2	—	—	—
1985	6	—	—	—
		14		
1986	1		14 + 15 = **29**	**29/8 = 3.625**
		15		
1987	5		15 + 16 = **31**	**31/8 = 3.875**
		16		
1988	3		16 + 17 = **33**	**33/8 = 4.125**
		17		
1989	2		17 + 18 = **35**	**35/8 = 4.375**
		18		
1990	2		18 + 19 = **37**	**37/8 = 4.625**
		19		
1991	6		19 + 20 = **39**	**39/8 = 4.875**
		20		
1992	4		20 + 21 = **41**	**41/8 = 5.125**
		21		
1993	8	—	—	—
1994	3	—	—	—

From the last columns of the table 1 and table 2, we notice that the 4-year centered moving average is equivalent to a 5-year weighted moving average with weights 1, 2, 2, 2, 1 respectively.

LINEAR TREND (BY THE METHOD OF LEAST-SQUARES)

14.16 METHOD OF LEAST SQUARES

This is the best method for obtaining the trend values. It provides a convenient basis for obtaining the line of best fit in a series. **Line of the best fit is a line from which the sum of the deviations of various points on either side is zero. Further the sum of the squares of these deviations would be the least as compared to the sum of squares of the deviations obtained by using other lines.** We know **that the sum of the deviations from the arithmetic mean is zero, therefore, the sum of the deviations from the line of the best fit is zero. For this reason the sum of the squares of the deviations of various points from the line of the Best Fit is the least.** That is why this method is known as the **Method of Least Squares. This method satisfies the following conditions.**

(*i*) **$\Sigma (Y - Y_C) = 0$, *i.e.*, the sum of deviations of the actual values of Y and computed values of Y is zero.**

(*ii*) **$\Sigma (Y - Y_C)^2$ is least, *i.e.*, the sum of the squares of deviations from the actual and the computed value of Y is least.**

The straight line trend has an equation of the type:

$Y = a + bX$, where Y represents the estimated values of the trend, X represents the deviations in time period; 'a' and 'b' are constants.

The values of two constants 'a' and 'b' are estimated by solving the following two normal equations.

$$\Sigma Y = Na + b\,\Sigma X$$
$$\Sigma XY = a\,\Sigma X + b\,\Sigma X^2,$$

where N represents number of years for which data is given.

The variable X can be measured from any point of time as origin. To make calculation simpler, it is better to take the mid-point of time as the origin because the negative values of first half of the time series will equalise the positive values in the second half of the series which symbolically gives $\Sigma X = 0$. **When $\Sigma X = 0$, the two normal equations for finding the constants 'a' and 'b' will be**

$$\Sigma Y = Na \Rightarrow a = \frac{\Sigma Y}{N} = \overline{Y}$$
$$\Sigma XY = b\,\Sigma X^2 \Rightarrow b = \frac{\Sigma XY}{\Sigma X^2}.$$

This provides that **the constant 'a' is simply equal to the mean of Y values and the constant 'b' gives the rate of change. The constant 'a' refers to the Y-intercept,** *i.e.*, the **difference between the point of origin and the point where the trend line touches the Y-axis. The constant 'b' refers to the slope of the line which indicates the change in Y for each unit change in X.**

MERITS AND LIMITATIONS OF THE LEAST SQUARES METHOD

Merits:

(i) ***This method gives the trend values for the entire time period.***

(ii) ***This method can be used to forecast future trend because trend line establishes a functional relationship between the value and the time.***

(iii) ***This method is a completely objective method.***

Limitations:

(i) ***It requires many calculations and is tedious and complicated.***

(ii) ***Future forecasts made by this method are based only on trend values, seasonal, cyclical or irregular variations are ignored.***

(iii) ***If even a single item is added to the series a new equation has to be formed.***

14.17 FITTING A STRAIGHT LINE TREND FOR ODD NUMBER OF YEARS.

Whe the number of years is odd, we have the following algorithm

ALGORITM:

Step 1. Denote time as the X variable and values as Y.

Step 2. Middle year is assumed as the period of origin and find out deviations.

Step 3. Square the time deviations and find X^2.

Step 4. Multiply the given value of Y by the respective deviation of X and find the total ΣXY.

Step 5. Find out the values of Y; get ΣY.

Step 6. The values so obtained are placed in the two equations

(1) $\Sigma Y = Na + b\Sigma X$. (2) $\Sigma XY = a\Sigma X + b\Sigma X^2$; find out the value of a and b.

Step 7. The calculated values of *a* and *b* are substituted and the trend value of Y_C are found for various values of *X*, where

$$Y_C = a + bX$$

Example 24. *Determine the equation of a straight line which best fits the following data:*

Year :	*1984*	*1985*	*1986*	*1987*	*1988*
Sales (in Rs. '000) :	*35*	*56*	*79*	*80*	*40*

Compute the trend values for all the years from 1984 to 1988.

Solution. Let the equation of the **straight line of best fit,** with the origin at the middle year 1986 and unit of *X* as 1 year, be

$$Y = a + bX. \quad \text{... (1)}$$

By the **Method of Least Squares,** the values of *a* and *b* are given by

$$a = \frac{\Sigma Y}{N} \text{ and } b = \frac{\Sigma XY}{\Sigma X^2} \quad \text{... (2)}$$

Here N = **number of years = 5.**

Table: Calculations for the Line of Best Fit

Year	*Sales (Rs. '000) Y*	*X*	X^2	*XY*
1984	35	– 2	4	– 70
1985	56	– 1	1	– 56
1986	79	0	0	0
1987	80	1	1	80
1988	40	2	4	80
Total	$\Sigma Y = 290$	**0**	$\Sigma X^2 = 10$	$\Sigma XY = 34$

Using (2), $a = \frac{\Sigma Y}{N} = \frac{290}{5} = 58$, and $b = \frac{\Sigma XY}{\Sigma X} = \frac{34}{10} = 3.4$

Substituting these values in (1), we get, the required equation of the best fitted straight line as:

$$Y = 58 + 3.4\, X., \text{ with } 1986 = 0$$

Table: Trend Values

Year	*X*	*Trend Values* ($Y = 58 + 3.4X$)
1984	– 2	58 + 3.4 × – (2) = **51.2**
1985	– 1	58 + 3.4 × – (1) = **54.6**
1986	0	58 + 3.4 × 0 = **58.0**
1987	1	58 + 3.4 × 1 = **61.4**
1988	2	58 + 3.4 × 2 = **64.8**

Example 25. *The annual production of a commodity is given as follows:*

Year:	*1990*	*1991*	*1992*	*1993*	*1994*	*1995*	*1996*
Production (in tonnes):	*70*	*80*	*90*	*95*	*102*	*110*	*115*

Fit a straight line trend by the method of least squares.

Solution.

Table: Fitting Straight line Trend

Year	*Production (Y)*	*Deviation from 1993 (X)*	*XY*	X^2
1990	70	– 3	– 210	9
1991	80	– 2	– 160	4
1992	90	– 1	– 90	1
1993	95	0	0	0
1994	102	+ 1	+ 102	1
1995	110	+ 2	+ 220	4
1996	115	+ 3	+ 345	9
$N = 7$	$\Sigma Y = 662$	$\Sigma X = 0$	$\Sigma XY = 207$	$\Sigma X^2 = 28$

Here $\Sigma X = 0$, $a = \dfrac{\Sigma Y}{N} = \dfrac{662}{7} = 94.6$

and $b = \dfrac{\Sigma XY}{\Sigma X^2} = \dfrac{207}{28} = 7.39.$

The **equation of the straight line trend is:** $Y = a + bX \Rightarrow Y = 94.6 + 7.39X.$

Hence, $Y = 94.6 + 7.39X$ is the required trend line.

Example 26. *Using 1991 as origin, obtain a linear trend equation by the method of least squares.*

Year :	*1987*	*1989*	*1990*	*1991*	*1992*	*1993*	*1994*
Value :	*140*	*144*	*160*	*152*	*168*	*176*	*180*

Find the trend value for the missing year 1988.

Solution. Let the **linear trend** be given by the equation

$Y = a + bX$, with origin July 1991 and unit = 1 year.

Table: Calculation For Fitting Linear Trend

Year	*Value*	*X = x – 1991*	X^2	*XY*
1987	140	– 4	16	– 560
1989	144	– 2	4	– 28
1990	160	– 1	1	– 160
1991	152	0	0	0
1992	168	1	1	168
1993	176	2	4	352
1994	180	5	25	90
Total	$\Sigma Y = 1120$	$\Sigma X = 1$	$\Sigma X^2 = 51$	$\Sigma XY = 412$

Hence $N = 7.$

The normal equations are

$\Sigma Y = Na + b\Sigma X$ ... (1); $\Sigma XY = a\Sigma X + b\Sigma X^2$ ... (2)

Substituting the values in (1) and (2), we get:

$1120 = 7a + b$ ... **(3)**; $412 = a + 51b.$ ... **(4)**

Solving (3) and (4), we get: $\boldsymbol{a = 159.29}$, $\boldsymbol{b = 4.96}$.

$\therefore$ The **linear trend equation** is: $\boldsymbol{Y = 159.29 + 4.96X}$.

with origin: July 1991 and Unit = 1 year.

Estimation For 1988: Here $X = 1988 - 1991 = -3$.

$\therefore \; Y_{1988} = 159.29 + 4.96 \times (-3) = \mathbf{144.41}$.

Example 27. *Given below are the figures of production (in lakh kg.) of a sugar factory*

Year :	*1971*	*1972*	*1973*	*1974*	*1975*	*1976*	*1977*
Production :	*40*	*45*	*46*	*42*	*47*	*50*	*46.*

Fit a straight line trend by the least squares method and tabulate the trend.

Solution.

Table for Computation of Trend Values by Least Squares Method

Year *t*	*Production in lakh kg.* *Y*	*t – 1974* *X*	X^2	*XY*	*Trend Values* $Y_C = 45.143 + 1.036X$
1971	40	– 3	9	– 120	**42.035**
1972	45	– 2	4	– 90	**43.071**
1973	46	– 1	1	– 46	**44.107**
1974	42	0	0	0	**45.143**
1975	47	+ 1	1	+ 47	**46.179**
1976	50	+ 2	4	+ 100	**47.215**
1977	46	+ 3	9	+ 138	**48.250**
N = 7	$\boldsymbol{\Sigma Y = 316}$	$\boldsymbol{\Sigma X = 0}$	$\boldsymbol{\Sigma X^2 = 28}$	$\boldsymbol{\Sigma XY = +29}$	$\boldsymbol{\Sigma Y_C = 316}$

$$a = \frac{\Sigma Y}{N} = \frac{316}{7} = \mathbf{45.143}; \; b = \frac{\Sigma XY}{\Sigma X^2} = \frac{29}{28} = \mathbf{1.036}.$$

$$\therefore \qquad Y_C = a + bX.$$

$$\Rightarrow \qquad Y_C = 45.143 + 1.036\,X.$$

When $X = -3$. $Y_C = 45.143 + [1.036 \times (-3)] = 42.035$.

Now we can add 1.036 to different values of Y_C for $X = -2, -1, 0, +1, +2, +3$ successively to get the trend values. Thus when $X = -2$, $Y_C = 42.035 + 1.036 = 43.071$. Similarly other values can be obtained.

Example 28. *Obtain the straight line trend equation and tabulate against each year after estimation of the trend and short-term fluctuations. Also draw the graph.*

Year	*1977*	*1978*	*1979*	*1980*	*1981*	*1982*	*1983*	*1984*	*1985*
Value	*380*	*400*	*650*	*720*	*690*	*620*	*670*	*950*	*1040*

Solution.

Table for Computation

Year	*Value* Y	*Deviation from 1981* X	X^2	XY	*Trend Value*	*Short-term Fluctuations* $(Y - Y_C)$
1977	380	– 4	16	– 1520	398.0	– 18.0
1978	400	– 3	9	– 1200	468.5	– 68.5
1979	650	– 2	4	– 1300	539.0	+ 111.0
1980	720	– 1	1	– 720	609.5	+ 110.5
1981	690	0	0	0	680.0	+ 10.0
1982	620	+ 1	1	+ 620	750.5	– 130.5
1983	670	+ 2	4	+ 1340	821.0	– 151.0
1984	950	+ 3	9	+ 2850	891.5	+ 58.5
1985	1040	+ 4	16	+ 4160	962.0	+ 78.0
$N = 9$	$\Sigma Y = 6120$	$\Sigma X = 0$	$\Sigma X^2 = 60$	$\Sigma XY = 4230$		

Now $Y_C = a + bX$.

$$a = \frac{\Sigma Y}{N} = \frac{6120}{9} = \mathbf{680}; \qquad b = \frac{\Sigma XY}{\Sigma X^2} = \frac{4230}{60} = \mathbf{70.5.}$$

Trend equation : $Y_C = 680 + 70.5X = 680 + 70.5X.$

$$\therefore \qquad Y_{1977} = 680 + (70.5)(-4) = 680 - 282 = \mathbf{398.}$$

Similarly, other trend values can be obtained by putting $X = -3, -2, -1, 0, 1, 2, 3, 4$.

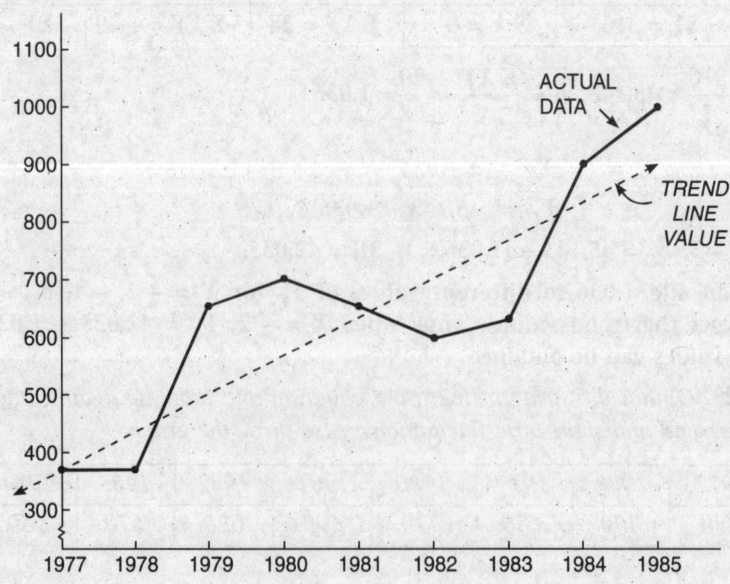

14.18 FITTING A STRAIGHT LINE TREND FOR EVEN NUMBER OF YEARS

The method of least-squares for even number of years is almost the same as that for odd years for computing the trend values.

If the number of years is even, there is no middle year and in this case the mid-point, which is taken as the origin, lies midway between the two middle years. In the following example, the mid-point (*i.e.*, the origin) lies midway between July 1, 1984 and July 1, 1985, which is January 1, 1985 (or December 31, 1984). To avoid fractions, the units of X are taken as $\frac{1}{2}$ year (or 6 months).

Example 29. *Fit a straight line trend equation by the method of least squares and estimate the trend values.*

Year:	*1981*	*1982*	*1983*	*1984*	*1985*	*1986*	*1987*	*1988*
Values :	*80*	*90*	*92*	*83*	*94*	*99*	*92*	*104*

Solution. Here N = Number of years = 8 which is even.

Let the straight line trend equation by the method of least squares with the origin at the mid-point of 1984 and 1985, and unit of $X = \frac{1}{2}$ year or half year be

$$Y = a + bX. \quad \text{... (1)}$$

Then a and b are given by the normal equations:

$$a = \frac{\Sigma Y}{N} \text{ and } b = \frac{\Sigma XY}{\Sigma X^2}. \quad \text{... (2)}$$

The midpoint = Jan 1, 1985 or 31st December 1984.

Table 7. Calculations for Fitting the Straight Line Trend

Year	*Values Y*	*X =* *(t — mid-point of X)*	X^2	*XY*
1981	80	– 7	49	– 560
1982	90	– 5	25	– 450
1983	92	– 3	9	– 276
1984	83	– 1	1	– 83
→				
1985	94	1	1	94
1986	99	3	9	297
1987	92	5	25	460
1988	104	7	49	728
Total	$\Sigma Y = 734$	**0**	$\Sigma X^2 = 168$	$\Sigma XY = 210$

Using (2), $a = \frac{\Sigma Y}{N} = \frac{734}{8} = 91.75$, and $b = \frac{\Sigma XY}{\Sigma X^2} = \frac{210}{168} = 1.25$.

∴ From (1), **the required equation of the straight line trend is:**

$$Y = 91.75 + 1.25X. \qquad \text{Unit: } \frac{1}{2} \text{ year}$$

Table: Trend Values

Year	*X*	*Trend Values* ($Y = 91.75 + 1.25X$)
1981	– 7	$91.75 + 1.25 \times (-7)$ = **83.0**
1982	– 5	$91.75 + 1.25 \times (-5)$ = **85.5**
1983	– 3	$91.75 + 1.25 \times (-3)$ = **88.0**
1984	– 1	$91.75 + 1.25 \times (-1)$ = **90.5**
1985	1	$91.75 + 1.25 \times 1$ = **93.0**
1986	3	$91.75 + 1.25 \times 3$ = **95.5**
1987	5	$91.75 + 1.25 \times 5$ = **98.0**
1988	7	$91.75 + 1.25 \times 7$ = **100.5**

Example 30. *Fit a straight line trend by the method of least squares to the following data. Assuming that the same rate of change continues what would be the predicted earnings for the year 1980?*

Year :	*1981*	*1982*	*1983*	*1984*	*1985*	*1986*	*1987*	*1988*
Earnings : (*Rs. lakhs*)	*38*	*40*	*65*	*72*	*69*	*60*	*87*	*95*

Solution. Let the oring = 1st July 1984

and unit = $\frac{1}{2}$ year.

Table for Computation of Straight Line Trend by the Method of Least Squares

Year (*t*)	*Earnings Rs. Lakhs* (*Y*)	*Deviations* (*t – 1984*)	*Deviations multiplie by 2.* (*X*)	*XY*	*Y²*
1981	38	– 3.5	– 7	– 266	49
1982	40	– 2.5	– 5	– 200	25
1983	65	– 1.5	– 3	– 195	9
1984	72	– 0.5	– 1	– 72	1
1985	69	+ 0.5	+ 1	+ 69	1
1986	60	+ 1.5	+ 3	+ 180	9
1987	87	+ 2.5	+ 5	+ 435	25
1988	95	+ 3.5	+ 7	+ 665	49
***N* = 8**	**$\Sigma Y = 526$**		**$\Sigma X = 0$**	**$\Sigma XY = 616$**	**$\Sigma X^2 = 168$**

$$a = \frac{\Sigma Y}{N} = \frac{526}{8} = \mathbf{65.75}; \qquad b = \frac{\Sigma XY}{\Sigma X^2} = \frac{616}{168} = \mathbf{3.667.}$$

$$Y_C = a + bX \Rightarrow Y_C = 65.75 + 3.667X.$$

For 1989, *X* will be + 9, unit = $\frac{1}{2}$ year

$$Y_C = 165.75 + 3.667X = 65.75 + 3.667\ (9)$$

$$\Rightarrow \qquad Y_C = 65.75 + 33.003 = \mathbf{98.753.}$$

The estimated earning for **the 1989 will be Rs. 98.753 lakhs.**

Example 31. *From the data given below fit a straight line trend by the method of least squares.*

Year	*1975*	*1976*	*1977*	*1978*	*1979*	*1980*	*1981*	*1982*
Sales (in lakhs)	*6.7*	*5.3*	*4.3*	*6.1*	*5.6*	*7.9*	*5.8*	*6.1*

[M.B.A; Anna University – Dec. 2002]

Solution. Let the origin = 1 st July 1978; and unit = half year.

Equation of a straight line $Y_C = a + bX$.

$$a = \frac{\Sigma Y}{N} = \frac{47.8}{8} = 5.975; \qquad b = \frac{\Sigma XY}{\Sigma X^2} = \frac{8.6}{168} = .05119.$$

Table for Computation of Straight Line Trend by The Method of Least Squares

Year (t)	*Sales (in lakhs) (Y)*	*Deviation (t-1978.5)*	*Deviation multiplied by 2 (X)*	*XY*	X^2	*Trend values* $Y_C = 5.975 + 0.05119X$
1975	6.7	– 3.5	– 7	– 46.9	49	5.61667
1976	5.3	– 2.5	– 5	– 26.5	25	5.71905
1977	4.3	– 1.5	– 3	– 12.9	9	5.82143
1978	6.1	– 0.5	– 1	– 6.1	1	5.92381
1979	5.6	+ 0.5	+ 1	+ 5.6	1	6.02619
1980	7.9	+ 1.5	+ 3	+ 23.7	9	6.12857
1981	5.8	+ 2.5	+ 5	+ 29.0	25	6.23095
1982	6.1	+ 3.5	+ 7	+ 42.7	49	6.3333
N = 8	**ΣY = 47.8**		**Σ X = 0**	**Σ XY = 8.6**	**Σ X² = 168**	

The equation of straight line would be Y_C **= 5.975 + .05119*X*; Unit =** $\frac{1}{2}$ **year.**

Thus for 1975 when $X = -7$, $Y_C = 5.975 + (-7 \times .05119)$
$= 5.975 - .35833 = \mathbf{5.61667}$.

Similarly, **for other trend values calculations can be made.**

Example 32. *Using 1974 as origin obtain a straight line trend equation by the method of least squares.*

Year	*1970*	*1972*	*1973*	*1974*	*1975*	*1976*	*1979*
Value	*140*	*144*	*160*	*152*	*168*	*176*	*180*

Find the Trend Value of missing year 1971.

Solution. Let us take 1974 as origin with unit = 1 year.

Table For Calculations

Year	*Value (Y)*	*Deviations from Year 1974 (X)*	*XY*	X^2
1970	140	– 4	– 560	16
1972	144	– 2	– 288	4
1973	160	– 1	– 160	1
1974	152	0	0	0
1975	168	+ 1	+ 168	1
1976	176	+ 2	+ 352	4
1979	180	+ 5	+ 900	25
N = 7	**ΣY = 1120**	**Σ X = 1**	**Σ XY = 412**	**Σ X² = 51**

The equation of the straight line trend is:

$$Y = a + bX.$$

For getting the value of the **constants 'a' and 'b'**, we have to solve the following two equations.

$$\Sigma Y = Na + b\,\Sigma X$$
$$\Sigma XY = a\,\Sigma X + b\,\Sigma X^2.$$

[If the $\Sigma X = 0$ then the value of 'a' and 'b' can be found out [$a = \dfrac{\Sigma Y}{N}$; $b = \dfrac{\Sigma XY}{\Sigma X^2}$.]

Since in this problem the $\Sigma X = 1$, we have to solve this question by solving these two equations to find the value of a and b

$$\Sigma Y = Na + b\,\Sigma X \quad \Rightarrow \quad 1120 = 7a + b \quad \ldots (1)$$
$$\Sigma XY = a\,\Sigma X + b\,\Sigma X^2 \quad \Rightarrow \quad 412 = a + 51b. \quad \ldots (2)$$

On solving the equations (1) and (2), we get : $\boldsymbol{a = 159.29, b = 4.955.}$

Putting these values **in equation of straight line trend,** we get

$$Y = a + bX = \mathbf{159.29 + 4.955X.}$$

The **Trend value for 1971 can be calculated by putting $X = -3$.**

$$\therefore \quad Y_{1971} = 159.29 + 4.955\,(-3) = 159.29 - 14.865 = \mathbf{144.425.}$$

Example 33. *Fit a straight line trend by the method of least squares to the following data and also forecast the earnings for the year 1985.*

Year :	*1974*	*1975*	*1976*	*1977*	*1978*	*1979*	*1980*
Earning : (Rs. Lakhs)	*15*	*14*	*18*	*20*	*17*	*24*	*27.*

Solution. Let us take 1977 as origin with unit = 1 year.

Table for Computation of Straight line Trend by the Least Squares Method.

Year	***Earnings (Rs. lakhs) (Y)***	***Deviations from year 1977 (X)***	***XY***	***X^2***
1974	15	− 3	− 45	9
1975	14	− 2	− 28	4
1976	18	− 1	− 18	1
1977	20	0	0	0
1978	17	+ 1	+ 17	1
1979	24	+ 2	+ 48	4
1980	27	+ 3	+ 81	9
$N = 7$	$\Sigma Y = 135$	$\Sigma X = 0$	$\Sigma XY = 55$	$\Sigma X^2 = 28$

The **equation of straight line trend is : $Y = a + bX$.**

Since $\boldsymbol{\Sigma X = 0}$, therefore,

$$a = \frac{\Sigma Y}{N} = \frac{135}{7} = \mathbf{19.29}; \qquad b = \frac{\Sigma XY}{\Sigma X^2} = \frac{55}{28} = \mathbf{1.96.}$$

Thus **substituting the value of 'a' and 'b' in the straight line of the trend,** we get

$$Y = a + bX = \mathbf{19.29 + 1.96\,X.}$$

Estimate of 1985 will be calculated on the basis of $X = +8$.

$$Y_{1985} = 19.29 + (1.96)\,(8) = 19.29 + 15.68 = \mathbf{34.97.}$$

Example 34. *Fit a straight line trend by the method of least squares (taking 1985 as year of origin) to the following data concening sales of a certain firm.*

Year :	*1981*	*1982*	*1983*	*1984*	*1985*	*1986*	*1987*	*1988*	*1989*
Sales : *Rs. ('000)*	*48*	*55*	*63*	*65*	*72*	*84*	*90*	*87*	*82*

Make an estimate of sales for the year 1990 from the trend line.

Solution. The equation of the straight line trend is : $Y = a + bX$.

The normal equations to find the value of a and b are

$$\Sigma Y = Na + b\,\Sigma X$$

$$\Sigma XY = a\,\Sigma X + b\,\Sigma X^2$$

When $\Sigma X = 0$, then

$$a = \frac{\Sigma Y}{N}; \text{ and } b = \frac{\Sigma XY}{\Sigma X^2}.$$

Table for Computing of Straight Line Trend

Year	*Production of wheat (Y)*	*Deviation from 1982 (X)*	*XY*	X^2
1981	48	– 4	– 192	16
1982	55	– 3	– 165	9
1983	63	– 2	– 126	4
1984	65	– 1	– 65	1
1985	72	0	0	0
1986	84	+ 1	+ 84	1
1987	90	+ 2	+ 180	4
1988	87	+ 3	+ 261	9
1989	82	+ 4	+ 328	16
$N = 9$	**$\Sigma Y = 646$**	**$\Sigma X = 0$**	**$\Sigma XY = 305$**	**$\Sigma X^2 = 60$**

Substituting these values, we get

$$a = \frac{\Sigma Y}{N} = \frac{646}{9} = 71.7778; \qquad b = \frac{\Sigma XY}{\Sigma X^2} = \frac{305}{60} = 5.0833.$$

Therefore trend line is $Y = a + bX = 71.7778 + 5.0833X$.

ESTIMATE OF SALES FOR 1990.

For 1990, $X = 5$, **so we have**

$$Y_{1990} = 71.7778 + 5.0833\,(5) = 71.7778 + 25.4165$$

$$= \mathbf{97.1943 \text{ thousand rupees.}}$$

Therefore, sale for the year 1990 is Rs. 97,194.

14.19 FITTING OF A SECOND DEGREE TREND OR A PARABOLIC TREND

The **equation of second degree or parabolic trend is**

$$Y = a + bX + cX^2$$

where a is the Y intercept, b is the slope of the curve at the origin and c is the rate of change in the slope. The values of a, b and c can be determined by solving the following three normal equations:

$$\Sigma Y = Na + b\,\Sigma X + c\,\Sigma X^2 \quad ...(1)$$
$$\Sigma XY = a\,\Sigma X + b\,\Sigma X^2 + c\,\Sigma X^3 \quad ...(2)$$
$$\Sigma X^2Y = a\,\Sigma X^2 + b\,\Sigma X^3 + c\,\Sigma X^4 \quad ...(3)$$

The first equation is obtained on summing both sides the equation of parabolic trend, the second is the summation of X multiplied by the equation of parabolic trend and the third is the summation of X^2 multiplied in the equation of parabolic trend.

We know that if the time origin is taken between the middle years ΣX would be zero. Applying this rule on the given above equations, we get

$$\Sigma Y = Na + c\Sigma X^2$$
$$\Sigma XY = b\Sigma X^2$$
$$\Sigma X^2Y = a\Sigma X^2 + c\Sigma X^4$$

Thus $a = \dfrac{\Sigma(Y) - c(\Sigma X^2)}{N}$; $b = \dfrac{\Sigma XY}{\Sigma X^2}$; $c = \dfrac{N(\Sigma X^2Y) - (\Sigma X^2)\Sigma(Y)}{N\Sigma(X^4) - (\Sigma X^2)^2}$ ... (4)

Parabolic trend line is considered to be an appro priate model for secular trend component of a time series when the data appear not to fall in a straight line.

Hence $Y_C = a + bX + cX^2$, where a, b, c are given by (4).

Example 35. *Fit a second degree parabola to the following data and also estimate the value for 1990 and give your comments.*

Year	*1955*	*1960*	*1965*	*1970*	*1975*	*1980*	*1985*
Production ('000 units)	*6*	*8*	*9*	*10*	*12*	*11*	*8*

Solution. The parabolic trend equation is

$$Y_C = a + bX + cX^2.$$

The values of a, b and c can be obtained by solving the following equations.

$$\Sigma Y = Na + b\,\Sigma X + c\,\Sigma X^2;$$
$$\Sigma XY = a\,\Sigma X + b\,\Sigma X^2 + c\,\Sigma X^3;$$
$$\Sigma X^2Y = a\,\Sigma X^2 + b\,\Sigma X^3 + c\,\Sigma X^4.$$

Since $\Sigma X = 0$ the equations change to

$\Sigma Y = Na + c\,\Sigma X^2$; $\Sigma XY = b\,\Sigma X^2$; $\Sigma X^2Y = a\,\Sigma X^2 + b\,\Sigma X^3 + c\,\Sigma X^4$.

Table for Computation

Year	*Production '000 units Y*	*X*	X^2	X^3	X^4	*XY*	X^2Y	*Trend Value* Y_C
1955	6	– 3	9	– 27	81	– 18	54	**5.57**
1960	8	– 2	4	– 8	16	– 16	32	**8.071**
1965	9	– 1	1	– 1	1	– 9	9	**9.786**
1970	10	0	0	0	0	0	0	**10.715**
1975	12	1	1	1	1	12	12	**10.858**
1980	11	2	4	8	16	22	44	**10.215**
1985	8	3	9	27	81	24	72	**8.786**
N = 7	**ΣY = 64**	**ΣX = 0**	**ΣX² = 28**	**ΣX³ = 0**	**ΣX⁴ = 196**	**ΣXY = 15**	**ΣX²Y = 223**	

Substituting the values, we get

$$64 = 7a + 28c \quad \text{... (1)}$$
$$15 = 28b \quad \text{... (2)}$$
$$223 = 28a + 196c. \quad \text{... (3)}$$

From (2) we get, $\boldsymbol{b} = \frac{15}{28} = \mathbf{0.536.}$

Multiply equation (1) by 4 and subtract equation (3) from (4) to get

$$256 = 28a + 112c \quad \text{... (4)}$$
$$223 = 28a + 196c$$

$$+33 = -84c \Rightarrow \boldsymbol{c} = \frac{33}{-84} = \mathbf{-0.393.}$$

Substituting the value of c in equation (1), we get

$$64 = 7a + 28(-0.393) \text{ or } 64 = 7a - 11.004$$

$$\Rightarrow 7a - 11.004 = 64 \Rightarrow 7a = 64 + 11.004$$

$$\Rightarrow \boldsymbol{a} = \frac{75.004}{7} = \mathbf{10.715}$$

Hence $Y_C = 10.715 + 0.536X - 0.393X^2$ $[\because Y_C = a + bX + cX^2]$

CALCULATION OF TREND VALUES

$Y_{1955} = 10.715 + [0.536 \times (-3)] + (0.393 \times 9) = 10.715 - 5.145 = \mathbf{5.57.}$

Similarly, the other trend values can be obtained by putting $X = -2, -1, 0$ etc.

For 1990, $X = 4$, $X^2 = 16$,

$\therefore\ Y_{1990} = 10.715 + (0.536 \times 4) + (-0.393 \times 16) = 10.715 + 2.144 - 6.288 = \mathbf{6.571.}$

Hence the estimated value for 1990 = 6.571.

Example 36. Fit a second degree parabola to the following data:

x:	*1*	*2*	*3*	*4*	*5*
y:	*1090*	*1220*	*1390*	*1625*	*1915*

Solution. **Computation Table**

X	Y	$X - 3 = u$	$\frac{Y - 1450}{5} = v$	u^2	u^4	u^2v	uv
1	1090	– 2	– 72	4	16	– 288	144
2	1220	– 1	– 46	1	1	– 46	46
3	1390	0	– 12	0	0	0	0
4	1625	1	35	1	1	35	35
5	1915	2	93	4	16	372	186
		0	– 2	10	34	73	411

If the parabola be $\boldsymbol{v = a + bu + cu^2}$, **then the normal equations are:**

$$-2 = 5a + 10c \quad \text{... (1)}$$
$$411 = 10b \quad \text{... (2)}$$
$$73 = 10a + 34c \quad \text{... (3)}$$

Solving (1), (2) and (3), we get

$a = -11.4.$ $b = 41.1,$ $c = 5.5$

$\therefore\ v = -11.4 + 41.1u + 5.5u^2$

$\Rightarrow y = 1024 + 40.5X + 27.5X^2$

Example 37. *The following data on production (in '000 units) of a commodity from the year 1990 to 1996.*

Year:	*1990*	*1991*	*1992*	*1993*	*1994*	*1995*	*1996*
Production (in '000 units):	6	7	5	4	6	7	5

Fit the trend of the type $Y = a + bx + cx^2$ to the above data

(Take 1993 as the year of origin)

Solution. Here **the trend is given by: $Y = a + bX + cX^2$.**

Also $N = 7$ and $x = 1993 - x$. We have the following table

Table: Calculation for fitting: $Y = a + bX + cX^2$

Year (x)	**Production (Y)**	**$X = x - 1993$ (1993 = 0)**	**X^2**	**X^3**	**X^4**	**XY**	**X^2Y**
1990	6	– 3	9	– 27	81	– 18	54
1991	7	– 2	4	– 8	16	– 14	28
1992	5	– 1	1	– 1	1	– 5	5
1993	4	0	0	0	0	0	0
1994	6	1	1	1	1	6	6
1995	7	2	4	8	16	14	28
1996	5	3	9	27	81	15	45
***n* = 7**	**40**	**0**	**28**	**0**	**196**	**– 2**	**166**

The **normal equations for $y = a + bX + cX^2$** are

$$\Sigma Y = Na + b\Sigma X + c\Sigma X^2$$
$$\Sigma XY = a\Sigma X + b\Sigma X^2 + c\Sigma X^3$$
$$\Sigma X^2Y = a\Sigma X^2 + b\Sigma X^3 + c\Sigma X^4$$

Putting the values from the above table in the normal equations, we find.

$$40 = 7a + 28c;$$
$$-2 = 28b;$$

and $$166 = 28a + 196c$$

Solving them, we get: $a = 5.429,$ $b = -0.071,$ $c = 0.071$

$\therefore$ **The required trend equation is: $Y = 5.429 - 0.071X + 0.71X^2$**

Example 38. *Given below is the time series data on production (in '000 units) of a certain firm:*

Year :	*1985*	*1986*	*1987*	*1988*	*1989*	*1990*	*1991*
Production ('000 units) :	42	49	62	75	92	122	158

(i) Compute the quadratic trend by least squares method.

(ii) Forecast the production for the year 1996.

Solution. Let the quadratic trend be: $y = a + bx + cx^2$

The normal equations are:

$$\Sigma = na + b\Sigma x + c\Sigma x^2$$

$$\Sigma xy = a\Sigma x + b\Sigma x^2 + c\Sigma x^3$$

and $\Sigma x^2y = a\Sigma x^2 + b\Sigma x^3 + c\Sigma x^4$

Let the **origin (or base year) = 1988**

TABLE: Calculation of Trend Curve: $y = a + bx + cx^2$

Year (X)	*Production* n (y)	*x = X – 1988*	x^2	x^3	x^4	*xy*	x^2y
1985	42	– 3	9	– 27	81	– 126	378
1986	49	– 2	4	– 8	16	– 98	196
1987	62	– 1	1	0	0	– 62	62
1988	75	0	0	0	0	0	0
1989	92	+ 1	1	1	1	92	92
1990	122	+ 2	4	8	16	244	488
1991	158	+ 3	9	27	81	474	1422
Total	**600**	**0**	**28**	**0**	**196**	**524**	**2638**

Substituting these values in normal equations, we have:

$$600 = 7a + 0b + 28c \quad ...(1)$$

$$524 = 0a + 28b + 0c \quad ...(2)$$

$$2638 = 28a + 0b + 196c \quad ...(3)$$

Solving (1), (2) and (3), we get: $\boldsymbol{a = 74.38, b = 18.71}$ **and** $\boldsymbol{c = 2.83.}$

$\therefore$ $\boldsymbol{y = 74.38 + 18.71x + 2.83x^2}$**, where** $\boldsymbol{x = X - 1988.}$

For the year 1996, x = 1996 – 1988 = **8.**

The forecast of the production for the year 1996 is given by:

$$y = 74.38 + 18.71 \times 8 + 2.83 \times 8^2 = \mathbf{405.18.}$$

$\therefore$ **The estimate for 1996 is 405180.**

Example 39. *Fit a non-linear trend of the form $Y = a + bX + cX^2$ given the plan following data.*

X′	*0*	*1*	*2*	*3*	*4*
Y	*1.0*	*1.5*	*1.5*	*2.5*	*3.5*

Solution. Let the non-linear trend equation be: $Y = a + bX + cX^2$.

X′	*X = (X′ – 2)*	*Y*	X^2	X^3	X^4	*XY*	X^2Y
0	– 2	1.0	4	– 8	16	– 2.00	4
1	– 1	1.5	1	– 1	1	– 1.5	1.5
2	0	1.5	0	0	0	0	0.0
3	+ 1	2.5	1	+ 1	1	2.5	2.5
4	+ 2	3.5	4	+ 8	16	7.0	14.0
	0	**10.0**	**10**	**0**	**34**	**6**	**22.00**

Substituting these values in the **normal equation,** we get

$$\Sigma Y = na + b\Sigma X + c\Sigma X^2 \Rightarrow 10 = 5a + 10c \quad ...(1)$$
$$\Sigma XY = a\Sigma X + b\Sigma X^2 + c\Sigma X^3 \Rightarrow 6 = 10b + 0 \quad ...(2)$$
$$\Sigma X^2Y = a\Sigma X^2 + b\Sigma X^3 + c\Sigma X^4 \Rightarrow 22 = 10a + 34c \quad ...(3)$$

Solving (1), (2) and (3), we get

$$a = \frac{12}{7} = 1.71; \quad b = \frac{6}{10} = 0.6; \quad c = \frac{1}{7} = 0.14.$$

∴ **Non-linear Trend** is given by:

$$Y = 1.71 + 0.6X + 0.14X^2$$

Example 40. *Compute the quadratic trend of the form $y = a + bx + cx^2$ for the data showing the production of an item over a 7 year period.*

Year	*1985*	*1986*	*1987*	*1988*	*1989*	*1990*	*1991*
Production (in '000 tonnes)	*10*	*11*	*12*	*9*	*10*	*13*	*11*

Solution. **Computation Table**

Year	***X* (*1988 = 0*)**	***Y***	***XY***	X^2	X^2Y	X^3	X^4
1985	– 3	10	– 30	9	90	– 27	81
1986	– 2	11	– 22	4	44	– 8	16
1987	– 1	12	– 12	1	12	– 1	1
1988	0	9	0	0	0	0	0
1989	1	10	10	1	10	1	1
1990	2	13	26	4	52	8	16
1991	3	11	33	9	99	27	81
$n = 7\ \Sigma \rightarrow$	**0**	**76**	**5**	**28**	**307**	**0**	**196**

Normal equations are:

$$\Sigma Y = na + b\Sigma X + \Sigma X^2 \Rightarrow 76 = 7a + 28c \quad ...(1)$$
$$\Sigma XY = a\Sigma X + b\Sigma X^2 + c\Sigma X^3 \Rightarrow 5 = 28b \quad ...(2)$$
$$\Sigma X^2Y = a\Sigma X^2 + b\Sigma X^3 + c\Sigma X^4 \Rightarrow 307 = 28a + 196c \quad ...(3)$$

Solving the three equations (1), (2) and (3), we get:

$a = 75/7 =$ **10.71**, $b = 5/28 = 0.178 \approx$ **0.18** and $c = \frac{3}{84} = 0.0357 \approx$ **0.036**

Hence **the quadratic trend is: $Y = a + bX + cX^2 = 10.71 + 0.18X + 0.036X^2$, where $X = 0$ for 1988.**

Example 41. *Fit a second degree parabola $Y = a + bX + cX^2$ to the following population data of a city.*

Year :	*1979*	*1980*	*1981*	*1982*	*1983*	*1984*	*1985*	*1986*	*1987*
Population : (in Lakhs)	*5*	*6*	*6*	*7*	*7*	*8*	*9*	*10*	*10*

(Take the year 1983 as the working origin)

Solution.

Computation Table

Year	*Population in Lakhs (Y)*	*Deviations from year 1983 (X)*	*XY*	X^2	X^2Y	X^3	X^4
1979	5	– 4	– 20	16	80	– 64	256
1980	6	– 3	– 18	9	54	– 27	81
1981	6	– 2	– 12	4	24	– 8	16
1982	7	– 1	– 7	1	7	– 1	1
1983	7	0	0	0	0	0	0
1984	8	+ 1	8	1	8	1	1
1985	9	+ 2	18	4	36	8	16
1986	10	+ 3	30	9	90	27	81
1987	10	+ 4	40	16	160	64	256
$N = 9$	$\Sigma Y = 68$	$\Sigma X = 0$	$\Sigma XY = 39$	$\Sigma X^2 = 60$	$\Sigma X^2Y = 459$	$\Sigma X^3 = 0$	$\Sigma X^4 = 708$

Substituting respective values in the normal equation

$$\Sigma Y = Na + b\Sigma X + c\Sigma X^2$$
$$\Sigma XY = a\Sigma X + b\Sigma X^2 + c\Sigma X^3$$
$$\Sigma X^2Y = a\Sigma X^2 + b\Sigma X^3 + c\Sigma X^4,$$

we get

$$68 = 9a + 60c \quad \text{... (i)}$$
$$39 = 60b \quad \text{... (ii)}$$
$$459 = 60a + 708c \quad \text{... (iii)}$$

From (*ii*), we get $\quad b = \dfrac{39}{60} = 0.65.$

On multiplying equation (*i*) by 60 and equation (*iii*) by 9, we get

$$4080 = 540\,a + 3600\,c$$
$$4131 = 540a + 6372c$$

$$-51 = -2772\,c \;\Rightarrow\; c = \frac{-51}{-2772} = 0.018$$

Substituting the value of c in equation (*i*) we get

$68 = 9a + 60 \times 0.018 \;\Rightarrow\; 68 = 9a + 1.08$

$\Rightarrow 9a = 68 - 1.08 = 66.92 \;\Rightarrow\; a = \dfrac{66.92}{9} = 7.435.$

Required second degree parabola is given by

$Y = a + bX + cX^2 = 7.435 + 0.65X + 0.018X^2.$

Example 42. *Compute a non-linear trend of the form* $Y = a + bX + cX^2$ *for the data showing the production of wheat (in '000 tonnes) during the years 1978 – 1986.*

Year :	*1978*	*1979*	*1980*	*1981*	*1982*	*1983*	*1984*	*1985*	*1986*
Production : of wheat ('000 tonnes)	9	10	12	15	13	10	8	16	15

(*Take the year 1982 as working origin*)

Solution.

Table for Computing a Non-Linear Trend

Year	*Production of wheat ('000 tonnes) (Y)*	*Deviation from 1982 (X)*	*XY*	X^2	X^2Y	X^3	X^4
1978	9	– 4	– 36	16	144	– 64	256
1979	10	– 3	– 30	9	90	– 27	81
1980	12	– 2	– 24	4	48	– 8	16
1981	15	– 1	– 15	1	15	– 1	1
1982	13	0	0	0	0	0	0
1983	10	+ 1	10	1	10	1	1
1984	8	+ 2	16	4	32	8	16
1985	16	+ 3	48	9	144	27	81
1986	15	+ 4	60	16	240	64	256
$N = 9$	$\Sigma Y = 108$	$\Sigma X = 0$	$\Sigma XY = 29$	$\Sigma X^2 = 60$	$\Sigma X^2Y = 723$	$\Sigma X^3 = 0$	$\Sigma X^4 = 708$

The equation of a non-linear trend is $Y = a + bX + cX^2$.

To find out the values of a, b, c we have to solve the following equation.

$$\Sigma Y = Na + b\Sigma X + c\Sigma X^2$$

$$\Sigma XY = a\Sigma X + b\Sigma X^2 + c\Sigma X^3$$

$$\Sigma X^2Y = a\Sigma X^2 + b\Sigma X^3 + c\Sigma X^4.$$

On substituting the values from the table, we get

$$108 = 9a + 60c \quad (i)$$

$$29 = 60b \quad (ii)$$

$$723 = 60a + 708c. \quad (iii)$$

From equation (*i*), (*ii*) and (*iii*), we get

$$a = 11935, \quad b = 0.4833 \quad \text{and} \quad c = 0.0097.$$

Whence **$Y = a + bX + cX^2 = 11.935 + 0.4833X + 0.0097X^2$ is the non-linear trend equation.**

Example 43. *From the given data*

Year :	*1974*	*1975*	*1976*	*1977*	*1978*	*1979*	*1980*
Earning : (Rs. lakhs)	*15*	*14*	*18*	*20*	*17*	*24*	*27*

Calculate the quadratic or parabolic trend by the least squares method and forecast the earning for the year 1985.

Solution.

Table for the Calculation of Quadratic or Parabolic Trend by the Method of Least Squares.

Year	*Earnings (Rs. lakhs) (Y)*	*Deviation from year 1977 (X)*	*XY*	X^2	X^2Y	X^3	X^4
1974	15	– 3	– 45	9	135	– 27	81
1975	14	– 2	– 28	4	56	– 8	16
1976	18	– 1	– 18	1	18	– 1	1
1977	20	0	0	0	0	0	0
1978	17	+ 1	+ 17	1	17	+ 1	1
1979	24	+ 2	+ 48	4	96	+ 8	16
1980	27	+ 3	+ 81	9	243	+ 27	81
$N = 7$	$\Sigma Y = 135$	$\Sigma X = 0$	$\Sigma XY = 55$	$\Sigma X^2 = 28$	$\Sigma X^2Y = 565$	$\Sigma X^3 = 0$	$\Sigma X^4 = 196$

The quadratic trend is given by the equation

$$Y = a + bX + cX^2$$

The normal equations for solving a, b, c are

$$\Sigma Y = Na + b\Sigma X + c\Sigma X^2$$
$$\Sigma XY = a\Sigma X + b\Sigma X^2 + c\Sigma X^3$$
$$\Sigma X^2Y = a\Sigma X^2 + b\Sigma X^3 + c\Sigma X^4.$$

Substituting the values of ΣY, N, ΣXY, ΣX^2, ΣX^2Y, ΣX^3 and ΣX^4, we get

$$135 = 7a + 28c \qquad \text{... (i)}$$
$$55 = 28b \qquad \text{... (ii)}$$
$$565 = 28a + 196c \qquad \text{... (iii)}$$

Solving equation (*i*) and (*iii*) and multiplying (*i*) by 4 we get

$$540 = 28a + 112c \qquad \text{... (iv)}$$
$$565 = 28a + 196c \qquad \text{... (v)}$$

Subtracting (*iv*) from (*v*), we have

$$25 = 84c \Rightarrow \boldsymbol{c} = \frac{25}{84} = \mathbf{0.298.}$$

Putting the value of c in equation (*i*), we get

$$135 = 7a + 28 \times 0.298$$

or $$135 = 7a + 8.344$$

$\Rightarrow$ $$7a = 135 - 8.344 = 126.656 \Rightarrow \boldsymbol{a} = \mathbf{18.095}.$$

Solving equation (*ii*), we get $\boldsymbol{b} = \frac{55}{28} = \mathbf{1.96.}$

Substituting these values, we get the **quadratic or parabolic trend**

$Y = a + bX + cX^2$

or $\boldsymbol{Y = 18.095 + 1.96X + 0.298X^2}$

Estimate for 1985 by taking $X = 8$, $X^2 = 64$

$Y_{1985} = 18.095 + (1.96)(8) + (0.298)(64) = 18.095 + 15.68 + 19.072 = \mathbf{52.847.}$

Example 44. *Fit a quadratic trend to the following data:*

Year	*1988*	*1989*	*1990*	*1991*	*1992*
Goods carried in Indian Railways (Lakhs of tonnes)	*180*	*192*	*195*	*204*	*202*

Solution. The quadratic trend is given by the equation

$$Y = a + bX + cX^2. \quad \text{... (A)}$$

The normal equations for solving *a, b, c* are

$$\Sigma Y = Na + b\Sigma X + c\Sigma X^2$$
$$\Sigma XY = a\Sigma X + b\Sigma X^2 + c\Sigma X^3$$
$$\Sigma X^2Y = 2\Sigma X^2 + b\Sigma X^3 + c\Sigma X^4.$$

Let 1990 be the origin year.

Table for Fitting of Quadratic Trend

Year	*Goods Carried (Y)*	*X*	*XY*	X^2	X^2Y	X^3	X^4
1988	180	– 2	– 360	4	720	– 8	16
1989	192	– 1	– 192	1	192	– 1	1
1990	195	0	0	0	0	0	0
1991	204	+ 1	+ 204	1	204	+ 1	1
1992	202	+ 2	+ 404	4	808	+ 8	16
$N = 5$	$\Sigma Y = 973$	$\Sigma X = 0$	$\Sigma XY = 56$	$\Sigma X^2 = 10$	$\Sigma X^2Y = 1924$	$\Sigma X^3 = 0$	$\Sigma X^4 = 34$

Substituting the values in normal equations, we get

$$973 = 5a + 10c \quad \text{... (1)}$$
$$56 = 10b \Rightarrow \mathbf{b = 5.6} \quad \text{... (2)}$$
$$1924 = 10a + 34c. \quad \text{... (3)}$$

From (1) and (3), we have

$$1946 = 10a + 20c \quad \text{... (4)}$$
$$1924 = 10a + 34c \quad \text{... (5)}$$

(4) and (5) $\Rightarrow 14c = -22 \Rightarrow c = \frac{-22}{14} = \mathbf{-1.57.}$

Putting the value of *c* in (1), we get

$5a - (10)(1.57) = 973 \quad \Rightarrow \quad 5a = 973 + 15.7 = 988.7$

$$\Rightarrow \quad a = \frac{988.7}{5} = \mathbf{197.74.}$$

Hence $\mathbf{Y = 197.74 + 5.6X - 1.57X^2}$. $[\because Y = a + bX + cX^2]$

Example 45. *Fit a parabolic curve of second degree to the data given below and estimate the value for 1989 and comment on it.*

Years :	*1983*	*1984*	*1985*	*1986*	*1987*
Sales (in Rs. '000) :	*10*	*12*	*13*	*10*	*8*

Solution. Let $Y = a + bX + cX^2$ ... **(1)**

be the equation of the parabolic curve to be fitted to the given data. By the method of least squares, the normal equations referred to mid-point of time as the origin are

$$\Sigma Y = Na + c\Sigma X^2,$$
$$\Sigma XY = b\Sigma X^2 \text{ and}$$
$$\Sigma X^2Y = a\Sigma X^2 + c\Sigma X^4.$$

Here $N = 5$.

Let 1985 be the origin year.

Table: Calculations for Fitting a Parabolic Curve

Years (*x*)	*Sales (in Rs. '000)* (*Y*)	*X = x – 1985*	X^2	X^3	X^4	*XY*	X^2Y
1983	10	– 2	4	– 8	16	– 20	40
1984	12	– 1	1	– 1	1	– 12	12
1985	13	0	0	0	0	0	0
1986	10	1	1	1	1	10	10
1987	8	2	4	8	16	16	32
Total	$\Sigma Y = 53$	$\Sigma X = 0$	$\Sigma X^2 = 10$	$\Sigma X^3 = 0$	$\Sigma X^4 = 34$	$\Sigma XY = -6$	**94**

Using the normal equations, we get

$$53 = 5a + c \times 10 \quad \text{...(2)}$$
$$-6 = b \times 10 \quad \text{...(3)}$$
$$94 = a \times 10 + c \times 34 \quad \text{...(4)}$$

From (3), we have: $b = \frac{6}{10} = -0.6$.

Multiplying equation (2) by 2 and then subtracting it from equation (4), we have:

$$-12 = 14c \Rightarrow c = -\frac{6}{7} = -0.857.$$

From equation we get: (2) $5a = 53 - 10c = 53 (-10) \times (-0.857) = \mathbf{61.57}$;

$\therefore$ $a = 12.314$.

Hence, **the required equation of the fitted parabolic curve is**

$$Y = 12.314 - 0.6X - 0.857X^2$$

with origin at the middle year 1985 and unit of *X* as 1 year.

In 1989, $X = 4$,

$\therefore$ $Y = 12.314 - 0.6 \times 4 - 0.857 \times 16 = 12.314 - 16.112 = -3.798$.

Negative trend value for the year 1989 indicates that the model (*i.e.*, fitting parabolic curve) is not suitable for the given data.

Thus under the circumstances it could be safely presumed that the sales of the company are likely to be discontinued very soon.

14.20 EXPONENTIAL TREND

Exponential Trend or Logarithmic Trend: The equation of the **exponential curve** is given by:

$$Y = a\,b^X, \qquad \text{... (1)}$$

where X = time, Y = time series data and a, b are constants to be obtained from normal equations. The values of Y obtained from (1) form a G.P., while the corresponding X-values form an A.P. For example, the data from the fields of biology (the growth of bacteria is exponential), banking and economics often exhibit exponential trend. The sales or earnings from business may grow exponentially over a short period. In the **exponential trend**, which is also known as **'Geometric Trend' the equal relative change in the variable is assumed for each time period.**

Taking logarithms of the both sides of (1), we get:

$$\log Y = \log a + X \log b \qquad \text{(Logarithmic Trend)}$$

When plotted on a **semi-logarithmic** graph, **the curve will be a straight line.** However, **when plotted on an arithmetic chart, the curve will give a non-linear trend.** The **value of the constants 'a' and 'b' are obtained by solving the following two normal equations (by using least squares method)** .

$$\Sigma Y = N(\log a) + (\log b)\,\Sigma X$$

$$\Sigma (X \log Y) = (\log a)\,\Sigma X + (\log b)\,\Sigma X^2.$$

where a = Y intercept, b = slope of the curve, and N = Total number of observations.

When deviations are taken from the middle year, then $\Sigma X = 0$, and the above two equations reduce to

$$\Sigma \log Y = N \log a \;\Rightarrow\; \log a = \frac{\Sigma \log Y}{N}$$

$$\text{and}\,\Sigma (X \log Y) = (\log b)\,\Sigma X^2 \;\Rightarrow\; \log b = \frac{\Sigma X \log Y}{\Sigma X^2}$$

$$\text{Thus } a = \text{antilog}\left(\frac{\Sigma \log Y}{N}\right); \qquad b = \text{antilog}\left[\left(\frac{\Sigma X \log Y}{\Sigma X^2}\right)\right]$$

14.21 WORKING RULE FOR FITTING THE EXPONENTIAL CURVE

Step 1. Obtain the deviation of each year from the middle year and denote these by 'X'.

Step 2. Calculate X^2 and find ΣX^2.

Step 3. Find log Y and Σ log Y.

Step 4. Find X log Y by multiplying log Y by the corresponding time deviations X. Also find ΣX log Y.

Step 5. Calculate $\log a = \dfrac{\Sigma \log Y}{N}$ *(Use Step 3)*

Step 6. Calculate $\log b = \dfrac{\Sigma (X \log Y)}{\Sigma X^2}$ *(Using STEP 2 and STEP 4), where log b is the rate of growth or the slope of the line.*

Step 7. Find trend ordinates in logarithms by substituting the values of log a and log b obtained in STEP 5 and STEP 6 respectively in log Y = log a + X log b

Step 8. Find the antilogs of these logs to get the actual trend values.

Example 46. *The sale of a company in thousands of rupees for the year 1980 through 1986 are given below.*

Year :	*1980*	*1981*	*1982*	*1983*	*1984*	*1985*	*1986*
Sales :	*32*	*47*	*65*	*92*	*132*	*190*	*275*

Estimate, the sales figure for the year 1987 using an equation of the form $Y = ab^X$, where X = years and Y = sales.

Solution. Taking origin at the middle year 1983, the normal equations to the exponential trend $Y = ab^X$ **are:**

$$\log a = \frac{\Sigma \log Y}{N}, \quad \log b = \frac{\Sigma(X \log Y)}{\Sigma X^2}.$$

Also the **trend equation is: $\log Y = \log a + X \log b$.**

Table: Computation of Fitting Exponential Trend

Year	*Sales*	*X*	*log Y*	X^2	*X log Y*	*log Y*	*Antilog Y*
1980	32	– 3	1.5051	9	– 4.5153	1.5084	**32.24**
1981	47	– 2	1.6721	4	– 3.3442	1.6624	**45.96**
1982	65	– 1	1.8129	1	– 1.8129	1.8164	**65.52**
1983	92	0	1.9638	0	0	1.9704	**93.42**
1984	132	+ 1	2.1206	1	+ 2.1206	2.1244	**133.10**
1985	190	+ 2	2.2788	4	+ 4.5576	2.2784	**189.00**
1986	275	+ 3	2.4398	9	+ 7.3119	2.4324	**270.60**
Total	**833**		**13.7926** $= \Sigma \log Y$	**28** $= \Sigma X^2$	**+ 4.3237** $= \Sigma X \log Y$		

Substituting the values of $\Sigma \log Y$, ΣX^2 and $\Sigma X \log Y$ in the normal equations, we get

$$\log a = \frac{\Sigma \log Y}{N} = \frac{13.7926}{7} = \mathbf{1.9704}$$

$$\Rightarrow \qquad \log b = \frac{\Sigma(X \log Y)}{\Sigma X^2} = \frac{\mathbf{4.3237}}{\mathbf{28}} = \mathbf{0.1544.}$$

From the trend equation: **$\log Y = \log a + X \log b = 1.9704 + 0.154\, X$**

ESTIMATION OF SALES FIGURES FOR 1987.

For the year 1987, X = 1987 – 1983 = 4.

$\therefore$ $\log Y = 1.9704 + 0.154 \times 4 = 2.5864$

$\therefore$ Y = antilog 2.5864 = 385.9.

Thus the estimated sale for 1987 is Rs. 385.9 lakhs.

Example 47. *The following table gives the total expenditure of the govt. of India during 1978-79 to 1984-85:*

Year :	*1978-79*	*1979-80*	*1980-81*	*1981-82*	*1982-83*	*1983-84*	*1984-85*
Expenditure : (in crores)	*177.2*	*185.0*	*224.9*	*254.0*	*304.9*	*359.9*	*438.8*

Fit an exponential trend curve $y = ab^x$ to the data.

Solution. We have to fit curve $y = ab^x$

Taking logarithm of both the sides, we get, $\log y = \log a + x \log b$

Year	*Expenditure (Y)*	*X*	X^2	*log Y*	*X log Y*
1978-79	177.2	– 3	9	2.2485	–6.7455
1979-80	185.0	– 2	4	2.2672	– 4.5344
1980-81	224.9	– 1	1	2.352	– 2.352
1981-82	254.0	0	0	2.4048	0
1982-83	304.9	1	1	2.4842	2.4842
1983-84	359.9	2	4	2.5562	5.1124
1984-85	438.8	3	9	2.6423	7.9269
	1944.7	**0**	**28**	**16.9552**	**1.8916**

$$\log a = \frac{\Sigma \log Y}{n} = \frac{16.9552}{7} = 2.4222 \Rightarrow a = \text{Antilog } 2.4222 = \mathbf{264.36}$$

$$\log b = \frac{\Sigma X \log Y}{\Sigma X^2} = \frac{\mathbf{1.8916}}{\mathbf{28}} = 0.06755 \Rightarrow b = \text{Antilog } 0.06755 = \mathbf{1.168}$$

Thus exponential trend curve is given by $y = ab^X = 264.36 \times (1.168)^X$

14.22 MODIFIED EXPONENTIAL TREND

The **modified exponential Trend** curve is:

$$Y = a + b^x. \quad \text{... (1)}$$

The **important characteristic of this curve is that the difference of the consecutive values of Y corresponding to the equidistant values of X change by a constant ratio.**

COWDEN'S METHOD OF FITTING THE MODIFIED EXPONENTIAL CURVE

In this case the given data are first plotted on ordinary, or semi-log graph paper and a trend line is drawn by the **freehand method.** Now for any 3 equidistant values of X, say for $X = 0, x, 2x$, we find three ordinates Y_0, Y_1, Y_2, where

$$Y_0 = a + bc^0 = a + b \quad \text{... (2)}$$

$$Y_1 = a + bc^x$$

$$Y_2 = a + bc^{2x}.$$

[$\because$ The points $(0, Y_0)$, (x, Y_1) and $(2x, Y_2)$ lie on (1).]

$\therefore$ $$Y_1 - Y_0 = bc^x - b = b\,(c^x - 1). \quad \text{... (3)}$$

Also $$Y_2 - Y_1 = bc^{2x} - bc^x = bc^x\,(c^x - 1). \quad \text{... (4)}$$

Dividing (4) by (3), we get:

$$\frac{Y_2 - Y_1}{Y_1 - Y_0} = c^x$$

Substituting the value of c^x in (3), we have:

$$Y_1 - Y_0 = b\left(\frac{Y_2 - Y_1}{Y_1 - Y_0} - 1\right) = b\left(\frac{Y_2 - 2Y_1 + Y_0}{Y_1 - Y_0}\right);$$

$$\Rightarrow \quad b = \frac{(Y_1 - Y_0)^2}{Y_2 - 2Y_1 + Y_0} \quad \text{... (5)}$$

Now (2) and (5)

$$\Rightarrow \qquad a = Y_0 - b = Y_0 - \frac{(Y_1 - Y_0)^2}{Y_2 - 2Y_1 + Y_0} = \frac{Y_0 Y_2 - Y_1^2}{Y_2 - 2Y_1 + Y_0}.$$

Thus the values of the constants *a, b, c* in (1) are determined. Substituting these values in (1), we get the equation of the modified exponential trend.

14.23 GROWTH-CURVES

The following two curves are most widely used in describing growth.

I. Logistic Curve.

II. Gompertz Curve.

I. LOGISTIC CURVE

The equation of the **Logistic curve** is given by:

$$\frac{1}{Y} = a + bc^X, \qquad \text{... (1)}$$

where *Y* represents time series values, *X* represents time and *a, b, c* are constants.

If we take $\frac{1}{Y} = y$, the equation (1) reduces to:

$$y = a + bc^X, \qquad \text{... (2)}$$

which is the modified exponential curve. Therefore, the differences of the consecutive values of $\frac{1}{Y}$ corresponding to equidistant values of *X* change by a constant ratio.

The equation (1) can be fitted to the given time series data by applying the method of three selected points as used in the Modified exponential curve.

USES

The logistic curve is generally used in population data of various kinds (both human and non-human). It is also used in economic series relating to industrial growth.

II. GOMPERTZ CURVE

The equation of **Gompertz Curve** is given by:

$$Y = a\, b^{c^{(X)}} \qquad \text{... (3)}$$

Taking logarithms of both sides of (3), we get:

$$\log Y = \log a + c^X \log b,$$

or, $$Y' = A + Bc^X \qquad \text{... (4)}$$

where $Y' = \log Y$; $\quad A = \log a$; $\quad B = \log b$.

This is similar to the equation of **Modified Exponential Curve with log *Y* instead of *Y*.** The curve (1) may be fitted by **Cowden's method with the modification that log *Y* is to be used in place of *Y*.**

USES

This curve is used in actuarial work and sometimes in business and population forecasting as a growth curve.

14.24 SELECTION OF THE TYPE OF TREND

The selection of a particular type of equation that best describes the data is a difficult job. **Croxton and Cowden states** that the following steps can be applied to the smoothe curve.

1. ***If the first differences are constant, use a straight line method.***
2. ***If the second differences are constant, use a second degree curve method***
3. ***If the trend is a straight line, use the straight line method.***

4. ***If the first differences of the logarithm are constant, use an exponential curve $Y = ab^X$.***
5. ***Use Gompertz curve if the first differences resemble a frequency curve.***
6. ***Use modified exponential Trend curve: $Y = a + bc^X$,***
   if the first differences trend to decrease by a constant percentage.

14.25 CHOICE OF THE TREND PERIOD

The trend period should be longer, only then the trend value will not be affected by cyclical or random influences. The period employed should encompass a number of business cycles and should begin and end in such a way that distortion is avoided.

14.26 CONVERSION OF TREND EQUATION

Trend equation depends on:

1. ***The origin of time reference.***
2. ***The units of time, viz., weekly, monthly, yearly etc.***
3. ***The time series values relate to annual figures or monthly averages.***

Trend values can be computed by the straight line method. For simplicity and convenience, the following methods are also used. They are

1. ***Shifting of origin***
2. ***Conversion of annual trend equation to monthly trend equation when the y – values are in annual totals and are given as monthly averages.***

14.27 SHIFTING OF ORIGIN

When we **calculate the trend value, the origin is generally taken in the middle year of the series.** But sometimes to compare the trend values or to convert a trend equation from an annual to half yearly, quarterly or monthly basis, it is necessary to change the origin of the trend equation to some other point in the equation. **It is used to know the seasonal or cyclical pattern.**

Let **the straight line annual trend equation be:**

$$Y_c = a + bX,$$

where X units: one year, Y units = Annual totals

Let the constant 'a' be the trend value at the year of origin, *viz.*, 1990, *i.e.*,

$$(Y_c)_{1990} = a \qquad (\because X = 0 \text{ in } 1990)$$

Now if we want to change the time series to have its origin say 1996, *i.e.*, we want to shift the new trend origin 6 years hence, **then the new trend equation is obtained by changing X to $(X + 6)$.** The new trend equation is

$$Y_c = a + b\,(X + 6) = (a + 6) + bX.$$

Thus, when we shift the origin, **it will affected the value of the constant 'a' in the equation, while the value of 'b' remains the same. Even though we change the origin, the slope of the trend line is the same.** The equation is

$Y = a + b\,(X + k)$, where k, is the number of time units shifted.

If we shift **the origin forward in time, k is positive; and backward in time, then k is negative.**

Example 48. *You are given the trend equation:*

$Y_t = 95 + 3X$; origin: 1985 time; unit : 1 year.

Shift the origin to 1993.

Solution. You are required to **shift the origin to 1993**, *i.e.*, (1993 – 1985) = 8 **year's forward.** Here $k = 8$. Using the formula for shifting of origin required equation is

$$Y_t = a + b\,(X + k)$$

or $$Y_t = 95 + 3\,(X + 8) = 95 + 24 + 3X \qquad (\because a = 95, b = 3, k = 8)$$

$\therefore$ $$Y = 119 + 3X \qquad \textbf{(Origin 1993, time unit = 1 year)}$$

Example 49. *Given the trend equation:* $Y_c = 56 - 3.5X$ *(Origin 1996, time unit = 1 year)* *Shift the origin to 1992.*

Solution. Changing the origin from 1996 to 1992 means going back by 4 years, so $k = -4$.

Using the formula: $Y_t = a + b\,(X + k)$, we have

$$Y_t = 56 - 3.5\,(X - 4) \qquad (\because a = 56, b = 3.5, k = -4)$$

$\Rightarrow$ $$Y_t = 56 + 14 - 3.5X = 70 - 30.5X$$

(Origin 1992; time unit = 1 year)

14.28 CONVERSION OF ANNUAL TREND EQUATION TO MONTHLY TREND EQUATION

Let $Y = a + bX$ be **the linear trend equation.** If Y units are annual totals and X units are in years, then the **Annual Trend Equation** is:

$$Y_t = a + bX.$$

where a = intercept on x-axis b = slope of the equation

This equation can be converted into monthly trend equation by dividing 'a' by '12' and 'b' by **12 × 12 = 144.** Thus the **Monthly Trend Equation is:**

$$Y_t = \frac{a}{12} + \frac{b}{144}X.$$

where **X = units are in months, and Y units are monthly totals.**

If the time units of X in the trend equation represents only 6 months, then 'b' is divided by 72. Then the **monthly Trend Equation is:**

$$Y_t = \frac{a}{12} + \frac{b}{72}X.$$

Example 50. *The Trend Equation:* $Y_t = 36 + 4X$, *where X units = 6 months, y-unit = annual production of computers convert this into a monthly trend equation.*

Solution. Here Y_t = annual production of computers

Time Unit = 6 months = $\frac{1}{2}$ year

$\therefore$ **Monthly Trend Equation:** $Y_t = \frac{a}{12} + \frac{b}{72}X$

or $$Y_t = \frac{36}{12} + \frac{4}{72}X = 3 + 0.55X.$$

Example 51. *You are given the following trend equation by the method of least squares of a company selling readymade garments.*

$Y = 480 + 36\,X$ *is origin 1988,*

$X =$ *Unit of one year,* Y = *number of units sold per year*

(*a*) *Convert the above trend equation into monthly trend equation*

(*b*) *Estimate the sale for the month of Oct. 1994.*

Solution. (*a*) Monthly Trend Equation: $Y = \frac{a}{12} + \frac{b}{144} X$

Here $a = 480, b = 36.$

$\therefore$ **Required Monthly Trend Equation is:** $Y = \frac{480}{12} + \frac{36}{144} X$

or $Y = 40 + 0.25 X$. Origin 1st July 1980

(*b*) X = Number of months from 1st July 1988 to 15th October 1994 = 6 years and $3\frac{1}{2}$ months

= 75.5 months

Now **the Monthly Trend Equation is: $Y = 40 + 0.25 X$**

[X = One Month, Y = Number of units sold per month]

Required Estimate sale for Oct. 1994:

$Y_{1994} = 40 + 0.25\ (75.5) = 58.87$ **units** $(\because X = 75.5)$

Example 52. *The trend of the annual sales of Steel Company is described by the following equation:*

$$Y = 12 + 0.7 X \quad \text{... (1)}$$

[Origin: (1st July) 1990 ; X unit = 1 year and Y unit = Annual production]

Step the equation down to a month to month basis and shift the origin to 1st January 1990.

Solution. The **monthly Trend Equation is:**

$$Y_c = \frac{a}{12} + \frac{b}{144} X.$$

The given equation (1), when reduced to monthly trend equation is:

$$Y_c = \frac{12}{12} + \frac{0.7}{144} X \quad (\because a = 12, b = 0.7)$$

or

$$Y_c = 1 + \frac{0.7}{144} X. \quad \text{... (2)}$$

[Origin: (1st July) 1990; X unit = 1 month; Y unit = Monthly production]

To shift the origin in (2) to 1st January, 1990, we have to step back 6 months. Hence, the new trend equation with 1st January 1990 as origin is obtained on changing X to $X - 6$ in (2) and is given by:

$$Y_c = 1 + \frac{0.7}{144}(X - 6) = 1 + 0.0048\,(X - 6)$$

or $\quad Y_C = 1 + 0.0048 X - 0.0288$

$\Rightarrow \quad Y_c = 0.9712 + 0.0048X.$

[Origin : 1st January 1990 ; X unit = 1 month].

14.29 SEASONAL VARIATIONS (*S*)

By seasonal variations in a time series we mean those variations which occur regularly and periodically with period less than one year. Seasonal movements affect most of the economic and business phenomena and unless a business executive knows about the nature of such variation he can not take policy decisions. ***The purpose of studying seasonal variation is two-fold:***

(*i*) **To isolate the effect of seasonal factors.**

(*ii*) **To eliminate these effects from the time series so that the values in a time series, without the effect of seasonal factors, may be computed.**

These values are useful for making forecasts over a long period of time. **The elimination of seasonal variations from a time series is known as the Process Of Deseasonalisation.** ***After a time series is deseasonalised it, is relatively easy to study the effects of cyclic and irregular variations. Thus, deseasonalisation helps in the process of decomposition of a time series into its various components,*** *viz*., **trend, seasonal variations, cyclic variations and irregular or random variations.**

It may be noted that for the study of seasonal variation in a time series, the data should be available seasonwise, weekly, monthly or quarterly. Seasonal effects are measured in two ways. In an **additive model** they **are expressed in terms of absolute values,** *i.e.*,

$$S = Y - (T + C + I).$$

In a **multiplicative model,** the seasonal effects are expressed as percentages or ratios and are given in multiplicative model, *i.e.*,

$$S = \frac{\text{TSCI}}{\text{TCI}} \times 100 \text{ or } \frac{\text{Y}}{\text{TCI}} \times 100.$$

14.30 METHODS FOR MEASURING SEASONAL VARIATIONS

The following are the important methods of studying seasonal variations

(I) Method of Simple Averages

(II) Ratio to Trend Method

(III) Ratio to Moving Average Method

(IV) Link Relative Method.

14.31 METHOD OF SIMPLE AVERAGES

This is the simplest and the easiest method of studying seasonal variations. The following steps are necessary to calculate the seasonal index:

(*i*) **Arrange the data by years and months or by years and quarters if data is given quarterly.**

(*ii*) **Find the sum of all figures relating to a month. It means add all values of January for all the years. Repeat the process for all the years. Repeat the process for all months.**

(*iii*) **Find the average of monthly figure,** ***i.e.,*** **divide the monthly total by the number of years.**

(*iv*) **Obtain the average of monthly averages by dividing the sum of averages by 12.**

(*v*) **Taking the average of monthly averages as 100, find out the percentages of monthly averages.**

For example for the average of the month January the percentage would be

$$\frac{\text{Average of January}}{\text{Average of monthly averages}} \times 100$$

If instead of the averages the monthly totals are taken into account the result would be the same.

From the table given on page 630, in which 5 years (monthly figures) have been given, we have first to find the totals of monthly figures in column 7. These totals have been converted into averages by dividing them by 5. These averages are given in column 8. The average of these averages comes out to be $\frac{2371.6}{12}$ or 197.6. This figure is taken as 100 and the monthly averages have been turned into percentages in column 9. Thus for the month of January the percentage or sesonal index is $\frac{197.0}{197.6} \times 100 = 99.7$. This figure can be found by using the total also *i.e.*, $\frac{985}{988} \times 100 = 99.7$. Similarly, the other figures have been calculated.

Illustration.

Table for Calculation of Seasonal Variations by Monthly Averages

Month	*Production of Commodity*					*Monthly Total for 5 years*	*Five Yearly Average*	*Percentage Seasonal Index*
	1988	**1989**	**1990**	**1991**	**1992**			
(1)	**(2)**	**(3)**	**(4)**	**(5)**	**(6)**	**(7)**	**(8)**	**(9)**
Jan.	131	145	208	238	263	985	197.0	99.7
Feb.	129	151	211	237	261	989	197.8	100.1
Mar.	121	149	207	225	250	952	190.4	96.3
Apr.	119	148	201	214	244	926	185.2	93.7
May	113	149	201	212	240	915	183.0	92.6
June	116	159	203	206	253	937	187.4	94.8
July	113	153	193	201	250	910	182.0	92.1
Aug.	123	165	205	210	255	958	191.6	97.0
Sep.	130	173	210	220	269	1002	200.4	101.4
Oct.	133	183	221	237	285	1059	211.8	107.2
Nov.	145	197	223	241	290	1096	219.2	110.9
Dec.	149	198	222	260	300	1129	225.8	114.2
					Total	**11858**	**2371.6**	**1200.0**
					Average	**988**	**197.6**	**100.0**

MERITS AND LIMITATIONS OF THE METHOD

The only merit that the method of simple averages possesses is its simplicity.

This method suffers from a **serious limitation** that it assumes that the data do not have any trend or cyclic or irregular variation or in any case their effect on the time series is negligible and can be ignored. Seasonal index computed by this method is actually an index of trends and seasons. ***This method tries to eliminate the effect of irregular and cyclic or random variables by averaging the monthly values over different years.*** This method serves the purpose because of its simplest form only where no definite trend exists.

Example 53. *Calculate the seasonal index for the following data by using the simple average method.*

Year	*1st Quarter*	*2nd Quarter*	*3rd Quarter*	*4th Quarter*
1974	*72*	*68*	*80*	*70*
1975	*76*	*70*	*82*	*74*
1976	*74*	*66*	*84*	*80*
1977	*76*	*74*	*84*	*78*
1978	*78*	*74*	*86*	*82*

Solution.

Table For Computation of Seasonal Indices

Year	*1st Quarter*	*2nd Quarter*	*3rd Quarter*	*4th Quarter*
1974	72	68	80	70
1975	76	70	82	74
1976	74	66	84	80
1977	76	74	84	78
1978	78	74	86	82
Total	376	352	416	384
Average	75.2	70.4	83.2	76.8
Seasonal index	**98.43**	**92.15**	**108.90**	**100.52**

Seasonal Index has been calculated as follows

$$\textbf{Average of Quarterly Averages} = \frac{75.2+70.4+83.2+76.8}{4} = \frac{305.6}{4} = \mathbf{76.4.}$$

$$\textbf{Seasonal Index for 1st Quarter} = \frac{\textbf{Average of 1st quarter}}{\textbf{Average of quarterly averages}} \times 100$$

$$= \frac{75.2}{76.4} \times 100 = \mathbf{98.43.}$$

$$\textbf{Seasonal Index for 2nd Quarter} = \frac{\textbf{Average of 2nd quarter}}{\textbf{Average of quarterly averages}} \times 100$$

$$= \frac{70.4}{76.4} \times 100 = \mathbf{92.15.}$$

$$\textbf{Seasonal Index for 3rd Quarter} = \frac{\textbf{Average of 3rd quarter}}{\textbf{Average of quarterly averages}} \times 100$$

$$= \frac{83.2}{76.4} \times 100 = \mathbf{108.90.}$$

$$\textbf{Seasonal Index for 4th Quarter} = \frac{\textbf{Average of 4th quarter}}{\textbf{Average of quarterly averages}} \times 100$$

$$= \frac{76.8}{76.4} \times 100 = \mathbf{100.52.}$$

Example 54. *Calculate seasonal indices for each quarter from the following percentages of whole sale price indices to their moving averages:*

	Quarter			
Year	*I*	*II*	*III*	*IV*
1986	—	—	*11.0*	*11.0*
1987	*12.5*	*13.5*	*15.5*	*14.5*
1988	*16.8*	*15.2*	*13.1*	*15.3*
1989	*11.2*	*11.0*	*12.4*	*13.2*
1990	*10.5*	*13.3*	—	—

Solution.

Table For Computation of Seasonal Indices.

Year	*Quarter* *I*	*II*	*III*	*IV*
1986	—	—	11.0	11.0
1987	12.5	13.5	15.5	14.5
1988	16.8	15.2	13.1	15.3
1989	11.2	11.0	12.4	13.2
1990	10.5	13.3	—	—
Total	**51.0**	**53.0**	**52.0**	**54.0**
Average	**12.75**	**13.25**	**13.0**	**13.5**

Average of Averages $= \frac{12.75+13.25+13.0+13.5}{4} = \frac{52.5}{4} = \mathbf{13.125}$.

SEASONAL INDICES

1st Quarter $= \frac{12.75}{13.125} \times 100 = \mathbf{97.143}$; **2nd Quarter** $= \frac{13.25}{13.125} \times 100 = \mathbf{100.952}$;

3rd Quarter $= \frac{13.0}{13.125} \times 100 = \mathbf{99.048}$; **4th Quarter** $= \frac{13.5}{13.125} \times 100 = \mathbf{102.852}$.

Seasonal indices are calculated by converting the respective quarterly averages on the basis of the average of averages = 100.

14.32 RATIO TO TREND METHOD

This method is an improvement over the simple average method of measuring seasonality. **It is based on the assumption that the seasonal fluctuations for any season (month, for monthly data and quarter for quarterly data) are a constant factor of the trend.** The following are the steps for measuring seasonal indices by this method.

(*i*) **Obtain the trend values (seasonwise) by the method of least squares or by parabolic curve or exponential curves.**

(*ii*) **Eliminate the Trend values. For eliminating trend in multiplicative model, we have**

$$\mathbf{Y = TSCI \times 100 \quad \text{or} \quad \frac{Y}{T} \times 100 = SCI \times 100.}$$

Now the percentages obtained by dividing the time series values by the trend values would contain S, C and I or seasonal, cyclic and irregular variations.

(*iii*) **To eliminate the cyclic and irregular variations from SCI, these values are arranged seasonwise for various years as is done in simple average method. These seasonwise figures for various years are averaged by using the median or arithmetic mean. These averages give the preliminary indices of seasonal variations.**

(*iv*) **These preliminary indices are adjusted to the total of 1200 for monthly data and to the total 400 for quarterly data by multiplying each one of them by a constant factor (c.f.) given by**

$$\text{c.f.} = \frac{1200}{\text{Sum of Averages of Monthly Indices}}$$

$$\text{or c.f.} = \frac{400}{\text{Sum of Averages of Quarterly Indices}}$$

These are the final indices of the seasonal variations which are nothing but the preliminary indices expressed as a percentage of their arithmetic mean.

Example 55. *Find seasonal variations by the Ratio to Trend method from the data given below:*

Year	*1st Quarter*	*2nd Quarter*	*3rd Quarter*	*4th Quarter*
1987	*30*	*40*	*36*	*34*
1988	*34*	*52*	*50*	*44*
1989	*40*	*58*	*54*	*48*
1990	*54*	*76*	*68*	*62*
1991	*80*	*92*	*86*	*82*

Solution. First, we have to determine the trend for yearly data by ***fitting a straight line trend by the method of least squares.***

Computation Table

Year	***Yearly Total***	***Quarterly Average (Y)***	***Deviation from mid year (X)***	***XY***	***X^2***	***Trend Values (Y_c)***
1987	140	35	– 2	– 70	4	32
1988	180	45	– 1	– 45	1	44
1989	200	50	0	0	0	56
1990	260	65	+ 1	+ 65	1	68
1991	340	85	+ 2	+ 170	4	80
		$\Sigma Y = 280$	$\Sigma X = 0$	$\Sigma XY = 120$	$\Sigma X^2 = 10$	

Equation of straight trend line: $Y = a + bX$.

Now $a = \frac{\Sigma Y}{N} = \frac{280}{5} = 56,$ $b = \frac{\Sigma XY}{\Sigma X^2} = \frac{120}{10} = 12.$

Therefore, **the Quarterly increment** = $\frac{12}{4} = 3.$

Now we have to calculate the quarterly trend value. Since there are four quarters in a year we have to start in between the 2nd and 3rd quarter each year. The trend values of each quarter can be calculated with the help of quarterly increment.

Quarterly trend Values

Year	***1st Quarter***	***2nd Quarter***	***3rd Quarter***	***4th Quarter***	***Trend***
1987	27.5	30.5	33.5	36.5	**32**
1988	39.5	42.5	44.5	48.5	**44**
1989	51.5	54.5	57.5	60.5	**56**
1990	63.5	66.5	69.5	72.5	**68**
1991	75.5	78.5	81.5	84.5	**80**

Next step is to present the quarterly trend values as percentage of the given trend values. Therefore, **for the 1st quarter of 1987 the percentage would be calculated as**

$\frac{30}{27.5} \times 100 = 1091$. Similarly, the other trend values can be expressed as percentages as compared with the given values in the time series. For the second quarter of 1990 the percentage would be $\frac{76}{55.5} \times 100 = 1143$.

Quantity Values as% of Quarterly Trend Values

Year	*1st Quarter*	*2nd Quarter*	*3rd Quarter*	*4th Quarter*
1987	109.1	131.1	107.5	93.1
1988	86.1	122.4	109.5	90.7
1989	77.7	106.4	93.7	70.3
1990	85.0	114.3	97.8	85.5
1991	106.0	117.1	105.5	97.0
Total	**463.9**	**591.3**	**514.6**	**445.6**
Average	**92.78**	**118.26**	**102.92**	**89.12**
Adjusted Seasonal index	**92.0**	**117.4**	**102.1**	**88.4**

Total of the Averages = 92.78 + 118.26 + 102.92 + 89.12 = 403.08

As the Total of the Average is more than 400 therefore adjustment of correction factor is to be made.

Correction factor = $\frac{400}{403.08}$.

To find the adjusted seasonal index each quarterly average is to be multiplied by the correction factor.

1st Quarter: $92.78 \times \frac{400}{403.08} = \mathbf{92.0}$

2nd Quarter: $118.26 \times \frac{400}{403.08} = \mathbf{117.4}$

3nd Quarter: $102.92 \times \frac{400}{403.08} = \mathbf{102.1}$

4th Quarter: $89.12 \times \frac{400}{403.08} = \mathbf{88.4}$.

Example 56. *Find seasonal variations by Ratio to Trend method from the data given below:*

Year	*1st Quarter*	*2nd Quarter*	*3rd Quarter*	*4th Quarter*
1987	*34*	*54*	*38*	*38*
1988	*36*	*60*	*52*	*48*
1989	*40*	*58*	*56*	*52*
1990	*52*	*76*	*64*	*58*
1991	*70*	*90*	*88*	*84*

Solution. For using the Ratio to Trend method of calculating seasonal variation, first we have to determine the trend for yearly data.

Table for Computation of Trend by the Least Squares Method

Year	*Yearly Totals*	*Quarterly Average (Y)*	*Deviation from midyear (X)*	*XY*	X^2	*Trend values*
1987	162	40.5	– 2	– 81.0	4	**37.60**
1988	196	49.0	– 1	49.0	1	**47.45**
1989	206	51.5	0	0	0	**57.30**
1990	250	62.5	+ 1	+ 62.5	1	**67.15**
1991	332	83.0	+ 2	+ 166.0	4	**77.00**
$N = 5$		$\Sigma Y = 286.5$	$\Sigma X = 0$	$\Sigma XY = 98.5$	$\Sigma X^2 = 10$	

The equation of the straight line trend is: Y = *a* + *bX*,

$$a = \frac{\Sigma Y}{N} = \frac{286.5}{5} = 57.3; \qquad b = \frac{\Sigma XY}{\Sigma X^2} = \frac{98.5}{10} = 9.85.$$

$\therefore$ **The equation of trend is: $Y = 57.3 + 9.85\,X$**

The Quarterly increment is = $\frac{9.85}{4} = 2.46.$

Now we have to calculate the quarterly trend value. Since there are four quarters in a year, so we have to start in between the 2nd and 3rd quarter each year. The trend values of each quarter can be calculated with the help of quarterly increment.

Quarterly Trend Values

Year	*1st Quarter*	*2nd Quarter*	*Trend value*	*3rd Quarter*	*4th Quarter*
1987	33.91	36.37	37.60	38.83	41.29
1988	43.76	46.22	47.45	48.68	51.14
1989	53.61	56.07	57.30	58.53	60.90
1990	63.46	65.92	67.55	68.38	70.84
1991	73.31	75.77	77.00	78.23	80.69

Next step is to present the Quarterly trend vlaues as percentage of the given trend values.

For **the First Quarter of 1987 the percentage would be calculated as** $\frac{34}{33.91} \times 100 = 100.30.$

Similarly, the other trend values can be expressed as percentage by comparing with the given values in the time series data.

Quarterly Value as % of Quarterly Trend Values

Year	*1st Quarter*	*2nd Quarter*	*3rd Quarter*	*4th Quarter*
1987	100.30	143.48	97.86	87.19
1988	82.28	129.81	106.82	93.86
1989	74.63	103.44	95.68	85.26
1990	81.95	115.30	93.60	81.88
1991	95.50	118.79	112.48	104.10
Total	**434.66**	**615.82**	**506.44**	**452.29**
Average	**86.93**	**123.16**	**101.29**	**90.46**
Adjusted seasonal index	$86.96 \times \frac{400}{401.84}$ **= 86.53**	$123.16 \times \frac{400}{401.84}$ **= 122.60**	$101.29 \times \frac{400}{401.84}$ **= 100.83**	$90.46 \times \frac{400}{401.84}$ **= 90.05**

The adjustment of seasonal index

Total of the averages = 86.93 + 123.16 + 101.29 + 90.46 = **401.84.**

As the total of the averages is more than 400 therefore adjustment with correction factors is to be made to the averages.

The **correction factor (c.f.)** $= \dfrac{400}{401.84}$.

The adjusted seasonal index is given by the last row of the table.

14.33 RATIO TO THE MOVING AVERAGES METHOD

This method is also known as **Percentage of Moving Average Method.** The steps involved in the computation of seasonal indices by this method are as follows:

(*i*) **Obtain the centered 12 months (4 quarter) moving averages values for the given series. This step eliminates the seasonal fluctuations if they are regular and constant. When we average the data by the moving averages method, the average would only have the trend and cyclical variation, the irregular variation would disappear.**

(*ii*) **Express each original value of the time series as a percentage of the trend value.**

(*iii*) **Arrange these percentages seasonwise for all the years and then calculate the average of these percentages, the resultant percentages would be seasonal indices.**

(*iv*) **The sum of these indices should be 1200 (for monthly data) or 400 (for quarterly data). But if it is not so, then it should be adjusted with the help of a correction factor (c.f.)**

$$\text{c.f.} = \frac{1200}{\text{Sum of monthly indices}} \qquad \textbf{[For monthy data]}$$

$$\text{or c.f.} = \frac{400}{\text{Sum of quarterly indices}}. \qquad \textbf{[For quarterly data]}$$

For adjusting with the help of c.f. the seasonal indices have to be multiplied by the c.f. The resultant ratio will be the Ratio to Moving Averages Indices.

MERITS AND LIMITATIONS

It is most widely used method for studying seasonal variations. This metod eliminates the trend and cyclical variations. If cycles are not regular then only in that case this method contains some effect of cyclical variation.

The limitation of this method is that there are no seasonal indices at the beginning and at the end of the series.

Example 57. *From the following data calculate seasonal indices by the Ratio to Moving Average method.*

Year	*1st Quarter*	*2nd Quarter*	*3rd Quarter*	*4th Quarter*
1981	*68*	*62*	*61*	*63*
1982	*65*	*58*	*56*	*61*
1983	*68*	*63*	*63*	*67*
1984	*70*	*59*	*56*	*62*
1985	*60*	*55*	*51*	*58*

[M.B.A; Anna University – Dec. 2002]

Solution.

Table For Computation of Seasonal Indices by Ratio to Moving Average

Year	*Quarter*	*Values*	*4 quarters moving totals*	*4 quarters moving totals centered*	*4 quarters moving average centered*	*Given vlaues as percentage of moving average*
1981	1	68	—	—	—	—
	2	62	—	—	—	—
			254	—	—	—
	3	61	—	**505**	63.1	**96.7**
—	—	—	251	—	—	—
	4	63	—	**498**	62.2	**101.3**
—	—	—	247	—	—	—
1982	1	65	—	**489**	61.1	**106.4**
—	—	—	242	—	—	—
	2	58	—	**482**	60.2	**96.3**
			240			
	3	56	—	**483**	60.4	**92.7**
—	—	—	243	—	—	—
	4	61	—	**491**	61.4	**99.3**
		—	248	—	—	—
1983	1	68	—	**503**	62.9	**108.2**
		—	255	—	—	—
	2	63	—	**516**	64.5	**97.7**
		—	261	—	—	—
	3	63	—	**524**	65.5	**96.2**
			263	—	—	—
	4	67	—	**522**	65.2	**102.8**
		—	259	--	—	—
1984	1	70	—	**511**	63.9	**109.5**
		—	252	—	—	—
	2	59	—	**499**	62.4	**94.6**
		—	247	—	—	—
	3	56	—	**484**	60.5	**92.6**
		—	237	—	—	—
	4	62	—	**470**	58.8	**105.4**
		—	233	—	—	—
1985	1	60	—	**461**	57.6	**104.2**
		—	228	—	—	—
	2	55	—	**452**	56.6	**97.3**
		—	224	—	—	—
	3	51	—	—	—	—
	4	58	—	—	—	—

Table for Computation of Seasonal Index.

Year	*Ind Quarter*	*2nd Quarter*	*3rd Quarter*	*4th Quarter*
1981	—	—	96.7	101.3
1982	106.1	96.3	92.7	99.3
1983	108.1	97.7	96.2	102.8
1984	109.5	94.6	92.2	105.4
1985	104.20	97.3	—	—
Total	**428.20**	**385.2**	**378.2**	**408.8**
Average	**107.10**	**96.5**	**94.5**	**102.2**
Adjusted Seasonal Index	**107.02**	**96.43**	**94.43**	**102.12**

The adjustment of seasonal index is required because the total of average of seasonal index is 107.1 + 96.5 + 94.5 + 102.2 = 400.3 which is more than 400.

Correction Factor (c.f.) = $\frac{\mathbf{400}}{\mathbf{400.3}}$

Each seasonal index average has to be adjusted by the c.f.

Adjusted Seasonal Index for 1st Quarter $= 107.1 \times \frac{400}{400.3} = \mathbf{107.02}$

Adjusted Seasonal Index for 2nd Quarter $= 96.5 \times \frac{400}{400.3} = \mathbf{96.43}$

Adjusted Seasonal Index for 3rd Quarter $= 94.5 \times \frac{400}{400.3} = \mathbf{94.43}$

Adjusted Seasonal Index for 4th Quarter $= 102.2 \times \frac{400}{400.3} = \mathbf{102.12.}$

Example 58. *Compute seasonal indices by the Ratio to Moving Average method from the following data.*

Year	*1st Quarter*	*2nd Quarter*	*3rd Quarter*	*4th Quarter*
1971	*75*	*60*	*54*	*59*
1972	*86*	*65*	*63*	*80*
1973	*90*	*72*	*66*	*85*
1974	*100*	*78*	*72*	*93*

Solution.

Table for Computation of Seasonal Indices by The Ratio to Moving Average Method

Year	Quarter (1)	Values (2)	4-Quarterly moving totals (3)	4-Quarterly moving totals centered (4)	4-Quarterly moving average Col.5 = $\frac{Col.4}{8}$ (5)	Given values as percentage of moving averages
1971	1	75	—	—	—	—
	2	60	—	—	—	—
		—	248	—	—	—
	3	54	—	507	**63.375**	**85.21**
			259	—	—	—
	4	59	—	523	**65.375**	**90.25**
		—	264	—	—	—
1972	1	86	—	537	**67.125**	**128.12**
		—	273	—	—	—
	2	65	—	567	**70.875**	**91.71**
		—	294	—	—	—
	3	63	—	592	**74.000**	**85.13**
		—	298	—	—	—
	4	80	—	603	**75.375**	**106.14**
		—	305	—	—	—
1973	1	90	—	613	**76.625**	**117.45**
		—	308	—	—	—
	2	72	—	621	**77.625**	**92.75**
		—	313	—	—	—
	3	66	—	636	**79.500**	**83.02**
		—	323	—	—	—
	4	85	—	652	**81.500**	**104.29**
		—	329	—	—	—
1974	1	100	—	664	**83.00**	**120.48**
		—	335	—	—	—
	2	78	—	678	**84.756**	**92.03**
		—	343	—	—	—
	3	72	—	—	—	—
	4	93	—	—	—	—

Table for Computation of Seasonal Indices

Year	1st Quarter	2nd Quarter	3rd Quarter	4th Quarter
1971	—	—	85.21	90.25
1972	128.12	91.71	85.13	106.14
1973	117.45	92.75	83.02	104.29
1974	120.48	92.03	—	—
Total	**366.05**	**276.49**	**253.36**	**300.68**
Average	**122.02**	**92.16**	**84.45**	**100.23**
Adjusted Seasonal Index	**122.37**	**92.42**	**84.69**	**100.52**

The adjustment of seasonal index is required when the total of the average of seasonal index is more or less than 400. In this case 122.02 + 92.16 + 84.45 + 100.23 is 398.86 is less than 400. The correction factor will be

$$\text{c.f.} = \frac{400}{398.86}.$$

Each seasonal index average has to be adjusted by c.f.

The Adjusted seasonal index are :

Adjusted Seasonal Index for 1st Quarter $= 122.02 \times \frac{400}{398.86} = \mathbf{122.37}$

Adjusted Seasonal Index for 2nd Quarter $= 92.16 \times \frac{400}{398.86} = \mathbf{92.42}$

Adjusted Seasonal Index for 3rd Quarter $= 84.45 \times \frac{400}{398.86} = \mathbf{84.69}$

Adjusted Seasonal Index for 4th Quarter $= 100.23 \times \frac{400}{398.86} = \mathbf{100.52.}$

14.34 LINK RELATIVE METHOD

The procedure for computing seasonal variation by Link relative method is as follows:

WORKING RULES:

Step 1. Calculate the Link relatives of the seasonal figures. Link Relative is the value of the given phenomenon in any season expressed as a percentage of its value in the preceding season. Symbolically.

$$\textbf{Link Relatives (L.R.)} = \frac{\textbf{Current Season's figure}}{\textbf{Previous Season's figures}} \times \mathbf{100.}$$

Step 2. Calculate the average of the link relatives for each season by using arithmetic mean or median.

Step 3. Convert these averages into chain relatives on the basis of the first season.

Step 4. Calculate the chain relatives of the first season on the basis of the last season.

Step 5. For correction, the chain relative of the first season calculated by the first method is subtracted from the chain relative of the first season calculated by the second method. The difference is divided by the number of seasons. The resulting figure multiplied by 1, 2, 3, etc. are deducted respectively from chain relatives of 2nd, 3rd 4th etc. seasons.

Step 6. The seasonal indices are available when we express the corrected chain relatives as percentage of their respective averages.

Example 59. *Apply the method of link relatives to the following data and calculate the seasonal indices.*

	Price of Rice (in Rupees per 10 kg)			
Quarter ↓ Year →	*1980*	*1981*	*1982*	*1983*
1	*75*	*86*	*90*	*100*
2	*60*	*65*	*72*	*78*
3	*54*	*63*	*66*	*72*
4	*59*	*80*	*82*	*93*

Solution.

Computation of Seasonal Indices by Link Relative Method.

Year	*Link Relatives*			
	1st Quarter	*2nd Quarter*	*3rd Quarter*	*4th Quarter*
1980	—	80.00	90.00	109.26
1981	145.76	75.58	96.82	126.98
1982	112.50	80.00	91.67	128.79
1983	117.65	78.00	92.31	129.17
Total of Link Relatives	**375.91**	**313.58**	**370.90**	**494.20**
Average of Link Relatives	**125.303**	**78.395**	**92.725**	**123.55**
Chain Relatives	**100**	$\frac{78.395\times100}{100}$ **= 78.395**	$\frac{78.395\times92.725}{100}$ **= 72.690**	$\frac{123.55\times72.69}{100}$ **= 89.810**
Adjusted Chain Relatives	**100**	78.395 – 3.135 **= 75.26**	72.690 – 6.270 **= 66.42**	89.810 – 9.405 **= 80.41**
Seasonal Indices	**124.20**	**93.47**	**82.49**	**99.87.**

WORKING NOTES :

Chain Relative of the 1st Quarter (On the Basis of 1st Quarter) = 100.

Chain Relative of the 1st Quareter (On the Basis of Last Quarter)

$$= \frac{125.303\times89.81}{100} = \mathbf{112.54.}$$

Difference per Quarter = $d = \frac{1}{4}$ **(New C.R. of 1st Quarter – 100)**

$$= \frac{1}{4}(112.54-100) = \frac{1}{4}\times12.54 = \mathbf{3.135.}$$

Adjusted chain relatives for 2nd, 3rd and 4th Quarters are obtained on subtracting *d*, 2*d* and 3*d* from the corresponding chain relatives.

Sum of adjusted Chain Relatives = 100 + 75.26 + 66.42 + 80.41 = **322.09.**

Indices of seasonal variations are obtained on adjusting these adjusted chain relatives to a total of 400 by multiplying each one of them with a constant factor

$$\text{c.f.} = \frac{400}{\text{Sum of adjusted Chain Relatives}} = \frac{400}{322.09}.$$

Example 60. *The data below give the average quarterly prices of a commodity for five years. Calculate seasonal indices by the method of link relatives:*

Quarters	*Years*				
	1	*2*	*3*	*4*	*5*
1	*30*	*35*	*31*	*31*	*34*
2	*26*	*28*	*29*	*31*	*36*
3	*22*	*22*	*28*	*25*	*26*
4	*31*	*36*	*32*	*35*	*33*

Solution. Table for calculations of seasonal indices by the Method of Link Relatives.

Years	Q_1	Q_2	Q_3	Q_4
1		86.7	84.6	140.9
2	112.9	80.0	78.6	163.6
3	86.1	93.5	96.6	114.3
4	96.9	100.0	80.6	140.0
5	97.1	105.9	72.2	126.9
Total	**393**	**466.1**	**412.6**	**685.7**
A.M.	**98.25**	**93.22**	**82.52**	**137.14**
Chain values	**100**	**93.22**	$\frac{93.22 \times 82.52}{100}$ **= 76.93**	$\frac{76.93 \times 137.14}{100}$ **= 105.5**
Adjusted chain relatives	**100**	**93.22 – *d* = 92.32**	**76.93 – 2*d* = 75.13**	**105.5 – 3*d* = 102.8**
Seasonal indices	**108.1**	**99.8**	**81.22**	**111.13**

Chain relative of Q_4 with respect to $Q_1 = \frac{105.5 \times 98.25}{100} = \mathbf{103.6}$

Difference per Quarter $= \frac{1}{4}(103.6 - 100) = \frac{3.6}{4} = \mathbf{0.9}$

Average of adjusted chain relatives $= \frac{100 + 92.32 + 75.13 + 102.58}{4} =$ **92.5075**

Season Indices $= \frac{\textbf{Adjusted chain relative}}{\textbf{92.5075}} \times \mathbf{100}$

Example 61. *Calculate seasonal indices by the Link Relative method from the data given below:*

Output of Rice in Million Tonnes

Year	*1st Quarter*	*2nd Quarter*	*3rd Quarter*	*4th Quarter*
1987	*60*	*65*	*62*	*69*
1988	*62*	*68*	*65*	*68*
1989	*65*	*70*	*64*	*62*
1990	*70*	*75*	*68*	*67*
1991	*72*	*80*	*70*	*78*

Solution.

Table for Computation of Seasonal Indices by Link Relatives Method

	Link Relatives			
Year	*1st Quarter*	*2nd Quarter*	*3rd Quarter*	*4th Quarter*
1987		$\frac{65}{60}\times 100 = \mathbf{108.3}$	$\frac{62}{65}\times 100 = \mathbf{95.4}$	$\frac{69}{62}\times 100 = \mathbf{111.3}$
1988	$\frac{62}{69}\times 100 = \mathbf{90.0}$	$\frac{68}{62}\times 100 = \mathbf{109.7}$	$\frac{65}{68}\times 100 = \mathbf{95.6}$	$\frac{68}{65}\times 100 = \mathbf{104.6}$
1989	$\frac{65}{68}\times 100 = \mathbf{95.6}$	$\frac{70}{65}\times 100 = \mathbf{107.7}$	$\frac{64}{70}\times 100 = \mathbf{91.4}$	$\frac{62}{64}\times 100 = \mathbf{96.9}$
1990	$\frac{70}{62}\times 100 = \mathbf{112.9}$	$\frac{75}{70}\times 100 = \mathbf{107.1}$	$\frac{68}{75}\times 100 = \mathbf{90.7}$	$\frac{67}{68}\times 100 = \mathbf{98.5}$
1991	$\frac{72}{67}\times 100 = \mathbf{107.5}$	$\frac{80}{72}\times 100 = \mathbf{111.1}$	$\frac{70}{80}\times 100 = \mathbf{87.5}$	$\frac{78}{70}\times 100 = \mathbf{111.4}$
Total of Link Relatives	**406.0**	**543.9**	**460.6**	**522.7**
Arithmetic Mean	**101.50**	**108.78**	**92.12**	**104.54**
Chain Relatives	**100**	$\frac{108.78\times 100}{100}$ **= 108.78**	$\frac{108.78\times 92.12}{100}$ **= 100.208**	$\frac{100.208\times 104.54}{100}$ **= 104.757**
Adjusted Chain Relatives	**100**	108.78 – 1.582 **= 107.198**	100.208 – 3.164 **= 97.044**	104.757 – 4.746 **= 100.011**
Seasonal Index	**100**	$107.198\times\frac{400}{404.253}$ **= 106.07**	$97.044\times\frac{400}{404.253}$ **= 96.02**	$100.011\times\frac{400}{404.253}$ **= 98.96.**

Example 62. *Distinguish between Historigram and Histogram.*

Solution. Historigram: ***Time series is a record of the values of a variate (i.e., quantity) during a particular period taken at successive intervals of time.*** When the values of the variate are plotted against time on a graph paper, and the points so obtained are joined by straight line segments, a graphical record of the behaviour of the variate becomes available.

(In this graph the time is shown on X-axis whereas the variable, i.e., quantity is shown on the Y-axis. The values of the variable are plotted against corresponding time period and the points are joined by straight lines).

The graph of the time series so obtained is called a Historigram of the time series. A historigram gives a rough idea about the nature of changes in the value of the variate with time.

Histogram. ***A histogram is graphic or diagrammatic representation of a continuous grouped frequency distribution. It is a set of vertical bars whose areas are proportional to the frequency of the corresponding class.*** In this case, the data is shown in the form of rectangles. The class intervals are shown on the *x*-axis whereas frequencies against the class intervals are shown on the y-axis. Thus there are as many rectangles as there are class intervals. In other words, **a histogram is a graph containing a set of adjacent rectangles, each being constructed to represent the size of class interval by its width and the frequency of the class interval by its height.** The altitude of the rectangles vary with the frequency of the interval and the bases may be of equal or unequal width according as the frequency distribution consists of equal or unequal class intervals. Thus there are two types of histograms: **(*a*) Histogram with equal width, (*b*) Histogram with unequal width. The rise or fall of the boundary lines shows the increase or decrease in the magnitude of the variable from interval to interval.**

The series of rectangles fairly shows the relative size of the various groups and the entire distribution of the frequencies among the several classes becomes at once visible. The area of a rectangle is proportional to the frequency of the corresponding class. **The total area under the histogram being proportional to the total frequencies of all the classes. Thus a histogram presents an accurate picture of the relative proportions of the total frequency from interval to interval (as the total area of the histogram equals the total number of frequencies).**

Histograms are used for the graphic location of the value of Mode.

A histogram has the following defects.

(i) Histogram cannot be constructed for frequency distribution with open end classes (i.e., for inclusive series).

(ii) Different groupings will give different shapes of histograms for the same data which are not comparable.

(iii) It is not possible to compare two distributions with reference to their histograms.

(iv) It does not show a continuous distribution of the values of the variable within a particular class interval and might lead one to conclude that all the variables in the group are situated at the mid-value of the group.

A histogram is also known as 'Block Diagram' or 'Staircase chart'.

Example 63. *Distinguish between Additive model and Multiplicative Model in the Analysis of Time Series.*

Solution. The analysis of a time series depends on the following four components.

(1) *Trend (T)*; **(2) *Seasonal variations (S)*;**

(3) *Cyclic fluctuations (C)*; **(4) *Irregular variations or fluctuations***

It may be noted that any one or all of these components may be present in any particular series. There are two models to represent the analysis of a time series, *viz.*,

(*i*) **Additive Model** and,

(*ii*) **Multiplicative Model.**

Additive Model. In this model the four components are assumed to be additive in effect. In this model original data *Y* is expressed as:

$$Y = T + C + S + I,$$

Each of these components can be eliminated by subtraction. The method of moving averages elminates *C, S, I* and thus determines the Trend (*T*). Then subtracting the estimated trend, we get the short time fluctuations as:

$$Y - T = C + S + I$$

If the values of *Y* are known for weeks, months or quarters of a number of years, then the moving average eliminates seasonal variation and gives us $T + C$. Subtracting this from the original series, we get the seasonal variation.

One disadvantage of additive model is that all the values are in absolute terms. Their relative movements are not comparable.

This disadvantage is avoided in the multiplicative model.

Multiplicative Model. In multiplicative model we have

$$Y = T \times S \times C \times I$$

Here *S*, *C* and *I* are expressed in ratios and percentages, whereas *T* is expressed in the units of original series.

This concept is generally more useful than that of the additive model.

The reason for this is **that the short time fluctuations *S*, *C* and *I* are nearly constant in magnitude relative to Trend rather than in absolute terms.** In addition, the movements of the variable are more meaningful when considered relative to each other than when considered in absolute terms. Here we have

$$(T \times C \times S \times I) \div S = T \times C \times I$$

and $$(T \times C \times I) \div T = C \times I.$$

When multiplied by 100, the first quotient gives **Trend-cycle as a percentage of Seasonal Fluctuations and the second quotient gives the Cyclical Movement as percentage of Trend.**

When the seasonal Index is very small, division by the same will increase the percentage Trend-cycle values sharply in the deseasonalised data. In such a situation, it is preferable to use additive model and eliminate the seasonal effect by subtraction.

In additive model, all units are expressed in units and in multiplicative model, the components *C*, *S* and *I* are expressed in relatives or percentages and Trend (*T*) is expressed in the units of the original series. In the additive model *S*, *C* and *I* are quantitatives deviations about trend that can be expressed as seasonal, cyclical and irregular in nature. In additive model only absolute values are added or deducted from the trend value to arrive at the observed value.

The Additive Model is based on the assumption that the different components viz., S, C, T and I are independent of each other.

In Multiplicative Model, it is presumed that the components S, C, T and I are dependent on each other, so that seasonal variations increase with the rising trend and that the ratio of cyclical fluctuations to trend also similarly remains constant.

Example 64. *Differentiate between Ratio to Moving Average Method and Ratio to Trend Method.*

Solution. Ratio to Moving Average Method: A seasonal index computed by **Ratio to Moving Average** method ordinarily **does not fluctuate so much as the index is based on straight line trends.** This is so because the 12-month moving average follows cyclical course of the actual data quite closely. Therefore, the index ratios obtained by this method are often more representative of the data which they are obtained (*i.e.*, true representative of data) than in the case of **Ratio of Trend method.** This method is more **flexible and satisfactory measuring seasonal index.** The index obtained by this method is less fluctuative than obtained from Ratio to Trend method. This method has a **draw back that seasonal indices cannot be obtained for each month for which data are available.** When a 12 month moving average is taken, then the six months in the beginning and the six months in the end are left out for which we cannot calculate seasonal indices.

Ratio To Trend Method. The **ratio to trend** method is similar to ratio to moving average method. The only difference lies in the way of obtaining **Trend values.** In case of **Ratio to Moving Average** method the trend *values are obtained by the method of moving average but in* **Ratio to Trend Method,** *the corresponding trend is obtained by the method of least squares.* This method has the advantage of simplicity and ease of interpretation. There is no loss of data as is occurred in ratio to moving average method. *It is a logical method of measuring seasonal variations.*

This method has the following limitations:

Limitations:

1. ***It may influenced by errors in the calculation of trend.***
2. ***It may be influenced by cyclical and erratic influences. But this error is eliminated by the selection of period of time in which depression is offset by prosperity.***
3. ***There will be more bias in a seasonal index computed by ratio to trend method.***

Example 65. *What are the merits and limitations of Ratio to Trend Method?*

Merits.

1. It gives a more representative value of seasonal variation than the simple average method.
2. It is easy to understand and simple to calculate.

Limitation.

This method does not follow the actual data as a 12 month moving average method does. So the seasonal index calculated by this method is more biased.

Example 66. *Define Trend and its types. What are its uses?*

Trend. The term trend (or Secular trend) is popularly used in business, economics and in our daily conversation on day to day subjects. When we speak of **rising trend** of prices or rising crimes, we mean the gradual increase in the prices or crimes over a peirod of time. Similarly, by **declining trend** of sales or production, we mean gradual decrease in the sales or production over a period of time.

Trend (or Secular Trend) is a smooth, regular and long term movement of a series showing continuous growth, stagnation or decline over a long period of time, Graphically, it shows general direction and shape of time series. The concept of trend does not include short range oscillations, but it referes to steady movements over a **long period of time. The various types of Trends are:**

1. **Linear or Straight line Trend.**
2. **Non-linear Trend.**

Uses of Trend.

1. ***The analysis helps us to compare two or more time series over different periods of time and to draw important conclusions about them.***
2. ***The study of data over a long period of time give us a general idea about the pattern of the behaviour of the phenomenon under consideration.***
3. ***It helps in business forecasting and planning future operations.***

Example 67. *Calculate the seasonal index for the following data by using the Average Method assuming trend is absent.*

Year	*1st Quarter*	*2nd Quarter*	*3rd Quarter*	*4th Quarter*
1987	*3.7*	*4.1*	*3.3*	*3.5*
1988	*3.7*	*3.9*	*3.6*	*3.6*
1989	*4.0*	*4.1*	*3.3*	*3.1*
1990	*3.3*	*4.4*	*4.0*	*4.0*

Solution.

Table for Computation of Seasonal Indices

Year	*1st Quarter*	*2nd Quarter*	*3rd Quarter*	*4th Quarter*
1987	3.7	4.1	3.3	3.5
1988	3.7	3.9	3.6	3.6
1989	4.0	4.1	3.3	3.1
1990	3.3	4.4	4.0	4.0
Total	14.7	16.5	14.2	14.2
Average	3.675	4.125	3.550	3.550
Seasonal index	$\frac{3.675}{3.725}\times 100$ **= 98.7**	$\frac{4.125}{3.725}\times 100$ **= 110.7**	$\frac{3.550}{3.725}\times 100$ **= 95.3**	$\frac{3.550}{3.725}\times 100$ **= 95.3**

Average of quarterly averages $= \frac{3.675+4.125+3.550+3.550}{4} = \mathbf{3.725.}$

Example 68. *Calculate seasonal indices for each quarter from the following percentage of wholesale price indices to their moving averages:*

Year	*1st Quarter*	*2nd Quarter*	*3rd Quarter*	*4th Quarter*
1985	—	—	*127*	*134*
1986	*130*	*122*	*122*	*132*
1987	*120*	*120*	*118*	*128*
1988	*126*	*116*	*121*	*130*
1989	*128*	*118*	—	—

Solution.

Table for Computation of Seasonal Indices

Year	*1st Quarter*	*2nd Quarter*	*3rd Quarter*	*4th Quarter*
1985	—	—	127	134
1986	130	122	122	132
1987	120	120	118	128
1988	126	116	121	130
1989	128	118	—	—
Total	**504**	**476**	**488**	**524**
Average	**126**	**119**	**122**	**131**
Adjusted seasonal index	$126\times\frac{400}{498}$ **= 101.20**	$119\times\frac{400}{498}$ **= 95.58**	$122\times\frac{400}{498}$ **= 97.99**	$131\times\frac{400}{498}$ **= 105.22**

The total of averages = 126 + 119 + 122 + 131 = **498.**

The correction factor (c.f.) $= \mathbf{\frac{400}{498}}$.

Example 69. *Calculate trend values for the following data using a second degree equation:*

Year :	*1984*	*1985*	*1986*	*1987*	*1988*	*1989*	*1990*
Output (thousand tons) :	*100*	*107*	*128*	*140*	*181*	*192*	*200*

Also make a forecast for 1993.

Solution. The second degree trend is given by the equation:

$$Y = a + bX + cX^2.$$

The normal equations for solving *a, b, c* are

$$\Sigma Y = Na + b\Sigma X + c\Sigma X^2$$
$$\Sigma XY = a\Sigma X + b\Sigma X^2 + c\Sigma X^3$$
$$\Sigma X^2 Y = a\Sigma X^2 + b\Sigma X^3 + c\Sigma X^4.$$

Table for Fitting of Quadratic Trend

Year	*Output (Y)*	*Deviations from 1987 (X)*	*XY*	X^2	X^2Y	X^3	X^4
1984	100	– 3	– 300	9	900	– 27	81
1985	107	– 2	– 214	4	428	– 8	16
1986	128	– 1	– 128	1	128	– 1	1
1987	140	0	0	0	0	0	0
1988	181	+ 1	+ 181	1	181	+ 1	1
1989	192	+ 2	+ 384	4	768	+ 8	16
1990	200	+ 3	+ 600	9	1800	+ 27	81
$N = 7$	$\Sigma Y = 1048$	$\Sigma X = 0$	$\Sigma XY = 523$	$\Sigma X^2 = 28$	$\Sigma X^2 Y = 4205$	$\Sigma X^3 = 0$	$\Sigma X^4 = 196$

Substituting the values in normal equations, we get:

$$1048 = 7a + 28c \quad \text{... (i)}$$
$$523 = 28b \quad \text{... (ii)}$$
$$4205 = 28a + 196c \quad \text{... (iii)}$$

Solving (*ii*) for value of *b*, we get $\mathbf{b} = \frac{523}{28} = \mathbf{18.68.}$

Solving (*i*) and (*iii*) for value of *a* and *c* by multiplying equation (*i*) by 4 and subtracting from it equation (*iii*), we get

$$4192 = 28a + 112c$$
$$4205 = 28a + 197c$$
$$\Rightarrow \quad \mathbf{-13 = -84c} \quad \Rightarrow \quad \mathbf{c} = \frac{13}{84} = \mathbf{0.15.}$$

Substituting the value of *c* in equation (*i*), we get

$$1048 = 7a + 28 \times 0.15 \Rightarrow 1048 - 28 \times 0.15 = 7a$$
$$\Rightarrow \quad 1048 - 4.2 = 7a \Rightarrow \mathbf{a} = \frac{1043.8}{7} = \mathbf{149.11.}$$

Hence $\mathbf{Y = 149.11 + 18.68X + 0.15X^2}$

Estimate for 1993; $X = +6, X^2 = 36$

$$Y_{1993} = 149.11 + 18.68 \times 6 + 0.15 \times 36 = 149.11 + 112.08 + 5.4$$
$$= \mathbf{266.69.}$$

Example 70. *The seasonal index of the rate of readymade garments in a store is given below:*

Quarter	*Seasonal index*
January to March	*98*
April to June	*90*
July to September	*82*
October to December	*130.*

If the total sales of garments in the first quarter is worth Rs. 1,00,000, determine how much worth of garments of the type should be kept in stock to meet the demand in each of the remaining quarters.

Solution. Total sales in the first quarter is Rs. 1,00,000. Since the seasonal index for the first quarter is 98, thus the seasonal effect is 98/100 or 0.98.

The deseasonalised sale for the first quarter would be = $\frac{100000}{98} \times 100 = 10{,}20{,}40.81.$

On the basis of the sale the estimated worth of garments for the remaining quarters would be

	Seasonal index: S.I.	**Estimated Sales**
April to June	**90**	$102040.81 \times \frac{90}{100} = \mathbf{91836.72}$
July to Sept.	**82**	$102040.81 \times \frac{82}{100} = \mathbf{83673.46}$
Oct to Dec.	**130**	$102040.81 \times \frac{130}{100} = \mathbf{132653.05.}$

Example 71. *Population figures for a city are as given below. Fit a curve of the type $Y = ab^X$ and estimate the population for 1987.*

Year :	*1981*	*1982*	*1983*	*1984*	*1985*
Population ('000) :	*132*	*142*	*157*	*170*	*191*

Solution.

Table for the Computation of Equation of the Form $Y = ab^x$

Year	*Population (Y)*	*Deviation from 1983 (X)*	*log Y*	*X log Y*	X^2
1981	132	– 2	2.1206	– 4.2412	4
1982	142	– 1	2.1523	– 2.1523	1
1983	157	0	2.1954	0	0
1984	170	+ 1	2.2307	+ 2.2307	1
1985	191	+ 2	2.2810	+ 4.5620	4
$N = 5$	$\Sigma Y = 792$	$\Sigma X = 0$	$\Sigma \log Y = 10.9800$	$\Sigma X \log Y = 0.3992$	$\Sigma X^2 = 10$

$$\log a = \frac{\Sigma \log Y}{N} = \frac{10.9800}{5} = 2.1960$$

$$\log b = \frac{\Sigma X \log Y}{\Sigma X^2} = \frac{0.3992}{10} = 0.03992.$$

The required curve is $\log Y = \mathbf{2.1960 + 0.03992X}$

or $Y = \textbf{antilog } (\mathbf{2.1960 + 0.03992\,X})$

For 1987; $X = 4$

$\log Y_{1987} = 2.1960 + 0.03992 \times 4 = 2.1960 + 0.15968 = \mathbf{2.35568.}$

$\therefore$ **Y_{1987} = Antilog (log Y)** = Antilog (2.35568) = **226.819 thousands**

Hence, **the estimated population for 1987 = 226819.**

Example 72. *A company estimates its sales for a particular year to be Rs. 24,00,000. The seasonal indices for sales are as follows:*

Month	*Seasonal index*	*Month*	*Seasonal index*
January	*75*	*July*	*102*
February	*80*	*August*	*104*
March	*98*	*September*	*100*
April	*128*	*October*	*102*
may	*137*	*November*	*82*
June	*119*	*December*	*73*

Using this information, calculate estimates of monthly sales of the company. Assume that there is no trend.

Solution. The total of seasonal indices is 1200 and information release to one year. Therefore the C and I are absen T. Further in this series there is no trend at all.

Now the yearly sales of Rs. 24,00,000 is to be distributed among 12 months i proportion to seasonal indices.

Month	***Sales***	***Month***	***Sales***
January	$24{,}00{,}000 \times \frac{75}{1200} = \mathbf{1{,}50{,}000}$	July	$24{,}00{,}000 \times \frac{102}{1200} = \mathbf{2{,}04{,}000}$
February	$24{,}00{,}000 \times \frac{80}{1200} = \mathbf{1{,}60{,}000}$	August	$24{,}00{,}000 \times \frac{104}{1200} = \mathbf{2{,}08{,}000}$
March	$24{,}00{,}000 \times \frac{98}{1200} = \mathbf{1{,}96{,}000}$	September	$24{,}00{,}000 \times \frac{100}{1200} = \mathbf{2{,}00{,}000}$
April	$24{,}00{,}000 \times \frac{128}{1200} = \mathbf{2{,}56{,}000}$	October	$24{,}00{,}000 \times \frac{102}{1200} = \mathbf{2{,}04{,}000}$
May	$24{,}00{,}000 \times \frac{137}{1200} = \mathbf{2{,}74{,}000}$	November	$24{,}00{,}000 \times \frac{82}{1200} = \mathbf{1{,}64{,}000}$
June	$24{,}00{,}000 \times \frac{119}{1200} = \mathbf{2{,}38{,}000}$	December	$24{,}00{,}000 \times \frac{73}{1200} = \mathbf{1{,}46{,}000}$

Chain Relative of the 1st Quarter (on the basis of 1st Quarter) = 100.

Chain Relative of the 1st Quarter (on the basis of last Quarter)

$$= \frac{101.48 \times 104.777}{100} = \mathbf{106.328.}$$

Difference Per Quarter = $d = \frac{1}{4}$ (New Chain Relative of 1st Quarter – 100)

$$= \frac{1}{4}(106.328 - 100) = \frac{6.328}{4} = \mathbf{1.582.}$$

Adjusted chain relatives for 2nd, 3rd and 4th quarters are obtained on substituting *d*, 2*d* and 3*d* form the corresponding chain relatives.

Sum of adjusted claim relatives; 100 + 107.198 + 97.044 + 100.031 = 404.273.

Indices of second variation are obtained by adjusting these adjusted chain relatives to a total of 400 by multiplying each one of them with a constant factor, *i.e.*,

$$\text{c.f.} = \frac{400}{\text{Sum of adjusted claim relatives}} = \frac{\mathbf{400}}{\mathbf{404.73}}.$$

Example 73. *Fit a straight line trend by least squares method to the data given below and estimate trend for 1983.*

Year :	*1977*	*1978*	*1979*	*1980*	*1981*	*1982*
Sales (in '000) :	*10*	*12*	*15*	*16*	*18*	*19*

Solution. Let us take the origin as 1st July 1979 with unit = 6 months.

Table for The Computation of Straight Line Trend by The Method of Least Squares

Year	*Sales Rs. (in '000) (Y)*	*Deviation from 1979.5*	*Deviations multiplied by 2 (X)*	X^2	*XY*
1977	10	– 2.5	– 5	25	– 50
1978	12	– 1.5	– 3	9	– 36
1979	15	– 0.5	– 1	1	– 15
1980	16	+ 0.5	+ 1	1	+ 16
1981	18	+ 1.5	+ 3	9	+ 54
1982	19	+ 2.5	+ 5	25	+ 95
$N = 6$	$\Sigma Y = 90$		$\Sigma X = 0$	$\Sigma X^2 = 70$	$\Sigma XY = 64$

The straight line trend is: $Y = a + bX$.

When $\Sigma X = 0$,

$$a = \frac{\Sigma Y}{N} = \frac{90}{6} = \mathbf{15;}$$

$$b = \frac{\Sigma XY}{\Sigma X^2} = \frac{64}{70} = \mathbf{0.914.}$$

Thus the equation will be $Y = 15 + 0.914\,X$.

The Trend for 1983 is based on $X = +7$.

Substituting this value we get

$Y = 15 + (0.914)(7) = 15 + 6.398 = \mathbf{21.398}$.

Example 74. *Fit a straight line trend to the following data by the least squares method after summing the given quarterly data to yearly data.*

Export of Cottom textiles (million Rs.)

Year	*1st Quarter*	*2nd Quarter*	*3rd Quarter*	*4th Quarter*
1979	*10*	*13*	*14*	*12*
1980	*12*	*14*	*15*	*13*
1981	*13*	*15*	*18*	*14*
1982	*15*	*19*	*21*	*18*
1983	*15*	*22*	*23*	*20*
1984	*20*	*21*	*25*	*20*

Also find out short period fluctuations for the given years using additive model.

Solution.

Table for Computation of Straight line Trend by The Method of Least Squares

Year	*Y*	*Deviation from 1981.5*	*Deviation multiplied by 2 X*	*XY*	X^2	Y_c	*Short Term Fluctuations* $(Y - Y_c)$
1979	49	– 2.5	– 5	– 245	25	47.30	**+ 1.7**
1980	54	– 1.5	– 3	– 162	9	55.19	**– 1.19**
1981	60	– 0.5	– 1	– 60	1	63.07	**– 3.0**
1982	73	+ 0.5	+ 1	+ 73	1	70.96	**+ 2.04**
1983	80	+ 1.5	+ 3	+ 240	9	78.84	**+ 1.16**
1984	86	+ 2.5	+ 5	+ 430	25	86.73	**– 0.70**
$N = 6$	$\Sigma Y =$ **402**		$\Sigma X = 0$	ΣXY **= 276**	ΣX^2 **= 70**		

The equation of the Trend line is: $Y_c = a + bX$

$$a = \frac{\Sigma Y}{N} = \frac{402}{6} = \mathbf{67}; \qquad b = \frac{\Sigma XY}{\Sigma X^2} = \frac{276}{70} = \mathbf{3.943.}$$

$\mathbf{Y_c = 67 + 3.943X}$

The Trend value of $Y_{1979} = 67 + (3.943)(-5) = 67 - 19.715 = 47.285 \simeq \mathbf{47.30}$.

Similarly, for all other years the trend vlaues can be computed. The Short Term function for year 1979 is (49 – 47.3) = 1.7.

EXERCISE – 14.1

1. What is a time series? What is the need to analyse the time series?
2. What are the components of a time series? How will you determine them?
3. Describe in brief the mathematical models of time series. Which of these models is more popular in practice and why?
4. Distinguish between trend, seasonal variations and cyclical fluctuations in a time series.
5. Explain briefly the various methods of determining trend in a time series.
6. Describe the different methods of measuring seasonal variations.
7. Distinguish between ratio to trend and ratio to moving average as methods of measuring seasonal variations. Which is better and why?

8. Why do we deseasonalize data?

9. Explain clearly what is meant by Time Series Analysis. Describe briefly the components of a time series.

10. Explain briefly, how the seasonal element in a time series data is isolated and eliminated.

11. Explain the methods of fitting quadratic and exponential curves. How would you use the fitted curves for forecasting?

12. Using three-year moving averages determine the trend and short-time fluctuations. Plot the originaland trend values on the same graph.

Year :	1968	1969	1970	1971	1972	1973	1974	1975	1976	1977
Production :	21	22	23	25	24	22	25	26	27	23
('000 units)										

13. Fit a straight line trend by the method of least squares to the following data.

Year :	1985	1986	1987	1988	1989	1990
Production :	7	10	12	14	17	24.
(Rs. in Crores)						

14. Obtain the straight line trend equation and tabulate against each year after estimating the trend and short term fluctuations.

Year :	1970	1971	1972	1973	1974	1975	1976	1977	1978
Value :	380	400	650	720	690	620	670	950	1040.

15. You are given the population figures of India as follows

Year :	1911	1921	1931	1941	1951	1961	1971
Production :	25.0	25.1	27.9	31.9	36.1	43.6	54.7.
(in crores)							

Estimate the population for the year 1981 using an equation of the form $y = ab^x$.

16. Using five year moving averages, determine the trend and short term fluctuations. Plot the original and trend values on the same graph paper.

Year :	*1968*	*1969*	*1970*	*1971*	*1972*	*1973*	*1974*	*1975*	*1976*	*1977*
Production :	*21*	*22*	*23*	*25*	*24*	*22*	*25*	*26*	*27*	*26.*
('000 units)										

17. Determine the equation of a straight line which best fits the following data

Year :	*1974*	*1975*	*1976*	*1977*	*1978*
Sales ('000 Rs.) :	*35*	*56*	*79*	*80*	*40.*

Compute the trend vlaues for all the years from 1974 to 1978 and estimate for 1980.

18. The prices of a commodity during 1980-85 are given below. Fit a parabola $y = a + bX + cX^2$ to this data. Estimate the price of commodity in 1986.

Year :	*1980*	*1981*	*1982*	*1983*	*1984*	*1985*
Price :	*100*	*107*	*128*	*140*	*181*	*292.*

Also plot the actual trend and given values on graph.

19. Compute the seasonal index numbers applying the simple average method for the following data :

Year	*Summer*	*Monsoon*	*Autumn*	*Winter*
1981	112	110	120	115
1982	80	145	105	90
1983	95	100	140	80
1984	110	90	130	100
1985	85	110	110	90
1986	92	120	100	85

[**Hint.** Computation of Seasonal Index Numbers by Simple Average Method

Year	*Summer*	*Monsoon*	*Autumn*	*Winter*
1981	112	110	120	115
1982	80	145	105	90
1983	95	100	140	80
1984	110	90	130	100
1985	85	110	110	90
1986	92	120	100	85
Total	**574**	**675**	**705**	**560**
Average Index	**95.67**	**112.5**	**117.5**	**93.33**
Seasonal Index	**91.33**	**107.40**	**112.17**	**89.10**

Average = 104.75.

$$\text{S.I. for 1st Qr.} = \frac{95.67}{104.75} \times 100 = \mathbf{91.33}$$

Similarly, **S.I. for 2nd Qr. = 107.40, 3rd Qr. = 112.17, 4th Qr. = 89.10]**

20. The seasonal indices of the sale of ready made garments of a particular type in a certain store are given below :

	Quarter	*Seasonal Index*
i.	Jan.-March	95
ii.	April-June	80
iii.	July-September	88
iv.	Oct.-December	120

If the total sales in the first quarter of a year be worth Rs. 60,000, determine how much worth of garments of this type should be kept in stock by the store to meet the demand in each of the remaining quarters.

21. Find the seasonal indices by the method of moving averages using additive model.

Year	*1st Quarter*	*2nd Quarter*	*3rd Quarter*	*4th Quarter*
1983	16.00	13.50	14.70	17.00
1984	15.90	12.20	15.60	18.00
1985	16.30	11.90	16.90	19.20
1986	17.10	13.20	15.00	18.70

22. Obtain the seasonal fluctuations from the following time series by the method of link relatives.

Year	*1st Quarter*	*2nd Quarter*	*3rd Quarter*	*4th Quarter*
1985	65	58	56	61
1986	68	63	63	67
1987	70	59	56	52
1988	60	55	51	53

23. From the following data calculate seasonal index by the method of Ratio to Trend.

Quarter \ *Year*	*1985*	*1986*	*1987*	*1988*
1	30	45	35	75
2	49	50	62	79
3	50	61	60	65
4	35	20	25	70

24. From the following data calculate the 4-yearly moving average and determine the trend values. Find the short-term fluctuations. Plot the original values and the trend on a graph paper.

Year	*Value*	*Year*	*Value*
1983	50.0	1988	38.1
1984	36.5	1989	32.6
1985	43.0	1990	41.7
1986	44.5	1991	41.1
1987	38.9	1992	33.8

25. Construct indices of seasonal variation from the following time series data on consumption of cold drinks on 1,000 bottles:

Quarter	*I*	*II*	*III*	*IV*
1989	90	75	87	70
1990	75	80	78	75
1991	80	75	75	72
1992	85	82	80	81

26. Deseasonalise the following production data by the method of moving averages:

Quarters	*Quarterly Output ('000 tons)*			
	1981	*1982*	*1983*	*1984*
I	30	49	35	75
II	49	50	62	79
III	50	61	60	65
IV	35	20	25	70

27. The following data gives the annual cement production of a factory

Year :	1986	1987	1988	1989	1990	1991	1991
Production (in m. tonnes)	12	10	14	11	13	15	16

Fit a trend line by the method of least squares and estimate the production for the year 1995.

[Hint. Fitting of the straight line trend

Year	*Production* Y	X	XY	X^2
1986	12	– 3	– 36	9
1987	10	– 2	– 20	4
1988	14	– 1	– 14	1
1989	11	0	0	0
1990	13	+ 1	+ 13	1
1991	15	+ 2	+ 30	4
1992	16	+ 3	+ 48	9
$N = 7$	$\Sigma Y = 91$	$\Sigma X = 0$	$\Sigma XY = 21$	$\Sigma X^2 = 28$

The equation of the straight line trend is: $\boldsymbol{Y_C = a + bX = 13 + 0.75X}$

$\therefore\ Y_{1995} = 13 + 0.75\ (6) = 17.5$ million tonnes with 1989 = 0]

28. For the following data:

Year :	1986	1987	1988	1989	1990	1991	1992
Labour force (million) :	79	81	83	86	81	89	89

Find the equation of linear trend using least squares method and estimate the value of 1994.

MULTIPLE CHOICE QUESTIONS

29. Seasonal variations repeat during a period of

(*a*) 5 years (*b*) One year
(*c*) 7 years (*d*) None of these

30. The most important factor causing seasonal variation is

(*a*) Depression in business (*b*) Growth of population
(*c*) Whether and social customs (*d*) None of these

31. If the trend is absent, the seasonal indices are known by

(*a*) Ratio to trend method (*b*) Simple average method
(*c*) Ratio to moving average method (*d*) None of these

32. The trend can be found by the method of least squares if the

(*a*) Trend is not clear (*b*) The trend is linear
(*c*) Trend is not linear (d) None of these

33. The trend is linear if

(*a*) Rate of growth is positive (*b*) Growth rate is constant
(*c*) Growth is not constant (*d*) None of these

34. The most widely used method of measuring seasonal variations is

(*a*) Ratio to trend (*b*) Link relative
(*c*) Ratio to moving average (*d*) None of these

35. Use the method of monthly averages to find the monthly indices for the following data of production of a commodity for the years 1979, 1980, 1981:

Production in Lakhs of Tonnes

Months	*1979*	*1980*	*1981*
January	12	15	16
February	11	14	15
March	10	13	14
April	14	16	16
May	15	16	15
June	15	15	17
July	16	17	16
August	13	12	13
September	11	13	10
October	10	12	10
November	12	13	11
December	15	14	15

[**Hints.** First find the total for each of the 12 months and divide each of these totals by 3 (no. of years) to find the monthly averages $\overline{X}_1, \overline{X}_2 ...., \overline{X}_{12}$ for 12 months. Then find the grand average $\overline{G}$ of these 12 monthly averages. The required monthly indices are $\overline{X}_1 - \overline{G}, \overline{X}_2 - \overline{G}, ..., \overline{X}_{12} - \overline{G}$.]

36. Find the forecast for different years using Mean Forecast, Naive Forecast. Linear Trend Forecast and Non-linear Trend Forecast from

Year :	1986	1987	1988	1989
Sales (*in million Rs.*)	30	40	45	65

37. A company estimates its average monthly sales in a particular year to be Rs. 2,00,000. The seasonal indices (S.I.) of the sales data are as follows:

Month :	Jan.	Feb.	Mar.	April	May	June
S.I. :	76	77	98	128	137	122

Assuming that there is no trend draw up a monthly sales budget for the months January to June.

38. An old forecast of 30 is given initially and the smoothing coefficient is 0.2. Using the method of exponential smoothing, find forecasts for the following sales data:

Time (Day) :	1	2	3	4	5	6	7	8
Sales ('00 Rs.) :	32	33	28	33	30	34	30	32

39. Calculate the quarterly seasonal indices in respect of the following data by using the simple average method:

Year	*Quarter I*	*Quarter II*	*Quarter III*	*Quarter IV*
1980	71	68	79	71
1981	76	69	82	74
1982	74	66	84	80
1983	76	73	84	78
1984	78	74	86	82

40. Fill in the blanks:

(*a*) In the trend equation $y = a + bx$, a is the __________ and b represents __________.

(*b*) The additive model of time series is expressed as __________.

(*c*) When the difference between successive observations of a series are constant (or nearly so) the __________ may be an appropriate representation of trend equation.

(*d*) The line obtained by the method of least squares is known as the line of __________.

(*e*) An overall rise or fall in a time series is called the __________.

(*f*) The link relative method is one of the methods for the estimation of__________ index in the context of analysis of time series.

ANSWERS

13. $Y = 14 + 3.086X$.

14. $Y = 680 + 70.5X$.

15. 57.23.

18. $Y = 126.55 + 18.04X + 1.386X^2$.

19. Ist Qr. = 92.54; IInd Qr. = 100.97; IIIrd Qr. = 111.73; IVth Qr. = 88.75.

20. 60.000; 50,526; 55,578.95; 75,789.47.

22. 109.33, 97.78, 93.89, 99.00.

24. 42.1, 40.9, 39.8, 38.1, 37.35.

25. 104.76, 99.05, 101.58, 94.6.

26. Deseasonalised data ('000 Tons): 28.28, 38.70, 39.28, 57.75, 47.28, 39.70, 50.28, 42.74, 33.28, 51.70, 49.28, 47.74, 73.28, 68.70, 54.28, 92.74.

27. $Y_C = 13 + .75X$; $Y_{1995} = 17.5$ million tonnes.

28. $Y_C = 84 + 1.471X$; $Y_{1994} = 91.355$,

29. (*b*).

30. (*c*). **31.** (*b*). **32.** (*c*). **33.** (*b*) **34.** (*c*).

35. (*i*) Using Additive Model: 0.666, – 0.334, – 1.334, 1.666, 2.000, 2.667, – 1.000, – 2.334, – 3.000, – 1.667, 1.000.

(*ii*) Using Multiplicative Model: 104.87, 97.56, 90.24, 112.19, 112.19, 114.63, 119.51, 92.68, 82.92, 78.05, 87.80, 107.32.

36. 52, 52, 52, 52, 52; ......, 40, 45, 65, 80; 27, 39.5, 52, 64.5.

37. Rs. 1,52,000; Rs. 1,54,000; Rs. 1,96,000; Rs. 2, 56, 000; Rs. 2,74,000; Rs. 2,44,000.

38. 30.4, 30.92, 30.34, 30.87, 30.70, 31.36, 31.09, 31.27.

39. – 1.25, – 6.25, 6.75, 0.75.

40. (*a*) Average of *Y*, Slope of trend line;

(*b*) $Y = T + S + C + I$; (*c*) Straight line; (*d*) best fit;

(*e*) Secular trend; (*f*) Seasonal variations.

14.35 MEASUREMENT OF CYCLICAL VARIATIONS (*C*)

Residual Method. ***The oscillatory movements in a time series with period of oscillation more than one year are called*** **cyclical variations** (*C*). These oscillatory movements in any business are due *to Business Cycles* which are four-phased cyclical, viz., **Prosperity (Boom), Recession, Depression and Recovery from time to time.**

In order to measure cyclical fluctuations, we first *find Trend* (*T*) *and Seasonal variations* (*S*) *by one of the methods discussed earlier. We eleminate Trend* (*T*) *and Seasonal variations* (*S*) *from the original data by using additive or Multiplicative model.*

We are now left with only **Cyclical Variations (*C*)** and **Irregular Variations (*I*)** in the following form;

$$C + I = Y - T - S \qquad \textbf{(Additive Model)}$$

$$CI = \frac{Y}{S \times T} \qquad \textbf{(Multiplicative Model)}$$

Irregular variation (*I*) is then removed from this data by using the method of moving average of appropriate period depending upon the average duration of irregular variations (*I*). leaving only the Cyclical variations. **This process of finding cyclical variations is called Residual Method.**

Limitation. ***It is based on the assumption that Trend and Seasonal variations can be accurately measured and, therefore, be removed atleast in large part.***

14.36 MEASUREMENT OF IRREGULAR VARIATIONS (I)

Irregular variations (*I*) can be obtained by removing Trend (*T*), Seasonal variations (*S*) and Cyclical variations (*I*) from the given (original) data, *i.e.*,

$$I = Y - T - S - C \qquad \textbf{(Additive Model)}$$

$$I = \frac{Y}{T \times S \times C} \qquad \textbf{(Multiplicative Model)}$$

In practice, ***Irregular variations, present in the given data are found to be of small magnitude.***

14.37 DESEASONALISATION OF DATA

The objective of studying seasonal variations is **(*i*) to measure them and (*ii*) to eliminate them from the given series.** ***Elimination of the seasonal effects from the given values is termed as*** **deseasonalisation of the data.** It helps to adjust the given time series from seasonal variations, thereby, leaving with trend, cyclical and irregular variations.

Deseasonalised data $= \dfrac{Y}{S} = \dfrac{TCSI}{S} = TCI.$ **[Multiplicative Model]**

Deseasonalised data $= Y - S = (T + S + C + I) - S = T + C + 1.$ **[Additive Model]**

Deseasonalisation is specially needed for the study of cyclic component. It helps the management in planning, forecasting and managerial control.

Example 75. *Deseasonalise the following data with the help of seasonal index given against*

Month	*Cash Balance ('000)*	*Seasonal index*
January	*360*	*120*
February	*400*	*80*
March	*550*	*110*
April	*360*	*90*
May	*350*	*70*
June	*550*	*100.*

Solution. Deseasonalised values are obtained on dividing the given time series values (Y) by the seasonal effect.

$$\text{Deseasonalised value} = \frac{Y}{\textbf{Seasonal effect}} = \frac{Y}{SI} \times 100$$

Table for Computation of Deseasonalised Values

Month	*Case Balance (Rs. '000) Y*	*Seasonal index SI*	*Deseasonalised Value* $= \frac{Y}{SI} \times 100$
January	360	120	$\frac{360}{120} \times 100 = \mathbf{300}$
Febraury	400	80	$\frac{400}{80} \times 100 = \mathbf{500}$
March	550	110	$\frac{550}{110} \times 100 = \mathbf{500}$
April	360	90	$\frac{360}{90} \times 100 = \mathbf{400}$
May	350	70	$\frac{350}{70} \times 100 = \mathbf{500}$
June	550	100	$\frac{550}{100} \times 100 = \mathbf{550}$

Example 76. *Deseanolise the following data using a multiplicative model:*

Quarter :	*1*	*2*	*3*	*4*
Sales (Rs. '000) :	*65.4*	*25.2*	*23.37*	*21.4*
Seasonal Index :	*148*	*124*	*78*	*50*

Solution. Deseasonalised data = $\dfrac{\textbf{Original data}}{\textbf{Seasonal index}} \times 100$

Table: Computation of Desceasonalised Data

Quarter	*Sale (Rs. '000)*	*Seasonal Index (S.I.)*	*Seasonal Effect = SI ÷ 100*	*Deseasonalisd Data*
1	65.4	148	1.48	65.4 ÷ 1.48 = **44.19**
2	25.2	124	1.24	25.2 ÷ 1.24 = **20.32**
3	23.7	78	0.78	23.7 ÷ 0.78 = **30.38**
4	21.4	50	0.50	21.4 ÷ 0.50 = **42.80**
Total	**135.7**	**400**	**4.00**	**137.69**

14.38 SMOOTHING METHODS IN TIME SERIES – EXPONENTIAL SMOOTHING

Exponential Smoothing. It is one of the forecasting techniques with varied applications Exponential smoothing is a term which is applied to a special type of weighted moving average useful in sales and inventory forecasting. Exponential smoothing overcomes almost all the problems associated with forecasting with a moving average. We know that a moving average gives equal weightage to all items (values) in a time series, whether these are more recent ones or old ones. But, ***under exponential smoothing the weights vary in geometric progression with a constant ratio*** α ***(0 <*** α ***< 1) in such a way that the weights get lower and lower as we go back in the distant past. It is also an average in the sense in which the most recent observation is assigned the highest weightage and it decreases in geometric progression as we move towards the older observation.*** This is, obviously, more reasonable because the most recent observation is more relevant for forecasting the future sales than the older observation. The **ratio of the geometric progression of the series of weights is** $(1-\alpha)$**, where** α **is called the smoothing coefficient.** A simple exponential smoothing equation is

$$S_t(x) = \alpha x_t + (1-\alpha) S_{t-1}(x).$$

or

$$S_t = ax_t + (1-\alpha) S_{t-1} \quad \text{... (1)}$$

where x_t = **Value of the series in the time period** t.

S_t or $S_t(x)$ = **Smoothed value (moving average) of the series in the time period** t,

S_{t-1} or $S_{t-1}(x)$ = **Smoothed Value (moving average) of the series in the time period just previous to the period** t,

α = **Smoothing constant known as Smoothing Coefficient.**

The repeated application of the recurrence relation (1) gives

$$S_{t-1} = \alpha x'_{t-1} + (1-\alpha) S_{t-2} \quad \text{... (2)}$$

Substituting the value of S_{t-1} from (2) in (1), we get

$$\Rightarrow \quad S_t = \alpha x_t + \alpha(1-\alpha) x_{t-1} + S_{t-2}$$

Repeated application of the result (1), gives

$$S_t = \alpha x_t + \alpha(1-\alpha) x_{t-1} + \alpha(1-\alpha)^2 x_{t-2} + \alpha(1-\alpha)^3 S_{t-3} + \alpha(1-\alpha)^4 S_{t-4} + \ldots\ldots$$

The various weights assigned to the observations at time t are given by the following table.

Observation at time t (A)	*Weightage Assigned* (B)	*Col. (A) × Col (B)* (C)
x_t	α	αx_t
x_{t-1}	$\alpha(1-\alpha)$	$\alpha(1-\alpha) x_{t-1}$
x_{t-2}	$\alpha(1-\alpha)^2$	$\alpha(1-\alpha)^2 x_{t-2}$
x_{t-3}	$\alpha(1-\alpha)^3$	$\alpha(1-\alpha)^3 x_{t-3}$
—	—	—
—	—	—
x_{t-n}	$\alpha(1-\alpha)^n$	$\alpha(1-\alpha)^n x_{t-n}$
—	—	—
—	—	—

∴ **Exponential Smoothing average**

$$= \frac{\alpha x_t + \alpha(1-\alpha)x_{t-1} + \alpha(1-\alpha)^2 x_{t-2} + \alpha(1-\alpha)^3 x_{t-3} + \ldots + \alpha(1-\alpha)^n x_{t-n} + \ldots}{\alpha + \alpha(1-\alpha) + \alpha(1-\alpha)^2 + \alpha(1-\alpha)^3 + \ldots + \alpha(1-\alpha)^n + \ldots} \quad \ldots \text{(B)}$$

But we know that

$\alpha + t(1-\alpha) + \alpha(1-\alpha)^2 t + \alpha(1-\alpha)^3 t + \ldots + \alpha(1-\alpha)^n t \ldots$ is an infinite geometric series whose common ratio is $(1-\alpha) < 1$ ($\because \alpha < 1$). Therefore **the sum of the weights**

$$= \alpha + \alpha(1-\alpha) + \alpha(1-\alpha)^2 + \ldots \alpha(1-\alpha)^n + \ldots = \frac{\alpha}{1-(1-\alpha)} = 1. \qquad \left[\because S_\infty = \frac{a}{1-r}, r < 1\right]$$

Hence, **the expression (*B*),** *i.e.*,

Exponential smoothing Average

$$= \alpha x_t + \alpha(1-\alpha)x_{t-1} + \alpha(1-\alpha)^2 x_{t-2} + \ldots + \alpha(1-\alpha)^n x_{t-n} + \ldots$$

In short, the Exponential Smoothing Average at time t is given by the relation (1), *viz.*,

$$S_t = \alpha x_t + (1-\alpha) S_{t-1}.$$

The above expression can also the written as

$$S_t = S_{t-1} + a(x_t - S_{t-1}) \quad \ldots (3)$$

The term $(x_t - S_{t-1})$ in the R.H.S. of (3) is the difference of latest observation and its forecasting raised a period earlier and is known as error. It is denoted by e_t.

Thus expression (2) becomes

$$S_t = S_{t-1} + \alpha e_t. \quad \ldots (4)$$

Value of the Smoothing Coefficient α

Case I. When α = 0.

Putting $\alpha = 0$ in (4), we get

$$S_t = S_{t-1} + 0.e_t \text{ or } S_t = S_{t-1}.$$

This implies that the earlier average stands and no account of any difference between the actual observation and its forecast is taken. **Thus α = 0 implies that the forecasting system does not react to the incoming observations.**

Case II. When α = 1.

In this case the correction is equal

$$\alpha e_t = e_t = (x_t - S_{t-1})$$

$$\therefore \quad S_t = S_{t-1} + e_t = S_{t-1} + (x_t - S_{t-1}) = x_t.$$

Thus **when α = 1, the observation itself is taken as the forecast for the next period.** The past series is completely ignored.

This suggests that a **low value of α gives more weightage to the past figures and less consideration to the incoming observation. Low Values are, therefore, to be used where the series is stable and Right Values are to be used where the series is rather fluctuating.**

Existence of Trend. The following equations may be used where the trend is required:

(*i*) $S_t = S_{t-1} + \alpha e_t$

(*ii*) $\lambda_t = \alpha(S_t - S_{t-1}) + (1-\alpha)\lambda_{t-1}$,

(*iii*) $S_t' = S_t + \left(\frac{1+\alpha}{\alpha} + 1\right)\lambda_t$, *where* S_t' *gives the necessary forecast for the next period.*

LIMITATIONS OF EXPONENTIAL SMOOTHING

1. *The method is useful for short-term forecasting.*
2. *It relies solely on the past history of the sales or production. There are cases where subjective estimates may provide better forecasts. There have been attempts to*

complement the two, i.e., forecasts from exponential smoothing and subjective estimates, successful to a great extent, but not fully satisfactory.

MERITS OF EXPONENTIAL SMOOTHING

1. ***Exponential smoothing is better than the moving average because of the better weightage scheme.***
2. ***Exponential smoothing provides a convenient, systematic method for revising the forecast for the next period, whenever discrepancy exists between the previous forecast sales for the current period and the actual demand for the current period.***
3. ***It is a very useful method for computer applications.***

METHOD TO PREPARE FORECAST BY EXPONENTIAL SMOOTHING

The smoothed values are calculated by the following result:

$$S_t = \alpha x_t + (1 - \alpha)\, S_{t-1},$$

where α is smoothing coefficient, and S_t and S_{t-1} denote smooth values at time t and $t - 1$ respectively.

Step I. Find the ΔS_t, the change in smoothed value, where

$$\Delta S_t = S_t - S_{t-1}.$$

Step II. Find the trend estimate T_t by the formula

$$T_t = \alpha\,(\Delta S_t) + (1 - \alpha)\, T_{t-1},$$

where T_t, T_{t-1} respectively denote the trend at time t and $t - 1$.

Step III. Prepare the forecast $\hat{x}_t$ for the period $(t + 1)$ as follows

$$\hat{x}_t = S_t + \frac{(1-\alpha)}{\alpha} T_t,$$

Note. In fact forecast $\hat{X}_t$, is for the period $(t + 1)$.

Step IV. Calculate the forecasting error e_1, where

$$e_t = \hat{S}_t - x_{t+1}.$$

Step V. Put the above steps in the following tabular form

Forecasting Table

Time period	*Actual Value*	*Smoothed Value*	*Changes in Smoothed Value*	*Trend Estimate*	*Forecast*	*Forecasting error*
(t)	(x_t)	(S_t)	(Δs_t)	(T_t)	$(\hat{x}_t)$	(e_t)

Example 77. *From the following values prepare forecast by the method of exponential smoothing taking initial estimates as 100, the value of α = 0.4 and initial trend value zero:*

Time period (t) :	*1983*	*1984*	*1985*	*1986*	*1987*	*1988*	*1989*	*1990*	*1991*	*1992*
Actual value (X) :	*104*	*108*	*118*	*115*	*120*	*122*	*123*	*125*	*128*	*130*

Solution. We have the following calculations

1. Calculation of S_t:

$$S_t = aX_t + (1 - \alpha)\, S_{t-1}.$$

S_{1984} = .4 (108) + .6 (100) = 43.2 + 60 = **103.2**

S_{1985} = .4 (118) + .6 (103.2) = 47.2 + 61.92 = **109.12**

S_{1986} = .4 (115) + .6 (109.12) = 46 + 65.47 = **111.47**

$S_{1987} = .4\,(120) + .6\,(111.47) = 48 + 66.88 = \mathbf{114.88}$

$S_{1988} = .4\,(122) + .6\,(114.88) = 48.8 + 68.93 = \mathbf{117.73}$

$S_{1989} = .4\,(123) + .6\,(117.73) = 49.2 + 70.64 = \mathbf{119.84}$

$S_{1990} = .4\,(125) + .6\,(119.84) = 50+71.90 = \mathbf{121.90}$

$S_{1991} = .4\,(128) + .6\,(121.9) = 51.2 + 73.14 = \mathbf{124.34}$

$S_{1992} = .4\,(130) + .6\,(124.34) = 52 + 74.60 = \mathbf{126.60}.$

2. Calculation of ΔS_t:

$$\Delta S_t = S_t - S_{t-1}.$$

$\Delta S_{1984} = S_{1984} - S_{1983} = 103.2 - 100 = \mathbf{3.2}$

$\Delta S_{1985} = S_{1985} - S_{1984} = 109.2 - 103.12 = \mathbf{5.92.}$

Similarly, $\Delta S_{1986} = \mathbf{2.35}$, $\Delta S_{1987} = \mathbf{3.41}$, $\Delta S_{1988} = \mathbf{2.85}$, $\Delta S_{1989} = \mathbf{2.11}$,

$\Delta S_{1990} = \mathbf{2.06}$, $\Delta S_{1991} = 2.44$, $\Delta S_{1992} = \mathbf{2.26}.$

3. Calculation of Trend Estimate $T_t = \alpha\,(\Delta S_t) + T_{t-1}$.

For 1984:. 4 (3.2) + .6 (0) = 1.28 + 0 = **1.28**

For 1985: .4 (5.92) + .6 (1.28) = 2.37+ .77 = **3.14**

For 1986: .4 (2.35) + .6 (3.14) = .94 + 1.88 = **2.82**

For 1987: .4 (3.41) + .6 (2.82) = 1.364 + 1.602 = **3.06**

For 1988: .4 (2.85) + .6 (3.06) = 1.14 + 1.836 = **2.98**

For 1989: .4 (2.11) + .6 (0.98) = 0.844 + 1.788 = **2.63**

For 1990: .4 (2.06) + .6 (2.63) = 0.824 + 1.678 = **2.40**

For 1991: .4 (2.44) + .6 (2.40) = 0.976 + 1.44 = **2.42**

For 1992: .4 (2.26) + .6 (2.42) = 0.904 + 1.452 = **2.36.**

4. Calculation for Forecast:

$$\hat{S}_t = S_t + \frac{(1-\alpha)}{\alpha} T_t = S_t + 1.5\,(T_t) \qquad \left(\because \frac{1-\alpha}{\alpha} = 1.5\right)$$

For 1984: 103.2 + 1.5 (1.28) = 103.2 + 1.92 = **105.12**

For 1985: 109.12 + 1.5 (3.14) = 109.12 + 4.71 = **113.83**

For 1986: 111.47 + 1.5 (2.82) = 111.47 + 4.23 = **115.70**

For 1987: 114.88 + 1.5 (3.06) = 114.88 + 4.59 = **119.47**

For 1988: 117.73 + 1.5 (2.98) = 117.73 + 4.47 = **122.20**

For 1989: 119.84 + 1.5 (2.63) = 119.84 + 3.945 = **123.79**

For 1990: 121.9 + (1.5) (2.4) = **125.50**

For 1991: 124.34 + (1.5) (2.42) = **127.97**

For 1992: 121.9 + (1.5) (2.36) = **130.14.**

5. Calculation for Forecasting Error: $e_t = \hat{S}_t - x_{t+1}$

For 1984: 105.12 – 118 = – **12.88**

For 1985: 113.83 – 115 = **– 1.17**

For 1986: 115.7 – 120 = – **4.30**

For 1987: 119.47 – 122 = – **2.53**

For 1988: 122.20 – 123 = – **0.80**

For 1989: 123.79 – 125 = – **1.21**

15

CHAPTER

NON-PARAMETRIC METHODS

15.1 INTRODUCTION

Most of the hypothesis-testing procedures discussed so far, such as z-test or t-test, are based on the assumption that **the random samples are selected from a normal population.** We have seen that most of these tests are still reasonably reliable for slight departures from normality particularly, when the sample size is large. Traditionally, these testing procedures have been known as **Parametric methods** as they depend on the parameters, *viz.*, mean or proportion or standard deviation etc. These parametric tests have used the parametric statistic of sample that comes from the population being tested.

The need for technique that apply more broadly has led to the development of **non-parametric methods. These do not require that the underlying population be normal — or indeed that they have any single mathematical form and some even apply to non-numerical data.** The nonparametric methods are also called **distribution-free methods** that assume no knowledge, what so ever, about the distributions of underlying populations, except perhaps that they **are continuous.**

15.2 TYPES OF NONPARAMETRIC TESTS

A large number of nonparametric tests exist, but in this chapter we shall discuss the following tests which are better known and widely used ones:

1. **Sign test for paired data and One Sample sign Test.** In these tests, positive or negative signs are substituted for quantitative values.
2. **Mann-Whitney U Test or a Rank sum Test.** This test is used to determine whether two independent samples have been drawn from the same population.
3. **Kruskal-Wallis Test.** It is another rank sum test which generalizes the analysis of variance, (discussed in the chapter on ψ^2-test) to enable us to dispense with the assumption that the population are normally distributed.
4. **One Sample Run Test.** It is a method to determine the randomness with which the sample items have been selected.
5. **Rank Correlation Test.** It is a method for finding correlation coefficient when the data available is not in numerical form but the information is sufficient to rank the data first, second, third, and so on.
6. **Kolmogorov-Smirnov Test.** It is another method of determining the goodness of fit between an observed sample and theoretical probability distribution.

15.3 ADVANTAGES OF NONPARAMETRIC METHODS

1. Nonparametric tests are generally simple to understand, quicker and easier to apply when the sample sizes are small.
2. They do not require lengthy and laborious calculations and hence they are less time consuming. If significant results are obtained no further work is necessary.
3. They do not require the assumption that a population is distributed in the shape of a normal curve or another specified shape.

4. Nonparametric tests can be applied to all types of data — qualitative (nominal scaling) data in rank form (ordinary scaling) as well as data that have been measured more precisely (internal or ratio scaling).
5. The chief advantage of nonparametric tests is their inherently greater range of applicability because of milder assumptions.
6. Nonparametric tests make fewer and less stringent assumptions (that are more easily met) than the parametric tests.

15.4 DIS ADVANTAGES OF NONPARAMETRIC METHODS

Distribution-free or nonparametric methods also have a number of important disadvantages which in practice often outweigh their advantages.

1. Nonparametric tests sometimes pay for freedom from assumptions by a marked lack of sensitivity.
2. Important effects can be entirely missed by a nonparametric technique because of its lack of power or because it produces confidence intervals that are too wide. To put it another way, some of these techniques are very blunt instruments. It may causes it unwise to substitute one of these techniques either because of computational ease or because the data do not quite satisfy some distributional assumption.
3. The competing distribution-tied procedure may be quiet robust to departures from that assumption.
4. An additional related disadvantage is that distribution-free techniques often disregard the actual scale of measurement and substitute either ranks or even a more gross comparison of relative magnitude. This is not invariably bad, but is often the basis for the loss of power or sensitivity referred to in the preceding paragraph.
5. The another disadvantage is one that is gradually being corrected by research in statistical methodology. At present, we have available a much larger body of techniques in distribution-free hypothesis testing than in distribution-free estimations. To some extent, this is an inevitable accompanying feature of techniques that disregard or weaken the underlying scale of measurement, because estimation is often tied to that scale, whereas hypothesis testing is concerned with selecting one of the two competing decisions or actions.
6. Nonparametric tests are often wasteful of information and thus less efficient than the standard techniques which they replace.

15.5 USES OF NONPARAMETRIC METHODS

There are four situations in which the use of a distribution-free or nonparametric technique is indicated:

1. When quick or preliminary data analysis is needed. May be you just want to see roughly how things are going with the data. May be somebody is about to leave for Atlantic City to present a paper tomorrow, and his or her chief thought it would be a nice idea to include some statistics in the paper, and you are feeling unusually tender hearted or browbeaten today. May be your computer is suffering an acute exacerbation of some rare electronic disease. May be your computer programmer is manifesting that chronic, epidemic, occupational disability: temporal disorientation with gross distortion of task-time relationships. May be you are doing research in a "primitive" setting without access to modern computational aids.
2. When the assumptions of a competing distribution-tied or parametric procedure are not satisfied and the consequences of this are either unknown or known to be serious. Notice

that violation of assumptions per se is not a sufficient reason to reject one of the classical procedures. As we have emphasised throughout this book, these procedures are often insensitive to such violations. For example, the *t*-tests and intervals for normal means are relatively insensitive to non-normality. We think that one of the most common errors made today is the under application of classical statistical methods because of the overzealous adherence to the letter of the law with respect to assumptions. Remember that assumptions will rarely (never ?) be exactly satisfied. It is, therefore, as important to have a grasp of the consequences of violations of assumptions are. A great deal of theoretical research in statistics today is directed toward investigating the robustness of statistical procedures.

3. When data are only roughly scaled; for example, when only comparative rather than absolute magnitudes are available. In dealing with clinical data, perhaps patients can only be classified as better, unchanged, or worse. Perhaps only ranks, that is, largest, second largest, ..., smallest, are available.
4. When the basic question of interest is distribution-free or nonparametric in nature. For example, do we have a random samples, or are these two samples drawn from populations with identical distributions?

EXERCISE – 15.1

1. What are nonparametric tests? Point out their advantages and disadvantages.
2. Explain briefly the various types of nonparametric tests and specify the situation in which they are applicable.
3. What are the limitations of nonparametric tests?
4. What are free distribution tests? In what ways are they different from parametric tests?
5. Discuss the usefulness of nonparametric tests in statistical analysis.
6. Discuss the uses of nonparametric tests.
7. Differentiate between parametric and nonparametric statistical tests.

15.6 'THE SIGN TEST FOR PAIRED DATA

The sign test is the easiest nonparametric test. Its name comes from the fact that it is based on the **direction (or signs for pluses or minuses) of a pair of observations** and not on their numerical magnitudes.

In such problems, **each pair of sample value is replaced by a plus sign** if **the difference between the paired observation is positive** (that is, if the first value exceeds the second value) **and by a minus sign if the difference between the paired observation is negative** (that is, if the first value is less than the second value) **and it is discarded if the difference is zero.**

For applying sign test to solve a problem, we count:

Number of +ve signs (+ signs)

Number of –ve signs (– signs)

Number of 0's (*i.e.*, which cannot be included either as positive or negative)

We take the **two hypothesis** as:

Null Hypothesis: $H_0 : p = 0.5$

Alternative Hypothesis: $H_1 : p \neq 0.5$. It is a case of **two tailed test.** If you look carefully at the above two hypothesis, you will see that the situation is similar to a fair coin toss. If we toss a fair coin 30 times, then probability 'p' would be **0.5** and we would expect about fifteen heads and fifteen tails. In that case, **we would use the binomial distribution** as the appropriate sampling distribution. We also know that when np **and** nq **are each atleast 5, we can use the normal distribution to**

approximate the binomial. Thus, we can apply **normal distribution test.** We calculate **the standard error of the proportion** p as given by:

Standard error of proportion p: S.E. (p): $\sigma_p = \sqrt{\frac{pq}{n}}$.

The two limits of acceptance region at 0.05 level of significance are:

$$p_{H_0} + 1.96\,\sigma_p \quad \text{and} \quad p_{H_0} - 1.96\,\sigma_p$$

It is important to note that if the conditions that both np and nq are not greater than 5, then we must use the binomial instead of normal distribution test.

Example 1. *Use the sign test to see if there is a difference between the number of days required to collect an account receivable before and after a new collection policy. Use the 0.05 significance level.*

Before : 33 36 41 32 39 47 34 29 32 34 40 42 33 36 27
After : 35 29 38 34 37 47 36 32 30 34 41 38 37 35 28

Solution. Table for Before and After Collection Policy

Before (1) :	33	36	41	32	39	47	34	29	32	34	40	42	33	36	27
After (2) :	35	29	38	34	37	47	36	32	30	34	41	38	37	35	28
Sign of Score : [(1) – (2)]	–	+	+	–	+	0	–	–	+	0	–	+	–	+	–

If we count the bottom row of above table, we get

Number of '+' signs = 6
Number of '–' signs = 7
Number of 0's = 2
Total sample size = 15.

When we use the sign test, we **exclude the evaluation 0's as we are testing perceivable differences.** We notice that we have **6 plus signs and 7 minus signs, for a total of 13 usable responses.** If there is no difference between the two types of collections, p (the probability that the first collection exceeds the second collection) would be 0.5 and we would expect to get equal number of plus and minus signs. We set up the hypothesis as :

1. **Null Hypothesis: $H_0 : p = 0.5$ ⇒ there is no difference between the two types of collections**

   Alternative Hypothesis: $H_1 : p \neq 0.5$ ⇒ there is a significant difference between the two types of collections ⇒ It is a case of **two tailed test.**

   Sample size : $n = 13$.

   Proportion : $p = \frac{6}{13} = 0.46$.

   $\therefore\ q = 1 - p = 0.54$. **Also $q = \frac{7}{13} = 0.54$.**

   Here $n\bar{q} = 13 \times \frac{6}{13} = 6$, and $n\bar{q} = 13 \times \frac{7}{13} = 7$ are both greater than 5, so **we use the normal distribution to approximate the binomial.**

2. **Level of significance:** $\alpha = 0.05$
3. **Z_α for two tailed test = 1.96 (from the tables)**

4. **Standard error of the proportion p : S.E. ($\bar{p}$) = $\sigma_p = \sqrt{\frac{\bar{p}\,\bar{q}}{n}}$**

$\therefore\ \sigma_p = \sqrt{\frac{0.5 \times 0.5}{13}} = \sqrt{0.02} = \mathbf{0.142.}$

5. **Limits of Acceptance Region.**

Upper Limit: $p_{H_0} + 1.96\,\sigma_p = 0.5 + (1.96) \times (0.142) = 0.78$

Lower Limit: $p_{H_0} - 1.96\,\sigma_p = 0.5 - (1.96) \times (0.142) = 0.22$

6. **Decision.** The two limits of acceptance region are 0.78 and 0.22. The sample proportion $\bar{p}$ = 0.46 lies within these two limits, so the sample proportion falls within the acceptance region for this hypothesis. We accept the null hypothesis $H_0 \Rightarrow$ there is no difference between the two types of collections at 0.05 level significance.

Example 2. *To determine the effectiveness of a new traffic control system, the number of accidents that occurred at 12 dangerous intersections during four weeks before and four weeks after the installation of the new system was observed, and the following data was obtained:*

3 and 1,	*5 and 2,*	*2 and 0,*	*3 and 2,*	*3 and 2,*	*3 and 0*
0 and 2,	*4 and 3,*	*1 and 3,*	*6 and 4,*	*4 and 1,*	*1 and 0*

Use the paired — simple sign test at α = 0.05 level of significance to test the null hypothesis that (μ_1), the new traffic-control system is only as effective as (μ_2), the earlier system (the population samples are continuous, but this does not matter so long as zero differences are discarded).

Solution.

1. **Setting up of Hypothesis:**

Null Hypothesis: $H_0 : \mu_1 = \mu_2$.

Alternative Hypothesis: $H_1 : \mu_1 > \mu_2$ $\Rightarrow$ It is a **case of one-tailed test.**

2. **Level of significance:** Here $\alpha = 0.05$.

3. **Test Statistic:** Use the test statistic X = the observed number of plus signs.

Replacing each positive difference of the two values by a plus sign and each negative difference by a minus sign, we get:

+ + + + + + – + – + + +

Here, the number of paired values **n = 12**

and the number of plus signs **X = 10**

It is **a case of Binomial distribution as the number of negative signs is only 2, which is less than 5.**

From the table 13, we find that for $n = 12$, $\theta = \frac{1}{2}$,

P **($X \geq$ 10)** = 0.0161 + 0.0029 + 0.0002 = **0.0192**

[Note: For α = 0.05, P ($X \geq$ 10) = P (X = 10) + P (X = 11) + P (X = 12) for n = 12

= 0.0161 + 0.0029 + 0.0002 = 0.0192]

4. **Decision:** Since P-value = 0.0192 is less than 0.05, so the null hypothesis must be rejected and we conclude that the **new traffic-control system is effective in reducing the number of accidents at dangerous crossings.**

15.7 ONE SAMPLE SIGN TEST

One-sample sign test is a nonparametric method which is used as an alternate to the one-sample t-test, where the null hypothesis is $\mu = \mu_0$ against a suitable alternative hypothesis. The only

assumptions underlying the sign test are: **the population sample is continuous and symmetrical.** We assume that the population is continuous so that there is zero probability of getting a value equal to μ_0, and we do not even need the assumption of symmetry if we change the null hypothesis to $\mu = \mu_0$, where μ is the population median.

In the sign test, we replace each sample value exceeding μ_0 with a plus sign and each value less than μ_0 with a minus sign. Then we test the null hypothesis H_0 that X, the number of plus sign is a value of a random variable having Binomial distribution with parametres n (the total number of plus or minus signs) **and probability $\theta = \frac{1}{2}$**. The two-sided alternative hypothesis $H_1 : \mu \neq \mu_0$ thus becomes $\theta \neq \frac{1}{2}$, and one-sided alternatives $\mu < \mu_0$ and $\mu > \mu_0$ becomes $\theta < \frac{1}{2}$ and $\theta > \frac{1}{2}$ respectively. **If a sample value equals μ_0**, (which can happen when we deal with rounded data even though the population is continuous), **we simply discard it.**

In order to perform a sign test, when **the sample size is very small, we use the Table of Binomial probabilities (see Table 13), and when the sample is large we use the normal approximation to the Binomial distribution with**

Mean = $n\theta$, and variance = $n\theta(1 - \theta)$

$$\therefore \qquad \textbf{S.E.}(\theta) = \sqrt{n\theta(1-\theta)}$$

The **test statistic** in this case is,

$$Z = \frac{x - n\theta}{\sqrt{n\theta(1-\theta)}}$$

with $\theta = \frac{1}{2}$, and x the number of plus signs.

Example 3. *The following are the measurements of breaking strength of a certain kind of 2-inch cotton ribbon in pounds:*

163	*165*	*160*	*189*	*161*	*171*	*158*	*151*	*169*	*162*
163	*139*	*172*	*165*	*148*	*166*	*172*	*163*	*187*	*173*

Use the sign test to test the null hypothesis $\mu = 160$ against the hypothesis $\mu > 160$ at the 0.05 level of significance.

Solution.

1. **Null Hypothesis: $H_0 : \mu = 160$.**

   Alternative Hypothesis: $H_1 : \mu > 160$ $\Rightarrow$ It is a case of **one tailed test.**

2. **Level of significance:** $\alpha = 0.05$.

3. **Test statistic: X = The observed number of plus signs.**

   Replacing each value exceeding 160 with a plus sign, each value less than 160 with a minus sign, and discarding the one value which equals 160, we get

   + + 0 + + + − − + + + − + + − + + + + +

   Here n = the total number of plus and minus signs = **19.**

   X = the total number of plus signs = **15.**

   It is a case of Binomial Distribution.

   From the **Binomial Table 13** given at the end, we find

   $P(X \geq 15) = 0.0095$ for $\theta = \frac{1}{2}$ and $n = 19$.

[For $\theta = 0.05$, $P(X \geq 15) = P(X = 15) + P(X = 16) + P(X = 17) + P(X = 18) + P(X = 19)$
$= 0.0074 + 0.0018 + 0.0003 + 0.0000 + 0.0000 = \mathbf{0.0095}$]

4. Decision: Since P-value (= 0.0095) is less than 0.05, so the null hypothesis must be rejected and we conclude that the **mean breaking strength of given kind of ribbon exceeds 160 pounds.**

Example 4. *The following data, in tons, are the amounts of sulfur oxides emitted by a large industrial plant in 40 days*

24	*15*	*20*	*29*	*19*	*18*	*22*	*25*	*27*	*9*
17	*20*	*17*	*6*	*24*	*14*	*15*	*23*	*24*	*26*
19	*23*	*28*	*19*	*16*	*22*	*24*	*17*	*20*	*13*
19	*10*	*23*	*18*	*31*	*13*	*20*	*17*	*24*	*14*

Use the sign test to test the null hypothesis $\mu = 21.5$ *against the alternative hypothesis* $\mu > 21.5$ *at the 0.01 level of significance.*

Solution.

1. Setting up the Hypothesis:

Null Hypothesis: $H_0 : \mu = 21.5$.

Alternative Hypothesis: $H_1 : \mu < 21.5 \Rightarrow$ **It is a case of one tailed test.**

2. Test Statistic:

Replacing each value exceeding 21.5 with a plus sign, each value less than 21.5 with a minus sign, and discarding the one value which equals 21.5, we get

$$+ \; - \; - \; + \; - \; - \; + \; + \; + \; - \; - \; - \; - \; - \; + \; - \; - \; + \; + \; +$$

$$- \; + \; + \; - \; - \; + \; + \; - \; - \; - \; - \; - \; + \; - \; + \; - \; - \; - \; + \; -$$

X = the number of plus signs = **16.**

n = **total** number of plus and minus signs = 16 + 24 = **40.**

As the sample size, n = 40 is very large, so we shall use the Normal approximation to Binomial distribution.

$$\therefore \textbf{ Test statistic: } Z = \frac{X - n\theta}{\sqrt{n\theta(1-\theta)}}$$

where $\theta = \frac{1}{2} = 0.5$, **$n$ = sample size.**

$$\therefore \qquad Z = \frac{16 - 40(0.5)}{\sqrt{40(0.5)(0.5)}} = \frac{-4}{3.16} = -1.26 \qquad [\because \; n = 40, \theta = 0.5]$$

$$\therefore \qquad |Z| = 1.26.$$

3. Level of significance: Here $\alpha = 0.05$

4. Critical values: The critical value $|Z_\alpha|$ for $\alpha = 0.05$ for one tailed test **is** 2.33.

Decision: Since $|Z| < |Z_\alpha|$ as 1.26 < 2.33, so we **accept the null hypothesis.**

EXERCISE – 15.2

1. Use the sign test to see if there is a difference between the number of days until collection of an account receivable before and after a collection policy. Use 0.05 significance level.

Before :	30	28	34	35	40	42	33	38	34	45	28	27	25	41	36
After :	34	29	33	32	47	43	40	42	37	44	27	33	30	38	36.

[**Ans.** There is no significant difference before and after new collection policy in the accounts receivable.]

2. The following data show employees' rates of defective work before and after a change in the wage incentive plan. Compare the two sets of data given below to see if the change lowered the defective units produced. Use the 0.10 level of significance.

Before :	8	7	6	9	7	10	8	6	5	8	10	8
After :	6	5	8	6	9	8	10	7	5	6	9	5.

3. Use the sign test on the data given below to determine whether there is a statistical increase in the values produced by treatment *B* over those produced by treatment *A*:

Subject :	1	2	3	4	5	6	7	8	9	10
Treatment *A* :	46	41	37	32	28	43	42	51	28	27
Treatment *B* :	52	43	37	32	31	39	44	53	26	31.

Use 0.05 level of significance.

4. When is the sign test used? The scores under two conditions *X* and *Y* obtained by the respondents are given below:

***X* :**	12	16	8	6	4	8
***Y* :**	7	12	17	5	12	11.

Apply the sign test and comment on your findings at 0.05 level of significance.

5. A company claims that if its product is added to an automobile's gasoline tank, the distance travelled in kilometres per litre will improve. To test the claim, 15 different automobiles are chosen and the distances with and without the additive are measured; the results are shown below. Assuming that the driving conditions are the same, determine whether there is a difference due to the additive at significance levels of (*a*) 0.05 and (*b*) 0.01.

With additive :	17.3	14.1	9.8	12.5	7.8	12.2	14.3	11.7	13.8	16.0	24.8	11.2	12.8	14.0	12.1
Without additive :	15.7	13.6	10.2	12.3	7.4	11.1	13.4	12.0	13.1	15.7	14.4	11.5	12.0	13.6	11.4

[**Ans.** **(*a*)** There is a difference at the 0.05 level of significance
(*b*) There is no difference at the 0.01 level of significance.]

6. Can one conclude at the 0.05 significance level that the distances achieved in Problem 5 are *better* with the additive than without it?

7. The *XY* Soda Company sets up "taste tests" in 27 locations around the country in order to determine the public's relative preference for two brands of cola, *A* and *B*. In eight locations brand *A* is preferred over brand *B*, in 17 locations brand *B* is preferred over brand *A*, and in the remaining locations there is indifference. Can one conclude at the 0.05 significance level that brand *B* is preferred over brand *A*?

[**Ans.** No]

8. A weight-loss club advertises that a special program that it has designed will produce a weight loss of at least 6% in 1 month if followed precisely. To test the club's claim, 36 adults undertake the program. Of these, 25 realize the desired loss, 6 gain weight, and the rest remain essentially unchanged. Determine at the 0.05 significance level whether the program is effective.

[**Ans.** The program is effective at the 0.05 level of significance.]

9. A training manager claims that by giving a special course to company sales personnel, the company's annual sales will increase. To test this claim, the course is given to 24 people. Of these 24, the sales of 16 increase, those of 6 decrease, and those of 2 remain unchanged. Test at the 0.05 significance level the hypothesis that the course increased the company's sales.

10. The following data represents the number of hours that a rechargeable hedge trimmer operates before a recharge is required.

1.5 2.2 0.9 1.3 2.0 1.6 1.8 1.5 2.6 1.2 1.7.

Use the sign test to test the hypothesis at 0.05 level of significance that this particular trimmer operates, on the average, 1.8 hours before requiring a recharge.

[M.B.A; Anna University, December 2002]

[**Ans.** Accept the null hypothesis $H_0 : \mu = 1.8$]

[**Hint:** $H_0 : \mu = 1.8$, $H_1 : \mu \neq 1.8$, we have $- + - - + - - + - -$, so, $n = 10$ and $X = 3$. Apply Binomial Table 13.]

11. On 12 visits to a doctor's clinic, a patient had to wait in minutes as under:

17 32 25 15 28 25 20 12 35 20 26 24

before being seen by the doctor. Use the sign test with $\alpha = 0.05$ to test the doctor's claim that, on the average, his patient do not wait more than 20 minutes before being examined by him.

[**Ans.** $H_0 : \mu = 20$ is accepted]

15.8 RANK SUM TESTS

Rank sum tests are a whole family of tests. We shall concentrate only on the following two members of this family:

1. **Mann-Whitney U Test**
2. **Kruskal-Wallis H Test.**

Mann-Whitney tests is used when there are only two populations whereas Kruskal-Wallis test is employed when more than two populations are involved. The use of these tests will enable us to determine whether independent samples have been drawn from the same population or different populations have the same distribution.

15.9 MANN-WHITNEY U-TEST

It is a **nonparametric method used to determine whether two independent samples have been drawn from populations with same distribution.** This test is also known as **U-Test.** This **test** enables us to test the null hypothesis $H_0 : \mu_1 = \mu_2$ without assuming whether the population samples have roughly the shape of a normal distribution.

This method helps us to determine whether the two **samples have come from identical populations.** If it is true that the samples **have come from the same population, it is reasonable to assume that the means of ranks assigned to the values of two sampls are more or less the same.** The **alternative hypothesis H_1** would be: **That the means of the populations are not equal, *i.e.*, $H_1 : \mu_1 \neq \mu_2$.** In this case, most of the smaller ranks will go to the values of one sample, while most of the higher ranks will go to the other sample.

WORKING METHOD

Step 1. Set the null hypothesis H_0 and alternative hypothesis H_1:

$H_0 : \mu_1 = \mu_2$

$H_1 : \mu_1 \neq \mu_2 \Rightarrow$ **a case of two tailed test**

Step 2. Combine all sample values in an array from smallest to the largest, and assign ranks to all these values. If two or more sample values are identical (*i.e.*, there are **tie scores**), the sample values are each assigned a rank equal to the **mean of the ranks** that would otherwise be assigned.

Step 3. Find the sum of the ranks for each of the samples. Let us denote these sums by R_1 and R_2. Also n_1 and n_2 are their respective sample sizes. For convenience, choose n_1 as the smaller size if they are unequal so that $n_1 \leq n_2$.

A significant difference between the rank sums R_1 and R_2, implies a significant difference between the samples.

Step 4. **Calculation of U to test the difference between the rank sums.**

U-Statistic: $$U = n_1 n_2 + \frac{n_1(n_1+1)}{2} - R_1 \quad \text{[Corresponding to Sample 1]}$$

or $$U = n_1 n_2 + \frac{n_2(n_2+1)}{2} - R_2 \quad \text{[Corresponding to Sample 2]}$$

The sampling distribution of U is symmetrical and has a mean and variance given by the formulae:

Mean: $\mu_U = \dfrac{n_1 n_2}{2}$

Variance: $\sigma_U^2 = \dfrac{n_1 n_2 (n_1 + n_2 + 1)}{12}$

If n_1 and n_2 are both atleast equal to 8, it turns out that the distribution of U is nearly normal and one could use the statistic Z, where

$$Z = \frac{U - \mu_U}{\sigma_U}$$

is normally distributed with mean 0 and variance 1.

Step 5. **Level of significance:** Take $\alpha = 0.05$

Step 6. **Critical Region:** Accept H_0 if $|Z| < |Z_\alpha|$, where Z_α is the tabled value of Z for the given level of significance α. The values of Z_α are given by the normal **Table 8** given at the end.

Example 5. *The nicotine contents of two brands of cigarettes, measured in milligrams, was found to be as follows:*

Brand A :	*2.1*	*4.0*	*6.3*	*5.4*	*4.8*	*3.7*	*6.1*	*3.3*		
Brand B :	*4.1*	*0.6*	*3.1*	*2.5*	*4.0*	*6.2*	*1.6*	*2.2*	*1.9*	*5.4*

Test the hypothesis, at the 0.05 level of significance, that the average nicotine contents of the two brands are equal against the alternative that they are unequal.

Solution.

1. Setting up of Hypothesis

Null Hypothesis: $H_1 : \mu_1 = \mu_2$, *i.e.*, the average nicotine contents of the two brands are equal.

Alternative Hypothesis: $H_1 : \mu_1 \neq \mu_2$, *i.e.*, the average nicotine contents of the two brands are not equal.

$\Rightarrow$ **It is a case of two tailed test.**

2. Level of significance: Here $\alpha = 0.05$.

3. Computation of U-statistic:

The observations are arranged in ascending order and ranks from 1 to 18 are assigned.

Original Data :	0.6	1.6	1.9	2.1	2.2	2.5	3.1	3.3	3.7	4.0	4.0	4.1	4.8	5.4	5.4	6.1	6.2	6.3
Rank	1	2	3	**4**	5	6	7	**8**	**9**	**10.5**	10.5	12	**13**	14.5	**14.5**	**16**	17	**18**

The ranks of the observation belonging to the smaller samples are underscored (put in bold form)

$R_1 = 4 + 8 + 9 + 10.5 + 13 + 14.5 + 16 + 18 = \mathbf{93}$

$R_2 = 1 + 2 + 3 + 5 + 6 + 7 + 10.5 + 12 + 14.5 + 17 = \mathbf{78.}$

Also, $n_1 = 8$; $n_2 = 10$.

$\therefore$ **U-Statistic:**
$$U = n_1 n_2 + \frac{n_1(n_1+1)}{2} - R_1$$
$$= 8 \times 10 + \frac{8 \times 9}{2} - 93 = 80 + 36 - 93 = \mathbf{23.}$$

$$\therefore \quad \textbf{Mean of U} = \mu_U = \frac{n_1 n_2}{2} = \frac{10 \times 8}{2} = \mathbf{40.}$$

$$\textbf{Variance of U} = \sigma_U^2 = \frac{(n_1 \times n_2)(n_1 + n_2 + 1)}{12}$$

$$= \frac{10 \times 8\,(10 + 8 + 1)}{12} = \frac{380}{3} = \mathbf{126.1 = 126.67.}$$

$$\therefore \quad \sigma_U = \sqrt{126.67} = \mathbf{11.25.}$$

Here $n_2 = 10$, so we can use the statistic

$$Z = \frac{U - \mu_U}{\sigma_U} = \frac{23 - 40}{11.25} = \frac{-17}{11.25} = \mathbf{-1.51.}$$

$$\therefore \quad |Z| = \mathbf{1.51.}$$

The tabled value of Z_α at $\alpha = 0.05$ is 1.96. **[From normal table 8]**

Now, $Z_\alpha \, |Z| < |Z_\alpha|$ **as 1.51 < 1.06, so we accept the null hypothesis H_0 and conclude that there is no significant difference in the average nicotine contents of the two brands of cigarettes.**

Example 6. *The following are the weight gains (in pounds) of two random samples of young Indians fed on two different diets but otherwise kept under identical conditions:*

Diet I :	*16.3*	*10.1*	*10.7*	*13.5*	*14.9*	*11.8*	*14.3*	*10.2*
	12.0	*14.7*	*23.6*	*15.1*	*14.5*	*18.4*	*13.2*	*14.0*
Diet II :	*21.3*	*23.8*	*15.4*	*19.6*	*12.0*	*13.9*	*18.8*	*19.2*
	15.3	*20.1*	*14.8*	*18.9*	*20.7*	*21.1*	*15.8*	*16.2.*

Use U test at 0.01 level of significance to test the null hypothesis that the two population samples are identical against the alternative hypothesis that on the average the second diet produces a greater gain in weight.

Solution.

1. **Setting up of Hypothesis:**
   Null Hypothesis: $H_0 : \mu_1 = \mu_2$
   Alternative Hypothesis: $H_1 : \mu_1 < \mu_2$.
2. **Level of significance:** $\alpha = 0.01$
3. **Test Statistic:** Ranking the data jointly according to size we find the values **first sample occupy the ranks:** 21, 1, 3, 8, 15, 4, 11, 2, 5.5, 13, 31, 16, 12, 22, 7 and 10 (the fifth and sixth values are both 12, so we assigned each the rank 5.5).

$\therefore$ **R_1 = The sum of the ranks of the first sample**

$= 21 + 1 + 3 + 8 + 15 + 4 + 11 + 2 + 5.5 + 13 + 31 + 16 + 12 + 22 + 7 + 10 = \mathbf{181.5.}$

Also **$n_1 = 16$, $n_2 = 16$.**

$$\therefore \quad \textbf{Statistics } U_1 = n_1 n_2 + \frac{n_1(n_1 + 1)}{2} - R_1$$

or

$$U = 16 \times 16 + \frac{6\,(16 + 1)}{2} - 181.5 = \mathbf{210.5.}$$

As the sample sizes n_1 and n_2 are both greater than 8, so the distribution of U is nearly normal with

$$\textbf{Mean: } \mu_U = \frac{n_1 n_2}{2} = \frac{16 \times 16}{2} = 128.$$

$$\textbf{Variance: } \sigma_U^2 = \frac{n_1 n_2 (n_1 + n_2 + 1)}{12} = \frac{16 \times 16 (16 + 17 + 1)}{12}$$

$$= \frac{16 \times 16 \times 33}{2} = 704.$$

$$\therefore \quad \text{S.E. (U)} = \sigma_U = \sqrt{704} = 26.53.$$

$$\textbf{Test Statistic: } Z = \frac{U - \mu_U}{\text{S.E.(U)}}$$

$$\text{or} \quad Z = \frac{210.5 - 128}{26.53} = \frac{82.5}{26.53} = 3.11.$$

Critical value. From the normal tables the value of Z for $\alpha = 0.01$ is **2.33.**

$$\therefore \quad | Z_\alpha | = 2.33.$$

4. **Decision:** Reject the hypothesis if the calculated value of $| Z |$ is more than the tabled value of $| Z_\alpha |$.

   Here $| Z | > | Z_\alpha |$ as $3.11 > 2.33$, so the null hypothesis is rejected and we conclude that **on the average the second diet produces a greater gain in weight.**

15.10 KRUSKAL-WALLIS TEST OR H-TEST

The Kruskal-Wallis test is a generalization of Mann-Whitney U-test to the case of $k > 2$ samples. This **test is also known as Kruskal-Wallis H-test.** It is used to test the null hypothesis H_0 that k independent samples are drawn from the identical population. **This test is an alternative nonparametric test to the F-test for testing the equality of means in the one factor analysis of variance when the experimenter wishes to avoid the assumption that the samples were selected from the normal populations.** Let n_i ($i = 1, 2, 3, \ldots\ldots, k$) be the number of observations in the ith sample. First we combine all k samples and arrange them to get $n = n_1 + n_2 + n_3 + \ldots\ldots + n_k$ observations in ascending order, substituting the appropriate rank from 1, 2, ..., n for each observation. In the case **of ties (identical observations), we follow the usual procedure of replacing the observations by the means of the ranks that the observations would have if they were distinguishable.** The sum of the ranks corresponding to the n_i observations in the ith sample is denoted by the random variable R_i. Now let us consider the statistic.

$$H = \frac{12}{n(n+1)} \sum_{i=1}^{k} \frac{R_i^2}{n_i} - 3(n+1).$$

which is approximated very well by a chi-square distribution with $k - 1$ degrees of freedom when H_0 is true and if each sample consists of at least five observations.

If H falls in the critical region $H > X_\alpha^2$ with $v = k - 1$ degrees of freedom, we reject H at the α level of significance; otherwise, we accept H.

Example 7. *A teacher wishes to test three different leading methods I, II and III. To do this, the teacher chooses at random three groups of five students each and teaches each group by a different method. The same examination is then given to all the students and the marks obtained are given below. Determine at α = 0.05 significance level whether there is a difference between the teaching methods.*

Method I :	*78*	*62*	*71*	*58*	*73*
Method II :	*76*	*85*	*77*	*90*	*87*
Method III :	*74*	*79*	*60*	*75*	*80*

Solution. Null Hypothesis: H_0 : there is difference between the teaching methods.

Alternative Hypothesis: H_1 : There is no difference between the three teaching methods.

Since there are three methods (I, II and III) and each group consists of 5 students, so we have $N_1 = N_2 = N_3 = 5$, and $N = N_1 + N_2 + N_3 = 5 + 5 + 5 =$ **15.** Arranging all these marks in increasing order of magnitude and assigning appropriate ranks, we get :

Marks:	58	60	62	61	73	74	75	76	77	78	79	80	85	87	90
Ranks:	**1**	**2**	**3**	**4**	**5**	**6**	**7**	**8**	**9**	**10**	**11**	**12**	**13**	**14**	**15**

Rank Table

Method 1	10	3	4	1	5	**23 = R_1**
Method 2	8	13	9	15	14	**59 = R_2**
Method 3	6	11	2	7	12	**38 = R_3**

$$\therefore \quad H = \frac{12}{N(N+1)}\left[\frac{R_1^2}{N_1} + \frac{R_2^2}{N_2} + \frac{R_3^2}{N_3}\right] - 3(N+1)$$

$$= \frac{12}{15 \times 16}\left[\frac{(23)^2}{5} + \frac{(59)^2}{5} + \frac{(38)^2}{5}\right] - 3 \times 16$$

$$H = \frac{1}{5}\,[529 + 3481 + 1444] - 48$$

$$= \frac{5454}{100} - 48 = \mathbf{6.54.}$$

Here **degrees of freedom = $k - 1$ = 3 – 1 = 2.**

Also **level of significance :** $\alpha = 0.05$

$\therefore$ **ψ^2 (for 2 d.f. and $\alpha = 0.05$) = $\psi_{0.05,\,2}$ = 5.991.** **[From ψ^2-tables]**

Decision: Reject H_0 if $H > \psi^2_{0.95}$

Now 6.54 > 5.991 $\Rightarrow$ $H > \psi^2_{0.05}$

$\Rightarrow$ **the null hypothesis H_0 is rejected and the alternative hypothesis H_1 is accepted. We conclude that there is no difference in teaching methods.**

Example 8. *During one semester a student received in various subjects the marks shown below. Test at 0.05 significance level whether there is a difference between the marks in these subjects.*

Mathematics	*72*	*80*	*83*	*75*	
Science	*81*	*74*	*77*		
English	*88*	*82*	*90*	*87*	*80*
Economics	*90*	*71*	*77*	*70*	

Solution. Null Hypothesis: H_0 **: There is a significant difference between the marks in these subjects.**

Alternative Hypothesis: H_1 **: There is no difference between the marks in these subjects.**

$\therefore \quad n = n_1 + n_2 + n_3 + n_4 = 4 + 3 + 5 + 4 = 16.$

Arranging these marks in increasing order of magnitude and assigning appropriate ranks, we have

Table of Ranks

Marks:	70	71	72	74	75	77	77	80	80	81	82	83	87	88	90	90
Ranks:	**1**	**2**	**3**	**4**	**5**	**6.5**	**6.5**	**8.5**	**8.5**	**10**	**11**	**12**	**13**	**14**	**15.5**	**15.5**

						Sum of Ranks
Maths	3	8.5	12	5		**28.5 =** R_1
Science	10	4	6.5			**20.5 =** R_2
English	14	11	15.5	13	8.5	**62 =** R_3
Economics	15.5	2	6.5	1		**25 =** R_4

Here $N_1 = 4, \quad N_2 = 3, \quad N_3 = 5, \quad N_4 = 4$

$\therefore \quad N = N_1 + N_2 + N_3 + N_4 = 4 + 3 + 5 + 4 = \mathbf{16}.$

Also $R_1 = 28.5, \quad R_2 = 20.5, \quad R_3 = 62, \quad R_4 = 25.$

$$\therefore \quad H = \frac{12}{N(N+1)}\left[\frac{R_1^2}{N_1} + \frac{R_2^2}{N_2} + \frac{R_3^2}{N_3} + \frac{R_4^2}{N_4}\right] - 3(N+1)$$

$$= \frac{12}{16 \times 17}\left[\frac{(28.5)^2}{4} + \frac{(20.5)^2}{3} + \frac{(62)^2}{5} + \frac{(25)^2}{4}\right] - 3 \times 17$$

$$= \frac{12}{16 \times 17}\left[\frac{812}{4} + \frac{420}{3} + \frac{3844}{5} + \frac{625}{4}\right] - 51$$

$$= \frac{12}{16 \times 17}[203 + 140 + 768.80 + 156.25] - 51$$

$$= \frac{12}{16 \times 17}(1268.05) - 51 = 55.94 - 51 = \mathbf{4.94}.$$

Degrees of freedom = $k - 1 = 4 - 1 = 3$

Level of significance : Here $\alpha = 0.05$

The value of ψ^2 for 3 degrees of freedom for $\boldsymbol{\alpha} = \mathbf{0.05}$ is **7.81** **[From ψ^2 table]**

Decision: Since the value of H is less than the tabled value of ψ^2 at $\alpha = 0.05$ for 3 degrees of freedom as $4.94 < 7.81$, **so we accept the null hypothesis** H_0 $\Rightarrow$ **there is a significant difference between the marks of these subjects.**

Example 9. *The following are the final examination of marks of three groups of students who were taught computer by three different methods.*

First Method: *94, 88, 91, 74, 87, 97*

Second Method: *85, 82, 79, 84, 61, 72, 80*

Third Method: *89, 67, 72, 76, 69.*

Use the H-test at the 0.05 level of significance to test the null hypothesis that the three methods are equally effective.

Solution. 1. Setting up of Hypothesis:

Null Hypothesis: $H_0 : \mu_1 = \mu_2 = \mu_3$.

Alternative Hypothesis: $H_1 : \mu_1, \mu_2$ and μ_3 are not all equal.

2. Level of significance: Here $\alpha = 0.05$.

3. Calculation of Test Statistic: Ranking the data jointly from 1 to 18, we find that

$$R_1 = 6 + 13 + 14 + 16 + 17 + 18 = \mathbf{84}$$

$$R_2 = 1 + 4.5 + 8 + 9 + 10 + 12 = \mathbf{55.5}$$

$$R_3 = 2 + 3 + 4.5 + 7 + 15 = \mathbf{31.5}$$

[∵ there is only one tie and the tied marks are each assigned the rank 5.5]

Also $\quad N_1 = 6, \quad N_2 = 7, \quad N_3 = 5$

∴ $\quad N = N_1 + N_2 + N_3 = 6 + 7 + 5 = \mathbf{18}.$

$$\therefore \quad \textbf{Test statistic : } H = \frac{12}{N(N+1)}\left[\frac{R_1^2}{N_1} + \frac{R_2^2}{N_2} + \frac{R_3^2}{N_3}\right] - 3(N+1)$$

or

$$H = \frac{12}{18 \times 19}\left[\frac{(84)^2}{6} + \frac{(55.5)^2}{7} + \frac{(31.5)^2}{5}\right] - 3 \times 19$$

$$= \frac{2}{57}\left[1176 + \frac{3080.25}{7} + \frac{992.25}{5}\right]$$

$$= \frac{2}{57}[1176 + 440.04 + 198.45] - 57$$

$$= \frac{2}{57} \times 1814.49 - 57 = 63.67 - 57 = \mathbf{6.67}$$

Degrees of Freedom $= k - 1 = 3 - 1 = 2$

∴ ψ^2 **(for 2 d.f. and $\alpha = 0.05$) = 5.991.** **[From ψ^2 table]**

Decision: Since $H = 6.67$ exceeds $\psi^2_{0.05,\,2} = 5.991$, so the null hypothesis is rejected. We **conclude that the three methods are not equally effective.**

EXERCISE – 15.3

1. What is Mann–Whitney U test? When is it used?
2. Explain the Mann–Whitney U test with the help of an example.
3. A farmer wishes to determine whether there is a difference in yields between two different varieties of wheat, I and II. The following data shows the production of wheat per unit area using the two varieties. Can the farmer conclude at signficance levels of (*a*) 0.05 and (*b*) 0.01 that a difference exists?

Wheat I	15.9	15.3	16.4	14.9	15.3	16.0	14.6	15.3	14.5	16.6	16.0
Wheat II	16.4	16.8	17.1	16.9	18.0	15.6	18.1	17.2	15.4		

4. Instructors *A* and *B* both teach a first course in chemistry at *XYZ* University. On a common final examination, their students received the marks shown as given below. Test at the 0.05 significance level the hypothesis that there is no difference between the two instructors' grades.

A	88	75	92	71	63	84	55	64	82	96				
B	72	65	84	53	76	80	51	60	57	85	94	87	73	61

5. A company wishes to determine whether there is a difference between two brands of gasoline, *A* and *B*. The following data shows the distances travelled in kilometres per litre for each brand. Can we conclude at the 0.05 significance level (*a*) that there is a difference between the brands and (*b*) that brand *B* is better than brand *A*?

A	15.2	14.4	14.6	16.3	15.9	14.7	15.4	15.6	15.3	16.0
B	16.7	14.9	15.0	15.8	16.9	15.5	15.7	14.8	16.4	16.5

6. Explain Kruskal–Wallis test with the help of an example.

7. In an experiment to determine which of three different missile systems is preferable, the propellant burning rates was measured. The data after coding is given below.

System I	24.0	16.7	22.8	19.8	18.9			
System II	23.2	19.8	18.1	17.6	20.2	17.8		
System III	18.4	19.1	17.3	17.3	19.7	18.9	18.8	19.3

Use the Kruskal–Wallis test and a significance level of $\alpha = 0.05$ to test the hypothesis that propellant burning rates are the same for the three missiles.

[**Hint.** Let $H_0 : \mu_1 = \mu_2 = \mu_2$, $\alpha = 0.05$.

Table : Ranks for Propellant Burning Rates

Missile System I	*Missile System II*	*Missile System III*
19	18	7
1	14.5	11
17	6	2.5
14.5	4	2.5
9.5	16	13
	5	9.5
		8
		12
$\mathbf{R_1 = 61.0}$	$\mathbf{R_2 = 63.5}$	$\mathbf{R_3 = 65.5}$

Also $n_1 = 5$, $n_2 = 6$, $n_3 = 8$ and $\boldsymbol{n = n_1 + n_2 + n_3 = 19}$

$$\therefore \qquad \boldsymbol{H = \frac{12}{n(n+1)}\left[\frac{R_1^{\,2}}{n_1} + \frac{R_2^{\,2}}{n_2} + \frac{R_3^{\,2}}{n_3}\right] - 3\,(n+1)}$$

$$= \frac{12}{19 \times 20}\left[\frac{(61.0)^2}{5} + \frac{(63.5)^2}{6} + \frac{(65.5)^2}{8}\right] - 3 \times 20 = \mathbf{1.66.}$$

Also, the value of $\psi^2_{0.05}$ **for 18 degrees of freedom at α = 0.05 is 28.869.** **[from ψ^2 tables]**

Since $H < \psi^2_{0.05,\,18}$ as $1.66 < 5.991$, so we **accept the null hypothesis H_0**]

8. The same mathematics papers were marked by three teachers *A*, *B* and *C*. The final marks were recorded as follows :

Teacher A	73	89	82	43	80	73	66	60	45	93	36	77
Teacher B	88	78	48	91	51	85	74	77	31	78	62	76
Teacher C	68	79	56	91	71	71	87	41	59	68	53	79

Use Kruskal–Wallis test, at the 0.05 level of significance to determine if the marks distributions given by the 3 teachers differ significantly.

9. The following data represent the operating times in hours for 3 types of scientific pocket calculators before a recharge is required.

Calculator		
A	*B*	*C*
4.9	5.5	6.4
6.1	5.4	6.8
4.3	6.2	5.6
4.6	5.8	6.5
5.3	5.5	6.3
	5.2	6.6
	4.8	

Use Kruskal–Wallis test, at the 0.01 level of significance, to test the hypothesis that the operating times for all three calculators are equal.

10. The following data represent the weights (in kg) of personal luggage carried on an aircraft by a member of 2 base ball clubs.

Club A	34	39	41	28	33	
Club B	36	40	35	31	39	36

Use Mann–Whitney U test with $\alpha = 0.05$ to test the hypothesis that the two clubs carry the same amount of luggage on the average.

11. Random samples of 4 brands of cigarettes were tested for tar content. The following figures show the milligrams of tar found in the 16 cigarettes tested.

Brand A	*Brand B*	*Brand C*	*Brand D*
14	16	16	17
10	18	15	20
11	14	14	19
13	15	12	21

Use the Kruskal–Wallis test, at the 0.05 level of significance, to test whether there is a significant difference in tar content among the 4 brands of cigarettes.

12. The following are the numbers of minutes it took random samples of 15 men and 12 women to complete a written test given for the renewal of their driver's licenses:

Men	9.9	7.4	8.9	9.1	7.7	9.7	11.8	9.2	10.0	10.2	9.5	10.8	8.0	11.0	7.5
Women	8.6	10.9	9.8	10.7	9.4	10.3	7.3	11.5	7.6	9.3	8.8	9.6			

Use the U test at the 0.05 level of significance to decide whether to accept the null hypothesis $\mu_1 = \mu_2$ or the alternative hypothesis $\mu_1 \neq \mu_2$, where μ_1 and μ_2 are the average amounts of time it takes men and women to complete the test.

13. The following are data on the breaking strength (in pounds) of random samples of two kinds of 2-inch cotton ribbons:

Type I ribbon	144	181	200	187	169	171	186	194	176	182	133	183	197	165	180	198
Type II ribbon	175	164	172	194	176	198	154	134	169	164	185	159	161	189	170	164

Use the *U* test at the 0.05 level of significance to test the claim that Type I ribbon is, on the average, stronger than Type II ribbon.

14. The following are the miles per gallon which a test driver got for 10 tankfuls of each of three kinds of gasoline:

Gasoline A	20	31	24	33	23	24	28	16	19	26
Gasoline B	29	18	29	19	20	21	34	33	30	23
Gasoline C	19	31	16	26	31	33	28	28	25	30.

Use the Kruskal–Wallis test at the 0.05 level of significance to test whether or not there is a difference in the actual average mileage yield of the three kinds of gasoline.

15. Three groups of guinea pigs were injected, respectively, with 0.5, 1.0 and 1.5 milligrams of a tranquilizer, and the following are the numbers of seconds it took them to fall asleep:

0.5-mg dose	8.2	10.0	10.2	13.7	14.0	7.8	12.7	10.9		
1.0-mg dose	9.7	13.1	11.0	7.5	13.3	12.5	8.8	12.9	7.9	10.5
1.5-mg dose	12.0	7.2	8.0	9.4	11.3	9.0	11.5	8.5.		

Use the H test at the 0.01 level of significance to test the null hypothesis that the difference in dosage have no effect on the length of time it takes guinea pigs to fall asleep.

16. To compare four bowling balls, a professional bowler bowls five games with each ball and gets the following results:

Ball A	208	220	247	192	229
Ball B	216	196	189	205	210
Ball C	226	218	252	225	202
Ball D	212	198	207	232	221.

Use the Kruskal–Wallis test at the 0.05 level of significance to test whether or not the bowler can expect to score equally well with the four bowling balls.

ANSWERS

1. (*a*) Yes (*b*) Yes.
4. There is no difference at 0.05 level of significance.
5. (*a*) Yes (*b*) Yes.
7. H_0 : The propellant burning rates are the same for the three missiles is accepted.
8. The marks given by the three teachers differ significantly.
9. The times for the three calculators are not equal.
10. Hypothesis is rejected.
11. There is a significant difference.
12. The null hypothesis is accepted.
13. The null hypothesis is accepted and $|Z| = 1.62$.
14. The null hypothesis cannot be rejected; $H = 0.86$.
15. The null hypothesis is accepted.
16. The null hypothesis is accepted.

15.11 THE ONE SAMPLE RUNS TEST

It is a nonparametric method to determine the randomness with which the samples items have been selected. **The runs test,** based on the order in which the sample observations are obtained, **is a useful technique for testing the null-hypothesis H_0 that the observations have indeed been drawn at random.** The runs test can also be used to detect departures in randomness of a sequence of quantitative measurements over time, caused by trends or periodicities.

RUN. A run is a subsequence of one or more identical symbols representing a common property of the data. In other words, a **run** is **defined as a set of identical (or related) symbols**

contained between two different symbols or no symbol (such as at the beginning or end of the sequence).

To **understand what a run is**, consider a sequence made up of two symbols, *A* and *B* such as;

$$AAA \mid BB \mid A \mid BBB \mid AAAA \mid BBBB \mid AA \mid \quad \text{... (I)}$$

In tossing a coin, for example, ***A*'s** could represent **'heads'** and ***B*'s** represent **'tails'.**

Proceeding from left to right in the above sequence (I), the first run indicated by a **vertical bar** | consists of three *A*'s, the second run consists of two *B*'s, the third run consists of one *A*, the fourth run consists of three *B*'s, the fifth run consists of four *A*'s, the sixth run consists of four *B*'s and the seventh run consists of two *A*'s. Thus, **there are seven runs in all.**

It seems that some relationship exists between randomness and the number of runs. Consider the another sequence

$$A \mid B \mid A \mid B \mid A \mid B \mid A \mid B \mid A \mid B \mid \quad \text{... (II)}$$

There is a **cyclic pattern**, in the above sequence in which we go from *A* to *B*, back to *A* again, etc., **which we could hardly believe to be random.** In such case, we have **too many runs** (in fact, we have the maxmum number possible for the given number of *A*'s and *B*'s).

On the other hand, the following sequence *AAAAAA* | *BBBB* | *AAAAA* | *BBB* | *AAAA* | *BB* | *AAA* | *B* | seems to be a **trend pattern**, in which *A*'s and *B*'s are grouped together in a pattern. In such there are **too few runs**, and **we could not consider the sequence to be random.**

We notice that: **"If there are too few runs, a definite grouping, or clustering or trend may be suspected. On the other hand, if there are too many runs some sort of repeated alternating pattern may be suspected."** Thus, **it may be possible to prove that too many or too few runs in a sample indicate something other than chance, (or randomness), when items were selected.**

SAMPLING DISTRIBUTION OF *V*-STATISTIC

The number of runs '*V*' is a statistic with a special sampling distribution. The mean μ_V of *V*-statistic is given by μ_V, where

Mean of *V*-Statistic: $\mu_V = \dfrac{2n_1n_2}{n_1+n_2} + 1$, **where**

n_1 = **the number of first response**

n_2 = **the number of second response**

The variance of *V*-statistic is σ_V^2, where

Variance of *V*-statistic: $\sigma_V^2 = \left[\dfrac{2n_1n_2\,(2n_1n_2 - n_1 - n_2)}{(n_1+n_2)^2\,(n_1+n_2-1)}\right]$

The standard error S.E. (*V*) of the *V*-statistic is given by σ_V.

The sampling distribution of *V*-statistic can be closely approximated by the normal distribution if either n_1 or n_2 is atleast equal to 8. In that case,

$$Z = \frac{V - \text{mean of } V \text{ statistic}}{\text{S.E.}(V)}$$

or

$$Z = \frac{r - \mu_V}{\sigma_V}$$

is normally distributed with mean 0 and variance 1 and thus normal Table can be used.

Example 10. *Twenty five individuals were sampled as to whether they like or did not like a product indicated by Y and N respectively. The resulting sample is shown by the following sequence:*

YY NNNN YYY N Y NN Y NNNNN YYYY NN.

(*a*) *Determine the number of runs, V.*

(*b*) *Test at 0.05 significance level whether the responses are random.*

Solution. (*a*) Using a vertical bar to indicate a run, we have

YY | *NNNN* | *YYY* | *N* | *Y* | *NN* | *Y* | *NNNNN* | *YYYY* | *NN* |

∴ **Number of runs** V **= 10.**

(*b*) Let n_1, n_2 respectively denote the number of *Y*'s and *N*'s.

∴ $n_1 = 11$ and $n_2 = 14$. Also $V = 10$.

1. **Setting of Hypothesis:**

   Null hypothesis : H_0 **:** The responses are random.

   Alternative Hypothesis : H_1 **:** The response are not random

   ⇒ It is a case of **two tailed test.**

2. **Computation of test statistic:**

$$\mu_V = \frac{2n_1n_2}{n_1+n_2} + 1 = \frac{2\times 11\times 14}{11+14} + 1 = \frac{308}{25} + 1 = 12.32 + 1 = \mathbf{13.32.}$$

$$\sigma_V^2 = \frac{2n_1n_2\,(2n_1n_2 - n_1 - n_2)}{(n_1+n_2)^2\,(n_1+n_2-1)} = \frac{2\times 11\times 14\,(2\times 11\times 14 - 11 - 14)}{(11+14)^2\,(11+14-1)}$$

$$= \frac{308\,(308-25)}{(25)^2\times 23} = \frac{308\times 283}{14375} = \frac{87164}{14375} = \mathbf{6.064.}$$

$$\therefore \quad \sigma_V = \sqrt{6.064} = \mathbf{2.462.}$$

$$\therefore \quad Z = \frac{V-\mu_V}{\sigma_V} = \frac{10-13.32}{2.462} = \frac{-3.32}{2.462} = \mathbf{-1.348.}$$

$$\therefore \quad |Z| = \mathbf{1.348.}$$

3. **Level of significance:** $\alpha = 0.05$.

4. **Critical value:** The value of Z_α at 5% level of significance for two tailed test is 1.96 **[From the table]**

   i.e., $|Z_\alpha| = \mathbf{1.96.}$

5. **Decision:** Since, the calculated value of $|Z|$ is less than the tabled value of $|Z_\alpha|$ as 1.348 < 1.96, so **we accept the null hypothesis** ⇒ **the responses are random.**

Example 11. *A machine is used to insert randomly one of the two types of toys A and B in each box. The manufactures choose a samples of 60 successive boxes to see if the machine is properly mixing the two types of toys. Test at 0.05 significance level whether the machine is inserting the toys in random order.*

B A BBB AAA BB A BBBB AAAA B A B AA BBB AA B A AAA BB A BB AAAA BB A BBBB AA BB A B AA BB.

Hint. Let n_1, n_2 respectively denote the number of boxes containing toy *A* and toy *B*.

Using a vertical bar to indicate a run, we have

B | *A* | *BBB* | *AAA* | *BB* | *A* | *BBBB* | *AAAA* | *B* | *A* | *B* | *AA* | *BBB* | *AA* | *B* | *AAAA* | *BB* | *A* | *BB* | *AAAA* | *BB* | *A* | *BBBB* | *AA* | *BB* | *A* | *B* | *AA* | *BB* |

∴ **The number of runs** V **= the number of vertical lines = 29.**

1. **Setting up of Hypothesis:**

   Null Hypothesis : H_0 : The toys are being inserted in the boxes in random order.

   Alternative Hypothesis : H_1 : The toys are not being inserted in the boxes in random order $\Rightarrow$ **It is a case of two tailed.**

2. **Computation of test statistic V:**

   Here n_1 **= 29,** n_2 **= 31 and** V **= 29.**

$$\textbf{Mean of } V\textbf{-Statistic} = \mu_V = \frac{2n_1n_2}{n_1+n_2}+1 = \frac{2\times 29\times 31}{29+31}+1 = \frac{1798}{60}+1$$

$$= 2997 + 1 = \mathbf{30.97.}$$

$$\textbf{Variance of } V\textbf{-Statistic} = \sigma_V{}^2 = \frac{2n_1n_2\,(2n_1n_2-n_1-n_2)}{(n_1+n_2)^2\,(n_1+n_2-1)}$$

$$= \frac{(2\times 29\times 31)\,(2\times 29\times 31-29-31)}{(29+31)^2\,(29+31-1)} = \frac{1798\times 1738}{(60)^2\times 59}$$

$$= \frac{3124924}{3600\times 59} = \frac{3124924}{212400} = \mathbf{14.712.}$$

$$\therefore \qquad \sigma_V = \sqrt{14.712} = \mathbf{3.84.}$$

$$\therefore \qquad \textbf{Stand Error of } V : \textbf{S.E. } (V) = \sigma_V = \mathbf{3.84.}$$

$$\textbf{Test Statistic : } Z = \frac{V-\mu_V}{\sigma_V} = \frac{29-30.97}{3.84} = -\frac{1.97}{3.84} = \mathbf{-\,0.513.}$$

$$\therefore \qquad |\,Z\,| = \mathbf{0.513.}$$

3. **Level of significance:** $\alpha = 0.05$

4. **Critical value:** The tabled value of $|\,Z_\alpha\,|$ at 5% level of significance for two tailed test is **1.96.**

5. **Decision:** Since, the calculated value of $|\,Z\,|$ is less than the tabelled value of $|\,Z_\alpha\,|$ as $0.513 < 1.96$, so **we accept the null hypothesis H_0 $\Rightarrow$ the toys are being inserted in the boxes in random order.**

Example 12. *In 30 tosses of a coin the following sequence of heads (H) and tails (T) is obtained*

H T T H T H H H T H H T T H T

H T H H T H T T H T H H T H T.

(*a*) *Determine the number of runs, V.*

(*b*) *Test at the 0.05 significance level whether the sequence is random.*

Solution. (*a*)

Using a vertical bar to indicate a run, we have :

H | T T | H | T | H H H | T | H H | T T | H | T |

H | T | H H | T | H | T T | H | T | H H | T | H | T |

$\therefore$ **The number of runs V = 22.**

Solution. (*b*)

1 Setting of Hypothesis :

Null Hypothesis : H_0 : The sequence is random.

Alternative Hypothesis : H_1 : The sequence is not random $\Rightarrow$ It is a case of **two tailed test.**

2. Calculation of Test Statistic: Let n_1, n_2 denote respectively the number of heads and tails. Then n_1 = 16 heads; n_2 = 14 tails.

Number of runs V= 22.

Now,
$$\mu_V = \frac{2n_1n_2}{n_1+n_2} + 1 = \frac{2\times16\times14}{16+14} + 1 = \mathbf{15.93.}$$

$$\sigma_V^2 = \frac{2n_1n_2\,(2n_1n_2 - n_1 - n_2)}{(n_1+n_2)^2\,(n_1+n_2-1)} = \frac{2\times16\times14\,(2\times16\times14-16-14)}{(16+14)^2\,(16+14-1)}$$

$$= \frac{448\,(448-30)}{900\times29} = \frac{187264}{26100} = \mathbf{7.175.}$$

$$\therefore \quad \sigma_V = \sqrt{7.175} = \mathbf{2.679.}$$

Test Statistic:
$$Z = \frac{V-\mu_V}{\sigma_V} = \frac{22-15.93}{2.679} = \frac{6.07}{2.679} = \mathbf{2.27.}$$

3. Level of Significance: $\alpha = 0.05$.

4. Critical value: The value of Z_α at 5% level of significance for two tailed test is 1.96.

[From the table]

5. Decision: Since, the calculated value of Z is greater than the tabled value of Z_α as 2.27 > 1.96, so **we reject the null hypothesis H_0 and accept the alternative hypothesis H_1** $\Rightarrow$ **the sequence is not random.**

Example 13. *The following is an arrangement of men, M, and women, W, lined up to purchase tickets for a rock concert:*

M W M W M M M W M W M M M W W M M M W W M W M

M M W M M M W W W M W M M M W M W M M M M W W M

Test for randomness of the arrangement at the 0.05 level of significance.

Solution.

1. Setting up of Hypothesis:

Null Hypothesis : H_0 : Arrangement is random.

Alternative Hypothesis : H_1 : Arrangement is not random. It is a case of two tailed test.

2. Level of Significance : $\alpha = 0.05$.

3. Test Statistic:
$$Z = \frac{V-\mu_V}{\sigma_V}, \text{ where}$$

$$\mu_V = \frac{2n_1n_2}{n_1+n_2} + 1$$

and
$$\sigma_V^2 = \frac{2n_1n_2\,(2n_1n_2-n_1-n_2)}{(n_1+n_2)^2\,(n_1+n_2-1)}.$$

Since, $n_1 = 30$, $n_2 = 18$ and $V = 27$, we get

$$\mu_V = \frac{2 \cdot 30 \cdot 18}{30 + 18} + 1 = \mathbf{23.5}$$

$$\sigma_V = \sqrt{\frac{2 \cdot 30 \cdot 18\,(2 \cdot 30 \cdot 18 - 30 - 18)}{(30 + 18)^2\,(30 + 18 - 1)}} = \mathbf{3.21}$$

and hence, $$Z = \frac{27 - 23.5}{3.21} = \mathbf{1.09.}$$

4. **Critical values:** The value of $|Z_\alpha|$ for $\alpha = 0.05$ is **1.96.** **[From normal tables]**
5. **Decision:** Since, $Z = 1.09$ falls between -1.96 and 1.96, so the null hypothesis cannot be rejected. In other words, there **is no real evidence to indicate that the arrangement is not random.**

15.12 RUNS ABOVE AND BELOW THE MEDIAN

The method discussed in the last section is not limited to tests of the randomness of series of attributes. Any sample which consists of numerical measurements or observations can be treated similarly by using the letters *a* and *b* to denote respectively, the values falling above and below the median of the sample. (Numbers equaling the median are omitted). The resulting series of *a*'s and *b*'s can then be tested for randomness on the basis of the total number of runs of *a*'s and *b*'s, namely, the total number of **runs above and below the median.**

Example 14. *The following are the speeds (in kilometre per hour) at which every fifth passnger car was timed at a certain checkpoint: 46, 58, 60, 56, 70, 66, 48, 54, 62, 41, 39, 52, 45, 62, 53, 69, 65, 65, 67, 76, 52, 52, 59, 59, 67, 51, 46, 61, 40, 43, 42, 77, 67, 63, 59, 63, 63, 72, 57, 59, 42, 56, 47, 62, 67, 70, 63, 66, 69 and 73. Test the null hypothesis of randomness at the 0.05 level of significance.* *[Given median speed = 59.5 km per hour]*

Solution.

1. **Setting the Hypothesis:**
   Null Hypothesis : H_0 : The sample is random.
   Alternative Hypothesis : H_1 : The sample is not random.
2. **Level of Significance:** $\alpha = 0.05$
3. **Test Statistic:**
   Here median speed = 59.5.
   Since, **the median of the speeds is 59.5, we get the following arrangement of *a*'s and *b*'s:**

   b b |a| b |a a| b b |a| b b b b |a| b |a a a a a| b b b b |a|
   b b |a| b b b |a a a| b |a a a| b b b b b |a a a a a a a|

   Then, since $n_1 = 25$, $n_2 = 25$ and $V = 20$, we get

$$\mu_V = \frac{2n_1 n_2}{n_1 + n_2} + 1$$

$$\Rightarrow \qquad \mu_V = \frac{2 \times 25 \times 25}{25 + 25} + 1 = \mathbf{26.}$$

$$\sigma_V^2 = \frac{2n_1 n_2\,(2n_1 n_2 - n_1 - n_2)}{(n_1 + n_2)^2\,(n_1 + n_2 - 1)}$$

$$\Rightarrow \quad \sigma_V^2 = \frac{2 \times 25 \times 25\,(2 \times 25 \times 25 - 25 - 25)}{(25+25)^2\,(25+25-1)} = \mathbf{12.2.}$$

$$\therefore \quad Z = \frac{V - \mu_V}{\sigma_V} = \frac{20-26}{\sqrt{12.2}} = -\ \mathbf{1.72.}$$

$$\therefore \quad |Z| = 1.72$$

4. **Critical value:** The value of $|Z_\alpha|$ for $\alpha = 0.05$ from the table is $|\mathbf{Z}_\alpha| = \mathbf{1.96.}$
5. **Decision:** Since the calculated value of $|Z| <$ Tabled value of $|Z_\alpha|$ as $1.72 < 1.96$, so the null hypothesis is accepted $\Rightarrow$ **the sample is random.**

EXERCISE – 15.4

1. Determine the number of runs, V, for each of these sequences. Also determine their mean and variance.
   (*i*) *HH TTT H T HHH TTTT H T H T TT H TT*
   (*ii*) *A B AA BBB A BB AA BBBB A B A BB AA BBB*
   (*iii*) *M W MM WW MM WWW M WW MM WW MMM*
   (*iv*) *A B A BB AAA BB A B.*
2. Use runs test on the sequence (*iii*) and (*iv*) of **question 1** and test their randomness at 5% level of significance.
3. A sample of 48 tools produced by a machine shows the following sequence of good (*G*) and defective (*D*) tools:

   G G G G G G D D G G G G G G G G
   G G D D D D G G G G G G D G G G
   G G G G G G D D G G G G G D G G.

   Test the randomness of the sequence at the 0.05 significance level.

   [**Hint.** Let n_1, n_2 denote the number of *D*'s and *G*'s respectively. Here $V = 11$.

   Let H_0 = the sequence is random.

   Now, $$\mu_V = \frac{2n_1n_2}{n_1+n_2} + 1 = \frac{2 \times 10 \times 38}{10+38} + 1 = \mathbf{16.83.}$$

   $$\sigma_V^2 = \frac{2n_1n_2\,(2n_1n_2 - n_1 - n_2)}{(n_1+n_2)^2\,(n_1+n_2-1)} = \frac{(2 \times 10 \times 38)(2 \times 10 \times 38 - 10 - 38)}{(10+38)^2\,(10+38-1)} = 4.997$$

   $$\Rightarrow \quad \sigma_V = \mathbf{2.235.}$$

   $$\therefore \quad Z = \frac{V - \mu_V}{\sigma_V} = \frac{11 - 16.83}{2.235} = -\ 2.61 \Rightarrow |Z| = \mathbf{2.61.}$$

   Also $|Z_\alpha|$ at $\alpha = 0.05 = 1.96$ [**from tables.**]

   Now $|Z| > |Z_\alpha|$ as $2.235 > 1.96 \Rightarrow H_0$ is rejected $\Rightarrow$ **the sequence is not random.**]
4. The following is the arrangement of 25 men (*M*) and 15 women (*W*) lined up to purchase tickets for a premier picture show:

   M | W W | M M M | W | M M | W | M | W | M | W W W | M M M |
   W | M M | W W W | M M M M M M | W W W | M M M M MM |

   Test for randomness at 5% level of significance.

   [**Hint.** Let H_0 : the arrangement is random.

   Here $n_1 = 25$, $n_2 = 15$, $V = 17$.

$$\therefore \quad \mu_V = \frac{2n_1n_2}{n_1+n_2} + 1 = \frac{2\times 25\times 15}{25\times 15} + 1 = \mathbf{19.75.}$$

$$\sigma_V^2 = \frac{2n_1n_2(2n_1n_2 - n_1 - n_2)}{(n_1+n_2)^2(n_1+n_2-1)} = \frac{(2\times 25\times 15)(2\times 25\times 15 - 25 - 15)}{(25+15)^2(25+15-1)}$$

$$= \frac{750\times 710}{1600\times 39} = \mathbf{8.533.}$$

$$\therefore \quad \sigma_V = \sqrt{8.533} = \mathbf{2.92.}$$

Test Statistic: $$Z = \frac{V-\mu_V}{\sigma_V} = \frac{17-19.75}{2.92} = \mathbf{-0.94.}$$

$$\therefore \quad |Z| = \mathbf{0.94.}$$

Also Z_α **at 5% = 1.96.**

Since, $|Z| < |Z_\alpha|$ at 5% as 0.94 < 1.96, so we accept the null hypothesis H_0 $\Rightarrow$ **the arrangement is random.]**

5. Form all possible sequence consisting of three *a*'s and two *b*'s and give the number of runs, *V*, corresponding to each sequence. Also obtain the sampling distribution of *V*.

[**Hint.** (*a*) The number of possible sequences consisting of three *a*'s and two *b*'s is $^5C_2 = \frac{5!}{2!\times 3!} = 10$. These sequences are shown in Table 1, along with the number of runs corresponding to each sequence.
(*b*) The sampling distribution of *V* is given in Table 2, where *V* denotes the number of runs and *f* denotes the frequency. For example, Table 2 shows that there is one 5, four 4's, etc.

Table 1

Sequence	*Runs (V)*
a a a b b	2
a a b a b	4
a a b b a	3
a b a b a	5
a b b a a	3
a b a a b	4
b b a a a	2
b a b a a	4
b a a a b	3
b a a b a	4

Table 2

V	*f*
2	2
3	3
4	4
5	1

6. Find mean and variance of **question number 5.**
[**Hint.** Here $n_1 = 3$, $n_2 = 2$.

$$\therefore \quad \mu_V = \frac{2n_1n_2}{n_1+n_2} + 1 = \frac{2\times 3\times 2}{3+2} + 1 = \frac{\mathbf{17}}{\mathbf{5}}.$$

$$\sigma_V^2 = \frac{2n_1n_2(2n_1n_2 - n_1 - n_2)}{(n_1+n_2)^2(n_1+n_2-1)} = \frac{2\times 3\times 2(2\times 3\times 2 - 3 - 2)}{(3+2)^2(3+2-1)} = \frac{\mathbf{21}}{\mathbf{25}}.]$$

7. What conclusion can you draw using runs test concerning the randomness of the following digits at $\alpha = 0.05$ significance level.
(*a*) $\sqrt{2} = 1.4142135623730950488$ ....

(*b*) $\sqrt{3}$ = 1.73205 08075 68877 2935 ....

(*c*) π = 3.14159 26535 89793 2643 ....

8. A machine is adjusted to dispence acrylic paint thinner into a container. Would you say that the amount of paint thinner being dispensed by this machine varies randomly if the contents of the next 15 containers are measured and found to be 3.6, 3.9, 4.1, 3.6, 3.8, 3.7, 3.4, 4.0, 3.8, 4.1, 3.9, 4.0, 3.8, 4.2 and 4.1 litres? Use 0.1 level of significance.

9. A random sample of 15 adults living in a small town are selected to estimate the proportion of voters favouring a certain candidate for mayor. Each individual was also asked if he or she was a college graduate. By letting *Y* and *N* designate the response of "yes" and "no" to the education question, the following sequence was obtained.

NNNN YY N YY N Y NNNN

Find the number of runs and use the runs test at the 0.1 level of significance if the sequence supports the contention that the sample was selected at random.

10. In an industry, items are inspected periodically for defectives. The following is a sequence of defective items, *D*, and nondefective items, *N*, produced by this industry.

DD NNN D NN DD NNNNN DDD NN D NNNN D N D

Use runs test to determine whether the defectives are occurring at random or not at 0.05 significance level.

11. In an industrial production line, items are inspected periodically for defectives. The following is a sequence of defective items, *D*, and nondefective items, *N*, produced by this production line:

D D N N N D N N D D N N N N
N D D D N N D N N N N D N D

Use the large-sample theory for the runs test, with the significance level of 0.05, to determine whether the defectives are occurring at random or not.

12. A random sample of 15 adults living in a small town are selected to estimate the proportion of voters favouring a certain candidate for mayor. Each individual was also asked if he or she was a college graduate. By letting *Y* and *N* designate the responses of "yes" and "no" to the education question, the following sequence was obtained:

N N N N Y Y N Y Y N Y N N N N.

Use the runs test at the 0.05 level of significance to determine if the sequence supports the contention that the sample was selected at random.

13. A silver-plating process is being used to coat a certain type of serving tray. When the process is in control, the thickness of the silver on the trays will vary randomly following a normal distribution with a mean of 0.02 millimetre and a standard deviation of 0.005 millimetre. Suppose that the next 12 trays examined show the following thickness of silver: 0.019, 0.021, 0.020, 0.019, 0.020, 0.018, 0.023, 0.021, 0.024, 0.022, 0.023, 0.022. Use the runs test to determine if the fluctuations in thickness from one tray to another is random. Let $\alpha = 0.05$.

14. Explain the working of one sample runs test with the help of an example. What are its limitations?

15. The following is the order in which red, *R*, and black, *B*, cards were dealt as a bridge player:

B B B R R R R R B B R R R

Test for randomness at the 0.05 level of significance.

16. The following is the order in which a broker received buy, *B*, and sell, *S*, orders for a certain stock:

B B B B B B B B S S B S S S S S S B B B B B

Test for randomness at the 0.05 level of significance.

17. The following are six year's quarterly sales (in millions of rupees) of a manufacturer of heavy machinery: 83.8, 102.5, 121.0, 90.5, 106.6, 104.8, 114.7, 93.6, 98.9, 96.9, 122.6, 85.6, 103.2, 96.9, 118.0, 92.1, 100.5, 92.9, 125.6, 79.2, 110.8, 95.1, 125.6 and 86.7. At the 0.05 level of significance, is there a real cyclical pattern?

18. The following arrangement indicates whether 60 consecutive cars that went by the toll booth of a bridge had local plates, L, or out-of-state plates, O:

$$L\ L\ O\ L\ L\ L\ L\ O\ O\ L\ L\ L\ L\ O\ L\ O\ O\ L\ L\ L\ L\ O\ L\ O\ O\ L\ L\ L\ L$$
$$O\ L\ L\ L\ O\ L\ O\ L\ L\ L\ O\ O\ L\ O\ O\ O\ O\ L\ L\ L\ L\ O\ L\ O\ O\ L\ L\ L\ O.$$

Test at the 0.05 level of significance whether this arrangement of L's and O's may be regarded as random.

19. A driver buys gasoline either at a Texaco station, T, or at a Mobil station, M, and the following arrangement shows the order of the stations from which she bought gasoline over a certain period of time:

$$T\ T\ T\ M\ T\ M\ T\ M\ M\ T\ T\ M\ T\ M\ T\ M\ T\ M\ M\ T\ M\ T$$

Test for randomness at the 0.05 level of significance.

20. The numbers of retail stores that opened for business and also quit business in the same year were 108, 103, 109, 107, 125, 142, 147, 122, 116, 153, 144, 162, 143, 126, 145, 129, 134, 137, 143, 150, 148, 152, 125, 106, 112, 139, 132, 122, 138, 148, 155, 146 and 158 during a period of 33 years. Making use of the fact that the medium is 138, test at the 0.05 level of significance whether there is a real trend.

21. The following are the numbers of students absent from school on 24 consecutive school days: 29, 25, 31, 28, 30, 28, 33, 31, 35, 29, 31, 33, 35, 28, 36, 30, 33, 26, 30, 28, 32, 31, 38 and 27. Test for randomness at the 0.01 level of significance.

ANSWERS

1. (*a*) **12** (*b*) **14** (*c*) **12** (*d*) **8.**

2. random; not random.

3. Not random.

4. The arrangement is random.

5. 10 ; 2, 4, 3, 5, 3, 4, 2, 4, 3, 4.

6. Mean = $\frac{17}{5}$; Variance = $\frac{21}{25}$.

7. (*a*) the digits are at random;

(*b*) the digits are at random;

(*c*) the digits are at random.

8. The sequence of measurement vary randomly.

9. $V = 7$; the sample is at random.

10. Not random.

11. The defective are not occurring at random.

12. $V = 7$. The sample was selected at random.

15. $V = 4$. The null hypothesis of randomness is accepted.

16. $V = 5$. The null hypothesis of randomness is accepted.

17. There is a cyclical pattern.

18. No, there is a cyclical pattern.

19. The null hypothesis of randomness is accepted.

20. The null hypothesis of real trend is accepted.

21. $V = 13$. The null hypothesis of randomness is accepted.

15.13 SPEARMAN RANK CORRELATION

15.13-1 Spearman Rank Correlation

Spearman rank correlation is a measure of the correlation that exists between the two sets of ranks, or it is a measure of degree of association between the variables that we would not have been able to calculate otherwise.

5.13-2 Rank Correlation Coefficient

In Chapter 7, we used the sample correlation coefficient r to measure the linear relationship etween two continuous variables X and Y. **If ranks 1, 2, ..., n are assigned to the X observations n order of magnitude and similarly to the Y observations, and if these ranks are then ubstituted for the actual numerical values into the formula for r, we obtain the onparametric counterpart of the conventional correlation coefficient.** A correlation oefficient calculated in this manner is known as the **Spearman rank correlation coefficient*** and **s denoted by r_s.** When there are no ties among either set of measurements, the formula for r_s educes to a much simpler expression, which is given below.

Rank Correlation Coefficient. A nonparametric measure of association between two ariables X and Y is given by the **rank correlation coefficient**

$$r_S = 1 - \frac{6\sum_{i=1}^{n} d_i^2}{n(n^2-1)}, \text{ where}$$

- **Σ = notation meaning "the sum of"**
- **d_i = the difference between the ranks assigned to x_i and y_i**
- **n = the number of paired observations**
- **r_s = coefficient of rank correlation**

[Notice that the subscripts, from Spearman distinguishes this r from sample correlation oefficient discussed in Chapter 7*]

In practice, the proceding formula is also used when there are ties either among x or y bservations. **The ranks for tied observations are assigned by averaging the ranks that would ave been assigned if the observations were distinguished.**

The value of r_s will usually be close to the value obtained by finding r based on numerical neasurements and is interpreted in much the same way. As before, **the values of r_s will range rom – 1 to + 1. A value of + 1 or – 1 indicates perfect association between X and Y, the plus ign occurring for identical rankings and the minus sign occurring for reverse rankings. Vhen r_s is close to zero, we would conclude that the variables are uncorrelated.**

5.13-3 Advantages of Rank Correlation Coefficient

1. We do not assume the underlying relationship between X and Y to be linear and, therefore, when the data possess a distinct curvilinear relationship, the rank correlation coefficient will likely be more reliable than the conventional measure.
2. A second advantage in using the rank correlation coefficient is the fact that no assumptions of normality are made concerning the distributions of X and Y.
3. The greatest advantage occurs when one is unable to make meaningful numerical measurements but nevertheless can establish rankings. Such is the case, for example, when different judges rank a group of individuals according to some attribute. The rank correlation coefficient can be used in this situation as a measure of the consistency of the two judges.

SOLVING A PROBLEM USING RANK CORRELATION

Example 15. *The following figures, released by the Federal Trade Commission, show the nilligrams of tar and nicotine found in 10 brands of cigarettes.*

* *For more details refer to Chapter 7, section 7.10.*

Cigarette Brand	*Tar Content*	*Nicotine Content*
Viceroy	*14*	*0.9*
Marlboro	*17*	*1.1*
Chesterfield	*28*	*1.6*
Kool	*17*	*1.3*
Kent	*16*	*1.0*
Raleigh	*13*	*0.8*
Old Gold	*24*	*1.5*
Philip Morris	*25*	*1.4*
Oasis	*18*	*1.2*
Players	*31*	*2.0*

Calculate the rank correlation coefficient to measure the degree of relationship between ta and nicotine content is cigarettes.

Solution. Let X and Y respectively represent the tar and nicotine contents. We assign rank to each set of measurements with the rank of 1 assigned to the lowest number in each set, the ran of 2 to the second lowest number in each set, and so forth, until the rank 10 is assigned to th largest number. The following table shows the individual rankings of the measurements and th differences in ranks for the 10 pairs of observations.

Table : Rankings for Tar and Nicotine Contents

Cigarette Brand	x_i	y_i	d_i	d_i^2
Viceroy	2	2	0	0
Marlboro	4.5	4	0.5	0.25
Chesterfield	9	9	0	0
Kool	4.5	6	– 1.5	2.25
Kent	3	3	0	0
Raleigh	1	1	0	0
Old Gold	7	8	– 1	1
Philip Morris	8	7	1	1
Oasis	6	5	1	1
Players	10	10	0	0
				$\Sigma d_i^2 = 5.5$

Substituting into the formula for r_s, we get

$$r_s = 1 - \frac{6 \Sigma d_i^2}{n(n^2 - 1)} = 1 - \frac{(6)(5.5)}{(10)(100-1)} = 0.97.$$

Hence, $r_s = 0.97$

Also $r_s = 0.97$ **indicates a high positive correlation between the amount of tar an nicotine found in cigarettes.**

Example 16. *The following table shows the weights of a sample of 12 fathers and their oldest dult sons. Find the coefficient of rank correlation.*

Weight of father (kg)	*63*	*65*	*67*	*64*	*62*	*68*	*70*	*66*	*67*	*68*	*69*	*71*
Weight of son (kg)	*66*	*68*	*68*	*65*	*66*	*69*	*68*	*65*	*67*	*71*	*68*	*70*

Solution. Arranging in the ascending order of magnitude, the **fathers' weights are**

62 63 64 65 66 67 67 68 68 69 71

Since, the sixth and seventh places in this array represent the same weight (67 kg), we assign mean rank $\frac{1}{2}(6 + 7) = 6.5$ to these places. Similarly, the eighth and ninth places are assigned the ınk $\frac{1}{2}(8 + 9) = 8.5$. Thus, the fathers' weights are assigned the ranks:

1 2 3 4 5 6.5 6.5 8.5 8.5 10 11 12

Similarly, arranging in the ascending order of magnitude, the **sons' weights are:**

65 65 66 66 67 68 68 68 68 69 70 71

nd since, the sixth, seventh, eighth and ninth places represent the same weight (68 kg), we assign e mean rank $\frac{1}{4}(6 + 7 + 8 + 9) = 7.5$ to these places. Thus, the **sons' weights are assigned the anks.**

1.5 1.5 3.5 3.5 5 7.5 7.5 7.5 7.5 10 11 12

We prepare the following table.

Table

Rank of father	4	2	6.5	3	8.5	1	11	5	8.5	6.5	10	12
Rank of son	7.5	3.5	7.5	1.5	10	3.5	7.5	1.5	12	5	7.5	11
D	– 3.5	– 1.5	– 1.0	1.5	– 1.5	– 2.5	3.5	3.5	– 3.5	1.5	2.5	1.0
D^2	12.25	2.25	1.00	2.25	2.25	6.25	12.25	12.25	12.25	2.25	6.25	100
												$\Sigma D^2 = 72.50$

$\therefore$ **Rank correlation coefficient** $r_s = 1 - \dfrac{6\,\Sigma D^2}{N(N^2-1)} = 1 - \dfrac{6\,(72.50)}{12\,(12^2-1)} = \mathbf{0.7465}$

5.14 TESTING HYPOTHESIS ABOUT RANK CORRELATION

To test the significance of the rank correlation coefficient, one needs to consider the istribution of the r_s values under the assumption that X and Y are independent. Critical values for $= 0.20, 0.10, 0.05, 0.02, 0.01$ and 0.002 have been calculated and are given in **Table 11 at the nd. This table is set up similar to the table of critical values for the *t* distribution except for he left column,** which **now gives the number of pairs of observations rather than the degrees f freedom.** Since, the distribution of the r_s values is symmetric about $r_s = 0$, the r_s value that aves an area of α to the left is equal to the negative of the r_s value that leaves an area of α to the ght. For a two-sided alternative hypothesis, the critical region of size α falls equally in the two ils of the distribution. For **a test in which the alternative hypothesis is negative, the critical egion is entirely in the left tail of the distribution, and when the alternative is positive, the ritical region is placed entirely in the right tail.**

SMALL SAMPLE TEST ($n \leq 30$):

For small values of (*n* less than or equal to 30), the distribution r_s is not normal, and nlike other small sample statistic, we have encountered, it is not appropriate to use the distribution. For testing hypothesis about the rank correlation coefficient we determine the cceptance and rejection regions for such hypothesis.

For a given level of significance α and the number of pairs of observation n, we reject th null hypothesis H_0, if

$$|\text{Calculated value of } r_s| > |\text{Tabled value of } r \text{ for given } n \text{ and } \alpha|$$

$$\Rightarrow \quad |r_s| > |r_\alpha|$$

LARGE SAMPLE TEST ($n > 30$):

If the sample size is greater than 30 (*i.e.*, $n > 30$), we can no longer use Table 7. However when n is greater than 30, the sampling distribution of r_s is approximately normal, with mean of zero and a standard deviation of $1/(\sqrt{n-1})$. Thus, the standard error of r_s is:

$$\text{Standard Error of } r_s\text{: S.E. }(r_s) = \sigma_s = \frac{1}{\sqrt{(n-1)}}$$

Consequently, when $n > 30$, one could test for a significant correlation by computing

$$z = \frac{r_s - 0}{1/\sqrt{n-1}} = r_s\sqrt{n-1}$$

and comparing with critical values obtained from the standard normal curve.

The test for independence of two continuous variables presented in this section is a simplifie alternative to the more cumbersome chi-square procedure using contingency tables. Unfortunately all too often the values of the two random variables must fall into certain established categories and therefore, cannot be measured on a continuous scale, thereby necessitating the more comple calculations associated with contingency tables.

Example 17. *Test the hypothesis that X and Y are independent against the alternative tha they are dependent if for a sample of size n = 50 pairs of observations we find that $r_s = -0.29$. Us $\alpha = 0.05$.*

Solution. Here $n = 50$. **It is case of large sample.**

1. **Setting up of Hypothesis:**

   Null Hypothesis : H_0 : X and Y are independent

   Alternative Hypothesis : H_1 : X and Y are dependent $\Rightarrow$ It is a **case of two taile test.**

2. **Level of Significance:** $\alpha = 0.05$
3. **Z_α for two tailed test:** $|Z_\alpha| = 1.96$ for $\alpha = 0.05$.
4. **Calculation of Z-statistic:**

$$\text{Standard Error of } r_s = \text{S.E. }(r_s) = \frac{1}{\sqrt{n-1}} = \frac{1}{\sqrt{50-1}} = \frac{1}{\sqrt{49}} = \frac{1}{7}.$$

Here rank correlation coefficient $r_s = -0.29$. **[give**

$$\text{Test Statistics} = \frac{r_s - 0}{1/(\sqrt{n-1})} = \frac{-0.29 - 0}{(1/7)} = 7\,(-0.29) = -2.03.$$

$$\therefore \quad |Z| = 2.03.$$

5. **Decision:** Since $|Z| > |Z_\alpha|$ as $2.03 > 1.96$, so we reject the null hypothesis H_0 and acce the alternative hypothesis H_1 $\Rightarrow$ ***X* and *Y* are dependent.**

Example 18. *Test the hypothesis that there is no correlation between air-quality levels an the incidence of pulmonary disease in the world's cities against the alternative that there association between them. For a sample size of n = 11 cities, $r_s = 0.736$. Use $\alpha = 0.05$.*

Solution. Here $r_s = 0.376$ and $\alpha = 0.15$.

1. **Setting up of Hypothesis:**

   Null Hypothesis : H_0 : This is no correlation between air-quality levels of incidence of pulmonary disease in the world's cities

   Alternative Hypothesis : H_1 : There is an association between air-quality levels and incidence of pulmonary disease in the world's cities

   $\Rightarrow$ It is a **case of two tailed test.**

2. **Level of Significance:** $\alpha = 0.05$.

3. **Critical values:** It is a case of two tailed test, so we look at **the Table 11** in the row for $n = 11$ (the number of cities) and the column for a significance level of $\alpha = 0.05$. There we find that the critical values of r_s are $\pm$ 0.6091, *i.e.*, the upper limit of acceptance region is 0.6091 and the lower limit of acceptance is – 0.6091.

   Decision: Since, the calculated value of correlation coefficient $r_s = 0.736$ lies outside the acceptance region $\pm$ 0.6091, so we reject the null hypothesis H_0 and accept the alternative hypothesis H_1 $\Rightarrow$ there **is an association between air-quality level and the incidence of pulmonary disease in the world's cities.**

15.15 KOLMOGOROV–SMIRNOV TEST

The statisticians A.N. Kolmogorov and N.V. Smirnov developed the Kolmogorov–Smirnov test. It is a simple **nonparametric test for testing whether there is a significant difference between an observed frequency distribution and a theoretical frequency distribution.** In short, this test is known as **K-S test.** The K-S test is, therefore, another measure of the **goodness of fit of a frequency distribution**, as was **the chi-square test.** Its basic advantages are:

1. **It is a more powerful test.**
2. **It is easier to use, since it does not require that the data be grouped in any way.**

10.15-1 K-S Statistic

The K-S statistic is the maximum absolute deviation of expected relative frequency F_e and the observed relative frequency F_0. It is denoted by D_n.

$\therefore$ **K-S statistic:** $D_n = \max | F_e - F_0 |$

The K-S test is always a one tailed test for a given α, the level of significance.

The critical values for D_n can be tabulated by using the table. We compare the calculated value of D_n with the critical value of D_n from the table.

If the tabled value for the chosen significance level is greater than the calculated value of D_n, then we will accept the null hypothesis H_0.

Example 19. *Below is the table of obsorved frequencies, along with the frequency to the observed under a normal distribution.*

(*a*) *Calculate the K-S statistic.*

(*b*) *Can we conclude that this distribution does infact follow a normal distribution? Use 0.10 level of significance.*

Test Score	*51–60*	*61–70*	*71–80*	*81–90*	*91–100*
Observed Frequency	*30*	*100*	*440*	*500*	*130*
Expected Frequency	*40*	*170*	*500*	*390*	*100*

Solution. Null Hypothesis : H_0 : This distribution follows a normal distribution.

Table: Calculation of K-S Statistic

Test Score	*Observed Frequency*	*Observed Cumulative Frequency*	*Observed Relative Frequency* (F_0)	*Expected Frequency*	*Expected Cumulative Frequency*	*Expected Relative Frequency* (F_e)	$D = \| F_e - F_0 \|$
51–60	30	30	$\frac{30}{1200} = 0.025$	40	40	$\frac{40}{1200} = 0.033$	0.008
61–70	100	130	$\frac{130}{1200} = 0.108$	170	210	$\frac{210}{1200} = 0.175$	0.067
71–80	440	570	$\frac{570}{1200} = 0.475$	500	710	$\frac{710}{1200} = 0.592$	**0.117**
81–90	500	1070	$\frac{1070}{1200} = 0.891$	390	1100	$\frac{1100}{1200} = 0.92$	0.029
91–100	130	1200	$\frac{1200}{1200} = 1$	100	1200	$\frac{1200}{1200} = 1$	0

(*a*) ∴ **K-S statistic:** $D_n = \max | F_e - F_0 | = 0.117$

(*b*) The tabulated value of D_n for $n = 5$ and $\alpha = 0.05$ is **0.510**. **[From table 12]**

Since, the table value of D_n (= 0.510) is greater calculated value of D_n (= 0.117), so we accept the null hypothesis H_0 ⇒ the distribution follows a normal distribution.

Example 20. *The following are the numbers of hours which 10 students studied for an examination and the scores they obtained:*

Number of hours studied (x)	*8*	*5*	*11*	*13*	*10*	*5*	*18*	*15*	*2*	*8*
Score (y)	*56*	*44*	*79*	*72*	*70*	*54*	*94*	*85*	*33*	*65*

Calculate r_s. Also test at 0.01 level of significance whether the value obtained $r_s = 0.98$ is significant.

Solution. Ranking the *x*'s and the *y*'s, and proceeding as in the following table, we get the following table for r_s.

Table: Calculation of r_s

Rank of x	*Rank of y*	*d*	d_2
6.5	7	– 0.5	0.25
8.5	9	– 0.5	0.25
4	3	1.0	1.00
3	4	– 1.0	1.00
5	5	0.0	0.00
8.5	8	0.5	0.25
1	1	0.0	0.00
2	2	0.0	0.00
10	10	0.0	0.00
6.5	6	0.5	0.25
	Total		3.00

Here $n = 10$.

$$\therefore\ r_s = 1 - \frac{6\,\Sigma d^2}{n\,(n^2 - 1)} = 1 - \frac{6 \times 3}{10\,(10^2 - 1)} = 1 - \frac{18}{10 \times 99} = 1 - 0.02 = \mathbf{0.98.}$$

TEST OF SIGNIFICANCE

1. **H_0 : There is no correlation.**
   H_1 : There is a correlation.
2. **Level of significance: $\alpha = 0.01$**
3. Reject the null hypothesis if $Z \le -2.575$ or $Z \ge 2.575$, where $Z = r_s\sqrt{n-1}$
4. Substituting $n = 10$ and $r_s = 0.98$, we get

$$Z = 0.98\sqrt{10-1} = 2.94.$$

5. Since $Z = 2.94$ exceeds 2.575, the null hypothesis must be rejected; we **conclude that there is a real (positive) relationship between study time and scores.**

EXERCISE – 15.5

1. The following table shows how 10 students, arranged in alphabetical order, were ranked according to their achievements in both the laboratory and lecture sections of a computer course. Find the coefficient of rank correlation.

Laboratory	8	3	9	2	7	10	4	6	1	5
Lecture	9	5	10	1	8	7	3	4	2	6

[**Hint.**

Difference of ranks (*D*)	– 1	– 2	– 1	1	–1	3	1	2	– 1	– 1
D^2	1	4	1	1	1	9	1	4	1	1
										$\Sigma D^2 = 24$

$$\therefore\ r_s = 1 - \frac{6\,\Sigma D^2}{N\,(N^2 - 1)} = 1 - \frac{6\,(24)}{10\,(10^2 - 1)} = \mathbf{0.8545.}]$$

2. In a contest, two judges were asked to rank eight candidates (numbered 1 through 8) in order of preference. The judges submitted their choices in the following manner.

First judge	5	2	8	1	4	6	3	7
Second judge	4	5	7	3	2	8	1	6

(*a*) Find the coefficient of rank correlation.

(*b*) Decide how well the judges agreed in their choices.

3. A consumer panel tested nine makes of electric heaters for overall quality. The ranks assigned by the panel and the suggested retail prices were as follows:

Manufacturer	*Panel Rating*	*Suggested Price in Rs.*
A	6	480
B	9	395
C	2	575
D	8	550
E	5	510
F	1	545
G	7	400
H	4	465
I	3	420

Is there a significant relationship between the quality and the price of a electric heaters?

4. The following data were obtained in a study of relationship between the weight and chest size of infants at birth.

Weight (kg) :	2.75	2.15	4.41	5.52	3.21	4.32	2.31	4.30	3.71
Chest size (cm) :	29.5	26.3	32.2	36.5	27.2	27.7	28.3	30.3	28.7

(*a*) Calculate the rank correlation coefficient.

(*b*) Test the hypothesis at the 0.025 level of significance that the rank correlation coefficient is zero against the alternative that it is greater than zero.

5. A plant supervisor ranked a sample of eight workers on the number of hours worked overtime and length of employment. Find the rank correlation between two measures. Is it significance at 0.01 level ?

Amount of overtime	:	5.0	8.0	2.0	4.0	3.0	7.0	1.0	6.0
Years of employment	:	1.0	6.0	4.5	2.0	7.0	8.0	4.5	3.0

6. Two judges at a college homecoming parade ranked 8 floats in the following order:

	Float							
	1	2	3	4	5	6	7	8
Judge A	5	8	4	3	6	2	7	1
Judge B	7	5	4	2	8	1	6	3

(*a*) Calculate the rank correlation coefficient.

(*b*) Test the hypothesis that the rank correlation coefficient is zero against the alternative hypothesis that it is greater than zero. Use $\alpha = 0.05$.

7. A social scientist randomly choses 32 couples to try to determine whether bright people tend to choose spouses who are also bright and found rank correlation coefficient $r_s = 0.809$. Test his hypothesis of no correlation at 0.01 level of significance.

8. Explain the rank correlation test with the help of an example.

9. Explain Kalmogorov-Smirnov test with the help of an example.

10. Kevin Morgan, national sales manager of an electronics firm, has collected the following salary statistics on his field salesforce earnings. He has both observed frequencies and frequencies expected if the distribution of salaries is normal. At the 0.10 level of significance, can Kevin conclude that the distribution of salesforce earnings is normal?

	Earnings in Thousands						
	25–30	31–36	37–42	43–48	49–54	55–60	61–66
Observed frequency	9	22	25	30	21	12	6
Expected frequency	6	17	32	35	18	13	4

11. Randall Nelson, salesman for the V-Star company, has seven accounts to visit per week. It is thought that the sales by Mr. Nelson may be described by the binomial distribution, with the probability of selling each account being 45. Examining the observed frequency distribution of Mr. Nelson's number of sales per week, determine whether the distribution does in fact correspond to the suggested distribution. Use the 0.05 significance level.

Number of sales per week	:	0	1	2	3	4	5	6	7
Frequency of the number of sales	:	25	32	61	47	39	21	18	12.

12. At the 0.05 level of significance, can we conclude that the following distribution follows a Poisson distribution with $\lambda = 3$?

Number of arrivals per day	:	0	1	2	3	4	5	6 or more
Number of days	:	6	18	30	24	11	2	9.

13. The following shows how a panel of nutrition experts and a panel of housewives ranked 15 breakfast foods on their palatability:

Breakfast food	Nutrition experts	Housewives
A	3	5
B	7	4
C	11	8
D	9	14
E	1	2
F	4	6
G	10	12
H	8	7
I	5	1
J	13	15
K	12	9
L	2	3
M	15	10
N	6	11
O	14	13

Calculate r_s as a measure of the consistency of the two rankings. Also test the null hypothesis of no correlation at the 0.05 level of significance.

14. Calculate r_s for the following data representing the statistics grades, x, and psychology grades, y, of 18 students:

x	y	x	y
78	80	97	90
86	74	74	85
49	63	53	71
94	85	58	67
53	55	62	64
89	86	74	69
94	90	74	71
71	84	70	67
70	71	74	71

Also test at the 0.05 level of significance whether the value obtained for r_s is significant.

ANSWERS

1. 0.845. **2.** (*a*) 0.67 (*b*) The judges did not agree too well in their choices.

3. $r_s = -0.47$; No significant relationship.

4. (*a*) $r_s = 0.72$ (*b*) Reject the null hypothesis that the rank correlation coefficient is zero.

5. $r_s = 0.185$. The rank correlation is not significant.

6. (*a*) $r_s = 0.71$ (*b*) Reject the null hypothesis

7. Reject the null hypothesis of no correlation and conclude that bright people tend to choose bright spouses.

10. $D_n = 0.064$. The data are well described by the suggested distribution.

11. $D_n = 0.1463$. The suggested distribution is unlikely to be correct.

13. $r_s = 0.54$, $z = 1.79$. The correlation is not significant.

14. $r_s = 0.86$; $z = 3.55$. The value obtaind for r_s is signifi cant.

16

CHAPTER

INDEX NUMBERS

16.1 INDEX NUMBERS

An index number *is a ratio or an average of ratios expressed as a percentage. In an index number two or more time periods are involved, one of which is the base time period. The value at the base time period serves as the standard point of comparison.*

In other words, **an index number is nothing more than a 'relative number' or relative which expresses the relationship between two figures, where one of the figures is used as base.** Wherever a comparative study has to be made, the index numbers are very handy. In present day, situations such as changes in production, consumption, exports, imports, national income, cost of living, incidence of crimes, number of road accidents, intra firm comparison and a very wide variety of other fields are studied with the help of index numbers. **Therefore, index numbers are supposed to be the barometers which measure the change in the level of a phenomena**.

There are two broad types : Simple or Composite. When the variation in the level of a single item is being studied, the index number is termed as **Simple or Univariate index** number. But when the changes in average level of a number of items is being studied collectively, this index number is termed as **Composite index number.**

An Index Number is a number which is used to measure the level of a certain phenomenon as compared to the level of the same phenomenon at some standard period. In other words, ***an index number is a number which is used as a device for comparison between the price, quantity or value of a group of articles (related variables) in different situations, e.g., at a certain place or a period of time and that of another place or period of time.*** When the comparison is in respect of prices, it is called an **Index Number of Prices**, when it is in respect of physical quantities, it is named as **Index Number of Quantities.** Other Index Numbers are defined in the similar manner. ***The Index numbers are meant for comparison of variations arising out of the difference in situations, e.g., change of time or change of place.*** In comparing the price in a given year with the price of a particular year called base year (or base period), the price of the base year is taken as 100 and the price at the given period of time (called current year), for which the Index Number is required, is expressed as percentage of the base year price.

Let P_0 be the price of any item in the base year and P_1 be the price of that item in the current year, then the corresponding

$$\textbf{Index Number} = \frac{P_1}{P_0} \times 100 = \frac{\textbf{Current price}}{\textbf{Base price}} \times 100.$$

16.2 RELATIVES OR PRICE RELATIVES

Relatives are derived because absolute numbers measured in some appropriate unit, are often of little importance and meaningless in themselves. If the meaning of a relative figure remains ambiguous, it is often necessary to know the absolute as well as the relative numbers.

A price relative or simply, a relative is the ratio of the price of a certain commodity at the current year to its price at the base year, expressed as a percentage.

$$\therefore \ \textbf{Price Relative} = \frac{\textbf{Current year's price}}{\textbf{Base year's price}} \times 100 = \frac{P_1}{P_0} \times 100.$$

It is generally denoted by *P*, where $P = \frac{P_1}{P_0} \times 100$.

Thus Price-Relatives are the same as Index Numbers for prices.

16.3 CHARACTERISTICS OF INDEX NUMBERS

1. **Index number is a special type of average.**
2. **Index numbers study the effect of such factors that cannot be measured directly. According to *Bowely:***

   "Index numbers are used to measure the changes in some quantity which we cannot observe directly"
3. **Index numbers are expressed in percentages.**
4. **Index numbers measure the relative change of differences from time to time or place to place.**

16.4 USES OF INDEX NUMBERS

1. **Index numbers help in studying trends. .**
2. **Index numbers help management in policy formulation.**
3. **Index numbers help in measuring the purchasing power of money.**
4. **Index numbers act as Economic Barometers.** ***Just as a barometer measures the atomospheric pressure, index numbers are helpful in measuring the pressure of economic and business behaviour of the economy of a country.***

   That's why index numbers are called Economic Barometers.
5. **Index numbers help in deflating various values.**

16.5 PROBLEMS IN THE CONSTRUCTION OF INDEX NUMBERS

The following are the problems in the construction of index numbers:

1. **Selection of Data.** For **calculating index number only standardized items are to be selected.**
2. **Selection of Base Period.** ***The base period of index numbers is the period against which comparisons are made.*** It may be a year, a month or a day. ***The Index for base period is always taken as 100.*** *Base period should be a normal one and it should not be too distant in the past.* ***There are two choices either to have a fixed base period or a chain base period.***
3. **Use of Average.** Out of ***these only Geometric mean is the best average for the construction of index number.***
4. **Selection of Weights.** *Weights are of two types–**quantity weight and value weight. A quantity weight means the amount of commodity produced, distributed or commenced during a period. A value weight means quantity produced multiplied by the price of the quantity.***
5. **Selection of Formula.** *The selection of a formula depends on the purpose of the index numbers and on the data available.* ***Fisher's method for constructing index number is regarded as ideal formula as it uses geometric mean for calculating average of data.***
6. **Choice of Variables. Index numbers are constructed with regard to price or quantity or any other measure. Thus it is important to decide the unit also.**

16.6 BASE YEAR OR BASE PERIOD

Out of the different periods selected for an index number **any one period is taken as base and is given the value of 100,** so that the **values of other time period can be calculated and compared with 100,** (*the value of the base period*). *The base selected in constructing index* *numbers should suffer no abnormalities, i.e., there is no war, famine, wide spread of a disease or earthquake or flood etc., causing abnormalities, (boom or depression) in prices in that period.*

Usually, bases are selected in four different ways. According to these four types of base periods, there are four methods of constructing **Index Numbers:**

(*i*) Fixed Base Method

(*ii*) Chain Base Method

(*iii*) Average Base Method

(*iv*) Laspeyre's Method

16.6.1 Fixed Base Method

In the ***fixed base method, the base period once selected remains fixed and does not change.*** The period may be any year (or a month or a day). The prices of a particular year selected as a base period are treated as equal to 100. The changes in the prices of consequent years are shown as percentages of the base year. **The price changes in the case of a fixed base are called Price Relatives** and are calculated as under:-

Price Relative : $$P = \frac{P_1}{P_0} \times 100$$

where P_1 = **price in the current year**

P_0 = **price in the base year.**

The Price Relatives of the other years are then the required Index Numbers.

16.6.2 Chain Base Method

When a year to year comparison is desired, the index numbers are calculated by Chain Base Method. In this method the base period continues to shift for every subsequent year. Each is taken as the base for immediately next year, the first year itself being its own base. **(For every current year its previous year is taken as its base and the price is taken as 100 for the base year). For example** if index numbers for 1990, 1991, 1992, 1993, 1994, 1995, and 1996 are to be constructed, the 1995 will be base year for 1996, 1994 for 1993, 1993 for 1994, 1990 for 1991 and 1990 and 1990 for itself for 1990.

The price changes under this method are called Link Relatives and are computed as under:

$$\text{Link Relatives} = \frac{P_1}{P_0} \times 100, \text{ where}$$

P_1 = **price in the current year**

P_0 = **price in the previous year.**

In **Chain Base Method** relative figures for each year are first expressed as *Link Relatives (percentages of the preceding years)*. These percentages or Link Relatives are then chained together by successive multiplication to form a **Chain Index.** Thus, ***this index, in addition to the information, which is provided by a fixed base method, also provides the information which can let us know the changes in relation to the year immediately preceding to the current year.***

16.6.3 Average Base Method

The ***average of a number of years may be used as base price in determining index numbers.*** This has the effect of minimising the abnormalities of any particular year.

Thus in this case

$$\textbf{Price Relative Index Number} = \frac{\textbf{Price in the current year under consideration}}{\textbf{Average price}}$$

16.6.4 Laspeyre's Method

If the base year quantities are used as weights, then Laspeyre's Index is given as under:

$$\textbf{Laspeyre's Index Number} = \frac{\Sigma p_1 q_0}{\Sigma p_0 q_0} \times 100, \text{ where}$$

q_0 = **quantity consumed in the base year**

p_0 = **price in the base year**

p_1 = **price in the current year.**

Example 1. *The following data relate to the price of rice per kg. in different years :*

Year:	*1998*	*1999*	*2000*	*2001*	*2002*	*2003*	*2004*	*2005*
Price in Rs.	*6*	*7*	*7*	*8*	*10*	*14*	*12*	*13*

Find out price relatives

(i) Taking 1998 as base

(ii) Taking 2002 as base

(iii) Taking average of 1998, 1999 and 2000 as the base.

Also *calculate the link relatives.*

Solution. Let p_0 be the price of rice in the base year 1998 and p_1 be the price in current year. Then

$$\textbf{Price Relatives} = \frac{p_0}{p_1} \times 100.$$

Table : Computation of Price Relatives

Year	*Price*	*Price Relatives*		*Average of 1998, 1999 and 2000* $= \frac{6+7+7}{3}$
	(In Rs)	***1998 (Base)***	***2002 (Base)***	***= 6.67 (Base)***
1998	6	100	$\frac{6}{10} \times 100 = 60$	$\frac{6}{6.67} \times 100 = \mathbf{89.96}$
1999	7	$\frac{7}{6} \times 100 = \mathbf{116.67}$	$\frac{7}{10} \times 100 = \mathbf{70}$	$\frac{7}{6.67} \times 100 = \mathbf{104.95}$
2000	7	$\frac{7}{6} \times 100 = \mathbf{116.67}$	$\frac{7}{10} \times 100 = \mathbf{70}$	$\frac{7}{6.67} \times 100 = \mathbf{104.95}$
2001	8	$\frac{8}{6} \times 100 = \mathbf{133.33}$	$\frac{8}{10} \times 100 = \mathbf{80}$	$\frac{8}{6.67} \times 100 = \mathbf{119.94}$
2002	10	$\frac{10}{6} \times 100 = \mathbf{166.67}$	**100**	$\frac{10}{6.67} \times 100 = \mathbf{149.93}$
2003	14	$\frac{14}{6} \times 100 = \mathbf{233.33}$	$\frac{14}{10} \times 100 = \mathbf{140}$	$\frac{14}{6.67} \times 100 = \mathbf{209.90}$
2004	12	$\frac{12}{6} \times 100 = \mathbf{200}$	$\frac{12}{10} \times 100 = \mathbf{120}$	$\frac{12}{6.67} \times 100 = \mathbf{179.91}$
2005	13	$\frac{13}{6} \times 100 = \mathbf{216.67}$	$\frac{13}{10} \times 100 = \mathbf{130}$	$\frac{13}{6.67} \times 100 = \mathbf{194.90}$

$$\textit{Link Relatives} = \frac{p_1}{p_0} \times 100,$$

where p_1 = price in the current year

p_0 = price in the previous year.

In the following table, **Link Relatives** have been calculated by taking the price of previous year as 100. **For example** for 1999, 1998 has been taken as 100, for 2000, 1999 has been taken as 100, and so on.

TABLE : Computation of Link Relatives

Year	*Price in (Rs)*	*Link Relatives (Chain Index)*
1998	**6**	**= 100**
1999	7	$\frac{7}{6} \times 100 = \mathbf{116.67}$
2000	7	$\frac{7}{7} \times 100 = \mathbf{100}$
2001	8	$\frac{8}{7} \times 100 = \mathbf{114.29}$
2002	10	$\frac{10}{8} \times 100 = \mathbf{125}$
2003	14	$\frac{14}{10} \times 100 = \mathbf{140}$
2004	12	$\frac{12}{14} \times 100 = \mathbf{85.71}$
2005	13	$\frac{13}{12} \times 100 = \mathbf{108.33}$

16.7 CONVERSION OF PRICE RELATIVES INTO LINK RELATIVES AND VICE-VERSA

The Price Relatives can be converted into Link Relatives by the following result :

$$\text{Link Relatives} = \frac{\text{Price Relative for the current year}}{\text{Price Relative for the previous year}} \times 100.$$

Similarly, the *Link Relatives can be converted into Price-Relatives by the following result :*

$$\text{Price Relative} = \frac{(\text{Link Relative for the current year}) \times (\text{Price Relative of the previous year})}{100}.$$

Example 2. *From the fixed base index numbers given below, calculate the Link Relatives:*

Year:	*2000*	*2001*	*2002*	*2003*	*2004*	*2005*
Price Relatives:	*188*	*199*	*204*	*190*	*196*	*200*

Solution. The **price relative of 2000 (*i.e.*, 188) will be taken as 100** and **the link relatives are** calculated by the result.

$$\text{Link Relative} = \frac{\text{Price Relative in the current year}}{\text{Price Relative in the previous year}} \times 100.$$

TABLE: Computation of Link Relatives

Year	*Price Relatives*	*Link Relatives*
2000	188	**100**
2001	199	$\frac{199}{188}\times 100 = \mathbf{105.85}$
2002	204	$\frac{204}{199}\times 100 = \mathbf{102.51}$
2003	190	$\frac{190}{204}\times 100 = \mathbf{93.14}$
2004	196	$\frac{196}{190}\times 100 = \mathbf{103.16}$
2005	200	$\frac{200}{196}\times 100 = \mathbf{102.04}$

Thus the respective Link Relatives are: 100, 105.85, 102.51, 93.14, 103.16, and 102.4.

Example 3. *An index is 100 in 1992. It rises 15% in 1993, falls 4% in 1994, falls 2% in 1995 and rises 10% in 1996. Calculate the index for 5 years with 1992 as the base.*

Solution. The **price relative for 1992 will be taken as 100.**

$$\textbf{Price Relative} = \frac{\textbf{Current year's Link Relative} \times \textbf{Previous year's Price Relative}}{\textbf{100}}.$$

Table : Computation of Price Relatives

Year	*Link Relative*	*Price Relative (1992 as Base)*
1992	100	100
1993	115	$\frac{115\times 100}{100} = \mathbf{115}$
1994	96	$\frac{96\times 115}{100} = \mathbf{110.40}$
1995	98	$\frac{98\times 110.40}{100} = \mathbf{108.19}$
1996	110	$\frac{110\times 108.19}{100} = \mathbf{119.01}$

Hence the Price Relatives are: 100, 115, 110.40, 108.19 and 119.01.

Example 4. *An index is at 100 in 2001. It rises 5% in 2002, falls 6% in 2003, falls 5% in 2004, rises 4% in 2005 and 7% in 2006. Calculate the index numbers for all these years with 2001 as base.*

Solution.

Table for Index Numbers

Year	*2001*	*2002*	*2003*	*2004*	*2005*	*2006*
Link relatives	100	105	94	95	104	107
Index Nos.	100	$\frac{105\times 100}{100}$	$\frac{94\times 105}{100}$	$\frac{98.7\times 95}{100}$	$\frac{93.8\times 104}{100}$	$\frac{97.6\times 107}{100}$
	= 100	**= 105**	**= 98.7**	**= 93.8**	**= 97.6**	**= 104.4**

Hence the index numbers for these years are: 100, 105, 98.7, 93.8, 97.6 and 104.4.

16.8 TYPES OF INDEX NUMBERS

Index numbers are classified on the basis of the phenomena whose changes they measure. In the economic and business, index numbers can be classified into the following four types:

I. Price Index numbers

II. Value Index numbers

III. Quantitative Index numbers

IV. Special Purpose Index numbers.

16.9 METHODS FOR CONSTRUCTION OF INDEX NUMBERS

The various methods of constructing index numbers can be classified in two groups:

(*a*) Simple or Unweighted Index Numbers

(*b*) Weighted Index Numbers.

In **Simple or unweighted index numbers each item is supposed to have the same weight but in weighted index numbers the weights are assigned to various items in accordance with their importance.**

Unweighted Index numbers can be further divided in two categories:

(*i*) SIMPLE AGGREGATIVE METHOD

(*ii*) SIMPLE AVERAGE OF RELATIVES METHOD.

Weighted index numbers can be divided into two categories:

(*i*) WEIGHTED AGGREGATIVE METHOD

(*ii*) WEIGHTED AVERAGE OF RELATIVES METHOD.

We can state the broad heads as follows:

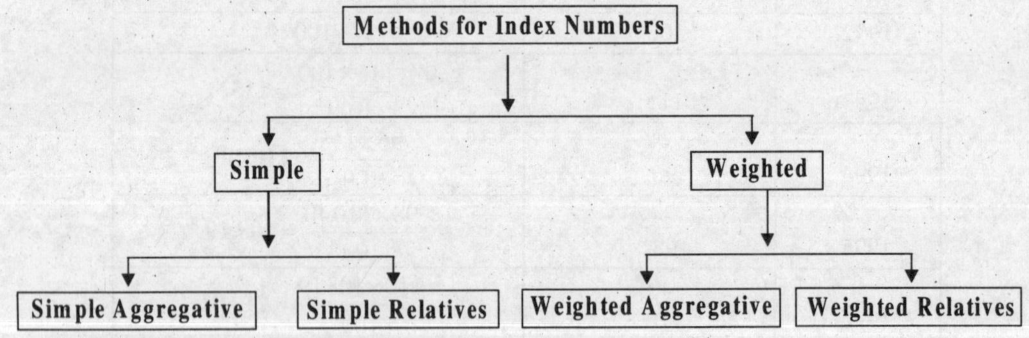

(*a*) SIMPLE OR UNWEIGHTED INDEX NUMBERS

(*i*) Simple Aggregative Method. This is a simple method of constructing index numbers and it consists in expressing the aggregate price of all commodities in the current year as a percentage of the aggregate price in the base year. Symbolically,

$$P_{01} = \frac{\Sigma P_1}{\Sigma P_0} \times 100$$

where ΣP_1 = **Total of current year's prices of all commodities**

ΣP_0 = **Total of base year's prices of all commodities.**

P_{01} = **Index number of current year.**

WORKING RULE :

Step 1. *Add the current year prices for all commodities to get* ΣP_1

Step 2. *Add the base year prices of all commodities to get* ΣP_0

Step 3. *Divide* ΣP_1 *by* ΣP_0 *and multiply the result by 100.*

This method has the following limitations.

LIMITATIONS :

(*i*) *This index is defective as it is affected by the units in which prices are quoted.*

(*ii*) *This index number is influenced by the magnitude of the prices, i.e., higher the price of a commodity the greater is its influence on the index number. Thus, high priced commodities receive greater weightage than low priced commodities.*

(*iii*) *The relative importance of various commodities is not taken into account in this index number as it is unweighted.*

Example 5. *From the following data construct the index number for the year 2004 on the basis of 2003.*

Items	*Units*	*Price (Rs.)*	
		(2003) P_0	*(2004)* P_1
Wheat	*Quintal*	*200*	*250*
Rice	*Quintal*	*300*	*400*
Pulses	*Quintal*	*400*	*500*
Milk	*Litre*	*2*	*3*
Clothing	*Meter*	*3*	*5*

Solution.

COMPUTATION TABLE

Items	*Units*	*Price (Rs.)*	
		2003 (P_0)	*2004* (P_1)
Wheat	Quintal	200	250
Rice	Quintal	300	400
Pulses	Quintal	400	500
Milk	Litre	2	3
Clothing	Meter	3	5
		$\Sigma P_0 = 905$	$\Sigma P_1 = 1158$

Index number of 2004 : $P_{01}\text{ L} = \frac{\Sigma P_1}{\Sigma P_0} \times 100 = \frac{1158}{905} \times 100 = 127.96.$

It means that prices in 2004 were 27.96% more than the prices in 2003.

(*ii*) Simple Average of Relatives method. Under this method **the prices of the commodities in the base year are assumed to be 100 and the current year's prices are expressed as percentage of the base year's prices.** Thus the current year's prices are expressed as a price relative to the base year's prices. These price relatives are then averaged to get the **index number.** The average used could be arithmetic mean, geometric mean or median.

INDEX NUMBER USING ARITHMETIC MEAN :

When **Arithmetic mean** is used, then the **formula** for construction of index number is

$$P_{01}\text{ (A.M.)} = \frac{\Sigma(P_1/P_0)\times 100}{N}.$$

INDEX NUMBER USING GEOMETRIC MEAN :

When **Geometric mean** is used, the formula for construction of index number

$$\log P_{01} = \frac{\Sigma\log(P_1/P_0)\times 100]}{N}$$

or $$P_{01}\text{ (G.M.)} = \text{Antilog}\left\{\frac{\Sigma\log(P_1/P_0)\times 100}{N}\right\},$$

where *N* is the number of items from which the index number is to be constructed.

ADVANTAGES :

This method has the following advantages:

(*i*) *It is not affected by the units in which prices are quoted. It is also not affected by absolute values of prices as prices are converted into price relatives.*

(*ii*) *It gives equal importance to all items and extreme items do not unduly affect the index number.*

(*iii*) *The index number calculated by this method satisfies the Unit Test.*

LIMITATIONS :

(*i*) *As it is an unweighted average the importance of all items is assumed to be the same.*

(*ii*) *The index number constructed by this method does not satisfy all the criteria laid down for an ideal index.*

Example 6. *From the following data construct the index number for the year 2002 taking 2001 as base by using (i) Arithmetic Mean (ii) Geometric Mean.*

Item	*Price (Rs.) (2001)*	*Price (Rs.) (2002)*
A	*6*	*10*
B	*2*	*2*
C	*4*	*6*
D	*10*	*12*
E	*8*	*12*

Solution.

TABLE : Computation of Index Number for the Year 2002 using Arithmetic Mean

Item	***Price 2001*** (*in Rs.*) P_0	***Price 2002*** (*in Rs.*) P_1	***Price Relatives*** $P = \frac{P_1}{P_0}\times 100$
A	6	10	166.7
B	2	2	100.0
C	4	6	150.0
D	10	12	120.0
E	8	12	150.0
			$\sum\frac{P_1}{P_0}\times 100 = 686.7$

$$P_{01} = \frac{\Sigma(P_1/P_0)\times 100}{N} = \frac{686.7}{5} = 137.34.$$

TABLE: Computation of Index Number for the year 2002 using Geometric Mean

Item	*Price 2001* (*in Rs.*) P_0	*Price 2002* (*in Rs.*) P_1	*Price Relatives* $P = \frac{P_1}{P_0}\times 100$	*logP*
A	6	10	166.7	2.2201
B	2	2	100.0	2.000
C	4	6	150.0	2.1761
D	10	12	120.0	2.0792
E	8	12	150.0	2.1761
				$\Sigma \log P = 10.6515$

$$P_{01} = \text{antilog}\left(\frac{\Sigma \log P}{N}\right) = \text{antilog}\left(\frac{10.6515}{5}\right) = \text{antilog } 2.1303 = \mathbf{134.99.}$$

It should be noted that since arithmetic mean gives greater weightage to bigger items, so the index number which uses arithmetic mean would be higher than the index number using geometric mean.

(B) Weighted Index Number Methods

To meet the weakness of the simple or unweighted method, we weigh the price of each commodity by a suitable factor often taken as the quantity or the volume of the commodity sold during the base year. In other words, **in this method appropriate weights are assigned to various commodities to reflect their relative importance in the group.** The weights can be production figures, consumption figures or distributive figures. For the **construction of the price index numbers, quantity weights are used. If *w* is the weight attached to a commodity, then the price index is** given by :

$$\text{Price Index: } P_{01} = \frac{\Sigma P_1 \times w}{\Sigma P_0 \times w}\times 100 .$$

Weighted index numbers can be classified into two broad groups:

(*i*) **Weighted Aggregative Index Numbers.**

(*ii*) **Weighted Average of Price Relatives.**

I. WEIGHTED AGGREGATIVE INDEX NUMBERS

In the weighted aggregative index numbers the weights are assigned to various items and the weighted aggregate of the prices are obtained. Weights are assigned in various ways and the weighted aggregates are obtained in different ways for the construction of index numbers. Some of the important methods of constructing weighted Aggregative Index Numbers are given below:

16.10 LASPEYRE'S PRICE INDEX

Laspeyre's method is based on fixed weights of the base year. Base year's quantities are used as weights. The formula given by **Laspeyre is given below:**

$$\text{LASPEYRE'S PRICE INDEX: } P_{01} = \frac{\Sigma P_1 q_0}{\Sigma P_0 q_0}\times 100,$$

where q_0 = quantity used or weight in the base year.

This formula was devised by **French Economist Laspeyre in 1871.**

16.11 PAASCHE'S METHOD

Paasche's method is based on current year's quantities. Current year's quantities are used as weights.

The formula given by **Paasche's** is given below :

PAASCHE'S PRICE INDEX: $P_{01} = \dfrac{\Sigma P_1 q_1}{\Sigma P_0 q_1} \times 100$.

where q_1 = quantity used or weight in the current year.

This formula was devised by **German Statistician Paasche in 1874.**

16.12 FISHER'S IDEAL INDEX METHOD

Fisher's ideal index number is the **Geometric mean of the Laspeyre's and Pasche's index numbers.** Fisher had examined more than 100 formulae for the construction of index numbers and ultimately derived this index number and calls it as **ideal index number.**

FISHER'S IDEAL INDEX NUMBER: $P_{01} = \sqrt{\dfrac{\Sigma P_1 q_0}{\Sigma P_0 q_0} \times \dfrac{\Sigma P_1 q_1}{\Sigma P_0 q_1}} \times 100$.

$$= \sqrt{\text{Laspeyre's Price Index} \times \text{Paasche's Price Index}}$$

This formula is based on geometric mean. This formula satisfies both the **Time Reversal Test as well as Factor Reversal Test.** It takes into account both **current year's as well as base year's weights.**

16.13 DORBISH AND BOWLEY'S METHOD

This method is a **combination of Laspeyre's and Paasche's method.** If we find the arithmetic average of Laspeyre's Index and the Paasche's Index, we get the index suggested by Dorbish and Bowley. This index number takes into account both the base year as well as the current year weights.

DORBISH AND BOWLEY INDEX NUMBER:

$$P_{01} = \frac{(\Sigma P_1 q_0 / \Sigma P_0 q_0) + \Sigma(P_1 q_1 / \Sigma P_0 q_1)}{2} \times 100.$$

16.14 MARSHALL-EDGEWORTH FORMULA

In this method both base year as well as current year prices and quantities are considered for calculating the index number. The formula for calculating the index number is given below:

MARSHALL EDGEWORTH INDEX NUMBER:

$$P_{01} = \frac{\Sigma(q_0 + q_1)P_1}{\Sigma(q_0 + q_1)P_0} \times 100 = \frac{\Sigma P_1 q_0 + \Sigma P_1 q_1}{\Sigma P_0 q_0 + \Sigma P_0 q_1} \times 100.$$

This method satisfies the Time Reversal Test. It gives a very close approximation to results of Fisher's ideal index number.

16.15 WALSCH INDEX NUMBER METHOD

Walsch suggested the geometrically crossed weighted aggregates for calculating the index number instead of arithmetically crossed weighted aggregates.

WALSCH'S INDEX NUMBER: $P_{01} = \dfrac{\Sigma P_1 \sqrt{q_0 q_1}}{\Sigma P_0 \sqrt{q_0 q_1}} \times 100.$

This **index also satisfies the Time Reversal Test.**

16.16 KELLY'S INDEX NUMBER METHOD

This formula, named after **Truman L. Kelly** requires the weights to be fixed for all the periods and is also sometimes known as aggregative index for all the periods and is also sometimes known as aggregative index weights and is given by the following formula :

KELLY'S METHOD

$$P_{01} = \frac{\Sigma P_1 q}{\Sigma P_0 q} \times 100, \quad \text{where } q = \frac{q_0 + q_1}{2}.$$

Example 7. *Compute Laspeyre's, Paasche's, and Fisher's Price index numbers for 2005 from the following data.*

Commodity	*2000*		*2005*	
	Price Rs.	*Quantity*	*Price*	*Quantity*
A	*15*	*15*	*22*	*12*
B	*20*	*5*	*27*	*4*
C	*4*	*10*	*7*	*5*

Solution.

Table for Computation of Index numbers.

Com-modity	*Price Rs.* (P_0)	*Quan-tity kg* (q_0)	*Price Rs.* (P_1)	*Quan-tity kg* (q_1)	$P_0 q_0$	$P_1 q_1$	$P_1 q_0$	$P_0 q_1$	$q_0 + q_1$	$q = \frac{q_0 + q_1}{2}$	$P_1 q$	$P_0 q$
A	15	15	22	12	225	264	330	180	27	13.5	297.0	202.5
B	20	5	27	4	100	108	135	80	9	4.5	121.5	90.0
C	4	10	7	5	40	35	70	20	15	7.5	52.5	30.0
					$\Sigma P_0 q_0 = 365$	$\Sigma P_1 q_1 = 407$	$\Sigma P_1 q_1 = 535$	$\Sigma P_0 p_1 = 280$			$\Sigma P_1 q = 471$	$\Sigma P_0 q = 322.5$

$$\textbf{Laspeyre's Price Index Number} = \frac{\Sigma P_1 q_0}{\Sigma P_0 q_0} \times 100 = \frac{535}{365} \times 100 = \mathbf{146.57}.$$

$$\textbf{Paasche's Price-Index Number} = \frac{407}{280} \times 100 = \mathbf{145.35}.$$

$$\textbf{Fisher's ideal Index Number} = \sqrt{\frac{\Sigma P_1 q_0}{\Sigma P_0 q_0} \times \frac{\Sigma P_1 q_1}{\Sigma P_0 q_1}} \times 100 = \sqrt{\frac{535}{365} \times \frac{407}{280}} \times 100$$

$$= \sqrt{1.4657 \times 1.4535} \times 100 = \sqrt{2.130395} \times 100 = 1.4596 \times 100 = \mathbf{145.96}.$$

II WEIGHTED AVERAGE OF PRICE RELATIVES

In weighted average of price relatives the price relatives for the current year are calculated on the basis of the base year prices. These price relatives are multiplied by the respective weights of the items. These products are added up and are divided by the sum of the weights. This gives the required index number.

$$P_{01}(\text{A.M.}) = \frac{\Sigma[(P_1/P_0)\times 100)]W}{\Sigma W} \quad \text{or} \quad P_{01} = \frac{\Sigma PW}{\Sigma W},$$

where P is the price relative $= \frac{P_1}{P_0}\times 100$ and W is the weight of the commodity concerned.

If the weighted geometric mean is used then the formula is

$$P_{01}\,(\text{G.M.}) = \text{Antilog}\left[\frac{\Sigma(\log P)W}{\Sigma W}\right].$$

Notes: (*i*) ***If base year's values are taken as weights, then weighted average of relative method gives us the Laspeyre's Price index number if arithmetic mean is used.***

(*ii*) ***If the weights $W = P_0q_0$ are base year prices multiplied by current year's weights, then the resultant weighted average of relative method gives us the same result as is obtained by Paasche's method.***

Example 8. *Compute price index for the following data by applying weighted average of price relative method using (i) Arithmetic and (ii) Geometric mean.*

Item	*Price in 1996 (Rs.)*	*Price in 1997 (Rs.)*	*Quantity in 1996*
Wheat	*2.00*	*2.50*	*40 kg.*
Sugar	*3.00*	*3.25*	*20 kg.*
Milk	*1.50*	*1.75*	*10 lit.*

Solution:

Table : Computation of Weighted Average of Price Relatives

Item	*Price in 1996* p_1	*Price in 1997* p_0	*Quantity 1996* q_0	$V = p_0q_0$	$P = \frac{P_1}{P_0}\times 100$	*PV*	*logP*	*V logP*
Wheat	2.00	2.50	40	80	125	10000	2.0969	167.7520
Sugar	3.00	3.25	20	60	108.33	6499.8	2.0346	122.0760
Milk	1.50	1.75	10	15	116.67	1750.05	2.0669	31.0065
Total				$\Sigma V = 155$		$\Sigma PV =$ **18249.85**		$\Sigma \log P =$ **320.8345**

(A) Weighted Average of Price Relatives (Using Arithmetic Mean)

$$P_{01} = \frac{\Sigma PV}{\Sigma V} = \frac{18249.85}{155} = 117.74.$$

(B) Weighted Average of Price Relatives (Using Geometric Mean)

$$P_{01} = \text{antilog}\left(\frac{\Sigma \log P}{\Sigma V}\right) = \text{antilog}\left(\frac{320.8345}{155}\right) = \text{antilog } 2.0699 = \mathbf{117.4}$$

Example 9. *Compute (i) Laspeyre's Price Index, (ii) Weighted Average of relatives Index number taking $W = p_0q_0$, (iii) Paasche's Index number, (iv) Weighted Average of relatives index number using $W = p_0q_1$.*

Commodity	*2004*		*2005*	
	Price	*Quantity*	*Price*	*Quantity*
A	*2*	*5*	*3*	*4*
B	*1*	*2*	*2*	*3*
C	*3*	*1*	*4*	*1*

Solution.

Table : Computation of Index Number

Com-modity	*Price (Rs.)* (P_0)	*Quantity (kg)* (q_0)	*Price (Rs.)* (P_1)	*Quantity (kg)* (q_1)	P_0q_0	P_1q_1	P_1q_0	P_0q_1	$P = \frac{p_1}{p_0} \times 100$
A	2	5	3	4	10	12	15	8	150
B	1	2	2	3	2	6	4	3	200
C	3	1	4	1	3	4	4	3	400/3
					$\Sigma P_0q_0 = 15$	$\Sigma P_1q_1 = 22$	$\Sigma P_1q_0 = 23$	$\Sigma P_0q_1 = 14$	

$$\textbf{Laspeyre's Method} = \frac{\Sigma P_1q_0}{\Sigma P_0q_0} \times 100 = \frac{23}{15} \times 100 = 153.33.$$

Weighted Average of Relatives using $W = P_0q_0$.

$$P_{01} = \frac{\Sigma[(P_1/P_0)\times 100]W}{\Sigma W} = \frac{\Sigma[(P_1/P_0)\times 100]P_0q_0}{\Sigma P_0q_0}$$

$$= \frac{150\times 10 + 200\times 2 + (400/3)\times 3}{15} = \frac{2300}{15} = 153.33.$$

$$\textbf{Paasche's Method} = \frac{\Sigma P_1q_1}{\Sigma P_0q_1} \times 100 = \frac{22}{14} \times 100 = 157.14.$$

Weighted Average of Relatives using $W = P_0q_1$.

$$P_{01} = \frac{\Sigma[(P_1/P_0)\times 100]W}{\Sigma W} = \frac{\Sigma[(P_1/P_0)\times 100]P_0q_1}{\Sigma P_0q_1}$$

$$= \frac{150\times 8 + 200\times 3 + (400/3)\times 3}{14} = \frac{2200}{14} = 157.14.$$

16.17 QUANTITY INDEX NUMBERS

The *quantity index numbers measure the average storage in quantities and enable us to compare changes in physical quantity of goods produced or sold.* These index numbers can also be simple or weighted. **Weights in quantity index number is price.** Therefore, **quantity index numbers can be easily obtained from Price Index numbers just by interchanging *P's* and *q's*.**

(A) SIMPLE OR UNWEIGHTED QUANTITY INDEX NUMBERS.

(*i*) Simple Aggregative Method: $Q_{01} = \frac{\Sigma q_1}{\Sigma q_0} \times 100$.

(*ii*) Simple Average of Relatives Method: $Q_{01} = \frac{\Sigma[(q_1/q_0)\times 100]}{N}$.

[Using Arithmetic Mean]

Simple Average of Relatives Method = $Q_{01} = \text{antilog}\left(\frac{\Sigma \log(q_1/q_0)\times 100}{N}\right).$

[Using Geometric Mean]

(B) WEIGHTED QUANTITY INDEX NUMBER.

(I) Weighted Aggregative Index Number:

(A) Laspeyre's Quantity Index Number: $Q_{01} = \frac{\Sigma q_1 P_0}{\Sigma q_0 P_0}\times 100.$

(B) Paascher's Quantity Index Number: $Q_{01} = \frac{\Sigma q_1 P_1}{\Sigma q_0 P_1}\times 100.$

(C) Fisher's Ideal Quantity Index Number: $Q_{01} = \sqrt{\frac{\Sigma q_1 P_0}{\Sigma q_0 P_0} + \frac{\Sigma q_1 P_1}{\Sigma q_0 P_1}}\times 100.$

(D) Dorbish and Bowleys Quantity Index numbers: $Q_{01} = \frac{\frac{\Sigma q_1 P_0}{\Sigma q_0 P_0} + \frac{\Sigma q_1 P_1}{\Sigma q_0 P_1}}{2}\times 100.$

(E) Marshall-Edgeworth Formula: $Q_{01} = \frac{\Sigma (P_0 + P_1) q_1}{\Sigma (P_0 + P_1) q_0}\times 100.$

(F) Kelly's Quantity Index Number: $Q_{01} = \frac{\Sigma q_1 P}{\Sigma q_0 P}\times 100$, **where** $P = \frac{P_0 + P_1}{2}.$

(II) Weighted Average of Relatives:

$$Q_{01} = \frac{\Sigma((q_1/q_0)\times 100)W}{\Sigma W}, \text{ where } W = q_0 P_0.$$

Example 10. *Compute Fisher's Ideal quantity index number from the following data:*

Commodity	*2004*		*2005*	
	Price (Rs.)	*Total value (Rs.)*	*Price*	*Total value (Rs.)*
A	*5*	*50*	*4*	*48*
B	*8*	*48*	*7*	*49*
C	*6*	*18*	*5*	*20*

Solution. Fisher's Ideal Quantity index number is given by

$$Q_{01} = \sqrt{\frac{\Sigma q_1 P_0}{\Sigma q_0 P_0} \times \frac{\Sigma q_1 P_1}{\Sigma q_0 P_1}}\times 100.$$

From the value given in the question, we can calculate the quantity for the year 2004 and year 2005.

Commodity	***A***	***B***	***C***
q_0 **(for 2004)** $= \frac{\textbf{Total value}}{\textbf{Price}}$	$\frac{50}{5} = \mathbf{10}$	$\frac{48}{8} = 6$	$\frac{18}{6} = \mathbf{3}$
q_1 **(for 2005)** $= \frac{\textbf{Total value}}{\textbf{Price}}$	$\frac{48}{4} = \mathbf{12}$	$\frac{49}{7} = \mathbf{7}$	$\frac{20}{5} = 4$

Table: Computation of Fisher's Quantity Index

	2004		*2005*					
Commodity	P_0	q_0	P_1	q_1	q_0P_0	q_1P_1	q_0P_1	q_1P_0
A	5	10	4	12	50	48	40	60
B	8	6	7	7	48	49	42	56
C	6	3	5	4	18	20	15	24
					Σq_0P_0 **= 116**	Σq_1P_1 **= 117**	Σq_0P_1 **= 97**	Σq_1P_0 **= 140**

Fisher's Quantity Index Number: $Q_{01} = \sqrt{\frac{\Sigma q_1 p_0}{\Sigma q_0 p_0} \times \frac{\Sigma q_1 p_0}{\Sigma q_0 p_0}} \times 100$

$$= \sqrt{\frac{140}{116} \times \frac{117}{97}} \times 100 = \mathbf{120.65}.$$

Example 11. *Compute Fisher's ideal price and quantity index numbers for the current year on the basis of the following data:*

Item	*Base Year*		*Current Year*	
	Price (in Rs.)	*Total Value (in Rs.)*	*Price (in Rs.)*	*Total Value (in Rs.)*
A	*6*	*300*	*10*	*560*
B	*2*	*200*	*2*	*240*
C	*4*	*240*	*6*	*360*
D	*10*	*300*	*12*	*288*
E	*8*	*320*	*12*	*432*

Solution.

Table: Computation of Price Index and Quantity Index Numbers

Item	P_0 *Price (in Rs.)*	p_0q_0 *Total Value (in Rs.)*	P_n *Price (in Rs.)*	P_nq_n *Total Value (in Rs.)*	q_0	q_n	P_nq_0	P_0q_n
A	6	300	10	560	50	56	500	336
B	2	200	2	240	100	120	200	240
C	4	240	6	360	60	60	360	240
D	10	300	12	288	30	24	360	240
E	8	320	12	432	40	36	480	288
	Total	**1360**		**1880**			**1900**	**1344**

Price Index $= \sqrt{\frac{\Sigma P_n q_0}{\Sigma P_0 q_0} \times \frac{\Sigma P_n q_n}{\Sigma P_0 q_n}} \times \mathbf{100} = \sqrt{\frac{1900}{1360} \times \frac{1880}{1344}} \times 100 = \sqrt{1.9954} \times 100 = \mathbf{139.79}$.

$$\textbf{Quantity Index} = \sqrt{\frac{\Sigma q_n P_0}{\Sigma q_0 P_0} \times \frac{\Sigma q_n P_n}{\Sigma q_0 P_n}} \times 100 = \sqrt{\frac{1344}{1360} \times \frac{1880}{1900}} \times 100 = \sqrt{0.9778} \times 100 = \mathbf{98.89}.$$

Example 12. *For the following compute the Laspeyre's and Paasche's price and quantity index number for 2005 with 1995 as the base year:*

	1995		*2005*	
Commodity	*Quantity*	*Value*	*Quantity*	*Value*
A	*50*	*350*	*60*	*420*
B	*120*	*600*	*140*	*700*
C	*30*	*330*	*20*	*200*
D	*20*	*360*	*15*	*300*
E	*5*	*40*	*5*	*50*

Solution.

Table for Computation

	1995		*2005*					
Com-modity	***Quantity*** (Q_0)	***Value*** $(V_0 = P_0 Q_0)$	***Quantity*** (Q_1)	***Value*** $(V_1 = P_1 Q_1)$	$P_0 = V_0 / Q_0$	$P_1 = V_1 / Q_1$	$P_1 Q_0$	$P_0 Q_1$
A	50	350	60	420	7	7	350	420
B	120	600	140	700	5	5	600	700
C	30	330	20	200	11	10	300	220
D	20	360	15	300	18	20	400	270
E	5	40	5	50	8	10	50	40
Total		**1680**		**1670**			**1700**	**1650**

PRICE INDEX NUMBERS:

$$\textbf{Laspeyre's: } \frac{\Sigma P_1 Q_0}{\Sigma P_0 Q_0} \times 100 = \frac{1700}{1680} \times 100 = \mathbf{101.19}.$$

$$\textbf{Paasche's: } \frac{\Sigma P_1 Q_1}{\Sigma P_0 Q_1} \times 100 = \frac{1670}{1650} \times 100 = \mathbf{101.21}.$$

QUANTITY INDEX NUMBERS:

$$\textbf{Laspeyre's: } \frac{\Sigma Q_1 P_0}{\Sigma Q_0 P_0} \times 100 = \frac{1650}{1680} \times 100 = \mathbf{98.21}.$$

$$\textbf{Paasche's: } \frac{\Sigma Q_1 P_1}{\Sigma Q_0 P_1} \times 100 = \frac{1670}{1700} \times 100 = \mathbf{98.24}.$$

16.18 VALUE INDEX NUMBERS

The value index numbers are very easy to calculate. **Value is the product of price and quantity. A simple value index number is equal to the value of the current year divided by the value of the base year. If this value is multiplied by 100, we get the value index number.** The formula, therefore, is

$$\textbf{VALUE INDEX NUMBER: } V = \frac{\Sigma P_1 q_1}{\Sigma P_0 q_0} \times 100.$$

Simple value Index Number: $V = \frac{\Sigma V_1}{\Sigma V_0}$,

where V_1 = value of the current year, V_0 = value of the base year.

Such index numbers are not weighted because they do not take into account both the prices and quantities. These index numbers are not very popular because the situation revealed by prices and quantities are not fully revealed by the values.

Example 13. *Calculate Paasche's quantity and Laspeyre's price index for the following data:*

Commodities	*Qty*	*Units*	*Value in Rs.*	
	1995	*1999*	*1995*	*1999*
A	100	150	500	900
B	80	100	320	500
C	60	72	120	360
D	30	33	360	397

Solution. We prepare the following table for calculating laspeyre's Price Index and Paasche's Quantity Index numbers.

Table : Computation of Laspeyre's Price and Paasche's Quantity Index

Commodities	*Qty.* q_0	q_1	p_0	p_1	p_1q_0	p_0q_0	q_1p_1	q_0p_1
A	100	150	5	6	600	500	900	600
B	80	100	4	5	400	320	500	400
C	60	72	2	5	300	120	360	300
D	30	33	12	9	270	360	297	270
					Σp_1q_0 **= 1570**	Σp_0q_0 **= 1300**	Σq_1p_1 **= 2057**	Σq_0p_1 **= 1570**

Laspeyre's Price Index : $P_{01} = \frac{\Sigma p_1q_0}{\Sigma p_0q_0} \times 100 = \frac{1570}{1300} \times 100 = \mathbf{120.77.}$

Paasche's Quantity Index : $Q_{01} = \frac{\Sigma q_1p_1}{\Sigma q_0p_1} \times 100 = \frac{2057}{1570} \times 100 = \mathbf{131.02.}$

Example 14. *Geometric mean of index numbers of Laspeyre and Paasche's is 229.5648 while the sum of Laspeyre and Paasche's index numbers is 480; find out Laspeyre and Paasche indices.*

Solution. We are given:

$$L + P = 480 \text{ and } \sqrt{L \times P} = 229.5648.$$

Now, $\quad (L + P)^2 - (L - P)^2 = 4\,(L \times P).$

$\therefore \quad (480)^2 - (L - P)^2 = 4 \times (229.5648)^2 = 4 \times 52700$

or $\quad (L - P)^2 = 230400 - 210800 = 19600$

or $\quad L - P = \sqrt{19600} = \mathbf{140.}$

Using $L + P = 480$ **and** $L - P = 140$**,** we get :

$$2L = 480 + 140 = 620 \text{ or } L = \frac{620}{2} = \mathbf{310.}$$

Now, making use of the value of L, we get :

$$P = 480 - 310 = \mathbf{170.}$$

Hence, Laspeyre's index is 310, while Paasche's index is 170.

Example 15. *Index numbers due to Paasche and Fisher are 144 and 132 respectively. Find index number of Laspeyre.*

Solution. Fisher's Index Number: $F = \sqrt{L \times P}$

$\therefore \quad 132 = \sqrt{L \times 144} \Rightarrow L = \frac{132 \times 132}{144} = \mathbf{121.}$

Example 16. *From the data given below construct Laspeyre's and Paasche's index numbers:*

Commodities	*Base Year*		*Current Year*	
	Price	*Expenditure*	*Price*	*Expenditure*
A	*2*	*40*	*5*	*75*
B	*4*	*16*	*8*	*40*
C	*1*	*10*	*2*	*24*
D	*5*	*25*	*10*	*60*

Solution.

Table : Computation of Laspeyre and Paasche Index

Commodity	p_0	q_0	p_1	q_1	p_0q_0	p_1q_0	p_1q_1	p_0q_1
A	2	20	5	15	40	100	75	30
B	4	4	8	5	16	32	40	20
C	1	10	2	12	10	20	24	12
D	5	5	10	6	25	50	60	30
Total					$\Sigma p_0q_0 = 91$	$\Sigma p_1q_0 = 202$	$\Sigma p_1q_1 = 199$	$\Sigma p_0q_1 = 92$

$$\textbf{Laspeyre's index} = \frac{\Sigma p_1q_0}{\Sigma p_0q_0} \times 100 = \frac{202}{91} \times 100 = \mathbf{221.98.}$$

$$\textbf{Paasche's index} = \frac{\Sigma p_1q_1}{\Sigma p_0q_1} \times 100 = \frac{199}{92} \times 100 = \mathbf{216.30.}$$

Example 17. *Construct index numbers of price from the following data applying:*

(*a*) *Laspeyre's method* (*b*) *Paasche's method.*

(*c*) *Fisher's ideal method*

	Base Year (2004)		*Current Year (2005)*	
	Kilo	*Rate (Rs)*	*Kilo*	*Rate (Rs)*
Wheat	*10*	*3*	*8*	*3.25*
Sugar	*20*	*15*	*15*	*20*
Rice	*2*	*25*	*3*	*23*

Solution.

Table : Computation of Index Numbers

	Base Year 2004		*Current Year 2005*		P_1q_0	P_0q_0	P_1q_1	P_0q_1
	Rate (Rs) P_0	*Kilo (kg)* q_0	*Rate (Rs)* P_1	*Kilo (kg)* q_1				
Wheat	3	10	3.25	8	32.50	30	26	24
Sugar	15	20	20.00	15	400.00	300	300	225
Rice	25	2	23.00	3	46.00	50	69	75
					ΣP_1q_0 = **478.50**	ΣP_0q_0 = **380**	ΣP_1q_1 = **395**	ΣP_0q_1 = **324**

(*a*) **Laspeyre's Index:**

$$P_{01}\ (La) = \frac{\Sigma P_1q_0}{\Sigma P_0q_0} \times 100 = \frac{478.50}{380} \times 100 = \mathbf{125.9.} \qquad \text{... (L)}$$

(*b*) **Paasche's Index:**

$$P_{01}\ (Pa) = \frac{\Sigma P_1 q_1}{\Sigma P_0 q_1} \times 100 = \frac{395}{324} \times 100 = \mathbf{121.9.} \qquad \text{... (P)}$$

Thus, Bowley's index is the arithmetic mean of Laspeyre's and Paasche's index numbers.

(*c*) **Fisher's Index:**

$$P_{01}\ (F) = \sqrt{L \times P} \times 100$$

or
$$P_{01}\ (F) = \sqrt{\frac{\Sigma P_1 q_0}{\Sigma P_0 q_0} \times \frac{\Sigma P_1 q_1}{\Sigma P_0 q_1}} \times 100 = \sqrt{\frac{478.50}{380} \times \frac{395}{324}} \times 100$$

$$= \sqrt{1.259 \times 1.219} \times 100 = 1.2388 \times 100 = \mathbf{123.88}.$$

Example 18. *If the index number of Laspeyre is 133.2 then find the value of x from the following data:*

Commodity	*Base Year*		*Current Year*
	Price	*Value*	*Price*
P	*5*	*12*	*8*
Q	*18*	*10*	*x*
R	*13*	*20*	*15*

Solution. **Table : Computation for x**

Commodity	***Base Year***		***Current Year***		
	Price p_0	***Quantity*** q_0	***Price*** p_1	$p_0 q_0$	$p_1 q_0$
P	5	12	8	60	96
Q	18	10	x	180	$10x$
R	13	20	15	260	300
				$\Sigma p_0 q_0$ **= 500**	$\Sigma p_1 q_0$ **= 10*x* + 396**

$$\therefore \qquad P_{01}^{La} = \frac{\Sigma p_1 q_0}{\Sigma p_0 q_0} \times 100 \Rightarrow 133.2 = \frac{10x + 396}{500} \times 100$$

or
$$666 = 10x + 396 \quad \Rightarrow \quad \mathbf{x = 27.}$$

EXERCISE – 16.1

1. What is an Index Number ? What are the characteristics of an Index Number ?
2. Discuss the steps involved in the construction of index numbers.
3. Discuss various problems that arise in connection with the construction of an index number.
4. Calculate the index number for 1998 with 1990 as base using the Weighted Average of Price Relative Method for the following data:

Commodity	*Weight*	*Price*	
		1990	*1998*
A	2	12	24
B	8	8	12
C	4	15	27
D	5	6	18
E	1	10	12

5. Compute price index by applying weighted average of price relatives:

Commodity	P_0 *(Rs.)*	q_0	P_1 *(Rs.)*
Sugar	*10*	*6 kg*	*15*
Rice	20	10 kg	25
Milk	10	8 lt.	14

6. Calculate Fisher's Ideal Index from the following data:

Commodity	*Base Year*		*Current Year*	
	Quantity	*Price*	*Quantity*	*Price*
A	15	4	10	6
B	20	3	25	4
C	10	6	20	5
D	30	5	25	5

7. From the data given below construct Fisher's quantity index numbers:

Commodities	*1994*		*1995*	
	Quantity	*Price*	*Quantity*	*Price*
A	2	4	6	18
B	5	3	2	2
C	7	8	4	24

8. Calculate the index number by the application of Laspeyre's formula and Paasche's formula from the following data: Also calculate Fisher's Ideal Index.

Commodity	*Units Consumed*		*Price per unit*	
	2004	*2006*	*2004*	*2006*
A	20	16	1.2	2.0
B	35	38	2.1	2.4
C	10	9	3.0	4.1
D	45	50	0.8	1.2

[**Hint.** $\Sigma P_1q_0 = 219$, $\Sigma p_0q_0 = 163.50$, $\Sigma P_1q_1 = 220.10$, $\Sigma P_0q_1 = 166$.

Laspeyre's Index $(P_{01}) = \frac{\Sigma P_1q_0}{\Sigma P_0q_0} \times 100 = \frac{219}{163.50} \times 100 = \mathbf{133.94.}$

Paasche's Index $= \frac{220.10}{166} \times 100 = \mathbf{132.59.}$

Fisher's Ideal Index $= \sqrt{\text{Laspeyre's Index} \times \text{Paasche's Index}}$

$= \sqrt{133.94 \times 132.59} = \mathbf{133.26.}$]

ANSWERS

4. $P_{01} = 197$. **5.** Price Index = 136.

6. Fisher's Ideal Index = 141.38. **7.** $Q_{01} = 87.34$.

8. Laspeyre's Index = 133.94; Paasche's Index = 132.59; Fisher's Ideal Index = 133.26.

APPENDIX – I

USE OF LOGARITHMIC TABLES

DEFINITION OF LOGARITHM

The logarithm of a number to a given base is the Index or the power to which the base must be raised to obtain that number. In other words, if $a^x = b$, where $b > 0$ and a is also a positive number greater than one (i.e., $a \neq 1, a > 0$), then the exponent x is called the logarithm of b to the base a. It is written as

$$x = \log_a b$$

It is to be noted that the two statements

$$a^x = b \qquad \textit{(Exponent Form)} \qquad \text{... (A)}$$

and $$b = \text{antilog } (a^x) \Rightarrow x = \log_a b \qquad \textit{(Logarithmic Form)} \qquad \text{... (B)}$$

are equivalent statements

Also the number b in (A) or (B) is called the **antilogarithm** of x to the base a and is written as

$$b = \text{antilog}_a (a^x) \qquad \textit{(Antilog Form)} \qquad \text{... (C)}$$

Remark. 1. Since $a^{\circ} = 1$, $\therefore$ $\log_a 1 = 0$.

In other words *logarithm of* 1 *to any base is zero.*

2. Since $a^{-\infty} = 0$, $(a > 1)$,

$\therefore$ $\log_a 0 = -\infty$

In other words ***logarithm of* 0 to any base greater than unity is – ∞**

3. Since $a^1 = a$.

$\therefore$ $\log_a a = 1$.

In other words, ***the logarithm of any number to the same base is always unity.***

LAWS OF LOGARITHMS

Logarithms have the following four fundamental rules. These rules are valid for any base a, $a > 0$, $a \neq 1$. (*i.e.*, a is a positive number greater than one).

(*i*) $\log_a mn = \log_a m + \log_a n$ **(Product formula)**

(*ii*) $\log_a (m/n) = \log_a m - \log_a n$ **(Quotient formula)**

(*iii*) $\log_a m^n = n \log_a m$ **(Power formula)**

(*iv*) $\log_a m = \log_n m \times \log_a n$ **(Base Changing Formula)**

SYSTEMS OF LOGARITHMS

There are two systems of logarithms, *viz*., **(*i*) Natural logarithms and (*ii*) Common logarithms.**

Natural logarithms. The logarithms to the base *e*, where e is the sum of the infinite series

$$1+\frac{1}{1!}+\frac{1}{2!}+\frac{1}{3!}+....... \quad = (2.718)$$

are called **natural logarithms**. They are used in theoretical calculations.

Common logarithms. ***Logarithms to the base 10 are called common logarithms.*** They are used in **numerical (*i.e.*, practical) calculations.** ***It is commonly used in calculations. Thus when no base is mentioned in numerical calculations, the base 10 is always understood.***

For example log 2, log 3 means $\log_{10} 2$ and $\log_{10} 3$ respectively. Hence forth by a logarithm we, shall mean a *common logarithm.*

Advantages of Common Logarithms. The followings are advantages of using 10 as the base of the system of logarithms.

(*i*) ***The characteristics of the logarithms of a number can be determined by inspection.***

(*ii*) ***The mantissa of the logarithms of all numbers consisting of the same digits in the same order (i.e, of numbers differing only in the position of the decimal point) is the same.***

CHARACTERISTIC AND MANTISSA

In common logarithms, the base is 10. As most numbers are not exact power of 10, so the *logarithm of such numbers consists of two parts:*

(*i*) Integral part and

(*ii*) Decimal or fractional part.

The integral part of the logarithm of a number (after expressing the decimal part as positive) is called the characteristic and the positive decimal part is called the mantissa.

For example. In log 20 = 1.3010, the characteristics is 1 and mantissa is 3010.

Example 1. ***The logarithm of a number is – 2.3010. Find the characteristic and mantissa.***

Solution. Let log x = – 2.3010. Here the decimal part – .3010, which is negative. ***Let us make it positive first by subtracting 1 from the integral part and adding one to the decimal part***

$\therefore$ – 2.3010 = – 2 – .3010 = (– 2 – 1) + (1 – .3010) = **– 3 + .6990.**

Now **the decimal part is positive, therefore, the characteristic = – 3 and mantissa = .6990.**

Note. (1) ***In practice write the logarithm of a number is such a way so that its decimal part is always positive.***

(2) ***– 3 + .6990. In short it is written as*** $\bar{3}.6990$.

The horizontal line called the bar, placed over 3 denotes that the integral part alone is negative, whereas the decimal part .6990 is positive. *Also* $\bar{3}$ *is read as 'Three bar.'*

Note. The mantissa is always positive.

RULES TO FIND THE CHARACTERISTIC. There are two rules to find the characteristic by inspection.

(*a*) **Rule 1 :** ***The characteristic of the logarithm of a number greater than unity is positive and is one less than the number of digits in the integral part of the numbers.***

Explanation

As we have $10^0 = 1$ $\therefore$ $\log_{10}1 = 0$

$10^1 = 10$ $\log_{10}10 = 1$

$10^2 = 100$ $\log_{10}100 = 2$

Example 2. ***Find the characteristics of the logarithms of (i) 5631 (ii) 5.631; (iii) 56.31.***

Solution: (*i*) The number 5631 is > 1 and the number of digits in the integral part is 4. Therefore, the **characteristic of the logarithm is 4 – 1 = 3.**

(*ii*) In this number 5.631, number of digits in the integral part = 1, therefore **characteristic of the logarithm is 1 – 1 = 0.**

(*iii*) In this case number of digits in the integral part of the number 56.32 is 2.

$\therefore$ **the characteristic of the logarithm = 2 – 1 = 1.**

Rule 2: ***The characteristic of the logarithm of a number less than 1 is negative and is numerically one more than the number of zeros immediately after the decimal point.***

Explanation: Since .1 = 10^{-1} $\therefore$ log .1 = – 1

.01 = 10^{-2} $\therefore$ log .01 = – 2

.001 = 10^{-3} $\therefore$ log .001 = – 3

and so on.

Example 3. ***Find the characteristic of the logarithm of*** **(*i*) 0.8253; (*ii*) 0.08253; (*iii*) 0.008253.**

Solution. (*i*) Since 0.8253 < 1 and there is no zero immediately after the decimal point, *i.e.*, number of **zero immediately after the decimal point is 0,**

$\therefore$ **the characteristic of the logarithm** = – (0 + 1) *i.e.*, = $\bar{\mathbf{1}}$.

(*ii*) Here the number of zeros immediately after the decimal point = 1.

Hence the characteristic of the logarithm is = – (1 + 1) = – 2 = $\bar{\mathbf{2}}$.

(*iii*) Here the number of zeros immediately after the decimal point = 2.

Hence the characteristic of the logarithm is = – (2 + 1) = – 3 = $\bar{\mathbf{3}}$.

RULE REGARDING THE MANTISSA

Rule: ***The mantissa of logarithms of all numbers consisting of the same digits in the same order are the same.***

This can be seen as follows:

Considers the numbers 125.4, 1.254, 0.01254. We shall show that the mantissa of log 125.4, log 1.254, log 0.01254 is the same.

$$\therefore \quad \textbf{log 125.4} = \log (100 \times 1.254) = \log 100 + \log 1.254$$
$$= \log 10^2 + \log 1.254 = \mathbf{2 + log\ 1.254.}$$
$$\textbf{log 0.01254} = \log \frac{1.254}{100} = \log 1.254 - \log 100 = \mathbf{log\ 1.254 - 2.}$$

So we find that the only difference lies in the characteristic, mantissae are the same.

DETERMINATION OF MANTISSA

The mantissa of the common logarithm of a number cannot be determined like its characteristics by inspecting the number. It is to be found from a table called **TABLE OF LOGARITHMS** or **LOG-TABLE.**

TABLE OF LOGARITHMS OR LOG – TABLES

In the extreme left hand column of the log table, we find figures ranging from 10 to 99 in bold-digits. The upper-most row contains the figures from 0 to 9. Below each of these figures we find numbers containing 4 digits (sometimes more also). These are the mantissae with decimal point at the beginning being dropped. Each mantissa is understood to have a decimal point preceding it. This forms the main table. There is also a side table on the right called Mean Difference Table.

USE OF LOG-TABLE TO FIND THE MANTISSA

Step I. ***Remove the decimal point from the given number.***

Step II. ***Look at the extreme left hand vertical column till you reach the first two digits from left of the number in STEP I.***

Step III. ***In the horizontal row beginning with above two figures, read the number under column headed by third digit (from the left) of the number.***

Step IV. ***To the number obtained in STEP III, add the number in the same horizontal line under the mean difference columns headed by fourth digit (from the left) of the number.***

Step V. ***Prefix the decimal point to the number obtained in STEP IV, which is the required mantissae.***

Explanation. Suppose we are to find log 46.98

$\therefore$ **Characteristics** = 2 – 1 = **1.** ... (1)

For Mantissa, in the positive decimal part

Remove decimal point, the number becomes 4698.

The first two digits are 46, the third is 9 and fourth is 8.

The figures at the intersection of row 46 and the column under 9 is 6712.

The figure at the cross of the same row and under mean difference column of 8 = 7.

Thus 6712 + 7 = 6719

∴ **Mantissa = 0.6719.**

∴ **log 46.98 = 1.6719.**

Remark 1. ***To find the mantissa of the logarithm of a number which contains less than 4 figures, add zeros to the right of the numbers until figures are 4.***

2. To find the Mantissa when the number contains five digits or more – When the number contains more than four significant digits, it can be approximated upto 4 significant digits for rough calculation. However, for more precise results, the method is illustrated below and it is called the Method of Interpolation. This method is based on an important principle of mathematics known as Principle of Proportional Part, which is stated below (without proof).

"If the value of a quantity depending on a variable quantity *x* be tabulated for different values of *x* at regular small intervals, then for a very small change in *x*, the corresponding small change in the tabulated value of the quantity is proportional to the change in *x*."

Example 4. ***Find log 3.5825***

Solution. Since 3.5825 lies between 3.582 and 3.583, therefore, from the tables,

$$\log 3.582 = 0.5541 \text{ and } \log 3.583 = 0.5543.$$

Now for a difference of (3.583 – 3.582) = 0.001,

the difference in the logarithm = 0.5543 – 0.5541 = 0.0002.

∴ For the difference of (3.5825 – 3.582) = 0.0005 in the number, the difference in the logarithm is

$$\frac{0.0002 \times 0.0005}{0.001} = 0.0001.$$

Adding this difference to log 3.582, we have

$$\log 3.5825 = 0.5541 + 0.0011 = 0.5542.$$

ANTILOGARITHM

Besides finding logarithm of a given number it is also necessary to reverse the process, that is, to find the number. This is called antilogarithm. Mathematically, **if log *x* = *y*, then *x* is called the antilogarithm of *y*.**

i.e., **x = antilog y.**

For example log 2 = 0.3010.

∴ **Antilog 0.3010 = 2.**

To find the antilogarithm, we use the tables of antilogarithms, whose arrangement is similar to that of logarithm.

READING THE ANTILOGARITHM TABLES. ***We shall explain it by the following example:***

Example 5. ***Find the number whose logarithm is 1.6078.***

Solution. The characteristic (= 1), which is one less than the number of digits in the integral part of the required number. Thus the number of digits in the integral part of the required number = 1 + 1 = 2.

(*ii*) Removing (the integral part) 1 from the given logarithm we get 0.6078. From the tables of antilogarithm, at the intersection of row against 0.60 and the column under 7, the value is 4046. In the row below the column of 8 under mean differences, we get 7.

∴ digits in antilog 1.6078 = 4046 + 7 = 4053.

Since the characteristic is 2, therefore, the required number is 40.53.

TABLE – 1

LOGARITHMS TABLES

LOGARITHMS

	0	1	2	3	4	5	6	7	8	9	Mean Differences								
											1	2	3	4	5	6	7	8	9
10	0000	0043	0086	0128	0170	0212	0253	0294	0334	0374	4	8	12	17	21	25	29	33	37
11	0414	0453	0492	0531	0569	0607	0645	0682	0719	0755	4	8	11	15	19	23	26	30	34
12	0792	0828	0864	0899	0934	0969	1004	1038	1072	1106	3	7	10	14	17	21	24	28	31
13	1139	1173	1206	1239	1271	1303	1335	1367	1399	1430	3	6	10	13	16	19	23	26	29
14	1461	1492	1523	1553	1584	1614	1644	1673	1703	1732	3	6	9	12	15	18	21	24	27
15	1761	1790	1818	1847	1875	1903	1931	1959	1987	2014	3	6	8	11	14	17	20	22	25
16	2041	2068	2095	2122	2148	2175	2201	2227	2253	2279	3	5	8	11	13	16	18	21	24
17	2304	2330	2355	2380	2405	2430	2455	2480	2504	2529	2	5	7	10	12	15	17	20	22
18	2553	2577	2601	2625	2648	2672	2695	2718	2742	2765	2	5	7	9	12	14	16	19	21
19	2788	2810	2833	2856	2878	2900	2923	2945	2967	2989	2	4	7	9	11	13	16	18	20
20	3010	3032	3054	3075	3096	3118	3139	3160	3181	3201	2	4	6	8	11	13	15	17	19
21	3222	3243	3263	3284	3304	3324	3345	3365	3385	3404	2	4	6	8	10	12	14	16	18
22	3424	3444	3464	3483	3502	3522	3541	3560	3579	3598	2	4	6	8	10	12	14	15	17
23	3617	3636	3655	3674	3692	3711	3729	3747	3766	3784	2	4	6	7	9	11	13	15	17
24	3802	3820	3838	3856	3874	3892	3909	3927	3945	3962	2	4	5	7	9	11	12	14	16
25	3979	3997	4014	4031	4048	4065	4082	4099	4116	4133	2	3	5	7	9	10	12	14	15
26	4150	4166	4183	4200	4216	4232	4249	4265	4281	4298	2	3	5	7	8	10	11	13	15
27	4314	4330	4346	4362	4378	4393	4409	4425	4440	4456	2	3	5	6	8	9	11	13	14
28	4472	4487	4502	4518	4533	4548	4564	4579	4594	4609	2	3	5	6	8	9	11	12	14
29	4624	4639	4654	4669	4683	4698	4713	4728	4742	4757	1	3	4	6	7	9	10	12	13
30	4771	4786	4800	4814	4829	4843	4857	4871	4886	4900	1	3	4	6	7	9	10	11	13
31	4914	4928	4942	4955	4969	4983	4997	5011	5024	5038	1	3	4	6	7	8	10	11	12
32	5051	5065	5079	5092	5105	5119	5132	5145	5159	5172	1	3	4	5	7	8	9	11	12
33	5185	5198	5211	5224	5237	5250	5263	5276	5289	5302	1	3	4	5	6	8	9	10	12
34	5315	5328	5340	5353	5366	5378	5391	5403	5416	5428	1	3	4	5	6	8	9	10	11
35	5441	5453	5465	5478	5490	5502	5514	5527	5539	5551	1	2	4	5	6	7	9	10	11
36	5563	5575	5587	5599	5611	5623	5635	5647	5658	5670	1	2	4	5	6	7	8	10	11
37	5682	5694	5705	5717	5729	5740	5752	5763	5775	5786	1	2	3	5	6	7	8	9	10
38	5798	5809	5821	5832	5843	5855	5866	5877	5888	5899	1	2	3	5	6	7	8	9	10
39	5911	5922	5933	5944	5955	5966	5977	5988	5999	6010	1	2	3	4	5	7	8	9	10
40	6021	6031	6042	6053	6064	6075	6085	6096	6107	6117	1	2	3	4	5	6	8	9	10
41	6128	6138	6149	6160	6170	6180	6191	6201	6212	6222	1	2	3	4	5	6	7	8	9
42	6232	6243	6253	6263	6274	6284	6294	6304	6314	6325	1	2	3	4	5	6	7	8	9
43	6335	6345	6355	6365	6375	6385	6395	6405	6415	6425	1	2	3	4	5	6	7	8	9
44	6435	6444	6454	6464	6474	6484	6493	6503	6513	6522	1	2	3	4	5	6	7	8	9
45	6532	6542	6551	6561	6571	6580	6590	6599	6609	6618	1	2	3	4	5	6	7	8	9
46	6628	6637	6646	6656	6665	6675	6684	6693	6702	6712	1	2	3	4	5	6	7	7	8
47	6721	6730	6739	6749	6758	6767	6776	6785	6794	6803	1	2	3	4	5	5	6	7	8
48	6812	6821	6830	6839	6848	6857	6866	6875	6884	6893	1	2	3	4	4	5	6	7	8
49	6902	6911	6920	6928	6937	6946	6955	6964	6972	6981	1	2	3	4	4	5	6	7	8
50	6990	6998	7007	7016	7024	7033	7042	7050	7059	7067	1	2	3	3	4	5	6	7	8

(Contd.)

LOGARITHMS

	0	1	2	3	4	5	6	7	8	9	Mean Differences								
											1	2	3	4	5	6	7	8	9
51	7076	7084	7093	7101	7110	7118	7126	7135	7143	7152	1	2	3	3	4	5	6	7	8
52	7160	7168	7177	7185	7193	7202	7210	7218	7226	7235	1	2	2	3	4	5	6	7	7
53	7243	7251	7259	7267	7275	7284	7292	7300	7308	7316	1	2	2	3	4	5	6	6	7
54	7324	7332	7340	7348	7356	7364	7372	7380	7388	7396	1	2	2	3	4	5	6	6	7
55	7404	7412	7419	7427	7435	7443	7451	7459	7466	7474	1	2	2	3	4	5	5	6	7
56	7482	7490	7497	7505	7513	7520	7528	7536	7543	7551	1	2	2	3	4	5	5	6	7
57	7559	7566	7574	7582	7589	7597	7604	7612	7619	7627	1	2	2	3	4	5	5	6	7
58	7634	7642	7649	7657	7664	7672	7679	7686	7694	7701	1	1	2	3	4	4	5	6	7
59	7709	7716	7723	7731	7738	7745	7752	7760	7767	7774	1	1	2	3	4	4	5	6	7
60	7782	7789	7796	7803	7810	7818	7825	7832	7839	7846	1	1	2	3	4	4	5	6	6
61	7853	7860	7868	7875	7882	7889	7896	7903	7910	7917	1	1	2	3	4	4	5	6	6
62	7924	7931	7938	7945	7952	7959	7966	7973	7980	7987	1	1	2	3	3	4	5	6	6
63	7993	8000	8007	8014	8021	8028	8035	8041	8048	8055	1	1	2	3	3	4	5	5	6
64	8062	8069	8075	8082	8089	8096	8102	8109	8116	8122	1	1	2	3	3	4	5	5	6
65	8129	8136	8142	8149	8156	8162	8169	8176	8182	8189	1	1	2	3	3	4	5	5	6
66	8195	8202	8209	8215	8222	8228	8235	8241	8248	8254	1	1	2	3	3	4	5	5	6
67	8261	8267	8274	8280	8287	8293	8299	8306	8312	8319	1	1	2	3	3	4	5	5	6
68	8325	8331	8338	8344	8351	8357	8363	8370	8376	8382	1	1	2	3	3	4	4	5	6
69	8388	8395	8401	8407	8414	8420	8426	8432	8439	8445	1	1	2	2	3	4	4	5	6
70	8451	8457	8463	8470	8476	8482	8488	8494	8500	8506	1	1	2	2	3	4	4	5	6
71	8513	8519	8525	8531	8537	8543	8549	8555	8561	8567	1	1	2	2	3	4	4	5	5
72	8573	8579	8585	8591	8597	8603	8609	8615	8621	8627	1	1	2	2	3	4	4	5	5
73	8633	8639	8645	8651	8657	8663	8669	8675	8681	8686	1	1	2	2	3	4	4	5	5
74	8692	8698	8704	8710	8716	8722	8727	8733	8739	8745	1	1	2	2	3	4	4	5	5
75	8751	8756	8762	8768	8774	8779	8785	8791	8797	8802	1	1	2	2	3	3	4	5	5
76	8808	8814	8820	8825	8831	8837	8842	8848	8854	8859	1	1	2	2	3	3	4	5	5
77	8865	8871	8876	8882	8887	8893	8899	8904	8910	8915	1	1	2	2	3	3	4	4	5
78	8921	8927	8932	8938	8943	8949	8954	8960	8965	8971	1	1	2	2	3	3	4	4	5
79	8976	8982	8987	8993	8998	9004	9009	9015	9020	9025	1	1	2	2	3	3	4	4	5
80	9031	9036	9042	9047	9053	9058	9063	9069	9074	9079	1	1	2	2	3	3	4	4	5
81	9085	9090	9096	9101	9106	9112	9117	9122	9128	9133	1	1	2	2	3	3	4	4	5
82	9138	9143	9149	9154	9159	9165	9170	9175	9180	9186	1	1	2	2	3	3	4	4	5
83	9191	9196	9201	9206	9217	9217	9222	9227	9232	9238	1	1	2	2	3	3	4	4	5
84	9243	9248	9253	9258	9263	9269	9274	9279	9284	9289	1	1	2	2	3	3	4	4	5
85	9294	9299	9304	9360	9315	9320	9325	9330	9335	9340	1	1	2	2	3	3	4	4	5
86	9345	9350	9355	9360	9365	9370	9375	9380	9385	9390	1	1	2	2	3	3	4	4	5
87	9395	9400	9405	9410	9415	9420	9425	9430	9435	9440	0	1	1	2	2	3	3	4	4
88	9445	9450	9455	9460	9465	9469	9474	9479	9484	9489	0	1	1	2	2	3	3	4	4
89	9494	9499	9504	9509	9513	9518	9523	9528	9533	9538	0	1	1	2	2	3	3	4	4
90	9542	9547	9552	9557	9562	9566	9571	9576	9581	9586	0	1	1	2	2	3	3	4	4
91	9590	9595	9600	9605	9609	9614	9619	9624	9628	9633	0	1	1	2	2	3	3	4	4
92	9638	9643	9647	9652	9657	9661	9666	9671	9675	9680	0	1	1	2	2	3	3	4	4
93	9685	9689	9694	9699	9703	9708	9713	9717	9722	9727	0	1	1	2	2	3	3	4	4
94	9731	9736	9741	9745	9750	9754	9759	9763	9768	9773	0	1	1	2	2	3	3	4	4
95	9777	9782	9786	9791	9795	9800	9805	9809	9814	9818	0	1	1	2	2	3	3	4	4
96	9823	9827	9832	9836	9841	9845	9850	9854	9859	9863	0	1	1	2	2	3	3	4	4
97	9868	9872	9877	9881	9886	9890	9894	9899	9903	9908	0	1	1	2	2	3	3	4	4
98	9912	9916	9921	9926	9930	9934	9939	9943	9948	9952	0	1	1	2	2	3	3	4	4
99	9956	9961	9965	9969	9974	9978	9983	9987	9991	9996	0	1	1	2	2	3	3	3	4

TABLE – 2

ANTILOGARITHMS

	0	1	2	3	4	5	6	7	8	9	Mean Differences								
											1	2	3	4	5	6	7	8	9
.00	1000	1002	1005	1007	1009	1012	1014	1016	1019	1021	0	0	1	1	1	1	2	2	2
.01	1023	1026	1028	1030	1033	1035	1038	1040	1042	1045	0	0	1	1	1	1	2	2	2
.02	1047	1050	1052	1054	1057	1059	1062	1064	1067	1069	0	0	1	1	1	1	2	2	2
.03	1072	1074	1076	1079	1081	1084	1086	1089	1091	1094	0	0	1	1	1	1	2	2	2
.04	1096	1099	1102	1104	1107	1109	1112	1114	1117	1119	0	1	1	1	1	2	2	2	2
.05	1122	1125	1127	1130	1132	1135	1138	1140	1143	1146	0	1	1	1	1	2	2	2	2
.06	1148	1151	1153	1156	1159	1161	1164	1167	1169	1172	0	1	1	1	1	2	2	2	2
.07	1175	1178	1180	1183	1186	1189	1191	1194	1197	1199	0	1	1	1	1	2	2	2	2
.08	1202	1205	1208	1211	1213	1216	1219	1222	1225	1227	0	1	1	1	1	2	2	2	3
.09	1230	1233	1236	1239	1242	1245	1247	1250	1253	1256	0	1	1	1	1	2	2	2	3
.10	1259	1262	1265	1268	1271	1274	1276	1279	1282	1285	0	1	1	1	1	2	2	2	3
.11	1288	1291	1294	1297	1300	1303	1306	1309	1312	1315	0	1	1	1	2	2	2	2	3
.12	1318	1321	1324	1327	1330	1334	1337	1340	1343	1346	0	1	1	1	2	2	2	2	3
.13	1349	1352	1355	1358	1361	1365	1368	1371	1374	1377	0	1	1	1	2	2	2	3	3
.14	1380	1384	1387	1390	1393	1396	1400	1403	1406	1409	0	1	1	1	2	2	2	3	3
.15	1413	1416	1419	1422	1426	1429	1432	1435	1439	1442	0	1	1	1	2	2	2	3	3
.16	1445	1449	1452	1455	1459	1462	1466	1469	1472	1476	0	1	1	1	2	2	2	3	3
.17	1479	1483	1486	1489	1493	1496	1500	1503	1507	1510	0	1	1	1	2	2	2	3	3
.18	1514	1517	1521	1524	1528	1531	1535	1538	1542	1545	0	1	1	1	2	2	2	3	3
.19	1549	1552	1556	1560	1563	1567	1570	1574	1578	1581	0	1	1	1	2	2	3	3	3
.20	1585	1589	1592	1596	1600	1603	1607	1611	1614	1618	0	1	1	1	2	2	3	3	3
.21	1622	1626	1629	1633	1637	1641	1644	1648	1652	1656	0	1	1	2	2	2	3	3	3
.22	1660	1663	1667	1671	1675	1679	1683	1687	1690	1694	0	1	1	2	2	2	3	3	3
.23	1698	1702	1706	1710	1714	1718	1722	1726	1730	1734	0	1	1	2	2	2	3	3	4
.24	1738	1742	1746	1750	1754	1758	1762	1766	1770	1774	0	1	1	2	2	2	3	3	4
.25	1778	1782	1786	1791	1795	1798	1803	1807	1811	1816	0	1	1	2	2	2	3	3	4
.26	1820	1824	1828	1832	1837	1841	1845	1849	1854	1858	0	1	1	2	2	3	3	3	4
.27	1862	1866	1871	1875	1879	1884	1888	1892	1897	1901	0	1	1	2	2	3	3	3	4
.28	1905	1910	1914	1919	1923	1928	1932	1936	1941	1945	0	1	1	2	2	3	3	4	4
.29	1950	1954	1959	1963	1968	1972	1977	1982	1986	1991	0	1	1	2	2	3	3	4	4
.30	1995	2000	2004	2009	2014	2018	2023	2028	2032	2037	0	1	1	2	2	3	3	4	4
.31	2042	2046	2051	2056	2061	2065	2070	2075	2080	2084	0	1	1	2	2	3	3	4	4
.32	2089	2094	2099	2104	2109	2113	2118	2123	2128	2133	0	1	1	2	2	3	3	4	4
.33	2138	2143	2148	2153	2158	2163	2168	2173	2178	2183	0	1	1	2	2	3	3	4	4
.34	2188	2193	2198	2203	2208	2213	2218	2223	2228	2234	1	1	2	2	3	3	4	4	5
.35	2239	2244	2249	2254	2259	2265	2270	2275	2280	2286	1	1	2	2	3	3	4	4	5
.36	2291	2296	2301	2307	2312	2317	2323	2328	2333	2339	1	1	2	2	3	3	4	4	5
.37	2344	2350	2355	2360	2366	2371	2377	2382	2388	2393	1	1	2	2	3	3	4	4	5
.38	2399	2404	2410	2415	2421	2427	2432	2438	2443	2449	1	1	2	2	3	3	4	4	5
.39	2455	2460	2466	2472	2477	2483	2489	2495	2500	2506	1	1	2	2	3	3	4	5	5
.40	2512	2518	2523	2529	2535	2541	2547	2553	2559	2564	1	1	2	2	3	4	4	5	5
.41	2570	2576	2582	2588	2594	2600	2606	2612	2618	2624	1	1	2	2	3	4	4	5	5
.42	2630	2636	2642	2649	2655	2661	2667	2673	2679	2685	1	1	2	2	3	4	4	5	6
.43	2692	2698	2704	2710	2716	2723	2729	2735	2742	2748	1	1	2	3	3	4	4	5	6
.44	2754	2761	2767	2773	2780	2786	2793	2799	2805	2812	1	1	2	3	3	4	4	5	6
.45	2818	2825	2831	2838	2844	2851	2858	2864	2871	2877	1	1	2	3	3	4	5	5	6
.46	2884	2891	2897	2904	2911	2917	2924	2931	2938	2944	1	1	2	3	3	4	5	5	6
.47	2951	2958	2965	2972	2979	2985	2992	2999	3006	3013	1	1	2	3	3	4	5	5	6
.48	3020	3027	3034	3041	3048	3055	3062	3069	3076	3083	1	1	2	3	4	4	5	6	6
.49	3090	3097	3105	3112	3119	3126	3133	3141	3148	3155	1	1	2	3	4	4	5	6	6
.50	3162	3170	3177	3184	3192	3199	3206	3214	3221	3228	1	1	2	3	4	4	5	6	7

(Contd.)

ANTILOGARITHMS

	0	1	2	3	4	5	6	7	8	9	Mean Differences								
											1	2	3	4	5	6	7	8	9
.51	3236	3243	3251	3258	3266	3273	3281	3289	3296	3304	1	2	2	3	4	5	5	6	7
.52	3311	3319	3327	3334	3342	3350	3357	3365	3373	3381	1	2	2	3	4	5	5	6	7
.53	3388	3396	3404	3412	3420	3428	3436	3443	3451	3459	1	2	2	3	4	5	6	6	7
.54	3467	3475	3483	3491	3499	3508	3516	3524	3532	3540	1	2	2	3	4	5	6	6	7
.55	3548	3556	3565	3573	3581	3589	3597	3606	3614	3622	1	2	2	3	4	5	6	7	7
.56	3631	3639	3648	3656	3664	3673	3681	3690	3698	3707	1	2	3	3	4	5	6	7	8
.57	3715	3724	3733	3741	3750	3758	3767	3776	3784	3793	1	2	3	3	4	5	6	7	8
.58	3802	3811	3819	3828	3837	3846	3855	3864	3873	3882	1	2	3	4	4	5	6	7	8
.59	3890	3899	3908	3917	3926	3936	3945	3954	3963	3972	1	2	3	4	5	5	6	7	8
.60	3981	3990	3999	4009	4018	4027	4036	4046	4055	4064	1	2	3	4	5	6	6	7	8
.61	4074	4083	4093	4102	4111	4121	4130	4140	4150	4159	1	2	3	4	5	6	7	8	9
.62	4169	4178	4188	4198	4207	4217	4227	4236	4246	4256	1	2	3	4	5	6	7	8	9
.63	4266	4276	4285	4295	4305	4315	4325	4335	4345	4355	1	2	3	4	5	6	7	8	9
.64	4365	4375	4385	4395	4406	4416	4426	4436	4446	4457	1	2	3	4	5	6	7	8	9
.65	4467	4477	4487	4498	4508	4519	4529	4539	4550	4560	1	2	3	4	5	6	7	8	9
.66	4571	4581	4592	4603	4613	4624	4634	4645	4656	4667	1	2	3	4	5	6	7	9	10
.67	4677	4688	4699	4710	4721	4732	4742	4753	4764	4775	1	2	3	4	5	7	8	9	10
.68	4786	4797	4808	4819	4831	4842	4853	4864	4875	4887	1	2	3	4	6	7	8	9	10
.69	4898	4909	4920	4932	4943	4955	4966	4977	4989	5000	1	2	3	5	6	7	8	9	10
.70	5012	5023	5035	5047	5058	5070	5082	5093	5105	5117	1	2	4	5	6	7	8	9	11
.71	5129	5140	5152	5164	5176	5188	5200	5212	5224	5236	1	2	4	5	6	7	8	10	11
.72	5248	5260	5272	5284	5297	5309	5321	5333	5346	5358	1	2	4	5	6	7	9	10	11
.73	5370	5383	5395	5408	5420	5433	5445	5458	5470	5483	1	3	4	5	6	8	9	10	11
.74	5495	5508	5521	5534	5546	5559	5572	5585	5598	5610	1	3	4	5	6	8	9	10	12
.75	5623	5636	5649	5662	5675	5689	5702	5715	5728	5741	1	3	4	5	7	8	9	10	12
.76	5754	5768	5781	5794	5808	5821	5834	5848	5861	5875	1	3	4	5	7	8	9	11	12
.77	5888	5902	5916	5929	5943	5957	5970	5984	5998	6012	1	3	4	5	7	8	10	11	12
.78	6026	6039	6053	6067	6081	6095	6109	6124	6138	6152	1	3	4	6	7	8	10	11	13
.79	6166	6180	6194	6209	6223	6237	6252	6266	6281	6295	1	3	4	6	7	9	10	11	13
.80	6310	6324	6339	6353	6368	6383	6397	6412	6427	6442	1	3	4	6	7	9	10	12	13
.81	6457	6471	6486	6501	6516	6531	6546	6561	6577	6592	2	3	5	6	8	9	11	12	14
.82	6607	6622	6637	6653	6668	6683	6699	6714	6730	6745	2	3	5	6	8	9	11	12	14
.83	6761	6776	6792	6808	6823	6839	6855	6871	6887	6902	2	3	5	6	8	9	11	13	14
.84	6918	6934	6950	6966	6982	6998	7015	7031	7047	7063	2	3	5	6	8	10	11	13	15
.85	7079	7096	7112	7129	7145	7161	7178	7194	7211	7228	2	3	5	7	8	10	12	13	15
.86	7244	7261	7278	7295	7311	7328	7345	7362	7379	7396	2	3	5	7	8	10	12	13	15
.87	7413	7430	7447	7464	7482	7499	7516	7534	7551	7568	2	3	5	7	9	10	12	14	16
.88	7586	7603	7621	7638	7656	7674	7691	7709	7727	7745	2	4	5	7	9	11	12	14	16
.89	7762	7780	7798	7816	7834	7852	7870	7889	7907	7925	2	4	5	7	9	11	13	14	16
.90	7943	7962	7980	7998	8017	8035	8054	8072	8091	8110	2	4	6	7	9	11	13	15	17
.91	8128	8147	8166	8185	8204	8222	8241	8260	8279	8299	2	4	6	8	9	11	13	15	17
.92	8318	8337	8356	8375	8395	8414	8433	8453	8472	8492	2	4	6	8	10	12	14	15	17
.93	8511	8531	8551	8570	8590	8616	8630	8650	8670	8690	2	4	6	8	10	12	14	16	18
.94	8710	8730	8750	8770	8790	8810	8831	8851	8872	8892	2	4	6	8	10	12	14	16	18
.95	8913	8933	8954	8974	8995	9016	9036	9057	9078	9099	2	4	6	8	10	12	15	17	19
.96	9120	9141	9162	9183	9204	9226	9247	9268	9290	9311	2	4	6	8	11	13	15	17	19
.97	9333	9354	9376	9397	9419	9441	9462	9484	9506	9528	2	4	7	9	11	13	15	17	20
.98	9550	9572	9594	9616	9638	9661	9683	9705	9727	9750	2	4	7	9	11	13	16	18	20
.99	9772	9793	9817	9840	9863	9886	9908	9931	9954	9977	2	5	7	9	11	14	16	18	20

TABLE 3
AREA OF A STANDARD NORMAL DISTRIBUTION

An entry in the table is the proportion under the entire curve which is between $z = 0$ and a positive value of z. Area for negative values of z are obtained by symmetry.

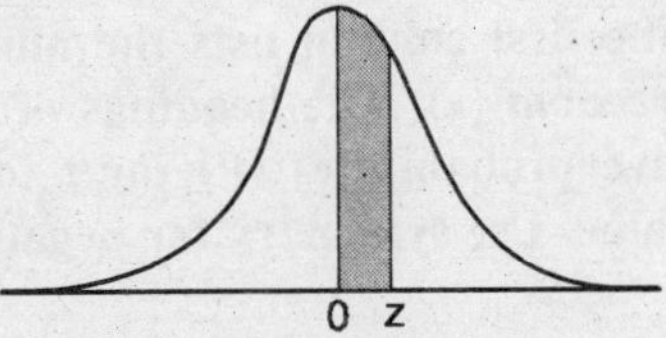

z	.00	.01	.02	.03	.04	.05	.06	.07	.08	.09
0.0	.0000	.0040	.0080	.0120	.0160	.0199	.0239	.0279	.0319	.0359
0.1	.0398	.0438	.0478	.0517	.0557	.0596	.0636	.0675	.0714	.0753
0.2	.0793	.8832	.0871	.0910	.0948	.0987	.1026	.1064	.1103	.1141
0.3	.1179	.1217	.1255	.1293	.1331	.1368	.1406	.1443	.1480	.1517
0.4	.1554	1591	.1628	.1664	.1700	.1736	.1772	.1808	.1844	.1879
0.5	.1915	.1950	.1985	.2019	.2054	.2088	.2123	.2157	.2190	.2224
0.6	.2257	.2291	.2324	.2357	.2389	.2422	.2454	.2486	.2517	.2549
0.7	.2580	.2611	.2642	.2673	.2703	.2734	.2764	.2794	.2823	.2852
0.8	.2881	.2910	.2939	.2967	.2995	.3023	.3051	.3078	.3106	.3133
0.9	.3159	.3186	.3212	.3238	.3264	.3289	.3315	.3340	.3365	.3389
1.0	.3413	.3438	.3461	.3485	.3508	.3531	.3554	.3577	.3599	.3621
1.1	.3643	.3665	.3686	.3708	.3729	.3749	.3770	.3790	.3810	.3830
1.2	.3849	.3869	.3888	.3907	.3925	.3944	.3962	.3980	.3997	.4015
1.3	.4032	.4049	.4066	.4082	.4099	.4115	.4131	.4147	.4162	.4177
1.4	.4192	.4207	.4222	.4236	.4251	.4265	.4279	.4292	.4306	.4319
1.5	.4332	.4345	.4357	.4370	.4382	.4394	.4406	.4418	.4429	.4441
1.6	.4452	.4463	.4474	.4484	.4495	.4505	.4415	.4525	.4535	.4545
1.7	.4554	.4564	.4573	.4582	.4591	.4599	.4608	.4616	.4625	.4633
1.8	.4641	.4649	.4656	.4664	.4671	.4678	.4686	.4693	.4699	.4706
1.9	.4713	.4719	.4726	.4732	.4738	.4744	.4750	.4756	.4761	.4767
2.0	.4772	.4778	.4783	.4788	.4793	.4798	.4803	.4808	.4812	.4817
2.1	.4821	.4826	.4830	.4834	.4838	.4842	.4846	.4850	.4854	.4857
2.2	.4861	.4864	.4868	.4871	.4875	.4878	.4841	.4884	.4887	.4890
2.3	.4893	.4896	.4898	.4901	.4904	.4906	.4909	.4911	.4913	.4916
2.4	.4918	.4920	.4922	.4925	.4927	.4929	.4931	.4932	.4934	.4936
2.5	.4938	.4940	.4941	.4943	.4945	.4946	.4948	.4949	.4951	.4952
2.6	.4953	.4955	.4956	.4957	.4959	.4960	.4961	.4962	.4963	.4964
2.7	.4965	.4966	.4967	.4968	.4969	.4970	.4971	.4972	.4973	.4974
2.8	.4974	.4975	.4976	.4977	.4977	.4978	.4979	.4979	.4980	.4981
2.9	.4981	.4982	.4982	.4983	.4984	.4984	.4985	.4985	.4986	.4986
3.0	.4987	.4987	.4987	.4988	.4988	.4989	.4989	.4989	.4990	.4990

TABLE 4
STUDENTS'S t DISTRIBUTION

The first column lists the number of degrees of freedom (v). The headings of the other columns give probabilities (P) for t to exceed the entry value. Use symmetry for negative t values.

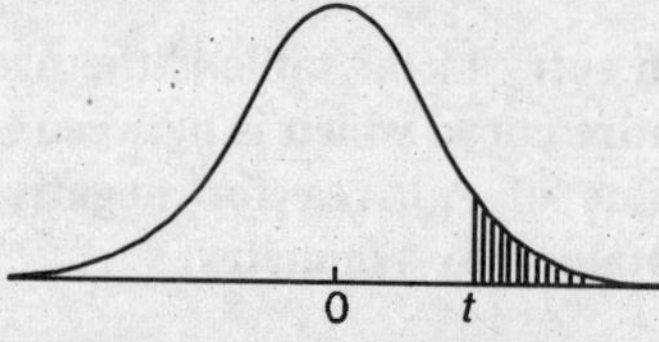

P/v	.10	.05	.025	.01	.005
1	3.078	6.314	12.706	31.812	63.657
2	1.886	2.920	4.303	6.965	9.925
3	1.638	2.353	3182	4.541	5.841
4	1.533	2.132	2.776	3.747	4.604
5	1.476	2.015	2.571	3.365	4.032
6	1.440	1.943	2.447	3.143	3.707
7	1.415	1.895	2.365	2.998	3.499
8	1.397	1.860	2.306	2.896	3.355
9	1.383	1.833	2.262	2.821	3.250
10	1.372	1.812	2.228	2.764	3.169
11	1.363	1.796	2.201	2.718	3.106
12	1.356	1.782	2.179	2.681	3.055
13	1.350	1.771	2.160	2.650	3.012
14	1.345	1.761	2.145	2.624	2.977
15	1.341	1.753	2.131	2.602	2.947
16	1.337	1.746	2.120	2.583	2.921
17	1.333	1.740	2.110	2.567	2.898
18	1.330	1.734	2.101	2.552	2.878
19	1.328	1.729	2.093	2.539	2.861
20	1.325	1.725	2.086	2.528	2.845
21	1.323	1.721	2.080	2.518	2.831
22	1.321	1.717	2.074	2.508	2.819
23	1.319	1.714	2.069	2.500	2.807
24	1.318	1.711	2.064	2.492	2.797
25	1.316	1.708	2.060	2.485	2.287
26	1.315	1.706	2.056	2.479	2.779
27	1.314	1.703	2.052	2.473	2.771
28	1.313	1.701	2.048	2.467	2.763
29	1.311	1.699	2.045	2.462	2.756
30	1.310	1.697	2.042	2.457	2.750
40	1.303	1.684	2.021	2.423	2.704
60	1.296	1.671	2.000	2.390	2.660
120	1.289	1.658	1.980	2.358	2.617
∞	1.282	1.645	1.960	2.326	2.576

TABLE 5
PERCENTAGE POINTS OF X^2 DISTRIBUTION

α / d.f.	.995	.990	.975	.950	.050	.025	.010	.005
1	392704×10^{-10}	157088×10^{-9}	982069×10^{-9}	393214×10^{-8}	3.84146	5.02389	6.63490	7.87944
2	.0100251	.0201007	.0506356	.102587	5.99147	7.37776	9.21034	10.5966
3	.0717212	.114832	.215795	.351846	7.81473	9.34840	11.3449	12.8381
4	.206990	.297110	.484419	.710721	9.48773	11.1433	13.2767	14.8602
5	.411740	.554300	.831211	1.145476	11.0705	12.8325	15.0863	16.7496
6	.675727	.872085	1.237347	1.63539	12.5916	14.4494	16.8119	18.5476
7	.989265	1.239043	1.68987	2.16745	14.0671	16.0128	18.4753	20.2777
8	1.344419	1.646482	2.17973	2.73264	15.5073	17.5346	20.0902	21.9550
9	1.734926	2.087912	2.70039	3.32511	16.9190	19.0228	21.6660	23.5893
10	2.15585	2.55821	3.24697	3.94030	18.3070	20.4831	23.2093	25.1882
11	2.60321	3.05347	3.81575	4.57481	19.6751	21.9200	24.7250	26.7569
12	3.07382	3.57056	4.40379	5.22603	21.0261	23.3367	26.2170	28.2995
13	3.56503	4.10691	5.00874	5.89186	22.3621	24.7356	27.6883	29.8194
14	4.07468	4.66043	5.62872	6.57063	23.6848	26.1190	29.1413	31.3193
15	4.60094	5.22935	6.2614	7.26094	24.9958	27.4884	30.5779	32.8013
16	5.14224	5.81221	6.90766	7.96164	26.2962	28.8454	31.9999	34.2672
17	5.69724	6.40776	7.56418	8.67176	27.5871	30.1910	33.4087	35.7185
18	6.26481	7.01491	8.23075	9.39046	28.2693	31.5264	34.8053	37.1564
19	6.84398	7.63273	8.90655	10.1170	30.1435	32.8523	36.1908	38.5822

TABLE 5 (*Continued*)

20	7.43386	8.26040	9.59083	10.8508	31.4104	34.1696	37.5662	39.9968
21	8.03366	8.89720	10.28293	11.5913	32.6705	35.4789	38.9321	41.4010
22	8.64272	9.54249	10.9823	12.3380	33.9244	36.7807	40.2894	42.7956
23	9.26042	10.19567	11.6885	13.0905	35.1725	38.0757	41.6384	44.1813
24	9.88623	10.8564	12.4011	13.8484	36.4151	39.3641	42.9798	45.5585
25	10.5197	11.5240	13.1197	14.6114	37.6525	40.6465	44.3141	46.9278
26	11.1603	12.1981	13.8439	15.3791	38.8852	41.9232	45.6417	48.2899
27	11.8076	12.8786	14.5733	16.1513	40.1133	43.1944	46.9630	49.6449
28	12.4613	13.5648	15.3079	16.9279	41.3372	44.4607	48.2782	50.9933
29	13.1211	14.2565	16.0471	17.7083	42.5569	45.7222	49.5879	52.3356
30	13.7867	14.9535	16.7908	18.4926	43.7729	46.9792	50.8922	53.6720
40	20.7065	22.1643	24.4331	26.5093	55.7585	59.3417	63.6907	66.7659
50	27.9907	29.7067	32.3574	34.7642	67.5048	71.4202	76.1539	79.4900
60	35.5346	37.4848	40.4817	43.1879	79.0819	83.2976	88.3794	91.9517
70	43.2752	45.4418	48.7576	51.7393	90.5312	95.0231	100.425	104.215
80	51.1720	53.5400	57.1532	60.3915	101.879	106.629	112.329	116.321
90	59.1963	61.7541	65.6466	69.1260	113.145	118.136	124.116	128.299
100	67.3276	70.0648	74.2219	77.9295	124.342	129.561	135.807	140.169

TABLE 6

Values of F for F Distribution at 5% Points

Degrees of freedom for numerator

	1	2	3	4	5	6	7	8	9	10	12	15	20	24	30	40	60	120	∞
1	161	200	216	225	230	234	237	239	241	242	244	246	248	249	250	251	252	253	254
2	18.5	19.0	19.2	19.2	19.3	19.3	19.4	19.4	19.4	19.4	19.4	19.4	19.4	19.5	19.5	19.5	19.5	19.5	19.5
3	10.1	9.55	9.28	9.12	9.01	8.94	8.89	8.85	8.81	8.79	8.74	8.70	8.66	8.64	8.62	8.59	8.57	8.55	8.53
4	7.71	6.94	6.59	6.39	6.26	6.16	6.09	6.04	6.00	5.96	5.91	5.86	5.80	5.77	5.75	5.72	5.69	5.66	5.63
5	6.61	5.79	5.41	5.19	5.05	4.95	4.88	4.82	4.77	4.74	4.68	4.62	4.56	4.53	4.50	4.46	4.43	4.40	4.37
6	5.99	5.14	4.76	4.53	4.39	4.28	4.21	4.15	4.10	4.06	4.00	3.94	3.87	3.84	3.81	3.77	3.74	3.70	3.67
7	5.59	4.74	4.35	4.12	3.97	3.87	3.79	3.73	3.68	3.64	3.57	3.51	3.44	3.41	3.38	3.34	3.30	3.27	3.23
8	5.32	4.46	4.07	3.84	3.69	3.58	3.50	3.44	3.39	3.35	3.28	3.22	3.15	3.12	3.08	3.04	3.01	2.97	2.93
9	5.12	4.26	3.86	3.63	3.48	3.37	3.29	3.23	3.18	3.14	3.07	3.01	2.94	2.90	2.86	2.83	2.79	2.75	2.71
10	4.96	4.10	3.71	3.48	3.33	3.22	3.14	3.07	3.02	2.98	2.91	2.85	2.77	2.74	2.70	2.66	2.62	2.58	2.54
11	4.84	3.98	3.59	3.36	3.20	3.09	3.01	2.95	2.90	2.85	2.79	2.72	2.65	2.61	2.57	2.53	2.49	2.45	2.40
12	4.75	3.89	3.49	3.26	3.11	3.00	2.91	2.85	2.80	2.75	2.69	2.62	2.54	2.51	2.47	2.43	2.38	2.34	2.30
13	4.67	3.81	3.41	3.18	3.03	2.92	2.83	2.77	2.71	2.67	2.60	2.53	2.46	2.42	2.38	2.34	2.30	2.25	2.21
14	4.60	3.74	3.34	3.11	2.96	2.85	2.76	2.70	2.65	2.60	2.53	2.46	2.39	2.35	2.31	2.27	2.22	2.18	2.13
15	4.54	3.68	3.29	3.06	2.90	2.79	2.71	2.64	2.59	2.54	2.48	2.40	2.33	2.29	2.25	2.20	2.16	2.11	2.07
16	4.49	3.63	3.24	3.01	2.85	2.74	2.66	2.59	2.54	2.49	2.42	2.35	2.28	2.24	2.19	2.15	2.11	2.06	2.01
17	4.45	3.59	3.20	2.96	2.81	2.70	2.61	2.55	2.49	2.45	2.38	2.31	2.23	2.19	2.15	2.10	2.06	2.01	1.96
18	4.41	3.55	3.16	2.93	2.77	2.66	2.58	2.51	2.46	2.41	2.34	2.27	2.19	2.15	2.11	2.06	2.02	1.97	1.92
19	4.38	3.52	3.13	2.90	2.74	2.63	2.54	2.48	2.42	2.38	2.31	2.23	2.16	2.11	2.07	2.03	1.98	1.93	1.88
20	4.35	3.49	3.10	2.87	2.71	2.60	2.51	2.45	2.39	2.35	2.28	2.20	2.12	2.08	2.04	1.99	1.95	1.90	1.84
21	4.32	3.47	3.07	2.84	2.68	2.57	2.49	2.42	2.37	2.32	2.25	2.18	2.10	2.05	2.01	1.96	1.92	1.87	1.81
22	4.30	3.44	3.05	2.82	2.66	2.55	2.46	2.40	2.34	2.30	2.23	2.15	2.07	2.03	1.98	1.94	1.89	1.84	1.78
23	4.28	3.42	3.03	2.80	2.64	2.53	2.44	2.37	2.32	2.27	2.20	2.13	2.05	2.01	1.96	1.91	1.86	1.81	1.76
24	4.26	3.40	3.01	2.78	2.62	2.51	2.42	2.36	2.30	2.25	2.18	2.11	2.03	1.98	1.94	1.89	1.84	1.79	1.73
25	4.24	3.39	2.99	2.76	2.60	2.49	2.40	2.34	2.28	2.24	2.16	2.09	2.01	1.96	1.92	1.87	1.82	1.77	1.71
30	4.17	3.32	2.92	2.69	2.53	2.42	2.33	2.27	2.21	2.16	2.09	2.01	1.93	1.89	1.84	1.79	1.74	1.68	1.62
40	4.08	3.23	2.84	2.61	2.45	2.34	2.25	2.18	2.12	2.08	2.00	1.92	1.84	1.79	1.74	1.69	1.64	1.58	1.51
60	4.00	3.15	2.76	2.53	2.37	2.25	2.17	2.10	2.04	1.99	1.92	1.84	1.75	170	1.65	1.59	1.53	1.47	1.39
120	3.92	3.07	2.68	2.45	2.29	2.18	2.09	2.02	1.96	1.91	1.83	1.75	1.66	1.61	1.55	1.50	1.43	1.35	1.25
∞	3.84	3.00	2.60	2.37	2.21	2.10	2.01	1.94	1.88	1.83	1.75	1.67	1.57	1.52	1.46	1.39	1.32	1.22	1.00

TABLE 7

Values oı F for *F* Distribution at 1% Points

Degrees of freedom for numerator

	1	2	3	4	5	6	7	8	9	10	12	15	20	24	30	40	60	120	∞
1	4,052	5,000	5,403	5,625	5,764	5,859	5,928	5,982	6,023	6,056	6,106	6,157	6,209	6,235	6,261	6,287	6,313	6,339	6,366
2	98.5	99.0	99.2	99.2	99.3	99.3	99.4	99.4	99.4	99.4	99.4	99.4	99.4	99.5	99.5	99.5	99.5	99.5	99.5
3	34.1	30.8	29.5	28.7	28.2	27.9	27.7	27.5	27.3	27.2	27.1	26.9	26.7	26.6	26.5	26.4	26.3	26.2	26.1
4	21.2	18.0	16.7	16.0	15.5	15.2	15.0	14.8	14.7	14.5	14.4	14.2	14.0	13.9	13.8	13.7	13.7	13.6	13.5
5	16.3	13.3	12.1	11.4	11.0	10.7	10.5	10.3	10.2	10.1	9.89	9.72	9.55	9.47	9.38	9.29	9.20	9.11	9.02
6	13.7	10.9	9.78	9.15	8.75	8.47	8.26	8.10	7.98	7.87	7.72	7.56	7.40	7.31	7.23	7.14	7.06	6.97	6.88
7	12.2	9.55	8.45	7.85	7.46	7.19	6.99	6.84	6.72	6.62	6.47	6.31	6.16	6.07	5.99	5.91	5.82	5.74	5.65
8	11.3	8.65	7.59	7.01	6.63	6.37	6.18	6.03	5.91	5.81	5.67	5.52	5.36	5.28	5.20	5.12	5.03	4.95	4.86
9	10.6	8.02	6.99	6.42	6.06	5.80	5.61	5.47	5.35	5.26	5.11	4.96	4.81	4.73	4.65	4.57	4.48	4.40	4.31
10	10.0	7.56	6.55	5.99	5.64	5.39	5.20	5.06	4.94	4.85	4.71	4.56	4.41	4.33	4.25	4.17	4.08	4.00	3.91
11	9.65	7.21	6.22	5.67	5.32	5.07	4.89	4.74	4.63	4.54	4.40	4.25	4.10	4.02	3.94	3.86	3.78	3.69	3.60
12	9.33	6.93	5.95	5.41	5.06	4.82	4.64	4.50	4.39	4.30	4.16	4.01	3.86	3.78	3.70	3.62	3.54	3.45	3.36
13	9.07	6.70	5.74	5.21	4.86	4.62	4.44	4.30	4.19	4.10	3.96	3.82	3.66	3.59	3.51	3.43	3.34	3.25	3.17
14	8.68	6.51	5.56	5.04	4.70	4.46	4.28	4.14	4.03	3.94	3.80	3.66	3.51	3.43	3.35	3.27	3.18	3.09	3.00
15	8.68	6.36	5.42	4.89	4.56	4.32	4.14	4.00	3.89	3.80	3.67	3.52	3.37	3.29	3.21	3.13	3.05	2.96	2.87
16	8.53	6.23	5.29	4.77	4.44	4.20	4.03	3.89	3.78	3.69	3.55	3.41	3.26	3.18	3.10	3.02	2.93	2.84	2.75
17	8.40	6.11	5.19	4.67	4.34	4.10	3.93	3.79	3.68	3.59	3.46	3.31	3.16	3.08	3.00	2.92	2.83	2.75	2.65
18	8.29	6.01	5.09	4.58	4.25	4.01	3.84	3.71	3.60	3.51	3.37	3.23	3.08	3.00	2.92	2.84	2.75	2.66	2.57
19	8.19	5.93	5.01	4.50	4.17	3.94	3.77	3.63	3.52	3.43	3.30	3.15	3.00	2.92	2.84	2.76	2.67	2.58	2.49
20	8.10	5.85	4.94	4.43	4.10	3.87	3.70	3.56	3.46	3.37	3.23	3.09	2.94	2.86	2.78	2.69	2.61	2.52	2.42
21	8.02	5.78	4.87	4.37	4.04	3.81	3.64	3.51	3.40	3.31	3.17	3.03	2.88	2.80	2.72	2.64	2.55	2.46	2.36
22	7.95	5.72	4.82	4.31	3.99	3.76	3.59	3.45	3.35	3.26	3.12	2.98	2.83	2.75	2.67	2.58	2.50	2.40	2.31
23	7.88	5.66	4.76	4.26	3.94	3.71	3.54	3.41	3.30	3.21	3.07	2.93	2.78	2.70	2.62	2.54	2.45	2.35	2.26
24	7.82	5.61	4.72	4.22	3.90	3.67	3.50	3.36	3.26	3.17	3.03	2.89	2.74	2.66	2.58	2.49	2.40	2.31	2.21
25	7.77	5.57	4.68	4.18	3.86	3.63	3.46	3.32	3.22	3.13	2.99	2.85	2.70	2.62	2.53	2.45	2.36	2.27	2.17
30	7.56	5.39	4.51	4.02	3.70	3.47	3.30	3.17	3.07	2.98	2.84	2.70	2.55	2.47	2.39	2.30	2.21	2.11	2.01
40	7.31	5.18	4.31	3.83	3.51	3.29	3.12	2.99	2.89	2.80	2.66	2.52	2.37	2.29	2.20	2.11	2.02	1.92	1.80
60	7.08	4.98	4.13	3.65	3.34	3.12	2.95	2.82	2.72	2.63	2.50	2.35	2.20	2.12	2.03	1.94	1.84	1.73	1.60
120	6.85	4.79	3.95	3.48	3.17	2.96	2.79	2.66	2.56	2.47	2.34	2.19	2.03	1.95	1.86	1.76	1.66	1.53	1.30
∞	6.63	4.61	3.78	3.32	3.02	2.80	2.64	2.51	2.41	2.32	2.04	2.04	1.88	1.79	1.70	1.59	1.47	1.32	1.00

TABLE 8
ORDINATES (*Y*) of the STANDARD NORMAL CURVE AT *Z*

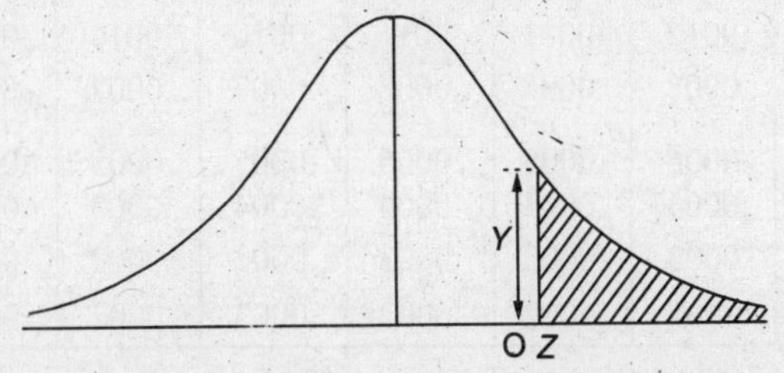

0	0.00	0.01	0.02	0.03	0.04	0.05	0.06	0.07	0.08	0.09
0.0	.3989	.3989	.3989	.3938	.3986	.3984	.3982	.3980	.3977	.3973
0.1	.3970	.3965	.3961	.3956	.3951	.3945	.3939	.3932	.3925	.3918
0.2	.3910	.3902	.3894	.3885	.3876	.3867	.3857	.3847	.3836	.3825
0.3	.3814	.3802	.3700	.3778	.3765	.3752	.3739	.3725	.3712	.3697
0.4	.3683	.3668	.3653	.3637	.3621	.3605	.3589	.3572	.3555	.3538
0.5	.3521	.3503	.3482	.3407	.3448	.3429	.3410	.3391	.3372	.3352
0.6	.3332	.3312	.3292	.3271	.3251	.3230	.3209	.3187	.3166	.3144
0.7	.3123	.3101	.3079	.3956	.3034	.3011	.2989	.2966	.2943	.2920
0.8	.2897	.2874	.2850	.2827	.2803	.2780	.2756	.2732	.2709	.2685
0.9	.2661	.2637	.2613	.2589	.2565	.2541	.2516	.2492	.2468	.2444
1.0	.2420	.2396	.2371	.2347	.2323	.2299	.2275	.2251	.2227	.2203
1.1	.2179	.2155	.2131	.2107	.2083	.2059	.2036	.2012	.1989	.1965
1.2	.1942	.1919	.1895	.1872	.1849	.1826	.1804	.1781	.1758	.1736
1.3	.1714	.1691	.1669	.1647	.1626	.1604	.1582	.1561	.1539	.1518
1.4	.1497	.1476	.1456	.1435	.1415	.1394	.1374	.1354	.1334	.1315
1.5	.1295	.1276	.1257	.1238	.1219	.1200	.1182	.1163	.1145	.1127
1.6	.1109	.1092	.1074	.1057	.1040	.1023	.1006	.0989	.0973	.957
1.7	.0940	.0925	.0909	.0893	.0878	.0863	.0848	.0833	.0818	.0804
1.8	.0790	.0775	.0761	.0748	.0734	.0721	.0707	.0694	.0681	.0669
1.9	.0656	.0644	.0632	.0620	.0608	.0596	.0584	.0573	.0562	.0551
2.0	.0540	.0529	.0519	.0508	.0498	.0488	.0478	.0468	.0459	.0449
2.1	.0440	.0431	.0422	.0413	.0404	.0396	.0387	.0379	.0371	.0363
2.2	.0355	.0347	.0339	.0332	.0325	.0317	.0310	.0303	.0297	.0290
2.3	.0283	.0277	.0270	.0264	.0258	.0252	.0246	.0241	.0235	.0229
2.4	.0224	.0219	.0213	.0208	.0203	.0198	.0194	.0189	.0184	.0180
2.5	.0175	.0171	.0167	.0163	.0158	.0154	.0151	.0147	.0143	.0139
2.6	.0136	.0132	.0129	.0116	.0122	.0119	.0116	.0113	.0110	.0107
2.7	.0104	.0101	.0099	.0096	.0093	.0091	.0088	.0086	.0084	.0081
2.8	.0079	.0077	.0075	.0073	.0071	.0069	.0067	.0065	.0063	.0061
2.9	.0060	.0058	.0056	.0055	.0053	.0051	.0050	.0048	.0047	.0046
3.0	.0044	.0043	.0042	.0040	.0039	.0038	.0037	.0036	.0035	.0034

3.1	.0033	.0032	.0031	.0030	.0029	.0028	.0027	.0026	.0025	.0025
3.2	.0024	.0023	.0022	.0022	.0021	.0020	.0020	.0019	.0018	.0018
3.3	.0017	.0017	.0016	.0016	.0015	.0015	.0014	.0014	.0013	.0013
3.4	.0012	.0012	.0012	.0011	.0011	.0010	.0010	.0010	.0009	.0009
3.5	.0009	.0008	.0008	.0008	.0008	.0007	.0007	.0007	.0007	.0006
3.6	.0006	.0006	.0006	.0005	.0005	.0005	.0005	.0005	.0005	.0004
3.7	.0004	.0004	.0004	.0004	.0004	.0004	.0003	.0003	.0003	.0003
3.8	.0003	.0003	.0003	.0003	.0003	.0002	.0002	.0002	.0002	.0002
3.9	.0002	.0002	.0002	.0002	.0002	.0002	.0002	.0002	.0001	.0001

TABLE 9

(VALUE OF e^{-m}); (0 < m < 1)

m	0	1	2	3	4	5	6	7	8	9
0.0	1.0000	.9900	.9802	.9704	.9608	.9512	.9418	.9324	.9231	.9139
0.1	0.9048	.8958	.8860	.8781	.8694	.8607	.8521	.8437	.8353	.8270
0.2	0.8178	.8106	.8025	.7945	.7866	.7788	.7711	.7634	.7558	.7483
0.3	0.7408	.7334	.7261	.7189	.7118	.7047	.6977	.6907	.6839	.6771
0.4	0.6703	.6636	.6570	.6505	.6440	.6376	.6313	.6250	.6188	.6126
0.5	0.6065	.6005	.5945	.5886	.5827	.5770	.5712	.5655	.5599	.5543
0.6	0.5488	.5434	.5379	.5326	.5278	.5220	.5160	.5117	.5066	.5016
0.7	0.4966	.4916	.4868	.4810	.4771	.4724	.4670	.4630	.4584	.4538
0.8	0.4493	.4449	.4404	.4360	.4317	.4274	.4232	.4190	.4148	.4107
0.9	0.4066	.4025	.3985	.3946	.3906	.3867	.3829	.3791	.3753	.3716

VALUE OF e^{-m} ; (m = 1, 2, 3, ..., 10)

m	1	2	3	4	5	6	7	8	9	10
e^{-m}	.36788	.13534	.04979	.01832	.006738	.002479	.000912	.000335	.000123	.000045

Note : To obtain value of e^{-m} for other values of m, use the laws of exponents.

Example : $e^{-2.35} = (e^{-2.00})\ (e^{-0.35}) = (0.13534)\ (0.7047) = 0.095374$

TABLE 10
CHI-SQUARE DISTRIBUTION

The following table provides the values of χ_α^2 that correspond to a given upper-tail area χ and a specified number of degrees of freedom.

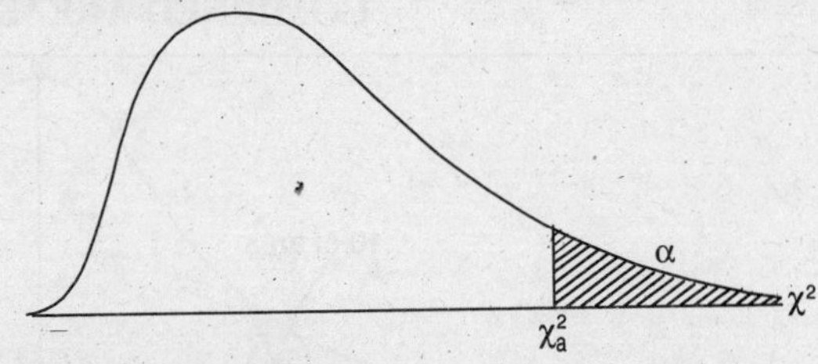

Degree of Freedom	Upper-Tail Area .20	.10	.05	.02	.01	.001
1	1.642	2.706	3.841	5.412	6.635	10.827
2	3.219	4.605	5.991	7.824	9.210	13.815
3	4.642	6.251	7.815	9.837	11.345	16.268
4	4.989	7.779	9.488	11.668	13.277	18.465
5	7.289	9.236	11.070	13.388	15.086	20.517
6	8.558	10.645	12.592	15.033	16.812	22.457
7	9.083	12.017	14.067	16.622	18.475	24.322
8	11.030	13.362	15.507	18.168	20.090	26.125
9	12.242	14.648	16.919	19.679	21.666	27.877
10	13.442	15.987	18.307	21.161	23.209	29.588
11	14.631	17.275	19.675	22.618	24.725	31.264
12	15.812	18.549	21.026	24.054	26.217	32.909
13	16.985	19.812	22.362	25.472	27.688	34.528
14	18.151	21.064	23.685	26.873	29.141	36.123
15	19.311	22.307	24.996	28.259	30.578	37.697
16	20.465	23.542	26.296	29.633	32.000	39.252
17	21.615	24.769	27.587	30.995	33.409	40.790
18	22.760	25.989	28.869	32.346	34.805	42.312
19	23.900	27.204	30.144	33.687	36.191	43.820
20	25.038	28.412	31.410	35.020	37.566	45.315
21	26.171	29.615	32.671	36.343	38.932	46.797
22	27.301	30.813	33.924	37.659	40.289	48.268
23	28.429	32.007	35.172	38.968	41.638	49.728
24	29.553	33.196	36.415	40.270	42.980	51.179
25	30.675	34.382	37.652	41.566	44.314	52.620
26	31.795	35.563	38.885	42.856	45.642	54.052
27	32.912	36.741	40.113	44.140	46.963	55.476
28	34.027	37.916	41.337	45.419	48.278	56.893
29	35.139	39.087	42.557	46.693	49.588	58.302
30	36.250	40.256	43.773	47.962	50.892	59.703

TABLE 11
VALUES FOR SPEARMAN'S RANK CORRELATION (r_S) FOR COMBINED AREAS IN BOTH TAILS

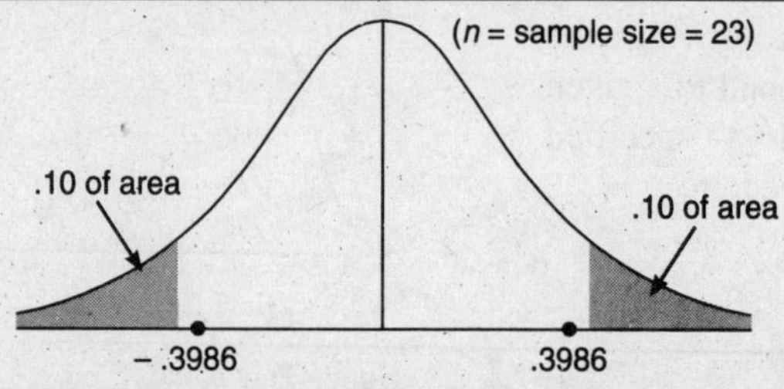

EXAMPLE: For a two-tailed test of significance at the 0.05 level, with n = 12, the appropriate value for r_s can be found by looking under the .20 column and proceeding down to the 12 row; there we find the appropriate r_s value to be 0.5804.

n	.20	.10	.05	.02	.01	.002
4	.8000	.8000				
5	.7000	.8000	.9000	.9000		
6	.6000	.7714	.8286	.8857	.9429	
7	.5357	.6786	.7450	.8571	.8929	.9643
8	.5000	.6190	.7143	.8095	.8571	.9286
9	.4667	.5833	.6833	.7667	.8167	.9000
10	.4424	.5515	.6364	.7333	.7818	.8667
11	.4182	.5273	.6091	.7000	.7455	.8364
12	.3986	.4965	.5804	.6713	.7273	.8182
13	.3791	.4780	.5549	.6429	.6978	.7912
14	.3626	.4593	.5341	.6220	.6747	.7670
15	.3500	.4429	.5179	.6000	.6536	.7464
16	.3382	.4265	.5000	.5824	.6324	.7265
17	.3260	.4118	.4853	.5637	.6152	.7083
18	.3148	.3994	.4716	.5480	.5975	.6904
19	.3070	.3895	.4579	.5333	.5825	.6737
20	.2977	.3789	.4451	.5203	.5684	.6586
21	.2909	.3688	.4351	.5078	.5545	.6455
22	.2829	.3597	.4241	.4963	.5426	.6318
23	.2767	.3518	.4150	.4852	.5306	.6186
24	.2704	.3435	.4061	.4748	.5200	.6070
25	.2646	.3362	.3977	.4654	.5100	.5962
26	.2588	.3299	.3894	.4564	.5002	.5856
27	.2540	.3236	.3822	.4481	.4915	.5757
28	.2490	.3175	.3749	.4401	.4828	.5660
29	.2443	.3113	.3685	.4320	.4744	.5567
30	.2400	.3059	.3620	.4251	.4665	.5479

TABLE 12
CRITICAL VALUES OF D_n IN THE KOLMOGOROV-SMIRNOV GOODNESS-OF-FIT TEST

Sample size	Level of significance for D_n = Maximum $\lvert F_e - F_o \rvert$				
(n)	.20	.15	.10	.05	.01
1	.900	.925	.950	.975	.995
2	.684	.726	.776	.842	.929
3	.565	.597	.642	.708	.828
4	.494	.525	.564	.624	.733
5	.446	.474	.510	.565	.669
6	.410	.436	.470	.521	.618
7	.381	.405	.438	.486	.577
8	.358	.381	.411	.457	.543
9	.339	.360	.388	.432	.514
10	.322	.342	.368	.410	.490
11	.307	.326	.352	.391	.468
12	.295	.313	.338	.375	.450
13	.284	.302	.325	.361	.433
14	.274	.292	.314	.349	.418
15	.266	.283	.304	.338	.404
16	.258	.274	.295	.328	.392
17	.250	.266	.286	.318	.381
18	.244	.259	.278	.309	.371
19	.237	.252	.272	.301	.363
20	.231	.246	.264	.294	.356
25	.21	.22	.24	.27	.32
30	.19	.20	.22	.24	.29
35	.18	.19	.21	.23	.27
Over 35	$\frac{1.07}{\sqrt{n}}$	$\frac{1.14}{\sqrt{n}}$	$\frac{1.22}{\sqrt{n}}$	$\frac{1.36}{\sqrt{n}}$	$\frac{1.63}{\sqrt{n}}$

TABLE 13
BINOMIAL PROBABILITIES

n	x	.05	.10	.15	.20	.25	.30	.35	.40	.45	.50
1	0	.9500	.9000	.8500	.8000	.7500	.7000	.6500	.6000	.5500	.5000
	1	.0500	.1000	.1500	.2000	.2500	.3000	.3500	.4000	.4500	.5000
2	0	.9025	.8100	.7225	.6400	.5625	.4900	.4225	.3600	.3025	.2500
	1	.0950	.1800	.2550	.3200	.3750	.4200	.4550	.4800	.4950	.5000
	2	.0025	.0100	.0225	.0400	.0625	.0900	.1225	.1600	.2025	.2500
3	0	.8574	.7290	.6141	.5120	.4219	.3430	.2746	.2160	.1664	.1250
	1	.1354	.2430	.3251	.3840	.4219	.4410	.4436	.4320	.4084	.3750
	2	.0071	.0270	.0574	.0960	.1406	.1890	.2389	.2880	.3341	.3750
	3	.0001	.0010	.0034	.0080	.0156	.0270	.0429	.0640	.0911	.1250
4	0	.8145	.6561	.5220	.4096	.3164	.2401	.1785	.1296	.0915	.0625
	1	.1715	.2916	.3685	.4096	.4219	.4116	.3845	.3456	.2995	.2500
	2	.0135	.0486	.0975	.1536	.2109	.2646	.3105	.3456	.3675	.3750
	3	.0005	.0036	.0115	.0256	.0469	.0756	.1115	.1536	.2005	.2500
	4	.0000	.0001	.0005	.0016	.0039	.0081	.0150	.0256	.0410	.0625
5	0	.7738	.5905	.4437	.3277	.2373	.1681	.1160	.0778	.0503	.0312
	1	.2036	.3280	.3915	.4096	.3955	.3602	.3124	.2592	.2059	.1562
	2	.0214	.0729	.1382	.2048	.2637	.3087	.3364	.3456	.3369	.3125
	3	.0011	.0081	.0244	.0512	.0879	.1323	.1811	.2304	.2757	.3125
	4	.0000	.0004	.0022	.0064	.0146	.0284	.0488	.0768	.1128	.1562
	5	.0000	.0000	.0001	.0003	.0010	.0024	.0053	.0102	.0185	.0312
6	0	.7351	.5314	.3771	.2621	.1780	.1176	.0754	.0467	.0277	.0156
	1	.2321	.3543	.3993	.3932	.3560	.3025	.2437	.1866	.1359	.0938
	2	.0305	.0984	.1762	.2458	.2966	.3241	.3280	.3110	.2780	.2344
	3	.0021	.0146	.0415	.0819	.1318	.1852	.2355	.2765	.3032	.3125
	4	.0001	.0012	.0055	.0154	.0330	.0595	.0951	.1382	.1861	.2344
	5	.0000	.0001	.0004	.0015	.0044	.0102	.0205	.0369	.0609	.0938
	6	.0000	.0000	.0000	.0001	.0002	.0007	.0018	.0041	.0083	.0156
7	0	.6983	.4783	.3206	.2097	.1335	.0824	.0490	.0280	.0152	.0078
	1	.2573	.3720	.3960	.3670	.3115	.2471	.1848	.1306	.0872	.0547
	2	.0406	.1240	.2097	.2753	.3115	.3177	.2985	.2613	.2140	.1641
	3	.0036	.0230	.0617	.1147	.1730	.2269	.2679	.2903	.2918	.2734
	4	.0002	.0026	.0109	.0287	.0577	.0972	.1442	.1935	.2388	.2734
	5	.0000	.0002	.0012	.0043	.0115	.0250	.0466	.0774	.1172	.1641
	6	.0000	.0000	.0001	.0004	.0013	.0036	.0084	.0172	.0320	.0547
	7	.0000	.0000	.0000	.0000	.0001	.0002	.0006	.0016	.0037	.0078
8	0	.6634	.4305	.2725	.1678	.1001	.0576	.0319	.0168	.0084	.0039
	1	.2793	.3826	.3847	.3355	.2670	.1977	.1373	.0898	.0548	.0312
	2	.0515	.1488	.2376	.2936	.3115	.2965	.2587	.2090	.1569	.1094
	3	.0054	.0331	.0839	.1468	.2076	.2541	.2786	.2787	.2568	.2188
	4	.0004	.0046	.0185	.0459	.0865	.1361	.1875	.2322	.2627	.2734

n	x	.05	.10	.15	.20	.25	.30	.35	.40	.45	.50
8	5	.0000	.0004	.0026	.0092	.0231	.0467	.0808	.1239	.1719	.2188
	6	.0000	.0000	.0002	.0011	.0038	.0100	.0217	.0413	.0703	.1094
	7	.0000	.0000	.0000	.0001	.0004	.0012	.0033	.0079	.0164	.0312
	8	.0000	.0000	.0000	.0000	.0000	.0001	.0002	.0007	.0017	.0039
9	0	.6302	.3874	.2316	.1342	.0751	.0404	.0207	.0101	.0046	.0020
	1	.2985	.3874	.3679	.3020	.2253	.1556	.1004	.0605	.0339	.0176
	2	.0629	.1722	.2597	.3020	.3003	.2668	.2162	.1612	.1110	.0703
	3	.0077	.0446	.1069	.1762	.2336	.2668	.2716	.2508	.2119	.1641
	4	.0006	.0074	.0283	.0661	.1168	.1715	.2194	.2508	.2600	.2461
	5	.0000	.0008	.0050	.0165	.0389	.0735	.1181	.1672	.2128	.2461
	6	.0000	.0001	.0006	.0028	.0087	.0210	.0424	.0743	.1160	.1641
	7	.0000	.0000	.0000	.0003	.0012	.0039	.0098	.0212	.0407	.0703
	8	.0000	.0000	.0000	.0000	.0001	.0004	.0013	.0035	.0083	.0176
	9	.0000	.0000	.0000	.0000	.0000	.0000	.0001	.0003	.0008	.0020
10	0	.5987	.3487	.1969	.1074	.0563	.0282	.0135	.0060	.0025	.0010
	1	.3151	.3874	.3474	.2684	.1877	.1211	.0725	.0403	.0207	.0098
	2	.0746	.1937	.2759	.3020	.2816	.2335	.1757	.1209	.0763	.0439
	3	.0105	.0574	.1298	.2013	.2503	.2668	.2522	.2150	.1665	.1172
	4	.0010	.0112	.0401	.0881	.1460	.2001	.2377	.2508	.2384	.2051
	5	.0001	.0015	.0085	.0264	.0584	.1029	.1536	.2007	.2340	.2461
	6	.0000	.0001	.0012	.0055	.0162	.0368	.0689	.1115	.1596	.2051
	7	.0000	.0000	.0001	.0008	.0031	.0090	.0212	.0425	.0746	.1172
	8	.0000	.0000	.0000	.0001	.0004	.0014	.0043	.0106	.0229	.0439
	9	.0000	.0000	.0000	.0000	.0000	.00001	.0005	.0016	.0003	.0098
	10	.0000	.0000	.0000	.0000	.0000	.0000	.0000	.0001	.0003	.0018
11	0	.5688	.3138	.1673	.0859	.0422	.0198	.0088	.0036	.0014	.0005
	1	.3293	.3835	.3248	.2362	.1549	.0932	.0518	.0266	.0125	.0054
	2	.0867	.2131	.2866	.2953	.2581	.1998	.1395	.0887	.0513	.0269
	3	.0137	.0710	.1517	.2215	.2581	.2568	.2254	.1774	.1259	.0806
	4	.0014	.0158	.0536	.1107	.1721	.2201	.2428	.2365	.2060	.1611
	5	.0001	.0025	.0132	.0388	.0803	.1321	.1830	.2207	.2360	.2256
	6	.0000	.0003	.0023	.0097	.0268	.0566	.0985	.1471	.1931	.2256
	7	.0000	.0000	.0003	.0017	.0064	.0173	.0379	.0701	.1128	.1611
	8	.0000	.0000	.0000	.0002	.0011	.0037	.0102	.0234	.0462	.0806
	9	.0000	.0000	.0000	.0000	.0001	.0005	.0018	.0052	.0126	.0269
	10	.0000	.0000	.0000	.0000	.0000	.0000	.0002	.0007	.0021	.0054
	11	.0000	.0000	.0000	.0000	.0000	.0000	.0000	.0000	.0002	.0005
12	0	.5404	.2824	.1422	.0687	.0317	.0138	.0057	.0022	.0008	.0002
	1	.3413	.3766	.3012	.2062	.1267	.0712	.0368	.0174	.0075	.0029
	2	.0988	.2301	.2924	.2835	.2323	.1678	.1088	.0639	.0339	.0161
	3	.0173	.0852	.1720	.2362	.2581	.2397	.1954	.1419	.0923	.0537
	4	.0021	.0213	.0683	.1329	.1936	.2311	.2367	.2128	.1700	.1208

(Contd.)

n	x	.05	.10	.15	.20	.25	.30	.35	.40	.45	.50
12	5	.0002	.0038	.0193	.0532	.1032	.1585	.2039	.2270	.2225	.1934
	6	.0000	.0005	.0040	.0155	.0401	.0792	.1281	.1766	.2124	.2256
	7	.0000	.0000	.0006	.0033	.0115	.0291	.0591	.1009	.1489	.1934
	8	.0000	.0000	.0001	.0005	.0024	.0078	.0199	.0420	.0762	.1208
	9	.0000	.0000	.0000	.0001	.0004	.0015	.0048	.0125	.0277	.0537
	10	.0000	.0000	.0000	.0000	.0000	.0002	.0008	.0025	.0068	.0161
	11	.0000	.0000	.0000	.0000	.0000	.0000	.0001	.0003	.0010	.0029
	12	.0000	.0000	.0000	.0000	.0000	.0000	.0000	.0000	.0001	.0002
13	0	.5133	.2542	.1209	.0550	.0238	.0097	.0037	.0013	.0004	.0001
	1	.3512	.3672	.2774	.1787	.1029	.0540	.0259	.0113	.0045	.0016
	2	.1109	.2448	.2937	.2680	.2059	.1388	.0836	.0453	.0220	.0095
	3	.0214	.0997	.1900	.2457	.2517	.2181	.1651	.1107	.0660	.0349
	4	.0028	.0277	.0838	.1535	.2097	.2337	.2222	.1845	.1350	.0873
	5	.0003	.0055	.0266	.0691	.1258	.1803	.2154	.2214	.1989	.1571
	6	.0000	.0008	.0063	.0230	.0559	.1030	.1546	.1968	.2169	.2095
	7	.0000	.0001	.0011	.0058	.0186	.0442	.0833	.1312	.1775	.2095
	8	.0000	.0000	.0001	.0011	.0047	.0142	.0336	.0656	.1089	.1571
	9	.0000	.0000	.0000	.0001	.0009	.0034	.0101	.0243	.0495	.0873
	10	.0000	.0000	.0000	.0000	.0001	.0006	.0022	.0065	.0162	.0349
	11	.0000	.0000	.0000	.0000	.0000	.0001	.0003	.0012	.0036	.0095
	12	.0000	.0000	.0000	.0000	.0000	.0000	.0000	.0001	.0005	.0016
	13	.0000	.0000	.0000	.0000	.0000	.0000	.0000	.0000	.0000	.0001
14	0	.4877	.2288	.1028	.0440	.0178	.0068	.0024	.0008	.0002	.0001
	1	.3593	.3559	.2539	.1539	.0832	.0407	.0181	.0073	.0027	.0009
	2	.1229	.2570	.2912	.2501	.1802	.1134	.0634	.0317	.0141	.0056
	3	.0259	.1142	.2056	.2501	.2402	.1943	.1366	.0845	.0462	.0222
	4	.0037	.0349	.0998	.1720	.2202	.2290	.2022	.1549	.1040	.0611
	5	.0004	.0078	.0352	.0860	.1468	.1963	.2178	.2066	.1701	.1222
	6	.0000	.0013	.0093	.0322	.0734	.1262	.1759	.2066	.2088	.1833
	7	.0000	.0002	.0019	.0092	.0280	.0618	.1082	.1574	.1952	.2095
	8	.0000	.0000	.0003	.0020	.0082	.0232	.0510	.0918	.1398	.1833
	9	.0000	.0000	.0000	.0003	.0018	.0066	.0183	.0408	.0762	.1222
	10	.0000	.0000	.0000	.0000	.0003	.0014	.0049	.0136	.0312	.0611
	11	.0000	.0000	.0000	.0000	.0000	.0002	.0010	.0033	.0093	.0222
	12	.0000	.0000	.0000	.0000	.0000	.0000	.0001	.0005	.0019	.0056
	13	.0000	.0000	.0000	.0000	.0000	.0000	.0000	.0001	.0002	.0009
	14	.0000	.0000	.0000	.0000	.0000	.0000	.0000	.0000	.0000	.0001
15	0	.4633	.2059	.0874	.0352	.0134	.0047	.0016	.0005	.0001	.0000
	1	.3658	.3432	.2312	.1319	.0668	.0305	.0126	.0047	.0016	.0005
	2	.1348	.2669	.2856	.2309	.1559	.0916	.0476	.0219	.0090	.0032
	3	.0307	.1285	.2184	.2501	.2252	.1700	.1110	.0634	.0318	.0139
	4	.0049	.0428	.1156	.1876	.2252	.2186	.1792	.1268	.0780	.0417
	5	.0006	.0105	.0449	.1032	.1651	.2061	.2123	.1859	.1404	.0916
	6	.0000	.0019	.0132	.0430	.0917	.1472	.1906	.2066	.1914	.1527
	7	.0000	.0003	.0030	.0138	.0393	.0811	.1319	.1771	.2013	.1964

(*Contd.*)

n	x	.05	.10	.15	.20	.25	.30	.35	.40	.45	.50
15	8	.0000	.0000	.0005	.0035	.0131	.0348	.0710	.1181	.1647	.1964
	9	.0000	.0000	.0001	.0007	.0034	.0116	.0298	.0612	.1048	.1527
	10	.0000	.0000	.0000	.0001	.0007	.0030	.0096	.0245	.0515	.0916
	11	.0000	.0000	.0000	.0000	.0001	.0006	.0024	.0074	.0191	.0417
	12	.0000	.0000	.0000	.0000	.0000	.0001	.0004	.0016	.0052	.0139
	13	.0000	.0000	.0000	.0000	.0000	.0000	.0001	.0003	.0010	.0032
	14	.0000	.0000	.0000	.0000	.0000	.0000	.0000	.0000	.0001	.0005
	15	.0000	.0000	.0000	.0000	.0000	.0000	.0000	.0000	.0000	.0000
16	0	.4401	.1853	.0743	.0281	.0100	.0033	.0010	.0003	.0001	.0000
	1	.3706	.3294	.2097	.1126	.0535	.0228	.0087	.0030	.0009	.0002
	2	.1463	.2745	.2775	.2111	.1336	.0732	.0353	.0150	.0056	.0018
	3	.0359	.1423	.2285	.2463	.2079	.1465	.0888	.0468	.0215	.0085
	4	.0061	.0514	.1311	.2001	.2252	.2040	.1553	.1014	.0572	.0278
	5	.0008	.0137	.0555	.1201	.1802	.2099	.2008	.1623	.1123	.0667
	6	.0001	.0028	.0180	.0550	.1101	.1649	.1982	.1983	.1684	.1222
	7	.0000	.0004	.0045	.0197	.0524	.1010	.1524	.1889	.1969	.1746
	8	.0000	.0001	.0009	.0055	.0197	.0487	.0923	.1417	.1812	.1964
	9	.0000	.0000	.0001	.0012	.0058	.0185	.0442	.0840	.1318	.1746
	10	.0000	.0000	.0000	.0002	.0014	.0056	.0167	.0392	.0755	.1222
	11	.0000	.0000	.0000	.0000	.0002	.0013	.0049	.0142	.0337	.0667
	12	.0000	.0000	.0000	.0000	.0000	.0002	.0011	.0040	.0115	.0278
	13	.0000	.0000	.0000	.0000	.0000	.0000	.0002	.0008	.0029	.0085
	14	.0000	.0000	.0000	.0000	.0000	.0000	.0000	.0001	.0005	.0018
	15	.0000	.0000	.0000	.0000	.0000	.0000	.0000	.0000	.0001	.0002
	16	.0000	.0000	.0000	.0000	.0000	.0000	.0000	.0000	.0000	.0000
17	0	.4181	.1668	.0631	.0225	.0075	.0023	.0007	.0002	.0000	.0000
	1	.3741	.3150	.1893	.0957	.0426	.0169	.0060	.0019	.0005	.0001
	2	.1575	.2800	.2673	.1914	.1136	.0581	.0260	.0102	.0035	.0010
	3	.0415	.1556	.2359	.2393	.1893	.1245	.0701	.0341	.0144	.0052
	4	.0076	.0605	.1457	.2093	.2209	.1868	.1320	.0796	.0411	.0182
	5	.0010	.0175	.0668	.1361	.1914	.2081	.1849	.1379	.0875	.0472
	6	.0001	.0039	.0236	.0680	.1276	.1784	.1991	.1839	.1432	.0944
	7	.0000	.0007	.0065	.0267	.0668	.1201	.1685	.1927	.1841	.1484
	8	.0000	.0001	.0014	.0084	.0279	.0644	.1134	.1606	.1883	.1855
	9	.0000	.0000	.0003	.0021	.0093	.0276	.0611	.1070	.1540	.1855
	10	.0000	.0000	.0000	.0004	.0025	.0095	.0263	.0571	.1008	.1484
	11	.0000	.0000	.0000	.0001	.0005	.0026	.0090	.0242	.0525	.0944
	12	.0000	.0000	.0000	.0000	.0001	.0006	.0024	.0081	.0215	.0472
	13	.0000	.0000	.0000	.0000	.0000	.0001	.0005	.0021	.0068	.0182
	14	.0000	.0000	.0000	.0000	.0000	.0000	.0001	.0004	.0016	.0052
	15	.0000	.0000	.0000	.0000	.0000	.0000	.0000	.0001	.0003	.0010
	16	.0000	.0000	.0000	.0000	.0000	.0000	.0000	.0000	.0000	.0001
	17	.0000	.0000	.0000	.0000	.0000	.0000	.0000	.0000	.0000	.0000

(*Contd.*)

n	x	.05	.10	.15	.20	.25	.30	.35	.40	.45	.50
18	0	.3972	.1501	.0536	.0180	.0056	.0016	.0004	.0001	.0000	.0000
	1	.3763	.3002	.1704	.0811	.0338	.0126	.0042	.0012	.0003	.0001
	2	.1683	.2835	.2556	.1723	.0958	.0458	.0190	.0069	.0022	.0006
	3	.0473	.1680	.2406	.2297	.1704	.1046	.0547	.0246	.0095	.0031
	4	.0093	.0700	.1592	.2153	.2130	.1681	.1104	.0614	.0291	.0117
	5	.0014	.0218	.0787	.1507	.1988	.2017	.1664	.1146	.0666	.0327
	6	.0002	.0052	.0301	.0816	.1436	.1873	.1941	.1655	.1181	.0708
	7	.0000	.0010	.0091	.0350	.0820	.1376	.1792	.1892	.1657	.1214
	8	.0000	.0002	.0022	.0120	.0376	.0811	.1327	.1734	.1864	.1669
	9	.0000	.0000	.0004	.0033	.0139	.0386	.0794	.1284	.1694	.1855
	10	.0000	.0000	.0001	.0008	.0042	.0149	.0385	.0771	.1248	.1669
	11	.0000	.0000	.0000	.0001	.0010	.0046	.0151	.0374	.0742	.1241
	12	.0000	.0000	.0000	.0000	.0002	.0012	.0047	.0145	.0354	.0708
	13	.0000	.0000	.0000	.0000	.0000	.0002	.0012	.0045	.0134	.0327
	14	.0000	.0000	.0000	.0000	.0000	.0000	.0002	.0011	.0039	.0117
	15	.0000	.0000	.0000	.0000	.0000	.0000	.0000	.0002	.0009	.0031
	16	.0000	.0000	.0000	.0000	.0000	.0000	.0000	.0000	.0001	.0006
	17	.0000	.0000	.0000	.0000	.0000	.0000	.0000	.0000	.0000	.0001
	18	.0000	.0000	.0000	.0000	.0000	.0000	.0000	.0000	.0000	.0000
19	0	.3774	.1351	.0456	.0144	.0042	.0011	.0003	.0001	.0000	.0000
	1	.3774	.2852	.1529	.0685	.0268	.0093	.0029	.0008	.0002	.0000
	2	.1787	.2852	.2428	.1540	.0803	.0358	.0138	.0046	.0013	.0003
	3	.0533	.1796	.2428	.2182	.1517	.0869	.0422	.0175	.0062	.0018
	4	.0112	.0798	.1714	.2182	.2023	.1491	.0909	.0467	.0203	.0074
	5	.0018	.0266	.0907	.1636	.2023	.1916	.1468	.0933	.0497	.0222
	6	.0002	.0069	.0374	.0955	.1574	.1916	.1844	.1451	.0949	.0518
	7	.0000	.0014	.0122	.0443	.0974	.1525	.1844	.1797	.1443	.0961
	8	.0000	.0002	.0032	.0166	.0487	.0981	.1489	.1797	.1771	.1442
	9	.0000	.0000	.0007	.0051	.0198	.0514	.0980	.1464	.1771	.1762
	10	.0000	.0000	.0001	.0013	.0066	.0220	.0528	.0976	.1449	.1762
	11	.0000	.0000	.0000	.0003	.0018	.0077	.0233	.0532	.0970	.1442
	12	.0000	.0000	.0000	.0000	.0004	.0022	.0083	.0237	.0529	.0961
	13	.0000	.0000	.0000	.0000	.0001	.0005	.0024	.0085	.0233	.0518
	14	.0000	.0000	.0000	.0000	.0000	.0001	.0006	.0024	.0082	.0222
	15	.0000	.0000	.0000	.0000	.0000	.0000	.0001	.0005	.0022	.0074
	16	.0000	.0000	.0000	.0000	.0000	.0000	.0000	.0001	.0005	.0018
	17	.0000	.0000	.0000	.0000	.0000	.0000	.0000	.0000	.0001	.0003
	18	.0000	.0000	.0000	.0000	.0000	.0000	.0000	.0000	.0000	.0000
	19	.0000	.0000	.0000	.0000	.0000	.0000	.0000	.0000	.0000	.0000
20	0	.3585	.1216	.0388	.0115	.0032	.0008	.0002	.0000	.0000	.0000
	1	.3774	.2702	.1368	.0576	.0211	.0068	.0020	.0005	.0001	.0000
	2	.1887	.2852	.2293	.1369	.0669	.0278	.0100	.0031	.0008	.0002
	3	.0596	.1901	.2428	.2054	.1339	.0716	.0323	.0123	.0040	.0011
	4	.0133	.0898	.1821	.2182	.1897	.1304	.0738	.0350	.0139	.0046

(*Contd.*)

n	x	.05	.10	.15	.20	.25	.30	.35	.40	.45	.50
20	5	.0022	.0319	.1028	.1746	.2023	.1789	.1272	.0746	.0365	.0148
	6	.0003	.0089	.0454	.1091	.1686	.1916	.1712	.1244	.0746	.0370
	7	.0000	.0020	.0160	.0545	.1124	.1643	.1844	.1659	.1221	.0739
	8	.0000	.0004	.0046	.0222	.0609	.1144	.1614	.1797	.1623	.1201
	9	.0000	.0001	.0011	.0074	.0271	.0654	.1158	.1597	.1771	.1602
	10	.0000	.0000	.0002	.0020	.0099	.0308	.0686	.1171	.1593	.1762
	11	.0000	.0000	.0000	.0005	.0030	.0120	.0336	.0710	.1185	.1602
	12	.0000	.0000	.0000	.0001	.0008	.0039	.0136	.0355	.0727	.1201
	13	.0000	.0000	.0000	.0000	.0002	.0010	.0045	.0145	.0366	.0739
	14	.0000	.0000	.0000	.0000	.0000	.0002	.0012	.0049	.0150	.0370
	15	.0000	.0000	.0000	.0000	.0000	.0000	.0003	.0013	.0049	.0148
	16	.0000	.0000	.0000	.0000	.0000	.0000	.0000	.0003	.0013	.0046
	17	.0000	.0000	.0000	.0000	.0000	.0000	.0000	.0000	.0002	.0011
	18	.0000	.0000	.0000	.0000	.0000	.0000	.0000	.0000	.0000	.0002
	19	.0000	.0000	.0000	.0000	.0000	.0000	.0000	.0000	.0000	.0000
	20	.0000	.0000	.0000	.0000	.0000	.0000	.0000	.0000	.0000	.0000

TABLE 14
POISSON PROBABILITIES

λ

x	0.1	0.2	0.3	0.4	0.5	0.6	0.7	0.8	0.9	1.0
0	.9048	.8187	.7408	.6703	.6065	.5488	.4966	.4193	.4066	.3679
1	.0905	.1637	.2222	.2681	.3033	.3293	.3476	.3595	.3659	.3679
2	.0045	.0164	.0333	.0536	.0758	.0988	.1217	.1438	.1647	.1839
3	.0002	.0011	.0033	.0072	.0126	.0198	.0284	.0383	.0494	.0013
4	.0000	.0001	.0002	.0007	.0016	.0030	.0050	.0077	.0111	.0153
5	.0000	.0000	.0000	.0001	.0002	.0004	.0007	.0012	.0020	.0031
6	.0000	.0000	.0000	.0000	.0000	.0000	.0001	.0002	.0003	.0005
7	.0000	.0000	.0000	.0000	.0000	.0000	.0000	.0000	.0000	.0001

λ

x	1.1	1.2	1.3	1.4	1.5	1.6	1.7	1.8	1.9	2.0
0	.3329	.3012	.2725	.2466	.2231	.2019	.1827	.1653	.1496	.1353
1	.3662	.3614	.3543	.3452	.3347	.3230	.3106	.2975	.2842	.2707
2	.2014	.2169	.2303	.2417	.2510	.2584	.2640	.2678	.2700	.2707
3	.0738	.0867	.0998	.1128	.1255	.1378	.1496	.1607	.1710	.1804
4	.0203	.0260	.0324	.0395	.0471	.0551	.0636	.0723	.0812	.0902
5	.0045	.0062	.0084	.0111	.0141	.0176	.0216	.0260	.0309	.0361
6	.0008	.0012	.0018	.0026	.0035	.0047	.0061	.0078	.0098	.0120
7	.0001	.0002	.0003	.0005	.0008	.0011	.0015	.0020	.0027	.0034
8	.0000	.0000	.0001	.0001	.0001	.0002	.0003	.0005	.0006	.0009
9	.0000	.0000	.0000	.0000	.0000	.0000	.0001	.0001	.0001	.0002

λ

x	2.1	2.2	2.3	2.4	2.5	2.6	2.7	2.8	2.9	3.0
0	.1225	.1108	.1003	.0907	.0821	.0743	.0672	.0608	.0550	.0498
1	.2572	.2438	.2306	.2177	.2052	.1931	.1815	.1703	.1596	.1494
2	.2700	.2681	.2652	.2613	.2565	.2510	.2450	.2384	.2314	.2240
3	.1890	.1966	.2033	.2090	.2138	.2176	.2205	.2225	.2237	.2240
4	.0992	.1082	.1169	.1254	.1336	.1414	.1488	.1557	.1622	.1680
5	.0417	.0476	.0538	.0602	.0668	.0735	.0804	.0872	.0940	.1008
6	.0146	.0174	.0206	.0241	.0278	.0319	.0362	.0407	.0455	.0504
7	.0044	.0055	.0068	.0083	.0099	.0118	.0139	.0163	.0188	.0216
8	.0011	.0015	.0019	.0025	.0031	.0038	.0047	.0057	.0068	.0081
9	.0003	.0004	.0005	.0007	.0009	.0011	.0014	.0018	.0022	.0027
10	.0001	.0001	.0001	.0002	.0002	.0003	.0004	.0005	.0006	.0008
11	.0000	.0000	.0000	.0000	.0000	.0001	.0001	.0001	.0002	.0002
12	.0000	.0000	.0000	.0000	.0000	.0000	.0000	.0000	.0000	.0001

λ

x	3.1	3.2	3.3	3.4	3.5	3.6	3.7	3.8	3.9	4.0
0	.0450	.0408	.0369	.0334	.0302	.0273	.0247	.0224	.0202	.0183
1	1397	.1304	.1217	.1135	.1057	.0984	.0915	.0850	.0789	.0733
2	.2165	.2087	.2008	.1929	.1850	.1771	.1692	.1615	.1539	.1465

	λ									
x	3.1	3.2	3.3	3.4	3.5	3.6	3.7	3.8	3.9	4.0
3	.2237	.2226	.2209	.2186	.2158	.2125	.2087	.2046	.2001	.1954
4	.1734	.1781	.1823	.1858	.1888	.1912	.1931	.1944	.1951	.1954
5	.1075	.1140	.1203	.1264	.1322	.1377	.1429	.1477	.1522	.1563
6	.0555	.0608	.0662	.0716	.0771	.0826	.0881	.0936	.0989	.1042
7	.0246	.0278	.0312	.0348	.0385	.0425	.0466	.0508	.0551	.0595
8	.0095	.0111	.0129	.0148	.0169	.0191	.0215	.0241	.0269	.0298
9	.0033	.0040	.0047	.0056	.0066	.0076	.0089	.0102	.0116	.0132
10	.0010	.0013	.0016	.0019	.0023	.0028	.0033	.0039	.0045	.0053
11	.0003	.0004	.0005	.0006	.0007	.0009	.0011	.0013	.0016	.0019
12	.0001	.0001	.0001	.0002	.0002	.0003	.0003	.0004	.0005	.0006
13	.0000	.0000	.0000	.0000	.0001	.0001	.0001	.0001	.0002	.0002
14	.0000	.0000	.0000	.0000	.0000	.0000	.0000	.0000	.0000	.0001

	λ									
x	4.1	4.2	4.3	4.4	4.5	4.6	4.7	4.8	4.9	5.0
0	.0166	.0150	.0136	.0123	.0111	.0101	.0091	.0082	.0074	.0067
1	.0679	.0630	.0583	.0540	.0500	.0462	.0427	.0395	.0365	.0337
2	.1393	.1323	.1254	.1188	.1125	.1063	.1005	.0948	.0894	.0842
3	.1904	.1852	.1798	.1743	.1687	.1631	.1574	.1517	.1460	.1404
4	.1951	.1944	.1933	.1917	.1898	.1875	.1849	.1820	.1789	.1755
5	.1600	.1633	.1662	.1687	.1708	.1725	.1738	.1747	.1753	.1755
6	.1093	.1143	.1191	.1237	.1281	.1323	.1362	.1398	.1432	.1462
7	.0640	.0686	.0732	.0778	.0824	.0869	.0914	.0959	.1002	.1044
8	.0328	.0360	.0393	.0428	.0463	.0500	.0537	.0575	.0614	.0653
9	.0150	.0168	.0188	.0209	.0232	.0255	.0280	.0307	.0334	.0363
10	.0061	.0071	.0081	.0092	.0104	.0118	.0132	.0147	.0164	.0181
11	.0023	.0027	.0032	.0037	.0043	.0049	.0056	.0064	.0073	.0082
12	.0008	.0009	.0011	.0014	.0016	.0019	.0022	.0026	.0030	.0034
13	.0002	.0003	.0004	.0005	.0006	.0007	.0008	.0009	.0011	.0013
14	.0001	.0001	.0001	.0001	.0002	.0002	.0003	.0003	.0004	.0005
15	.0000	.0000	.0000	.0000	.0001	.0001	.0001	.0001	.0001	.0002

	λ									
x	5.1	5.2	5.3	5.4	5.5	5.6	5.7	5.8	5.9	6.0
0	.0061	.0055	.0050	.0045	.0041	.0037	.0033	.0030	.0027	.0025
1	.0311	.0287	.0265	.0244	.0225	.0207	.0191	.0176	.0162	.0149
2	.0793	.0746	.0701	.0659	.0618	.0580	.0544	.0509	.0477	.0446
3	.1348	.1293	.1239	.1185	.1133	.1082	.1033	.0985	.0938	.0892
4	.1719	.1681	.1641	.1600	.1558	.1515	.1472	.1428	.1383	.1339
5	.1753	.1748	.1740	.1728	.1714	.1697	.1678	.1656	.1632	.1606
6	.1490	.1515	.1537	.1555	.1571	.1584	.1594	.1601	.1605	.1606
7	.1086	.1125	.1163	.1200	.1234	.1267	.1298	.1326	.1353	.1377
8	.0692	.0731	.0771	.0810	.0849	.0887	.0925	.0962	.0998	.1033
9	.0392	.0423	.0454	.0486	.0519	.0552	.0586	.0620	.0654	.0688
10	.0200	.0220	.0241	.0262	.0285	.0309	.0334	.0359	.0386	.0413
11	.0093	.0104	.0116	.0129	.0143	.0157	.0173	.0190	.0207	.0225
12	.0039	.0045	.0051	.0058	.0065	.0073	.0082	.0092	.0102	.0113
13	.0015	.0018	.0021	.0024	.0028	.0032	.0036	.0041	.0046	.0052
14	.0006	.0007	.0008	.0009	.0011	.0013	.0015	.0017	.0019	.0022
15	.0002	.0002	.0003	.0003	.0004	.0005	.0006	.0007	.0008	.0009
16	.0001	.0001	.0001	.0001	.0001	.0002	.0002	.0002	.0003	.0003
17	.0000	.0000	.0000	.0000	.0000	.0001	.0001	.0001	.0001	.0001

λ

x	6.1	6.2	6.3	6.4	6.5	6.6	6.7	6.8	6.9	7.0
0	.0022	.0020	.0018	.0017	.0015	.0014	.0012	.0011	.0010	.0009
1	.0137	.0126	.0116	.0106	.0098	.0090	.0082	.0076	.0070	.0064
2	.0417	.0390	.0364	.0340	.0318	.0296	.0276	.0258	.0240	.0223
3	.0848	.0806	.0765	.0726	.0688	.0652	.0617	.0584	.0552	.0521
4	.1294	.1249	.1205	.1162	.1118	.1076	.1034	.0992	.0952	.0912
5	.1579	.1549	.1519	.1487	.1454	.1420	.1385	.1349	.1314	.1277
6	.1605	.1601	.1595	.1586	.1575	.1562	.1546	.1529	.1511	.1490
7	.1399	.1418	.1435	.1450	.1462	.1472	.1480	.1486	.1489	.1490
8	.1066	.1099	.1130	.1160	.1188	.1215	.1240	.1263	.1284	.1304
9	.0723	.0757	.0791	.0825	.0858	.0891	.0923	.0954	.0985	.1014
10	.0441	.0469	.0498	.0528	.0558	.0588	.0618	.0649	.0679	.0710
11	.0245	.0265	.0285	.0307	.0330	.0353	.0377	.0401	.0426	.0452
12	.0124	.0137	.0150	.0164	.0179	.0194	.0210	.0227	.0245	.0264
13	.0058	.0065	.0073	.0081	.0089	.0098	.0108	.0119	.0130	.0142
14	.0025	.0029	.0033	.0037	.0041	.0046	.0052	.0058	.0064	.0071
15	.0010	.0012	.0014	.0016	.0018	.0020	.0023	.0026	.0029	.0033
16	.0004	.0005	.0005	.0006	.0007	.0008	.0010	.0011	.0013	.0014
17	.0001	.0002	.0002	.0002	.0003	.0003	.0004	.0004	.0005	.0006
18	.0000	.0001	.0001	.0001	.0001	.0001	.0001	.0002	.0002	.0002
19	.0000	.0000	.0000	.0000	.0000	.0000	.0000	.0001	.0001	.0001

λ

x	7.1	7.2	7.3	7.4	7.5	7.6	7.7	7.8	7.9	8.0
0	.0008	.0007	.0007	.0006	.0006	.0005	.0005	.0004	.0004	.0003
1	.0059	.0054	.0049	.0045	.0041	.0038	.0035	.0032	.0029	.0027
2	.0208	.0194	.0180	.0167	.0156	.0145	.0134	.0125	.0116	.0107
3	.0492	.0464	.0438	.0413	.0389	.0366	.0345	.0324	.0305	.0286
4	.0874	.0836	.0799	.0764	.0729	.0696	.0663	.0632	.0602	.0573
5	.1241	.1204	.1167	.1130	.1094	.1057	.1021	.0986	.0951	.0916
6	.1468	.1445	.1420	.1394	.1367	.1339	.1311	.1282	.1252	.1221
7	.1489	.1486	.1481	.1474	.1465	.1454	.1442	.1428	.1413	.1396
8	.1321	.1337	.1351	.1363	.1373	.1382	.1388	.1392	.1395	.1396
9	.1042	.1070	.1096	.1121	.1144	.1167	.1187	.1207	.1224	.1241
10	.0740	.0770	.0800	.0829	.0858	.0887	.0914	.0941	.0967	.0993
11	.0478	.0504	.0531	.0558	.0585	.0613	.0640	.0667	.0695	.0722
12	.0283	.0303	.0323	.0344	.0366	.0388	.0411	.0434	.0457	.0481
13	.0154	.0168	.0181	.0196	.0211	.0227	.0243	.0260	.0278	.0296
14	.0078	.0086	.0095	.0104	.0113	.0123	.0134	.0145	.0157	.0169
15	.0037	.0041	.0046	.0051	.0057	.0062	.0069	.0075	.0083	.0090
16	.0016	.0019	.0021	.0024	.0026	.0030	.0033	.0037	.0041	.0045
17	.0007	.0008	.0009	.0010	.0012	.0013	.0015	.0017	.0019	.0021
18	.0003	.0003	.0004	.0004	.0005	.0006	.0006	.0007	.0008	.0009
19	.0001	.0001	.0001	.0002	.0002	.0002	.0003	.0003	.0003	.0004
20	.0000	.0000	.0001	.0001	.0001	.0001	.0001	.0001	.0001	.0002
21	.0000	.0000	.0000	.0000	.0000	.0000	.0000	.0000	.0001	.0001

x	λ 8.1	8.2	8.3	8.4	8.5	8.6	8.7	8.8	8.9	9.0
0	.0003	.0003	.0002	.0002	.0002	.0002	.0002	.0002	.0001	.0001
1	.0025	.0023	.0021	.0019	.0017	.0016	.0014	.0013	.0012	.0011
2	.0100	.0092	.0086	.0079	.0074	.0068	.0063	.0058	.0054	.0050
3	.0269	.0252	.0237	.0222	.0208	.0195	.0183	.0171	.0160	.0150
4	.0544	.0517	.0491	.0466	.0443	.0420	.0398	.0377	.0357	.0337
5	.0882	.0849	.0816	.0784	.0752	.0722	.0692	.0663	.0635	.0607
6	.1191	.1160	.1128	.1097	.1066	.1034	.1003	.0972	.0941	.0911
7	.1378	.1358	.1338	.1317	.1294	.1271	.1247	.1222	.1197	.1171
8	.1395	.1392	.1388	.1382	.1375	.1366	.1356	.1344	.1332	.1318
9	.1256	.1269	.1280	.1290	.1299	.1306	.1311	.1315	.1317	.1318
10	.1017	.1040	.1063	.1084	.1104	.1123	.1140	.1157	.1172	.1186
11	.0749	.0776	.0802	.0828	.0853	.0878	.0902	.0925	.0948	.0970
12	.0505	.0530	.0555	.0579	.0604	.0629	.0654	.0679	.0703	.0728
13	.0315	.0334	.0354	.0374	.0395	.0416	.0438	.0459	.0481	.0504
14	.0182	.0196	.0210	.0225	.0240	.0256	.0272	.0289	.0306	.0324
15	.0098	.0107	.0116	.0126	.0136	.0147	.0158	.0169	.0182	.0194
16	.0050	.0055	.0060	.0066	.0072	.0079	.0086	.0093	.0101	.0109
17	.0024	.0026	.0029	.0033	.0036	.0040	.0044	.0048	.0053	.0058
18	.0011	.0012	.0014	.0015	.0017	.0019	.0021	.0024	.0026	.0029
19	.0005	.0005	.0006	.0007	.0008	.0009	.0010	.0011	.0012	.0014
20	.0002	.0002	.0002	.0003	.0003	.0004	.0004	.0005	.0005	.0006
21	.0001	.0001	.0001	.0001	.0001	.0002	.0002	.0002	.0002	.0003
22	.0000	.0000	.0000	.0000	.0001	.0001	.0001	.0001	.0001	.0001

x	λ 9.1	9.2	9.3	9.4	9.5	9.6	9.7	9.8	9.9	10
0	.0001	.0001	.0001	.0001	.0001	.0001	.0001	.0001	.0001	.0000
1	.0010	.0009	.0009	.0008	.0007	.0007	.0006	.0005	.0005	.0005
2	.0046	.0043	.0040	.0037	.0034	.0031	.0029	.0027	.0025	.0023
3	.0140	.0131	.0123	.0115	.0107	.0100	.0093	.0087	.0081	.0076
4	.0319	.0302	.0285	.0269	.0254	.0240	.0226	.0213	.0201	.0189
5	.0581	.0555	.0530	.0506	.0483	.0460	.0439	.0418	.0398	.0378
6	.0881	.0851	.0822	.0793	.0764	.0736	.0709	.0682	.0656	.0631
7	.1145	.1118	.1091	.1064	.1037	.1010	.0982	.0955	.0928	.0901
8	.1302	.1286	.1269	.1251	.1232	.1212	.1191	.1170	.1148	.1126
9	.1317	.1315	.1311	.1306	.1300	.1293	.1284	.1274	.1263	.1251
10	.1198	.1210	.1219	.1228	.1235	.1241	.1245	.1249	.1250	.1251
11	.0991	.1012	.1031	.1049	.1067	.1083	.1098	.1112	.1125	.1137
12	.0752	.0776	.0799	.0822	.0844	.0866	.0888	.0908	.0928	.0948
13	.0526	.0549	.0572	.0594	.0617	.0640	.0662	.0685	.0707	.0729
14	.0342	.0361	.0380	.0399	.0419	.0439	.0459	.0479	.0500	.0521
15	.0208	.0221	.0235	.0250	.0265	.0281	.0297	.0313	.0330	.0347
16	.0118	.0127	.0137	.0147	.0157	.0168	.0180	.0192	.0204	.0217
17	.0063	.0069	.0075	.0081	.0088	.0095	.0103	.0111	.0119	.0128
18	.0032	.0035	.0039	.0042	.0046	.0051	.0055	.0060	.0065	.0071
19	.0015	.0017	.0019	.0021	.0023	.0026	.0028	.0031	.0034	.0037
20	.0007	.0008	.0009	.0010	.0011	.0012	.0014	.0015	.0017	.0019
21	.0003	.0003	.0004	.0004	.0005	.0006	.0006	.0007	.0008	.0009
22	.0001	.0001	.0002	.0002	.0002	.0002	.0003	.0003	.0004	.0004
23	.0000	.0001	.0001	.0001	.0001	.0001	.0001	.0001	.0002	.0002
24	.0000	.0000	.0000	.0000	.0000	.0000	.0000	.0001	.0001	.0001

	λ									
x	11	12	13	14	15	16	17	18	19	20
0	.0000	.0000	.0000	.0000	.0000	.0000	.0000	.0000	.0000	.0000
1	.0002	.0001	.0000	.0000	.0000	.0000	.0000	.0000	.0000	.0000
2	.0010	.0004	.0002	.0001	.0000	.0000	.0000	.0000	.0000	.0000
3	.0037	.0018	.0008	.0004	.0002	.0001	.0000	.0000	.0000	.0000
4	.0102	.0053	.0027	.0013	.0006	.0003	.0001	.0001	.0000	.0000
5	.0224	.0127	.0070	.0037	.0019	.0010	.0005	.0002	.0001	.0001
6	.0411	.0255	.0152	.0087	.0048	.0026	.0014	.0007	.0004	.0002
7	.0646	.0437	.0281	.0174	.0104	.0060	.0034	.0018	.0010	.0005
8	.0888	.0655	.0457	.0304	.0194	.0120	.0072	.0042	.0024	.0013
9	.1085	.0874	.0661	.0473	.0324	.0213	.0135	.0083	.0050	.0029
10	.1194	.1048	.0859	.0663	.0486	.0341	.0230	.0150	.0095	.0058
11	.1194	.1144	.1015	.0844	.0663	.0496	.0355	.0245	.0164	.0106
12	.1094	.1144	.1099	.0984	.0829	.0661	.0504	.0368	.0259	.0176
13	.0926	.1056	.1099	.1060	.0956	.0814	.0658	.0509	.0378	.0271
14	.0728	.0905	.1021	.1060	.1024	.0930	.0800	.0655	.0514	.0387
15	.0534	.0724	.0885	.0989	.1024	.0992	.0906	.0786	.0650	.0516
16	.0367	.0543	.0719	.0866	.0960	.0992	.0963	.0884	.0772	.0646
17	.0237	.0383	.0550	.0713	.0847	.0934	.0963	.0936	.0863	.0760
18	.0145	.0256	.0397	.0554	.0706	.0830	.0909	.0936	.0911	.0844
19	.0084	.0161	.0272	.0409	.0557	.0699	.0814	.0887	.0911	.0888

TABLE 15
CRITICAL VALUES FOR THE *U* – TEST

Values of $U_{.10}$

n_1 \ n_2	2	3	4	5	6	7	8	9	10	11	12	13	14	15
2				0	0	0	1	1	1	1	2	2	3	3
3		0	0	1	2	2	3	4	4	5	5	6	7	7
4		0	1	2	3	4	5	6	7	8	9	10	11	12
5	0	1	2	4	5	6	8	9	11	12	13	15	16	18
6	0	2	3	5	7	8	10	12	14	16	17	19	21	23
7	0	2	4	6	8	11	13	15	17	19	21	24	26	28
8	1	3	5	8	10	13	15	18	20	23	26	28	31	33
9	1	4	6	9	12	15	18	21	24	27	30	33	36	39
10	1	4	7	11	14	17	20	24	27	31	34	37	41	44
11	1	5	8	12	16	19	23	27	31	34	38	42	46	50
12	2	5	9	13	17	21	26	30	34	38	42	47	51	55
13	2	6	10	15	19	24	28	33	37	42	47	51	56	61
14	3	7	11	16	21	26	31	36	41	46	51	56	61	66
15	3	7	12	18	23	28	33	39	44	50	55	61	66	72

Values of $U_{.05}$

n_1 \ n_2	2	3	4	5	6	7	8	9	10	11	12	13	14	15
2							0	0	0	0	1	1	1	1
3				0	1	1	2	2	3	3	4	4	5	5
4			0	1	2	3	4	4	5	6	7	8	9	10
5		0	1	2	3	5	6	7	8	9	11	12	13	14
6		1	2	3	5	6	8	10	11	13	14	16	17	19
7		1	3	5	6	8	10	12	14	16	18	20	22	24
8	0	2	4	6	8	10	13	15	17	19	22	24	26	29
9	0	2	4	7	10	12	15	17	20	23	26	28	31	34

10	0	3	5	8	11	14	17	20	23	26	29	30	36	39
11	0	3	6	9	13	16	19	23	26	30	33	37	40	44
12	1	4	7	11	14	18	22	26	29	33	37	41	45	49
13	1	4	8	12	16	20	24	28	30	37	41	45	50	54
14	1	5	9	13	17	22	26	31	36	40	45	50	55	59
15	1	5	10	14	19	24	29	34	39	44	49	54	59	64

Values of $U_{.02}$

n_1 \ n_2	**2**	**3**	**4**	**5**	**6**	**7**	**8**	**9**	**10**	**11**	**12**	**13**	**14**	**15**
2												0	0	0
3						0	0	1	1	1	2	2	2	3
4				0	1	1	2	3	3	4	5	5	5	7
5			0	1	2	3	4	5	6	7	8	9	10	11
6			1	2	3	4	6	7	8	9	11	12	13	15
7		0	1	3	4	6	7	9	11	12	14	16	17	19
8		0	2	4	6	7	9	11	13	15	17	20	22	24
9		1	3	5	7	9	11	14	16	18	21	23	26	28
10		1	3	6	8	11	13	16	19	22	24	27	30	33
11		1	4	7	9	12	15	18	22	25	28	31	34	37
12		2	5	8	11	14	17	21	24	28	31	35	38	42
13	0	2	5	9	12	16	20	23	27	31	35	39	43	47
14	0	2	6	10	13	17	22	26	30	34	38	43	47	51
15	0	3	7	11	15	19	24	28	33	37	42	47	51	56

Values of $U_{.01}$

n_1 \ n_2	**3**	**4**	**5**	**6**	**7**	**8**	**9**	**10**	**11**	**12**	**13**	**14**	**15**
3							0	0	0	1	1	1	2
4				0	0	1	1	2	2	3	3	4	5
5			0	1	1	2	3	4	5	6	7	7	8
6		0	1	2	3	4	5	6	7	9	10	11	12
7		0	1	3	4	6	7	9	10	12	13	15	16
8		1	2	4	6	7	9	11	13	15	17	18	20
9	0	1	3	5	7	9	11	13	16	18	20	22	24
10	0	2	4	6	9	11	13	16	18	21	24	26	29
11	0	2	5	7	10	13	16	18	21	24	27	30	33
12	1	3	6	9	12	15	18	21	24	27	31	34	37
13	1	3	7	10	13	17	20	24	27	31	34	38	42
14	1	4	7	11	15	18	22	26	30	34	38	42	46
15	2	5	8	12	16	20	24	29	33	37	42	46	51

TABLE 16
CRITICAL VALUES FOR THE RUNS TEST

Values of $u'_{.025}$

n_1 \ n_2	**2**	**3**	**4**	**5**	**6**	**7**	**8**	**9**	**10**	**11**	**12**	**13**	**14**	**15**
2											2	2	2	2
3					2	2	2	2	2	2	2	2	2	3
4				2	2	2	3	3	3	3	3	3	3	3
5			2	2	3	3	3	3	3	4	4	4	4	4
6		2	2	3	3	3	3	4	4	4	4	5	5	5
7		2	2	3	3	3	4	4	5	5	5	5	5	6
8		2	3	3	3	4	4	5	5	5	6	6	6	6

9		2	3	3	4	4	5	5	5	6	6	6	7	7
10		2	3	3	4	5	5	5	6	6	7	7	7	7
11		2	3	4	4	5	5	6	6	7	7	7	8	8
12	2	2	3	4	4	5	6	6	7	7	7	8	8	8
13	2	2	3	4	5	5	6	6	7	7	8	8	9	9
14	2	2	3	4	5	5	6	7	7	8	8	9	9	9
15	2	3	3	4	5	6	6	7	7	8	8	9	9	10

Values of $U_{.02}$

n_1 \ n_2	4	5	6	7	8	9	10	11	12	13	14	15
4		9	9									
5	9	10	10	11	11							
6	9	10	11	12	12	13	13	13	13			
7		11	12	13	13	14	14	14	14	15	15	15
8		11	12	13	14	14	15	15	16	16	16	16
9			13	14	14	15	16	16	16	17	17	18
10			13	14	15	16	16	17	17	18	18	18
11			13	14	15	16	17	17	18	19	19	19
12			13	14	16	16	17	18	19	19	20	20
13				15	16	17	18	19	19	20	20	21
14				15	16	17	18	19	20	20	21	22
15				15	16	18	18	19	20	21	22	22

Values of $u'_{.005}$

n_1 \ n_2	3	4	5	6	7	8	9	10	11	12	13	14	15
3										2	2	2	2
4						2	2	2	2	2	2	2	3
5				2	2	2	2	3	3	3	3	3	3
6			2	2	2	3	3	3	3	3	3	4	4
7			2	2	3	3	3	3	4	4	4	4	4
8		2	2	3	3	3	3	4	4	4	5	5	5
9		2	2	3	3	3	4	4	5	5	5	5	6
10		2	3	3	3	4	4	5	5	5	5	6	6
11		2	3	3	4	4	5	5	5	6	6	6	7
12	2	2	3	3	4	4	5	5	6	6	6	7	7
13	2	2	3	3	4	5	5	5	6	6	7	7	7
14	2	2	3	4	4	5	5	6	6	7	7	7	8
15	2	3	3	4	4	5	6	6	7	7	7	8	8

Values of $U_{.005}$

n_1 \ n_2	5	6	7	8	9	10	11	12	13	14	15
5		11									
6	11	12	13	13							
7		13	13	14	15	15	15				
8		13	14	15	15	16	16	17	17	17	
9			15	15	16	17	17	18	18	18	19
10			15	16	17	17	18	19	19	19	20
11			15	16	17	18	19	19	20	20	21
12				17	18	19	19	20	21	21	22
13				17	18	19	20	21	21	22	22
14				17	18	19	20	21	22	23	23
15					19	20	21	22	22	23	24

TABLE 17 — Powers, Roots and Reciprocals

n	n^2	n^3	$\sqrt{n}$	$\sqrt[3]{n}$	$\sqrt{10n}$	$\sqrt[3]{10n}$	$\sqrt[3]{100n}$	$\frac{1}{n}$
1	1	1	1	1	3.162	2.154	4.642	1
2	4	8	1.414	1.260	4.472	2.714	5.848	0.5000
3	9	27	1.732	1.442	5.477	3.107	6.694	0.3333
4	16	64	2	1.587	6.325	3.420	7.368	0.2500
5	25	125	2.236	1.710	7.071	3.684	7.937	0.2000
6	36	216	2.449	1.817	7.746	3.915	8.434	0.1667
7	49	343	2.646	1.913	8.367	4.121	8.879	0.1429
8	64	512	2.828	2.000	8.944	4.309	9.283	0.1250
9	81	729	3.000	2.080	9.487	4.481	9.655	0.1111
10	100	1000	3.162	2.154	10.000	4.642	10.000	0.1000
11	121	1331	3.317	2.224	10.488	4.791	10.323	0.09091
12	144	1728	3.464	2.289	10.954	4.932	10.627	0.08333
13	169	2197	3.606	2.351	11.402	5.066	10.914	0.07692
14	196	2744	3.742	2.410	11.832	5.192	11.187	0.07143
15	225	3375	3.873	2.466	12.247	5.313	11.447	0.06667
16	256	4096	4.000	2.520	12.649	5.429	11.696	0.06250
17	289	4913	4.123	2.571	13.038	5.540	11.935	0.05882
18	324	5832	4.243	2.621	13.416	5.646	12.164	0.05556
19	361	6859	4.359	2.668	13.784	5.749	12.386	0.05263
20	400	8000	4.472	2.714	14.142	5.848	12.599	0.0500
21	441	9261	4.583	2.759	14.491	5.944	12.806	0.04762
22	484	10648	4.690	2.802	14.832	6.037	13.006	0.04545
23	529	12167	4.796	2.844	15.166	6.127	13.200	0.04348
24	576	13824	4.899	2.884	15.492	6.214	13.389	0.04167
25	625	15625	5.000	2.924	15.811	6.300	13.572	0.04000
26	676	17576	5.099	2.962	16.125	6.383	13.751	0.03846
27	729	19683	5.196	3.000	16.432	6.463	13.925	0.03704
28	784	21952	5.292	3.037	16.733	6.542	14.095	0.03571
29	841	24389	5.385	3.072	17.029	6.619	14.260	0.03448
30	900	27000	5.477	3.107	17.321	6.694	14.422	0.03333
31	961	29791	5.568	3.141	17.607	6.768	14.581	0.03226
32	1024	32768	5.657	3.175	17.889	6.840	14.736	0.03125
33	1089	35937	5.745	3.208	18.166	6.910	14.888	0.03030
34	1156	39304	5.831	3.240	18.439	6.980	15.037	0.02941
35	1225	42875	5.916	3.271	18.708	7.047	15.183	0.02857
36	1296	46656	6.000	3.302	18.974	7.114	15.326	0.02778
37	1369	50653	6.083	3.332	19.235	7.179	15.467	0.02703
38	1444	54872	6.164	3.362	19.494	7.243	15.605	0.02632
39	1521	59319	6.245	3.391	19.748	7.306	15.741	0.02564
40	1600	64000	6.325	3.420	20.000	7.368	15.874	0.02500
41	1681	68921	6.403	3.448	20.248	7.429	16.005	0.02439
42	1764	74088	6.481	3.476	20.494	7.489	16.134	0.02381
43	1849	79507	6.557	3.503	20.736	7.548	16.268	0.02326
44	1936	85184	6.633	3.530	20.976	7.606	16.386	0.02273
45	2025	91125	6.708	3.557	21.213	7.663	16.510	0.02222
46	2116	97336	6.782	3.583	21.448	7.719	16.631	0.02174
47	2209	103823	6.856	3.600	21.679	7.775	16.751	0.02178
48	2304	110592	6.928	3.634	21.909	7.830	16.869	0.02083
49	2401	117649	7.000	3.659	22.136	7.884	16.985	0.02041
50	2500	125000	7.071	3.684	22.361	7.937	17.100	0.02000

TABLE 17 (Contd.) — Powers, Roots and Reciprocals

n	n^2	n^3	$\sqrt{n}$	$\sqrt[3]{n}$	$\sqrt{10n}$	$\sqrt[3]{10n}$	$\sqrt[3]{100n}$	$\frac{1}{n}$
51	2601	132651	7.141	3.708	22.583	7.990	17.213	0.01961
52	2704	140608	7.211	3.733	22.804	8.041	17.325	0.01923
53	2809	148877	7.280	3.756	23.022	8.093	17.435	0.01887
54	2916	157464	7.348	3.780	23.238	8.143	17.544	0.01852
55	3025	166375	7.416	3.803	23.452	8.193	17.652	0.01818
56	3136	175616	7.483	3.826	23.664	8.243	17.758	0.01786
57	3249	185193	7.550	3.849	23.875	8.291	17.863	0.01754
58	3364	195112	7.616	3.871	24.083	8.340	17.967	0.01724
59	3481	205379	7.681	3.893	24.290	8.387	18.070	0.01695
60	3600	216000	7.746	3.915	24.495	8.434	18.171	0.01667
61	3721	226981	7.810	3.936	24.698	8.481	18.272	0.01639
62	3844	238328	7.874	3.958	24.900	8.527	18.371	0.01613
63	3969	250047	7.937	3.979	25.100	8.573	18.469	0.01587
64	4096	262144	8.000	4.000	25.298	8.618	18.566	0.01562
65	4225	274625	8.062	4.021	25.495	8.662	18.663	0.01538
66	4356	287496	8.124	4.041	25.690	8.707	18.758	0.01515
67	4489	300763	8.185	4.062	25.884	8.750	18.852	0.01493
68	4624	314432	8.246	4.082	26.077	8.794	18.945	0.01471
69	4761	328509	8.307	4.102	26.268	8.837	19.038	0.01449
70	4900	343000	8.367	4.121	26.458	8.879	19.129	0.01429
71	5041	357911	8.426	4.141	26.646	8.921	19.220	0.01408
72	5184	373248	8.485	4.160	26.833	8.963	19.310	0.01389
73	5329	389017	8.544	4.179	27.019	9.004	19.399	0.01370
74	5476	405224	8.602	4.198	27.203	9.045	19.487	0.01351
75	5625	421875	8.660	4.217	27.386	9.086	19.574	0.01333
76	5776	438976	8.718	4.236	27.568	9.126	19.661	0.01316
77	5929	456533	8.775	4.254	27.749	9.166	19.747	0.01299
78	6084	474552	8.832	4.273	27.928	9.205	19.832	0.01282
79	6241	493039	8.888	4.291	28.107	9.244	19.916	0.01266
80	6400	512000	8.944	4.309	28.284	8.283	20.000	0.01250
81	6561	531441	9.000	4.327	28.460	9.322	20.083	0.01235
82	6724	551368	9.055	4.344	28.636	9.360	20.165	0.01220
83	6889	571787	9.110	4.362	28.810	9.398	20.247	0.01205
84	7056	592704	9.165	4.380	28.983	9.435	20.328	0.01190
85	7225	614125	9.220	4.397	29.155	9.473	20.408	0.01176
86	7396	636056	9.274	4.414	29.326	9.510	20.488	0.01163
87	7569	658503	9.327	4.431	29.496	9.546	20.567	0.01149
88	7744	681472	9.381	4.448	29.665	9.583	20.646	0.01136
89	7921	704969	9.434	4.465	29.833	9.619	20.724	0.01124
90	8100	729000	9.487	4.481	30.000	9.655	20.801	0.01111
91	8281	753571	9.539	4.498	30.166	9.691	20.878	0.01099
92	8464	778688	9.592	4.514	30.332	9.726	20.954	0.01087
93	8649	804357	9.644	4.531	30.496	9.761	21.029	0.01075
94	8836	830584	9.695	4.547	30.659	9.796	21.105	0.01064
95	9025	857375	9.747	4.563	30.822	9.830	21.179	0.01053
96	9216	884736	9.798	4.579	30.984	9.865	21.253	0.01042
97	9409	912673	9.849	4.595	31.145	9.899	21.327	0.01031
98	9604	941192	9.899	4.610	31.305	9.933	21.400	0.01020
99	9801	970299	9.950	4.626	31.464	9.967	21.472	0.01010
100	10000	1000000	10.000	4.642	31.623	10.000	21.544	0.0100

TABLE 18 – Constants of Control Chart

Sample size	*Meant chart* Factors for control limits			*Standard deviation chart* Factor for central line	Factors for control limits				*Range chart* Factors for central line	Factors for control limits			
n	A	A_1	A_2	c_2	B_1	B_2	B_3	B_4	d_2	D_1	D_2	D_3	D_4
2	2.121	3.76	1.88	0.5642	0	1.843	0	3.267	1.128	0	3.686	0	3.267
3	1.732	2.384	1.023	7236	0	1.858	0	2.566	1.693	0	4.358	0	2.575
4	1.500	1.88	0.729	0.7979	0	1.808	0	2.266	2.059	0	4.698	0	2.282
5	1.342	1.596	0.577	0.8407	0	1.756	0	2.089	2.326	0	4.918	0	2.115
6	1.225	1.41	0.483	0.8686	0.026	1.711	0.08	1.97	2.531	0	5.078	0	2.004
7	1.131	1.277	0.419	0.8882	0.105	1.672	0.118	1.882	2.704	0.205	5.203	0.076	1.924
8	1.061	1.175	0.373	0.9027	0.167	1.638	0.185	1.815	2.817	0.387	5.307	0.136	1.864
9	1.000	1.091	0.337	0.9139	0.219	1.609	0.239	1.761	2.970	0.546	5.394	0.184	1.816
10	0.949	1.028	0.308	0.9227	0.262	1.531	0.284	1.816	3.078	0.687	5.469	0.223	1.747
11	0.905	0.973	0.235	0.939	0.299	1.561	0.321	1.679	3.173	0.812	5.534	0.256	1.774
12	0.866	0.925	0.265	0.9359	0.331	1.511	0.354	1.646	3.258	0.924	5.592	0.284	1.716
13	0.832	0.881	0.249	0.941	0.359	1.523	0.382	1.618	3.336	1.026	5.646	0.308	1.692
14	0.802	0.848	1.235	0.9453	0.384	1.507	0.406	1.594	3.407	1.21	2.593	0.329	1.671
15	0.775	0.816	0.223	0.949	0.406	1.492	0.428	1.572	3.472	1.207	5.737	0.348	1.652
16	0.750	0.788	0.212	0.9523	0.427	1.478	1.448	1.552	3.532	1.285	5.779	0.365	1.636
17	0.728	1.762	0.203	0.9551	0.415	1.465	0.466	1.531	3.588	1.359	5.817	0.379	1.621
18	0.707	0.738	0.194	0.9575	0.461	1.454	0.482	1.518	3.640	1.426	5.854	0.392	1.608
19	0.688	0.717	0.187	0.9599	0.477	1.443	0.497	1.503	3.689	1.490	5.888	0.404	1.596
20	0.671	0.697	0.18	0.9619	0.401	1.433	0.510	1.490	3.735	1.548	5.922	0.414	1.586
21	0.655	0.679	0.173	0.9633	0.504	1.424	0.523	1.471	3.778	1.606	5.950	0.425	1.575
22	0.640	0.662	0.167	0.9555	0.516	1.415	0.534	1.466	3.819	1.659	5.972	0.434	1.566
23	0.626	0.647	0.162	0.9670	0.527	1.407	0.515	1.455	3.858	1.710	6.006	0.443	1.557
24	0.612	0.632	0.157	0.9684	0.533	1.399	0.555	1.445	3.395	1.759	6.031	0.452	1.548
25	0.600	0.619	0.153	0.9695	0.548	1.392	0.565	1.435	3.931	1.804	6.058	0.459	1.541

TABLE 19 – CONVERSION OF PEARSONS INTO CORRESPONDING FISHER'S Z COEFFICIENT*

$$Z = \frac{1}{2}\log\left(\frac{1+r}{1-r}\right) = 1.513 \quad \log_{10}\left(\frac{1+r}{1-r}\right)$$

r	z	r	z	r	z	r	z	r	z	r	z
0.25	0.26	0.40	0.42	0.55	0.62	0.70	0.87	0.85	1.26	0.950	1.83
0.26	0.27	0.41	0.44	0.56	0.63	0.71	0.89	0.86	1.29	0.955	1.89
0.27	0.28	0.42	0.45	0.57	0.65	0.72	0.91	0.87	1.33	0.960	1.95
0.28	0.29	0.43	0.46	0.58	0.66	0.73	0.93	0.88	1.38	0.970	2.01
0.29	0.30	0.44	0.47	0.59	0.68	0.74	0.95	0.89	1.42	0.975	2.09
0.30	0.31	0.45	0.48	0.60	0.69	0.75	0.97	0.90	1.47	0.980	2.18
0.31	0.32	0.46	0.50	0.61	0.71	0.76	1.00	0.905	1.50	0.985	2.30
0.32	0.33	0.47	0.51	0.62	0.73	0.77	1.02	0.910	1.53	0.990	2.44
0.33	0.34	0.48	0.52	0.60	0.74	0.78	1.05	0.915	1.56	0.995	2.65
0.34	0.35	0.49	0.54	0.64	0.76	0.79	1.07	0.920	1.59	1.00	2.99
0.35	0.37	0.50	0.55	0.65	0.78	0.8	1.10	0.925	1.62		
0.36	0.38	0.51	0.56	0.66	0.79	0.81	1.13	0.930	1.66		
0.37	0.39	0.52	0.58	0.67	0.81	0.82	1.16	0.935	1.70		
0.38	0.40	0.53	0.59	0.68	0.83	0.83	1.19	0.940	1.74		
0.39	0.41	0.54	0.60	0.69	0.85	0.84	1.22	9.945	1.78		

* r's under 0.25 may be taken as equivalent to z's.

TABLE 20 – CRITICAL VALUES OF n FOR THE SIGN TEST

n	α(two sided) 0.10 α(one-sided) 0.05	0.05 0.025	0.02 0.01	0.01 0.005
1				
2				
3				
4				
5	0,5			
6	0,6	0,8		
7	0,7	0,8	0,7	
8	1,8	1,8	0,8	0,8
9	1,9	1,8	0,9	0,9
10	2,9	1,9	0,10	0,10
11		1,10	1,10	0,11
12	2,10	2,10	1,11	1,11
13	3,10	2,11	1,12	1,12
14	3,11	2,22	2,12	1,13
15	3,12	3,12	2,13	2,13
16	4,12	3,13	2,14	2,14
17	4,12	4,13	3,14	2,15
18	5,13	4,14	3,15	2,15
19	5,14	4,15	4,15	3,16
20	5,15	5,15	4,16	3,17
21	6,15	5,16	4,17	4,17
22	6,16	5,17	5,17	4,18
23	7,16	6,17	5,18	4,19
24	7,17	6,18	5,19	5,19
25	7,18	7,18	6,19	5,20
26	8,18	7,19	6,20	6,20
27	8,19	7,20	7,20	6,21
28	9,19	8,20	7,21	6,22
29	9,20	8,21	7,22	7,22
30	10,20	9,21	8,22	7,23
31	10,21	9,22	8,23	7,24
32	10,22	9,23	8,24	8,24
33	11,22	10,23	9,24	8,25
34	11,23	10,24	9,24	8,25
35	12,23	11,24	10,25	9,26
36	12,24	11,25	10,26	9,27
37	13,34	12,25	10,27	10,27
38	13,25	12,26	11,27	10,28
39	13,26	12,57	11,28	11,28

TABLE 20 (Contd.) – CRITICAL VALUES OF n FOR THE SIGN TEST

n	α(two sided) 0.10 α(one-sided) 0.05	0.05 0.025	0.02 0.01	0.01 0.005
40	14,26	13,27	12,28	11,29
41	14,27	13,28	12,29	11,30
42	14,27	14,28	13,20	12,30
43	15,28	14,29	13,30	12,31
44	16,28	15,29	13,31	13,31
45	16,29	15,30	14,31	13,32
46	16,30	15,31	14,32	13,33
47	17,30	16,31	15,32	14,33
48	17,31	16,32	15,33	14,34
49	18,31	17,32	15,34	15,34
50	18,32	17,33	16,34	15,35
51	19,32	18,33	16.35	15,36
52	19,33	18,34	17,35	16,36
53	20,33	18,35	17,36	16,37
54	20,34	19,35	18,36	17,37
55	20,35	19,36	18,37	17,38
56	21,35	20,36	18,38	17,39
57	21,36	20,37	19,38	18,39
58	22,36	21,37	19,39	18,40
59	22,37	21,38	20,39	19,40
60	22,37	21,39	20,40	19,41
61	23,38	22,39	20,41	20,41
62	24,38	22,40	21,41	20,42
63	34,39	23,40	21,42	20,43
64	24,40	23,41	22,42	21,43
65	25,40	24,41	22,43	21,44
66	25,41	24,42	23,43	22,44
67	26,41	25,42	23,44	22,45
68	26,42	25,43	23,45	22,46
69	27,42	25,44	24,45	23,46
70	27,43	26,44	24,46	23,47
71	28,43	26,45	25,46	24,47
72	28,44	27,45	25,47	24,48
73	28,45	27,46	26,47	25,48
74	29,45	28,46	26,48	25,49
75	29,46	28,47	26,49	25,50

TABLE 20 (Contd.) – CRITICAL VALUES OF n FOR THE SIGN TEST

n	α(two sided) 0.10 α(one-sided) 0.05	0.05 0.025	0.02 0.01	0.01 0.005
76	30,46	28,48	27,49	26,50
77	30,47	29,48	27,50	26,51
78	31,47	29,49	28,50	27,51
79	31,48	30,49	28,51	27,52
80	32,48	30,50	29,51	28,52
81	32,49	31,50	29,52	28,53
82	33,49	31,51	30,52	28,54
83	33,50	32,51	30,53	39,54
84	33,51	32,52	30,54	29,55
85	34,51	32,53	31,54	30,55
86	34,52	33,53	31,55	30,56
87	35,52	33,54	32,55	31,56
88	35,53	34,54	32,56	31,57
89	36,53	34,55	33,56	31,58
90	36,54	35,55	33,57	32,58
91	37,54	35,57	33,58	32,59
92	37,55	36,56	34,58	33,59
93	38,55	36,57	34,59	33,60
94	38,56	37,57	35,59	34,60
95	38,57	37,58	35,60	34,61
96	39,57	37,59	36,60	34,62
97	39,58	28,59	36,61	35,62
98	40,58	38,60	37,61	35,63
99	40,59	37,62	37,62	37,63
100	41,59	39,61	37,63	36,64

TABLE 21 – WILCOXON SIGNED RANK TEST
n = number of pairs

	Critical Values			
n	α ≦ 0.10	α ≦ 0.05	α ≦ 0.02	α ≦ 0.01
1				
2				
3				
4				
5	0,15			
6	2,19	0,21		
7	3,25	2,26	0,28	
8	5,31	3,33	1,35	0,36
9	8,37	5,40	3,42	1,44
10	10,45	8,47	5,50	3,52
11	13,53	10,56	7,59	5,61
12	17,61	13,65	9,69	7,71
13	21,70	17,74	12,79	9,82
14	25,80	21,84	15,90	12,93
15	30,90	25,95	19,101	15,105
16	35,101	29,107	23,113	19,117
17	41,112	34,119	28,125	23,130
18	47,124	40,131	32,139	27,144
19	53,137	46,144	37,153	33,158
20	60,150	52,158	43,167	37,173
21	67,164	58,173	49,182	42,189
22	75,178	66,187	55,198	48,205
23	83,193	73,203	62,214	54,222
24	91,209	81,210	69,231	61,239
25	100,225	89,236	76,249	68,257

TABLE 22 – DISTRIBUTION-FREE CONFIDENCE LIMITS FOR THE MEDIAN OF A CONTINUOUS DISTRIBUTION

n	0.95	0.99	*n*	0.95	0.99	*n*	0.95	0.99	*n*	0.95	0.99
6	1-6		31	10-22	8-24	56	21-36	18-39	81	32-50	29-53
7	1-7		32	10-23	9-24	57	21-37	19-39	82	32-51	29-54
8	1-8	1-8	33	11-23	9-25	58	22-37	19-40	83	33-51	30-54
9	2-8	1-9	34	11-24	10-25	59	22-38	20-40	84	33-52	30-55
10	2-9	1-10	35	12-24	10-26	60	22-39	20-41	85	33-53	31-55
11	2-10	1-11	36	12-25	10-27	61	23-39	21-41	86	34-53	31-56
12	3-10	2-11	37	13-25	11-27	62	23-40	21-42	87	34-54	32-56
13	3-11	2-12	38	13-26	11-28	63	24-40	21-43	88	35-54	32-57
14	3-12	2-13	39	13-27	12-28	64	24-41	22-43	89	35-55	32-58
15	4-12	3-13	40	14-27	13-29	65	25-41	22-44	90	36-55	32-59
16	4-13	3-14	41	14-28	12-30	66	25-42	23-44	91	36-56	33-59
17	5-13	3-15	42	15-28	13-30	67	26-42	23-45	92	37-56	34-59
18	5-14	4-15	43	15-29	13-31	68	26-43	23-46	93	37-57	34-60
19	5-15	4-16	44	16-29	14-31	69	26-44	24-46	94	38-57	35-60
20	6-15	4-17	45	16-30	14-32	70	27-44	24-47	95	38-58	35-61
21	6-16	5-17	46	16-31	14-33	71	27-45	25-47	96	38-59	35-62
22	6-17	5-18	47	17-31	15-33	72	28-45	25-18	97	39-59	36-62
23	7-17	2-19	48	17-32	15-34	73	38-46	26-48	98	39-60	36-63
24	7-18	6-19	49	18-32	16-34	24	29-46	26-49	99	40-60	37-63
25	8-18	6-20	50	18-33	16-35	75	29-47	26-50	100	40-61	37-64
26	8-19	7-20	51	19-33	10-36	76	29-48	27-50			
27	8-20	7-21	52	19-34	17-36	77	30-48	27-51			
28	9-20	7-22	53	19-35	17-37	78	30-49	28-51			
29	9-21	8-22	54	20-35	18-37	79	31-49	28-52			
30	10-21	8-23	55	20-36	18-38	80	31-50	29-52			

TABLE-23 – RUNS ABOVE AND BELOW THE MEDIAN*

n'	Critical Values $\alpha = 0.10$	$\alpha = 0.05$	$\alpha = 0.02$	$\alpha = 0.01$
1				
2				
3				
4	2,8			
5	3,9	2,10	2,10	
6	3,11	3,11	2,12	2,12
7	4,12	3,13	3,13	3,13
8	5,13	4,14	4,14	3,15
9	6,14	5,15	4,16	4,16
10	6,16	6,16	5,17	5,17
11	7,17	7,17	6,18	5,19
12	8,18	7,19	7,19	6,20
13	9,19	8,20	7,21	7,21
14	10,20	9,21	8,22	7,23
15	11,21	10,22	9,23	8,24
16	11,23	11,23	10,24	9,25
17	12,24	11,25	10,26	10,26
18	13,25	12,26	11,27	11,27
19	14,26	13,27	12,28	11,29
20	15,27	14,28	13,29	12,30
21	16,28	15,29	14,30	13,31
22	17,29	16,30	14,32	14,32
23	17,31	16,32	15,33	14,34
24	18,32	17,33	16,34	15,35
25	19,33	18,34	17,25	16,36
26	20,24	19,35	18,36	17,37
27	21,35	20,36	19,37	18,38
28	22,36	21,37	19,39	18,40
29	23,37	22,38	20,40	19,41
30	24,38	22,40	21,2[illegible]	20,42
31	25,39	23,41	22,42	21,23
32	25,41	24,42	23,43	22,44
33	26,42	25,43	24,44	23,45
34	27,43	26,44	24,46	23,47
35	28,44	27,45	25,47	24,48
36	29,45	28,46	26,48	25,49
37	30,46	29,47	27,49	26,50
38	31,47	30,48	28,50	27,51
39	32,48	30,50	29,51	28,52
40	33,49	31,51	30,52	29,53
41	34,50	32,52	31,53	29,55

TABLE-23 (CONTD.) – RUNS ABOVE AND BELOW THE MEDIAN

n'	Critical Values $\alpha = 0.10$	$\alpha = 0.05$	$\alpha = 0.02$	$\alpha = 0.01$
42	35,51	33,53	31,54	30,56
43	35,53	34,54	32,56	31,57
44	36,54	35,55	33,57	32,58
45	37,55	36,56	34,58	33,59
46	38,56	37,57	35,59	34,60
47	39,57	38,58	36,60	35,61
48	40,58	38,60	37,61	36,63
49	41,59	39,61	38,62	36,64
50	42,60	40,62	38,64	37,65
51	43,61	41,63	39,65	38,66
52	44,62	42,64	40,66	39,67
53	45,63	43,65	41,67	40,68
54	45,65	44,66	42,68	41,69
55	46,66	45,67	43,69	42,70
56	47,67	46,68	44,70	42,72
57	48,68	47,69	45,71	43,73
58	49,69	47,71	46,72	44,74
59	50,70	48,72	46,74	45,75
60	51,71	49,73	47,75	46,76
61	52,72	50,74	48,76	47,77
62	53,73	51,75	49,77	48,78
63	54,74	52,76	50,78	49,79
64	55,75	53,77	51,79	49,81
65	56,76	54,78	52,80	50,82
66	57,77	55,79	53,81	51,83
67	58,78	56,80	54,82	52,84
68	58,80	57,81	54,84	53,85
69	59,81	58,82	55,85	54,86
70	60,82	58,84	56,86	55,87
71	61,83	59,85	57,87	56,88
72	62,84	60,86	58,88	57,89
73	63,85	61,87	59,89	57,91
74	64,86	62.88	60,90	58,92
75	65,87	63,89	61,91	59,93
76	66,88	64,90	62,92	60,94
77	67,89	65,91	63,93	61,95
78	68,90	66,92	64,94	62,96
79	69,91	67,93	64,96	63,97
80	70,92	68,94	65,97	64,98
81	71,93	69,95	66,98	65,99
82	71,95	69,97	67,99	65,100

TABLE-23 (CONTD.) – RUNS ABOVE AND BELOW THE MEDIAN

n'	Critical Values			
	$\alpha = 0.10$	$\alpha = 0.05$	$\alpha = 0.02$	$\alpha = 0.01$
83	72,96	70,98	68,100	66,102
84	73,97	71,99	69,101	67,103
85	74,98	72,100	70,102	68,104
86	75,99	73,101	71,103	69,105
87	76,100	74,102	72,104	70,106
88	77,101	75,103	73,105	71,107
89	78,102	76,104	74,106	72,108
90	79,103	77,105	74,108	73,109
91	80,104	78,106	75,109	74,110
92	81,105	79,107	76,110	75,111
93	82,106	80,108	77,111	75,113
94	83,107	81,109	78,112	76,114
95	84,108	82,110	79,113	77,115
96	85,109	82,112	80,114	78,116
97	86,110	83,113	81,115	79,117
98	87,111	84,114	82,116	80,118
99	87,113	85,115	83,117	81,119
100	88,114	86,116	84,118	82,120